SQUIRRELS OF THE WORLD

Squirrels of the World

Richard W. Thorington, Jr., John L. Koprowski, Michael A. Steele, and James F. Whatton

THE JOHNS HOPKINS UNIVERSITY PRESS BALTIMORE

This book was brought to publication through the generous contributions of the Smithsonian Institution, Wilkes University, and the University of Arizona.

 Published 2012
Printed in China on acid-free paper
9 8 7 6 5 4 3 2 1

The Johns Hopkins University Press
2715 North Charles Street
Baltimore, Maryland 21218-4363
www.press.jhu.edu

Library of Congress Cataloging-in-Publication Data

Squirrels of the world / by Richard W. Thorington, Jr. . . . [et al.].
p. cm.
Includes bibliographical references and index.
ISBN-13: 978-1-4214-0469-1 (hdbk. : alk. paper)
ISBN-10: 1-4214-0469-9 (hdbk. : alk. paper)
1. Sciuridae. I. Thorington, Richard W.
QL737.R68S68 2012
599.36—dc23 2011030931

A catalog record for this book is available from the British Library.

The Johns Hopkins University Press uses environmentally friendly book materials, including recycled text paper that is composed of at least 30 percent post-consumer waste, whenever possible.

Title Page Illustrations: (*left to right*) *Tamiops mcclellandii,* photo courtesy Nick Baker, www.ecologyasia.com; *Sciurus granatensis,* photo courtesy Caroline M. Thorington; *Marmota caligata,* photo courtesy U.S. Fish and Wildlife Service; *Sciurus niger cinereus,* photo courtesy Cynthia C. Winter.

CONTENTS

PREFACE

Our objective in writing *Squirrels of the World* is to provide a book that exhibits the great diversity of squirrels, describes some of their biology, and is illustrated with photographs and distribution maps for all species. Unfortunately, photographs were not available for all of the squirrel species, despite the efforts of our many photographic collaborators, who contributed images of 231 out of the total of 285 species that we recognize in this book. In preparing the maps, we discovered that many of the published distribution maps are inaccurate. Accordingly, we carefully checked the described distributions of each species and based the maps on documented records of occurrence. Our knowledge of species and distributions continues to change, but we have endeavored to provide the most accurate assessments available.

Our format is straightforward, starting with the scientific name, followed by the name of the person who described it and the date of the description, the common name, a brief description of the species, including some measurements, and its overall distribution. In the section on geographic variation, we provide the scientific names of the subspecies we recognize, their geographic range, and characteristics used to distinguish them. In some species, the subspecies are remarkably distinct from one another. Our goal is that our readers will be able to identify the species and subspecies of the squirrels that they observe. We then comment on the conservation status, although frequently the abundance of a species is poorly known. Finally, we supply a brief description of the preferred habitat of the species, and a summary of what is known about its natural history. We hope that the extreme brevity of natural history information for many species will serve to stimulate further research.

This book is a compilation from many sources. In order to facilitate reading, we do not include citations to our sources within the text. However, at the end of each species account we supply a reference section, giving a brief citation for the publications we have used, which are subsequently cited in full in the bibliographic section at the back of the book. The online International Union for Conservation of Nature (IUCN) *Red List of Threatened Species,* version 2010.3, provided information for the conservation status of each species and many of the range maps. We include the IUCN online reference for an individual species account in our book only if we used information from it beyond the conservation status and the map. This format should allow interested readers access to the rich literature on squirrels.

Normally a mammalian account includes illustrations of the skull. Instead of this we provide illustrations of the skulls of most genera of squirrels in an appendix. We hope this will assist in the identification of museum specimens, field discoveries, and fossils.

We are grateful for our assistants who have helped us prepare this book, particularly Paula W. Bohaska of the Smithsonian Institution, Maressa Takahashi of Columbia University, Jennie Miller of Yale University, Hsiang Ling Chen and Katharine Derrick of the Conservation Research Laboratory at the University of Arizona, and Melissa Bugdal of Wilkes University.

Abbreviations

HB	head and body length measurement
HF	hind foot measurement
IUCN	International Union for Conservation of Nature
T	tail length measurement
TL	total length

SQUIRRELS OF THE WORLD

Taxonomic Introduction

Squirrels constitute a family, the Sciuridae, of the order Rodentia, all being descendants of a common ancestor that lived some 30–40 million years ago. All share certain anatomical features of their teeth and jaw muscles by which they can be recognized, but distinguishing different groups within the Sciuridae and discerning interrelationships among them have been difficult. A particularly interesting question has been the relationship of flying squirrels to the others. Beginning in the nineteenth century and continuing still, there has been debate about whether flying squirrels evolved from tree squirrels or represent an independent lineage with a much longer history and thus are not closely related to tree squirrels. Comparative anatomists usually considered them to be true squirrels, whereas paleontologists contended that they were the sister group of all the other squirrels, one which evolved independently at about the same time as the others. Molecular data have recently been used in this debate, and the evidence adduced from the DNA clearly supports the hypothesis that living flying squirrels are most closely related to the Sciurini (consisting of the American and the northern Eurasian tree squirrels), and thus that flying squirrels are not an independent lineage. However, this conclusion leaves paleontologists with the problem of determining what the fossil squirrels are, since those occur-

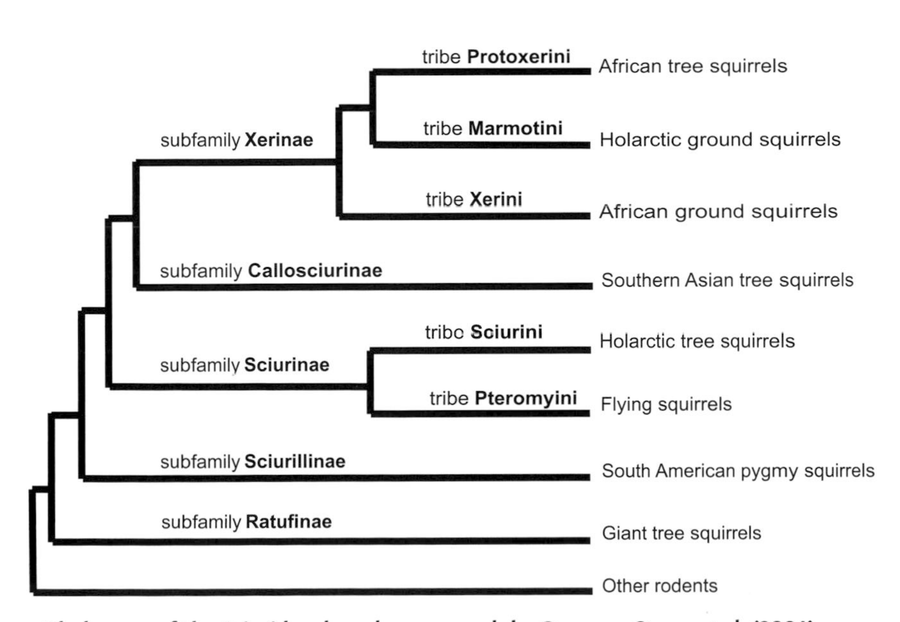

Phylogeny of the Sciuridae, based on research by Steppan, Storz, et al. (2004).

Table 1.1
Classification of the Sciuridae (58 genera, 285 species), based on Thorington and Hoffmann (2005), Helgen et al. (2009)

family **Sciuridae** Fischer de Waldheim, 1817

- subfamily **Ratufinae** Moore, 1959 (1 genus, 4 species)
    - *Ratufa* Gray, 1867 (4)
- subfamily **Sciurillinae** Moore, 1959 (1 genus, 1 species)
    - *Sciurillus* Thomas, 1914 (1)
- subfamily **Sciurinae** Fischer de Waldheim, 1817 (20 genera, 81 species)
    - tribe **Sciurini** Fischer de Waldheim, 1817 (5 genera, 37 species)
        - *Microsciurus* J. A. Allen, 1895 (4)
        - *Rheithrosciurus* Gray, 1867 (1)
        - *Sciurus* Linnaeus, 1758 (28)
        - *Syntheosciurus* Bangs, 1902 (1)
        - *Tamiasciurus* Trouessart, 1880 (3)
    - tribe **Pteromyini** Brandt, 1855 (15 genera, 44 species)
        - *Aeretes* G. M. Allen, 1940 (1)
        - *Aeromys* Robinson and Kloss, 1915 (2)
        - *Belomys* Thomas, 1908 (1)
        - *Biswamoyopterus* Saha, 1981 (1)
        - *Eoglaucomys* A. H. Howell, 1915 (1)
        - *Eupetaurus* Thomas, 1888 (1)
        - *Glaucomys* Thomas, 1908 (2)
        - *Hylopetes* Thomas, 1908 (9)
        - *Iomys* Thomas, 1908 (2)
        - *Petaurillus* Thomas, 1908 (3)
        - *Petaurista* Link, 1795 (9)
        - *Petinomys* Thomas, 1908 (8)
        - *Pteromys* G. Cuvier, 1800 (2)
        - *Pteromyscus* Thomas, 1908 (1)
        - *Trogopterus* Heude, 1898 (1)
- subfamily **Callosciurinae** Pocock, 1923 (14 genera, 67 species)
    - *Callosciurus* Gray, 1867 (14)
    - *Dremomys* Heude, 1898 (6)
    - *Exilisciurus* Moore, 1958 (3)
    - *Funambulus* Lesson, 1835 (5)
    - *Glyphotes* Thomas, 1898 (1)
    - *Hyosciurus* Archbold and Tate, 1935 (2)
    - *Lariscus* Thomas and Wroughton, 1909 (4)
    - *Menetes* Thomas, 1908 (1)
    - *Nannosciurus* Trouessart, 1880 (1)
    - *Prosciurillus* Ellerman, 1947 (7)
    - *Rhinosciurus* Blyth, 1856 (1)
    - *Rubrisciurus* Ellerman, 1954 (1)
    - *Sundasciurus* Moore, 1958 (17)
    - *Tamiops* J. A. Allen, 1906 (4)
- subfamily **Xerinae** Osborn, 1910 (22 genera, 132 species)
    - tribe **Xerini** Osborn, 1910 (3 genera, 6 species)
        - *Atlantoxerus* Forsyth Major, 1893 (1)
        - *Spermophilopsis* Blasius, 1884 (1)
        - *Xerus* Hemprich and Ehrenberg, 1833 (4)
    - tribe **Protoxerini** Moore, 1959 (6 genera, 31 species)
        - *Epixerus* Thomas, 1909 (1)
        - *Funisciurus* Trouessart, 1880 (10)
        - *Heliosciurus* Trouessart, 1880 (6)
        - *Myosciurus* Thomas, 1909 (1)
        - *Paraxerus* Forsyth Major, 1893 (11)
        - *Protoxerus* Forsyth Major, 1893 (2)
    - tribe **Marmotini** Pocock, 1923 (13 genera, 95 species)
        - *Ammospermophilus* Merriam, 1892 (5)
        - *Callospermophilus* Merriam, 1897 (3)
        - *Cynomys* Rafinesque, 1817 (5)
        - *Ictidomys* J. A. Allen, 1877 (3)
        - *Marmota* Blumenbach, 1779 (15)
        - *Notocitellus* A. H. Howell, 1938 (2)
        - *Otospermophilus* Brandt, 1844 (3)
        - *Poliocitellus* A. H. Howell, 1938 (1)
        - *Sciurotamias* Miller, 1901 (2)
        - *Spermophilus* F. Cuvier, 1825 (15)
        - *Tamias* Illiger, 1811 (25)
        - *Urocitellus* Obolenskij, 1927 (12)
        - *Xerospermophilus* Merriam, 1892 (4)

ring in the Eocene and identified as "flying" squirrels are as old as any fossils identified as tree squirrels.

How to subdivide the family Sciuridae into subfamilies was not at all clear until the 1920s, when Reginald Pocock, of the London Zoo, conducted a study of the bacula (the penis bones) of squirrels, which are quite different among the various subfamilies. Based on this work, he recognized six subfamilies, in addition to the flying squirrels, thus demonstrating significant differences among the tree squirrels of Southeast Asia, those of the Americas and northern Eurasia, and those of Africa. However, it was unclear how these subfamilies were related to one another.

Pocock's conclusions served as the basis of squirrel classifications until 2003, when molecular studies based on the DNA of several genes supported a classification of five subfamilies, modifying the Pocock arrangement in some interesting ways. Our classification in this book is based on the DNA evidence. Two of the subfamilies appear to be the most ancient lineages within the Sciuridae: the Ratufinae, which includes the giant tree squirrels of southern Asia (four species within the genus *Ratufa*), and the Sciurillinae, which is made up of the South American pygmy squirrels (one or perhaps several species of the genus *Sciurillus*). A third subfamily, the Sciurinae, consists of the tree squirrels of the Americas and northern Eurasia (tribe Sciurini) and its sister group, the flying squirrels (tribe Pteromyini), as mentioned above. A fourth subfamily, the Callosciurinae, contains the Southeast Asian tree squirrels recognized by Pocock, plus the small striped squirrels (*Funambulus*) of the Indian subcontinent. The fifth subfamily, the Xerinae, combines three distinct groups: the African ground squirrels (tribe Xerini); the African tree and bush squirrels (tribe Protoxerini); and the other ground squirrels, chipmunks, and marmots of North America and northern Eurasia (tribe Marmotini). This classification is summarized in table 1.1 and provides the organization of this book.

Paleontology

Squirrels are not well known in the fossil record. However, paleontologists have worked assiduously, screen washing and extracting fossil teeth from sediment, which has resulted in many finds showing the distinctive features of squirrel teeth. From these, a general picture of squirrel evolution has emerged.

One of the earliest fossil squirrels, dating from the late Eocene of North America, 36 million years ago (MYA), is Jefferson's squirrel, *Douglassciurus jeffersoni*. Remarkably, a complete skeleton of the species was found. Jefferson's squirrel was clearly a tree squirrel, with its body proportions and most details of its postcranial anatomy extremely similar to that of modern eastern fox squirrels (*Sciurus niger*). However, the rostrum was shorter than in other squirrels, and the jaw musculature was extremely primitive, with the masseter muscle originating only from the ventral surface of the zygomatic arch under the eye (protrogomorphy) and not from the zygomatic plate and the side of the rostrum in front of the eye (sciuromorphy), as it does in other modern and fossil squirrels. Because the more anterior origin of the masseter musculature has been the main defining character of the Sciuridae, there was some initial reluctance on the part of paleontologists to consider *Douglassciurus* to be a sciurid, but in all other respects it is very much a squirrel. It probably looked like a snub-nosed modern tree squirrel. It is closely related to *Protosciurus*, found from the early Oligocene (33 MYA) to the early Miocene (23 MYA) in North America. *Protosciurus* differs in having at least the beginning of the typical squirrel masseter muscle arrangement (sciuromorphy, instead of protrogomorphy). Presumably they also looked very much like modern North American tree squirrels. *Miosciurus* (approximately 23 MYA), in turn, is closely related to *Protosciurus*, but *Miosciurus* is suggested to be close to the divergence of tree squirrels and ground squirrels, although it is not ancestral to either. Tree squirrels attributed to the modern genus *Sciurus*, which was probably derived from *Protosciurus*, are found from the Miocene (approximately 16 MYA) to the present in North America. They are also found from the late Pliocene (3 MYA) to the present in Europe, and from the middle Pleistocene (0.7 MYA) to the present in Asia. They probably reached South America at the time that the Panamanian isthmus formed (3 MYA). Other New World tree squirrels are known only from Pleistocene fossils (the North American red squirrel, *Tamiasciurus*) or not at all as fossils (the dwarf and pygmy squirrels of Central and South America, *Microsciurus* and *Sciurillus*). Similarly, the Bornean *Rheithrosciurus* is not known as a fossil.

Asian tree squirrels are very rare as fossils. In Europe, a large tree squirrel in the early Miocene (23 MYA) is attributed to the genus *Ratufa*. This genus includes the four species of giant tree squirrels presently living in southern Asia, but no Recent (~10,000 years ago to the present) species in Europe. Other paleontologists have questioned this attribution and have considered this fossil to be "a large generalized tree squirrel" of unknown affiliation. *Ratufa* is known with more certainty from the middle Miocene (16 MYA) of southern Asia. Late Miocene (11 MYA) fossils from Asia are attributed to the callosciurine squirrels, which are now the dominant tree squirrels in Southeast Asia. These include *Dremomys*, *Callosciurus*, and *Tamiops*, which are known with more certainty from the early to the mid-Pleistocene, less than 2 MYA (*Dremomys*); the early Pleistocene (*Callosciurus*); and the late Pliocene to the middle Pleistocene, less than 3 MYA (*Tamiops*). All of the other genera of callosciurine squirrels in the Recent fauna are not known as fossils.

***Douglassciurus jeffersoni* skeleton at Smithsonian Institution's National Museum of Natural History.** Photo courtesy James Di Loreto, Smithsonian Institution.

The xerine ground squirrels presently occur mostly in Africa (five species) but are also found in southwest Asia (*Spermophilopsis*). Fossil forms include *Heteroxerus,* known from the early Oligocene to the late Miocene (26-5 MYA) in Europe. The 8-10 recognized species are more common in southwestern Europe than in central Europe. Another genus, *Aragoxerus,* has been described from the early Miocene (23-16 MYA) in Spain. A third genus, *Atlantoxerus,* with more robust teeth than *Heteroxerus,* includes five species, although the number may be reduced with further study. *Atlantoxerus* is rare, but it is found from the early Miocene to the early Pliocene (23-3 MYA) in Eurasia and from the middle Miocene (16 MYA) onward in Africa, with one species still living in northwest Africa.

Other African squirrels are also rare as fossils. *Vulcanisciurus,* of unknown affinities within the Sciuridae, is known from the early to late Miocene (23-5 MYA) in Africa. The giant palm squirrel, *Kubwaxerus,* is known from the late Miocene in Kenya, and the African sun squirrel, *Heliosciurus,* is known from the late Pleistocene (< 1 MYA) to the Recent in Africa, with six living species.

The marmotine ground squirrels (including chipmunks, marmots, and North American ground squirrels) are better represented as fossils, particularly in North America. In Europe the reputed ground squirrel, *Palaeosciurus,* is attributed to this group. It is a common fossil, comprising four species found from the early Oligocene to the early Miocene (33-16 MYA)—more frequent in central Europe than in southwestern Europe—and in the early Miocene (23-16 MYA) in Asia. Chipmunks occur early in the fossil record in North America. *Nototamias* is reported in the late Oligocene to the middle Miocene (24-12 MYA), and *Tamias* from the early Miocene (23 MYA) to the Recent. In Europe, three species of *Tamias* are recognized from the early Miocene to the early Pliocene (18-3 MYA), mostly restricted to southeastern Mediterranean Europe, and the genus is known from the middle Miocene (16 MYA) to the Recent in Asia. *Sinotamias* is known from the middle to the late Miocene (16-5 MYA) in Asia, and the Chinese rock squirrel, *Sciurotamias,* is known from the late Miocene (12 MYA) and intermittently to the Recent in Asia.

Protospermophilus occurs in the late Oligocene to the late Miocene (24-5 MYA) in North America. It retained the tooth morphology of tree squirrels, although it is found in situations that suggest it was a terrestrial squirrel. One supposition is that it was still a seed feeder, like the tree squirrels, and thus could not compete with the ground squirrels, which were better adapted to grasslands. *Miospermophilus,* which occurs from the late Oligocene to the middle Miocene (24-12 MYA) in North America, has been suggested to be ancestral to *Spermophilus,* which has become the dominant ground squirrel of North America and northern Eurasia and is now divided into eight distinct genera (*Notocitellus, Otospermophilus, Callospermophilus, Spermophilus, Ictidomys, Poliocitellus, Xerospermophilus,* and *Urocitellus*). Another possibility is that modern ground squirrels are derived from *Spermophilinus,* which is found from the early Miocene to the early Pliocene (18-3 MYA) in central and southwestern Europe, with four species being recognized; possibly in the late Miocene in western Asia; and more certainly in the early Pliocene (5-3 MYA) in eastern Asia. *Spermophilus* itself occurs from the middle Miocene (16 MYA) to the Recent in North America; in the late Miocene (6 MYA) in Europe, where it is still extant; and in the early Pleistocene (2 MYA), the middle Pleistocene, and the Recent in Asia. In North America, antelope ground squirrels, *Ammospermophilus,* are known from the late Miocene (11 MYA) to the Recent; and prairie dogs, *Cynomys,* are known from the late Pliocene (2 MYA) to the Recent, with a dubious record in the middle Miocene.

Three genera—*Paenemarmota, Palaearctomys,* and *Arcto-*

myoides—developed marmotlike characteristics in the late Miocene (12-5 MYA). They are probably derived from early ground squirrels of the *Otospermophilus* group. Similarly, *Marmota* seems to have evolved from this group in the late Miocene. Marmots are known from the late Miocene (12 MYA) to the Recent in North America, and in the Pleistocene and the Recent in Europe and Asia. There are now six species in North America and eight species in Eurasia.

Hesperopetes thoringtoni (36 MYA), slightly older than *Douglassciurus*, and hence the oldest known fossil squirrel, is known only from teeth and is attributed to the "flying" squirrel group, a group distinguished by having wrinkled or crenulated enamel on their molars. Squirrels of this group are represented by fossils from late Eocene until the Pliocene (36-2 MYA) and, if they are related to true flying squirrels, are still extant. *Hesperopetes* includes three species in North America from the late Eocene (36 MYA) and the late Oligocene (30 MYA). *Oligopetes*, with two species in Europe, is an early Oligocene (33 MYA) "flying" squirrel. However, molecular data suggest that modern flying squirrels diverged from the Sciurini tree squirrels in the early Miocene (23 MYA). Crenulated molars did not evolve in this lineage until the middle or late Miocene (10 MYA), which is discordant with the dates of these fossils reported from the Eocene and the Oligocene. Unfortunately, there are no postcranial bones of these purported "flying" squirrels to show whether these animals were gliders, like modern flying squirrels.

Within this "flying" squirrel group there are early Miocene fossils of four genera: *Petauristodon, Aliveria, Miopetaurista*, and *Blackia*. *Petauristodon*, with six species existing in North America from the early to the late Miocene (19-9 MYA), probably did not glide and most likely was not related to modern flying squirrels (tribe Pteromyini). *Aliveria* consists of two species in the early Miocene (18-16 MYA) and is considered to be a reasonable ancestor of flying squirrels, together with the similar *Shuanggouia* of the middle Miocene and *Albanensia* (with four species) in the middle and the late Miocene (15-11 MYA). *Miopetaurista*, with six species ranging from the early Miocene to the Pliocene (22-3 MYA), has smooth enamel but is otherwise similar to *Aliveria*. *Blackia*, with two species ranging from the late Oligocene to the late Pliocene (30-2 MYA), was a small squirrel with extreme pitting of the enamel and small cusps. No other squirrels exhibit this pattern, suggesting that *Blackia* was not ancestral to any other fossil "flying" squirrels or modern squirrels.

Other middle Miocene "flying" squirrels include *Forsythia*, a single species in the middle Miocene (14-11 MYA) in Europe, and *Meinia* and *Parapetaurista* in Asia. *Meinia* was described as being similar to *Blackia*. *Parapetaurista* is considered to be similar to *Miopetaurista* or *Albanensia*. *Pliopetaurista* (with four species) and *Pliosciuropterus* are late Miocene and Pliocene (11-2 MYA) genera, with similarities to modern giant flying squirrels (*Petaurista*). However, one species of *Pliopetaurista* (*P. meini*) appears to be more similar to *Callosciurus*.

Three species from the early Miocene to the early Pliocene (23-3 MYA) in Europe are attributed to the genus *Hylopetes*, congeneric with species in the modern fauna of Southeast Asia.

Flying squirrels of the Pliocene and the Pleistocene are attributed to four of the genera described above (*Blackia, Miopetaurista, Pliopetaurista*, and *Pliosciuropterus*) or to modern genera.

It is doubtful whether one can recognize true flying squirrels (tribe Pteromyini) in the fossil record by means of the pitting and crenulations on the teeth. These features are found on the teeth of 3 of the 15 genera of living flying squirrels, but also on 7 genera of tree squirrels. In fact, other characteristics of the teeth of flying squirrels that are touted as diagnostic are found among tree squirrels as well, leading to the suggestion that the gliding habits of fossil squirrels can be recognized only by characteristics of their postcranial skeleton, which is unfortunately seldom preserved. In the Pliocene and the Pleistocene, however, the close similarity of the teeth of some fossils to those of modern genera makes it highly probable that these were indeed true flying squirrels.

Anatomy: Form and Function

The largest squirrels living today are the marmots (*Marmota*) of North America and Asia, including the well-known yellow-bellied marmot of the Rocky Mountains and the woodchuck of the eastern USA and Canada (the "groundhog" of February 2). One of the largest is the gray marmot, found in the mountains of Kazakhstan. Marmots put on weight before they enter hibernation, and may even double their weight, so the animals are heaviest at the end of the summer. At this time, the largest gray marmots may weigh more than 8 kg (18 lb). The largest tree squirrels, the giant tree squirrels of Southeast Asia (*Ratufa*), are not nearly as big as marmots, but they are still quite large—ranging from 2 kg (4.4 lb) up to 3 kg (6.6 lb). With their beautiful long tails and striking coloration, these squirrels are an impressive sight bounding through the trees.

In contrast, the smallest squirrels are the pygmy tree squirrels of western Africa (*Myosciurus*) and Southeast Asia (*Exilisciurus*), which are smaller than some mice. The smallest adults of both genera average approximately 14 or 15 g (roughly 0.5 oz).

Rodents have jaws that function in two different ways, for gnawing and for chewing. In gnawing, the jaw is moved forward so that the incisors meet. In chewing, the jaw is moved backward so that the upper and lower molars come in contact with each other, and the upper and lower incisors do not. The muscles that pull the jaw forward are arranged differently in various groups of rodents. However, all squirrels have one of the jaw muscles (the anterior deep masseter) originating on the side of the face in front of the eye and then passing beneath the eye to insert onto the lower jaw. This muscle appears both to move the jaw forward and to strengthen the gnawing capabilities of squirrels, for which they are well known.

The incisor teeth are similar in all squirrels, although in some they are stouter and in others they are more gracile. There are also differences in how far they protrude (performing an almost forcepslike action) or in how vertically they are oriented (for efficient gnawing). All incisors continue to grow throughout the life of the squirrel, unlike the remaining teeth. There is much greater variance in the cheek teeth, depending on the diet of the species. Those squirrels that feed on animal matter and soft vegetable matter have the simplest teeth, with few cones and few ridges on the premolar and molar surfaces. Those that feed on leaves and other items that need a substantial amount of chewing have more complex ridges and cones on their tooth surfaces.

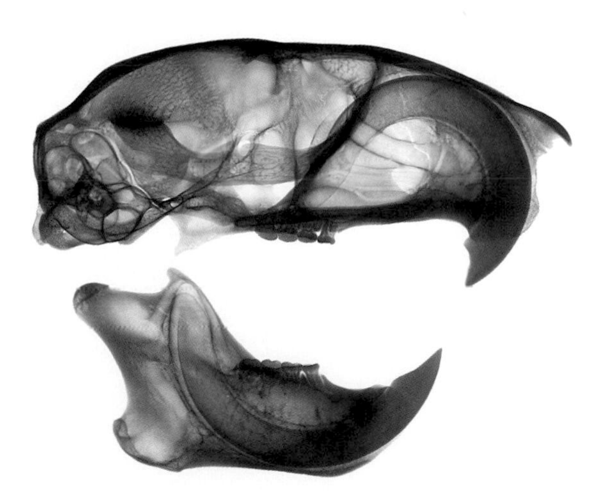

X-ray of a *Rheithrosciurus* skull, showing the incisors and molars.

The body proportions of squirrels vary in several ways, most notably in limb lengths and tail length relative to body size. The shortest legs and tails are found in ground squirrels. Tree squirrels have longer legs and tails, and the longest of both are found in large flying squirrels. The greatest elongation of the limbs is in the forearm of flying squirrels, which supports the wing tip. Presumably, shorter limbs and tails are more functional in burrowing animals, and the tail may even be used for sensing the way when the animal is moving backward. Longer limbs allow tree squirrels to have a more extensive reach around a tree and a greater jumping ability. The longer tail is obviously used for balance when jumping, among other functions. The limb elongation of flying squirrels permits these animals to have a larger "wing." The total wing is almost square (or even wider than long), with an aspect ratio (width/length) commonly between 1.0 and 2.0, which gives it good stability and enables the squirrel to land at low speeds. Flying squirrels clearly use their longer tails for balance and for stability while gliding. Among arboreal squirrels there is a great deal of flexion at the base of the tail, which allows the tail to the moved as a unit for balance. The short vertebrae at the base of the tail result in more intervertebral joints, which is where the flexion occurs. The flexibility at the base of the tail enables squirrels to hold their tails in their very distinctive way.

On many squirrels the hairs of the tail are arranged in a distichous fashion, with the lateral hairs being longer than the hairs on the top and the bottom. This featherlike arrangement is presumably excellent for steering and balancing in small tree squirrels and flying squirrels. Large flying squirrels (> 1000 g) and giant tree squirrels have long tails with hairs that are the same length on the top, the sides, and the bottom, which suggests that distichous tails are not effective on larger animals. In these animals, their tails appear to be used less for steering and more for stability, like a kite tail.

The ankle morphology of squirrels is also very distinctive, and it enables tree squirrels to rotate their hind feet 180 degrees, so that a squirrel coming down a tree trunk headfirst can have its hind feet pointed back up the tree. Most of this rotation occurs between the two major bones, the astragalus and the calcaneus, with their broad joint facets.

Flying squirrels evolved from tree squirrels, and they have developed a number of anatomical specializations associated with their gliding flight. The most obvious is the extension of the skin to form a "wing" (the patagium), stretching from the forelimb to the hind limb. The normal skin musculature of tree squirrels is incorporated into this membrane and enables the flying squirrel to gather the patagium against its body when it is not gliding. Another skin muscle extends from the wrist to the ankle and stabilizes the outer edge of the patagium when the animal is gliding, like the leech line on the sail of a sailboat. On large flying squirrels there is usually an extension of the wing (the uropatagium) between the hind limb and the tail. One of the hamstring muscles of the thigh supports this, and it also serves as a leech line, becoming disassociated from the other thigh muscles when the uropatagium is large.

Flying squirrels also have a specialized wrist. One of the sesamoid bones of the wrist in tree squirrels becomes elongated, forming the styliform process at the heel of the hand in flying squirrels. It is hinged so that it can be held against the forearm when the squirrel is not gliding, but it can be extended at 90 degrees from the wrist, supporting the wing tip of the patagium when the animal is gliding. When gliding, the squirrel holds its wrist in such a way that the wing tip is elevated above the rest of the patagium, which presumably serves one or two functions: reducing the drag of the wing and/or increasing stability. The musculature that enables the styliform process to be extended is remarkable. The styliform process is on the little-finger side of the wrist,

A Congo rope squirrel (*Funisciurus congicus*), showing ankle rotation. Photo courtesy Jon Hall, www.mammalwatching.com.

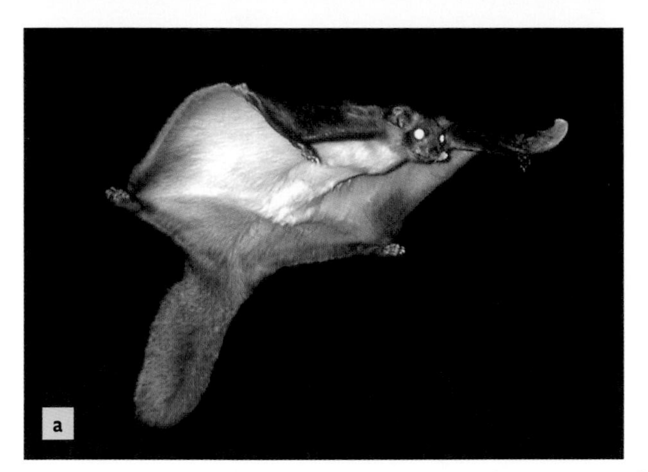

(a) A Japanese giant flying squirrel (*Petaurista leucogenys*) in flight, showing both the patagium and the uropatagium. Photo courtesy Takeo Kawamichi. **(b) Red giant flying squirrel (*Petaurista petaurista*) flight sequence.** Photo courtesy David Bakewell.

but the muscle (the *abductor pollicis*) that extends it is connected to the thumb, which is minuscule in flying squirrels. There is a ligament that reaches across the hand from the thumb to the base of the styliform process in such a way that when the squirrel contracts the *abductor pollicis* and dorsiflexes the wrist, the styliform process and the wing tip are extended and elevated.

The forearm of flying squirrels is notable in one other way. It appears that there is decreased mobility between three carpal bones of the wrist (the triquetrum, the scapholunate, and the pisiform), probably to provide greater stability for the wing tip. In addition, a number of flying squirrels have the two bones of the forearm (the radius and the ulna) tightly bound together with ligaments, presumably for the same reason. In most mammals this would preclude rotation of the hand (pronation and supination), but flying squirrels are able to manipulate their food easily, pronating and supinating their hands with ease. The rotation does not occur at the wrist, however, but rather at the elbow, with the ulna rocking on its articulation with the humerus, which is an extraordinary way to maintain manual dexterity.

Modern ground squirrels probably evolved from tree squirrels, and the anatomy of a tree squirrel seems to require little change for the animal to adapt to terrestrial habitats. Cheek pouches are one distinctive feature of the tribe Marmotini—including marmots, chipmunks, antelope ground squirrels, and other ground squirrels of this group. These enable the squirrels to carry more food in their mouths than they would be able to otherwise. Various muscles are associated with the cheek pouches, such as the one that helps them empty the pouch. This originates behind the upper incisors, where there is a fairly distinct muscle scar, which has enabled paleontologists to determine when cheek pouches evolved in the fossil record of ground squirrels.

Another interesting change that occurred in only some of the Marmotini—marmots, ground squirrels, and prairie dogs—is the development of the deltoid muscle, which normally extends from the clavicle to the upper half of the hu-

An eastern chipmunk (*Tamias striatus*) with full cheek pouches. Photo courtesy Jen Goellnitz.

merus, as it does in humans. In these burrowing squirrels there is an extra part of the deltoid (the clavobrachialis) that extends onto the forearm and assists the other forearm flexors (the biceps and the brachialis). It would be interesting to know exactly when it is used—during walking and running, or more specifically during digging activities. In other ways, these ground squirrels seem just like short-tailed tree squirrels, with short, stout legs.

Squirrels have very good eyesight and exceptional focusing ability. They have sharp vision across the entire retina, like the vision humans have in their *fovea centralis,* the small area of the retina where the cones are most densely packed and thus where a person's eyesight is most acute. This allows a motionless squirrel, without moving its head, to see clearly what is next to it and above it at the same time. Many squirrels have yellow-tinted eye lenses: ground squirrels have dark yellow lenses, and tree squirrels have paler yellow ones. These yellow lenses, much like sunglasses, reduce glare from bright light and increase the contrast between colors, giving the squirrel sharper vision. Flying squirrels, however, have clear lenses. Since they are nocturnal and seldom encounter bright light, they have no need for a tinted lens.

As in most mammals, a squirrel's retina contains both rods and cones. Rods are specially adapted cells that allow vision in low light, and cones are specifically designed cells for daylight vision, color vision, and the discrimination of detail. The retinas of diurnal squirrels contain rods, because the animals need to be able to see at dusk, dawn, and in shaded areas. Ground squirrels, such as prairie dogs, are superbly adapted to bright light and have many more cones than rods. In fact, it was once thought that they had no rods at all. The retinas of nocturnal flying squirrels, on the other hand, have mostly rods and only a few cones, which give them excellent night vision.

Ecology, Behavior, and Conservation

Many consider squirrels to be model mammals for the study of behavior and ecology. Although some may disagree, we strongly support this assertion, and submit the more than five decades of ecological and evolutionary studies on squirrels as evidence for their importance to such key research areas as mammalian social behavior, kin selection, foraging ecology, food hoarding, and seed dispersal.

Their relevance to the study of behavior and ecology is due in part to the simple fact that squirrels occupy a variety of niches throughout the temperate, subtropical, and tropical regions of the world. With the exception of flying squirrels, most squirrels are diurnal and easily observed; in fact, they are one of the few mammals that can be readily marked and directly studied for extended periods of time. In the case of social ground squirrels and prairie dogs, for example, nearly every individual in a colony can be marked, observed, and followed to measure characteristics such as dispersal, longevity, relatedness, and lifetime reproductive success. Such opportunities are clearly lacking in many other systems. Moreover, squirrels often function as keystone herbivores, seed predators, or seed dispersers; as such, they may provide critical services that are essential to ecosystem function. Our current state of knowledge is based upon temperate and holarctic species, and many new discoveries probably await us in the grasslands and forests of the tropics. Here, we provide a brief overview of the ecology and behavior of squirrels by highlighting some of the critical areas in which they have been studied.

Population Ecology

The modest body sizes of the Sciuridae—ranging from the 15 g African pygmy squirrel (*Myosciurus pumilio*) to the 8000 g Tarbagan marmot (*Marmota sibirica*)—in combination with high levels of aboveground activity and their often diurnal habits, have made squirrels the focus of many important studies in population ecology. Most of the Sciuridae mature in their first year of life; however, some of the larger-bodied ground squirrels, such as marmots (*Marmota*) and prairie dogs (*Cynomys*), often have delayed maturity that extends until 2 to 4 years of age, when first reproduction takes place. Hibernation of up to eight months a year occurs only in the Marmotini and, importantly, constrains resource use to the brief active season. High-latitude marmots, prairie dogs, ground squirrels, and chipmunks cannot complete reproduction late in this short season, whereas tree, flying, and tropical ground squirrels do not have such severe temporal constraints and often rely on cached food items and fat stores to remain active overwinter, with only occasional bouts of short-duration torpor.

A few significant patterns emerge from the recent phylogenetic assessments by V. Hayssen in characterizing the reproductive performance of the Sciuridae, despite variations such as litter sizes that range from 1.0 to 9.7 young (mean = 3.8 young) in some antelope ground squirrels (*Ammospermophilus*). Gestation (pregnancy) extends from 22 to 80 days (mean = 34.6 days), and females nurse from 21 to 105 days (mean = 45.0 days). Tree squirrels invest more into reproduction than do most ground squirrels and often have multiple modest-sized (two to five young) litters each year, especially in the tropics, as evidenced from the few species that have been studied. Flying squirrels have smaller litter sizes (one to four young) than most tree squirrels and ground squirrels. Marmotine ground squirrels typically produce a single large litter of small young and thus have a modest annual investment, despite a sizeable energetic investment in this single large litter, and some large-bodied ground squirrels

A litter of round-tailed ground squirrels (*Xerospermophilus tereticaudus*). Photo courtesy Karen Munroe.

do not reproduce each year. Non-marmotine ground squirrels have a reproductive pattern similar to that of tree squirrels, although typically with smaller litter sizes and slower growth.

The ecological longevity of squirrels is typically 5–10 years; however, in captivity (physiological longevity) many squirrels can survive 10–20 years. Survivorship in the wild is poor during the first year of life but increases after natal dispersal and upon reaching adulthood. Studies on natal dispersal in yellow-bellied marmots (*Marmota flaviventris*) suggest that survival during dispersal may not be as meager as once hypothesized. Classic studies conducted on several ground squirrels (Belding's ground squirrels [*Urocitellus beldingi*] and Columbian ground squirrels [*U. columbianus*]) and yellow-bellied marmots tell us that attaining an appropriate body mass is a key proximate cue that triggers natal dispersal; ultimately, however, inbreeding avoidance and reproductive opportunity appear to play an important role in dispersal from the natal area. Sex differences in patterns of space use, reproductive success, and annual mortality (greater for males than for females) are paralleled by sexual dimorphism in body size in some of the ground squirrels (particularly the marmotine ground squirrels, including *Marmota* and *Cynomys*), whereas most tree and flying squirrels demonstrate little if any dimorphism in body size but often exhibit these same ecological differences between the sexes. Chipmunks (*Tamias*), however, demonstrate an intriguing pattern of reversed sexual dimorphism, with the females being larger than the males, which is rare among mammals; the cause of this pattern is unclear, but females accrue advantages in dominance and resource acquisition.

Communication

Individuals communicate using several sensory modalities. The excellent vision of most squirrels allows them to use visual cues. Posture, foot stamping, and rapid movements of the tail appear to convey information on dominance status between individuals, as well as to indicate the level of alarm among conspecifics and heterospecifics. Unfortunately, we know very little about the messages conveyed by such visual displays in any species of squirrel. As in many mammals, chemical communication is a particularly important means of passing along information. Most squirrels have glands that produce chemicals (believed to be pheromones) in their cheeks, back, and flanks. Oral-nasal contact (often termed greeting or kissing) is a common means of interacting among the most social species, such as when individuals approach each other and touch nose to cheek. Even in less social species, the glands of the oral and cheek regions are used to deposit scent-marks. Rubbing other parts of the body and applying urine to structures and substrates also convey data on occupancy and perhaps dominance, because scent-marking increases with dominance in some species, such as *Sciurus niger*.

The calls of the Sciuridae are quite variable, from high-pitched single notes and prolonged squeals to low-pitched "chucks" and guttural tones. Some nestlings produce ultrasonic calls, and at least some species of *Sciurus* are capable of hearing in this range; however, the ultrasonic frequencies have not received much attention. Many nocturnal species of flying squirrel appear to be relatively silent. Studies by D. T. Blumstein on the alarm calls of squirrels have served as models for similar research in other taxa and have generated and tested important hypotheses. For example, they have shown that diurnality, and not sociality, is the most important factor in the evolution of rodent alarm calls. They have also begun to unravel the influence of physical and social environments on vocalizations. The alarm calls of some species of ground and tree squirrels appear to be relatively simple and intuitively related to the acoustics of their habitat; however, recent work demonstrates that calls by Gunnison's prairie dogs (*Cynomys gunnisoni*) encode a great deal of information. C. N. Slobodchikoff and his collaborators show that the calls of this social species distinguish not only the type of predator (terrestrial versus aerial)—a finding that has been described for other marmotine ground squirrels—but also unique characteristics of the potential predator, including its color.

Social Behavior

The interspecific variation in social and mating systems makes the Sciuridae excellent model organisms for the study of social evolution. Early models recognized relatively asocial species (such as many of the chipmunks [*Tamias*], small-bodied ground squirrels, and tree squirrels) and highly social species with overlapping generations of primarily female relatives living together (such as many of the marmots [*Marmota*], prairie dogs [*Cynomys*], African ground squirrels [*Xerus*], and eastern gray squirrels [*Sciurus carolinensis*]). The ability to achieve independence within the first year of life appears to be an important correlate of sociality; species in which juveniles are able to reach adult size and disperse by the time they are 1 year old are rarely social. Sociality occurs primarily when female kin remain to form kin clusters within local populations, and kin-differential behaviors often take place within these female kin clusters.

A round-tailed ground squirrel (*Xerospermophilus tereticaudus*) with a rattlesnake. Photo courtesy Karen Munroe.

Communal nesting, however, is among the most common behaviors documented not only within highly social species, but also in ground squirrels, flying squirrels, and tree squirrels. Although sociality has been well described for many temperate-zone squirrels, nocturnal flying squirrels and many tropical tree and ground squirrels remain relatively unstudied; most likely they will provide excellent tests of existing hypotheses on the evolution of sociality.

Kin Recognition, Kin Selection, Alarm Calling, and Infanticide

The overlapping generations of kin within kin clusters provide unique opportunities for the study of kin recognition and kin selection. The importance of nepotistic behaviors in nonhuman systems was exemplified by classic work on the ground squirrels (*Urocitellus beldingi, U. columbianus, Xerospermophilus tereticaudus*, and *Cynomys ludovicianus*). More recent studies suggest that such forms of conduct might also exist in the marmots, tree squirrels, and flying squirrels, especially with regard to communal nesting in challenging winter climates. Kinship influences many actions by ground squirrels, including nesting associations, alarm calling, and the distribution of amicable and agonistic behaviors. For example, Belding's ground squirrels with relatives nearby are more likely to call and risk detection by a predator. Detailed studies on Belding's ground squirrels and golden-mantled ground squirrels (*Callospermophilus lateralis*) demonstrate that recognition of close and distant relatives occurs at least in part through olfactory cues; however, differential treatment of individuals based on kinship only occurs in *U. beldingi*, whereas *C. lateralis* treat unrelated and related individuals similarly and agonistically. Studies to assess the costs and benefits of kin recognition in a number of other species are likely to produce similarly fruitful results.

A potential cost of living in such groups is the likelihood of negative actions, such as infanticide. Infanticide has been reported in a number of ground squirrels, and it is probably more common than has been observed. Initial reports suggested that it was an aberration; however, recent studies in the marmotines (*Cynomys* and *Urocitellus beldingi*) indicate that this behavior is most likely a strategy of females. In Belding's ground squirrels, infanticide appears to be a means of increasing genetic benefits by eliminating nonrelatives. In black-tailed prairie dogs, lactating females direct their infanticidal efforts at the offspring of related lactating females, and the benefits appear to be increased immediate nutrition (from eating the babies that are killed), reduced competition for resources, and potential cooperation in other duties—defending territory, watching for predators, and maintaining burrows—from marauded females (i.e., those whose offspring have been killed, and who thus no longer have infants to care for). Again, kinship seems to play an important role.

Foraging Ecology

Seed Predation and Herbivory

Although it would not be incorrect to suggest that squirrels eat everything, it would be quite misleading. Indeed, most species are opportunistic and typically include a variety of plant as well as animal material in their diets. Such diet breadth varies across and within taxa, and even across geography or season for the same species. Some species, however, are dietary specialists, such as the woolly flying squirrel (*Eupetauras cinereus*) from Pakistan, India, and Tibet, which feeds almost entirely on conifer needles, and the shrew-faced squirrel (*Rhinosciurus laticaudatus*) from the Malay Peninsula, which specializes on ants and termites. Others are bark gleaners (pygmy squirrels: *Nannosciurus, Exilisciurus, Myosciurus*, and *Sciurillus*), and at least a few species of *Tamiops* and *Funambulus* are reported to feed on nectar and, in the process, may serve as pollinators. Most squirrels are also known to include animal material in their diets: arthropods, bird eggs, baby birds, adult birds or mammals, and even conspecifics. Such carnivorous behavior and cannibalism, however, is usually restricted to periods of food shortages. Limiting nutrients (i.e., those that are less readily obtained, such as calcium and sodium) are also sought from various sources, such as hypogeous fungi, antlers, and bone.

Despite these specializations, many generalities exist. Tree squirrels, for example, are primarily granivores or frugivores, feeding mostly on seeds, nuts, and fruits. But within this general context many species show remarkable specializations, depending on the biomes or forest types in which they are found. In conifer forests, for example, tree squirrels will often concentrate on the seeds from pines, spruce, fir, and other evergreens. They do this by systematically stripping the bracts off of cones to get to the seeds at the base of the bracts, thereby acting predominantly (but not always) as seed predators in these forests. Many species have been shown to systematically remove, eat, and larder-hoard the entire cone crops of conifers.

Squirrels are highly selective with regard to feeding efficiency in these forests. For example, one of us has shown that fox squirrels (*Sciurus niger*) will systematically sample one or a few cones from most of the longleaf pines (*Pinus palustris*) within their home range and then systematically select the most profitable trees in which to concentrate their feeding activity. Similar patterns of cone selection have been reported in other tree squirrels that reside predominantly in conifer forests. Such patterns of selection often result in those trees that invest the most in female cone production experiencing the highest rates of seed predation. In

(a) A Siberian flying squirrel (*Pteromys volans*) foraging on European aspen (*Populus tremula*) leaves. Photo courtesy Marko Schrader. **(b) An Indian palm squirrel (*Funambulus palmarum*) feeding on nectar.** Photo courtesy Natasha Mhatre.

mixed conifer stands, C. C. Smith has noted that *Tamiasciurus* systematically remove and larder-hoard the cones from the most profitable species of trees, then concentrate on the next most useful species in the forest, and so on, until cones are harvested from all species.

Clearly the efficiency with which tree squirrels select and harvest cones places strong selection pressure on the trees, as the tree characteristics do on the squirrels. C. C. Smith first suggested that attributes such as cone hardness, most likely the result of serotiny, place strong selection on squirrels for a larger body size and more robust jaw musculature to handle the cones. However, such adaptations in squirrels also probably influence tree and cone characteristics, such as the orientation and strength of the point of attachment to the branch, as well as the number of seeds per cone, all of which help to reduce squirrel feeding efficiency and deter cone predation. Recent studies by C. W. Benkman and associates show how these evolutionary interactions between squirrels and conifers may have resulted in mosaics of selection across large geographic regions.

Tree squirrels that reside in hardwood forests and in many parts of the subtropical and tropical regions of the world also are seed predators, but they often double as both seed predators and seed dispersers. In these environments squirrels certainly exert selective pressures as seed predators, which is evidenced by the heavy outer seed coats of many nuts (e.g., *Carya* and *Juglans*). Indeed, for some tree species squirrels are their exclusive seed predators, suggesting a long evolutionary history between squirrels and these trees. For many genera of trees, however, tree squirrels also function as a primary agent of seed dispersal. This complicates the relationship; in a number of cases squirrels are agents of dispersal and forest regeneration, even though they depend on these seeds as a food resource (see below).

Squirrels are equally efficient and well adapted for herbivory. Although granivorous tree squirrels seasonally shift to herbivory only during periods of low food availability, ground squirrels and prairie dogs have mastered the art of specializing on plant material. However, at least one species of *Sciurus*, Abert's squirrel (*S. aberti*), has perfected the art of herbivory. Found throughout the range of ponderosa pine (*Pinus ponderosa*) in the southwestern USA, *S. aberti* is almost entirely dependent on this conifer species for its survival. The squirrel nests in the ponderosa and feeds on the immature female cones. But, after the cones mature and the seeds disperse, the squirrel shifts its diet to the nutrient-poor cambium and phloem of the terminal twigs of the tree. As shown by M. A. Snyder, Abert's squirrels are highly selective with respect to the trees in which they feed, eating the cambium of the same trees each year. This selectivity, based on the chemical characteristics of the xylem and phloem and the concentration of limiting nutrients (e.g., sodium in the phloem), results in repeated defoliation and strong selective pressure on the tree.

Scatter-Hoarding and Larder-Hoarding

Most species of squirrels, especially in temperate regions, store energy to make it through periods of food shortages and harsh conditions. Whereas hibernating squirrels usually store their energy in the form of body fat that enables them to reduce their metabolic rates and activity during these periods, many of the tree squirrels, in contrast, store their food and remain active. In North America, numerous studies on *Tamiasciurus* and *Sciurus* have been devoted to research on two alternative strategies for storing food: larder-hoarding and scatter-hoarding. At one extreme, larder-hoarding involves the placement of all food reserves in a single location that is then vigorously defended against all other competitors. *Tamiasciurus hudsonicus* and *T. douglasii* are the squirrels best studied and most well known for this. Both species construct extensive larders (middens) containing hundreds of cones that ensure their survival through periods of deep snowfall and low temperatures.

In contrast, many species of *Sciurus*—as well as other genera (e.g., *Sciurotamias*)—in temperate, subtropical, and even tropical regions engage in scatter-hoarding. This behavior, in its most extreme case, involves the placement of each of many items in individual, widely dispersed cache sites. Scatter-hoarding is thought to decrease the detection of food stores by competitors and thereby reduce the energy needed to defend the stores. The tradeoff, of course, is the extent to which the seeds can be spread before the costs of recovery become prohibitive. Scatter-hoarding squirrels, such as the eastern gray squirrel (*S. carolinensis*), appear to have mastered this technique.

As was first suggested by C. C. Smith for *Sciurus* and *Tamiasciurus* in North America, the type of plant food directly influences a squirrel's food-storage strategy, and this, in turn, can have a cascading effect on other aspects of squirrel biology, such as social behavior and even population dynamics. In coniferous forests, for example, cones are best stored in central middens, where they are protected from other squirrels and where cool, moist conditions prevent them from opening and releasing seeds. However, these central larder-hoards must be vigorously defended by an ever-vigilant owner. In *Tamiasciurus* this has led to a social system involving extreme territoriality, and a spatial system in which the habitat is divided evenly among the squirrels. Mothers often bequeath these territories and middens to their daughters. In deciduous forests, where several species of *Sciurus* reside, a different behavior is seen. Here squirrels scatter-hoard seeds and nuts in shallow pits just below the leaf litter. Scatter-hoards generally require no defense. Consequently, many species of *Sciurus* show overlapping home ranges and do not exhibit territoriality; their social system involves a dominance hierarchy in which dominance is often established at dense patches of food. Interestingly, where *Tamiasciurus* is found in mixed forests, it often shifts its behavior and scatter-hoards. However, in the coniferous forests of North America where *Sciurus* is found, these squirrels generally neither scatter-hoard nor larder-hoard. Hence the relationship between

(a) A red squirrel (*Tamiasciurus hudsonicus*) midden, mainly of spruce cones, in the White Mountains, Arizona. Photo courtesy Melissa Merrick. **(b) Another red squirrel (*Tamiasciurus hudsonicus*) midden, Mount Graham, Arizona.** Photo courtesy Claire Zugmeyer.

habitat and social structure is a strong but not a universal pattern.

Social ground squirrels and prairie dogs provide a completely different perspective regarding food type and social behavior, but one that is still consistent with the view above. These species generally feed on nondefensible food items (other than secondary metabolites) that do not store well. Instead, their strategy involves fat storage. However, some species (e.g., *Tamias*) use a combination of both food storage and the accumulation of body fat to overcome periods of food shortages and harsh environmental conditions.

Seed Dispersal

In the process of scatter-hoarding, squirrels—primarily *Sciurus*, but also members of several subtropical and tropical genera (e.g., *Sciurotamias*)—will store hundreds if not thousands of nuts or fruits per individual squirrel. Early observations led to the long-held notion that squirrels store more than they need and then forget where many seeds and nuts are cached, allowing some of them to germinate and become established. More recent research, however, suggests a far more complicated picture: one in which the squirrels are considerably more efficient at nut recovery, and one in which germination and establishment occur under far more limited circumstances than previously thought. This is not to say that scatter-hoarding squirrels are not critical for seed dispersal—they are. In fact, in many deciduous forests systems, as well as some subtropical and tropical forests, they may serve as keystone agents of seed dispersal. Experimental evidence strongly suggests that individual squirrels remember precise locations of their stored nuts, most likely based on spatial information, as is also shown for their avian counterparts, the scatter-hoarding corvids (rooks, jays, and crows). This is not to say that they don't steal from one another; pilfering is common and sometimes even extreme. Yet these scatter-hoarders may have domain over their own scatter-hoards in many situations.

The process by which scatter-hoarding squirrels disperse seeds, nuts, and fruits is rather involved. It is perhaps best understood in the eastern gray squirrel, which resides predominantly in oak forests throughout much of the central and eastern USA. Numerous experimental studies now demonstrate that this species is highly selective with respect to the nuts that are eaten and those that are stored. Acorns of red oak species, for example, are significantly preferred for scatter-hoarding over those of white oak. Typically, the acorns of red oak species are higher in fat content and tannin levels (which reduce the palatability and digestibility of plant foods), but they also exhibit dormancy prior to germination. White oak acorns, in contrast, have lower tannin and fat levels and germinate rapidly in the autumn, sometimes while still attached to the tree. Behavioral experiments by M. Steele, P. Smallwood, and others show that in the autumn, gray squirrels selectively consume white oak acorns, but at the same time they also selectively cache red oak acorns, primarily because of their delayed germination schedules and reduced perishability in the cache. Tannins and fats secondarily influence the squirrels' eating preferences, but overall the primary determinant of their fall caching decisions is the germination pattern.

Indeed, early germination in white oak acorns appears to have exerted a strong selective pressure on eastern gray squirrels. Several decades ago J. Fox demonstrated that when faced with heavy crops of white oak acorns, gray squirrels cache these more perishable food stores, but, when doing so, they carefully excise the small embryo at the apical end of the acorn with a few quick scrapes of their incisors. Interestingly, this embryo-excision of white oak acorns has been demonstrated recently in at least two other species of *Sciurus* (fox squirrels [*S. niger*] and Mexican gray squirrels [*S. aureogaster*]) from North America and at least one species of *Sciurotamias* (*S. davidianus* [Père David's rock squirrel]) from China. Moreover, in experiments with captive-raised eastern gray squirrels having no previous experience with acorns, they also attempt embryo-excision, but often perform it incorrectly or on the wrong part of an acorn. This suggests that the behavior may be largely innate. Although numerous other rodents—such as mice (*Peromyscus*), chipmunks (*Tamias striatus*), and southern flying squirrels (*Glaucomys volans*)—selectively cache red oak acorns, these species do not appear to perform embryo removal. Hence this excision behavior may be unique to one or a few lineages of squirrels that regularly scatter-hoard acorns.

Numerous other nut characteristics (e.g., nut size, nut mass, insect infestation, and food abundance) also influence the food-hoarding decisions of scatter-hoarding squirrels in many predictable ways. Ultimately, many of these behavioral decisions in turn influence where nuts are cached and the likelihood that the nuts will germinate and establish if they are not recovered. For example, we currently know that the sites frequently selected by eastern gray squirrels for scatter-hoarding are often coincidentally optimal for germination, because these are also the sites where acorns store well. We also know that the probability of seedling establishment increases during bumper-crop (or mast) years, when animals store large quantities of seeds and nuts and some are not recovered. What we don't yet understand is how other processes, such as interactions with competitors (potential cache pilferers) and predators, indirectly influence cache recovery and seedling establishment. Despite this fascinating interrelationship between squirrels and trees, we still have a great deal to learn before we fully understand how scatter-hoarding squirrels contribute to forest regeneration. That they do this is undeniable. How, when, and the precise mechanisms by which they scatter-hoard are still largely unanswered. However, research to date provides good evidence that squirrels are critical agents of seed dispersal for at least nine genera of hardwood trees (*Aesculus, Carya, Castanea, Castanopsis, Corylus, Fagus, Juglans, Lithocarpus*, and *Quercus*) and two species of conifer (*Pinus koraiensis* and *P. jeffreyi*).

Dispersal of Fungi

Squirrels frequently supplement their diets with a variety of fungi. When these fungi include hypogeous, endorhyzal species, squirrels insert themselves into a three-way mutualism that is often crucial for forest function and regeneration. Endorhyzal fungi are closely associated with the roots of many tree species: the tree provides critical metabolic byproducts for fungal growth, whereas the fungi sequester vital nutrients from the soil that are essential for tree growth. The fungi clearly cannot live without the tree, and the tree's growth is greatly inhibited without its association with the fungi. However, these fungi fruit underground, and their spores appear to be solely dependent on animals (primarily rodents) for their dispersal. Rodents, especially squirrels, feast on these truffles when other foods are not available or when critical nutrients (e.g., calcium) are needed. The fungal spores are then readily dispersed in the rodents' feces. The most thoroughly studied relationship is that of the sciurids of the Pacific Northwest in the USA and the moist, temperate, conifer forests in which they reside. The northern flying squirrel (*Glaucomys sabrinus*), in particular, is a truffle specialist and, as a consequence, an essential contributor to forest function. A similar relationship has been suggested for the southeastern fox squirrel (*Sciurus niger niger*) in the longleaf pine forests of the southeastern coastal plain of the USA. Although this relationship between squirrels, fungi, and trees has only been explored in a few systems, it is likely to exist in numerous forests throughout the world.

Ecosystem Services and Conservation

The closely knit relationships between squirrels and plants demonstrate the substantial ecosystem services they provide. A myriad of values are ascribed to squirrels. While

A Himalayan striped squirrel (*Tamiops mcclellandii*, foreground) and Finlayson's squirrels (*Callosciurus finlaysonii*) at a market in Thailand. Photo courtesy Israel Didham.

squirrels can cause tree damage, depredate crops, gnaw electrical and telephone wires, burrow beneath human structures, pilfer seeds, and serve as vectors for disease, their significance in forest and grassland ecosystems undoubtedly outweighs their potential costs. Squirrels perform considerable ecosystem services—such as seed planting and dispersal, pollination, and fungal spore dispersal—and they may serve as indicator species for the health of some ecosystems.

Some squirrels are seen as keystone species, due to their disproportionate influence on their ecosystems, often because of their role as ecosystem engineers that structurally change the environment. Examples include marmots (*Marmota*), prairie dogs (*Cynomys*), and some ground squirrels that create burrows and maintain shortgrass habitats that are used by many other species. Red squirrels (*Tamiasciurus hudsonicus*) and Douglas's squirrels (*T. douglasii*) form large piles of conifer-cone debris and cache newly harvested cones within these middens, and many animals use the resulting structures and food concentrations.

Squirrels also form a significant prey base for a multitude of other species. Subsistence and sport hunting of squirrels, as well as trapping for pelts, occur in many areas, ranging from the use of *Callosciurus* for food in Southeast Asia to the millions of pelts of *Sciurus* and *Marmota* harvested annually throughout Eurasia. Additionally, squirrel byproducts are collected for their potential medicinal value, such as the droppings of *Trogopterus* (in China) and *Eupetaurus* (in Pakistan and India), and the oils of *Marmota* in Asia.

Squirrels serve as models for human functions, pathologies, and diseases: *Funambulus* (for metabolism and reproduction), *Sciurus niger* (for metabolic disorders), *S. carolinensis* (for the lens and color vision in the human eye), *Marmota monax* (for the hepatitis virus and related carcinomas, and for atherosclerosis), and *Ictidomys tridecemlineatus* (for cholesterol-induced pathology and atherosclerosis). Moreover, numerous species provide examples for environmental education. In some countries, encouraging recreational observations of squirrels has also generated a lucrative industry (examples of some consumer products are nest boxes and squirrel feeders) and that, along with the legal pet trade, provides a substantial economic impact. As a result, squirrels can produce significant financial benefits.

Meanwhile, in terms of conservation, most major ecosystems have been fragmented through human activities during the last 200 years. In particular, the forested and grassland ecosystems inhabited by the squirrels have undergone excess levels of loss and fragmentation, with worldwide deforestation rates of 0.33 percent per year and grassland system reductions at a rate of 0.08 percent per year—a combined habitat loss of over 4 percent each decade at the current rates.

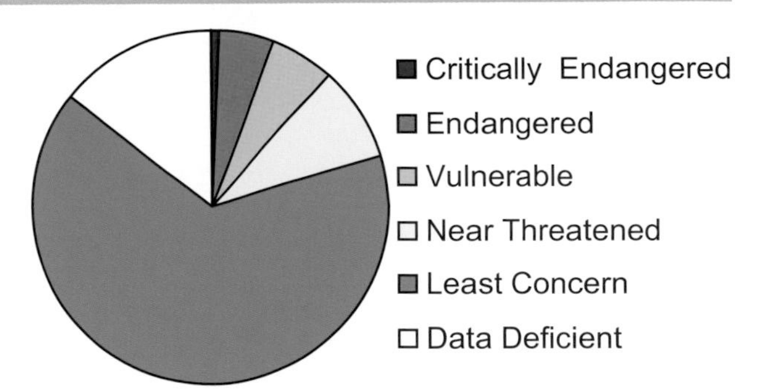

IUCN level of imperilment for all species of the Sciuridae.

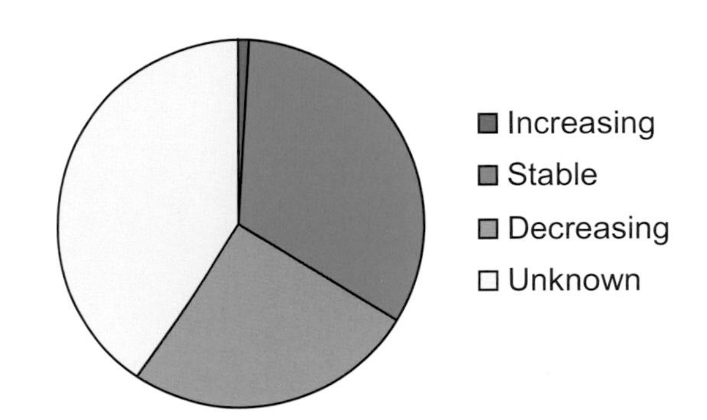

IUCN assessment of the population trends for all species of the Sciuridae.

Habitat fragmentation has important consequences for the biotic components of ecosystems, often changing genetic variation, population densities, home range size, reproductive performance, and biodiversity. The consequences of habitat loss and fragmentation can be profound.

Awareness of the conservation status of squirrels thus takes on great importance when viewed in the context of the ecological significance of the taxa. We know that the major threats to their persistence are habitat loss and fragmentation, overexploitation and persecution, and invasive species. When we assess the level of imperilment of squirrels using the IUCN *Red List* database, we see that most species are considered to be of least concern with respect to their conservation status. This presents a positive outlook for the conservation of squirrels. However, about one in five species is either of elevated conservation risk or so poorly known that an assessment cannot be rendered.

The state of our knowledge of the population trends for the members of the Sciuridae is of additional concern. According to the IUCN, more than 60 percent of the species are either decreasing (~ 25%) or the trend is unknown (~ 35%); only just over 35 percent of the species are considered to be stable or increasing. Clearly there is much work to be done toward the conservation of squirrels.

SPECIES ACCOUNTS

Subfamily Ratufinae, Moore, 1959

This subfamily—comprising the giant tree squirrels (*Ratufa*) of southern Asia—includes four species of squirrels: two of these are restricted to India and Sri Lanka, and the other two range from Nepal and southern China to the islands of Borneo, Java, and Sumatra. These giant tree squirrels were recognized as a separate tribe within the family Sciuridae on the basis of their morphology, and they are considered one of the two most identifiable genera of squirrels. Molecular evidence confirmed their distinctness and led to them being treated as the subfamily Ratufinae.

Ratufa Gray, 1867

This genus contains four species. They are canopy dwellers in tropical rainforests, nesting and foraging high in the trees and feeding on fruits, seeds, and leaves.

Ratufa affinis (Raffles, 1821)
Pale Giant Squirrel

DESCRIPTION: This is the brown giant tree squirrel of the Sunda Shelf. It is pale on the abdomen and darker on the back, at least in the midline.

SIZE: Female—HB 342.2 mm (n = 24); T 423.6 mm (n = 10); Mass 1236 g (n = 17).

Male—HB 335.2 mm (n = 30); T 409.3 mm (n = 50); Mass 1064.4 g (n = 31).

Sex not stated—HB 337.6 mm (n = 34); T 411.1 mm (n = 13); Mass 1120.7 g (n = 30).

DISTRIBUTION: This species is found through the Malay Peninsula, Sumatra (Indonesia), the Island of Borneo (divided among Malaysia, Brunei Darussalam, and Indonesia), and adjacent small islands.

GEOGRAPHIC VARIATION: Nine subspecies are recognized.

R. a. affinis—peninsular Thailand, peninsular Malaysia, and Singapore. It is pale below and uniformly brown above, with paler animals in the southern part of its range. None are as dark as the animals from the island of Borneo.

R. a. bancana—Bangka Island (Indonesia). This subspecies resembles *R. a. polia*, but it has lighter forefeet, less white on the head, and smaller teeth.

R. a. baramensis—Sabah, Sarawak, and Banggi Island (Malay-

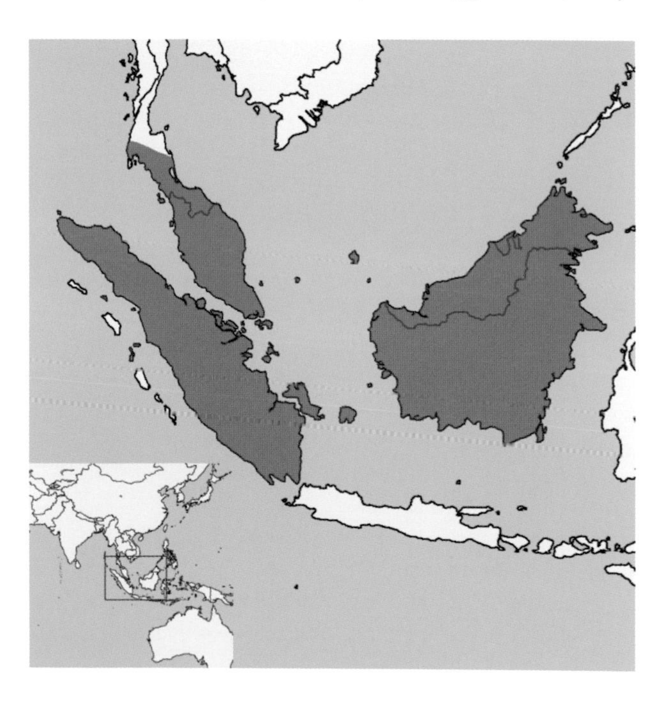

***Ratufa affinis*.** Photo courtesy Nick Baker, www.ecologyasia.com.

***Ratufa affinis*.** Photo courtesy Morten Strange.

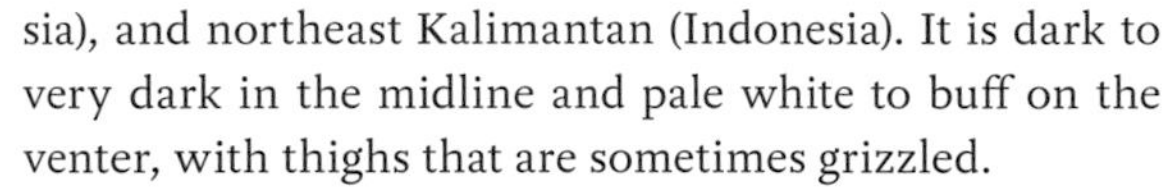

sia), and northeast Kalimantan (Indonesia). It is dark to very dark in the midline and pale white to buff on the venter, with thighs that are sometimes grizzled.

R. a. bunguranensis—Bunguran Island (Indonesia), in the northern Natuna Islands, between the island of Borneo and the Malay Peninsula. It resembles the Malayan forms, with a uniform brown dorsal pelage.

R. a. cothurnata—West Kalimantan (Indonesia). This form resembles *R. a. baramensis*, but with pale ungrizzled thighs.

R. a. ephippium—southeastern Kalimantan (Indonesia). This subspecies is very dark in the dorsal midline; otherwise it is similar to *R. a. cothurnata*.

R. a. hypoleucos—Sumatra (Indonesia). The top of the head is mixed with gray; the nose and forehead are gray; the throat, sides of the head, cheeks, and ventral surface of the body are white; and the dorsal surface of the tail is darker, with a yellowish tip.

R. a. insignis—Riau Islands (Indonesia). The upperparts are burnt umber; the underparts and inner surface of the limbs are cream buff tinged with brownish yellow, particularly on the front legs, and are lined by a light tawny ochraceous coloration.

R. a. polia—Billiton (= Belitung) Island (Indonesia). This subspecies resembles *R. a. ephippium*, but it has a distinctly gray or dirty white head, with more grizzled and less reddish sides.

CONSERVATION: IUCN status—near threatened. Population trend—decreasing.

HABITAT: This species is found in the evergreen broadleaf dipterocarp and lower montane forests of the Sunda Shelf.

NATURAL HISTORY: The pale giant tree squirrel lives high in the canopy, commonly between 20 and 40 m, where it makes a spherical nest of twigs and small branches in the crown of a tall tree. It feeds on seeds; occasionally on fruit pulp, sap, or bark; and uncommonly on flowers and leaves. It rarely descends to the ground and only very infrequently enters plantations.

GENERAL REFERENCES: J. B. Payne 1980.

Ratufa bicolor (Sparrman, 1778)
Black Giant Squirrel

DESCRIPTION: This is a black-and-white giant tree squirrel, with a white or buff bib and white or buff on the side of the face. The buff throat and cheeks are separated by a black mustache mark. On the back the hair is black or (in some subspecies) frosted. On the ventral surface the hair color ranges from creamy buff to reddish yellow.

SIZE: For most subspecies:

Female—HB 365.4 mm (n = 24); T 423.6 mm (n = 10); Mass 1236 g (n = 17).

Male—HB 335.2 mm (n = 30); T 409.3 mm (n = 50); Mass 1064.4 g (n = 31).

Sex not stated—HB 337.6 mm (n = 34); T 411.1 mm (n = 13); Mass 1120.7 g (n = 30).

For the diminutive subspecies *R. b. condorensis*:

Sex not stated—HB 304.8 mm (n = 6); T 324 mm (n = 6).

DISTRIBUTION: This species is found through eastern Nepal and southeast Tibet to southern Yunnan and Hainan (China), Assam (India), Myanmar, Thailand, Laos, Cambodia, and Vietnam, and south through the Malay Peninsula to Java and Bali (Indonesia).

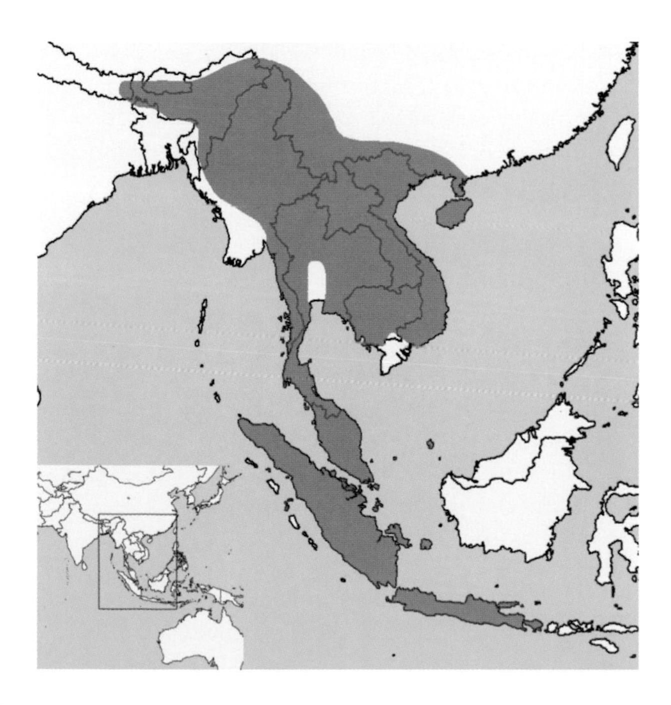

GEOGRAPHIC VARIATION: Eleven subspecies are recognized.

R. b. bicolor—Java and Bali (Indonesia). It lacks ear tufts, and it has a flash pattern (i.e., an extension of the buff coloration of the inner surface of the forearm onto the dark dorsal surface) on the forearms. This subspecies has very distinctive pelage, with pale-tipped hairs on the dorsum and completely pale hairs on the tail, making the tail appear white. It tends to be paler on the head and shoulders, darker on the back, and lacks a black mustache line.

R. b. angusticeps—Myanmar and Thailand south through peninsular Malaysia and the Natuna Islands (Indonesia). It is virtually indistinguishable from *R. b. leucogenys*, *R. b. melanopepla*, and *R. b. phaeopepla*. It lacks ear tufts, and it exhibits a flash pattern on the forearm. Island forms on the western sides of Myanmar, Thailand, and Malaysia are darker black on the back and darker reddish yellow on the belly, in contrast with the mainland forms.

R. b. condorensis—Côn Son Island (Vietnam), south of the mainland. It is distinctive because of its small size. Otherwise it is virtually identical to the mainland form, *R. b. leucogenys*.

R. b. felli—small area in northern Myanmar, south and east of the Chindwin River. It differs from *R. b. gigantea* in lacking ear tufts, as well as in having a flash pattern on the forearms. It differs from all neighboring subspecies in having pale-tipped hairs on the back.

R. b. gigantea—widespread through the northern part of the range. It is distinguished by its ear tufts and its lack of a flash pattern on the forearm. On the back these animals are more brown than black. On the venter they range from creamy buff to yellow cream.

R. b. hainana—Hainan Island (China). This form is similar to *R. b. gigantea*, but it is darker on the back, almost black in color; it is also darker (a reddish yellow color) on the abdomen.

R. b. leucogenys—Myanmar and Thailand south through peninsular Malaysia and the Natuna Islands (Indonesia). This subspecies is virtually indistinguishable from *R. b. angusticeps*, *R. b. melanopepla*, and *R. b. phaeopepla*. It lacks ear tufts, and it exhibits a flash pattern on the forearm. Island forms on the western sides of Myanmar, Thailand, and Malaysia are darker black on the back and darker reddish yellow on the belly, in contrast with the mainland forms.

R. b. melanopepla—Myanmar and Thailand south through peninsular Malaysia and the Natuna Islands (Indonesia). It is virtually indistinguishable from *R. b. angusticeps*, *R. b. leucogenys*, and *R. b. phaeopepla*. It lacks ear tufts, and it exhibits a flash pattern on the forearm. Island forms on

Ratufa bicolor. Photo courtesy Pitchaya and Rattapon Kaichid.

the western sides of Myanmar, Thailand, and Malaysia are darker black on the back and darker reddish yellow on the belly, in contrast with the mainland forms.

R. b. palliata—Sumatra (Indonesia). It is similar to *R. b. bicolor*, but the tail hairs are only pale-tipped, and the head and shoulders are darker than the back.

R. b. phaeopepla—Myanmar and Thailand south through peninsular Malaysia and the Natuna Islands (Indonesia). It is virtually indistinguishable from *R. b. angusticeps*, *R. b. leucogenys*, and *R. b. melanopepla*. It lacks ear tufts, and it exhibits a flash pattern on the forearm. Island forms on the western sides of Myanmar, Thailand and peninsular Malaysia are darker black on the back and darker reddish yellow on the belly, in contrast with the mainland forms.

R. b. smithi—western Vietnam. It has ear tufts, a flash pattern on the forearm, and frosted dorsal pelage.

CONSERVATION: IUCN status—near threatened. Population trend—decreasing.

HABITAT: This species prefers the evergreen and semi-evergreen broadleaf forests of Indochina and the Sunda Shelf.

NATURAL HISTORY: During a study in Kuala Lompat (Malaysia), *Ratufa bicolor* lived in the middle and upper canopy, seldom below 20 m. Based on feeding time, the squirrels fed extensively on seeds (61%) and pulp (20%), with figs accounting for half of the pulp. The rest of their diet included flowers, leaves, and sap/bark, but no insects. They did not store food. The median rate of travel for animals when foraging and feeding was 30 m/hour, ranging from a median 65 m/hour in the morning to 10 m/hour at midday and 50 m/hour at the end of the day. The estimated median for total distance traveled was 315 m/day. The population density of the squirrels was 11 individuals/100 ha. Nests (dreys) were constructed with long leafy twigs, more than 31 m above the ground in the canopy, commonly in the outer branches of *Koompassia* or *Shorea* trees. These squirrels usually exited their nests in the morning, between 6 and 8 a.m., but after rainy nights they might exit as late as 10 a.m. They were most active between dawn and 10 a.m., with a lesser peak of activity between 2 and 4 p.m. The median time at which they reentered their nests was 6:01 p.m. When attacked by a Crested Serpent Eagle (*Spilornis cheela*), a squirrel ran about in the tree calling loudly, with its hair piloerected, as if to intimidate the bird. Breeding follows a mating chase, apparently involving one male and the estrous female. Litter size averages 1.44, and females may have two litters per year. On peninsular Malaysia, pregnant *Ratufa* were found between April and September, suggesting that parturition occurs at the beginning or during the peak of the fruiting season.

GENERAL REFERENCES: Moore and Tate 1965; J. B. Payne 1980.

Ratufa indica (Erxleben, 1777)
Indian Giant Squirrel

DESCRIPTION: This dramatic giant squirrel has large red or maroon ear tufts that are approximately 20 mm long, with the back and tail predominantly maroon and black.

SIZE: Female—HB 365.4 mm (n = 45); T 462.7 mm (n = 45); Mass 1807.5 g (n = 8).

Male—HB 362.2 mm (n = 51); T 447.3 mm (n = 50); Mass 1678.3 g (n = 15).

Sex not stated—HB 366.8 mm (n = 44); T 456.7 mm (n = 11); Mass 1442.0 g (n = 33).

DISTRIBUTION: This species occurs through central and southern India, excluding the central lowlands.

GEOGRAPHIC VARIATION: Four subspecies are recognized.

R. i. indica—Western Ghats, south of Mumbai (India). It is maroon on the head, body, sides, and tail, with a pale tail tip.
R. i. centralis—eastern and central India. It has red ear tufts; red on the back, sides, and hind legs; and black on the forelimbs, shoulders, and tail.
R. i. dealbata—Dangs region, north of Mumbai (India). It is a very pale form, with a white tail.
R. i. maxima—southwestern tip of India. This subspecies is similar to *R. i. indica,* except that it is black across the shoulders and on the rump and tail.

CONSERVATION: IUCN status—near threatened. Population trend—decreasing. *Ratufa indica dealbata* may be extinct.

HABITAT: This species prefers the evergreen and semi-evergreen broadleaf forests of peninsular India.

NATURAL HISTORY: The Indian giant tree squirrel is solitary, territorial, and a facultative frugivore. When fruit is rare, it feeds on leaves, flowers, seeds, and bark. At one site (Magod), its intake of mature leaves ranged from 0 percent of the diet when fruit was readily available to 62.9 percent when fruit was scarce. At Bhimashankar, mature leaves constituted 4.5 to 18.8 percent of their diet. In a low-diversity seasonal cloud forest in the Western Ghats (India), these squirrels collected hard nuts of six species of trees and one species of vine, which they stored in a larder-hoard within their nests. At the Parambikulam Wildlife Sanctuary in Kerala, the squirrels are principally seed feeders, consuming the seeds of 10 different species of trees. The Indian giant tree squirrel builds nests made of twigs and leaves high in the canopy; in semi-deciduous woodlands it may choose vine-covered trees in riparian locations. Population densities are estimated to vary from 2.4 individuals/km^2 at Bandipur to 12.3 individuals/km^2 at Lakkavalli, although local densities at Magod appear to have been much higher, with 11 animals in an area of 25 ha and territory sizes of 1.2 ha for

Ratufa indica. Photo courtesy Sudhir Shivaram.

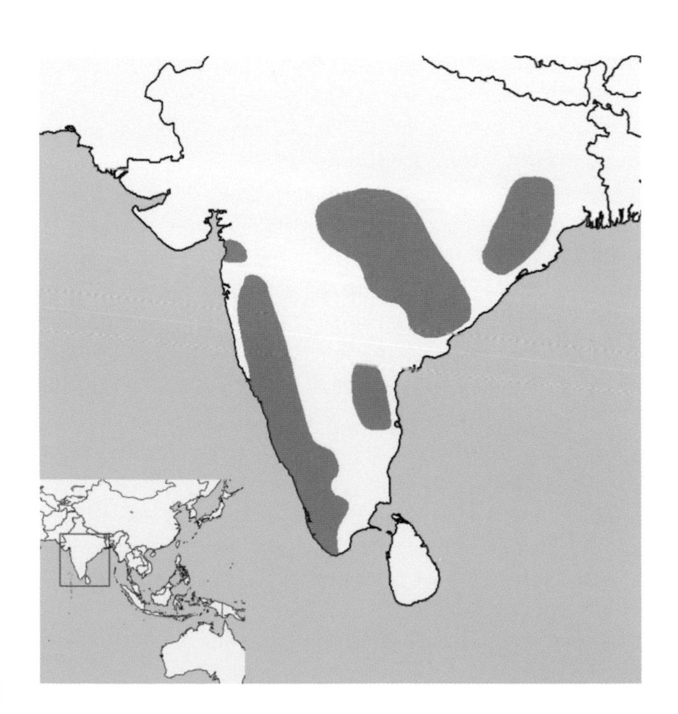

males and 1.1 ha for females. This species is preyed upon by a number of mammals and birds. Researchers have observed attempts at or successful predation by lion-tailed macaques (*Macaca silenus*), leopards (*Panthera pardus*), martens (*Martes*), Forest Eagle Owls (*Bubo nipalensis*), Black Eagles (*Ictinaetus malayensis*), Crested Serpent Eagles (*Spilornis cheela*), and Crested Hawk-Eagles (*Nisaetus cirrhatus*). Birds of prey may be mobbed by the squirrels, and in areas where aerial predators are particularly common, the squirrels may rest during the middle of the day within their nests, which is suggested to be an antipredator strategy. Litter size is normally one (n = 2), and gestation length is reported to be 2.5 months (n = 1).

GENERAL REFERENCES: Borges 1990, 1992, 1993, 1998, 2006; Borges et al. 2006.

Ratufa macroura (Pennant, 1769)
Sri Lankan Giant Squirrel

DESCRIPTION: This species of giant squirrel has a black or blackish crown, with a pale band between the ears separating the crown from the nape of the neck. The digits are also black or blackish; and the dorsum of the hands plus the feet, forearms, and ankles are yellowish. It has small short ear tufts.

SIZE: Female—HB 338.1 mm (n = 22); T 350.3 mm (n = 9); Mass 1600.0 g (n = 10).

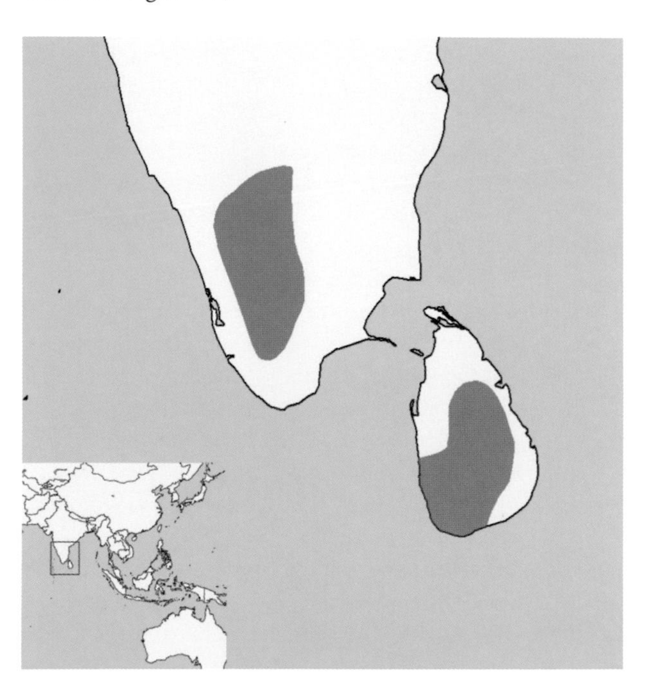

Ratufa macroura**.** Photo courtesy Sudhir Shivaram.

Male—HB 335.3 mm (n = 21); T 360.5 mm (n = 10); Mass 1610.0 g (n = 5).

Sex not stated—HB 362.0 mm (n = 1).

DISTRIBUTION: This species is found in southern India and Sri Lanka.

GEOGRAPHIC VARIATION: Three subspecies are recognized.

R. m. macroura—central Sri Lanka. It is distinguished by its black tail with white hair tips, and by a gray band separating the black back from the yellowish venter.

R. m. dandolena—southern India and northcentral and eastern Sri Lanka. This subspecies has a black crown; the shoulders and dorsal midline are variably black, but the lower back is brown. It has long pale hairs on the tail, making the tail appear grayer than in the other subspecies.

R. m. melanochra—southwestern Sri Lanka. It has a black back and tail, but it lacks the white hair tips and gray line separating the back from the yellowish venter.

CONSERVATION: IUCN status—near threatened. Population trend—decreasing.

***Ratufa macroura*.** Photo courtesy James Eaton/Birdtour Asia.

HABITAT: This species favors the evergreen broadleaf forests of southern India and Sri Lanka.

NATURAL HISTORY: On the eastern side of the Western Ghats (in Kerala, Tamil Nadu, and Karnataka), the Sri Lankan giant squirrel lives in riparian forests dominated by *Terminalia, Tamarindus, Mangifera, Pongamia, Albizia,* and *Syzygium*. It feeds on the young leaves of *Tamarindus,* and on the young leaves, pollen, and bark of *Albizia*.

GENERAL REFERENCES: Karthikeyan et al.1992; Ramachandran 1989.

Subfamily Sciurillinae Moore, 1959

This subfamily has just one genus.

Sciurillus Thomas, 1914

This genus contains a single species of pygmy squirrel.

Sciurillus pusillus (É. Geoffroy, 1803) Neotropical Pygmy Squirrel

DESCRIPTION: Neotropical pygmy squirrels have a gray dorsum suffused with yellow. They are cinnamon to red on the head; they have prominent white to buff postauricular patches, and black-tipped ears. The venter is gray to buff.

SIZE: Both sexes—HB 89-115 mm; T 89-120 mm; Mass 33-45 g.

DISTRIBUTION: This species is found in Peru, Colombia, Venezuela, Guyana, Suriname, French Guiana, and Brazil.

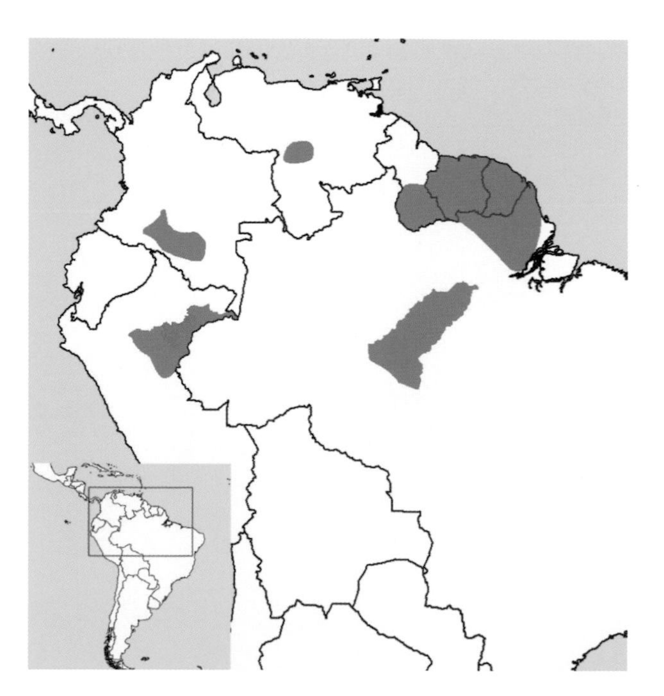

GEOGRAPHIC VARIATION: Three subspecies are recognized.

S. p. pusillus—Guyana, Suriname and French Guiana. This form is slightly more buff or brown, and the head is more reddish.

S. p. glaucinus—lower Amazon of Brazil. This is a paler form.

S. p. kuhlii—upper Amazon of Peru. This subspecies typically lacks postauricular patches, and it is more buff to brown.

***Sciurillus pusillus*.** Photo courtesy Edward Massiah.

CONSERVATION: IUCN status—data deficient. Population trend—no information.

HABITAT: Neotropical pygmy squirrels prefer mature evergreen lowland Amazon forests.

NATURAL HISTORY: This species is diurnal. Pregnant and lactating females have been documented in June with a litter size of one or two. Solitary individuals or family groups are seen feeding at nearly every layer of the canopy, but *S. pusillus* appears to be common around 10 m. Some groups contain more than one adult. Agonistic chases are common. Energetic and excitable, adult and juvenile neotropical pygmy squirrels quickly dart through the canopy and ascend and descend large trees. A major food resource is the sap and other exudates of the large trees that these squirrels frequent on a regular basis, and food trees are often distinguished by an accumulation of bark chips at the base. Neotropical pygmy squirrels tolerate modest levels of forest fragmentation. Their alarm call is given at varying levels of intensity, but it sounds much like a cricket.

GENERAL REFERENCES: Anthony and Tate 1935; Dalecky et al. 2002; Hafner et al. 1994; Heymann and Knogge 1997.

Subfamily Sciurinae Fischer de Waldheim, 1817

This subfamily contains two tribes, the Sciurini and the Pteromyini.

Tribe Sciurini Fischer de Waldheim, 1817

This tribe includes 20 genera.

Microsciurus J. A. Allen, 1895

This genus contains four species of dwarf squirrel.

Microsciurus alfari (J. A. Allen, 1895)
Central American Dwarf Squirrel

DESCRIPTION: Central American dwarf squirrels have a dull-colored back that is olivaceous brown to olivaceous black, sometimes with a reddish tinge. The venter is dull buff to gray to pale orange buff. Often there are white spots at the base of the ears. The tail is dark olivaceous brown at the core, but frosted with orange. The limbs are relatively long for the body length.

SIZE: Both sexes—HB 108-146 mm; T 80-130 mm; Mass 72-105 g.

DISTRIBUTION: This species is found in far southern Nicaragua, Costa Rica, Panama, and Colombia.

GEOGRAPHIC VARIATION: Six subspecies are currently recognized. However, this is probably in need of revision.

M. a. alfari—northeast Costa Rica, in the vicinity of the Turrialba Volcano. It is more rufescent than *M. a. venustulus.*
M. a. alticola—northcentral Costa Rica. This highland form has long dense fur and a modestly large size.
M. a. browni—southwest Costa Rica and northwest Panama. This subspecies is a low-elevation long-tailed variant.
M. a. fusculus—extreme southeastern Panama and Colombia. The upperparts are blackish brown; the face, sides, and chest are tawny or tawny ochraceous.
M. a. septentrionalis—northcentral Costa Rica and marginally in Nicaragua. This form is less rufescent on the dorsum than neighboring subspecies.
M. a. venustulus—central Panama and the Caribbean coast of western Panama. This subspecies is less rufescent than

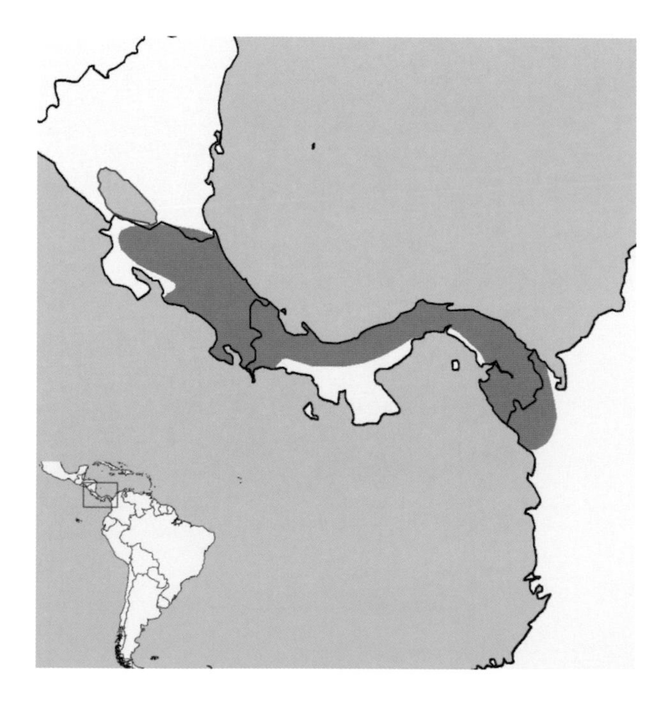

Microsciurus alfari. Photo courtesy Rhett Butler.

M. a. alfari: the upperparts are brownish black, grizzled with cinnamon rufous or rusty red, and it is more cinnamon rufous on the head and cheeks, with dark rusty brownish feet. The underparts are grayish brown.

CONSERVATION: IUCN status—least concern. Population trend—stable.

HABITAT: Central American dwarf squirrels can be common in dense evergreen forests, including cloud forests, at elevations up to about 2600 m. They are sometimes seen in disturbed forests.

NATURAL HISTORY: This species is diurnal. It forages on the ground and in the canopy, eating plant sap and exudates, fruits, and insects. *M. alfari* appears to be seasonally monoestrus, and these squirrels are occasionally seen in pairs. Central American dwarf squirrels scurry quickly across the ground, through the canopy, and along lianas or tree trunks. They are graceful and bold in their movements. *M. alfari* is rather silent but will make a series of "chatters" and a high-pitched trill.

GENERAL REFERENCES: J. A. Allen 1915; Emmons and Feer 1990; Fleming 1973; Giacalone et al. 1987; F. A. Reid 1997; R. W. Thorington and Heaney 1981; R. W. Thorington and Santana 2007.

Microsciurus flaviventer (Gray, 1867) Amazon Dwarf Squirrel

DESCRIPTION: Amazon dwarf squirrels have a dark brown dorsum suffused with reddish to olivaceous tones, and a pale yellow postauricular patch. The venter is grayish washed with orange to striking pale or deep orange. The tail is grizzled brown to black with a slight frosting of steel gray.

SIZE: Both sexes—HB 120-160 mm; T 96-150 mm; Mass 60-128 g.

DISTRIBUTION: This species is found in the Amazon Basin of Colombia, Ecuador, Peru, Brazil (west of the Negro and Juruá rivers), and marginally in Bolivia.

GEOGRAPHIC VARIATION: Eight subspecies are recognized.

M. f. flaviventer—eastern portion of the species' distribution in Brazil. The pelage is blackish olive, with the middle of the dorsum being blacker. The venter and the inner side

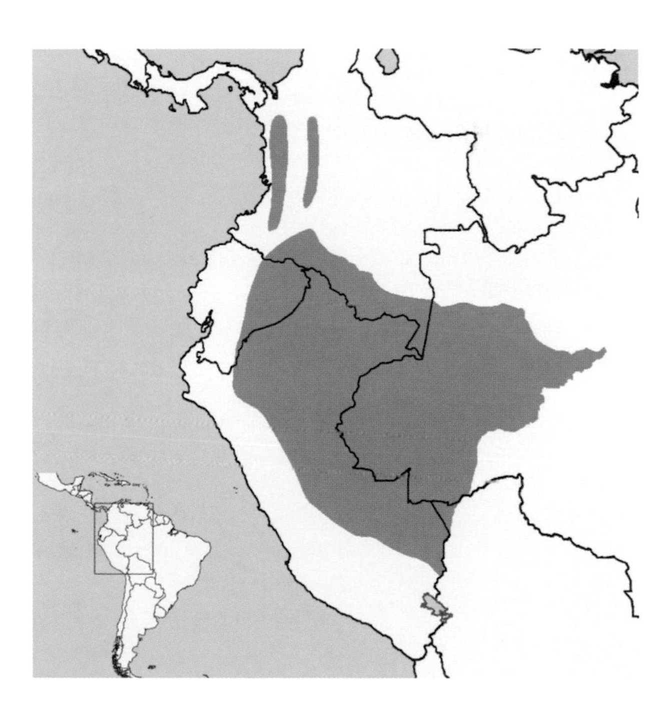

Microsciurus flaviventer. Photo courtesy Kenneth Chiou.

of the limbs are reddish yellow, and the tail is washed with yellow.

M. f. napi—Ecuador and southern Colombia, east of the Andes. This subspecies is similar to *M. f. peruanus* in coloring and postauricular patches, but it can be distinguished by the more faintly yellowish rufous color of the venter.

M. f. otinus—mountains of central and northern Colombia. The ears are tipped with white, and the tail is frosted whitish.

M. f. peruanus—northwestern Peru. This subspecies is yellowish rufous below, with large white postauricular patches.

M. f. rubrirostris—central Peru. This is a large form, with a long tail frosted in yellow, and an orange ochraceous venter.

M. f. sabanillae—southern Ecuador. This is a large form, with an ochraceous venter, and a lack of postauricular patches.

M. f. similis—southern Colombia. This subspecies is orange rufous below, and it lacks patches on or behind the ears.

M. f. simonsi—central Ecuador. This dark unspotted form has a yellow eye ring, as well as a fulvous venter.

CONSERVATION: IUCN status—data deficient. Population trend—no information.

HABITAT: Amazon dwarf squirrels inhabit upper Amazonian evergreen tropical rainforests, at elevations less than 2000 m.

NATURAL HISTORY: This form is diurnal. It may nest in typical leaf nests (dreys)—composed of woven leaves lined with fibers—in palm trees. Litter size has been reported to be two young. Occasionally these squirrels are seen in pairs; they forage on the ground and throughout all levels of the multistoried canopy. They are not found in flooded forests, but are more common in terra firma forests. They rarely use treefalls, gaps, or liana forests, but they are common within the high forest, usually foraging at heights of 1-5 m. They feed while sitting on logs or tree boles. Amazon dwarf squirrels make use of very small and very large trees as they hurriedly move through the forest, and they appear to actively search for insects and other arthropods as they go. Amazon dwarf squirrels follow mixed flocks of insectivorous birds in the morning, in order to enhance their own success in gleaning insects from the forest. Individuals gnaw on tree branches and trunks in apparent attempts to extract exudates. *M. flaviventer* is occasionally hunted along roads. These squirrels are not considered a nuisance.

GENERAL REFERENCES: J. A. Allen 1914, 1915; Buitrón-Jurado and Tobar 2007; Emmons and Feer 1990; Haugaasen and Peres 2005; Mena-Valenzuela 1998; Youlatos 1999.

Microsciurus mimulus (Thomas, 1898)
Western Dwarf Squirrel

DESCRIPTION: Western dwarf squirrels have a grizzled brown dorsum suffused with pale yellow or orange buff, sometimes with a black median line. The venter is a pale to

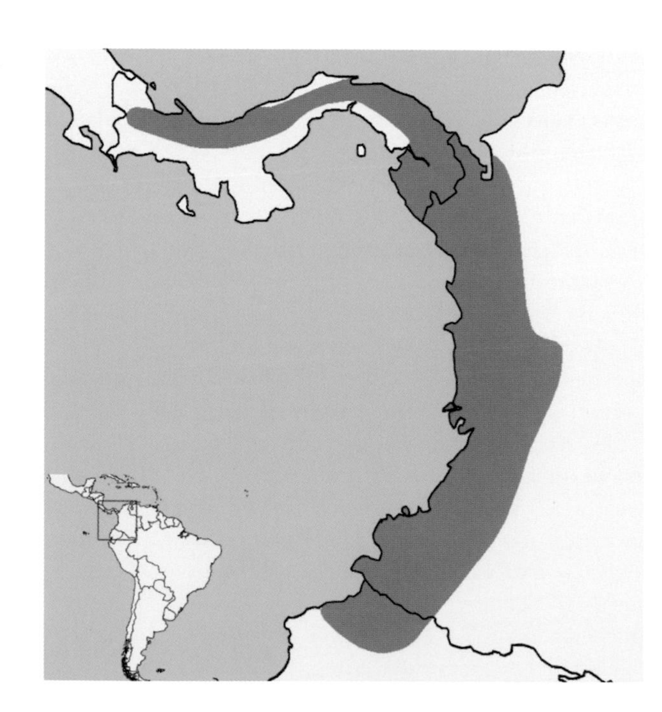

Microsciurus mimulus. Photo courtesy Luis G. Olarte.

red orange. The tail is grizzled black brown with a slight frosting of yellow or gray (in darker forms). The limbs are relatively elongated.

SIZE: Both sexes—HB 135-148 mm; T 94-116 mm; Mass 120 g.

DISTRIBUTION: This species is distributed in Panama, western Colombia, and northwestern Ecuador.

GEOGRAPHIC VARIATION: Three subspecies are recognized.

M. m. mimulus—southern half of the range, from northern Ecuador into Colombia. This form sometimes shows a black median dorsal line.

M. m. boquetensis—nearly all of Panama, and extending into northern Colombia. It is somewhat more pale dorsally, and ochraceous rufous ventrally.

M. m. isthmius—primarily in the central portion of the range, in extreme southwestern Panama and Colombia. This subspecies is more yellowish on the dorsum, with a reddish color apparent near the posterior.

CONSERVATION: IUCN status—least concern. Population trend—stable.

HABITAT: Western dwarf squirrels are found in evergreen foothills and highland forests, at elevations above 800 m in Panama, but at lower elevations in Colombia and Ecuador.

NATURAL HISTORY: This species is diurnal. It is occasionally seen in pairs. Western dwarf squirrels forage rapidly along the ground and in the lower forest canopy, perhaps foraging for insects. Where it occurs in sympatry with *M. alfari*, *M. mimulus* is found in higher-elevation forests; but where it is found with *Sciurus pucheranii*, it seems to move to lower elevations. Deforestation is rapidly affecting the regions inhabited by this dwarf squirrel.

GENERAL REFERENCES: J. A. Allen 1914, 1915; Emmons and Feer 1990; F. A. Reid 1997; R. W. Thorington and Santana 2007.

Microsciurus santanderensis (Hernandez-Camacho, 1957) Santander Dwarf Squirrel

DESCRIPTION: Santander dwarf squirrels have a dark brownish dorsum, with a grayish venter.

SIZE: Both sexes—TL 272-308 mm; T 136-152 mm.

DISTRIBUTION: This species is found in the state of Santander (and perhaps surrounding states) in Colombia, between the Magdalena River and the Cordillera Oriental and west toward the upper portion of the Cordillera Occidental.

GEOGRAPHIC VARIATION: None.

CONSERVATION: IUCN status—data deficient. Population trend—no information.

HABITAT: This species is found in marshlands and montane forests within its restricted range.

NATURAL HISTORY: *M. santanderensis* is diurnal. Once considered a subspecies of *Sciurus pucheranii*, this dwarf squirrel is poorly known. It is reported from forests, at elevations of 100-1000 m and 2700-3800 m.

GENERAL REFERENCES:

Alberico and Rojas-Dias 2002; Alberico et al. 2000; Eisenberg 1989; Leonard et al. 2009.

Rheithrosciurus Gray, 1867

This genus contains a single species.

Rheithrosciurus macrotis (Gray, 1856)
Tufted Ground Squirrel

DESCRIPTION: Tufted ground squirrels have a distinctive longitudinal groove on their incisors. The dorsum is brown with a slightly grizzled reddish tone. The shoulders and front limbs are sometimes more grizzled with a grayish wash. The hips, hind limbs, and base of the tail are a reddish bay. A white to buff to yellow stripe runs longitudinally along the flank, sometimes in tandem with a dark brownish stripe. The ears are large, with exaggerated red to dark brown tufts. The venter is pale gray to buff. The tail is grizzled charcoal heavily frosted with white.

SIZE: Both sexes—HB 335-352 mm; T 299-342 mm; Mass 1170-1280 g.

DISTRIBUTION: This species is found on the island of Borneo (divided among Malaysia, Brunei Darussalam, and Indonesia).

GEOGRAPHIC VARIATION: None.

CONSERVATION: IUCN status—vulnerable. Population trend—decreasing, totally protected in Sarawak (Malaysia).

HABITAT: Tufted ground squirrels are only recorded from lowland primary forest, where they are found on hillsides, at elevations less than 1100 m; however, survey efforts in other habitat types have not focused on this species. On occasion it can be seen in orchards and secondary forests.

Rheithrosciurus macrotis. Photo courtesy Wildlife Conservation Society-Malaysia Program.

NATURAL HISTORY: This species is diurnal. Tufted ground squirrels are rare and at low density; however, opportunistic sightings and camera traps suggest that this species forages on the ground and in the lower canopy, most often with the tail held high overhead. They will also climb high into the canopy for food. Their diet includes fruits, thick-shelled nuts and seeds, and insects. Tufted ground squirrels demonstrated little fear of macaques (*Macaca*) and often foraged in the same tree with these primates. Hunting of this relatively large-bodied squirrel undoubtedly occurs and is legal (with a license) in Sabah (Malaysia). Tufted ground squirrels cause damage in forest gardens and are considered destructive by some local residents. *R. macrotis* is probably extremely sensitive to logging.

GENERAL REFERENCES: Ambu et al. 2009; Azlan and Lading 2006; Blate et al. 1998; Corlett 1998; Kurland 1973; MacKinnon 1996; Meijaard and Sheil 2008; J. Payne and Francis 1985; Salafsky 1993.

Sciurus Linnaeus, 1758

This genus is the most speciose, with 28 members.

Sciurus aberti (Woodhouse, 1853)
Abert's Squirrel

DESCRIPTION: *S. aberti* is a large tree squirrel with a gray to charcoal dorsum, often with a medial reddish band; a white venter; a slate tail, frosted with white; and conspicuous ear tufts that are especially prominent in winter. Melanism is common, especially in the northern portions of the range. *S. a. kaibabensis* has an entirely charcoal body with a white tail.

SIZE: Female—HB 262.3 mm; T 221.7 mm; Mass 618.6 g.
Male—HB 274.0 mm; T 208.0 mm; Mass 593.8 g.

DISTRIBUTION: This species is found in Wyoming south through the Rocky Mountains in Utah, Colorado, Arizona, and New Mexico (USA), and the Sierra Madre in Sonora, Chihuahua, and Durango (México). This species' range has been increased within Arizona and New Mexico by purposeful introductions.

GEOGRAPHIC VARIATION: Six subspecies are recognized.

S. a. aberti—central Arizona (south of the Grand Canyon), southeastern Utah, southwestern Colorado, and western New Mexico (USA). It has a steel gray back, usually with a narrow medial reddish band and a lateral black band. The venter is normally white (rarely black), and the tail is white below and frosted black above. Abert's squirrels in the Hualapai Mountains of western Arizona, as well as the Pinaleño and Santa Catalina mountains of southern Arizona, are introduced populations of *S. a. aberti*.

S. a. barberi—western Chihuahua (México). It is the largest of the subspecies. It has a steel gray back with a lateral black band and (usually) a mid-dorsal reddish patch. The body and tail are white ventrally; and the tail is gray black on the dorsal side, with scattered white hairs.

S. a. chuscensis—northern part of the Arizona-New Mexico line (USA). It has a steel gray back, with a mid-dorsal reddish patch and a black lateral band. The body and tail are white ventrally, and the tail black dorsally, with frosting.

S. a. durangi—western Chihuahua and Durango (México). It is the smallest of the subspecies, with the tail relatively long compared with body length. The head and body are steel gray dorsally, usually with a mid-dorsal reddish patch and a lateral black band. The ventral side of both

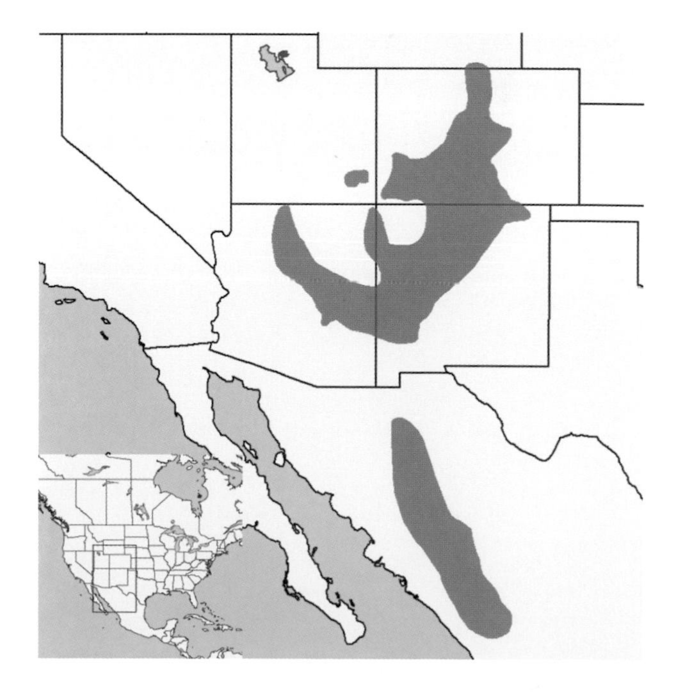

Sciurus aberti. Photo courtesy Randall D. Babb.

Sciurus aberti kaibabensis. Photo courtesy Randall D. Babb.

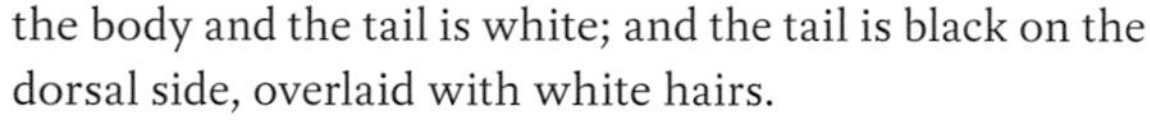

the body and the tail is white; and the tail is black on the dorsal side, overlaid with white hairs.

S. a. ferreus—extreme southcentral Wyoming and Colorado, and northern New Mexico east of the Río Grande (USA). This subspecies is small, with a relatively short tail and variable coloration, including an all black phase, a dark brown phase, and a normal phase. The normal phase consists of a steel gray back, usually a mid-dorsal reddish patch, and a black lateral band. The body and tail are white ventrally, and the tail black dorsally, with scattered white hairs.

S. a. kaibabensis—Kaibab Plateau, north of the Grand Canyon (USA). It is of medium size, with relatively the shortest tail of any of the *S. aberti* subspecies. It is steel gray dorsally, usually with a reddish dorsal patch. The venter is black or blackish gray; and the tail white dorsally and ventrally, with a few black dorsal hairs.

CONSERVATION: IUCN status—least concern. Population trend—stable.

HABITAT: Abert's squirrels are often suggested to be obligates of ponderosa pine (*Pinus ponderosa*) forests that contain abundant conifer seeds and moderately closed canopies for efficient foraging and protection from predators. Areas used by Abert's squirrels typically have denser growth than other portions of the forest. Recent evidence suggests that *S. aberti* uses and can persist in more diverse coniferous forests. Mosaic, heterogeneous, and unevenly aged forests are particularly favored.

NATURAL HISTORY: This species is diurnal. *S. aberti* demonstrates a strongly seasonal reproductive pattern, with males possessing scrotal testes in winter through summer, after which the testes are withdrawn into the abdominal cavity. Females are in estrus for less than one day, when males are attracted to and compete for access to them through a linear dominance hierarchy. Females mate with multiple males and solicit copulations from some low-ranking males. Females produce a single litter (averaging 3.2–3.4 young) each year, typically in late spring or early summer, after a gestation of 46 days. However, a female may produce a second litter in some localities. Paternal care is not provided, and the young remain with their mother until natal dispersal, at about 10 weeks of age. Most individuals are able to reproduce by the time they are 1 year old. Perhaps the most striking and popularized aspect of the biology of Abert's squirrels is their dependence, sometimes considered obligatory, on a single tree species—*Pinus ponderosa*. While this dependence on ponderosa pines is overstated, Abert's squirrels clearly are unique in their ability to utilize phloem as an important food source during periods of winter food shortages, in part due to a lengthened digestive tract. Individuals clip 2-5 cm lengths of small twigs from the distal portion of branches, strip the bark, and eat the phloem. The completely stripped clippings are discarded, and they have been used to estimate the density and document the presence or absence of this species. Squirrels select trees with reduced chemical defenses when feeding on phloem. Abert's squirrels also consume a wide variety of tree seeds—including those of pine (*Pinus*), true firs (*Abies*), Douglas fir (*Pseudotsuga menziesii*), and oaks (*Quercus*)—in addition to buds, fungi, insects, bone, and (occasionally) soil. *S. aberti* does not cache seeds. However, they are known to pilfer from other species that do store seeds, such as *Tamiasciurus hudsonicus*.

Abert's squirrels overlap considerably in space with their neighbors. Home ranges are relatively large, often greater than 20 ha, with male ranges sometimes larger than those of females, perhaps due to mate searching. *S. aberti* nests primarily in dreys, constructed of leaves and twigs within the forest canopy. The dense growth of twigs (known as witch's brooms) caused by parasitic mistletoes are commonly used to anchor and obscure nests. Much less frequently, Abert's squirrels employ cavities as nests. Both nest types are used to rear the young and serve as nocturnal nests. Nests are located in larger and taller trees with canopies connected to adjacent trees. The nests are found within more dense stands of trees, rather than being situated at random in the surrounding area. Individuals change nests regularly and maintain several active nests simultaneously. Communal nesting in mixed-sex assemblages occurs frequently in winter but is rare in the warmer months. Abert's squirrels are not highly vocal, but they do use a gruff low-pitched barking alarm call when startled or after detecting predators such as raptors, foxes, or bobcats (*Lynx rufus*). This species is hunted for food and sport throughout its range. Due to the low human population densities within the range of this species, *S. aberti* is of little economic value or detriment; however, their practice of clipping branch tips can significantly reduce cone availability. The introduction of *Sciurus aberti* to isolated mountain ranges in Arizona may have negatively impacted populations of endemic native tree squirrels, such as *Sciurus arizonensis* and *Tamiasciurus hudsonicus grahamensis*.

GENERAL REFERENCES: R. Davis and Brown 1988; Edelman and Koprowski 2005, 2006; Lamb et al. 1997; M. A. Snyder 1992, 1993; Yensen and Valdés-Alarcón 1999.

Sciurus aestuans (Linnaeus, 1766)
Guianan Squirrel

DESCRIPTION: The Guianan squirrel has an olivaceous gray dorsum grizzled with brown and yellow. The eye ring is pale yellowish brown, and sometimes there is a pale buffy yellow postauricular patch. The upper surface of the tail is the same color as the dorsum, with a rufous and more variable underside. The venter is white to cream to buff to yellowish. Albinism and melanism occur but are not common.

SIZE: Both sexes—HB 160-186 mm; T 163-250 mm; Mass 160-380 g.

DISTRIBUTION: This species can be found throughout much of Brazil, as well as in Colombia, Venezuela, Guyana, Suriname, French Guiana, and Argentina.

GEOGRAPHIC VARIATION: Ten subspecies are recognized.

S. a. aestuans—north of the Amazon, in Colombia, Venezuela, Guyana, Suriname, French Guiana, and Brazil. This form

Sciurus aestuans. Photo courtesy Patrícia Alexandre Formozo.

has white postauricular patches, and a reddish wash to the body and feet.

S. a. alphonsei—northern coast of Brazil. This form has a pale yellowish dorsum, with a grayish venter.

S. a. garbei—Espirito Santo and Bahia (Brazil). This form has chestnut ochraceous upperparts and orange ochraceous underparts, with a paler throat region.

S. a. georgihernandezi—northwestern portion of the range, including Colombia.

S. a. henseli—far southern region of Brazil, and northeast Argentina. The sides are ashy colored, and the venter is white.

S. a. ingrami—eastern and southern coasts of Brazil. This subspecies has an olivaceous dorsum, with a white to buff venter.

S. a. macconnelli—mountains of southern Venezuela, Guyana, and probably northern Brazil. This form is more brownish olivaceous.

S. a. quelchii—southern Guyana and northcentral Brazil. This form is olivaceous, with a yellow belly.

S. a. sebastiani—restricted to Sebastian Island (Brazil). This large form has a stronger darker brown color tone than the mainland varieties. The tail color is reddish brown, whereas the tails of those on the mainland are gray brown.

S. a. venustus—near Mount Duida (Venezuela). This is a diminutive subspecies, with coloration similar to *S. a. aestuans*.

CONSERVATION: IUCN status—least concern. Population trend—no information.

HABITAT: *S. aestuans* is commonly found in tropical rainforests, swamps and wet forests, Atlantic gallery forests, secondary forests, gardens, and plantations. It can also be found in urban parks.

NATURAL HISTORY: This species is diurnal. Guianan squirrels are found in all forest levels, but most commonly in the midstory, at 5–12 m. They make nests of leaves in the canopy of trees. Guianan squirrels appear to be solitary. Exclusive use of territories does not occur, and home range overlap is considerable. The average male home range (6.5 ha) is twice the size of that of the female (3.1 ha) in fall, and it gradually decreases in winter after mating. Males track and sniff females, and eventually several males chase one female through the canopy to mate. Females are pregnant in winter and summer, suggesting two annual mating seasons. Primarily herbivorous, *S. aestuans* feeds predominately on seeds and fruits of the highly diverse palms in neotropical forests, oaks (*Quercus*), a variety of other trees, and up to five shrubs. Fungi are eaten in winter and spring. Guianan squirrels demonstrate a consistent palm nut handling behavior for each palm species in order to obtain the protected seed endosperm; the behavior is actually learned, beginning as a juvenile. Guianan squirrels serve as an important seed disperser by carrying and caching fruit in autumn and winter, sometimes storing it at heights of more than 30 m. Within forest fragments of the Atlantic forest in southeastern Brazil, *S. aestuans* accounts for 96 percent of the seed predation for some species. *S. aestuans* are probably prey for many forest-dwelling carnivores—such as felids, procyonids, canids, mustelids, primates, snakes, lizards, and raptors—but documented predators are ocelots (*Leopardus pardalis*), margays (*Leopardus wiedii*), and capuchin monkeys (*Cebus*). *S. aestuans* has a varied vocal repertoire that includes alarm calls of sharp "chucks" or "chatters," soft single "chips," and high-pitched twanging whines. When alarmed, Guianan squirrels ascend a tree and chatter while running to hide, or depart through the canopy. *S. aestuans* is hunted for food in much of its range.

GENERAL REFERENCES: Alvarenga and Talamoni 2006; Bordignon and Monteiro-Filho 1999, 2000; Cullen et al. 2001; Fagundes et al. 2003; Galetti 1990; Galetti et al. 1992; Grelle 2003; Müller and Vesmanis 1971; Paschoal and Galetti 1995; Souza 2000.

Sciurus alleni (Nelson, 1898)
Allen's Squirrel

DESCRIPTION: Allen's squirrels have yellow brown backs grizzled with black and gray; the head is generally darker; and the sides are lighter. The eye ring is white to buff. The legs are white to buff, with a dorsal grizzling of black. The venter is white. The tail is black, sometimes suffused with faint buff or yellow, and frosted with white. Melanism has been reported.

SIZE: Female—HB 220-254 mm; T 217-247 mm; Mass 461 g (345-510 g).

Male—HB 220-254 mm; T 217-247 mm; Mass 408 g (290-491 g).

DISTRIBUTION: This species can be found in southeastern Coahuila through central Nuevo León, and south through western Tamaulipas to extreme northern San Luis Potosí (México).

GEOGRAPHIC VARIATION: None.

CONSERVATION: IUCN status—least concern. Population trend—decreasing.

HABITAT: Allen's squirrel is found primarily in oak (*Quercus*) and oak-pine (*Quercus, Pinus*) forests in montane environments. However, it does extend into similar habitats in coastal plains.

NATURAL HISTORY: This species is diurnal. Allen's squirrels are active on the ground and in the canopy. In mixed forests, *S. alleni* selects oak (*Quercus*) and pine (*Pinus*) forests during all seasons. Pregnancy and lactation have been reported in all months; litter size varies from one to four young. Nests are sited in natural cavities in trees or are built of leaves and sticks placed on tree branches. Primarily herbivores, Allen's squirrels feed upon the seeds and fruits of native trees, as well as on peanuts, corn (maize), oats, apples, peaches, mangos, plums, grapes, and tomatoes. They will also consume insects (adults and larvae) and anurans. *S. alleni* occasionally enters cornfields and can cause considerable damage to the ripening crop. Logging, burning, and the conversion of forests to agriculture are the primary threats to its conservation.

Sciurus alleni. Photo courtesy Gabriel Bojórquez.

GENERAL REFERENCES: Best 1995a; Guevara 1998; Jimenez-Guzman and Guerrero-Vazquez 1992; Leopold 1959; Morales 1985.

Sciurus anomalus (Gmelin, 1778) Caucasian Squirrel

DESCRIPTION: Caucasian squirrels have a dorsum that is generally grizzled gray to black, often suffused with buff to yellow to tawny rust. A pale buff to yellow eye ring is present. The venter ranges from a chestnut gray buff to pale grizzled gray to a bright orange. The tail varies from light brownish yellow to strong rust above with grayish yellow below.

SIZE: Female—HB 213 mm (198-235 mm); T 147 mm (120-162 mm); Mass 349 g (274-410g).

Male—HB 216 mm (192-248 mm); T 143 mm (120-157 mm); Mass 336 g (250-401 g).

DISTRIBUTION: This species is found on the Isle of Lesbos (Greece) and in Turkey, Transcaucasia (Georgia, Azerbaijan, and Armenia), Syria, Lebanon, Jordan, Iraq, Iran, and Israel.

GEOGRAPHIC VARIATION: Three subspecies are recognized.

S. a. anomalus—Isle of Lesbos (Greece), Transcaucasia, Turkey, and possibly Kurdistan and northern Iraq. This form has a chestnut gray-buff venter.

S. a. pallescens—Iraq, the Zagros Mountains (western Iran), and the Fars District (southern Iran). It is characterized by a venter that is pale grizzled gray.

S. a. syriacus—Lebanon, Syria, Jordan, and Israel. This subspecies possesses a bright yellow to golden venter.

CONSERVATION: IUCN status—least concern. Population trend—decreasing.

Sciurus anomalus. Photo courtesy Süleyman Ugar.

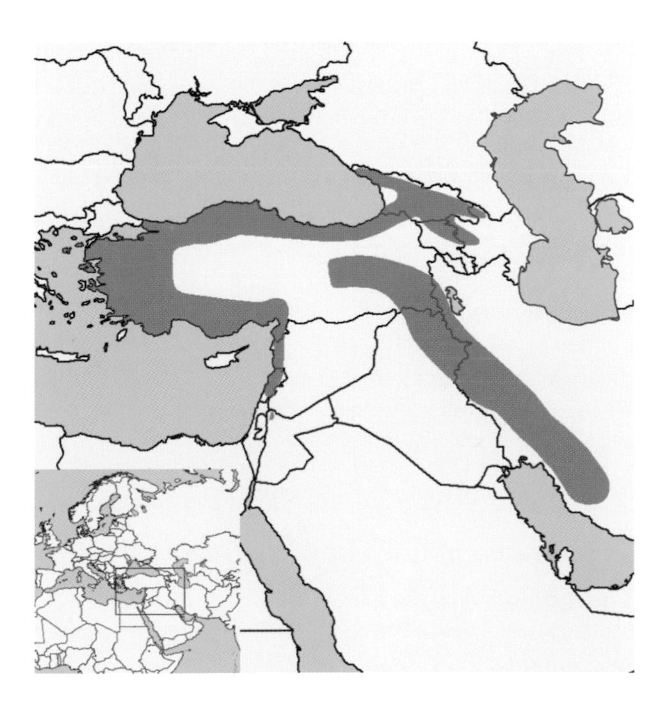

HABITAT: Caucasian squirrels inhabit mixed and deciduous forests, although they also occur in coniferous forests and in rocky outcrops. They can be found in chestnut, walnut, almond, and olive plantations.

NATURAL HISTORY: This species is diurnal. Caucasian squirrels forage and move nimbly on the ground and in the canopy throughout the year. Mating occurs twice each year: spring mating takes place in April to late May, and summer mating occurs in August and September. A litter averaging three (range = 1-7) young is born in a nest. Male parental care does not occur, and females nurse their young for five to six weeks. The young mature at 5-6 months of age. Females born in winter or early spring can breed in late summer or early autumn, whereas males probably do not breed until their second year of life. Caucasian squirrels nest in tree hollows lined with moss and dry leaves; however, nests may also be found between stones or tree roots, or on buildings. Nests may contain more than one adult. *S. anomalus* feeds on a variety of coniferous and deciduous tree seeds, nuts, and fruits, as well as on mushrooms; animal matter (insects, bird eggs, and nestlings) is occasionally eaten. Seeds with low perishability can be larder-hoarded within tree cavities or scatter-hoarded on the ground in shallow storage locations. The vocalization is a high-pitched metallic-sounding "chit-chit-chit," used in response to potential predators. Wild and domestic felids, canids, mustelids, and raptors are known to prey upon these squirrels. Their pelts are harvested for fur clothing and ornamentation; however, the quality of *S. anomalus* pelts is viewed as considerably less than that of *S. vulgaris*. Caucasian squirrels are hunted for food in much of their range and are considered pests in some regions with nut and fruit orchards. Young squirrels are sold as pets. Habitat loss due to timber harvesting and the conversion of forests to agriculture remains a threat.

GENERAL REFERENCES: Albayrak and Arslan 2006; Amr 2000; Demirsoy et al. 2006; Gavish 1993; D. L. Harrison and Bates 1991; Qumsiyeh 1996.

Sciurus arizonensis (Coues, 1867)
Arizona Gray Squirrel

DESCRIPTION: *S. arizonensis* has a silvery gray dorsum, occasionally with a faint medial brownish band; a white to cream venter; and a gray tail frosted with white. A cream to white eye ring is prominent.

Sciurus arizonensis. Photo courtesy Nichole Cudworth.

SIZE: Female—HB 259.1 mm; T 254.5 mm; Mass 667 g.
Male—HB 248.2 mm; T 245.8 mm; Mass 736 g.

DISTRIBUTION: This species is found in Arizona and extreme western New Mexico (USA) south into Sonora (México).

GEOGRAPHIC VARIATION: There is disagreement as to whether *S. arizonensis* is monotypic or whether three subspecies should be recognized. Currently, this species is considered to be monotypic.

CONSERVATION: IUCN status—data deficient. Population trend—no information. It is considered to be threatened in México.

HABITAT: Arizona gray squirrels are found in forests, ranging from low-elevation Madrean forests with a mixture of pine (*Pinus*) and oak (*Quercus*) to higher-elevation mixed conifer forests. Riparian areas with large cottonwoods (*Populus*) and sycamores (*Platanus*) also harbor high densities of Arizona gray squirrels.

NATURAL HISTORY: This species is diurnal. The ecology of Arizona gray squirrels is poorly known, despite their original description more than 140 years ago. *S. arizonensis* demonstrates a seasonal reproductive pattern. There is a short day-long estrus in spring (February-May), when multiple males pursue females, followed by the birth of a single litter of two to four young (mean = 3.1) in late spring or summer. Males possess scrotal testes from winter through summer. The testes are withdrawn into the abdominal cavity during the remaining months. Arizona gray squirrels nest in dreys composed of sticks and leaves within the forest canopy, as well as in cavities within large-diameter trees. Nests are used to rear the young and as nocturnal rest sites. Nest trees tend to be in the largest and most interconnected trees at a site. Communal nesting is known. Male home ranges are extensive, 113 ha as compared with those of females, which are 14 ha. Males appear to maximize overlap with females, particularly during the breeding season. *S. arizonensis* does not hibernate. Large raptors, foxes, and bobcats (*Lynx rufus*) are probably their major predators. Arizona gray squirrels are generally silent unless alarmed, when they will bark and "chuck" from elevated locations. This species is hunted for food and sport throughout its restricted range. Due to low human population densities within the range of this species, *S. arizonensis* is of little economic value or detriment. Habitat loss and degradation due to human development, timber harvests, and catastrophic wildfires pose significant risks to populations. Introductions of *Sciurus aberti* to portions of the Arizona gray squirrels' and other species' ranges are believed to negatively impact populations of the native species. *S. arizonensis* forages extensively on the ground and in the forest canopy for tree seeds and flowers, in addition to fungi. Seeds from the cones of pines (*Pinus*), Douglas fir (*Pseudotsuga menziesii*), and true firs (*Abies*) are consumed by removing individual cone scales. Acorns and walnuts are also eaten when available, along with a variety of other tree seeds. Arizona gray squirrels will rarely cache seeds by scatter-hoarding in leaf litter and topsoil.

GENERAL REFERENCES: Best and Riedel 1995; D. E. Brown 1984; Cudworth and Koprowski 2010; Hoffmeister 1986; Yensen and Valdés-Alarcón 1999.

Sciurus aureogaster (Cuvier, 1829) Red-Bellied Squirrel

DESCRIPTION: Red-bellied squirrels have a frosted pale to dark grizzled gray dorsum. Patches of varying sizes and colors may occur on the nape, shoulders, rump, and sides. The venter ranges from white through orange to chestnut. The tail can be a variegated grayish buff when the underparts are pale to orange red, or chestnut when the venter is deep orange. Partial melanism and complete melanism are common.

SIZE: Both sexes—HB 232-310 mm; T 215-284 mm; Mass 375-680 g.

DISTRIBUTION: This species is found from Nayarit, Guanajuato, and Nuevo León (México) to central and southwestern Guatemala. It was introduced to Elliott Key, Florida (USA).

GEOGRAPHIC VARIATION: Two subspecies are recognized.

S. a. aureogaster—eastern México, at elevations below 1525 m, with the exception of a population at Pinal de Amoles in the mountains of northeastern Querétaro, at an elevation of about 2440 m. This subspecies has an orange red to chestnut venter, small inconspicuous postauricular patches, a less patchy dorsum, and a white tail suffused with black and sometimes chestnut.

S. a. nigrescens—Pacific coastal plain and into the mountains of central and southeastern México. This form has a white to chestnut venter, often conspicuous postauricular patches, a sharp patchy dorsum, and a gray to chestnut tail.

CONSERVATION: IUCN status—least concern. Population trend—stable.

HABITAT: Red-bellied squirrels inhabit a variety of forests, especially dry open woodlands, including pine-oak (*Pinus, Quercus*) and thorn scrub. It can be found in secondary forests, plantations, and urban areas.

NATURAL HISTORY: This species is diurnal. Red-bellied squirrels are adept at moving through closed canopies; how-

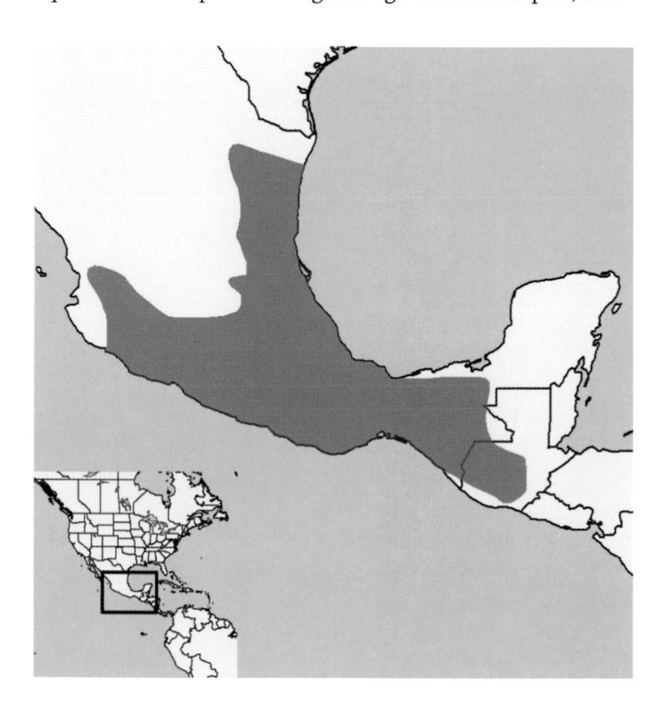

Sciurus aureogaster. Photo courtesy Juan Cruzado.

ever, they also forage on the ground, especially in open forests. They are active throughout the year. Mating chases with multiple male participants can be seen most often from January to May; however, the breeding season seems prolonged, and lactating females have been reported in most months. A litter of one to four is born in a leaf nest or den cavity, both located in the largest trees in relatively dense stands. The young are weaned and disperse from their natal area soon after emergence from the den. Individuals are not territorial, and home ranges overlap. Males range over 2.3 (0.9–4.7) ha, and females over 0.9 (0.5–1.2) ha. Individuals are usually solitary, except when attracted to an abundant food resource while foraging. *S. aureogaster* feeds predominantly on the tree seeds of oak (*Quercus*) and pine (*Pinus*), but it also eats flowers, buds, fruits, and fungi. Red-bellied squirrels excise the embryos of white oak acorns and also remove the embryo from germinated red oaks prior to caching the seeds, behaviors common in congeners from temperate forests. *S. aureogaster* is generally quiet and calls only when distressed. Two calls are used in this context: a low-pitched barking call emitted when the animals are apprehensive about a distant threat, and a high-pitched barking call given when danger is nearby. Their predators include raptors, felids, procyonids, and snakes. Red-bellied squirrels are viewed as pests in cornfields—as well as in cacao, papaya, mango, and coconut plantations—and they are hunted in some portions of their ranges as a source of meat and pelts. These colorful squirrels are occasionally sold as pets, but they may also have value to ecotourists. Habitat loss and fragmentation remain an issue as forests are harvested and converted to agriculture.

GENERAL REFERENCES: L. N. Brown and McGuire 1975; Estrada and Coates-Estrada 1985; Musser 1968; Ramos-Lara and Cervantes 2007; Romero-Balderas et al. 2006; Steele, Turner, et al. 2001.

Sciurus carolinensis (Gmelin, 1788) Eastern Gray Squirrel

DESCRIPTION: Eastern gray squirrels have a grizzled pale to slate gray dorsum that is sometimes washed with cinnamon, especially on the hips. Postauricular patches are white to buff and most prominent in the winter. A white eye ring is usually visible. The venter is white to buff to cinnamon. The tail color is the same as the dorsum, with a light frosting of white. Melanism is common in northern populations, where an energetic advantage occurs in winter. White and albino populations are known from several locations in the midwestern and eastern USA.

SIZE: Both sexes—HB 200–315 mm; T 150–250 mm; Mass 300–710 g.

DISTRIBUTION: This species is found from Saskatchewan (Canada) south to eastern Texas (USA) and east through Canada and the USA to the Atlantic coast. It was introduced into Britain, Scotland, Ireland, Italy, South Africa, Australia,

Sciurus carolinenesis. Photo courtesy Caroline M. Thorington.

México, and various localities in western Canada, Hawaii, and the western continental USA. Introductions to Australia, Hawaii, and perhaps México have failed.

GEOGRAPHIC VARIATION: Five subspecies are recognized.

S. c. carolinensis—most of the southern portion of the range, except for the Appalachian Mountains (USA). This moderately sized form has the coloration described above.

S. c. extimus—lower half of Florida (USA). This subspecies is the smallest and darkest.

S. c. fuliginosus—southern half of Louisiana, and extreme southern Mississippi and Alabama (USA). This form has rich dark colors, with the venter often a shade of cinnamon.

S. c. hypophaeus—northwestern portion of the range, including most of Minnesota, Wisconsin, and the Upper Peninsula of Michigan (USA). This form is the largest of the subspecies; and it often has a reduced area of white on the venter, with an increased encroachment of the dorsal coloration.

S. c. pennsylvanicus—northeastern portion of the range, the Appalachian Mountains, and extending from Iowa across the southern and eastern Great Lakes states (USA). This is another large subspecies, but it lacks the encroachment of dorsal coloration to the venter.

CONSERVATION: IUCN status—least concern. Population trend—stable.

HABITAT: Eastern gray squirrels are found in hardwood or mixed forests, especially with a diversity of mature to overly mature trees that provide abundant supplies of hard mast (acorns, walnuts, and hickory nuts) and nest cavities. It can be found in urban and suburban areas, even when trees are somewhat sparse.

NATURAL HISTORY: This species is diurnal. Eastern gray squirrels move readily in the canopy or on the ground throughout the year. Mating activity peaks twice annually: December-February and May-June. Males follow females to assess the latter's reproductive condition by sniffing genitalia, and they repeatedly visit when a female approaches estrus. Estrus lasts less than one day during a breeding season, and males congregate at the nest of the female, often at first light. Up to 34 males will pursue a female during her mating bout. Dominant males are able to mate by actively pursuing the female; however, lower-ranking males, typically younger individuals, are able to copulate by waiting on the periphery of the bout and locating the female when she evades the ultra-aggressive dominant males. Copulation lasts less than 30 seconds. A copulatory plug forms within the female from the coagulation of semen. Females will remove the plugs and mate with several males. After a 44-day gestation, a litter averaging two to four (maximum of eight) young is born in a tree cavity or leaf nest. Leaf nests are large spherical structures constructed in the canopy, using twigs and leaves, with a central nest chamber often lined with shredded bark and leaves. A few females may produce two litters in a year; however, most adult females only produce a single litter

Sciurus carolinenesis. Photo courtesy Brian E. Kushner, Brian KushnerPhoto.com.

***Sciurus carolinenesis*, black morph.** Photo courtesy Laura Finnegan.

each year. The young emerge from the nest after 7–8 weeks and are weaned beginning at 10 weeks. Adult size is reached by 8–9 months of age. Most individuals do not reproduce until at least their yearling year; precocious breeding at 5.5 months by females is known when there is a superabundance of food. Natal dispersal occurs soon after weaning, with males leaving their natal area and some females remaining to form matrilines. Groups of related females and, occasionally, unrelated males will nest together each night, often in tree cavities. The nest sites are aggressively defended from a nongroup member attempting to enter. Scent-marking with cheek glands is used to mark trees, and these glands are smelled when individuals greet each other. Traditional scent-marking sites are found in protected areas, such as under low branches or on the underside of a slanted tree trunk; primarily males gnaw the location and then wipe their cheeks back and forth to deposit scent, and sometimes they urinate to scent-mark.

Eastern gray squirrels feed heavily on tree seeds, but they also eat fruits, shoots, flowers, fungi, and occasionally animal matter (such as insects, bird eggs, and nestlings). A conspicuous behavior is the annual scatter-hoarding of hard low-perishability nuts and acorns each fall to serve as the winter food store. The squirrels collect tree seeds near the parent tree, disperse to greater distances, and bury them just under the surface. These behaviors all serve to decrease the loss of this critical food source to seed predators. To prevent white oak-group acorns from germinating once buried, eastern gray squirrels excise the embryo to ensure that the seed's nutrients remain available until recovered. Caches are recovered using a combination of spatial memory and olfactory cues. Reproduction is strongly tied to hard mast production; litter size and the prevalence of lactation decrease in poor food years. Overlap in home ranges is considerable, but females often defend exclusive-use cores, especially during lactation. Home ranges vary from 0.5 ha to 20.5 ha, are larger in males, and decrease with density and with decreasing habitat fragment size. First-year survivorship is only about 25 percent, but then it increases to more than 50 percent throughout adult life. Record longevity in the field for males is more than 9.0 years, for females 12.5 years, and as great as 20 years in captivity. Predators of *S. carolinensis* are numerous and include mustelids, procyonids, felids, canids, raptors, and snakes. Alarm calls are a series of medium-pitched barks, often followed by long high-pitched whines.

Eastern gray squirrels are frequently valued in parks and cities, and they are also hunted as game animals. *S. carolinensis* is considered a pest in orchards, gardens, agricultural fields, and forest plots. Having been introduced to England, this species continues to spread with deleterious consequences: the stripping of bark (which kills trees), and the rapid displacement of native Eurasian red squirrels (*Sciurus vulgaris*). In the state of Washington (USA), introductions of this species are believed to play a role in the reduction of western gray squirrels (*Sciurus griseus*). Habitat loss and frag mentation remain the most significant threats to eastern gray squirrels in their native range.

GENERAL REFERENCES: Bertolino 2009; Hopewell and Leaver 2008; Hopewell et al. 2008; Koprowski 1994a, 1996; Steele and Koprowski 2001; Steele, Manierre, et al. 2006; Wauters, Tosi, et al. 2002.

Sciurus colliaei (Richardson, 1839) Collie's Squirrel

DESCRIPTION: Collie's squirrel is a gray-colored squirrel of moderate size, and it is sometimes referred to as a "gray squirrel" in México, along with *S. aureogaster*. However, the dorsum is usually dark, grizzled gray with a yellowish wash that continues to the base of the tail. The sides are light gray. The venter is usually white, but it can sometimes be light orange. The dorsal surface of the tail, other than the base, is black with a white wash. The ventral surface of the tail is grizzled gray, or dark gray and yellow, with white along its edges. This species has two upper premolars, except in the northern parts of its range, where one is missing in some individuals.

SIZE: Female—HB 243.4 mm; T 260.4 mm; Mass 440.8 g.
Male—HB 248.6 mm; T 243.2 mm; Mass 335.2 g.
Sex not stated—HB 266.1 mm; T 274.1 mm.

DISTRIBUTION: This species occupies a narrow area along the westcentral coast of México, from the states of Sonora and Chihuahua to Sinaloa, Durango, Nayarit, Jalisco, and Colima.

GEOGRAPHIC VARIATION: Four subspecies are recognized.

S. c. colliaei—Nayarit (México). See description above.

S. c. nuchalis—Jalisco and Colima (México). This subspecies has a yellowish nape, a black rump, and ears that are more yellow or rust colored than those of *S. c. colliaei.*

S. c. sinaloensis—Sinaloa (México). This subspecies has a larger skull than *S. c. colliaei.*

S. c. truei—Sonora, Chihuahua, and Sinaloa (México). The skull is broader and the cranium is flatter than in *S. c. colliaei.*

CONSERVATION: IUCN status—least concern. Population trend—no information.

HABITAT: This species is found in thick tropical and subtropical vegetation in the coastal plain of the central Pacific coast of México, where it is reported to occupy a number of forest types, especially in Jalisco. It sometimes occurs in canyons and at lower elevations (up to 2000 m). It is often associated with figs (*Ficus*) and palms (e.g., *Orbignya cohune*), upland oak (*Quercus*) forests, and riparian or tropical dry forests (i.e., arroyo forests) of the Pacific coast. In the northern part of its range, Collie's squirrel often occupies subtropical canyons.

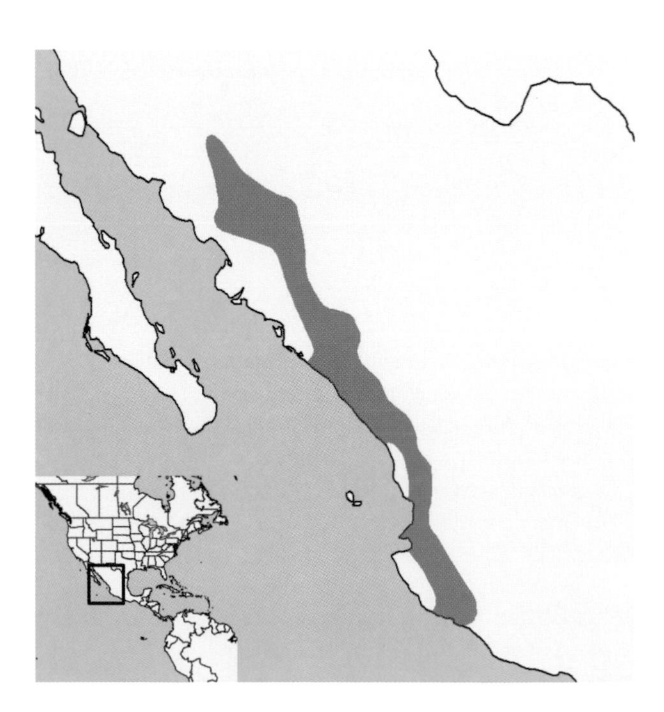

Sciurus colliaei. Photo courtesy Petr Myska, vivanatura.org.

NATURAL HISTORY: Collie's squirrel appears to specialize on the fruits and nuts of palms, figs (*Ficus*), and possibly oaks (*Quercus*). Forest fragmentation seems to reduce the distribution of this species, as measured by patterns of fruit predation in continuous and fragmented forests; in fragmented forests only 34 percent of *Ceiba aesculifolia* trees sustain fruit damage by *S. colliaei*, whereas 100 percent of these trees experience squirrel damage in continuous forests. Collie's squirrel is primarily (but not exclusively) arboreal. It is diurnal: active mostly in the morning (just after sunrise) and again late in the day (until sunset). This species is reported to build dreys on outer tree branches, but also to form nests in cavities in tree trunks and in abandoned arboreal termite nests. This species appears to breed in March and April, and its young have been reported in April in two parts of its range (Jalisco and Sonora). However, two females have been reported with embryos (three each): one in late May (Chihuahua), and another in mid-June (Durango). Hence more detail is needed on the reproductive biology of this species. The shape and size of the baculum are similar to the bacula of several other species of Mexican *Sciurus* (e.g., *S. aureogaster, S. variegatoides,* and *S. yucatanensis*). *S. aureogaster* is closely related to *S. colliaei*, and the others were considered part of the same fragmented species complex in the past, based on similarities in pelage, cranial features, and ecology. Collie's squirrel is allopatric with *S. aberti* and *S. nayaritensis* where their ranges overlap in the north (Sonora), but sympatric with *S. aureogaster* and *S. nayaritensis* throughout other parts of the range. Three species of lice (*Enderleinellus mexicanus, E. pratti,* and *Neohaematopinus sciurinus*) and no endoparasites are reported from Collie's squirrel.

GENERAL REFERENCES: Best 1995b; de Grammont et al. 2008; Herrerías-Diego et al. 2008; Peterson et al. 1999.

Sciurus deppei Peters, 1863
Deppe's Squirrel

DESCRIPTION: A small member of the genus, Deppe's squirrel shows considerable individual variation in color, from a gray to yellowish brown or rust-colored brown. The legs are dark gray or rust colored. The dorsal surface of the tail is black, interspersed with white hair. The ventral surface of the tail is yellowish orange to rust colored. The hairs on the tip of the tail are white. The face is gray. This species has two upper premolars. Its range overlaps with that of at least seven other species of *Sciurus* (*S. alleni, S. aureogaster, S. granatensis, S. oculatus, S. richmondi, S. variegatoides,* and *S. yucatanensis*). All of these, except *S. richmondi,* are larger than *S deppei. S. richmondi* is about the same size, but it has one upper premolar.

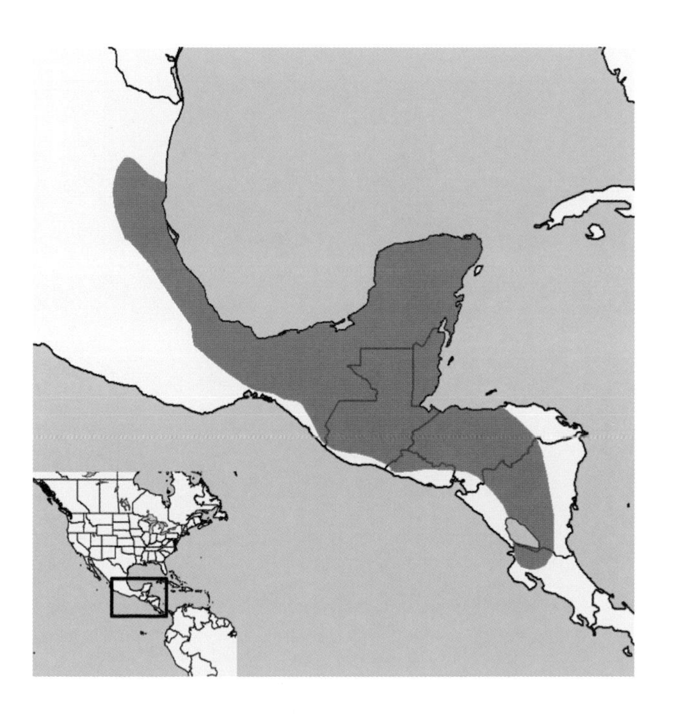

SIZE: Female—HB 210.2 mm; T 169.4 mm; Mass 287.3 g.
Male—HB 207.2 mm; T 176.0 mm; Mass 268.3 g.
Sex not stated—HB 197.4 mm; T 181.7 mm; Mass 203.3 g.

DISTRIBUTION: Deppe's squirrel ranges from the eastern coast of México—in the states of Tamaulipas through Chiapas and the Yucatán Peninsula—to northwestern Costa Rica. The range of *S. deppei* overlaps with that of *S. alleni, S. aureogaster, S. granatensis, S. oculatus, S. richmondi, S. variegatoides,* and *S. yucatanensis.*

GEOGRAPHIC VARIATION: Five subspecies are recognized.

S. d. deppei—northern Veracruz (México) to western Honduras. See description above.
S. d. matagalpae—western Honduras, and across Nicaragua. This form may be differentiated by the yellowish brown color of the upperparts and yellow underparts.
S. d. miravallensis—southern terminus of the species' range in Costa Rica. The underparts are gray and slightly washed with ochraceous orange. The upperparts, including the legs and feet, are dark grizzled yellowish brown, slightly darker along the middle and top of the head. The tail has a white border.
S. d. negligens—Tamaulipas, San Luis Potosí, Hidalgo, and northern Veracruz (México). This subspecies is distinguished from *S. d. deppei* by its longer ears, a more gray brown pelage on its upperparts, and a lighter ventral pelage.
S. d. vivax—Yucatán Peninsula. In comparison with *S. d. deppei, S. d. vivax* has a pale rusty dorsal pelage, and a larger rostrum.

CONSERVATION: IUCN status—least concern. Population trend—stable.

HABITAT: Deppe's squirrel is a habitat generalist. This species is typically associated with dense vegetation in damp tropical forests, usually in lowland areas at elevations between 200 and 1500 m. However, it is reported to occur as high as 3000 m in some parts of its range (e.g., Guatemala). It is also found in cloud forests, drier oak (*Quercus*) woodlands, pine-oak (*Pinus, Quercus*) forests throughout México, dry subtropical forests, and lowland riparian zones. It appears to tolerate some level of forest loss, but not severe clearing.

NATURAL HISTORY: Deppe's squirrel is primarily granivorous and frugivorous, feeding on a variety of fruits, nuts, and seeds across its range, including palm nuts, acorns, and

Sciurus deppei. Photo courtesy Tracey Dixon, www.trp.dundee.ac.uk/~bl/.

berries. It appears to function predominantly as a seed predator, although studies on scatter-hoarding and seed dispersal are generally lacking. It has been reported to possibly aid in the first stage of a two-stage dispersal process of *Guarea glabra* and *G. kunthiana* in Costa Rica by removing the arils of their seeds and then dropping them below the parent trees, where they may be further dispersed and cached by other rodents. At least one study, however, reports a limited consumption of leaves, and others note the use of fungi and insects. This species is also known to damage crops (e.g., corn/maize). *S. deppei* is diurnal and shows a bimodal pattern of activity, with peaks in early morning and just before sunset. It is frequently observed foraging on the ground, often in dense vegetation; yet this species is arboreal and relies on both cavities and dreys (made of leaves and twigs) for nesting. It is agile and moves easily through all parts of the arboreal habitat. While Deppe's squirrel is not considered social, it is often observed in small groups. It is reported to vocalize under some circumstances, although it is also considered secretive. This species' vocalizations are reported to be high-pitched trills that are often accompanied by rapid tail flicks. Based on anecdotal observations of lactating females, embryos, and nestlings, it appears that reproduction in this species can occur throughout the year, but at least one author reports that reproduction happens most often at the end of the dry season. Litter sizes range from two to eight, and average about four. At least two subspecies of *S. deppei* have three pairs of functional mammae, instead of four as in all other species of *Sciurus*. Densities have been reported as low as 2.2 and as high as 100 animals/km^2. Estimates of home range size average 1.5 ha.

This species appears to be relatively tolerant of human activity. In a study in Guatemala that compared mammal and bird densities in sites heavily visited by tourists and in control sites, *S. deppei* was one species showing higher densities in tourist sites (32.3/km^2 versus 2.2/km^2 in tourist versus control sites). This species is hunted and sold for food in relatively low numbers in markets in Chiapas (México). The range of Deppe's squirrel overlaps with that of *S. aureogaster*, although these two species seem to show signs of microhabitat segregation. It is also occasionally found in the same forests with *S. variegatoides* and *S. yucatanensis*. Compared with all three of these species, *S. deppei* spends more time on the ground, perhaps because it resides in more dense vegetation. Only ectoparasites have been reported from *S. deppei*. These include one tick, (*Ixodes tamaulipas*), one chigger (*Eutrombicula alfreddugesi*), two lice (*Enderleinellus deppei* and *E. extremus*), and four fleas (*Kohlsia graphis*, *Orchopeas howardi*, *Plusaetis dolens*, and *Trichopsylla graphis*).

GENERAL REFERENCES: Barragán et al. 2007; Best 1995c; Eckerlin 2005; Guzmán-Cornejo et al. 2007; Hidinger 1996; Koprowski, Roth, Woodman, et al. 2008; Sánchez-Cordero and Martínez-Gallardo 1998; Wenny 1999.

Sciurus flammifer (Thomas, 1904) Fiery Squirrel

DESCRIPTION: The pelage of fiery squirrels is variable, with complete melanism as well as partial albinism documented.

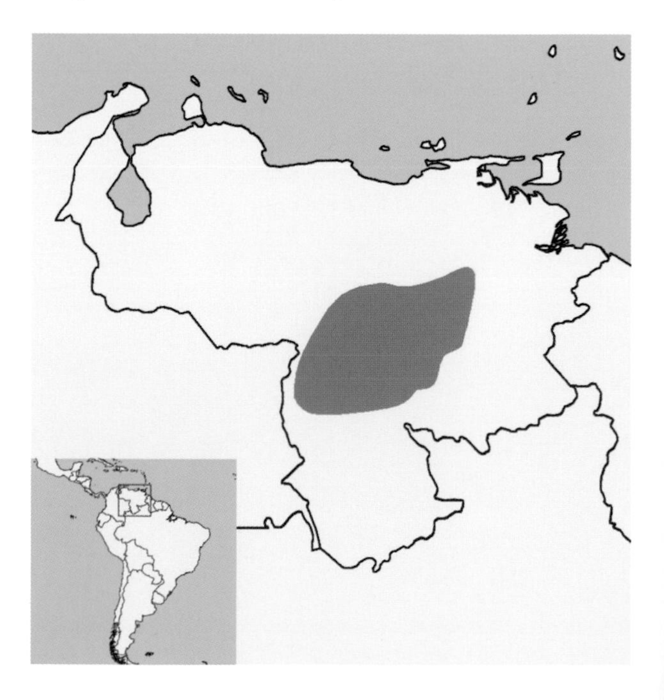

The dorsum is usually grizzled black to brown, suffused with yellow. It has a rufous tinge on the head, ears, and rump. The cheeks and chin are washed with yellow to orangish. The venter is white to cream and offset from the dorsal colors by an orange lateral line. The tail is orange to orange red, except near the base, where the coloration is black frosted with fiery orange red.

SIZE: Both sexes—HB 272-303 mm; T 242-310 mm.

DISTRIBUTION: It can be found in Venezuela south of the Orinoco River, extending from the Colombian border to Cuidad Bolívar.

GEOGRAPHIC VARIATION: None.

CONSERVATION: IUCN status—data deficient. Population trend—no information.

HABITAT: Fiery squirrels inhabit tropical lowlands and semideciduous and evergreen forests; they are not known to occur in secondary or disturbed forests.

NATURAL HISTORY: This species is diurnal. Fiery squirrels are strongly arboreal and are difficult to observe as they forage, which is typically high in the canopy of large trees with dense cover, and often in low humid marshy palm groves. They feed heavily on tree seeds, nuts, and fruits, especially palms. *S. flammifer* nests in leaf nests constructed high in the canopy. Fiery squirrels are occasionally hunted, but habitat loss is probably the major current threat to this species, given their apparent sensitivity to forest disturbance.

GENERAL REFERENCES: J. A. Allen 1915; Eisenberg and Redford 1999; Linares 1998.

Sciurus gilvigularis (Wagner, 1842) Yellow-Throated Squirrel

DESCRIPTION: *S. gilvigularis* has a short thin pelage, with a dorsal coloration grizzled with ochraceous buff and black. A narrow pale buff eye ring is present. No postauricular patches are evident. The ventral pelage is a dark ochraceous orange that is darkest on the chest and upper abdomen and paler on the throat and lower abdomen. The tail is grizzled with buff and black and edged in buff, and it may exhibit faint banding.

SIZE: Both sexes—HB 155-177 mm; T 165-195 mm; Mass 150-165 g.

DISTRIBUTION: The poorly known distribution of this species is often described as disjunct through northcentral and northeastern South America (Venezuela, Guyana, and northern Brazil).

GEOGRAPHIC VARIATION: Two subspecies are recognized.

S. g. gilvigularis—northern portion of the range. The tail is washed with fulvous.

S. g. paraensis—southern portion of the Brazilian range. The tail is washed with white.

CONSERVATION: IUCN status—data deficient. Population trend—no information.

HABITAT: Yellow-throated squirrels inhabit coastal and evergreen forests dominated by palms, lianas, and rattans. *S. gilvigularis* is not found in secondary or logged forests.

NATURAL HISTORY: This species is diurnal. Yellow-throated squirrels feed primarily on the seeds and fruit of palms. Fruit from *Maximiliana martiana* is a dominant food item, as are a diversity of small tree seeds. *S. gilvigularis* eats while sitting on a branch or rattan near a palm; the sound of teeth grinding and nibbled rinds that drop from trees indicate their presence. In Venezuela, *S. gilvigularis* is rare, restricted to low and humid wooded zones.

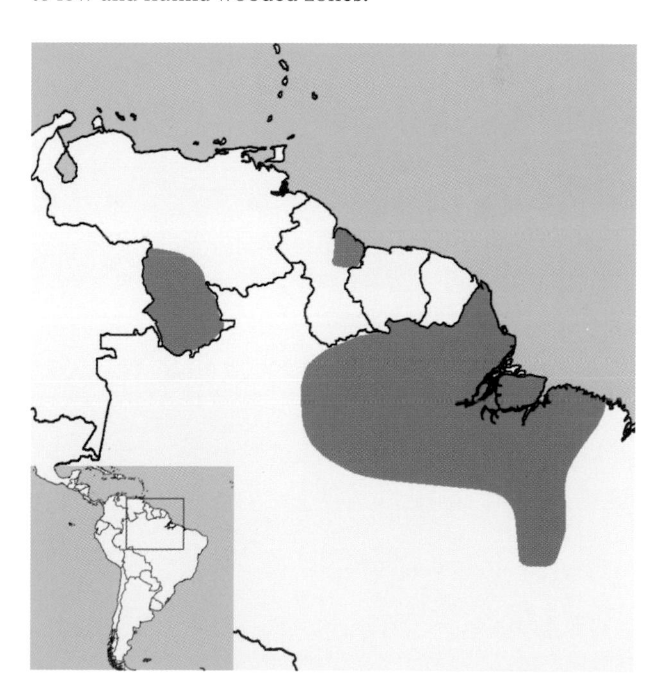

GENERAL REFERENCES: J. A. Allen 1915; Eisenberg 1989; Emmons 1984; Ghilardi and Alho 1990; Linares 1998; Tate 1939.

Sciurus granatensis (Humboldt, 1811) Red-Tailed Squirrel

DESCRIPTION: Red-tailed squirrels possess an extremely variable pelage. The dominant color in the dorsum is often a deep red. Variations range from a grizzled black and yellow to charcoal with a tinge of yellow; black can take the form of a median longitudinal stripe. A faint buff to ochre eye ring is sometimes visible. Melanism is also known. The venter is nearly always lighter, but it can be white to yellow to orange red. The upper side of the tail is red to rust, often grizzled with black and sometimes with a black tip; the lower surface is yellowish brown to charcoal to black, often frosted with reddish hues. Populations east of the Andes most often are white-bellied, Venezuelan populations often contain more brownish variants, and coastal Ecuador and Colombia have more blackish forms with a black-tipped red tail.

SIZE: Both sexes—HB 200-285 mm; T 140-280 mm; Mass 212-520 g.

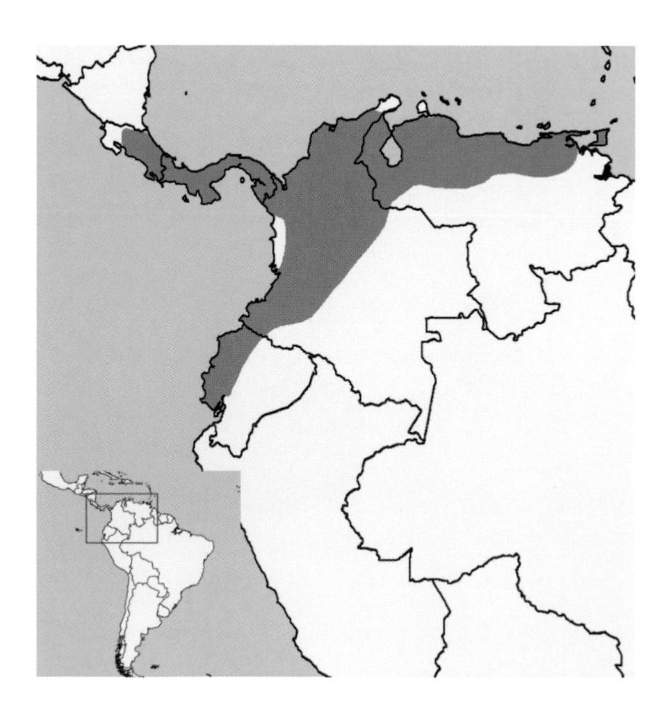

DISTRIBUTION: This species can be found in Costa Rica, Panama, Colombia, Venezuela (including Margarita Island), Trinidad and Tobago, and Ecuador.

GEOGRAPHIC VARIATION: This is an extremely variable form that probably needs a thorough assessment to more clearly define evolutionary relationships, as indicated by the great deal of overlap in the characteristics of the subspecies. Here we recognize 32 subspecies.

- *S. g. granatensis*—near the city of Cartagena (Colombia). The dorsum is orange to orange red, with a sharply contrasting white venter.
- *S. g. agricolae*—near Magdalena (Colombia). It has a brilliant orange dorsum with a white underside, and a burnt sienna tail.
- *S. g. bondae*—near Magdalena (Colombia). It has a light reddish dorsum with reduced black grizzling, and the underparts are whitish.
- *S. g. candelensis*—Huila (Colombia). It is a darker and more olivaceous form, with fulvous inner legs.
- *S. g. carchensis*—western slope of the Andes (Ecuador). This form has a grizzled ochraceous buff and black dorsum with a whitish venter; the tail is the same color as the

Sciurus granatensis. Photo courtesy Caroline M. Thorington.

Sciurus granatensis. Photo courtesy Cecil Schwalbe.

dorsum at its base, with the rest of the tail ochraceous orange for most of its length and black tipped.

S. g. chapmani—Trinidad and Tobago. It is a small yellowish-reddish brown form.

S. g. chiriquensis—Panama. It is a medium-sized brownish form, with a yellow to orange red venter.

S. g. chrysuros—near the city of Santa Fe de Bogotá (Colombia). It is a large yellowish to reddish brown form, with a yellow reddish venter and a black tip on the tail.

S. g. ferminae—Napo (eastern Ecuador). This form resembles *S. g. griseogena*, but the upperparts tend more toward brown and less toward olivaceous. The postauricular spots are small and well defined. The tail is very red.

S. g. gerrardi—near the city of Medellín (Colombia). It has red upperparts and underparts, with a black median band. The tail has a black base and a red distal portion.

S. g. griseimembra—eastern slope of the Andes (Colombia). It is a darker and more olivaceous form, with a gray inner surface on the legs.

S. g. griseogena—coast of Venezuela. This form is characterized by a red tail with a black tip.

S. g. hoffmanni—near the city of San José (Costa Rica). It is a medium-sized brownish form, with a yellow to orange red venter.

S. g. imbaburae—Imbabura (Ecuador). The dorsum is ochraceous tawny grizzled with black, the venter is orange, and the tail has a black tip.

S. g. llanensis—Los Llanos region (Venezuela). This form is distinguished from *S. g. griseogena* by its larger size and by its orange front legs, shoulders, and underparts.

S. g. manavi—Manabi (Ecuador). It is a medium-sized brownish form, with a reddish venter and a black base to the tail.

S. g. maracaibensis—near Lake Maracaibo (Venezuela). This form has a dark dorsum, mostly black, with only the shoulders orange rufous. The underparts are orange rufous.

S. g. meridensis—near the city of Mérida (Venezuela). This form is characterized by a red tail with a black tip.

S. g. morulus—area now occupied by the Panama Canal (Panama). It has a reddish dorsum, with a black-tipped tail.

S. g. nesaeus—Margarita Island (Venezuela). This subspecies has a light ochraceous tail and dorsum, with a bright orange rufous venter.

S. g. norosiensis—Bolivar (Colombia). It is a bright agouti form, with a paler and less blackish dorsum and a red venter.

S. g. perijae—Magdalena (Colombia). This form is mixed orange and black on the posterior of the dorsum, with a black median stripe, orange shoulders, and a white venter.

S. g. quindianus—Valle del Cauca (Colombia). This is a large form, with a dark black dorsal stripe and a black tip on the tail.

S. g. saltuensis—Magdalena (Colombia). This subspecies has a light yellowish red dorsum grizzled with black, and a whitish venter.

S. g. soederstroemi—near the city of Quito (Ecuador). It is a small brownish form, with ochraceous underparts.

S. g. splendidus—near Magdalena (Colombia). This form has a dark red dorsum and a white venter.

Sciurus granatensis. Photo courtesy Wilmer Quiceno, Medellín (Colombia).

S. g. sumaco—Mount Sumaco (Oriente, in eastern Ecuador). This form is more gray, and the tail is black and buff, not red.

S. g. tarrae—near Santander (Colombia). It is a very dark form that is uniformly dark agouti above, with deep red underparts.

S. g. valdiviae—Antioquia (Colombia). This is a subspecies with a blackish yellow dorsum and a reddish orange venter; the tail is black at the base, with a reddish tip.

S. g. variabilis—near Magdalena (Colombia). It is a large-bodied form that is blackish and reddish on the dorsum, with a sharply defined white venter.

S. g. versicolor—northern Ecuador. It is red above and below, with a black median stripe the length of the dorsum, and a tail that is black at the base and the tip.

S. g. zuliae—Venezuela, near the border with Colombia. This is a reddish form, with a very black tip on the tail.

CONSERVATION: IUCN status—least concern. Population trend—stable.

HABITAT: Red-tailed squirrels inhabit many types of forests, from sea level to montane, including secondary forests and picnic areas and parks.

NATURAL HISTORY: This species is diurnal. Red-tailed squirrels are specialists on large tropical tree seeds, and they can be heard gnawing through the thick protective seed coats while sitting motionless on a forest perch. *S. granatensis* plays a major role in the evolution and dispersal of seeds within tropical forests. A number of other seeds, fruits, leaves, flowers, fungi, plant saps and exudates, and even animal materials are eaten opportunistically. Foods are scatter-hoarded on the ground and hidden in branch forks or lianas in the canopy. Red-tailed squirrels can be found foraging in multiple layers of forest canopies throughout the year. Most often found in the midcanopy and the upper canopy when seeds ripen, *S. granatensis* is less commonly seen on the ground. From November to August, mating chases occur where several to many males will follow and chase the female while grunting and snorting in an effort to gain access during a single day of estrus. Copulations last only about 10 seconds, and males lose interest within 30 minutes after copulation. Gestation is probably about 45 days, and a litter of one to three young is born in a leaf nest within the canopy. As many as three litters may be produced during what appears to be a prolonged breeding season. The young continue to nurse for up to two months, and all disperse from their natal areas. Individuals live a solitary life, although feeding aggregations may occur at seasonal food sources. Female home ranges are relatively exclusive and defended from other females, and they cover about 0.64 ha; male home ranges are larger, at about 1.5 ha, with considerable interindividual overlap. Annual survivorship is 50–64 percent; preweaning juvenile survival is low. Maximum longevity in the wild is at least 7 years. Their major predators are monkeys, mustelids, procyonids, felids, raptors, and a multitude of snakes. *S. granatensis* will produce a series of hoarse "chucks" when alarmed; however, it usually remains relatively silent. Red-tailed squirrels are considered pests throughout much of their range in places where they come into contact with agriculture, as this species will eat a diversity of fruit and nut crops; harvested cuts, orchards, and the edges of pastures and urban areas harbor these squirrels. *S. granatensis* is hunted as a pest and for food throughout its range.

GENERAL REFERENCES: Adler et al. 1997; Carvajal and Adler 2008; Heaney and Thorington 1978; Mena-Valenzuela 1998; Nitikman 1985.

Sciurus griseus (Ord, 1818)
Western Gray Squirrel

DESCRIPTION: Western gray squirrels have a slate to silver gray dorsum that sometimes appears finely grizzled with black, and a white to buff eye ring. The venter is white, sometimes with a faint wash of buff. The tail is colored similarly

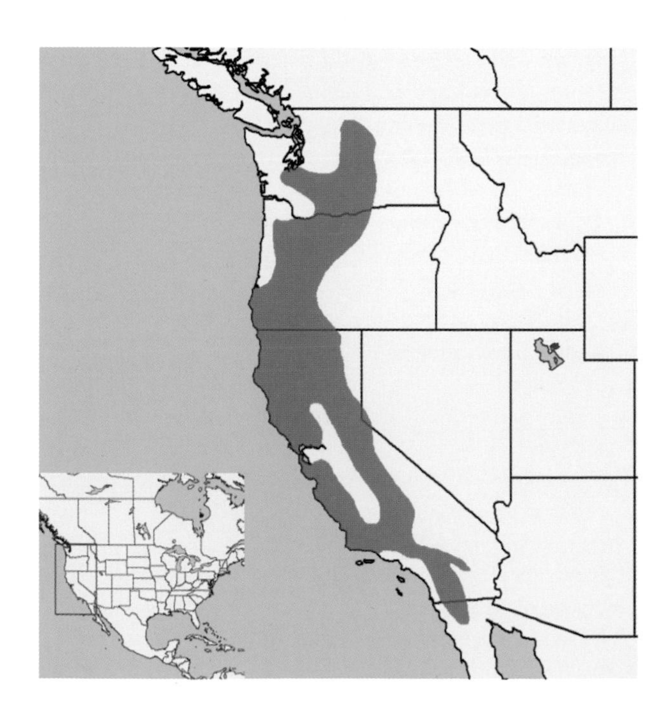

to the dorsum, but it is sometimes darker at the core and has a frosting of white to silver gray. The ears are a steel or silver gray and quite prominent, nearly double their relative length in many other squirrels.

SIZE: Both sexes—HB 265-323 mm; T 240-309 mm; Mass 520-942 g.

DISTRIBUTION: This species is found in central Washington, western Oregon, and California (USA) to Baja California (México).

GEOGRAPHIC VARIATION: Three subspecies are recognized.

S. g. griseus—Washington, Oregon, northern California, and most of the Sierra Nevada (USA). This form is a silvery gray color on the dorsum and limbs.

S. g. anthonyi—extreme southern portion of California (USA), from north of Los Angeles southward into Baja California (México). This subspecies is intermediate in coloration.

S. g. nigripes—Coast Range, from San Francisco Bay southward to north of Los Angeles (USA). This form is much darker on the dorsum and feet, approaching slate gray in many individuals.

Sciurus griseus. Photo courtesy Larry McCombs/Flickr.

CONSERVATION: IUCN status—least concern. Population trend—no information. *S. griseus* is listed as a threatened species in the state of Washington and as sensitive in Oregon (USA).

HABITAT: Western gray squirrels occur in oak-conifer woodlands and mixed conifer forests on the mesic western side of the mountains. Open pine (*Pinus*) forests are frequented on the more xeric eastern side of the mountains. This species is also found in cities and, occasionally, in nut orchards, but it does not seem to thrive in such close proximity to humans.

NATURAL HISTORY: This species is diurnal. Western gray squirrels are often seen foraging on the ground during all seasons, and they are less apt to be found moving in the canopy than most tree squirrels. When in the canopy, this large-bodied squirrel appears less agile and rarely leaps between branches, as do other tree squirrels. Males chase a female through the canopy and across the ground on her single day of estrus, in an attempt to mate; aggression levels between males are considerable. The breeding season ranges from late December to June or July. After a gestation of about 44-45 days, a litter of one to six is born in a nest cavity or leaf nest. Cavities may be preferred for rearing offspring. Most of the young leave the nest by May, and all disperse from their natal area. Sexual maturity is reached at 10-11 months of age; however, young females have less success in producing a litter than older females. Adult individuals are solitary and are not territorial. Home ranges cover from less than 5 ha for both sexes in some parklands, to 73 ha for males and 22 ha for females elsewhere, with considerable overlap between individuals. Males overlap with many females, especially during the breeding season. Nests are in large trees within reasonably dense and diverse forest sites. *S. griseus* is primarily herbivorous, feeding on the seeds of oaks (*Quercus*) and a variety of conifers. Epigeous (aboveground) and hypogeous (underground) fungi are also important foods and may make up most of the diet in some forests. In addition, western gray squirrels will feed on flowers, fruits, forbs, sap, and cambium. When disturbed or threatened, *S. griseus* usually ascends a tree and faces the threat while flagging its tail and projecting an alarm call in a series of moderately pitched "chucks," often increasing in length as the end of the series approaches. Major predators include raptors, bobcats (*Lynx rufus*), domestic cats, coyotes (*Canis latrans*), foxes, domestic dogs, and mustelids. Western gray squirrels feed on a variety of nut crops and are considered pests in some areas. This species is hunted outside of the state of Washington. Major threats to persistence are habitat loss and fragmentation, introduced eastern gray squirrels

(*S. carolinensis*) and eastern fox squirrels (*S. niger*), and poor regeneration of oak woodlands.

GENERAL REFERENCES: Bayrakçi et al. 2001; Carraway and Verts 1994; Garrison et al. 2005; Gregory et al. 2010; Linders et al. 2004; Mellink and Contreras 1993; Ryan and Carey 1995.

Sciurus ignitus (Gray, 1867)
Bolivian Squirrel

DESCRIPTION: Bolivian squirrels have an agouti brown gray dorsum, with protruding ears and a postauricular buffy patch. The venter is whitish buff to pale orange. The tail is the same as the dorsum, but often frosted in yellow to pale orange.

SIZE: Both sexes—HB 140-202 mm; T 150-199 mm; Mass 183-242 g.

DISTRIBUTION: This species is found in Peru, Brazil, Bolivia, and Argentina.

GEOGRAPHIC VARIATION: Five subspecies are recognized.

S. i. ignitus—east of the Andes, in Bolivia and adjacent Peru. This form has an ochraceous buff venter.

Sciurus ignitus. Photo courtesy Fabrice Schmitt.

S. i. argentinius—from Bolivia into extreme northwestern Argentina. This subspecies is distinguished by a yellowish venter and a pinkish to reddish tinge to the ears.

S. i. boliviensis—much of Bolivia. This is a large-bodied form, with a white venter.

S. i. cabrerai—Brazil. The dorsal surface is dark chestnut grizzled by reddish hair tips. The ears are edged in orange hairs, with the remainder of the ears fulvous. There is a distinct postauricular tuft of fulvous fur. The upper surface of the hands and feet are reddish.

S. i. irroratus—Peru, east of the Andes. This subspecies has a yellow venter.

CONSERVATION: IUCN status—data deficient. Population trend—no information.

HABITAT: Bolivian squirrels are found primarily in evergreen lowland tropical forests, which are transitional between humid tropical forests and dry tropical forests. *S. ignitus* can also be found in disturbed forests.

NATURAL HISTORY: This species is diurnal. Bolivian squirrels use all but the highest strata of the forest and will travel on the ground, often in dense vegetation. *S. ignitus* is solitary, and it nests high in the canopy (6-10 m), in bolus-shaped leaf nests hidden in vines or palms. It forages in cover for seeds, nuts, fruits, and fungi, and it will eat insects. When alarmed, this squirrel gives a series of slow soft "chucks." It is rarely hunted and not considered a pest.

GENERAL REFERENCES: S. Anderson 1997; Emmons and Feer 1990; Haugaasen and Peres 2005; Woodman et al. 1995.

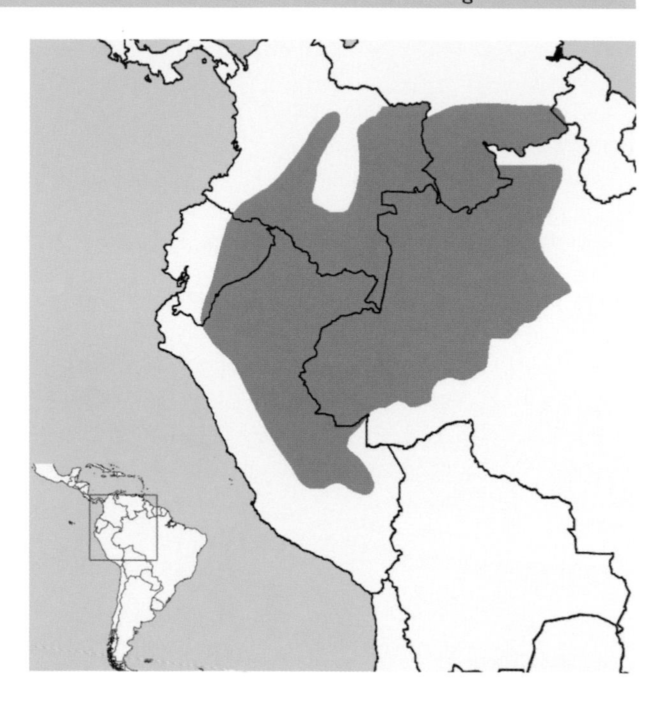

Sciurus igniventris (Wagner, 1842) Northern Amazon Red Squirrel

DESCRIPTION: Northern Amazon red squirrels have a dark chestnut red or rusty orange yellow dorsal pelage grizzled with black. The ears are thinly haired and extend above a crown that is often black; postauricular patches are absent or poorly defined and slightly yellow. The feet are bright red or orange, without black. The ventral pelage sharply contrasts with the dorsal pelage and has a pale orange, red, or white coloration, sometimes offset from the dorsum by a black lateral line. The tail is very bushy and black at the base, and orange or rusty at the distal end. Melanism occurs occasionally.

Sciurus igniventris. Photo courtesy Geoffrey Palmer.

SIZE: Both sexes—HB 240-295 mm; T 240-305 mm; Mass 500-900 g.

DISTRIBUTION: This species is found in Colombia, Venezuela, Ecuador, Peru, and Brazil.

GEOGRAPHIC VARIATION: Two subspecies are recognized.

S. i. igniventris—eastern part of the range, in Venezuela, Brazil, and eastern Peru. This form has an ochraceous reddish dorsum, with a ferruginous venter.

S. i. cocalis—western part of the range, in Colombia, Ecuador, and northern Peru. This subspecies has a dark blackish median band from head to tail, with a pale ochraceous buff venter.

CONSERVATION: IUCN status—least concern. Population trend—no information.

HABITAT: Northern Amazon red squirrels inhabit the lowland forests of the Amazon.

NATURAL HISTORY: This species is diurnal. Northern Amazon red squirrels are specialists on large tree seeds with extremely thick hard endocarps. The squirrel's body size and skull characteristics are well adapted to the challenge of opening these large protected food sources. *S. igniventris* is common in mature and disturbed rainforests, and it is often seen on the ground, in low undergrowth, and in the vicinity of palm trees. Individuals live and forage alone, only occasionally interacting with others at concentrated food sources. Open areas are not often used; instead, foraging takes place in all levels of the canopy, with about 10 percent of the time spent on the ground. Travel and feeding most often occur in the canopy. Primarily herbivores, northern Amazon red squirrels feed on the nuts and fruits of palms and other trees; however, they will eat insects gleaned from among lianas. *S. igniventris* is often the major seed predator of a tree. Seeds are collected as soon as they are ripe, and the supply is used by these squirrels until it is exhausted. Their loud gnawing to reach the endosperm through its thick coverings is an excellent indicator of their presence. Northern Amazon red squirrels are easily flushed, and they run away through the undergrowth along the ground when alarmed. They rarely call, but when threatened they produce an alarm call that is a low-frequency short series of "chatters" and "chucks." *S. igniventris* populations decrease in areas where it is hunted for food. The desirability of northern Amazon red squirrels as a food source varies considerably with location; in some regions its meat is highly valued, while in other areas squirrels are not viewed as food. In many regions northern Amazon red squirrels are valued as pets and sold in the pet trade.

GENERAL REFERENCES: Emmons 1984; Emmons and Feer 1990; Mena-Valenzuela 1998; Patton 1984; Silvius 2002; Youlatos 1999; Zapata-Ríos 2001.

Sciurus lis (Temminck, 1844) Japanese Squirrel

DESCRIPTION: Japanese squirrels have a brown dorsum that is heavily suffused with red. Patches of orange occur low on the sides and on the shoulders and hips. A faint white to buff eye ring is often visible. The venter is white. The tail is similar in color to the dorsum, with a light frosting of white to buff sometimes in evidence. The winter pelage is much grayer on the dorsum and tail.

SIZE: Both sexes—HB 160-220 mm; T 130-170 mm; Mass 250-310 g.

DISTRIBUTION: This species was previously found on the islands of Honshu, Shikoku, and Kyushu (Japan). It is believed to be extirpated from the island of Kyushu, however.

GEOGRAPHIC VARIATION: None.

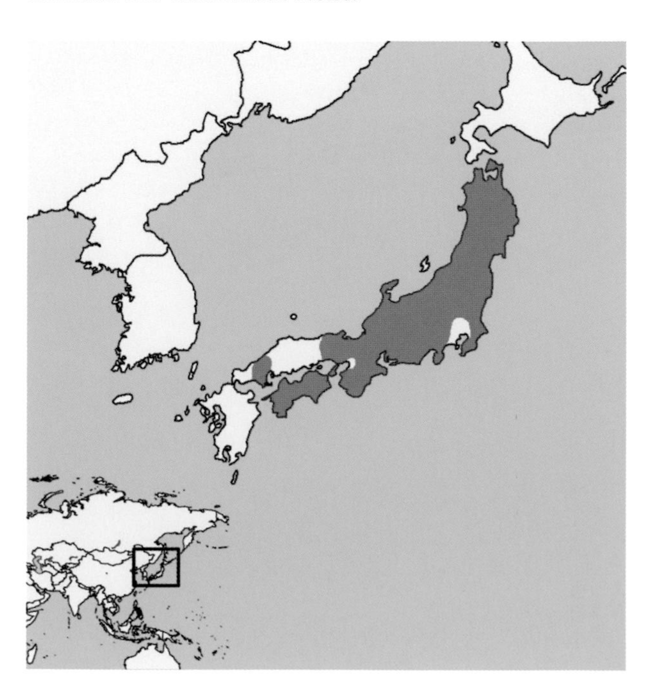

Sciurus lis. Photo courtesy Mikio Okubo, Sciurus lis of Nagano Pre. Japan.

CONSERVATION: IUCN status—least concern. Population trend—stable. Japan's *Red Data Book* lists *Sciurus lis* as a locally threatened population for populations in the Chugoku and Kyushu districts.

HABITAT: Japanese squirrels are found in lowland to subalpine primary and secondary pine (*Pinus*) and mixed forests, and they can also be found in suburban forests.

NATURAL HISTORY: This species is diurnal. Japanese squirrels are active throughout the year and can often be seen foraging on the ground or in the canopy. A group of males follows the estrous female on her single day of receptivity during a breeding season. Breeding occurs in two pulses: February-March and May-June. After a gestation of 39-40 days, a litter of two to six young are born in a leaf nest, den cavity, or burrow. The young are weaned within a few weeks of exiting the nest and disperse from their natal areas soon afterward. Adults are solitary, but occasionally they have been observed to nest with other adults, especially in winter. Male home ranges are between 4 and 30 ha, whereas female home ranges are from 4 to 17 ha; overlap between individuals is modest. Home ranges expand in heavily fragmented environments, because individuals must incorporate poor-quality marginal habitat. Japanese squirrels are primarily herbivores, feeding on seeds, buds, flowers, and fruits, particularly those of trees. Insects and epigeous (aboveground) fungi are also consumed when available. *S. lis* scatter-hoards thick-shelled nonperishable foods (such as walnuts) against winter food shortages. Annual mortality has been estimated at 63 percent. Predators include martens (*Martes*), foxes, domestic cats, domestic dogs, and a variety of raptors and crows. Although *S. lis* was hunted in the past, this has been illegal since 1994. Habitat loss, habitat fragmentation, and forest degradation due to disease are major challenges for their conservation.

GENERAL REFERENCES: Kataoka and Tamura 2005; Tamura 2004; Tamura and Hayashi 2007, 2008.

Sciurus nayaritensis (J. A. Allen, 1890) Mexican Fox Squirrel

DESCRIPTION: *S. nayaritensis* has a grizzled brown dorsum with a yellow to rufous venter, and a charcoal tail frosted with white. Two molts occur each year; the winter pelage is more rufous, and the scrotum is often ringed with white.

SIZE: Female—HB 259.1 mm; T 254.5 mm; Mass 667g.
Male—HB 248.2 mm; T 245.8 mm; Mass 736 g.

DISTRIBUTION: This species is found in the Chiricahua Mountains in extreme southeastern Arizona (USA) and south along the Sierra Madre Occidental into Jalisco (México).

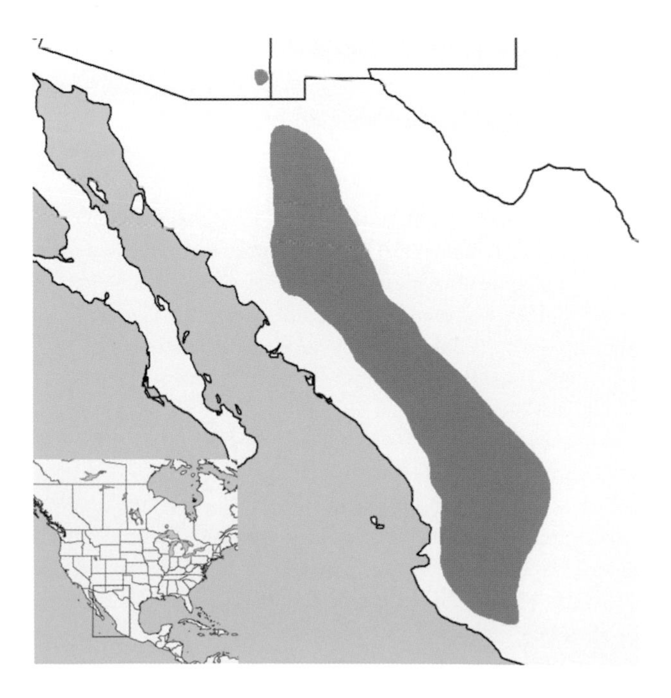

Sciurus nayaritensis. Photo courtesy Bruce D. Taubert.

GEOGRAPHIC VARIATION: Three subspecies are recognized.

S. n. nayaritensis—southern portion of the range (México). This is a smaller and more yellow subspecies.
S. n. apache—northern and middle portion of the range (México). This form is intermediate in size and coloration.
S. n. chiricahuae—Chiricahua Mountains of southeastern Arizona (USA). An endemic form, it is characterized by a more reddish coloration throughout the dorsal pelage and venter.

CONSERVATION: IUCN status—least concern. Population trend—no information. *S. n. chiricahuae* is a species of special concern in the USA.

HABITAT: Mexican fox squirrels are found in forests, ranging from low-elevation (> 1500 m) Madrean forests with a mixture of pine (*Pinus*) and oak (*Quercus*) to higher-elevation (< 2700 m) mixed conifer forests. Riparian areas with large cottonwoods (*Populus*) and sycamores (*Platanus*) often harbor the highest densities of these squirrels.

NATURAL HISTORY: This species is diurnal. The ecology of Mexican fox squirrels is poorly known, although densities are often low in the open arid forests that are inhabited by this species. *S. nayaritensis* typically produces a single small litter of one or two young in late spring or summer, after a short day-long estrus in spring or early summer (February-July), when multiple males pursue females. Males possess scrotal testes in winter through summer; the testes are withdrawn into the abdominal cavity during the remaining months. Mexican squirrels nest almost exclusively in dreys composed of sticks and leaves within the forest canopy; cavities within large-diameter trees are occasionally used by each sex and may be favored by nursing females. Communal nesting is known. *S. nayaritensis* does not hibernate. Home ranges of males average 35 ha, whereas females occupy 15 ha on average. Males maximize their overlap with females, especially during the breeding season. Large raptors, canids, felids, procyonids, and snakes are probably this species' major predators. Mexican fox squirrels are notably silent and appear to prefer to seek cover and remain motionless unless they are startled; then they will bark and "chuck" from elevated locations. *S. nayaritensis* forages extensively on the ground and in the forest canopy for tree seeds and flowers, in addition to fungi. Seeds from the cones of pines (*Pinus*), Douglas firs (*Pseudotsuga menziesii*), and true firs (*Abies*) are consumed by removing individual cone scales to excise the seed. Acorns and walnuts are also eaten when available, along with a variety of other tree seeds. In addition, hypogeous (underground) and occasionally epigeous (aboveground) fungi, as well as insects, are consumed by this opportunistic feeder. Mexican fox squirrels occasionally cache large storable seeds by scatter-hoarding them in leaf litter and topsoil. Mexican fox squirrels are hunted for food and sport throughout their range. Due to the low human population densities within the range of this species, *S. nayaritensis* is of little economic value or detriment.

GENERAL REFERENCES: Best 1995d; D. E. Brown 1984; Hoffmeister 1986; Kneeland et al. 1995; Koprowski and Corse 2001, 2005; Pasch and Koprowski 2006a, 2006b; Yensen and Valdés-Alarcón 1999.

Sciurus niger (Linnaeus, 1758)
Eastern Fox Squirrel

DESCRIPTION: The dorsal pelage of eastern fox squirrels is variable. *S. niger* from the western and northern portions of the range are grizzled with a suffusion of buff to orange; the venter is white to cinnamon, but usually rufous. In the southeastern USA, the dorsum is grizzled buff to gray to agouti to black, with a white or cream nose, ears, and feet and a black crown and nape; the venter is white to tan to rust. Fox squirrels in the northeastern coastal portion of the range are silvery gray washed with buff to a reddish tone on the hips, feet, and head; the tail is pale gray; the venter is white to pale gray, sometimes cinnamon. Melanism is common, especially in the south.

SIZE: Both sexes—HB 260-370 mm; T 200-330 mm; Mass 507-1361 g.

DISTRIBUTION: This species can be found from Saskatchewan and Manitoba (Canada) south to Texas (USA) and adjacent México and east to the Atlantic coast. In addition, it has been introduced to Ontario (Canada) and Washington, Oregon, California, Idaho, Montana, Colorado, Arizona, and New Mexico (USA).

GEOGRAPHIC VARIATION: Ten subspecies are recognized.

S. n. niger—southeastern USA (North and South Carolina, Georgia, and the panhandle of Florida). This is a highly variable subspecies, with buff to gray to black forms; a white snout, ears and feet; and (usually) a white to cream venter.

S. n. avicennia—throughout the Everglades and Big Cypress regions in the southern tip of Florida (USA). This is a highly variable subspecies, ranging from dark brown to tawny to cinnamon forms, with a white mask around the snout and cheeks. The feet and ears are white.

S. n. bachmani—Mississippi and Alabama (USA). It is characterized by a grizzled brown dorsum with an orangish venter, and a black head with varying amounts of white to buff on the ears and snout.

S. n. cinereus—Delmarva Peninsula, encompassing Delaware and portions of Maryland and Virginia (USA), in the extreme northeast part of the range. This endangered subspecies is silver gray grizzled finely with black. The venter is white to buff.

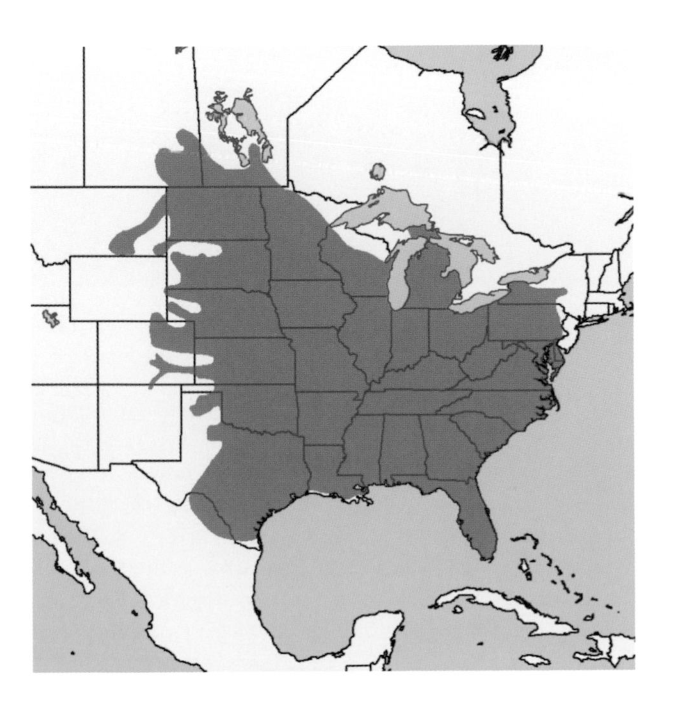

S. n. limitis—central Texas (USA) and northern México. This is a modest-sized cinnamon form with a grizzled dorsum and cinnamon below.

S. n. ludovicianus—eastern Texas and western Louisiana (USA). This is a large subspecies, with a light reddish tone to the dorsum and venter.

S. n. rufiventer—west of the Appalachian Mountains, from Oklahoma, Arkansas, and Tennessee (USA) northward to Saskatchewan and Manitoba (Canada). This is the most common subspecies, characterized by an orange to rust-colored venter.

S. n. shermani—northern and central peninsular Florida (USA). The dorsum is nearly all black to silvery gray grizzled with black. The head is often black; the ears and muzzle are often white. The venter is tawny. The tail is the same color as the dorsum and frosted with white.

S. n. subauratus—western bank of the Mississippi River, extending from central Arkansas south to Louisiana (USA). This is a smallish subspecies, with a rich dark reddish color to the dorsum and venter.

S. n. vulpinus—eastern and southern Pennsylvania, West Virginia, Maryland, and northern Virginia (USA). This subspecies is a grayish form washed with orange reddish on the dorsum, sides, limbs, and underside of the tail. The venter is white to cream.

CONSERVATION: IUCN status—least concern. Population trend—stable. *S. n. cinereus* is listed as endangered in the USA. There is a nonessential experimental population in Sussex County, Delaware (USA). *S. n. avicennia, S. n. cinereus, S. n. shermani,* and *S. n. vulpinus* are of special concern within their distributions.

HABITAT: Fox squirrels prefer open woodland habitats, with scattered trees and an open understory. In the western portion of the range, this includes riparian areas and fence-rows. In the north and midwest, the habitat is open hardwood forests and savannas. In the southeast, this species' habitat is pine (*Pinus*) woodlands, and even scattered swamp-bound stands of bald cypress (*Taxodium*) and pine.

NATURAL HISTORY: This species is diurnal. Eastern fox squirrels favor movement on the ground, but they will forage in the canopy throughout the year. Mating activity peaks twice annually: November-February (with a December peak) and April-July (with a June peak). Males track the females' reproductive condition by following females and sniffing their genitalia. Estrus lasts less than one day during a breeding season. Males congregate at the nest of the female,

Sciurus niger shermani. Photo courtesy Bo Chambers.

often at first light, or—in low-density populations—they may sleep in the female's nest the night prior to estrus. Up to 11 males will pursue a female during her mating bout. Dominant males are able to mate by pursuing the female; however, lower-ranking males, typically young individuals, are able to copulate by waiting on the periphery of the bout and tracking the female when she evades the aggression of dominant males. Copulation lasts less than 30 seconds. A copulatory plug forms within the female from the coagulation of semen. Females will remove the plugs and mate with several males. After a gestation of 44 or 45 days, a litter averaging two to three (maximum of seven) young is born in a tree cavity or leaf nest. Leaf nests are large spherical structures constructed in the canopy using twigs and leaves, with a central nest chamber often lined with shredded bark and leaves. A few females produce two litters in a year; adult females typically produce a single litter each year. The young emerge after 6-8 weeks, and weaning is not completed un-

Sciurus niger niger. Photo courtesy Daniela Duncan.

til as late as 12 weeks afterward. Adult size is not reached until after 1 year of age. Most individuals do not reproduce until at least their yearling year; precocious breeding at 8.0 months by females has occurred. Natal dispersal takes place soon after weaning, and all males and females leave their natal area. Groups of unrelated males and (rarely) females nest together at night. Scent-marking with cheek glands is used to mark trees, and these glands are smelled when individuals greet each other. Traditional scent-marking sites are found in protected areas, such as under low branches or on the underside of a slanted tree trunk; primarily males gnaw the location and then wipe their cheeks back and forth to deposit scent, and they sometimes urinate to scent-mark.

Eastern fox squirrels feed heavily on tree seeds, but they also eat fruits, buds, flowers, fungi, herbs, and (occasionally) animal matter such as insects, bird eggs, nestlings, fish, and conspecifics. A conspicuous behavior of *S. niger* is the annual scatter-hoarding of hard low-perishability nuts and acorns each fall to serve as their winter food store. The squirrels collect tree seeds near the parent tree, disperse to greater distances, and bury them just under the surface. These behaviors all serve to decrease the loss of this critical food source to other seed predators. Caches are recovered using a combination of spatial memory and olfactory cues. The home ranges of individuals overlap considerably, but females may defend exclusive-use cores, especially during lactation. Resident adult females are important in regulating ingress and settlement. Home ranges vary from 0.9 ha to

Sciurus niger avicennia. Photo courtesy Keith Bradley.

Sciurus niger cinereus. Photo courtesy Cynthia C. Winter.

Sciurus niger rufiventer. Photo courtesy Lowell B. Johnson, www.flickr.com/lbjbirds Lowell Johnson.

42.8 ha, are larger for males, and decrease with habitat fragment size. Annual adult survivorship is high, at greater than 60 percent throughout adult life. Longevity in the field for males is 8.3 years, and for females 12.6 years; in captivity, it is at least 13.0 years. Predators of eastern fox squirrels include mustelids, procyonids, felids, canids, opossums, raptors, and snakes. Their alarm calls are a series of medium-pitched barks, often followed by long high-pitched whines. Eastern fox squirrels are a game animal throughout their range, and they are highly prized as a food. These squirrels are also often valued in parks and cities as wildlife that is easily watched, and they present an educational opportunity for studies of ecology and behavior. Furthermore, through the burial of seeds, this species plays an integral role in the succession of grasslands to mature forests. *S. niger* is considered a pest in gardens, fields of agricultural crops (corn/maize, oats, wheat, soybeans, and sorghum), and orchards (apples, oranges, blueberries, and cherries), especially where these squirrels were introduced in California. Habitat loss and fragmentation remain the most significant threats to eastern fox squirrels.

GENERAL REFERENCES: Adam 1984; Bertolino 2009; Derge and Steele 1999; Geluso 2004; Koprowski 1994b, 1996; Kotler et al. 1999; McCleery et al. 2008; K. A. Schmidt 2000; Steele and Koprowski 2001; Steele and Weigl 1992; Thorson et al. 1998; Wrigley et al. 1991.

Sciurus oculatus (Peters, 1863)
Peters's Squirrel

DESCRIPTION: Peters's squirrels are large tree squirrels. They have a grizzled gray dorsum, sometimes with a median black band. The eye ring is white to buff. The venter is white suffused with pale yellow to ochraceous buff. The tail is black suffused with white above, whereas below a yellow cast is sometimes apparent.

SIZE: Both sexes—TL 508-543 mm; T 256-269 mm.

DISTRIBUTION: This species is found in the Mexican states of San Luis Potosí, Guanajuato, Querétaro, Hidalgo, México D.F., Puebla, and Veracruz.

GEOGRAPHIC VARIATION: Three subspecies are recognized.

S. o. oculatus—eastern portion of the range. This form has a black median band along the back.
S. o. shawi—northern portion of the range. This subspecies has a pale gray dorsum; the venter is pinkish cinnamon.
S. o. tolucae—western portion of the range. This form has a median band of gray along the dorsum, with a grayish white or sometimes buffy venter.

CONSERVATION: IUCN status—least concern. Population trend—no information. *S. oculatus* is considered endangered by the Mexican government.

HABITAT: Peters's squirrel inhabits pine (*Pinus*) and oak (*Quercus*) forests in arid mountains. It can sometimes be observed in deep arroyos in nearby valleys.

NATURAL HISTORY: This species is diurnal. Little is known about the ecology of Peters's squirrels. They move well in the canopy and clear gaps of greater than 2 m. The breeding season is apparently prior to July and August. Peters's squirrels are primarily herbivorous and are known to feed on tree seeds and fruits, such as almonds, acorns, and wild figs. *S. oculatus* is listed as a fragile species in México, in large part because about 56 percent of its potential habitat has been transformed.

GENERAL REFERENCES: Best 1995e; Sánchez-Cordero et al. 2005.

Sciurus pucheranii (Fitzinger, 1867) Andean Squirrel

DESCRIPTION: Andean squirrels have a reddish brown dorsal pelage, sometimes with a dark longitudinal midline, and a gray or yellow venter. The dorsal and ventral hairs of the tail are black with white tips and gray to black with white tips, respectively. Some individuals have a patch of black fur on the posterior portion of the crown. The body fur is soft and thick, and the ears are sparsely furred.

SIZE: Both sexes—HB 140-184 mm; T 119-160 mm; Mass 100-146 g.

DISTRIBUTION: This species is found in the Colombian Andes.

GEOGRAPHIC VARIATION: Three subspecies are recognized.

S. p. pucheranii—the type specimen is from near the city of Bogotá, and this subspecies is found in the eastern Andes

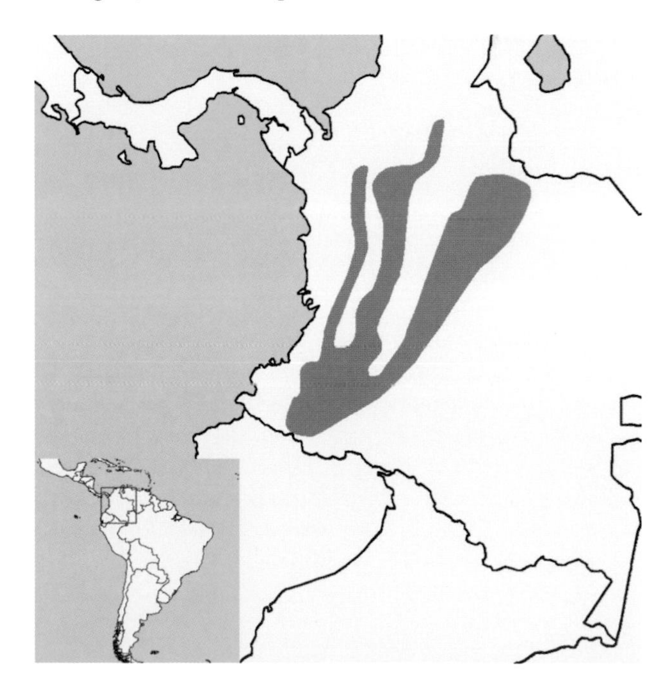

Sciurus pucheranii. Photo courtesy Martin Reid, www.martin reid.com.

of Colombia. It usually lacks a midline and possesses brownish gray underparts, with the pectoral region tinged in buff.

S. p. caucensis—western Andes of Colombia. It usually lacks a midline and possesses brownish underparts washed with ochraceous buff.

S. p. medellinensis—the type specimen is from near the city of Medellín (Colombia). This form usually has a blackish midline and a venter washed with white.

CONSERVATION: IUCN status—data deficient. Population trend—no information.

HABITAT: Andean squirrels inhabit sub-Andean and Andean forests, at elevations from 2200 to 3500 m.

NATURAL HISTORY: This species is diurnal. Nothing is known about the ecology of Andean squirrels. Their dense soft fur reflects the high elevations and cold, moist environments to which *S. pucheranii* is exposed. Forests at these elevations are still dense and tall (35 m). The larger red-tailed squirrels (*Sciurus granatensis*) repeatedly chased Andean squirrels away from eating sap. Deforestation and fragmentation are increasing in the Colombian Andean Cordilleras; reforestation programs are underway. Given their small size, Andean squirrels are probably not subjected to significant hunting pressure.

GENERAL REFERENCES: J. A. Allen 1915; Hernandez-Camacho 1960; Hershkovitz 1977; Leonard et al. 2009.

Sciurus pucheranii. Photo courtesy Martin Reid, www.martin reid.com.

Sciurus pyrrhinus (Thomas, 1898) Junin Red Squirrel

DESCRIPTION: Junin red squirrels have a lush dark red dorsal pelage, sometimes with flecks of white. The venter is strikingly colored, from orange to cream to white, sometimes with patches of each. The tail is orangish red at the tip and is decidedly more chestnut to black toward the base.

SIZE: Both sexes—HB 240-280 mm; T 208-240 mm.

DISTRIBUTION: This species is found on the eastern slopes of the Andes in Peru.

GEOGRAPHIC VARIATION: None.

CONSERVATION: IUCN status—data deficient. Population trend—no information. *S. pyrrhinus* is classified as vulnerable within Peru.

HABITAT: Junin red squirrels are found in montane forests, at elevations from 600 to 2500 m.

NATURAL HISTORY: This species is diurnal. The ecology and evolutionary relationships of *S. pyrrhinus* are very poorly known. A lactating female was taken in January. These squirrels are commonly observed visiting clay licks, both to eat clay individually and in small groups, suggesting some intraspecific tolerance. Individuals were reported to play during the visits. The Junin red squirrel is considered a rare endemic in Peru.

GENERAL REFERENCES: J. A. Allen 1915; Eisenberg and Redford 1999; Emmons and Feer 1990; Hammer and Tatum-Hume 2003; Pacheco 2002; Pacheco, Cadenillas, et al. 2009.

Sciurus richmondi (Nelson, 1898) Richmond's Squirrel

DESCRIPTION: Richmond's squirrel has a dorsum that is nearly uniform dark brown washed with ochraceous. The venter is yellow buff or yellow orange. The tail is the same color as the dorsum above and tawny ochraceous below.

SIZE: Both sexes—HB 160-218 mm; T 130-184 mm; Mass 235-268 g.

DISTRIBUTION: This species occurs in Nicaragua.

GEOGRAPHIC VARIATION: None.

CONSERVATION: IUCN status—near threatened. Population trend—no information.

HABITAT: Richmond's squirrels are found in lowland mature primary and gallery forests that are often associated with riparian zones. They can also be found in plantations and secondary forests.

NATURAL HISTORY: This species is diurnal. Richmond's squirrels forage on the ground and in the lower branches of trees and on tree trunks, rarely in the canopy. Apparently this species is solitary. The breeding season is prolonged, extending from February to September; litters of two to three young are produced in nests. Due both to their small range and to significant forest loss and fragmentation, Richmond's squirrels may face a considerable conservation challenge.

GENERAL REFERENCES: J. K. Jones and Genoways 1971, 1975; Pine 1971; F. A. Reid 1997; Ulmer 1995.

Sciurus sanborni (Osgood, 1944)
Sanborn's Squirrel

DESCRIPTION: Sanborn's squirrel has a uniform olive brown dorsum, with a pale buff eye ring, mouth, and postauricular patches. The venter is white to yellow to pale orange. The tail is the same color as the dorsum, with black banding sometimes evident in its slightly more grizzled appearance.

SIZE: Both sexes—HB 152-175 mm; T 164-184 mm.

DISTRIBUTION: This species is found in the department of Madre de Dios, Peru.

GEOGRAPHIC VARIATION: None.

CONSERVATION: IUCN status—data deficient. Population trend—no information. *S. sanborni* is classified as vulnerable within Peru.

HABITAT: Sanborn's squirrels are found within a restricted range in lowland Amazonian rainforest, at elevations of 300-580 m.

NATURAL HISTORY: This species is diurnal. The ecology and evolutionary relationships of *S. sanborni* are very poorly known. Sanborn's squirrel is considered a rare endemic in Peru. Due to its small size, this species is unlikely to be hunted.

Sciurus sanborni. Photo courtesy Tor Egil Hogsås.

GENERAL REFERENCES: J. A. Allen 1915; Eisenberg and Redford 1999; Emmons and Feer 1990; Pacheco 2002; Pacheco, Cadenillas, et al. 2009.

Sciurus spadiceus (Olfers, 1818)
Southern Amazon Red Squirrel

DESCRIPTION: Southern Amazon red squirrels have a dark chestnut red or rusty orange dorsal pelage mixed with black. The ears are thinly haired and protrude above the crown, which is often black or with a less-defined dark cap; postauricular patches are absent or poorly defined. The feet are dark red mixed with black or solid black. The ventral pelage is pale orange, white, or yellowish, sharply contrasting with the dorsal coloration. The tail is very bushy and black at the base, and orange or rusty at the distal end. Melanism occurs occasionally.

SIZE: Both sexes—HB 240-290 mm; T 235-340 mm; Mass 570-660 g.

DISTRIBUTION: This species is found in Colombia, Ecuador, Peru, Brazil, and Bolivia.

GEOGRAPHIC VARIATION: Three subspecies are recognized. The range and subspecies in Colombia are uncertain.

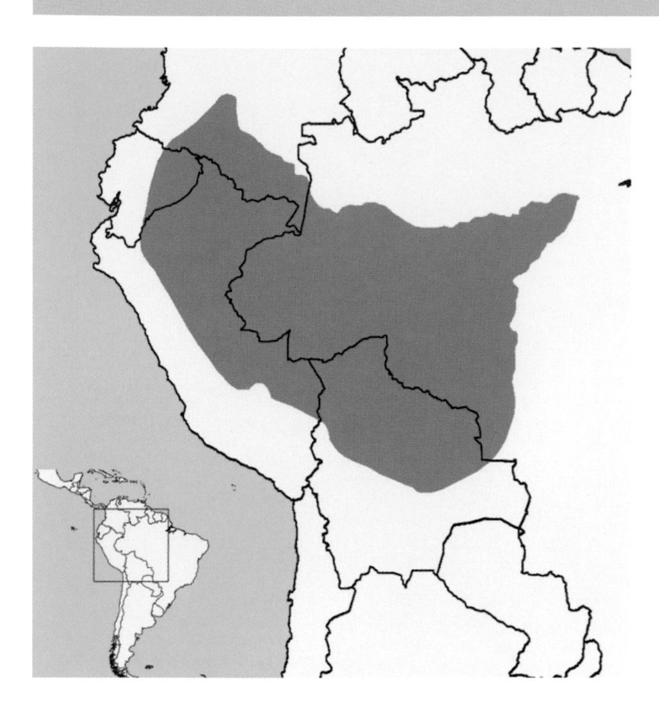

S. s. spadiceus—Brazil. The dorsum is grizzled pale yellowish and dusky. The underparts are ochraceous buff, washed with reddish orange on the cheeks and head.

S. s. steinbachi—Bolivia. This is a large form. It is paler, with the upperparts grizzled pale yellowish and dusky, and the underparts ochraceous buff. It is washed with yellowish on the cheeks and head.

S. s. tricolor—Ecuador and Peru. This is a form with a dark brown to blackish dorsum washed with ochraceous; the venter is pale yellow.

CONSERVATION: IUCN status—least concern. Population trend—no information.

HABITAT: Southern Amazon red squirrels occur in lowland forests of the Amazon and in the foothills of the Andes.

NATURAL HISTORY: This species is diurnal. Southern Amazon red squirrels are common in mature and disturbed rainforests; they are often seen on the ground, in low undergrowth, or in palm trees. Primarily herbivores, *S. spadiceus* feeds on the nuts and fruits of palms and other trees. In the western Amazon region, southern Amazon red squirrels specialize on large nuts with thick hard endocarps; the fruits from three large-seeded genera (*Astrocaryum, Dipteryx,* and *Scheelea*) provide 98 percent of their diet. Their loud gnawing to reach the endosperm through the thickened coverings is an excellent indicator of the presence of southern Amazon red squirrels. Nuts are often cached in the ground, and several individuals can easily remove hundreds of nuts from a palm. Southern Amazon red squirrels do not travel in the forest canopy, but instead run away in the undergrowth when alarmed. They rarely call, but when threatened they produce an alarm call that sounds much like a sneeze, followed by a series of "chatters" and "chucks." Ocelots (*Leopardus pardalis*) and jaguars (*Panthera onca*) are known predators. Dogs are trained to hunt *S. spadiceus*. Although southern Amazon red squirrels are not a preferred item for hunters, the numbers of *S. spadiceus* decrease demonstrably in areas where this large-bodied squirrel is hunted for food.

Sciurus spadiceus. Photo courtesy William E. Quatman.

GENERAL REFERENCES: S. Anderson 1997; Bodmer 1995; Brigido et al. 2004; Eisenberg and Redford 1999; Emmons 1984, 1987; Emmons and Feer 1990; Jernigan 2009; Mena-Valenzuela 1998; Patton 1984; Peres and Baider 1997.

Sciurus stramineus (Eydoux and Souleyet, 1841) Guayaquil Squirrel

DESCRIPTION: Guayaquil squirrels show two morphs, but much variation exists within a population. One morph has a black head, ears, and feet; the dorsum is a grizzled black and white that grades to a pale orange near the hips and the base of the tail. The venter is cream to tan to faint rust. In the second morph, the body is charcoal to black frosted with white, which yields a pale gray agouti effect. The hips are tinged with buff to orange; the top of the neck behind the skull has a white to buff patch. The ears and feet are black, and the venter is grayish. The tail of both morphs is black frosted with white.

SIZE: Both sexes—HB 180-320 mm; T 250-330 mm; Mass 460-495 g.

DISTRIBUTION: This species exists in southwestern Ecuador and extreme northwestern Peru, in the area surrounding the Gulf of Guayaquil. It was introduced to Lima, Peru.

GEOGRAPHIC VARIATION: None.

CONSERVATION: IUCN status—least concern. Population trend—no information.

HABITAT: Guayaquil squirrels inhabit wet evergreen forests and dry tropical forests, including secondary forests. Their tolerance for human disturbance is evident, in that *S. stramineus* can also be found in coffee plantations and urban areas.

NATURAL HISTORY: This species is diurnal. The squirrels forage on the ground and in all levels of the canopy for seeds, nuts, and fruits. Their nests are large bolus-shaped leaf nests in the canopy. Guayaquil squirrels are considered rare in Peru, in part because their range just reaches the northwestern corner of the country. *S. stramineus* has been introduced and has spread throughout portions of Lima (Peru), where individuals can be seen walking on telephone wires to cross streets. An individual in captivity lived to 7.3 years. Because of their colorful pelage, this species is often trapped and sold in the pet trade. Guayaquil squirrels may serve as important indicators of biological change: this species has been used to model future change in forested habitat. Habitat loss and fragmentation in the heavily populated range of this species may put its conservation at risk.

Sciurus stramineus. Photo courtesy Melissa J. Merrick.

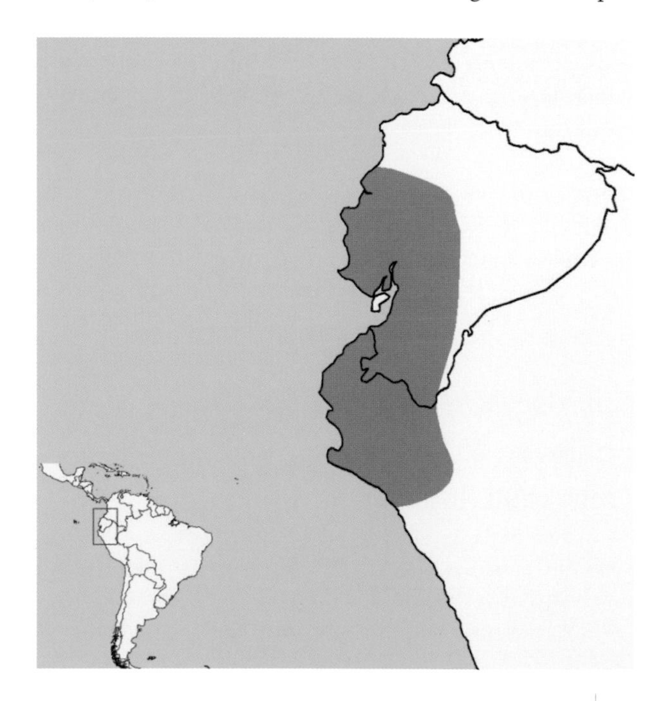

GENERAL REFERENCES: Eisenberg and Redford 1999; Emmons and Feer 1990; Grijalva and Villacis 2009; Jessen et al. 2010; Pacheco 2002; Pacheco, Cadenillas, et al. 2009; Peralvo et al. 2007; R. Weigl 2005.

Sciurus variegatoides Ogilby, 1839
Variegated Squirrel

DESCRIPTION: The variegated squirrel is so named because of the tremendous variation in its pelage characteristics: from almost completely steel gray, to grizzled gray with a yellow wash, to nearly all black. In some areas, animals may have a distinct dorsal patch of brown, or be completely reddish, or have a rust-colored neck, sides, and legs. The tail is either black with interspersed white-tipped hairs or solid black edged with white hairs. The tail color usually distinguishes it from other tree squirrels. In Costa Rica, for example, *S. granatensis* is smaller and has a reddish tail. Light-colored feet distinguish *S. variegatoides* from *S. aureogaster* in México and Guatemala.

SIZE: Female—HB 260.6 mm; T 278.7 mm; Mass 468.8 g.
Male—HB 255.9 mm; T 262.8 mm; Mass 536.9 g.
Sex not stated—HB 263.5 mm; T 268.9 mm; Mass 576.8 g.

DISTRIBUTION: This species ranges from southern Chiapas (México) across Central America to Panama.

GEOGRAPHIC VARIATION: Fifteen subspecies are recognized in this highly geographically variable species.

Sciurus variegatoides rigidus. Photo courtesy Randall D. Babb.

S. v. variegatoides—tropical forest highlands of Guatemala (along the borders with El Salvador and Honduras) south through eastern Honduras to the Pacific coast in eastern El Salvador. All of the upperparts and the base of the tail are a dingy yellowish gray. The underparts and feet are dull ochraceous. The backs of the ears are pale rusty, with basal patches of dull buff. The dorsal surface of the tail is black washed with white, while the ventral surface is colored like the back but bordered in black and edged in whitish gray.

S. v. adolphei—tropical lowlands of Realego (= Realejo) (Nicaragua). The top of the head and the nape are iron gray, but paler than the back. The rest of the upperparts and the base of the tail are dark grayish brown. The rest of the tail is washed with white. The underparts and the inside of the limbs are reddish chestnut brown, sometimes with patches of white.

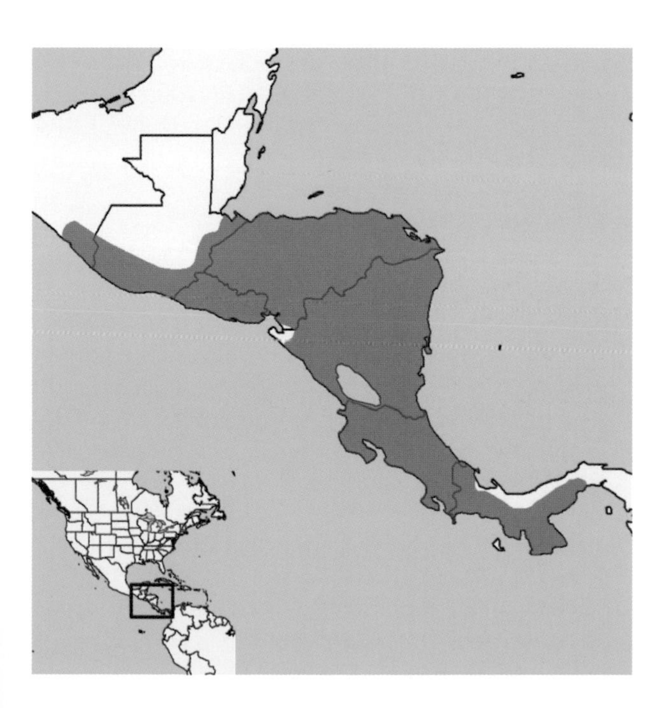

S. v. atrirufus—Tambor (Nicoya Peninsula, Costa Rica). There is a broad black band on the back, and the cinnamon rufus coloring on the underparts extends well up on the sides, without any trace of a lateral stripe. The top of the head is cinnamon rufous, and the ears are bordered in black. There are black stripes over the eyes that meet between the ears, extend down the back, and narrow toward the tail. The tail is black above and cinnamon rufous below.

S. v. bangsi—eastern part of the Pacific coast in Guatemala to the western portion of El Salvador. This form is most similar to *S. v. goldmani,* but it is lighter and grayer than the grayest extreme of that subspecies, with no rusty wash on the dorsal pelage. The ears are edged in black and sometimes have a buffy or rusty ear tuft. A pure white basal car patch is present. The feet are light gray, with white toes. The underparts are white.

S. v. belti—humid tropical forests, ranging from the western coast of Honduras and Nicaragua inland to Yoro (Honduras) and Chontales and Río San Juan (Nicaragua). The underparts are uniformly rusty rufous, and the feet are grizzled with rusty yellowish.

S. v. boothiae—tropical coastal forests of the northern region of Honduras (around the city of San Pedro Sula), southeasterly through central Honduras to the city of San Juan de Murra (= Murra) in Nicaragua. The back is dark grayish brown washed with shining black, and the feet are

Sciurus variegatoides. Photo courtesy Randall D. Babb.

blackish. The venter is white; the ears are black, edged with dark rusty patches at the base. The lateral line on the flank is dull red.

S. v. dorsalis—comparatively arid regions around Lake Nicaragua (Nicaragua) and the northwest area of Costa Rica, including the Nicoya Peninsula. The colors on this form are brighter than on *S. v. adolphei*. The midline of the dorsal surface is blackish brown or grizzled yellowish brown, surrounded by lighter pelage ranging from buffy to white. The underparts are white, buffy yellow, or bright rufous. The head coloration is lighter and paler than that of the dorsal surface.

S. v. goldmani—arid tropical forests in the southeastern coastal band in Guatemala, extending north over the border to Finca Esperanza (Chiapas, México). The upperparts are iron gray, and the underparts are white. The ears are black rimmed, rufous tufted, and have a white patch behind the base. The tail is black on top, washed with white.

S. v. helveolus—Panama, along the Pacific coast from Panama City toward the city of Santiago. This form is similar to *S. v. dorsalis* and *S. v. variegatoides*, but the limbs and underparts are paler.

S. v. loweryi—Valle del General and Valle de Coto Brus (Costa Rica). The upperparts are predominately black, intermixed with tawny. The lateral and underparts are tawny, as are the postauricular patches. The dorsal surface of the tail is black and white. The ventral surface of the tail has hairs that are brown at the base, followed by bands of tawny and black, and tipped with white. This subspecies may be distinguished from *S. v. dorsalis* by *S. v. loweryi*'s smaller size and the presence of tawny (never white) patches on the dorsum.

Sciurus variegatoides thomasi. Photo courtesy Alex Vargas, www.pbase.com/alex_vargas.

S. v. managuensis—humid tropical forests along the Managua River in a small area around Quiraguá (Guatemala), by the border with Honduras. This form is smaller than *S. v. boothiae*. The dorsal surface is blackish yellow, and the ventral surface is buffy yellow.

S. v. melania—Pacific coastal lowlands of the eastern portion of Panama. This form, as indicated by the subspecies name, possesses pelage of a uniformly polished black color.

S. v. rigidus—central Costa Rica, extending to the coast of the Gulf of Nicoya. The black hairs are ringed once with a rust or ochre color. The upperparts are black, sprinkled with rusty ochre yellow or white. The underparts are rusty, and whitish toward the midline. The chin and the patches at the base of the ears are also whitish. The extremities are rust colored, mixed with black.

S. v. thomasi—humid tropical forests in the central and Caribbean coastal regions of Costa Rica. All of the upperparts are black, with underlying dark yellow or ferrugi-

***Sciurus variegatoides*.** Photo courtesy Alex Vargas, www.pbase.com/alex_vargas.

nous brown shining through. The underparts are a deep ferruginous color. The ears are blackish, with black tufts and ferruginous basal patches. White patches occasionally occur on the underparts.

S. v. underwoodi—from El Caliche Cedros (Honduras) south to the city of Matagalpa (Nicaragua); also in Port Parker Bay (= Golfo Santa Elena) in Costa Rica. The underparts are white. The postauricular spots are buffy, and the feet are dark ochraceous buffy or black. This form has a dark ochraceous buff lateral line. The upperparts are paler than in *S. v. boothiae*.

CONSERVATION: IUCN status—least concern. Population trend—stable.

HABITAT: *S. variegatoides* is found in dry tropical forests, generally consisting of varying proportions of deciduous and evergreen species, usually at lower elevations (< 1800 m). However, it is reported from elevations as high as 2600 m in Costa Rica. It is often associated with pine-oak (*Pinus*, *Quercus*) or oak habitats in Honduras, Nicaragua, and Coast Rica, but not in México and Guatemala, where it occupies wet tropical lowland forests. It is also frequently observed in plantations, such as cacao farms in Costa Rica.

NATURAL HISTORY: *S. variegatoides* is primarily granivorous but sometimes is frugivorous, consuming nuts, some fruits, and the seeds of many fruits. However, it tends to be highly selective and avoids hard-shelled seeds and nuts, though acorns are frequently consumed. The tree species from which reproductive tissues are eaten by these squirrels vary with habitat and are described in detail in the literature. Animal material (e.g., insects and nestling birds) is also consumed. Dietary differences between young animals and adult squirrels may relate more to accessibility and the importance of cover than to actual food preferences. This species is primarily a seed predator, as observations of these squirrels dispersing and storing seeds are lacking. Additionally, this species is rarely seen on the ground, which allows little opportunity for scatter-hoarding. *S. variegatoides* is diurnal; it resides primarily in leaf nests that are constructed near the trunks of trees, although the use of tree cavities is also reported. Deep vocalizations, interpreted as alarms calls, are also noted but not well studied. This species is abundant and locally common throughout its range, yet nothing definitive is reported about its breeding and reproductive behavior, or its population ecology and genetics. Pelage and cranial features suggest a close relationship with *S. aureogaster*, as well as the possibility that *S. variegatoides* is part of the same species complex as *S. colliaei* and *S. yucatanensis*. Where its range overlaps with that of *S. granatensis* in eastern Costa Rica and Panama, *S. variegatoides* is more of a habitat generalist and will frequently occupy dry deciduous forests. This species is also observed to be sympatric with *Syntheosciurus brochus*, *Microsciurus alfari*, and *Sciurus granatensis* on the Poas Volcano (Costa Rica). Extensive observations over several years in Costa Rica indicate that the variegated squirrel is frequently preyed upon by white-faced capuchins (*Cebus capucinus*), which may engage in cooperative hunting to aid in capturing the squirrels. Two ectoparasites—one chigger (*Microtrombicula nicaraguae*) and one louse (*Enderleinellus hondurensis*)—are the only parasites recorded from this host.

GENERAL REFERENCES: Best 1995f; Caro et al. 2001; Ceballos, Arroyo-Cabrales, et al. 2002; Gálvez and Jansen 2007; E. R. Hall and Kelson 1959; Koprowski, Roth, Reid, et al. 2008; Koster 2008; Monge and Hilje 2006; Nelson 1899; Rose 1997, 2001; Shelley and Blumstein 2005.

Sciurus vulgaris (Linnaeus, 1758) Eurasian Red Squirrel

DESCRIPTION: The Eurasian red squirrel has a uniformly dark dorsum, with the color varying from dark red to black to brown to gray to bluish. The ear tufts are pronounced in winter and reduced in summer. The venter is white to cream. The tail is most often the same color as the dorsum, but it can be lighter in shade or nearly white in some subspecies. Melanism is common, especially in mainland, boreal, and montane populations.

SIZE: Both sexes—HB 206-250 mm; T 150-205 mm; Mass 235-480 g.

DISTRIBUTION: This species is found in forested regions of the Palearctic, from Great Britain south to the Iberian Peninsula and then east to the Mediterranean and Black seas, the Kamchatka Peninsula and Sakhalin Island (Russia), the island of Hokkaido (Japan), northern Mongolia, and western and northeastern China.

GEOGRAPHIC VARIATION: Evolutionary relationships are sorely in need of investigation across the massive geographic range of *S. vulgaris*. There are 22 subspecies.

- *S. v. vulgaris*—southern Scandinavia. This smallish subspecies has a reddish dorsum and lacks a dark phase.
- *S. v. alpinus*—Iberian Peninsula, southern France, and all of Italy. This form is an intense russet brown over the dorsum.
- *S. v. altaicus*—Altai (eastern Russia) and northern Mongolia. This is a highly variable form that includes extremely dark variants.
- *S. v. anadyrensis*—far northern Russia and across Siberia. During winter this form has a tinge of chestnut suffusion in its grayish dorsal pelage.

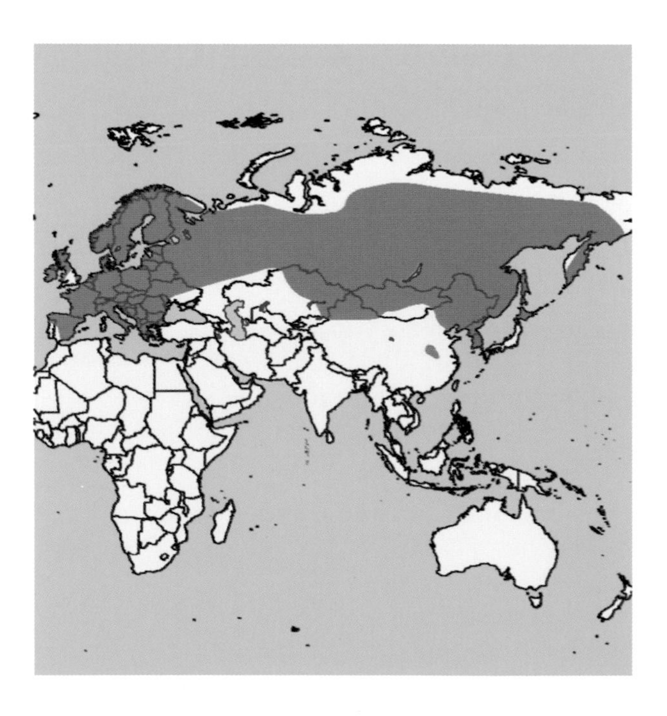

Sciurus vulgaris. Photo courtesy A. Coke Smith, www.cokesmithphototravel.com.

- *S. v. arcticus*—boreal Siberia to the Lena River. This form has a reddish to bay dorsum, with a white venter.
- *S. v. balcanicus*—eastern Balkan Peninsula (southeastern Europe). This is a more brownish form.
- *S. v. chiliensis*—eastern China. This form is black across the dorsum and tail, with varying amounts of reddish.
- *S. v. dulkeiti*—Amur region of far eastern Siberia including the Shantar Islands (Russia). The form with a black or black chestnut dorsum and a black tail throughout the year is common.
- *S. v. exalbidus*—western Siberia, east of the Ob River (Russia). This form is characterized by a light coloration that is pale buff gray in winter with a light gray tail.
- *S. v. fedjushini*—Belarus and vicinity. This subspecies has a rich dark reddish brown dorsum, with a dark chestnut or red tail.
- *S. v. formosovi*—northwestern Russia. The winter pelage is gray, with a chestnut tail.
- *S. v. fuscoater*—central Europe. This is a bright red form, with a white venter.
- *S. v. fusconigricans*—northern Mongolia and China. This subspecies is usually dark chestnut or black chestnut in summer and dark black to bluish gray in winter. The venter is white over a small area.
- *S. v. leucourus*—British Isles. It is characterized by a reddish dorsum, with a tail that often fades to near white in summer.

S. v. lilaeus—western Balkan peninsula. It is a generally brownish form.

S. v. mantchuricus—"Manchuria" (northeastern China) and the Korean Peninsula. This subspecies has a black tail year-round, and a dark gray winter pelage.

S. v. martensi—western Siberia, to west of the Ob River (Russia). This form has a chestnut tail, with a pale gray winter pelage and yellow buff on the hips in the summer pelage.

S. v. ognevi—western Russia. This smallish form is highly variable in coloration, but the winter pelage often is light grayish with a tinge of buff.

S. v. orientis—island of Hokkaido (Japan). This endemic form is gray, with a white venter.

S. v. rupestris—Sakhalin Island (far eastern Siberia). This has a black or black chestnut dorsum, with a black tail throughout the year.

S. v. ukrainicus—Ukraine. This form has a dorsum and tail that retain much of their reddish brown color in the winter pelage.

S. v. varius—northern Scandinavia. This form has a light bluish gray winter dorsum, with a dark chestnut to red tail.

CONSERVATION: IUCN status—least concern. Population trend—decreasing. The subspecies *S. v. leucurus* (in the British Isles) is considered to be in critical decline. *S. vulgaris* is considered near threatened in Mongolia and Croatia and vulnerable in Poland.

HABITAT: Eurasian red squirrels are most abundant in large tracts of coniferous forest, but they also occur in deciduous and mixed forests and parklands, and range from lowlands to subalpine forests within the Alps. They can be found in urban and suburban environs, conifer plantations, and secondary forests.

NATURAL HISTORY: This species is diurnal. Eurasian red squirrels forage in the canopy and on the ground throughout the year. Mating activity begins as early as December and can continue until August. Males track the females' reproductive condition by following and sniffing their genitalia. Estrus lasts less than one day during a breeding season. Males congregate, and more than 10 males can pursue a female during her mating bout. Dominant males are able to obtain the majority of matings, which occur along tree trunks, in branches, on the ground, or in dens. Copulation lasts less than 30 seconds. A copulatory plug forms within the female from the coagulation of semen. Females will remove the plugs and mate with multiple males. A litter averaging one to six young is born in a tree cavity or leaf nest. Leaf nests are large spherical structures constructed in the canopy using twigs and leaves. They have a central nest chamber, which is often lined with shredded bark, mosses, and leaves. Peak seasons of birth are February-April and May-August. Both body condition and dominance rank positively influence the reproductive success of males and females. A few females produce two litters in a year; adult females typically produce a single litter each year. The young emerge after 6-7 weeks and weaning is completed between 8 and 12 weeks. Adult size is reached at about 1 year. Most individuals do not reproduce until at least 10-11 months of age, and many will not do so until they are 2 years old. The lifetime reproductive success for adult females averages about five young per lifetime (range = 1-11). Natal dispersal occurs soon after weaning, and all males and females leave their natal area. Eurasian red squirrels are relatively solitary; however, several adults will share a nest at night, especially in cold months during the winter and spring. They will scent-mark trees with their cheek glands, often at traditional sites; they also seem to use urine for scent-marking.

Eurasian red squirrels feed heavily on tree seeds, especially those of conifers, but they also eat fruits, buds, flowers, shoots, herbs, fungi, bark, and lichens. Occasionally animal matter (including insects, bird eggs, and nestlings) is consumed. A conspicuous behavior is the annual scatter-hoarding of hard low-perishability acorns, other nuts, and cones each fall to serve as the winter food store. Hoarding activities are more common in deciduous forests, in comparison with coniferous forests. Squirrels collect tree seeds near the parent tree, disperse to considerable distances, and bury them just under the surface. These behaviors serve to decrease loss of this critical food source to seed predators. Caches are recovered using a combination of spatial memory and olfactory cues; however, the spatial memory of *S. vulgaris* seems to be less acute than that of introduced eastern gray squirrels (*S. carolinensis*). Fungi are also cached in the branches of trees. Success in the recovery of caches increases the probability of survival and augments reproductive success. Home range overlap is modest, especially among females that defend core areas, such as during lactation. Home ranges vary from 2.4 to 19.7 ha in females and 6.2 to 31.4 ha in males, and can be up to 47.0 ha in high-elevation forests. Survivorship and reproduction are positively correlated with the seed crop of trees. The first-year survival rate is low, often less than 25 percent. Annual adult survivorship is high, at greater than 50 percent throughout adult life; longevity in the field is greater than 7.0 years and, in captivity, greater than 10.0 years.

Predators of Eurasian red squirrels include mustelids, felids, canids, and raptors. Predation accounts for 16-61 percent of mortalities. Their alarm calls are a series of medium-pitched barks or "chucks," often followed by a long higher-pitched whine. Eurasian red squirrels are a game animal throughout their range. They are valued as food and are hunted and trapped in much of their range. Furthermore, their pelts are greatly desired, and a considerable market for them exists, one that has led to serious conservation concerns for populations in countries such as Mongolia. *S. vulgaris* is often valued in parks and cities as an appealing wildlife species. At a local scale, Eurasian red squirrels are also considered to be a pest in gardens and orchards; their habit of stripping bark to obtain cambium can result in a diminution of tree growth and vitality. Eastern gray squirrels (*S. carolinensis*) have spread since they were introduced to England, with a concomitant decline in the range and abundance of Eurasian red squirrels. Eastern gray squirrels occur at higher densities in broadleaf forests and form kin clusters of related females. They nest in larger groups, digest acorns more efficiently, and accumulate fat more quickly than Eurasian red squirrels in some habitats. *S. carolinensis* pilfers the caches of *S. vulgaris*; eastern gray squirrels also decrease summer breeding and limit juvenile recruitment of the imperiled native species. Furthermore, eastern gray squirrels harbor parapoxvirus, which is fatal to Eurasian red squirrels but not to the invasive species. In addition, Eurasian red squirrels appear very sensitive to fragmentation of their habitat. Thus habitat loss and fragmentation constitute a significant threat to *S. vulgaris* in many regions. Long-term conservation strategies will require wise and large-scale forest management efforts.

GENERAL REFERENCES: Grill et al. 2009; Lurz et al. 2005; Ognev 1963; Wauters, Bertolino, et al. 2005; Wauters, Tosi, et al. 2002; Wauters, Vermeulen, et al. 2007.

Sciurus yucatanensis (Allen, 1877)
Yucatan Squirrel

DESCRIPTION: The Yucatan squirrel has a black-and-white to gray grizzled dorsum, often with a wash of olive brown or tawny that appears rough in texture. A buffy postauricular patch is often present. The feet are dark brown to charcoal to black. The venter is buff to gray to grizzled black and white. The tail is black at the core and frosted with white to buff.

SIZE: Both sexes—HB 200-322 mm; T 194-271 mm; Mass 341-475 g.

DISTRIBUTION: This species is found in the Yucatán Peninsula (México, northern and southwestern Belize, and northern Guatemala).

GEOGRAPHIC VARIATION: Three subspecies are recognized.

S. y. yucatanensis—northern end of the Yucatán Peninsula (México). This subspecies is the lightest in its dorsal and ventral coloration.

S. y. baliolus—midsection of the Yucatán Peninsula (southern Campeche, México). This is a lighter form, with a buffy venter and a lack of distinct black patches at the base of the ears.

S. y. phaeopus—southern Yucatán Peninsula (México, Belize, and Guatemala). This is a dark form with black legs, black patches at the base of the ears, and a grayish venter.

CONSERVATION: IUCN status—least concern. Population trend—stable.

HABITAT: The Yucatan squirrel occurs in deciduous and evergreen forests, semiarid pine-oak (*Pinus, Quercus*) woodlands, and secondary forests; it also ranges into mangrove swamps. *S. yucatanensis* can be found in coffee and cacao plantations.

NATURAL HISTORY: This species is diurnal. Yucatan squirrels are primarily arboreal; they are often seen moving

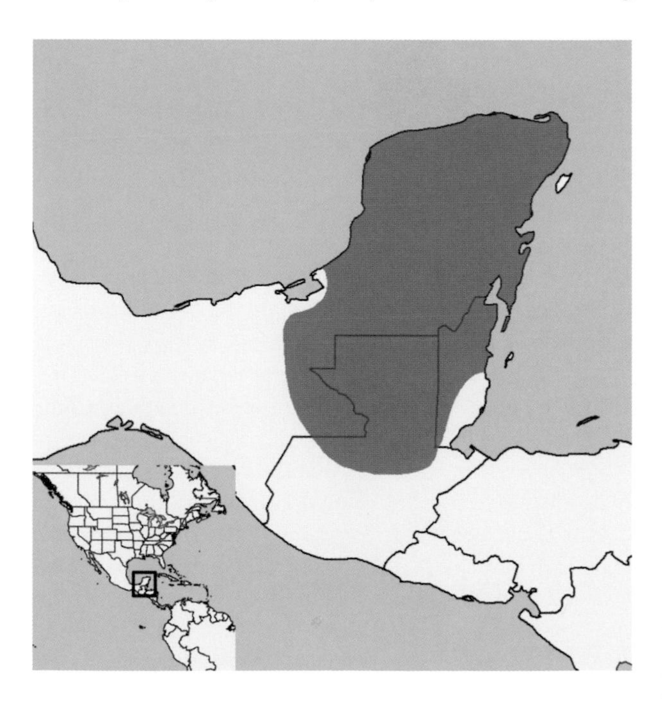

Sciurus yucatensis. Photo courtesy Jim McCulloch.

through the canopy in the early morning or resting on a branch during the heat of the day. Leaf nests are built on branches high in the canopy. Primarily herbivorous, *S. yucatanensis* feeds on soft fruits, nuts, seeds, flowers, buds, and shoots. A litter of two to three young is produced during the dry season (April-July). Raptors are probably a major predator, along with felids, canids, procyonids, primates, and snakes. Deforestation is the major threat to this species; however, Yucatan squirrels are hunted, especially in Yucatán and Quintana Roo (México), and exploited populations may be reduced.

GENERAL REFERENCES: Best, Ruiz-Piña, et al. 1995; Faller-Menéndez et al. 2005; Gerhardt et al. 1993; León and Montiel 2008; Pozo de la Tijera and Escobedo Cabrera 1999; F. A. Reid 1997.

Syntheosciurus Bangs, 1902

This genus contains a single species.

Syntheosciurus brochus (Bangs, 1902)
Bangs's Mountain Squirrel

DESCRIPTION: Bangs's mountain squirrels have a grizzled brown olivaceous dorsum, continuing onto the tail. The underside of the tail is charcoal to black, and the tip is black. The ventral surface of the body is grayish to a pale orange.

SIZE: Both sexes—HB 150-185 mm; T 120-160 mm.

DISTRIBUTION: This species is found in Costa Rica and northern Panama.

GEOGRAPHIC VARIATION: None.

CONSERVATION: IUCN status—near threatened. Population trend—no information.

HABITAT: Bangs's mountain squirrels are found in montane cloud forests, evergreen forests, secondary forests, and pasture edges, at elevations from 1900 to 2300 m.

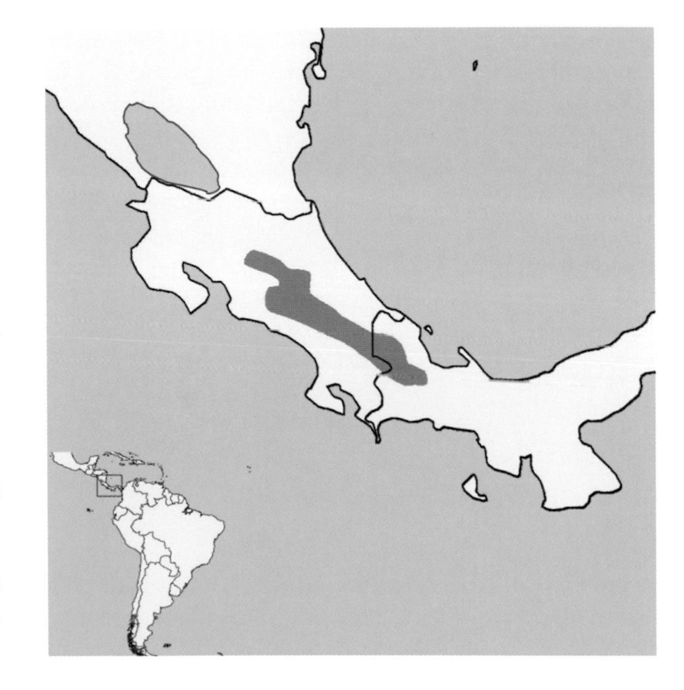

Syntheosciurus brochus. Photo courtesy Gregory E. Willis.

NATURAL HISTORY: This species is diurnal and crepuscular. Bangs's mountain squirrels can be seen on the ground, but they are secretive, remaining in heavy cover; this species uses all forest levels frequently. *S. brochus* nests in cavities and will forage in the upper canopy for flowers, fruits, sap, and bark. Mating occurs in the spring, when six to eight males will chase the female throughout the canopy. Litter sizes are two to five young. Individuals appear to remain in family groups for extended periods and will forage alone or with a partner. Males and females will travel together, rest during the day next to each other, and sleep in den cavities together, suggesting a level of pair-bonding that is extremely uncommon among tree squirrels.

GENERAL REFERENCES: Emmons and Feer 1990; Enders 1953, 1980; Giacalone et al. 1987; F. A. Reid 1997; Wells and Giacalone 1985.

Tamiasciurus Trouessart, 1880

This genus contains three species.

Tamiasciurus douglasii (Bachman, 1839)
Douglas's Squirrel

DESCRIPTION: Douglas's squirrel is a small tree squirrel. Its dorsal coloration varies from olivaceous gray to gray brown, often with a dark or chestnut midline band. In the summer it has a prominent lateral black stripe (which is faint or absent in the winter) separating the dorsal coloration from the underparts, which vary from white or pale buff to yellowish or reddish orange. The tail is similar in color to the back, but with more black in it, and is fringed with yellow- or white-tipped hairs. It has distinct white eye rings and slight ear tufts that are most evident in the winter.

SIZE: Female—HB 185.0 mm (n = 66); T 145.0 mm (n = 7); Mass 199.3 g (n = 115).

Male—HB 181.0 mm (n = 2); T 119.0 mm (n = 2); Mass 206.9 g (n = 1).

Sex not stated—HB 189.7 mm (n = 3); T 126.3 mm (n = 2); Mass 242.5 g (n = 4).

DISTRIBUTION: This species is found in the Coast and Cascade ranges and the Sierra Nevada, from southwestern British Columbia (Canada, but excluding Vancouver Island) to southern California (USA).

GEOGRAPHIC VARIATION: Two subspecies are recognized in *Tamiasciurus douglasii.*

T. d. douglasii—immediate vicinity of the Pacific coast in Washington and Oregon (USA). It can be recognized by a yellowish fringe on its tail.

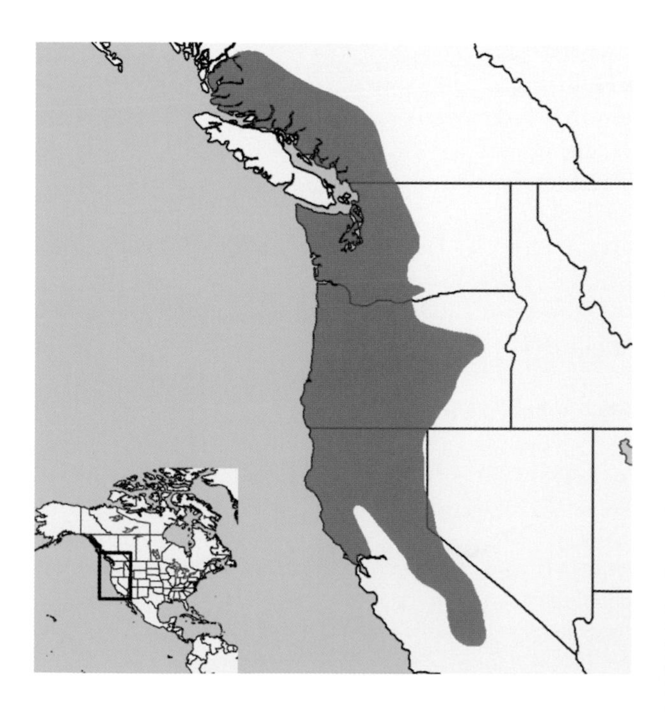

Tamiasciurus douglasii. Photo courtesy A. Coke Smith, www.cokesmithphototravel.com.

T. d. mollipilosus—more inland than the other subspecies in Washington and Oregon (USA), as well as along the coast in British Columbia (Canada) and northern California (USA), and in the Sierra Nevada of northern and western California (USA). It has a whitish fringe on its tail.

CONSERVATION: IUCN status—not listed. Population trend—no information.

HABITAT: *T. douglasii* occurs in single and mixed stands of conifers: pine (*Pinus*), fir (*Pseudotsuga* and *Abies*), spruce (*Picea*), and hemlock (*Tsuga*). Douglas's squirrels are considered an old growth-dependent species, although a few studies suggest that the abundance of *T. douglasii* may not vary considerably with forest age. Research on the effects of forest fragmentation shows that fragment size has little effect on *T. douglasii*, but that the percentage of primary or old secondary growth in the surrounding matrix, and its proximity to primary forests and riparian corridors, has a strong influence.

NATURAL HISTORY: Douglas's squirrel is diurnal and active throughout the year. Their activity is bimodal from spring to autumn, with peaks occurring in the morning and late afternoon. In winter, their activity is restricted to a single midday peak. When active, this species is generally quite visible and audible, as a result of the frequent interspecific interactions necessary to maintain territory boundaries. "Rattle" calls, detectable at well over 100 m, are often followed by vigorous chases of territorial intruders. The breeding season of *T. douglasii* begins in March and continues for several months. This extended breeding season results from asynchrony in estrus, which may reduce competition among dispersing young. A second breeding season is common when food is abundant and the climate is mild, although reproductive failure can occur when cone crops fail. Female Douglas's squirrels are in estrus for a single day during the breeding season, within which time several males converge on the female's territory and the female mates with one or more of these males. *T. douglasii* is a spontaneous ovulator, resulting in immediate conception. Litter sizes range between four and eight (average = 5.36), although limited data are available. Gestation and development are poorly documented for this species, but are probably similar to *T. hudsonicus*. Individuals reach sexual maturity at 10–12 months; they can reproduce in the first year, but they usually do not.

T. douglasii defends exclusive territories established around food supplies, nest sites, and the midden (a single larder in the center of the territory). Territory size and shape are both determined by the type, availability, and distribution of food; territory size is inversely related to cone crops. Throughout the year, *T. douglasii* aggressively defends territory boundaries from conspecifics via vocalizations and chases. Territories are often bequeathed to their young upon weaning. The evolution of territoriality in the genus appears to be primarily related to the availability of food for larder-hoarding (i.e., cones), the storability of cones in the damp middens, and the efficiency with which the midden can be defended. *T. douglasii* feeds heavily on tree seeds, mushrooms, and other fungi as available throughout the year, but it is also opportunistic when its primary diet items are scarce. Foods reported in the diet include the seeds of conifers, the seeds and fruits of angiosperms, tree buds, cambium, fruiting bodies of epigeous (aboveground) and hypogeous (underground) fungi, and numerous secondary diet items. These studies also give the energetic value of some of these items. Significant shifts in diet occur through the year. The mean stomach content consisted primarily of conifer cones in autumn, but it dropped through the winter to only 3 percent by spring, during which time the animals fed extensively on the cambium of terminal buds and shoots of conifers (50% by volume). Fungi, especially hypogeous (underground) truffles, accounted for 90 percent of their diet in summer. Cambium and fungi are low in calories but are important sources of protein, limiting minerals, and water. Population density can vary with the size of the cone crops, although the density of *T. douglasii* is to some extent limited by its territorial behavior. Population densities across forest types ranged from 0.03 to 0.9 animals/ha both between and within the same forest types. Annual variation in abundance is due primarily to variation in juvenile recruitment, which appears to be linked directly to annual cone production. The frequency and mass of *T. douglasii* in the diet of northern goshawks (*Accipiter gentilis*) increased during both the squirrel's breeding season and annually in response to cone crops.

GENERAL REFERENCES: J. B. Buchanan et al. 1990; Carey 1995; Flyger and Gates 1982; Ingles 1965; Keane et al. 2006; Koford 1982; Lair 1985; Larsen and Boutin 1994; Lomolino and Perault 2001; McKeever 1964; Perault and Lomolino 2000; Ransome 2001; Sanders 1983; C. C. Smith 1965, 1968, 1978, 1981; W. P. Smith, Anthony, et al. 2003; Steele 1998, 1999; Steele and Koprowski 2001; Steele, Wauters, et al. 2005; Sullivan and Sullivan 1982; Yahner 2003.

Tamiasciurus hudsonicus (Erxleben, 1777) Red Squirrel

DESCRIPTION: *T. hudsonicus* is recognized by its small size (200–250 g), reddish dorsum, white venter, and white eye ring. The dorsal surface is reddish, ferruginous brown, or olivaceous gray, usually with a reddish median band. *T. hudsonicus* is distinguished from *T. douglasii* by its venter, which is white or yellowish; that of *T. douglasii* is rust-colored or reddish, often with a blackish wash. The tail hairs of *T. hudsonicus* have yellowish to rusty tips with a black band, whereas those of *T. douglasii* have yellowish to white tips.

SIZE: Female—HB 189.7 mm (n = 191): T 123.6 mm (n = 139); Mass 213.0 g (n = 455).

Male—HB 187.0 mm (n = 177); T 123.7 mm (n = 177); Mass 194.0 g (n = 227).

Sex not stated—HB 186.6 mm (n = 9); T 123.4 mm (n = 9); Mass 195.0 g (n = 2).

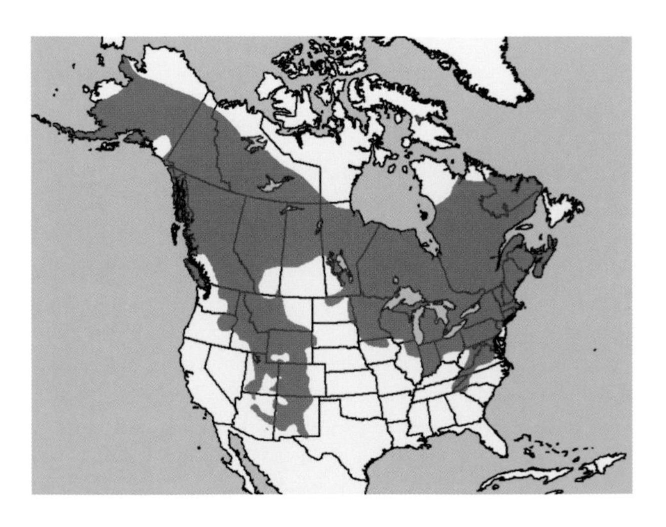

DISTRIBUTION: This species is found in Alaska (USA); throughout Canada (south of the tundra), including Vancouver Island; the western USA (in the mountain states); and the northeastern USA south to northwestern South Carolina. *T. hudsonicus* has been introduced to Newfoundland (Canada), as well as to numerous islands along the Pacific coast of Canada and Alaska (USA).

GEOGRAPHIC VARIATION: Twenty-five subspecies are recognized in this wide-ranging species.

The eastern subspecies include the following:

T. h. hudsonicus—Manitoba and Ontario (Canada), and northern Minnesota and northern Michigan (USA). This form is as described above, but with the dorsal surface rusty red from head to tail.

T. h. abieticola—southern Appalachians (USA). It is the darkest subspecies of *T. hudsonicus*, with darker and deeper red coloration dorsally than *T. h. loquax*.

T. h. gymnicus—Vermont, New Hampshire, and Maine (USA), and north in Canada to the St. Lawrence River and Nova Scotia. This form is slightly smaller and darker than *T. h. hudsonicus*, being rich rusty red above with gray underparts in winter, and duller red above with white underparts in summer.

T. h. laurentianus—southeastern Québec (Canada), mostly north of the mouth of the St. Lawrence River. This form is smaller and duller in color than *T. h. ungavensis*.

T. h. loquax—southern Michigan and Indiana east to New York state and southern New England (USA), as well as adjoining Ontario and Québec (Canada), and south to Virginia (USA). This form is brighter red above, with the tail less black in summer and winter than *T. h. hudsonicus*.

T. h. minnesota—eastern North Dakota, southern Minnesota, Wisconsin, northern Iowa, and northern Illinois (USA). This form is paler in coloration than *T. h. loquax*.

T. h. pallescens—North Dakota (USA), plus Saskatchewan and the Manitoba border area (Canada). It is similar to *T. h. dakotensis*, but is more gray and less buffy.

T. h. regalis—Isle Royale in Lake Superior (USA). This form is paler and less reddish than *T. h. loquax*.

T. h. ungavensis—northern Québec and Labrador (Canada). This form is a much deeper red color dorsally in the winter than *T. h. hudsonicus*, with the tail edged with ochraceous buff.

The western subspecies include the following:

T. h. baileyi—northcentral Montana to southcentral Wyoming (USA). It is larger than *T. h. hudsonicus*, and darker and more olivaceous above.

T. h. columbiensis—southern Yukon to central British Columbia and southwestern Alberta (Canada). This form is paler in color than *T. h. streatori,* with less black and red on the tail.

T. h. dakotensis—southeastern Montana, northeastern Wyoming, and western South Dakota (USA). It is larger and paler than *T. h. hudsonicus,* with the underparts light yellowish rufous and paler yellow in summer.

T. h. dixiensis—southwestern Utah (USA). This form is a large subspecies. It is dark in color, with little yellow suffusion of the pelage.

T. h. fremonti—Colorado (USA). In winter it is gray with a rusty yellow cast on the back and grayish white underparts; in summer it is yellowish gray above and whitish to grayish below.

T. h. grahamensis—Mount Graham, in southern Arizona (USA). This form is similar to *T. h. mogollonensis,* but with more yellow and less red above in its summer pelage.

T. h. kenaiensis, T. h. petulans, T. h. picatus, and *T. h. lanuginosus* (subspecies listed north to south)—along coastal Alaska (USA) and British Columbia (Canada), including Vancouver Island. These are similar to *T. h. hudsonicus,* but differ variously in being brighter or darker dorsally and more yellowish below.

T. h. lychnuchus—central New Mexico (USA). This form is similar to *T. h. fremonti,* but more red or yellow red dorsally.

T. h. mogollonensis—Arizona and New Mexico (USA). This form is large, with yellowish rufous coloration above.

T. h. preblei—south of the tundra in Alaska (USA) and northern Canada, as well as in Alberta and Saskatchewan (Canada). It is larger than *T. h. hudsonicus,* with paler coloration above and below.

T. h. richardsoni—southeastern British Columbia (Canada), and western Oregon, central Idaho, and western Montana (USA). It is large, with a dark upper surface on the tail and body and white underparts.

T. h. streatori—southern British Columbia (Canada), and eastern Washington and northern Idaho (USA). This form differs from *T. h. richardsoni* by a more olivaceous coloration above in summer and less black on the tail.

T. h. ventorum—eastern Wyoming, extending west into Idaho, north into Montana, and south into Utah (USA). This form is darker and more olivaceous above than *T. h. baileyi,* as well as blacker on the upper side of the tail and grayer on the lower surface of the tail.

CONSERVATION: IUCN status—least concern. Population trend—stable. The subspecies *T. h. grahamensis* is listed as endangered under the U.S. Endangered Species Act, and as critically endangered by the IUCN. Other subspecies are not listed by the IUCN.

Tamiasciurus hudsonicus. Photo courtesy Bruce D. Taubert.

HABITAT: Red squirrels in North America (also known as pine squirrels) are most common in boreal coniferous forests with abundant conifer seeds, interlocking canopies for efficient foraging and escape from predators, and cool moist conditions, which facilitate fungal growth and cone storage in larders (middens). In the southeastern portion of its range, *T. hudsonicus* also occupies a variety of other, less optimal forest types, including mixed coniferous-deciduous forests, hedgerows, parks, and even hardwood forests with scattered conifers. In these more marginal habitats, the primary determinants of trap success were herbaceous density, canopy cover, and proximity to tree cavities, underground burrows, and logs. In the southwestern portion of the range, the animal occupies subalpine forests down through mixed conifer forests and occasionally the transition zone between ponderosa pine (*Pinus ponderosa*) and Douglas fir (*Pseudotsuga menziesii*) forests.

NATURAL HISTORY: Red squirrels are diurnal; they show bimodal periods of activity in spring and summer, but in winter the animals are active only in midday. Activity outside the nest is devoted almost entirely to food gathering, maintenance behavior (resting and grooming), and feeding. Their activity is reduced by severe weather or abundant food sources. Red squirrels are primarily granivorous, but they are opportunistic feeders in the absence of mast foods. Primary diet items vary with habitat and include the seeds of conifers, oak (*Quercus*), hickory (*Carya*), beech (*Fagus*), walnut (*Juglans*), yellow poplar (*Liriodendron tulipifera*), sycamores (*Platanus*), and maple (*Acer*). Both epigeous (aboveground) and hypogeous (underground) mycorrhizal fungi are also important, especially in summer and during other periods of food shortages. Additional secondary food items (in approximate order of relative importance) include tree buds and flowers, fleshy fruits, tree sap, bark, insects, and other animal material. During winter, spring, and early summer, *T. hudsonicus* can cause serious damage to trees by stripping bark and girdling the trees to consume phloem and cambium. This species hoards food, primarily conifer cones, in a larder (midden) that is usually established in the center of the range near large trees, dense understory, and fallen logs. Secondary caches also may be common throughout the territory. The midden, critical for overwinter survival, provides a moist environment for maintaining conifer seeds, which prevents the cones from opening and decreases their pilferage by competitors. High rates of mutual pilferage of stored cones have been reported in a Vermont population of *T. hudsonicus*. C. C. Smith argued that the storability of conifer cones and the defensibility of the midden specifically led to the evolution of territoriality in the genus, and this social structure, in turn, probably limits population densities and the carrying capacity in conifer forests.

T. hudsonicus is highly systematic and selective when harvesting cones and provisioning the midden. Individual squirrels retrieve cones throughout their territory and may stockpile as many as 16,000 cones in a single year. *T. hudsonicus* tends to harvest cones first from the species with the highest energy profitability (i.e., seed density per cone) and then move to other species in order of declining profitability. In mixed stands in the Pacific Northwest, *T. hudsonicus* harvests cones in the following order: Pacific silver fir (*Abies amabilis*), Douglas fir (*Pseudotsuga menziesii*), Engelmann spruce (*Picea engelmannii*), and then western hemlock (*Tsuga heterophylla*). Squirrels appear to select individual trees only after cones have been harvested from all but the species with the least available energy per cone. *T. hudsonicus* selects individual cones based on seed number, the ratio of seed mass to cone mass, cone hardness, the orientation of the cone to the branch, and the distance between the point of harvest and the midden. Although some authors argue that *Tamiasciurus* is unlikely to exert a strong selective force on cone characteristics, several recent studies suggest otherwise.

This species shows two annual molts: one in spring through summer, and another during the autumn. The tail molt occurs only once annually. The structure of the male and female reproductive tracts of *Tamiasciurus* differs significantly from those of other tree squirrels. The penis is long, slender, and symmetrical. The Cowper's gland is reduced, the bulbar gland is absent, and the baculum is vestigial. The coiled vagina is unique to *Tamiasciurus*. Seasonal changes in the structure of male and female reproductive tracts are detailed by J. N. Layne. Among North American squirrels, the smaller body size of red squirrels may result from selection for increased foraging ability and agility in the smaller branches of conifer trees. Body temperature in this species is unusually high for mammals. Vascular bundles involved in countercurrent heat exchange may be present in the base of the tail. When extended over the body, the tail increases the insulation of the pelt by 13.7 percent. The tail also reduces the rate of heat gain by more than 40 percent, through shading and the dissipation of heat.

Red squirrels are spontaneous ovulators and are in estrus for only one day during the breeding season. Gestation averages 33 days and litter sizes (summarized from seven studies) range from 3.2 to 5.4. *T. hudsonicus* typically reproduces once annually, but females can produce two litters per year when food is abundant, especially in the eastern portion of the range. Seasonal variation in reproductive anatomy was reported for more than 550 animals in New York state (USA). The young are born pink and hairless; their eyes open between 26 and 35 days; the pelage is fully developed by 40 days; and the remainder of their development, including dentition, is complete by 125 days. The young are active outside the nest by 7 weeks of age, weaned by 7 weeks, and fully independent a few weeks after weaning. Maximum longevity is 10 years. Male-biased adult sex ratios may result from the differential mortality of the sexes. *T. hudsonicus* shows a Type III survivorship curve, with more than 60 percent mortality occurring in the first year of life, more than 80 percent by the second, and more than 90 percent by the third. Squirrels born in a year of abundant seed crops may show higher survivorship, but food supplementation often fails to improve survival. The annual population cycle is characterized by a peak in density during the summer following recruitment, and a gradual decline through autumn and winter. Red squirrels are probably prey for snakes, owls, hawks, and a number of carnivorous mammals. Numerous comprehensive surveys of the parasites of red squirrels are available.

GENERAL REFERENCES: Benkman and Siepielski 2004; Bertolino 2009; G. A. Kemp and Keith 1970; Larsen and Boutin 1994; Layne 1954; Rusch and Reeder 1978; C. C. Smith 1965, 1968, 1970, 1978, 1981; Steele 1998.

Tamiasciurus mearnsi (Townsend, 1897) Mearns's Squirrel

DESCRIPTION: Mearns's squirrels are gray on the dorsum and head, with a faint brownish tinge extending from the pelage behind the ears down the middle of the back. A gray to charcoal lateral stripe borders the white venter. The feet are a pale yellowish white to gray, and the ear tufts are composed of dense blackish tufts. A whitish eye ring is present. The tail consists primarily of black hairs tipped with white to gray to yellow.

SIZE: Both sexes—HB 201.0 mm ± 13.2 mm.

DISTRIBUTION: This species is endemic to about 40,000 ha in the highest elevation of the Sierra de San Pedro Mártir in northern Baja California (México). It is only known from three locations.

GEOGRAPHIC VARIATION: None.

CONSERVATION: IUCN status—endangered. Population trend—decreasing. *T. mearnsi* is considered threatened in México.

HABITAT: This species is restricted to limited high-elevation pine (*Pinus*) and fir (*Abies*) forests.

NATURAL HISTORY: The three sites from which Mearns's squirrels are known are within 10 km of each other, at elevations of 2100 m, 2500 m, and 2750 m. In stark contrast to its two congeners (*Tamiasciurus hudsonicus* and *T. douglasii*), *T. mearnsi* does not larder-hoard cones in middens or build dreys in the canopy, instead relying on cavities for nests. Males observed in May displayed scrotal testes, indicative of reproductive activity. A mating bout was observed in May as well, with as many as three males pursuing the female that had a swollen pink vulva. Mearns's squirrels feed upon current-year cones of Jeffrey pine (*Pinus jeffreyi*) and white fir (*Abies concolor*), branch tips of white fir, and the basidiomycete fungus veiled polypore (*Cryptoporus volvatus*), found on the upper trunk of white firs. Three vocalizations are known: a call similar to the territorial "rattle" of congeners but of seemingly higher pitch, a "chirp" call when aggravated, and a woofing bark or "buzz" call when startled or mildly aggravated. Coyotes (*Canis latrans*), bobcats (*Lynx rufus*), and raptors are probably this species' major predators.

Tamiasciurus mearnsi. Photo courtesy Bret Pasch.

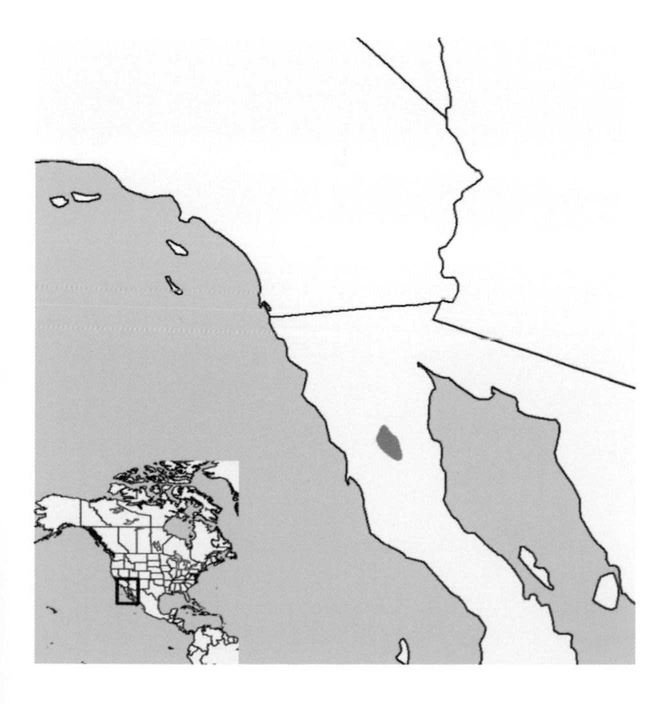

Original biological assessments of Sierra de San Pedro Mártir described *T. mearnsi* as very common among other small mammals; however, recent assessments found that Mearns's squirrel populations are usually low. The introduction of eastern gray squirrels (*Sciurus carolinensis*) in 1946 may be a serious threat; however, whether or not eastern gray squirrels remain in the Sierra de San Pedro Mártir is not known.

GENERAL REFERENCES: Koprowski, Ramos, et al. 2006; Yensen and Valdés-Alarcón 1999.

Tribe Pteromyini Brandt, 1855

This tribe includes 15 genera.

Aeretes G. M. Allen, 1940

This genus has a single species. The wrist anatomy of *Aeretes* suggests a close relationship with *Petaurista*.

Aeretes melanopterus (Milne-Edwards, 1867) Northern Chinese Flying Squirrel

DESCRIPTION: This is a medium-sized flying squirrel, with a bushy flattened tail. The dorsum is sandy brown, but it is trimmed in black along the edges of the patagium; the feet are black. The dorsal hairs are long and soft; the ventral hairs are short. The venter is white or pale. The face and throat are grayish. *Aeretes* is distinguished from *Petaurista* and other genera of flying squirrels by its grooved upper incisors.

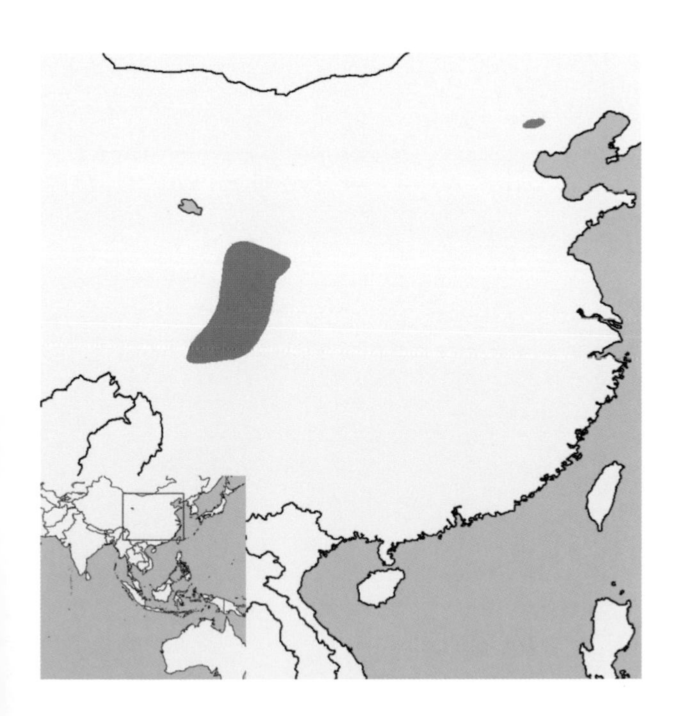

SIZE: Female—HB 350.0 mm; T 310.0 mm.
Sex not stated—HB 305.0 mm; T 336.5 mm.

DISTRIBUTION: Endemic to central China, this species is represented by two isolated populations: one with a range extending from southern Gansu to central and northeastern Sichuan, and the other farther to the northeast, in Hebei.

GEOGRAPHIC VARIATION: Two subspecies are recognized.

A. m. melanopterus—Hebei (China). This form is smaller, with the pelage being duller sandy buff above and dull white, washed slightly with buff, below.

A. m. szechuanensis—Gansu and Sichuan (China). This form is larger, with a darker pelage. The upperparts are dark grayish drab, and the underparts are a duller buffy white.

CONSERVATION: IUCN status—near threatened. Population trend—declining.

HABITAT: The northern Chinese flying squirrel (also known as the groove-toothed flying squirrel) is found in montane forests, at elevations below 3000 m. This species is assumed to be threatened because of habitat loss within its range.

NATURAL HISTORY: Virtually nothing is known about the biology of *A. melanopterus*.

GENERAL REFERENCES: Lee and Liao 1998; A. T. Smith and Johnston 2008a; A. T. Smith and Xie 2008; Tong 2007; Y. Z. Wang et al. 1966; Xiang et al. 2004; Yu 2002.

Aeromys Robinson and Kloss, 1915

This genus contains two species of flying squirrel. Like other large flying squirrels, *Aeromys* experiences high wing loading and must glide faster than smaller flying squirrels. *Aeromys* has an uropatagium (an extension of the "wing") that reaches from the knee to the tail, as in *Petaurista*. *Aeromys* possesses the largest uropatagium among the flying squirrels. This structure increases the patagial ("wing") area, helps overcome increased loading on the patagial membrane, and allows these larger squirrels to have more maneuverability during landing. Contrary to previous arguments, analysis of the wrist anatomy suggests that *Aeromys* does not belong to the *Petinomys* group.

Aeromys tephromelas (Günther, 1873)
Black Flying Squirrel

DESCRIPTION: This species is a large flying squirrel, with naked ears and a body mass comparable to *Petaurista elegans*. One subspecies is all black, and the other is orange red.

SIZE: Female—HB 375.4 mm; T 443.5 mm; Mass 1253.8 g.
Male—HB 385.0 mm; T 395.0 mm; Mass 1068.0 g.
Sex not stated—HB 384.4 mm; T 466.7 mm; Mass 1092.7 g.

DISTRIBUTION: The black flying squirrel is distributed across southern Thailand and peninsular Malaysia, including Penang Island. It also occurs in Sumatra (Indonesia) and on the island of Borneo (divided among Malaysia, Brunei Darussalam, and Indonesia). Reports of *A. tephromelas* in northern Thailand are not confirmed.

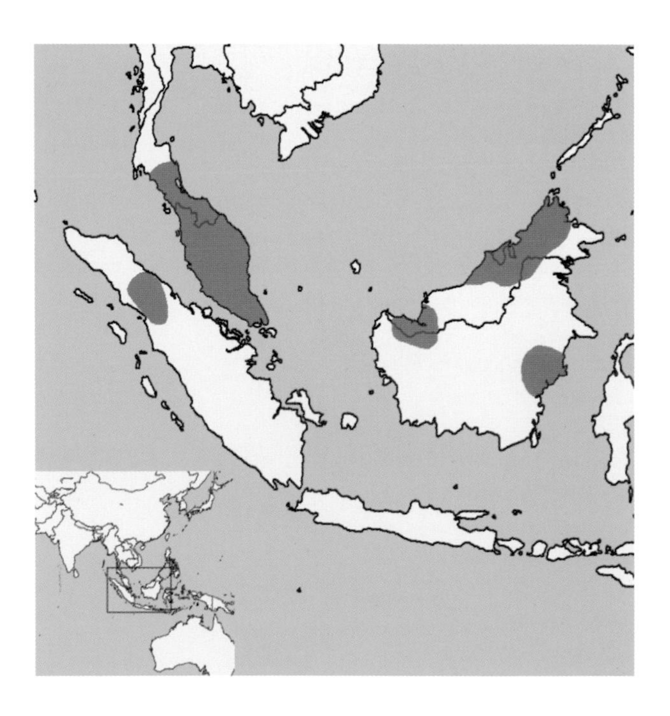

GEOGRAPHIC VARIATION: Two subspecies are recognized.

A. t. tephromelas—See distribution above, except for Kalimantan (Indonesia). It is a totally black squirrel.
A. t. phaeomelas—Kalimantan (Indonesia). This is the orange red subspecies.

CONSERVATION: IUCN status—data deficient. Population trend—no information.

HABITAT: *A. tephromelas* is found in both primary and secondary tropical lowland forests, often near villages and, less often, in more dense and isolated forests.

NATURAL HISTORY: The black flying squirrel appears to be opportunistic and flexible in the range of habitats it occupies, as well as in its diet, which includes seeds, fruits, and other plant material. This species is nocturnal and nests in tree cavities when available. It is widespread in the region but occurs in relatively low numbers.

GENERAL REFERENCES: Aplin et al. 2008; Muul and Liat 1971; Scheibe et al. 2007; R. W. Thorington and Darrow

Aeromys tephromelas. Photo courtesy Jon Hall, www.mammalwatching.com.

2000; R. W. Thorington and Heaney 1981; R. W. Thorington, Pitassy, et al. 2002; R. W. Thorington and Santana 2007.

Aeromys thomasi (Hose, 1900)
Thomas's Flying Squirrel

DESCRIPTION: *A. thomasi* is uniformly dark brown or black. It is larger than its congener *A. tephromelas*, but similar in coloration.

SIZE: Female—HB 343.3 mm; T 410.0 mm; Mass 1117.0 g.
Male—HB 300.0 mm; T 370.0 mm.
Sex not stated—HB 363.3 mm; T 380.0 mm; Mass 1423.3 g.

DISTRIBUTION: This species occurs in most of the island of Borneo (divided among Malaysia, Brunei Darussalam, and Indonesia), except the southeastern portion of the island.

GEOGRAPHIC VARIATION: None.

CONSERVATION: IUCN status—data deficient. Population trend—no information.

HABITAT: *A. thomasi* is found in lowland to midaltitude forests in montane regions on the island of Borneo. Although no reliable information is available, this species is likely to be threatened by habitat loss due to human activities.

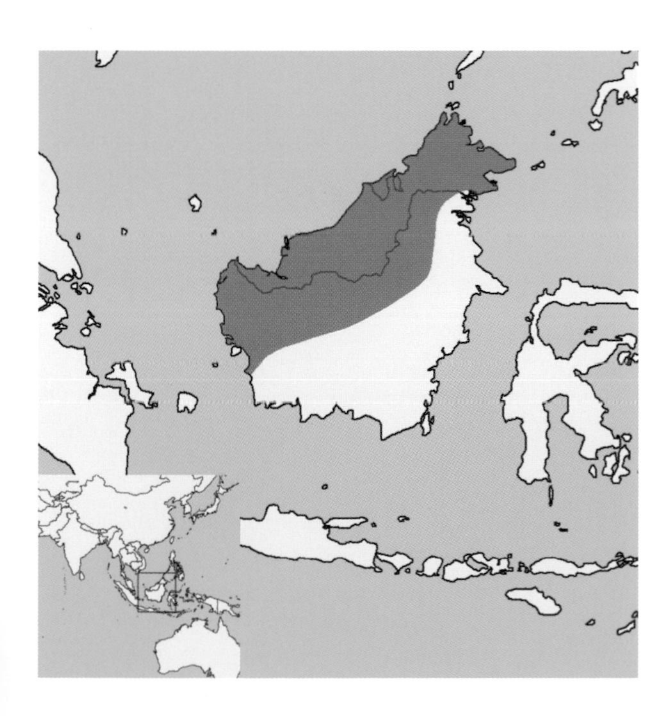

***Aeromys thomasi*.** Photo courtesy David Bakewell.

NATURAL HISTORY: A nocturnal frugivore, Thomas's flying squirrel may be relatively common, but no data are available on the population biology of this species. Information on its ecology and behavior is also limited. Recent camera-trap photographs taken in Kalimantan (Indonesia) were erroneously identified as a new species of carnivore, but the animals were subsequently identified as *A. thomasi*.

GENERAL REFERENCES: W. P. Harris 1944; Meijaard et al. 2006; Ruedas et al. 2008a.

Belomys Thomas, 1908

This genus contains a single species. Sequences of the mitochondrial cytochrome *b* gene suggest that *Belomys* diverged early from several other flying squirrel genera (*Glaucomys, Hylopetes, Petaurista, Petinomys,* and *Pteromys*); based on morphological characteristics, it appears to be most closely related to *Aeromys, Pteromyscus,* and *Trogopterus.*

Belomys pearsonii (Gray, 1842)
Hairy-Footed Flying Squirrel

DESCRIPTION: This is a small flying squirrel, with large ears and long black hairs at the base of the ears. The dorsal pelage is reddish brown, and the dorsal surface of the patagium ("wing") is blackish. The tail is short, bushy, and light brown. The belly is dark gray, and the ventral surface of the patagium is dark with a yellowish or orange wash. This species is most closely related to *Trogopterus xanthipes,* which is larger and more uniformly reddish in color.

SIZE: Female—HB 190.9 mm; T 169.3 mm.
Male—HB 189.2 mm; T 161.0 mm; Mass 155.9 g.
Sex not stated—HB 214.5 mm; T 135.8 mm.

DISTRIBUTION: This species' range extends from eastern Nepal, Bhutan, and India (Sikkim to Assam), through China (Guizhou, Yunnan, Guangxi, Guangdong, Hainan, and possibly Sichuan and Hunan), and Taiwan, and southward through Myanmar, Thailand, Laos, and Vietnam (see Agrawal and Chakraborty). However, this distribution is extremely fragmented, warranting additional research.

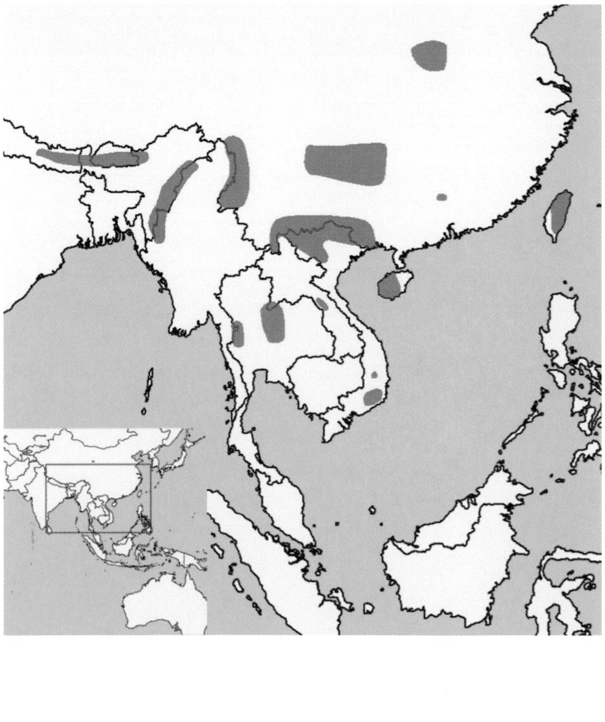

GEOGRAPHIC VARIATION: As many as four subspecies are recognized. Distribution records from Sichuan and Hunan (China) were identified only at the species level.

B. p. pearsonii—Nepal, India (Sikkim to Assam), and Bhutan. See description above. It is characterized by its rich ochraceous coloration.

B. p. blandus—Yunnan, Guizhou, Guangxi, Guangdong, and Hainan (China), Vietnam, Laos and Thailand. This subspecies is smaller than *B. p. pearsonii* and has rufescent underparts.

B. p. kaleensis—Taiwan. This subspecies is similar in appearance to *B. p. pearsonii* and *B. p. blandus,* but it has larger teeth than the latter.

B. p. trichotis—Yunnan (China) and Myanmar. This subspecies has white underparts and narrower tooth rows.

CONSERVATION: IUCN status—data deficient. Population trend—no information.

HABITAT: The hairy-footed flying squirrel is found in primary forests. In the northern part of its range it occurs in temperate deciduous or mixed coniferous-deciduous forests, and in the southern part it is found in subtropical forests or mixed stands, at elevations between 800 and 2400 m.

Belomys pearsonii. Photo courtesy Shih-Wei Chang.

NATURAL HISTORY: This nocturnal species is reported to feed on the leaves or needles of oak, cedar, and pine, and on various fruits. It nests in tree dens and rock crevices. In Taiwan, a two-year survey resulted in only 14 sightings. This and previous surveys have led to the conclusion that this species is found throughout the Central Mountain Range of this country, but occurs at low densities. Despite numerous studies of *B. pearsonii*, little else is known about its ecology, behavior, and population biology.

GENERAL REFERENCES: Agrawal and Chakraborty 1979; Chaimanee and Jaeger 2000; Duckworth and Molur 2008; Lee 1998; Lunde et al. 2003; Mishra et al. 2006; Oshida 2006; Oshida, Lin, Masuda, et al. 2000; Oshida, Lin, Yanagawa, et al. 2002; Oshida, Shafique, et al. 2004; Scheibe et al. 2007; A. T. Smith and Xie 2008; Srinivasulu et al. 2004; R. W. Thorington and Darrow 2000; R. W. Thorington, Pitassy, et al. 2002; Yu 2002.

Biswamoyopterus Saha, 1981

This genus contains a single species.

Biswamoyopterus biswasi Saha, 1981
Namdapha Flying Squirrel

DESCRIPTION: This large flying squirrel is red, grizzled with white, on the upperparts. The crown of the head has a patch of pale violet gray pelage. The patagium ("wing") is a glossy mahogany red above. The tail beyond the interfemoral membrane is proximally pale smoky gray, changing distally to vinaceous rufous to russet to clove brown near the tip. The pale smoky gray region is also washed with red. The pelage of the muzzle is vinaceous red, changing to mahogany red around the eyes; there is a black line across the nasal bridge. The body underparts are white with grayish tones. The patagium below is washed faintly orange rufous. The interfemoral membrane underneath has a band of pale red, grizzled with gray and white near the margin. The underarm is red.

SIZE: Sex not stated—HB 405.0 mm; TL 1010 mm.

DISTRIBUTION: *B. biswasi* is known only from the type locality: the western slope of the Patkai Range in northeastern India.

GEOGRAPHIC VARIATION: None.

CONSERVATION: IUCN status—critically endangered. Population trend—decreasing.

HABITAT: The Namdapha flying squirrel is reported to occupy tropical dry deciduous montane forests in riparian zones, at elevations of 100-350 m. This species may occupy a single valley of less than 100 km^2 in Namdapha National Park, Tirap District, Arunachal Pradesh (northeastern India). Extensive surveys in the area have not produced sightings.

NATURAL HISTORY: *B. biswasi* is considered to be represented by extremely small populations that are probably decreasing and susceptible to a range of threats, including logging, agriculture, poaching for food, and extreme weather (e.g., flooding). The species is thought to be crepuscular, but virtually no data are available on its distribution, densities, behavior, and ecology. No specific conservation measures are in place.

GENERAL REFERENCES: Kumar 1998; Lee and Liao 1998; Meijaard and Groves 2006; Molur 2008a; Molur et al. 2005; Saha 1981.

Eoglaucomys A. H. Howell, 1915

This genus contains a single species of squirrel. Previously considered to be a subgenus of *Hylopetes*, *Eoglaucomys* was recently determined to be a distinct clade, based on the identification of five derived dental, skeletal, and muscular characteristics.

Eoglaucomys fimbriatus (Gray, 1837) Kashmir Flying Squirrel

DESCRIPTION: This is a relatively large gray or brown flying squirrel, with a round or half-round tail that is flat on the underside. The distal third to half of the tail is commonly black. The feet are usually black. Melanistic individuals are known.

SIZE: Female—HB 290.2 mm; T 290.1 mm; Mass 560.1 g.
Male—HB 288.5 mm; T 289.0 mm; Mass 733.6 g.
Sex not stated—HB 271.0 mm.

DISTRIBUTION: The Kashmir flying squirrel is found in the mountains of northeastern Afghanistan (in the provinces of Badakshan and Nuristan); in northern Pakistan to northern India (in the disputed areas of Jammu and Kashmir); and in the states of Himachal Pradesh, Punjab, Uttarakhand, and Uttar Pradesh (India).

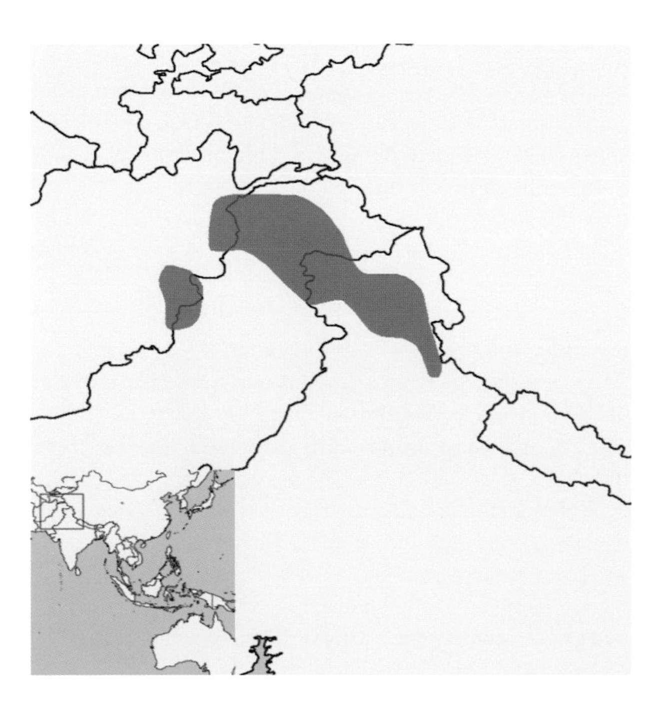

GEOGRAPHIC VARIATION: Two subspecies of *Eoglaucomys fimbriatus* are recognized, but there is a question about the validity of *E. f. baberi*; its identity is based on dental characteristics, which may be merely the milk dentition of the animal.

E. f. fimbriatus—see description above.
E. f. baberi—this form is externally the same as *E. f. fimbriatus*, and distinguished on the basis of dental characteristics, as noted above.

CONSERVATION: IUCN status—least concern. Population trend—no information.

HABITAT: This species is found in moist temperate coniferous forests, usually at elevations between 1600 and 3600 m.

NATURAL HISTORY: The diet of *E. fimbriatus* includes the seeds of at least 22 conifer and deciduous tree species, staminate cones, fruits, flowers, buds, and shoots. Leaves, lichens, moss, and bark are also consumed, but at a lower frequency. This species is more granivorous and relies far less on leaves than *Petaurista petaurista*, which is sometimes sympatric and feeds in the same tree species. Two breeding seasons per year are reported; litter sizes range from two to four. Populations are predicted to be declining, but additional

***Eoglaucomys fimbriatus*.** Photo courtesy Tatsuo Oshida.

data are needed to further document the status of this species. Although numerous potential threats are reported within its range (e.g., hunting for fur, the illegal pet trade, timber harvesting), this species appears relatively secure.

GENERAL REFERENCES: Molur 2008b; Nowak 1999; Oshida, Shafique, et al. 2004; Shafique et al. 2006; R. W. Thorington, Musante, et al. 1996.

Eupetaurus Thomas, 1888

This genus contains a single species.

Eupetaurus cinereus Thomas, 1888
Woolly Flying Squirrel

DESCRIPTION: This is a large flying squirrel with a gray dorsal pelage. A dark chocolate brown pelage has also been described.

SIZE: Female—HB 419.1 mm; T 381.0 mm.
Sex not stated—HB 499.7 mm; T 406.4 mm.

DISTRIBUTION: *E. cinereus* is found at high elevations, from northern Pakistan and Kashmir to Sikkim (India), to Tibet, and possibly to Yunnan (China), although only the Pakistan locations are fully confirmed.

GEOGRAPHIC VARIATION: None.

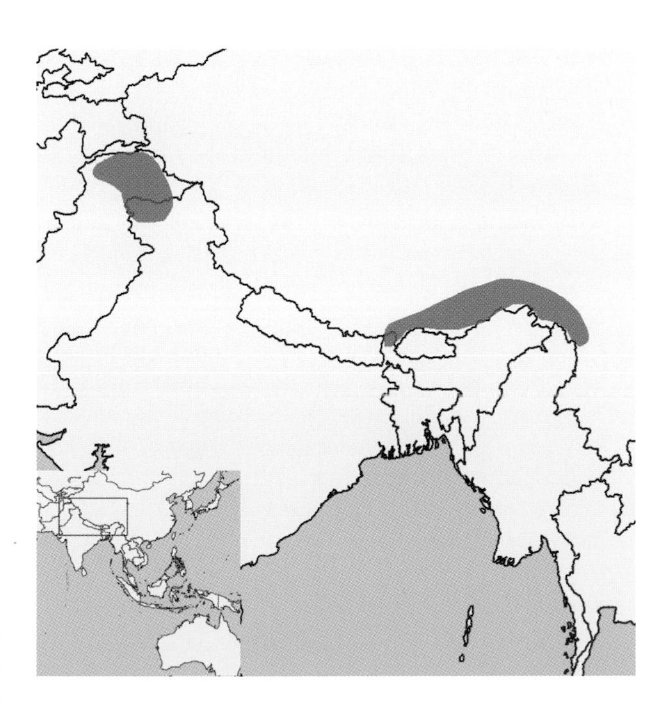

CONSERVATION: IUCN status—data deficient. Population trend—no information.

HABITAT: The woolly flying squirrel resides on rock faces and in caves, at elevations greater than 2000 m. The surrounding forest cover is described as "high, cold desert dominated by *Artemisia* and *Juniperus*" as well as *Pinus* (Zahler).

NATURAL HISTORY: One of the largest of all sciurids, the woolly flying squirrel is believed to be solitary, nocturnal, and active throughout the year. Other behavioral observations suggest a slow-moving squirrel that spends most of its time on cliffs and rock faces. Observations of glide events and comparisons of glide statistics with other mammalian gliders dismiss arguments that *E. cinereus* is too large for efficient gliding. The vocalizations of one mature female were described as "soft grunts and a quiet chirr" (Zahler). Body parts found near the roost of an Eagle Owl (*Bubo bubo*) implicate this avian species as a predator of woolly flying squirrels. Fecal analyses of four squirrels indicate a nearly exclusive diet of pine needles, which probably explains its hypsodont dentition—a characteristic not reported for any other sciurid.

***Eupetaurus cinereus*.** Photo courtesy Peter Zahler.

Local people in Pakistan have noted a decline in this species over the past few decades, probably resulting from the growing human population, a lack of forest regeneration due to overgrazing by sheep and goats, and illegal deforestation.

GENERAL REFERENCES: Yu, Yu, McGuire, et al. 2004; Zahler 1996, 2001; Zahler and Khan 2003.

Glaucomys Thomas, 1908

This genus contains two species of flying squirrels.

Glaucomys sabrinus (Shaw, 1801) Northern Flying Squirrel

DESCRIPTION: *G. sabrinus,* usually weighing more than 100 g, is larger than *G. volans,* the only other flying squirrel in North America. Its fur is thick, and the hairs on the ventral surface are lead colored, which help distinguish it from the white or creamy white venter of *G. volans.* The tail is dark gray; the skull length is greater than 36 mm, whereas that of *G. volans* is always less than 36 mm. Where hybridization of the two species occurs in the northeast, hybrids are intermediate in size and pelage characteristics.

SIZE: Female—HB 168.9 mm; T 129.4 mm; Mass 141.3 g.
Male—HB 158.8 mm; T 135.4 mm; Mass 141.9 g.
Sex not stated—HB 159.0 mm; T 133.5 mm; Mass 120.8 g.

DISTRIBUTION: This species ranges through much of Alaska (USA) and Canada. It extends southward into the USA along the Pacific coast and the Rocky Mountains, with isolated populations in southern California. In the USA, *G. sabrinus* is also found from the Black Hills of western South Dakota east to the northeastern states and southward to the southern Appalachian Mountains, although its range becomes highly fragmented between New York state and Georgia.

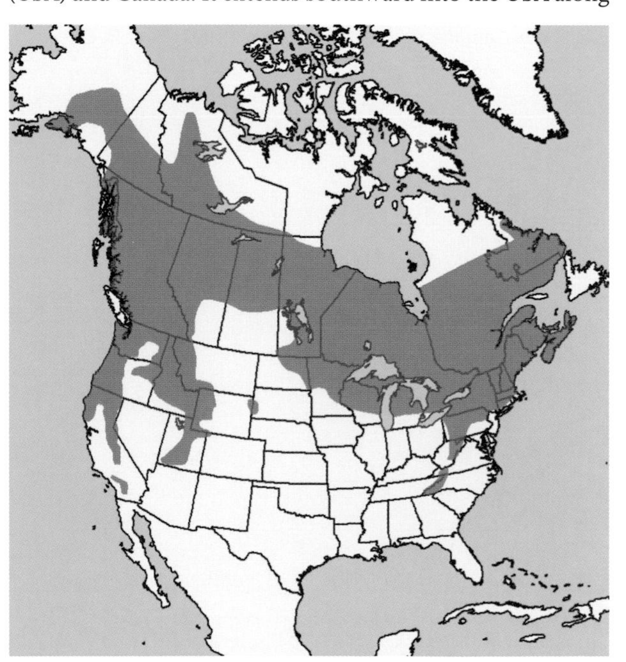

GEOGRAPHIC VARIATION: Twenty-five subspecies are recognized.

G. s. sabrinus—interior of Canada, "from Fort Simpson (possibly Fort Anderson), Mackenzie [River], and lower Churchill River, west side of Hudson Bay, south to northern Minnesota, extreme northwestern Wisconsin, southern Ontario (vicinity of Lake Nipissing), and southern Québec (Lake Edward)" (Howell). This squirrel is also known as the Hudson Bay flying squirrel. The upperparts are vinaceous cinnamon; the sides of the head are smoke gray; the eye ring and the dorsal surface of the tail are fuscous; the ventral surface of the tail is shaded with brown; and the underparts are soiled white, washed with pale yellowish and shaded with drab.

G. s. alpinus—interior of British Columbia, extending slightly over the northern border into the Yukon and over the eastern border into southwestern Alberta (Canada). Known as the Richardson flying squirrel, this form is grayer and less vinaceous above. The tail is darker, wood brown mixed with fuscous on both the dorsal and ventral surfaces.

G. s. bangsi—USA, in the "mountains of central Idaho, eastern Oregon, southwestern Montana," and northwestern Wyoming (Howell). Known as the Bangs flying squirrel, this form has more drab upperparts, grayer feet, and pinkish cinnamon underparts.

G. s. californicus—USA, in the "San Bernardino and San Jacinto Mountains, California" (Howell). Known as the San Bernardino flying squirrel, this form is paler and grayer than *G. s. lascivus,* and there is a grayish wash on the front of the face. The upperparts are light drab to yellowish wood brown, and the underparts are soiled white washed with buffy.

G. s. canescens—northeastern portion of North Dakota (USA), extending over the border into Minnesota (USA) and Manitoba (Canada). This is the palest of the subspecies. It resembles *G. s. macrotis*, but *G. s. canescens* has a grayer head. The upperparts and feet are pale, and the underparts are whiter than in *G. s. sabrinus*.

G. s. coloratus—small region straddling the border of North Carolina and Tennessee (USA). The dorsum is between cinnamon brown and mikado brown (from Ridgway color standards), and the underparts are pale yellow orange to ochraceous buff. The underside of the tail is suffused with orange, with the tail sooty along the edges of the distal half and at the tip. The sides of the face are gray washed with buff.

G. s. columbiensis—Canada and USA, in the "interior valleys and foothills of southern British Columbia and northern Washington, from Shuswap Lake and Cranbrook, British Columbia, south to Lake Chelan, Washington" (Howell). This form is known as the Okanagan flying squirrel. The upperparts are vinaceous cinnamon to vinaceous fawn. It is paler above and below than *G. s. oregonensis*.

G. s. flaviventris—uppermost strip of northern California (USA), from the Trinity Mountains in Siskiyou and Trinity counties east to the Warner Mountains in Modoc County. Known as the yellow-bellied flying squirrel, this form is smaller than *G. s. klamathensis*, with more yellow on the feet and underparts, which are white washed with pale greenish yellow, merging into wood brown along the sides.

G. s. fuliginosus—Canada and USA, in the "Cascade Range, from southern British Columbia south through Washington and Oregon to the Siskiyou Mountains, California" (Howell). Known as the Cascade flying squirrel, this form resembles *G. s. columbiensis*, but it is browner above, darker below, and less brown on the tail. *G. s. fuliginosus* is less rufescent than *G. s. oregonensis*.

G. s. fuscus—West Virginia (USA). This form is smaller and darker, with clear gray cheeks.

G. s. goodwini—tip of the Québec peninsula, north of New Brunswick, on the Gulf of St. Lawrence (Canada). This form has a "sayal brown to avellaneous" back (Howell). The underparts are light pinkish cinnamon, and the feet are dark fuscous. It is grayer on the face and cheeks than *G. s. gouldi*, with a darker and duller winter coat. It is paler than *G. s. makkovikensis*. The tail is smoke gray and slightly darker brown toward the tip, with the underside of the tail smoke gray to dull cinnamon buff.

G. s. gouldi—Prince Edward Island and Nova Scotia (Canada). The dorsal area is buffy brown to dull orange cinnamon. The underparts are white, with the hairs slightly buffy at the tips. The dorsal surface of the tail is dull cinnamon, and the ventral side is pinkish buff. The feet are a pale fuscous. The coat color is brighter and the tail is paler (above and below) than in both *G. s. macrotis* and *G. s. goodwini*.

G. s. griseifrons—Lake Bay, in Prince of Wales Island, Alaska (USA). This form is similar to *G. s. zaphaeus*, but the upperparts of *G. s. griseifrons* are slightly darker colored, including the tail, hind feet, and gliding membrane. The underparts are more whitish and less buffy, and the sides of the head and neck are more extensively grayish.

G. s. klamathensis—central Oregon, east of the Cascade Mountains (USA). Known as the Klamath flying squirrel, this form resembles *G. s. fuliginosus*, but *G. s. klamathensis* is grayer on the upperparts.

G. s. lascivus—USA, in the "Sierra Nevada Range and northward to eastern Shasta Co., California" (Howell). Known as the Sierra flying squirrel, this form is similar to *G. s. flaviventris*, but smaller and without a yellow suffusion on the underparts. Instead, the underparts are grayish white with a faint wash of light pink cinnamon.

G. s. latipes—Canada and USA, in the "Selkirk Range, and other ranges in southeastern British Columbia, higher mountains of northern Idaho and northwestern Montana; south to Mullan and Orofino, Idaho" (Howell). Known as the broad-footed flying squirrel, this form is relatively large. It is similar to *G. s. fuliginosus*, though *G. s. latipes* is darker and grayer, and the upperparts are drab mixed with brown.

G. s. lucifugus—Utah (USA). This form is most similar to *G. s. bangsi*, but *G. s. lucifugus* is paler, with less red and more gray, especially on the face. The feet are drabber; the underparts are whitish, lacking the pinkish cinnamon tone.

G. s. macrotis—Canada and USA, in "New Brunswick, Maine, New Hampshire, Vermont, northern Pennsylvania, southern Ontario, northern part of Michigan, and northeastern Wisconsin; west to Elk River, Minnesota" (Howell). Known as the Mearns flying squirrel, this form tends to be on the smaller side. The ears are slightly longer than typical, the upperparts and hind feet are pale cinnamon, and the underparts are white with irregular washes of light pinkish cinnamon.

G. s. makkovikensis—Canada, in the "coast region of Labrador and eastern Québec; exact limits unknown" (Howell). Known as the Labrador flying squirrel, this form is slightly larger than typical. It has darker upperparts, feet, and a darker face and tail.

G. s. murinauralis—southwestern region of Utah (USA). This form resembles *G. s. lucifugus*, but *G. s. murinauralis* has upperparts that are more yellowish orange. The head, face, and hind feet are grayer; the dorsal surface of the

Glaucomys sabrinus. Photo courtesy Phil Myers, Animal Diversity Web, animaldiversity.org.

tail is uniformly gray, without rufescence, and darker on the apical third of the tail.

G. s. oregonensis—coastal regions of southern British Columbia (Canada) and Washington and Oregon (USA). Known as the Bachman flying squirrel, this form is redder than *G. s. zaphaeus*, with the upperparts being dark reddish brown, and the underparts cinnamon or buff.

G. s. reductus—midcoastal area of British Columbia, near the Koeye River, and inland toward Wisteria and Chezacut (Canada). Known as the Atnarko flying squirrel, this form is paler, less red on the dorsal surface, and whiter on the ventral surface. It has clear gray cheeks.

G. s. stephensi—USA, in the "coast region of northern California; limits of range unknown" (Howell). Known as the California coast flying squirrel, this form's reddish upperparts are intermediate between *G. s. lascivus* and *G. s. oregonensis*. More precisely, the upperparts are wood brown, and the underparts are whitish with irregular washes of light pinkish cinnamon.

G. s. yukonensis—interior of the Yukon, near the Yukon River region (Canada), and west toward Tanana, Alaska (USA). Known as the Yukon flying squirrel, this form is larger than the typical *G. sabrinus*, with a longer tail and a broader hind foot. The upperparts are cinnamon, pinkish to vinaceous.

G. s. zaphaeus—Canada and USA, in the "coast region of southeastern Alaska and northern British Columbia; limits of range unknown" (Howell). Known as the Alaska coast flying squirrel, this form resembles *G. s. alpinus*, but *G. s. zaphaeus* has browner upperparts and darker underparts. The eye ring is blackish.

CONSERVATION: IUCN status—least concern. Population trend—stable.

G. s. coloratus is listed as endangered by the U.S. Endangered Species Act.

HABITAT: The northern flying squirrel is found in a variety of forest types, but it is most often associated with boreal conifer forests or mixed stands with a significant conifer component. Although previously considered a specialist of primary forests, there is increasing evidence that this species is a habitat generalist. In the Appalachian Mountains (from New York state to northwestern Georgia [USA]), *G. sabrinus* is often associated with spruce (*Picea*), fir (*Abies*), and hemlock (*Tsuga*); but it can also be found in mature deciduous forests of yellow birch (*Betula alleghaniensis*), sugar maple (*Acer saccharum*), beech (*Fagus*), and oak (*Quercus*). Numerous reports suggest that there is considerable variation and flexibility in the habitat types used by this species throughout its range.

NATURAL HISTORY: Although the northern flying squirrel consumes seeds and nuts of both conifers and hardwoods, it is considered less granivorous than *G. volans* and may be an obligate mycophagist in many parts of its range, often consuming large quantities of hypogeous (underground) fungi, especially at certain times of the year. As a mycophagist, it may serve a keystone role in some forests by dispersing fungal spores and inoculating tree roots. *G. sabrinus* also consumes tree buds, catkins, staminate cones, tree sap, and animal material (such as songbirds, bird eggs, and insects), as available. The northern flying squirrel is generally nocturnal, but it is also active either shortly before or after dawn and immediately after sunset. This species moves terrestrially and by gliding short distances, usually less than 20 m. Details of its gliding performance and kinematics are reported in the literature. *G. sabrinus* is active throughout the winter and often nests with conspecifics to conserve energy, but it does not appear to show the same physiological adaptations for dealing with extreme cold that are observed in *G. volans*. Vocalizations are common when interacting with conspecifics or when disturbed, but details on the context and function of its vocalizations are not available. This species builds nests in cavities, but it also constructs dreys of most available materials: twigs, bark, moss, lichens, and roots. Cavities are preferred in colder climates. The insides of nests are lined with finely shredded material to provide insulation, typically allowing an occupied nest to be maintained above 27°C, even when outside temperatures are as low as 4°C.

Details of courtship and copulation are reported in the literature and appear similar to those observed in *G. volans*. Gestation is 37-42 days; litters range in size from one to six, but usually average two to four. As many as three litters per year have been reported, although, as in *G. volans,* it is unlikely that the same female produces more than one litter annually. The young are 5-6 g at birth; details of their development are available in the literature. Population densities typically average 0.25 to 2.5 animals/ha, but have been observed to reach 12/ha. In the Oregon Cascades (USA), *G. sabrinus* averages about 2 animals/ha in both secondary and primary Douglas fir (*Pseudotsuga menziesii*) stands. Home range sizes vary considerably with habitat and location; reports typically differ across studies—from 1 to 7 ha per animal—but their ranges have been observed to reach 35 ha in the east. Supplementation experiments show that in the Pacific Northwest (USA), food (but not nest cavities) limits the densities of northern flying squirrels. Where this species is sympatric with *G. volans,* it may be excluded from nest cavities. In Pennsylvania (USA), *G. sabrinus* and *G. volans* converge on stands of hemlock (*Tsuga*) for nesting, and the two species have even been observed to nest together. Here, *G. volans* spends most of its time foraging in adjacent hardwoods, whereas *G. sabrinus* restricts nearly all of its activity to the hemlocks.

A number of avian and mammalian predators are reported to take northern flying squirrels, and this species is a preferred food item of the Northern Spotted Owl (*Strix occidentalis*) in the Pacific Northwest (USA). *G. sabrinus* hosts relatively high numbers of both ecto- and endoparasites for a sciurid. More ectoparasites are found on males than on females, and they more heavily infest smaller males than larger ones. At least 9 flea species, 3 louse species, and 10 mite and tick species are reported. Endoparasites include 2 species of *Eimeria* (protozoans), 6 nematodes, and 4 species of cestodes. The nonoverlapping geographic ranges of *G. volans* and *G. sabrinus* may result from parasite-mediated competitive exclusion, wherein the nematode *Strongyloides robustus*, a common parasite of *G. volans,* is potentially lethal to *G. sabrinus.* Fossils of this species are reported since the late Pleistocene. Cytochrome *b* sequence data from mitochondrial DNA is significantly more variable in *G. sabrinus* than in *G. volans,* and it shows two separate lineages in *G. sabrinus,* suggesting the possibility of a third species of *Glaucomys.* Loss of genetic variation and genetic differentiation are evident in populations on coastal islands in the Pacific Northwest (USA). Despite previous evidence of differences in chromosome number and bacular structure, the two species of *Glaucomys* have been reported to hybridize, producing viable offspring that are intermediate in pelage characteristics and size. Such hybridization in the northeastern part of the range may be due to climate change, habitat loss, or a combination of the two, which has increased sympatry between the two species.

GENERAL REFERENCES: Anthony 1928; Arbogast 1999; Arbogast et al. 2005; Bakker and Hastings 2002; Bidlack and Cook 2001; Bradley and Marzluff 2003; Carey 2000, 2001, 2002; Carey, Horton, et al. 1992; Carey, Kershner, et al. 1999; Demboski et al. 1998; Garroway et al. 2010; Hackett and Pagels 2003; D. S. Hall 1991; Holloway and Malcolm 2007; Howell 1915a; Krichbaum et al. 2010; A. W. Linzey and Hammerson 2008a; Loeb and Tainter 2000; Meyer, North, et al. 2005; D. Mitchell 2001; Odom et al. 2001; J. L. Payne et al. 1989; Perez-Orella and Schulte-Hostedde 2005; Pyare and Longland 2001; Ransome and Sullivan 1997, 2002, 2003, 2004; Raphael 1984; Rosenberg and Anthony 1992; Scheibe et al. 2007; W. P. Smith 2007; W. P. Smith, Gende, et al. 2004; Vernes 2001; Vernes et al. 2004; P. D. Weigl 2007; Wells-Gosling and Heaney 1984; Witt 1991, 1992; Zittlau et al. 2000.

Glaucomys volans (Linnaeus, 1758)
Southern Flying Squirrel

DESCRIPTION: *G. volans* is a small flying squirrel, usually weighing approximately 60 g. It is easily identified by its grayish brown dorsum; white, cream or yellowish venter; and the dark brownish edge of the dorsal and ventral surfaces of the patagium ("wing"). The head is gray, and the cheeks are white. The dorsal pelage consists of fine dense fur. *G. volans* is distinguished from the only other flying squirrel in North America—*G. sabrinus*—by its smaller size and the whitish base of the ventral hairs (which are dark gray in *G. sabrinus*). Despite differences in the bacula and

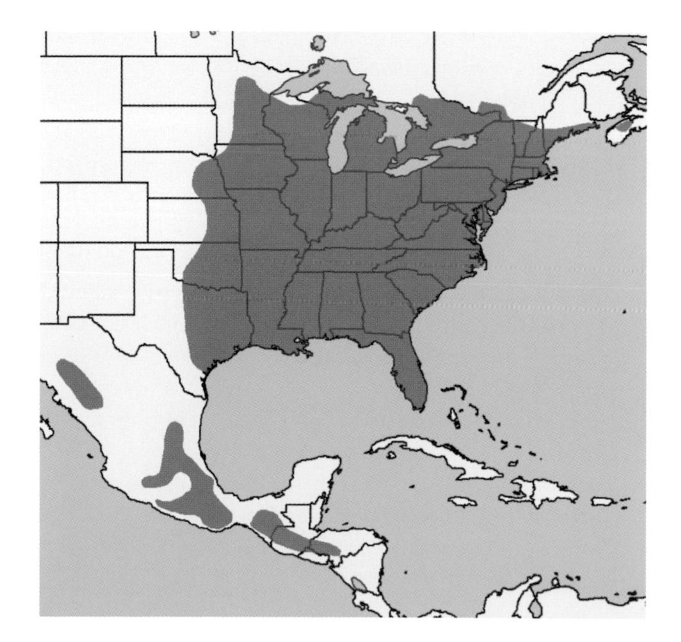

reports of chromosome differences, a zone of hybridization between *G. volans* and *G. sabrinus* has been reported in Ontario (Canada) and Pennsylvania (USA). The hybrids are intermediate in size and pelage characteristics.

SIZE: Female—HB 132.5 mm; T 103.0 mm; Mass 57.6 g.
Male—HB 131.1 mm; T 103.1 mm; Mass 53.2 g.
Sex not stated—HB 134.0 mm; T 100.3 mm; Mass 70.0 g.

DISTRIBUTION: This species ranges from Nova Scotia (Canada) southward across the eastern half of the USA. It also occurs on isolated mountains from northwestern México to Honduras.

GEOGRAPHIC VARIATION: Eleven subspecies are recognized.

G. v. volans—Canada and USA, in the "northeastern USA and extreme southern Canada, from central Minnesota, Wisconsin, and Michigan, southern Ontario, northern New York state (Lewis County), and southern New Hampshire south to North Carolina (Raleigh), Tennessee (Nashville), and northern Arkansas and Oklahoma (Boston Mountains); west to eastern Nebraska (Otoe and Nemaha counties) and eastern Kansas (Douglas and Woodson counties)" (Howell). This subspecies is known as the small eastern flying squirrel. See description above.

G. v. chontali—Santo Domingo Chontecomatlán, Yautepec District (Oaxaca, México). This form is the most brightly colored, with its upperparts ochraceous tawny and the dorsal surface of the patagium blackish brown. The dorsal surface of the tail is similar in color to the back, but darker. The underparts are white, with a pinkish buff tinge on the underside of the limbs. The ventral surface of the tail is a clear cinnamon buff.

G. v. goldmani—mountain slopes of Chiapas (México). This form is relatively large, with sayal brown upperparts. The underparts are creamy white irregularly washed with pinkish buff.

G. v. guerreroensis—Sierra Madre del Sur (Guerrero, México). This form is the darkest of the Mexican and Central American flying squirrels. It is dark brownish dorsally, blackish on the upper patagium, and creamy white on the venter.

G. v. herreranus—mountains of southern México, in southern Tamaulipas, southeastern San Luis Potosí, Querétaro, Michoacán, Veracruz, and Oaxaca. It is similar in coloring to *G. v. goldmani*, but *G. v. herreranus* is darker, the darkest of the subspecies. The upper part of the patagium is a deep glossy black.

G. v. madrensis—Sierra Madre Occidental (Chihuahua, México). This form is similar in coloring to *G. v. texensis*, but *G. v. madrensis* is paler, with the upperparts overlain with buffy gray. The underparts are white, tinged with pale buff on the underside of the patagium.

G. v. oaxacensis—Pacific side of southern México, in the highlands of Oaxaca. It is similar to *G. v. goldmani*, but *G. v. oaxacensis* is paler dorsally, with a pinkish buff tone.

G. v. querceti—peninsular Florida and the coastal region of Georgia (USA). Known as the Florida flying squirrel, this form is darker than *G. v. volans*, but lighter than *G. v. saturatus*. The toes are not conspicuously white.

G. v. saturatus—in the "southeastern United States (excepting peninsular Florida and the coast region of Georgia) from South Carolina and western North Carolina west to central Oklahoma and north in the Mississippi Valley to southwestern Kentucky" (Howell). Known as the southeastern flying squirrel, its upperparts are darker, snuff brown to hair brown; the toes are not conspicuously white.

G. v. texensis—eastern Texas (USA). Known as the Texas flying squirrel, this form has more ochraceous upperparts, wood brown with yellowish tinge varying to drab, that are less dark than in *G. v. saturatus*.

G. v. underwoodi—mountains of southern Guatemala and Honduras. It is similar in coloring to *G. v. goldmani*, but *G. v. underwoodi* is paler, with a pale sayal brown coloring.

CONSERVATION: IUCN status—least concern. Population trend—stable.

HABITAT: Throughout much of its range, *G. volans* is associated with oak-hickory (*Quercus, Carya*) forests, and the hard mast forms an important part of its diet. In the northern part of its range, this species also occurs in mixed deciduous-conifer forests. In the southern Appalachian Mountains, it can be found as high as 1500 m. From México southward, it usually inhabits oak-conifer associations in the mountains, at elevations of 1200-3000 m. This species is also common in coastal pine-oak (*Pinus, Quercus*) associations of the eastern and southeastern USA. Studies of its microhabitat use show a preference for areas with higher densities of shrubs, reduced understory cover, and more tree cavities.

NATURAL HISTORY: This species feeds mostly on nuts, seeds, and fruits, but it also includes fungi, lichens, moss, and bark in its diet. *G. volans* is considered by some to be highly carnivorous, as it consumes animal material when it is available. Nuts and seeds are both scatter-hoarded on the ground

Glaucomys volans. Photo courtesy Brian E. Kushner, BrianKushnerPhoto.com.

and larder-hoarded in trees or in the nest. Tree cavities are preferred for nesting, and two or more nests are usually maintained simultaneously, although one normally serves as the primary nest. *G. volans* often competes for nest structures and can exclude some species (e.g., Red-Cockaded Woodpeckers [*Picoides borealis*]) from nest hollows. One avian cavity nester, the Great Crested Flycatcher (*Myiarchus crinitus*), lines its nests with a piece of snakeskin, which appears to deter predation by *G. volans*. Nest aggregations as high as dozens of animals occur primarily in northern regions, for thermoregulatory conservation. Such nest aggregations often consist of close relatives, which may reduce the costs of intraspecific competition between nestmates. *G. volans* also competes with tree squirrels for nest structures and, where the two *Glaucomys* species are sympatric, it may exclude *G. sabrinus* from tree cavities. This species shows a resting metabolic rate (RMR) that is 33-38 percent lower than predicted by its body mass. However, less than 20 percent of this reduction is compensated for by the thermoregulatory advantages gained from communal nesting. Instead, *G. volans* relies on nonshivering thermogenesis (NST), which is negatively correlated with ambient temperatures and peaks in the winter. The higher energetic costs, coupled with a reduced food supply, probably influence the northern limit of this species' range.

Courtship and mating are described in detail in the literature. Although two breeding seasons are reported, the same females probably do not regularly breed during both seasons. Parturition occurs in the spring (April-May) and late summer (August-September); gestation is 40 days; the young are hairless and weigh 3-5 g at birth. Litter sizes vary between two and seven, but usually average about four. The young may stay with the mother beyond the normal 8 weeks if conditions warrant it. Details of their development are reported in the literature. *G. volans* may live as long as 10 years in captivity. Population densities vary from 2 to 12 animals/ha. Two studies listed mean home range sizes of 2.45 and 7.8 ha for adult males, and 1.95 and 3.8 ha for adult females. One study noted a mean home range size of 0.61 animals/ha for juveniles. At least two studies reported on the homing ability of *G. volans*, citing homing distances of 1.0-1.6 km. This species will frequently move through tree canopies by means of gliding. Glides of up to 90 m are possible; details of their gliding mechanics and performance are available in the literature.

This species is exclusively nocturnal and, as such, is primarily vulnerable to predation by owls and snakes. *G. volans* hosts a number of ectoparasites and, because of its highly carnivorous diet, a number of internal parasites as well. At least six species of fleas, three lice, four mites, three protozoans, one acanthocephalan, one cestode, and five nematodes have been reported. The southern flying squirrel is also a reservoir host for ectoparasites that carry *Rickettsia prowazekii*, which causes epidemic typhus. *G. volans* has been used to study potential interactions between *Strongyloides robustus* and *Capillaria americana*. It has been hypothesized that the generally nonoverlapping geographic ranges of *G. volans* and *G. sabrinus* may result from parasite-mediated competitive exclusion, wherein the nematode *Strongyloides robustus*, a common parasite of *G. volans*, is potentially lethal to *G. sabrinus*. For example, recent parasite surveys from the northeastern portion of the two species' ranges show that where *G. sabrinus* and *G. volans* are sympatric, they both host *S. robustus*; but 100 km farther north, where only *G. sabrinus* is found, the parasite appears to be absent.

GENERAL REFERENCES: Bendel and Gates 1987; Bishop 2006; Bowman et al. 2005; COSEWIC 2006; Day and Benton 1977; Dolan and Carter 1977; Garroway et al. 2010; E. R. Hall 1981; Harlow and Doyle 1990; Healy and Brooks 1988; D. J. Holmes and Austad 1994; Krichbaum et al. 2010; Lavers et al. 2006; Laves and Loeb 1999; Leung and Cheng 1997; D. W. Linzey and Linzey 1979; Mahan et al. 2010; Medlin and Risch 2006; Merritt et al. 2001; Patrick 1991; Risch and Brady 1996; Risch and Loeb 2004; Sawyer and Rose 1985; Stapp 1992; Stapp et al. 1991; Stone, Heidt, Baltosser, et al. 1996; Stone, Heidt, Caster, et al. 1997; K. K. Thorington et al. 2010; Tyler and Donelson 1996; P. D. Weigl 1968, 1978.

Hylopetes Thomas, 1908

This genus contains nine species of flying squirrels. Despite previous suggestions, morphometric analyses of the bacula, foot pads, musculature, teeth, wrist, and ankles validate three distinct genera (clades) for *Eoglaucomys, Glaucomys,* and *Hylopetes*. Based on five species of *Hylopetes,* one of which was *H. platyurus,* these analyses identified eight derived characteristics of the genus. Both morphological characteristics and recent phylogenetic analyses inferred from mitochondrial cytochrome *b* gene sequences in this and six other genera of flying squirrels indicate a close relationship between *Hylopetes* and New World flying squirrels, *Glaucomys*. These cytochrome *b* data also show these two genera to be closely related to *Petinomys*; the data suggest that *Glaucomys* and *Hylopetes* diverged approximately 28.6 MYA, and that *Glaucomys* and *Petinomys* diverged 29.2 MYA. *Petinomys* and *Hylopetes* are sympatric throughout their ranges, morphologically quite similar (being distinguished by the number of septa in the auditory bullae), and estimated to have diverged during the Pleistocene (2.2 MYA) in Southeast Asia.

Earlier studies considered *Eoglaucomys* to be a subgenus of *Hylopetes,* while others suggested that *Glaucomys sabrinus* was more closely related to *Hylopetes* than to *G. volans*. More recent morphological studies (e.g., bacula, foot pads, musculature, crania, and postcrania) indicate three distinct clades for *Eoglaucomys, Glaucomys,* and *Hylopetes*. Although *Hylopetes* is considered to be declining in many parts of its range, details on its population biology are not available.

Hylopetes alboniger (Hodgson, 1836)
Particolored Flying Squirrel

DESCRIPTION: *H. alboniger* is a relatively small flying squirrel. The dorsal hair is a medium brown; the sides, dorsal patagium ("wing"), and legs are dark gray to black. The venter is white to cream colored. The cheeks and throat are white; the ventral surface of the tail is dark.

SIZE: Female—HB 223.6 mm; T 202.5 mm.
Male—HB 214.5 mm; T 196.1 mm; Mass 269.3 g.
Sex not stated—HB 214.6 mm; T 187.7 mm; Mass 240.0 g.

DISTRIBUTION: The particolored flying squirrel's range extends from Nepal and Assam (India) to the Chinese provinces of Yunnan, Sichuan, Guizhou, Guangxi, Hainan, and Zhejiang, and then southward through Indochina to southern Vietnam.

GEOGRAPHIC VARIATION: Four subspecies are recognized.

H. a. alboniger—the type locality is the central and northern regions of Nepal. See description above.

H. a. chianfengensis—island of Hainan (China). We have not located the description.

H. a. leonardi—northwest Yunnan (China). This form is a small subspecies. It is blackish buffy above, with pinkish

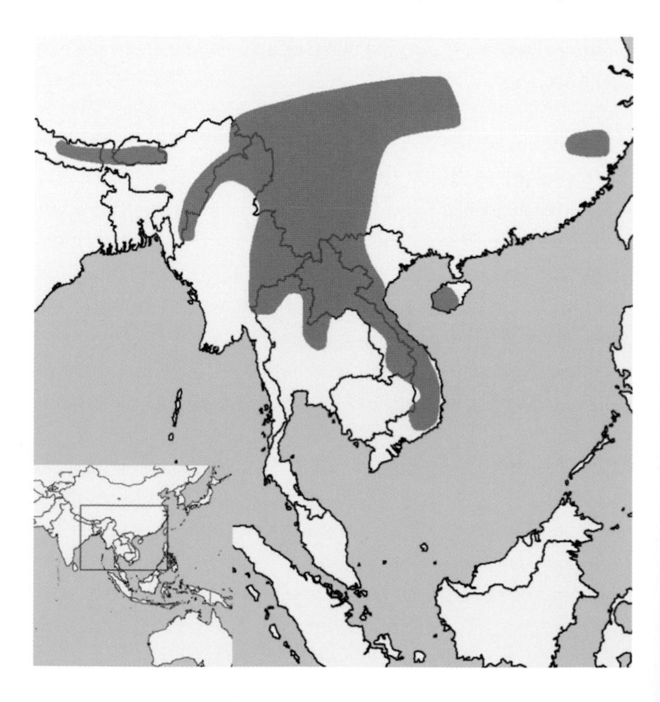

Hylopetes alboniger. Photo courtesy Randall D. Babb.

buffy underparts but a white chest and axillae. The face and sides of the neck are buff. The ears are large and black at the base; the hands and feet are blackish.

H. a. orinus—Yunnan, Sichuan, Guizhou, Guangxi, and Zhejiang (China). This form is a large subspecies. The terminal three-quarters of the tail is black.

CONSERVATION: IUCN status—least concern. Population trend—decreasing.

HABITAT: Throughout its range, the particolored flying squirrel is found at midelevations in montane regions that are classified as either tropical or subtropical. In China this species is found most often in oak (*Quercus*) and rhododendron (*Rhododendron*) forests, at elevations up to 3500 m.

NATURAL HISTORY: *H. alboniger* feeds primarily on a combination of fruits and nuts, as well as on leaves and other plant parts. Breeding takes place in the spring, from April to mid-June; the litter size is two to three. This species is nocturnal, and it nests in tree cavities. In China, *H. alboniger* is often sympatric with one or more species of *Petaurista*. This small flying squirrel produces high-pitched vocalizations that can be used to locate it in night surveys. *H. alboniger* is reported to be one of 134 species of wild animals harvested for meat by the indigenous people of northeast India.

GENERAL REFERENCES: Duckworth, Tizard, et al. 2008; Hilaluddin et al. 2005; A. T. Smith and Xie 2008; Thomas 1921; R. W. Thorington, Musante, et al. 1996.

Hylopetes bartelsi (Chasen, 1939) Bartels's Flying Squirrel

DESCRIPTION: This species is similar to *H. vordermanni*, but the whiskers and ear tufts are large and the tail is markedly bicolored at the base. This species was formerly included in *Petinomys*.

SIZE: Male—HB 197.0 mm.
Sex not stated—HB 139.0 mm.

DISTRIBUTION: This species is endemic to Java (Indonesia). It is known only from the type locality: on Mount Pangrango, in western Java.

GEOGRAPHIC VARIATION: None.

CONSERVATION: IUCN status—data deficient. Population trend—no information.

HABITAT: This species is thought to occupy both lowland and montane subtropical and tropical primary forests.

NATURAL HISTORY: Nothing is known about the biology of this species.

GENERAL REFERENCES: Corbet and Hill 1992; Duckworth and Hedges 2008a.

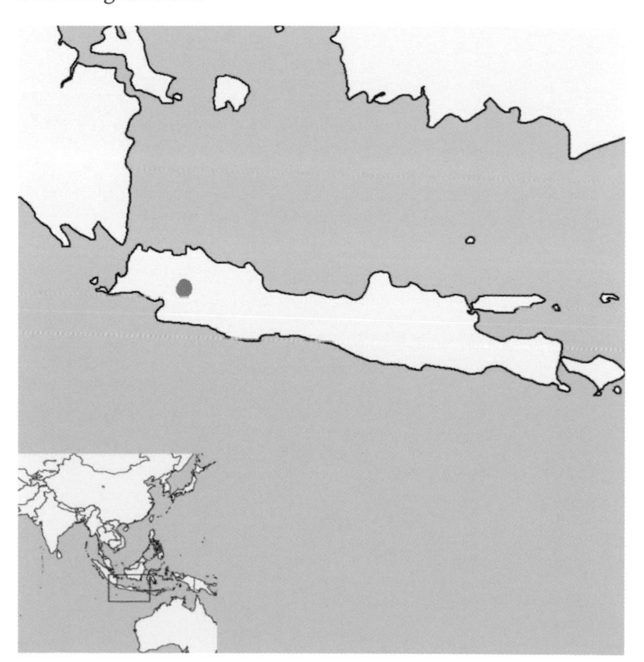

Hylopetes lepidus (Horsfield, 1822) Gray-Cheeked Flying Squirrel

DESCRIPTION: *H. lepidus* is a small flying squirrel, similar in size and pelage coloration to *H. platyurus* and *H. spadiceus*. These three species have long been considered problematic, because of their overlapping characteristics. Recent morphometric studies, however, show that the three can be distinguished by a multivariate analysis of their external measurements and craniodental dimensions, with *H. lepidus* intermediate in size between *H. platyurus* and *H. spadiceus*. The cheeks and ventral base of the tail of *H. lepidus* are pinkish and intermediate in coloration to the other two species; *H. platyurus* is mostly gray in color, and *H. spadiceus* is a reddish orange.

SIZE: Female—HB 122.9 mm; T 123.3 mm; Mass 43.3 g.
Male—HB 115.1 mm; T 91.5 mm; Mass 38.8 g.
Sex not stated—HB 127.1 mm; T 102.3 mm.

DISTRIBUTION: The gray-cheeked flying squirrel is native to the island of Java and Bangka Island (Indonesia). Although the full extent of its range is not well known, it appears to be represented by two widely separated populations.

GEOGRAPHIC VARIATION: None. This species is monotypic. This species was previously called *Hylopetes sagitta* Linnaeus, 1766. It also formerly included *H. platyurus*. More information is needed to clarify the relationship between *H. lepidus* and *H. platyurus*.

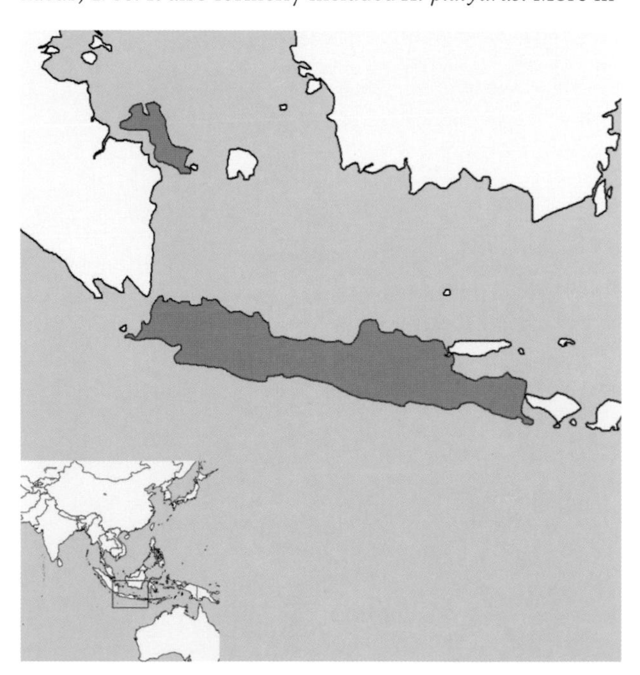

CONSERVATION: IUCN status—data deficient. Population trend—no information.

HABITAT: The specific habitat requirements of the gray-cheeked flying squirrel are not well known. In western Malaysia *H. lepidus* (but probably *H. platyurus* instead) is reported to be relatively abundant in older stands and secondary forests at elevations below 3000 m, but it is less common in disturbed forests or plantations.

NATURAL HISTORY: Similarly, little is known about the natural history of this species, and the species accounts may be confused between *H. lepidus* and *H. platyurus*. Reproduction may occur irregularly, but pregnant females have been reported in both February-March and July-August. One species of sucking louse is reported from this host.

GENERAL REFERENCES: Duckworth and Hedges 2008b; Durden and Adams 2005; Muul and Liat 1971, 1974; Nor 2001; Rasmussen and Thorington 2008; R. W. Thorington, Musante, et al. 1996.

Hylopetes nigripes (Thomas, 1893) Palawan Flying Squirrel

DESCRIPTION: This is a large member of the genus. It is grizzled grayish brown dorsally; is whitish to grayish white on the throat, chest, and abdomen; and has large ears that are thinly haired and blackish toward the tip. The tail is thickly furred and indistinctly distichous, with the proximal hairs broadly tipped with brown, and the more distal hairs fully dark brown.

SIZE: Female—HB 283.2 mm; T 313.9 mm.
Male—HB 264.1 mm; T 314.6 mm.
Sex not stated—HB 270.0 mm; Mass 534.0 g.

DISTRIBUTION: Endemic to the Philippines, *H. nigripes* is found only on the islands of Palawan and Bancalan in the Palawan Faunal Region.

GEOGRAPHIC VARIATION: Two subspecies are recognized.

H. n. nigripes—Palawan (Philippines). See description above.
H. n. elassodontus—Bancalan (Philippines). This is a smaller subspecies.

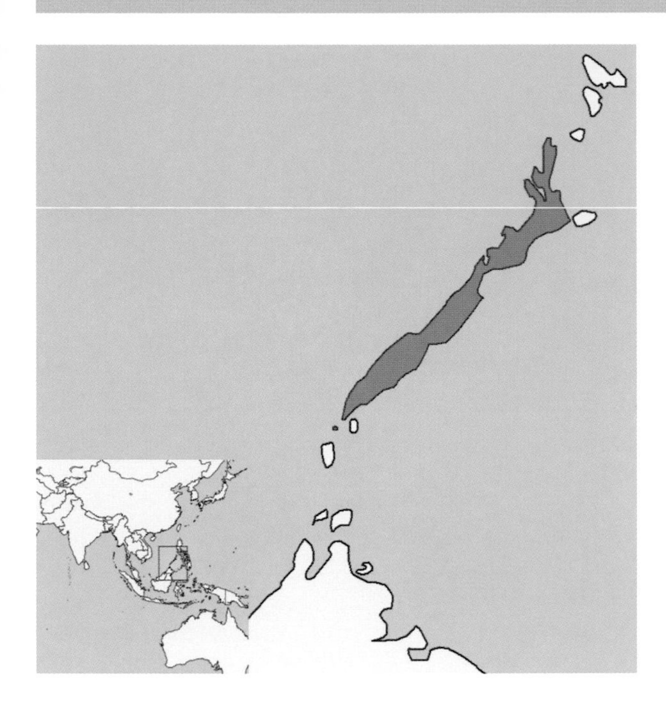

CONSERVATION: IUCN status—near threatened. Population trend—decreasing.

HABITAT: The Palawan flying squirrel is found at lower elevations in both primary and secondary forests. It appears to withstand some degree of habitat disturbance, such as selective logging, but more information is needed on its susceptibility to illegal logging and clearing for agriculture.

NATURAL HISTORY: This nocturnal species appears to be relatively common in older forests, possibly because of its preferences for large mature trees with cavities. It is often hunted for food and captured for the pet trade.

GENERAL REFERENCES: Esselstyn et al. 2004; Ong et al. 2008; Thomas 1893; Timm and Birney 1980.

Hylopetes phayrei (Blyth, 1859) Indochinese Flying Squirrel, Phayre's Flying Squirrel

DESCRIPTION: This is the smallest flying squirrel in China. It has a flattened tail, a reddish brown dorsal pelage, white cheeks, and a whitish ventral pelage with a faint yellow wash.

Hylopetes phayrei. Photo courtesy Pitchaya and Rattapon Kaichid.

SIZE: Female—HB 158.3 mm; T 143.7 mm; Mass 113.4 g.
Male—HB 162.3 mm; T 146.7 mm.
Sex not stated—HB 170.8 mm; T 143.4 mm; Mass 171.0 g.

DISTRIBUTION: *H. phayrei* can be found in Myanmar, Thailand, Laos, Vietnam, and southern China (Hainan Island, Guizhou, Guangxi, and Fujian).

GEOGRAPHIC VARIATION: Two subspecies are recognized.

H. p. phayrei—mainland China. See description above.
H. p. electilis—Hainan Island (China). The dorsal surface is a uniform pale cinnamon.

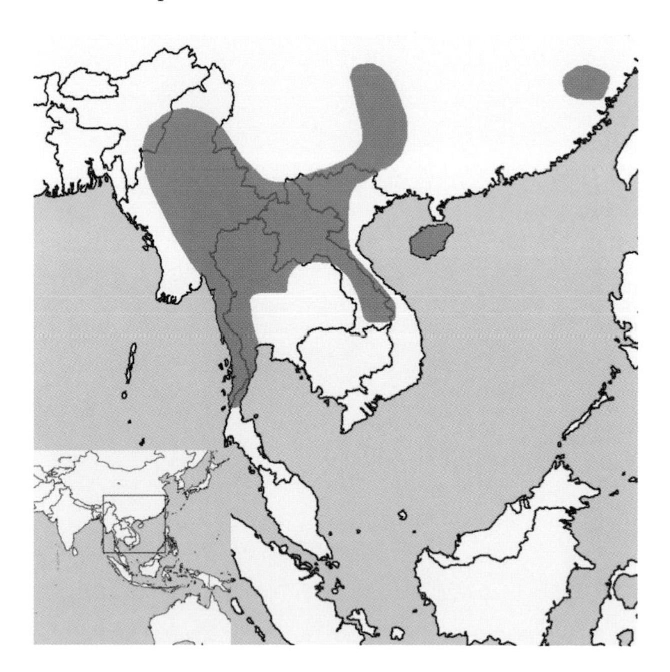

CONSERVATION: IUCN status—least concern. Population trend—stable.

HABITAT: In southern China, this species is found in mixed deciduous and lowland montane forests. It is sympatric with *H. alboniger* on Hainan Island (China).

NATURAL HISTORY: *H. phayrei* is nocturnal, and it nests in tree cavities. Little additional information is available on this species.

GENERAL REFERENCES: A. T. Smith and Xie 2008.

Hylopetes platyurus (Jentink, 1890) Jentink's Flying Squirrel

DESCRIPTION: Taxonomically confused with *H. spadiceus* and *H. lepidus* in the past, *H. platyurus* is distinctly smaller in its overall body dimensions and in at least 21 craniodental measures than these other two species. In *H. platyurus* the cheeks and the base of the tail are gray, as compared with those of *H. spadiceus* and *H. lepidus*, which are more pinkish, although overlap occurs between the three species. The tail of *H. lepidus* is also shorter and narrower than that of the other two species.

SIZE: Female—HB 130.0 mm; T 100.0 mm.

DISTRIBUTION: Jentink's flying squirrel is found only in northern Sumatra (Indonesia), peninsular Malaysia, and the island of Borneo (divided among Malaysia, Brunei Darussalam, and Indonesia). Neither the South China Sea nor the Malacca Strait has been a barrier for *H. platyurus*.

GEOGRAPHIC VARIATION: None.

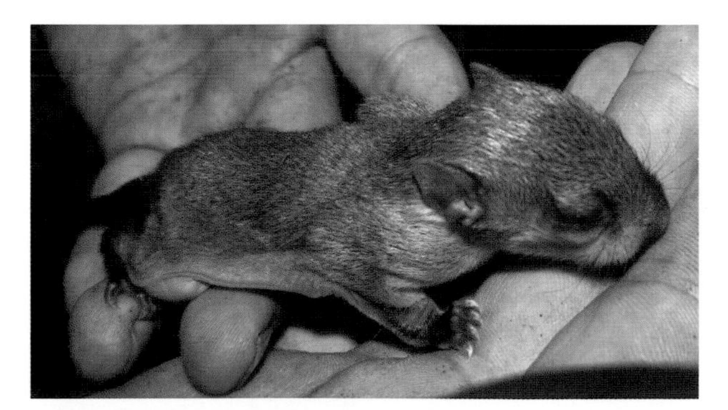

Hylopetes platyurus. Photo courtesy Andy Boyce.

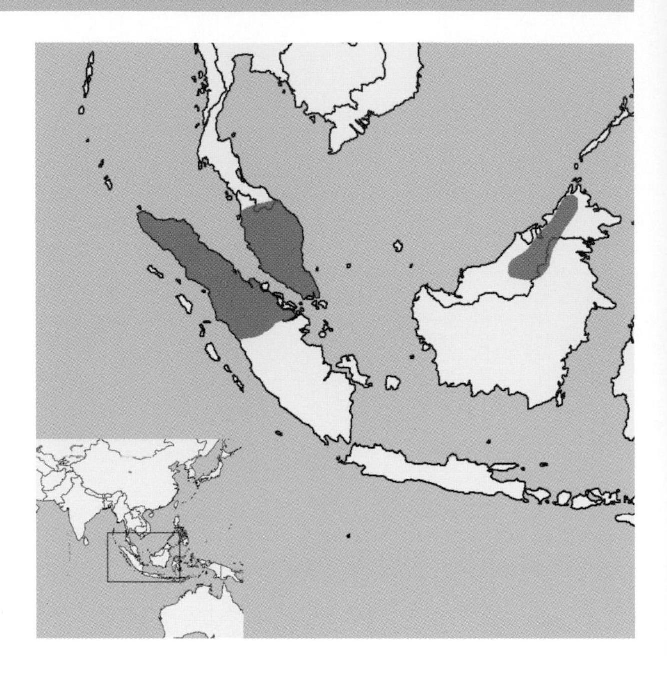

CONSERVATION: IUCN status—data deficient. Population trend—no information.

HABITAT: Little is known about this species' preferred habitat. It is reported from rubber forests, forest edges, and near human activities; however, it has not been observed in primary forests.

NATURAL HISTORY: Little is known about the distribution, ecology, and behavior of this species.

GENERAL REFERENCES: Meijaard 2003a; Muul and Liat 1971; Rasmussen and Thorington 2008; Thorington and Heaney 1981; Thorington, Musante, et al. 1996.

Hylopetes platyurus. Photo courtesy James Eaton/Birdtour Asia.

Hylopetes sipora Chasen, 1940
Sipora Flying Squirrel

DESCRIPTION: The tail is almost completely dull brownish black, paler at the base and white at the tip, and not markedly bicolored. This species has dark brown hands and feet. The ventral surface is buff to orange buff, and there is a white patch on the chest. Chasen did not describe the dorsal coloration.

SIZE: Sex not stated—HB 140.0 mm; Mass 89.2 g.

DISTRIBUTION: The Sipora flying squirrel is found only on Sipora Island (Indonesia), southwest of Sumatra in the Mentawai Archipelago.

HABITAT: *H. sipora* is a lowland species that resides in primary tropical and subtropical forests.

GEOGRAPHIC VARIATION: None.

CONSERVATION: IUCN status—endangered. Population trend—decreasing.

NATURAL HISTORY: Populations of *H. sipora* are reported to be declining, primarily as a result of deforestation for agriculture and wood products. Little is known about the precise distribution and ecology of this species.

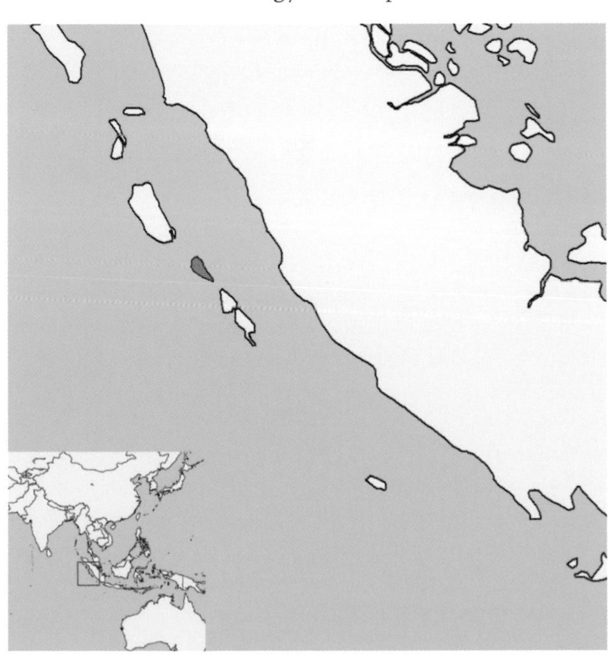

GENERAL REFERENCES: Baillie and Groombridge 1996; Heaney 2008; Meijaard 2003a; Ruedas et al. 2008b; R. W. Thorington and Hoffmann 2005.

Hylopetes spadiceus (Blyth, 1847)
Red-Cheeked Flying Squirrel

DESCRIPTION: This small flying squirrel is orange brown on the head and back; the cheeks are orange; the throat is white. The dorsal surface of the gliding membrane is black; the tail is blackish, with an orange brown or buff base. The feet are reddish brown, and the ventral surface is creamy white.

Hylopetes spadiceus. Photo courtesy Nick Baker, www.ecologyasia.com.

Hylopetes spadiceus. Photo courtesy Nick Baker, www.ecologyasia.com.

SIZE: Female—HB 146.4 mm; T 129.1 mm; Mass 78.0 g.
Male—HB 142.0 mm; T 118.1 mm; Mass 70.9 g.
Sex not stated—HB 147.9 mm; T 125.8 mm; Mass 75.9 g.

DISTRIBUTION: The red-cheeked flying squirrel extends south from Myanmar to Thailand, peninsular Malaysia, Indonesia—on Sumatra (including Bangka Island), Kundur Island, and Bunguran Island (in the northern South China Sea)—and the island of Borneo (divided among Malaysia, Brunei Darussalam, and Indonesia). It also occurs in Laos and South Vietnam. This species has a relatively wide distribution and is thought to be more common than records indicate, especially on the island of Borneo.

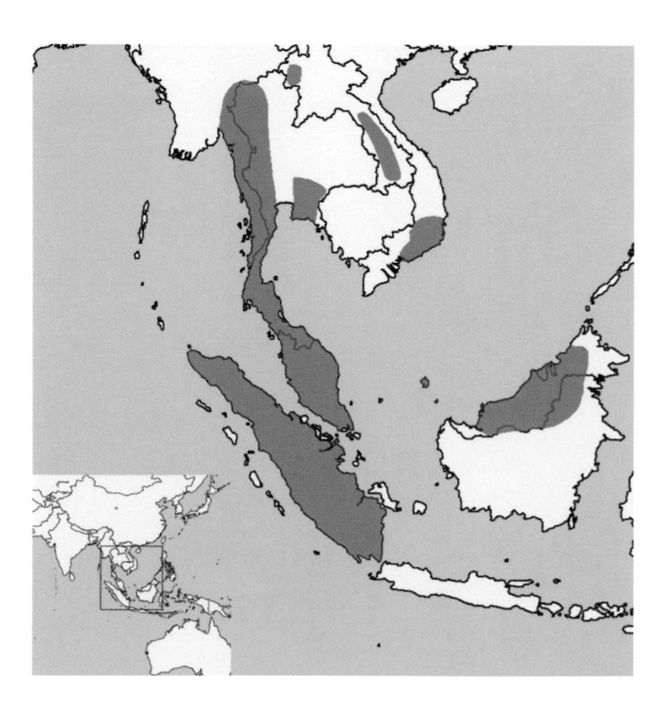

GEOGRAPHIC VARIATION: Three subspecies are recognized.

H. s. spadiceus—peninsular Malaysia.
H. s. everetti—island of Borneo (divided among Malaysia, Brunei Darussalam, and Indonesia).
H. s. sumatrae—Sumatra (Indonesia).

CONSERVATION: IUCN status—least concern. Population trend—no information.

HABITAT: This species is found in a wide range of habitats—from primary forests to degraded and cultivated forests—at elevations below 1500 m. It occurs in several protected areas.

NATURAL HISTORY: Little is known about the population ecology and behavior of *H. spadiceus*. An extended history of inconsistencies regarding the nomenclature of *H. spadiceus* with *H. lepidus* and *H. platyurus* has resulted in confusion over both the number of species of *Hylopetes* and the biogeographical factors that influenced these three species of *Hylopetes* in the region of Sumatra, Java, the island of Borneo,

Hylopetes spadiceus. Photo courtesy Nick Baker, www.ecologyasia.com.

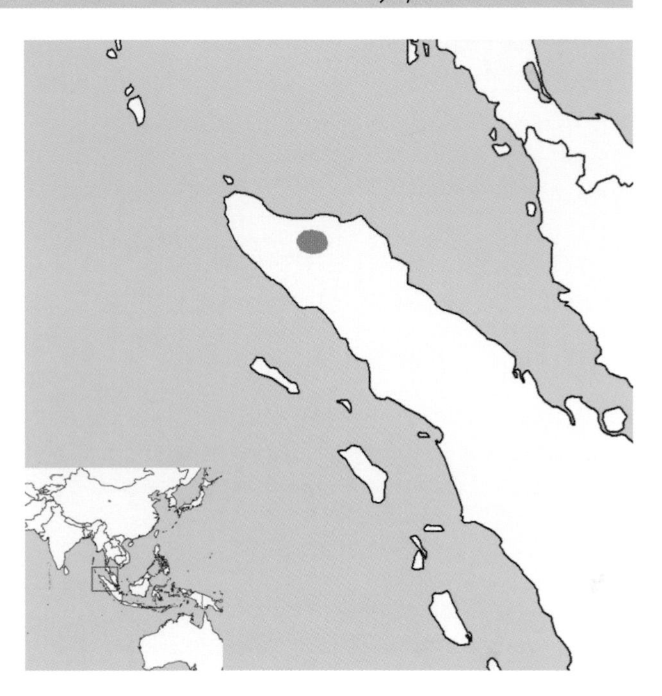

and Malaysia. Recent analyses of the pelage and craniodental characteristics, however, indicate a clear differentiation of the three species. This taxonomic clarification implies that the Sunda Strait, between Java and southern Sumatra, was a barrier preventing colonization of Java by *H. spadiceus*. This is not unique to this species: among 25 species of sciurids found on Sumatra, only nine also occur on Java. In contrast, the Malacca Strait, between Sumatra and the Malay Peninsula, appears not to have been a barrier either for *H. spadiceus* or for 19 other sciurids found on Sumatra. Likewise, neither the Isthmus of Kra on the Malay Peninsula, nor the South China Sea, separating the Malay Peninsula and the island of Borneo, has served as a dispersal barrier for *H. spadiceus*. One unidentified nematode and *Capillaria hepatica* are reported from *H. spadiceus*.

GENERAL REFERENCES: Ahl 1987; Corbet and Hill 1992; Duckworth and Hedges 2008c; Dunn et al. 1968; Liat et al. 1977; Meijaard 2003a; Rasmussen and Thorington 2008; R. W. Thorington and Hoffmann 2005.

Hylopetes winstoni (Sody, 1949)
Sumatran Flying Squirrel

DESCRIPTION: This is a small flying squirrel, with tufts of stiff hairs—directed upward and backward—in front of and above the ears and on the cheeks.

SIZE: Male—HB 142.0 mm; T 143.0 mm.
Sex not stated—HB 142.0 mm.

DISTRIBUTION: The Sumatran flying squirrel is known only from the type locality: "Baleq, E. Atjeh [= Aceh], in northern Sumatra, 1200 m" (Indonesia).

GEOGRAPHIC VARIATION: None.

CONSERVATION: IUCN status—data deficient. Population trend—no information.

HABITAT: The preferred habitat for this species is reported to be primary forest, at elevations of 1000 to 1500 m, but details on its habitat preferences are lacking.

NATURAL HISTORY: Information on this critically endangered species, found only on Sumatra, is based on a single specimen. No information is available on its population biology or ecology, but this species is considered highly threatened by habitat loss, due to deforestation. More work is required to clarify the distinction between *H. winstoni* and *H. bartelsi*.

GENERAL REFERENCES: Ruedas et al. 2008c.

Iomys Thomas, 1908

This genus contains two species. Based on both morphological and sequence data, flying squirrels are considered to have diverged into two main clades. *Iomys* is a member of the *Glaucomys* clade, which includes *Eoglaucomys, Glaucomys, Hylopetes, Iomys, Petaurillus,* and *Petinomys.*

Iomys horsfieldii (Waterhouse, 1838) Javanese Flying Squirrel

DESCRIPTION: This flying squirrel is brown on the back, and the tail is reddish. It has a black eye ring with a short black stripe extending to the base of the whiskers. The ventral surface is orange buff. (The description is based on a specimen from the island of Borneo.)

SIZE: Female—HB 189.6 mm; T 179.6 mm; Mass 209.8 g.
Sex not stated—HB 191.8 mm; T 176.5 mm; Mass 165.3 g.

DISTRIBUTION: The range of the Javanese flying squirrel extends across peninsular Malaysia to Sumatra and Java (Indonesia) and the island of Borneo (divided among Malaysia, Brunei Darussalam, and Indonesia). It is also found in Tioman and Penang islands (Malaysia), but it may now be extinct in Singapore.

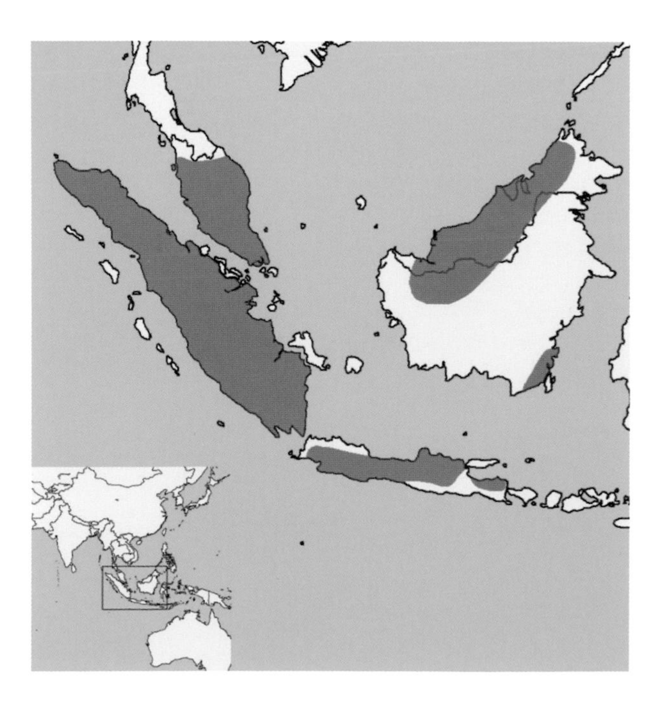

GEOGRAPHIC VARIATION: Four subspecies are recognized.

I. h. horsfieldii—Java and Sumatra (Indonesia). The upperparts are rufous brown, and the underparts are yellow or yellowish white. The underside of the tail is bright rust. The margin of the flank skin is reddish yellow, as are the sides of the face below the eyes.

I. h. davisoni—peninsular Malaysia and Singapore. The color on the back is dark slaty gray with rufous orange hair tips, and the edges of the sides are lined with rich orange. The tail is rich brown above and rich rufous orange below.

I. h. penangensis—Penang Island (Malaysia). Similar to *I. h. davisoni*, but *I. h. penangensis* is lighter and brighter on the dorsal surface, hands, feet, and tail.

I. h. thomsoni—island of Borneo (divided among Malaysia, Brunei Darussalam, and Indonesia). The upperparts and head are a dark smoky brown grizzled with dull buffy white. The undersides are whitish or pale fulvous.

CONSERVATION: IUCN status—least concern. Population trend—stable.

HABITAT: This species uses a broad range of habitats, including forest stands of all ages, disturbed and scrub vegetation, forest edges, and a variety of plantations, where it is sometimes considered a pest. It appears to be less common in dense forests, and it generally occurs at elevations below 1000 m.

Iomys horsfieldii. Photo courtesy Norman Lim.

NATURAL HISTORY: The Javanese flying squirrel is reported to feed on fruits, but details of its diet throughout the year are unknown. This nocturnal species relies on nest cavities and, although *I. horsfieldii* is common in parts of its range, little is known about this species' behavior and ecology. Studies of limb allometry show that the limbs of *I. horsfieldii* deviate significantly from regressions based on data from numerous rodent species, suggesting specific adaptations for locomotion. The tail length, relative limb length, and overall wing loading are lowest in smaller-bodied flying squirrels like the Javanese flying squirrel. Hence smaller flying squirrels show greater maneuverability when gliding.

GENERAL REFERENCES: Aplin and Lunde 2008; Bou et al. 1987; Corlett 1992; W. P. Harris 1944; Lim et al. 1999; Medway 1966; Muul and Liat 1971; R. W. Thorington and Heaney 1981; R. W. Thorington and Santana 2007; Waterhouse 1838.

Iomys sipora (Chasen and Kloss, 1928) Mentawai Flying Squirrel

DESCRIPTION: This species is dark brown on the back, with an even darker (almost black) tail. It is paler ventrally. The ears are naked. Females have two pairs of inguinal nipples.

SIZE: Female—HB 179.5 mm; T 179.5 mm.
Male—HB 196.0 mm; T 175.0 mm.

DISTRIBUTION: The Mentawai flying squirrel is found only in a few locations on two islands in the Mentawai Archipelago of Indonesia: Sipora and North Pagai.

GEOGRAPHIC VARIATION: None.

CONSERVATION: IUCN status—endangered. Population trend—decreasing.

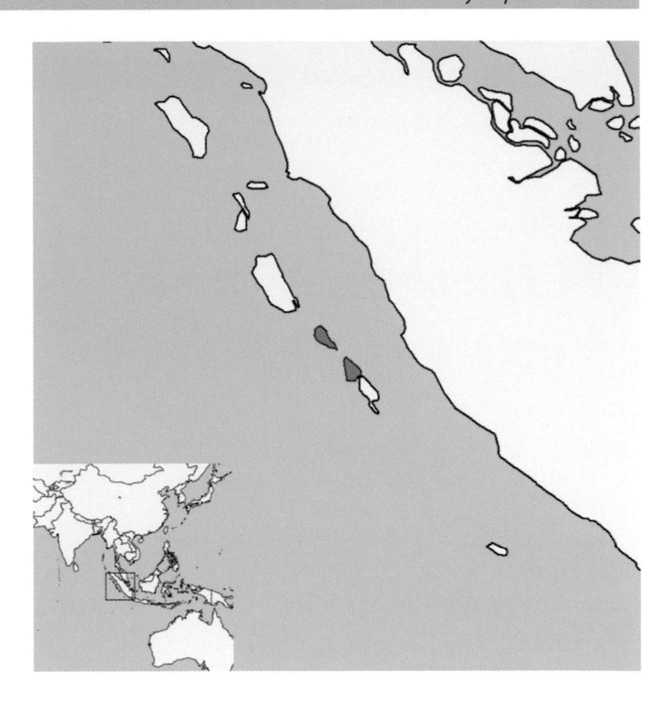

HABITAT: *I. sipora* resides in mature subtropical and tropical forests, at elevations below 500 m.

NATURAL HISTORY: Recent surveys have failed to detect this species, and it is assumed to be on the decline, due to extensive habitat loss as a result of various forms of agriculture, human development, and timber harvests. Nothing is known about the behavior, ecology, and reproduction of this species.

GENERAL REFERENCES: Mercer and Roth 2003; Ruedas et al. 2008d; R. W. Thorington and Darrow 2000; R. W. Thorington, Pitassy, et al. 2002.

Petaurillus Thomas, 1908

This genus contains three species of pygmy flying squirrels.

Petaurillus emiliae Thomas, 1908
Lesser Pygmy Flying Squirrel

DESCRIPTION: This is the smallest flying squirrel, and possibly the smallest sciurid in the world—the lesser pygmy flying squirrel is the size of a small mouse. It is only known from the type specimen. Further research on *P. emiliae* is needed to determine if this species is distinct from *P. hosei.*

SIZE: Sex not stated—HB 70.0 mm; T 64.5 mm; Mass 13.5 g.

DISTRIBUTION: *P. emiliae* is reported only from Sarawak (Malaysia).

GEOGRAPHIC VARIATION: None.

CONSERVATION: IUCN status—data deficient. Population trend—no information.

HABITAT: The lesser pygmy flying squirrel is assumed to be an arboreal forest resident at risk of habitat loss, due to clearing for timber and agriculture. However, no information is available on its habitat use.

NATURAL HISTORY: Nothing is known about the biology of this species.

GENERAL REFERENCES: Duckworth and Francis 2008a; Hayssen 2008a; R. W. Thorington and Ferrell 2006.

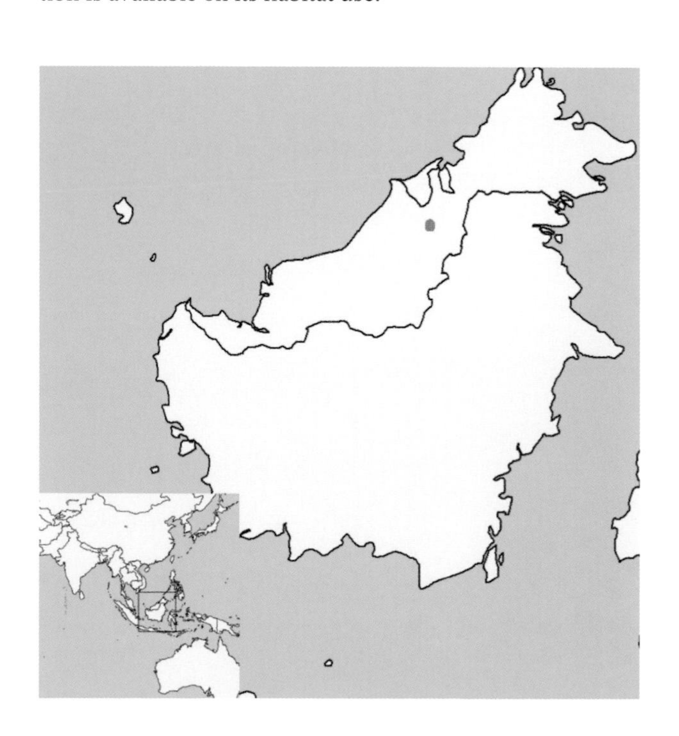

Petaurillus hosei (Thomas, 1900)
Hose's Pygmy Flying Squirrel

DESCRIPTION: *P. hosei* is a small flying squirrel, but it is approximately twice the mass of *P. emiliae*. It has a dark dorsum, a whitish belly, and a gray to dark-colored tail, sometimes with a white tip.

SIZE: Female—HB 87.0 mm; T 98.0 mm.
Sex not stated—HB 83.0 mm; T 88.8 mm; Mass 31.1 g.

DISTRIBUTION: Hose's pygmy flying squirrel is found only on the northern area of the island of Borneo, in Sabah and Sarawak (Malaysia) and Brunei Darussalam.

GEOGRAPHIC VARIATION: None.

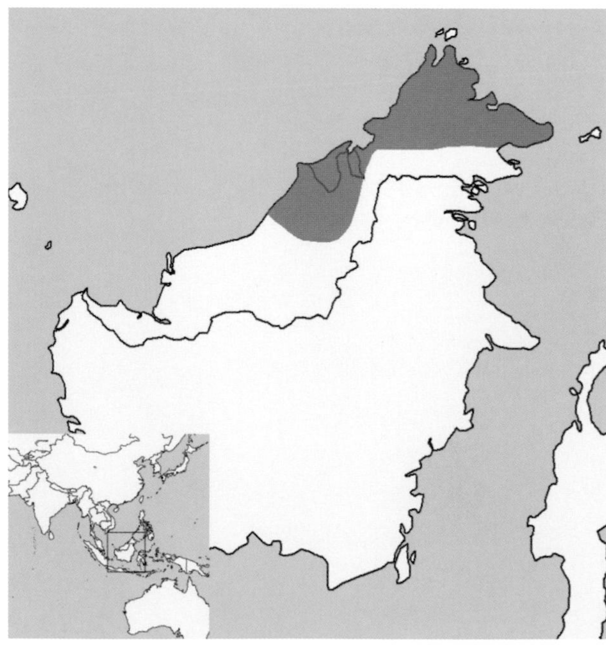

CONSERVATION: IUCN status—data deficient. Population trend—no information.

HABITAT: *P. hosei* is found in lowland forests consisting of numerous species of the Dipterocarpaceae. It is now considered extremely threatened by habitat loss, as a result of clearing for agriculture and timber harvests. Nothing is known about its specific habitat requirements.

NATURAL HISTORY: Like other flying squirrels, this species is arboreal and nocturnal. Because of their small size, *P. hosei* and the other two members of this genus have low wing loading (30% of that of the largest flying squirrels), which probably allows these diminutive flying squirrels to glide more slowly and with greater maneuverability than larger flying squirrel species.

GENERAL REFERENCES: Duckworth and Francis 2008b; R. W. Thorington and Heaney 1981.

Petaurillus kinlochii (Robinson and Kloss, 1911)
Selangor Pygmy Flying Squirrel

DESCRIPTION: The Selangor pygmy flying squirrel is brown on the head and body. It has white patches behind the ears; wing membranes that are dark with a black edge; and a tail that is buffy at the base, gradually becoming darker distally, but with a white tip. It is intermediate in size between *P. emiliae* and *P. hosei.*

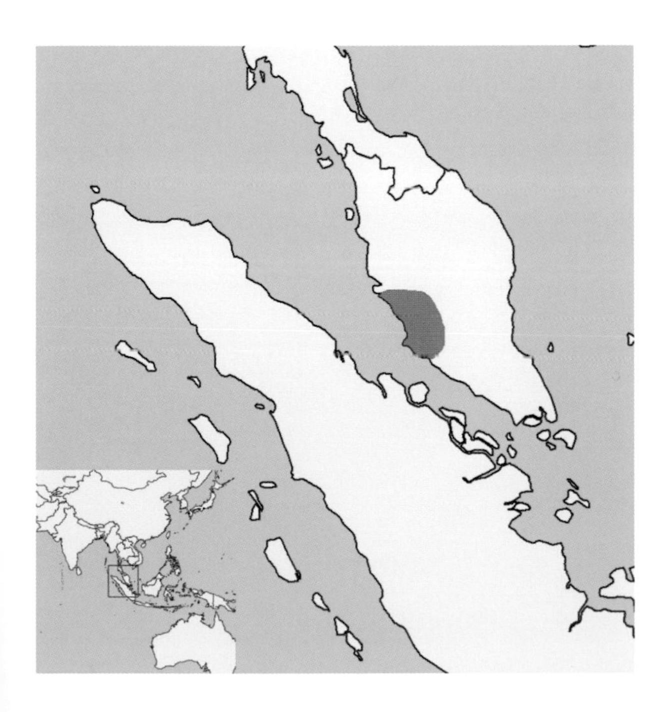

Petaurillus kinlochii. Photo courtesy Gareth Jones.

SIZE: Female—HB 90.0 mm; T 83.0 mm; Mass 28.0 g.
Male—HB 88.0 mm; Mass 19.8 g.

DISTRIBUTION: *P. kinlochii* is reported to occur only in Selangor (peninsular Malaysia). Early observations of this species in Selangor were in an area that has since been deforested.

GEOGRAPHIC VARIATION: None.

CONSERVATION: IUCN status—data deficient. Population trend—no information.

HABITAT: This species has been found in both natural forests and plantations, but it may have a broader distribution across peninsular Malaysia than current records indicate. Detailed surveys are needed to determine the distribution and habitat requirements for this species. This species is probably at risk because of deforestation, and it may already have been driven to local extinction in several localities.

NATURAL HISTORY: Virtually nothing is known about the ecology, behavior, population biology, or reproductive biology of this species. It has been suggested that *P. kinlochii* may communicate by means of tail flicking, but this is unconfirmed. Studies of dwarfism in flying squirrels suggest that, in contrast with the smallest tree squirrels, *P. kinlochii* and its congeners exhibit mandible characteristics that are highly evolved, and their eyes are set more anteriorly, which probably aids stereoscopic vision and depth perception.

GENERAL REFERENCES: Corbet and Hill 1992; Francis and Duckworth 2008a; Hautier et al. 2009; Scheffer et al. 1948; R. W. Thorington and Ferrell 2006.

Petaurista Link, 1795

This genus contains nine species of flying squirrel, many of them quite large.

Petaurista alborufus (Milne-Edwards, 1870) Red-and-White Giant Flying Squirrel

DESCRIPTION: The largest species in the genus, it has a white throat, a pinkish brown venter, and a speckled dorsum, with white or light maroon hairs. The tail has a whitish or pinkish brown ring at the base.

SIZE: Female—HB 383.5 mm; T 474.3 mm; Mass 1454.3 g.
Male—HB 421.2 mm; T 433.3 mm; Mass 1529.0 g.
Sex not stated—HB 496.0 mm; T 438.0 mm.

DISTRIBUTION: The red-and-white giant flying squirrel ranges from a portion of Myanmar through southcentral China. It also occurs in Taiwan.

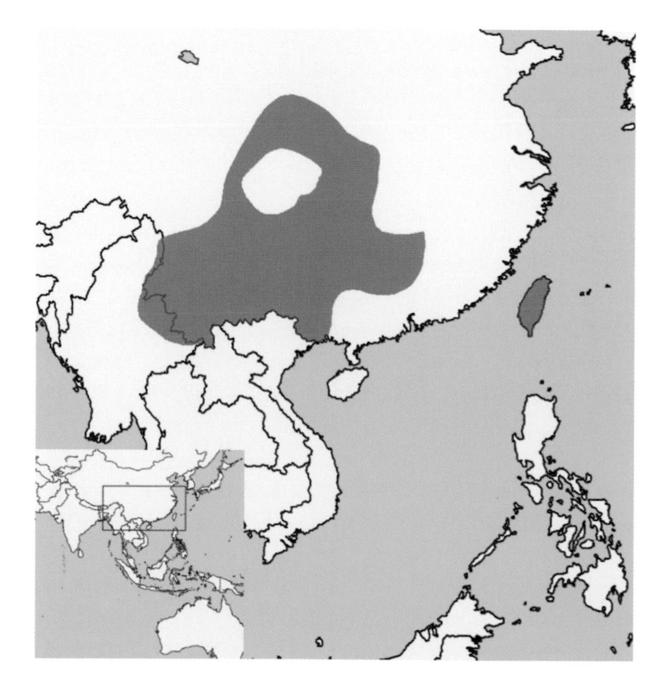

GEOGRAPHIC VARIATION: Five subspecies are recognized.

P. a. alborufus—western Sichuan and Gansu (China). The feet have reddish hairs.
P. a. castaneus—eastern Sichuan, Yunnan, Guizhou, and Hubei (China). It has black hair on the feet, and there are distinct rings at the base of the tail.
P. a. lena—island of Taiwan. It is the smallest subspecies.
P. a. leucocephalus—Tibet. The whole head, starting from the posterior margin of the ear, is white. It has a white line down the nape of the neck. The back, chin, throat, shoulders, and the outer side of the upper arm to the wrist joint, extending onto the patagium ("wing"), are also white.
P. a. ochraspis—Yunnan and Guangxi (China) into Myanmar. This subspecies has reddish hair on the feet, white rings at the base of the tail, and black on much of the rest of the tail.

CONSERVATION: IUCN status—least concern. Population trend—no information.

HABITAT: This species is typically found in dense forests of both hardwoods and conifers, generally at elevations of 2000-3000 m, but it has also been reported at lower eleva-

***Petaurista alborufus*.** Photo courtesy Wu Chi Ying.

tions. Hardwood forests are clearly preferred over those of conifers. Based on spotlight counts, it is estimated that *P. alborufus* is significantly more abundant in hardwood forests than in conifer forests, and that it is less abundant than *P. petaurista* in conifer stands. There is also significant seasonal variation in the abundance of *P. alborufus* in hardwood forests, with the highest densities occurring in the fall (0.44 animals/ha) and the lowest observed in winter (0.22 animals/ha). In forests where *P. alborufus* is sympatric with *P. petaurista,* both use the same tree species, but *P. alborufus* appears to reside higher in the canopy.

NATURAL HISTORY: Little information is available on this species. It is known to nest in hollow trees and cliff crevices. *P. alborufus* is nocturnal, and it is reported to glide over 400 m. Although some consider it to be a folivore, its diet is highly omnivorous and includes seeds, nuts, and fruits; leaves and other vegetation; and insects and other animal material. This species is reported to be sympatric with several other sciurids, including *Hylopetes alboniger, Petaurista philippensis, P. petaurista,* and *Trogopterus xanthipes.* In Taiwan, where two of these species are often found in the same forests, it is suggested that differences in body size (*P. petaurista* = 989-1597 g; *P. alborufus* = 1223-1930 g), along with abundant food resources, allow the two to coexist, partitioning nest cavities and resting sites and thus avoiding competitive interactions, which are reported to be uncommon. In most sightings in trees, these squirrels were found at heights of more than 10 m, and none were sighted at less than 3 m above the ground. The mating behavior of *P. alborufus* was observed only in June. Vocalizations by this species, described as high-pitched whistles, were produced less frequently and were qualitatively different from those made by *P. petaurista,* which were described as low-pitched sounds. Their reproductive rate is assumed to be low (less than two young per litter), as in other members of the genus.

GENERAL REFERENCES: Lee 1998; Lee, Progulske, et al. 1986, 1993; Oshida, Lin, Masuda, et al. 2000; A. T. Smith and Xie 2008.

Petaurista caniceps (Gray, 1842) Gray-Headed Flying Squirrel

DESCRIPTION: *Petaurista caniceps* differs from *P. elegans* by the absence of dorsal spots. This species has a grayish dorsum, a white throat, and a whitish brown venter. The base of the ears is light brown, and the sides of feet are orange.

SIZE: Sex not stated—HB 300-370 mm; T 360-400 mm.

DISTRIBUTION: This species is found in Nepal, Bhutan, and from India into Myanmar and southcentral China.

GEOGRAPHIC VARIATION: None. There is some question whether this is a distinct species, or just a subspecies of *P. elegans.*

CONSERVATION: IUCN status—least concern. Population trend—stable.

HABITAT: Little information is available for this species. *P. caniceps* is found in montane oak-rhododendron (*Quercus, Rhododendron*) and conifer forests, at elevations usually between 2100 and 3600 m.

NATURAL HISTORY: These squirrels are reported to feed on rhododendron leaves, buds, and conifer cones. They nest in tree cavities in rhododendron trees and conifers, but they are also known to construct leaf nests from ferns. Vocalizations, described as long cries, should aid in locating the animals in surveys. Although little is known about its reproductive behavior, litter sizes are assumed to be one to two young, as in other members of the genus. A new genus and species of flea (*Smitipsylla maseri*) has been collected from *P. caniceps* in Nepal. This species is also known to host larval ticks.

GENERAL REFERENCES: Hoogstraal and Mitchell 1971; R. E. Lewis 1971; A. T. Smith and Xie 2008.

Petaurista elegans (Müller, 1840) Spotted Giant Flying Squirrel

DESCRIPTION: A relatively small member of the genus, *P. elegans* is distinguished by the presence of white dorsal spots against a dark gray, yellowish gray, or brown dorsum. The tail is similar in color to the dorsal pelage, but not spotted. The patagium ("wing") is reddish orange, and the ventral pelage is orange to brown. The rump and base of the tail are reddish brown.

SIZE: Female—HB 332.8 mm; T 363.3 mm; Mass 759.8 g.
Male—HB 330.3 mm; T 355.0 mm; Mass 948.0 g.
Sex not stated—HB 346.7 mm; T 366.4 mm; Mass 1040.0 g.

DISTRIBUTION: The spotted giant flying squirrel occurs in Nepal, Sikkim (India), Bhutan, Yunnan and Guangxi (China), extreme northeastern India, northern and western Myanmar, Laos, Vietnam, Thailand, peninsular Malaysia, Sumatra and Java (Indonesia), and Sabah and Sarawak (Malaysia) on the island of Borneo. The current distribution, similar to that of *P. petaurista*, is considered to be a result of lower sea levels (200 m) and tectonic events during the Pliocene-Pleistocene that allowed dispersal throughout the Sundaic region.

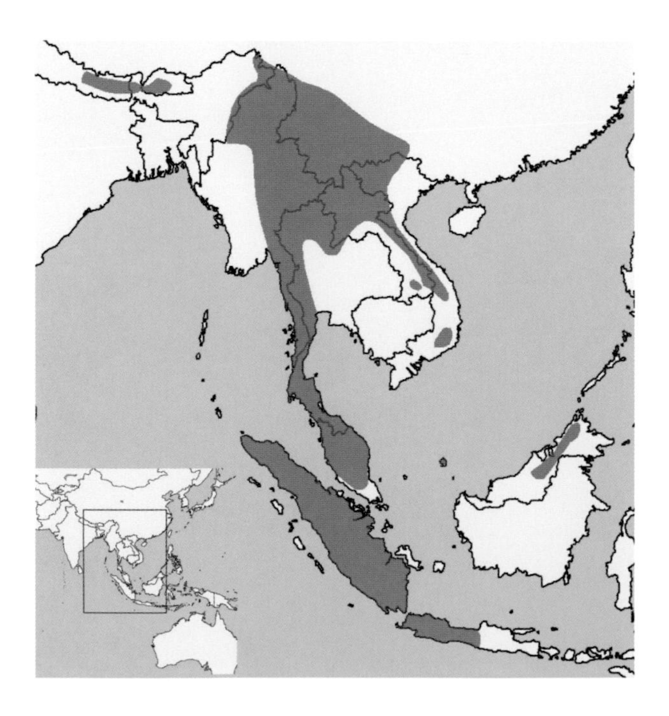

GEOGRAPHIC VARIATION: Eight subspecies are recognized here. Distribution records from Laos and Vietnam were identified only at the species level.

P. e. elegans—Java (excluding Mount Slamat). See description above.

P. e. banksi— Sabah and Sarawak (Malaysia). The top of the head, the neck, and the back are glossy black flecked with white. The tail is black and chestnut. The patagium is deep chestnut with tawny rufous below. The underparts are paler, an ochraceous rufous. The edges of the feet are black. There are dark rings around the eyes, and there is a small black spot on the chin.

P. e. caniceps— Nepal, Sikkim (India), and Bhutan. There is some question whether this is a distinct species (see *Petaurista caniceps*), or a subspecies.

P. e. marica—Yunnan and Guangxi (China) and northern Thailand. It has light upperparts with a few white spots on the back, an orange brown tail, and brown feet.

P. e. punctatus—southern part of Thailand, on the Malay Peninsula. It has dark upperparts with many white spots, a black tail, and black feet.

P. e. slamatensis—Mount Slamat (Central Java). The black grizzled color on the upper head, neck, and back continues into a black line down the back to the base of the tail. The feet are reddish brown. The throat is lighter in color than in the other subspecies. There is a black band over the toes, proximal of the nails.

P. e. sumatrana—Pandang Highlands (West Sumatra, Indonesia). This form is similar to *P. e. punctatus*, but *P. e. sumatrana* is flecked with far fewer white spots. The spots are almost absent on the head, neck, shoulders, rump, thighs, and basal part of the tail.

P. e. sybilla—India, west of Kindat (Myanmar). This form is similar to *P. e. marica*, but *P. e. sybilla* is smaller. It has been considered to be a distinct species.

CONSERVATION: IUCN status—least concern. Population trend—stable.

HABITAT: The spotted giant flying squirrel is found in temperate montane coniferous forests with an understory of rhododendron (*Rhododendron*), as well as in tropical evergreen forests. In the western part of its range it is found at elevations of 3000-4000 m. In western Malaysia it often occurs at much lower elevations (from 200 to 3000 m) in primary forests and those in which some timbering has occurred. It appears to be less common near human settlements.

NATURAL HISTORY: Common and locally abundant, the spotted giant flying squirrel nests in hollows or on rocky ledges,

Petaurista elegans. Photo courtesy Chan Kwok Wai.

where it breeds prior to the rainy season. Little else is known about its behavior, ecology, and population biology. Gliding dynamics have been indirectly investigated in this and related species. *P. elegans* appears to show one of the highest wing loadings among all the flying squirrels, which probably requires a faster glide to maintain the glide ratio (horizontal distance/vertical height). Like other members of the genus, *P. elegans* is reported to have small litters (one young per litter). Frequent vocalizations are reported, although details are not available. A new genus and species of flea (*Smitipsylla maseri*) and a new genus and species of sucking louse (*Atopophthirus emersoni*) have been reported from *P. elegans*. The microfilariae stage of some nematodes has been detected from blood samples taken from spotted giant flying squirrels. At least one species of *Eimeria* is also reported from this host. Although one study suggests that *P. elegans* is most closely related to *P. alborufus castaneus*, another indicates that the phylogenetic relationship of *P. elegans* to other members of the genus is not yet resolved.

GENERAL REFERENCES: Chowattukunnel and Esslinger 1979; Kim 1977; Lee, Progulske, et al. 1993; R. E. Lewis 1971; Lin et al. 1985; Muul and Liat 1971; Muul and Lim 1978; Muul et al. 1973; Oshida, Ikeda, et al. 2001; Scheibe et al. 2007; A. T. Smith and Xie 2008; R. W. Thorington and Darrow 2000; R. W. Thorington and Heaney 1981; Walston, Duckworth, and Molur 2008a; Yu, Yu, Pang, et al. 2006.

Petaurista leucogenys (Temminck, 1827) Japanese Giant Flying Squirrel

DESCRIPTION: This species shows considerable variation in its dorsal pelage coloration, from nearly black to dark reddish brown to grayish yellow. The ventral surface also varies, from yellowish brown to white. This species is one of the larger members of the genus.

SIZE: Female—HB 465.5 mm; T 314.8 mm.
Male—HB 367.8 mm; T 363.3 mm.
Sex not stated—HB 375.0 mm; T 345.0 mm; Mass 1178.9 g.

DISTRIBUTION: Except for the island of Hokkaido, this species is endemic to Japan. It also occurs in South Korea.

GEOGRAPHIC VARIATION: Four subspecies are recognized.

P l. leucogenys—islands of Shikoku and south Kiusu (= Kyushu) (Japan). The color on the upperparts is intermediate brown. The patch below the eye is grayish brown, succeeded behind by a fairly prominent light cheek patch, and by a dull fulvous one below and behind the ear.

P. l. hintoni—Seoul (South Korea). This form is similar to *P. l. leucogenys*, but the fur on the back is tinged with rufous brown. The tail is shorter, with a paler dorsal surface. The underparts are whiter, tinged with pale vinous in the center and becoming wine rufous on the marginal membrane.

P. l. nikkonis—Nikko, central region of the island of Hondo (Japan). The fur is particularly long, and the tail is bushy. The general color is paler than *P. l. leucogenys*, more grayish brown or drab. The underparts are white. The muzzle is whitish. It is grayish brown below the eye. The cheek patches are prominently white.

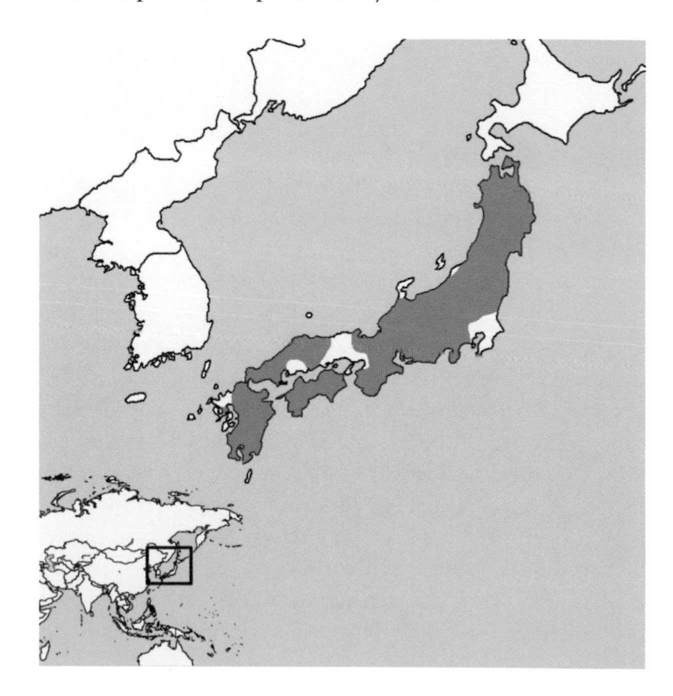

Petaurista leucogenys. Photo courtesy Takeo Kawamichi.

P. l. oreas—Wakayama, southern peninsula of the island of Hondo (Japan). The upperparts are a rich brown, suffused with rufous on the head. The underparts are washed with buffy. The patch below the eye is dark rufous, as is the forehead. The cheek patch is light dull grayish washed with buffy. The tail is dull cinnamon.

CONSERVATION: IUCN status—least concern. Population trend—stable.

HABITAT: *P. leucogenys* is found in mature stands of either primary or secondary forests. Few other details are available on its habitat.

NATURAL HISTORY: This species shows seasonal shifts in its diet (from folivory to granivory), feeding on the parts of at least 45 species of trees, yet it is reported to be highly selective when foraging. Overall, its diet is estimated to consist primarily of leaves (26%-40%), seeds (34%), and the staminate cones of various conifers (12%). However, even pine needles (*Pinus*) and cedar foliage (*Cryptomeria japonica*) are included in small percentages. Although animal material is occasionally consumed, this species is an obligate folivore, especially at certain times of the year. It is also exclusively arboreal and never hoards food. A nine-year study on the diet of *P. leucogenys* revealed strong seasonal patterns of food use: flowers and leaves constitute the majority of the diet in winter, spring, and early summer; seeds make up most of the diet in late summer, fall, and early winter. Mature leaves are consumed most often during February and again in July, at the same time that females are pregnant and the availability of other foods is lowest. During lactation, however, leaf consumption ceases. Their herbivorous diet is aided by their ability to remain in a sitting position on small branches while extending their forepaws to reach branches with buds and leaves. Compared with smaller squirrels, *P. leucogenys* spends more time foraging among smaller branches, and it also spends more time feeding while perched on vertical surfaces and smaller supports relative to its body size. Possibly because its diet often includes a high percentage of plant material, feeding bouts are relatively long, but they occur rather infrequently throughout the day. This species is reported to have a large caecum.

Under natural conditions, *P. leucogenys* usually becomes active about 30 minutes after sunset. Individuals show considerable overlap in their home ranges. For nesting, they often converge on mature forest patches, where nest cavities are available. One study reports that home range use is not homogenous, but instead is concentrated in patches of secondary forests, where food is abundant. Two mating seasons are reported: one from mid-November to mid-January, and another from mid-May to mid-June. Gestation is 74 days. The young begin feeding independently at 80 days postpartum, and they are usually weaned by day 91. Studies on their glide performance have reported average glide ratios (horizontal distance/vertical height) of 1.87, with upper limits of 3.5. The range of air speed was 4.39-9.47 m/second, with 3.3-7.0 m/second recorded in two studies. Horizontal glide distances in four study areas ranged between 10 and 100 m and averaged 17.1-33.1 m. Phylogenetic analyses based on partial mitochondrial cytochrome *b* sequences indicate that *P. leucogenys* and other members of the genus are more closely related to *Pteromys* than to four other genera: *Belomys*, *Glaucomys*, *Hylopetes*, and *Petinomys*. A study of genetic variation in *P. leucogenys* across Japan identifies three primary lineages: one from the island of Kyushu, a second consisting of haplotypes from the islands of Kyushu and Honshu, and a third defined by haplotypes from the islands of Honshu and Shikoku. The genetic differences do not correspond with geographic distances, but instead suggest recent range expansions (since the late Pleistocene).

GENERAL REFERENCES: Andō and Imaizumi 1982; Andō and Shiraishi 1993; Andō et al. 1984, 1985a, 1985b; Baba et al. 1982; Hiroyuki 1999; Ishii and Kaneko 2008a; T. Kawamichi 1997a, 1997b, 1998, 1999; Miayo 1972; Mori and Takatori 2006; Nakano et al. 2004; Oshida 2006; Oshida, Hachiya, et al. 2000; Oshida, Hiraga, et al. 2000; Oshida, Ikeda, et al. 2001; Oshida, Lin, Masuda, et al. 2000; Oshida and Obara 1993; Oshida and Yoshida 1999; Shafique et al. 2006; Stafford et al. 2002, 2003;

R. W. Thorington and Stafford 2001; Thorington, Darrow, and Anderson 1998; Wheeler and Myers 2004.

Petaurista magnificus (Hodgson, 1836) Hodgson's Giant Flying Squirrel

DESCRIPTION: This large flying squirrel is readily distinguished from other species in the genus by the wide dark brown or black dorsal stripe extending from the nose to the base of the tail. This stripe contrasts significantly with the light yellowish brown pelage on the sides and legs, and with the yellow shoulder patches. The anterior dorsal patagium ("wing") is reddish brown. The tail is dark at the base, lighter and more reddish brown for most of its length, and black at the tip. This species is often sympatric with *P. elegans*, from which it can be distinguished by its larger size (25%), its dorsal stripe (in contrast with the dorsal spots on *P. elegans*), and the lighter pelage of its sides and tail.

SIZE: Female—HB 442.5 mm; T 497.5 mm; Mass 1800.0 g.
Male—HB 382.3 mm; T 451.7 mm.
Sex not stated—HB 413.2 mm; T 480.0 mm.

DISTRIBUTION: This species' range extends from Nepal through Sikkim (India), Bhutan, southern Tibet, and western and northern Myanmar.

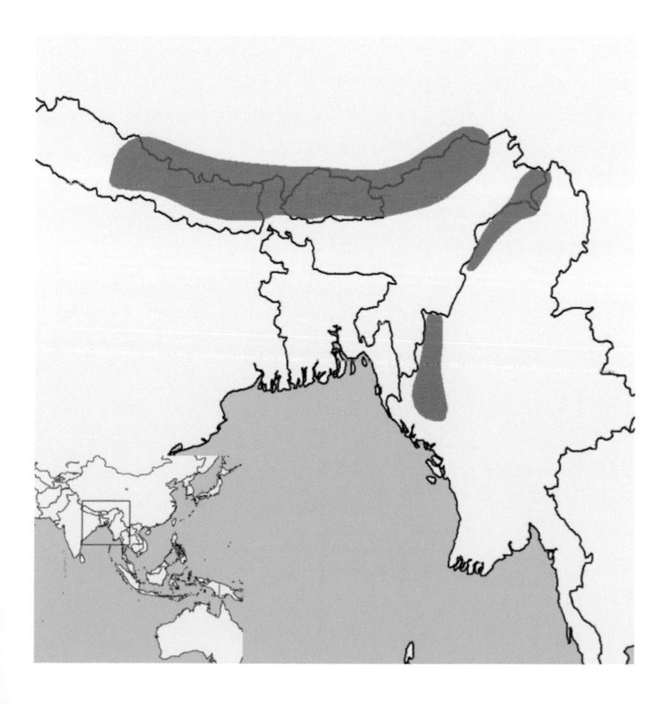

Petaurista magnificus. Photo courtesy David Cahlander.

GEOGRAPHIC VARIATION: None.

CONSERVATION: IUCN status—near threatened. Population trend—decreasing.

HABITAT: Hodgson's giant flying squirrel is found in tropical and subtropical evergreen and broadleaf deciduous forests, such as oak (*Quercus*) forests. It occurs from lowland areas to elevations of about 3000 m.

NATURAL HISTORY: Primarily herbivorous, Hodgson's giant flying squirrel feeds on grasses, buds, flowers, and leaves. It is nocturnal, and it tends to stay high in the trees. Little is known about its population and ecology; in part this may be due to its naturally low numbers. This species is reported to produce small litters (~ 1 per litter), similar to other members of the genus. It is also considered to be declining in many parts of its range, because of an increasing loss of habitat due to human development, fire, the conversion of forests to agriculture, and hunting pressure. This species is known for its deep calls, which are given at the onset of activity around sunset, when it can be observed gliding for distances of up to 100 m.

GENERAL REFERENCES: Chabaud and Bain 1976; Chatterjee and Majhi 1975; Chowattukunnel and Esslinger 1979; S. Dasgupta et al. 1978; Fan and Jiang 2009; Hayssen 2008b; Lee and Liao 1998; Lin et al. 1985; Mackerras 1962; R. M. Mitchell 1979; Molur 2008c; Muul et al. 1973; A. T. Smith and Xie 2008; Spratt and Varughese 1975; Srinivasulu et al. 2004; R. W. Thorington and Heaney 1981; Yamaguti 1941; Yorke and Maplestone 1926.

Petaurista nobilis (Gray, 1842)
Bhutan Giant Flying Squirrel

DESCRIPTION: This squirrel has upperparts that are bright chestnut brown with yellow tips to some of the hairs; and underparts that are pale rufous. It is pale fulvous on the top of the head, on the shoulders, and in a narrow streak down the middle of the upper part of the back.

SIZE: Female—HB 427.0 mm; T 522.9 mm.
Male—HB 417.8 mm; T 468.3 mm.
Sex not stated—HB 490.0 mm; T 460.0 mm; Mass 2710.0 g.

Petaurista nobilis. Photo courtesy Umesh Srinivasan.

DISTRIBUTION: This species' narrow range extends from central Nepal through Sikkim (India) to Bhutan.

GEOGRAPHIC VARIATION: Two subspecies are recognized.

P. n. nobilis—Nepal and India. See description above.

P. n. singhei—northeastern India and Bhutan. *P. n. singhei* is described as possessing thick ("woolly") orange buff or brownish buff hair; a dark dorsal saddle with shoulder patches; a black tipped tail; a yellow, orange or brownish patagium ("wing"); and a pale or yellowish venter.

CONSERVATION: IUCN status—vulnerable. Population trend—decreasing.

HABITAT: *P. n. singhei* is reported at elevations between 1500 and 3000 m in subtropical forests and temperate broadleaf forests, but it is also expected to occur in mixed broadleaf and coniferous forests.

NATURAL HISTORY: Less is known about this species than others in the genus. Descriptions of specimens and observations of live individuals of *P. n. singhei* document this subspecies in Arunachal Pradesh (northeastern India), 100 km east of its previously known range. Field observations there indicate that this species is nocturnal but, unlike most other flying squirrels, it is frequently sighted on the ground, often seeking mineral licks along newly cleared roads. It is often active at dawn or dusk. *P. nobilis* is known by the locals in northeastern India as *khiaw,* where it appears to be threatened by habitat loss and hunting.

GENERAL REFERENCES: Choudhury 2002; Ghose and Saha 1981; Saha 1977; Thapa et al. 2008.

Petaurista petaurista (Pallas, 1766)
Red Giant Flying Squirrel

DESCRIPTION: The dorsal color of *P. petaurista* appears to be reddish brown, but the hairs are gray at the base. The face and lower jaw are a darker brown. The ventral surface is light brown. Although similar in size to *P. alborufus*, *P. petaurista* is consistently smaller in body mass (85%) and other body measurements, including tail length. Females tend to be larger than males. *P. petaurista* seems to be closely related to *P. philippensis*. Some observers consider the subspecies *P.*

philippensis grandis, which occurs on Taiwan, to be a subspecies of *P. petaurista*.

SIZE: Female—HB 380.3 mm; T 493.3 mm; Mass 1405.3 g.
Male—HB 381.2 mm; T 432.2 mm; Mass 1264.3 g.
Sex not stated—HB 378.9 mm; T 459.1 mm; Mass 2004.0 g.

DISTRIBUTION: This species' range extends from extreme eastern Afghanistan, northern Pakistan, and northern India (in Kashmir and Punjab) along the edge of the Himalayas and then from eastern Nepal and eastern India (Assam) to Tibet and the southern Chinese provinces of Yunnan, Sichuan, Guangxi, Guangdong, and Fujian, possibly the island of Taiwan, and south through Myanmar, Thailand, peninsular Malaysia, Sumatra and Java (Indonesia), and the island of Borneo (divided among Malaysia, Brunei Darussalam, and Indonesia).

GEOGRAPHIC VARIATION: Eighteen subspecies are recognized. Records for extreme eastern Afghanistan, and for the area extending from northern Pakistan and northern India (in Kashmir and Punjab) along the edge of the Himalayas and then from eastern Nepal and eastern India (Assam) to Tibet, are identified only at the species level.

P. p. petaurista—Preanger regencies, West Java (Indonesia). See description above.

P. p. albiventer—Pakistan and Yunnan (China). This form has a well-defined blackish eye ring, white throat, and black color on the distal third of the tail. The rest of the tail is bright bay. The cheeks are rufous. The underparts are ochraceous buff, deepening to rufous on the patagium ("wing").

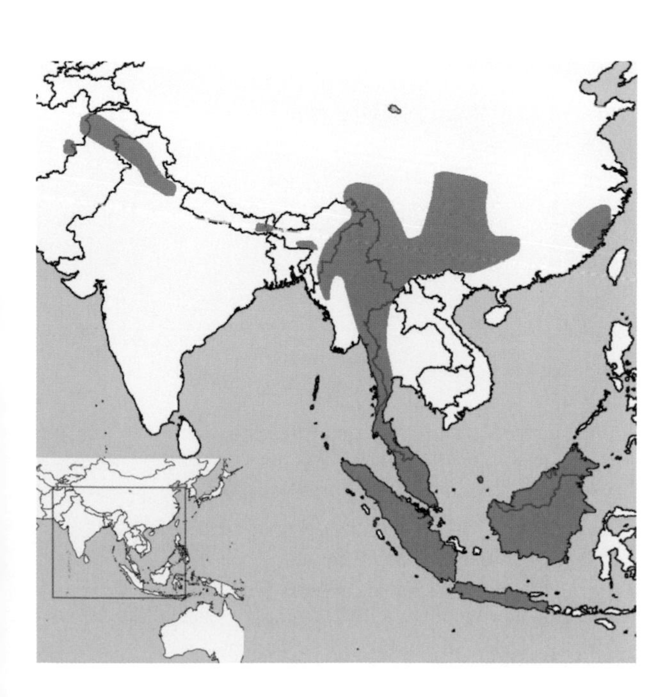

Petaurista petaurista. Photo courtesy Kalyan Varma.

P. p. batuana—Batu Islands, Sumatra (Indonesia). This form is similar to *P. p. nitidula*, but *P. p. batuana* is slightly larger. Its total length is 825 mm, and the head and body measurement is 405 mm.

P. p. candidula—Kindat (upper Myanmar). This form is paler, and the back is heavily washed with white.

P. p. cicur—northeastern Malay Peninsula. This subspecies is similar to other forms found on the Malay Peninsula, but the black areas of pelage are more extensive, and the hairs on the back have a distinct black tip.

P. p. interceptio—Mount Tjerimai, West Java (Indonesia). This form is similar to *P. p. petaurista*, but *P. p. interceptio*'s back is more rufous and less buff. The upper surface of the tail is pure rufous at the base, with the distal end being black.

P. p. lumholtzi—central region of Kalimantan (Indonesia), on the island of Borneo. The general color is deep ferruginous bay, though this is much lighter on the posterior half of the body. Beside and below the eye it is distinctively black. Above the eyes it is a pinkish cinnamon that continues backward, with a deepening color to the ears. The black tuft behind the ears is small and ill defined. The underparts are bright cinnamon rufous.

P. p. marchio—Sumatra (Indonesia). This form is similar to *P. p. nitidula*, but *P. p. marchio* has a slightly larger ear, and the reddish brown upperparts are paler. The dorsal color is relatively dark, a deep rufous chestnut. The dorsal hairs have black tips.

P. p. melanotus—China, possibly Laos, and the southern Malay Peninsula (Thailand and Malaysia). This form is similar to *P. p. cicur*, but on *P. p. melanotus* the black-tipped

Petaurista petaurista. Photo courtesy David Bakewell.

dorsal hairs are reduced in number and confined to the mid-dorsal line down the shoulders and back. The upperparts are rufous or burnt sienna.

P. p. nigrescens—Sabah (Malaysia), from the forests around Sandakan Bay north of Sungai (= stream) Kinabatangan. This form can be distinguished from *P. p. rajah* and *P. p. lumholtzi* by its much darker dull coloration. The nose tip, chin, eye ring, postauricular patch, extremities, and tip of the tail are black; the rest of the upperparts are dark cinnamon brown. The underparts are dull cinnamon brown. The proximal portion of the tail is dull cinnamon brown.

P. p. nigricaudatus—East Java (Indonesia). The general color of the tail is black, with the subbasal portions of the tail hairs being ferruginous maroon. The ears are dark chestnut, and it has an extensive black orbital eye ring. The feet and hands are black, with the black color continuing along the margins of the uropatagium (an extension of the "wing").

P. p. nitidula—Natuna Islands (Indonesia). This form is similar to *P. p. petaurista*, but *P. p. nitidula* is smaller in size.

P. p. penangensis—Telok Bahang, Penang Island (Malaysia). This form is similar to *P. p. melanotus*, but *P. p. penangensis* is smaller in size.

P. p. rajah—eastern Sarawak (Malaysia). The pelage color is intermediate between *P. p. melanotus* and *P. p. marchio*, with black tips on the dorsal hairs. It has dark orbital rings. The ears are relatively shorter, with the backs of the ears heavily tufted with black hairs, forming a conspicuous black patch on each side.

P. p. rufipes—Sichuan, Guangxi, Guangdong, and Fujian (China). The dorsal surface and the tail are a tawny or ferruginous glossy color, and there is a slight darkening on the nape and mid-dorsal area, caused by a slight blackening of the hair tips. It has a narrow black eye ring and a small dull brown spot on the chin. The entire underpart of the body is pinkish rufous or ochraceous salmon, deepening to "tawny" at the border of the patagium.

P. p. stellaris—Bontang (= Bintan) Island (Rhio [= Riau] Archipelago, Indonesia). This form is similar to *P. p. batuana*, but *P. p. stellaris* is less blackened on the upperparts, and paler on the cheeks and muzzle. The tail and underparts are paler, less rufous and more cinnamon.

P. p. taylori—southern Tenasserim (Myanmar). The tail is dirty whitish.

P. p. terutaus—Pulo Terutau (= Pulau Tarutao), off the west coast of peninsular Thailand. This form is similar to *P. p. nitidula*, but in *P. p. terutaus* the top of the head has a grayish wash, there is a slight buffy wash on the sides of body, and there is a more extensive black portion on the distal end of the tail.

CONSERVATION: IUCN status—least concern. Population trend—decreasing.

HABITAT: The red giant flying squirrel is found in a variety of forests—wet tropical lowlands (southern India), montane temperate forests (Pakistan), and hardwood forests (Taiwan)—but it is often reported to prefer evergreen broadleaf and coniferous forests. It also frequently occurs in plantations and orchards. This species appears fairly resilient to forest fragmentation, and it tends to show higher densities in smaller fragments than in larger stands.

Petaurista petaurista. Photo courtesy David Bakewell.

NATURAL HISTORY: Like several other members of the genus, *P. petaurista* is highly folivorous, but it is also known to eat seeds, fruits (e.g., figs of *Ficus*), flowers, buds, bark, and lichens. In northern Pakistan, *P. petaurista* is reported to feed on the leaves of 27 tree species. Bark stripping is reported in Taiwan. This nocturnal species gives a species-specific call, often at dusk. Leaf nests are occasionally used, but *P. petaurista* usually nests in tree cavities and rock crevices. In Taiwan, two reproductive peaks occur: one in spring, and another in late autumn. This species is often observed in pairs. Males enter reproductive conditions earlier than females. About 50 percent of the females reproduce during each season, and it is not known if a female can have more than one litter in a year. In Taiwan, litters range from one to two but are usually just one. *P. petaurista* is hunted for food and medicinal purposes in various parts of its range. In northeastern India, for example, the red giant flying squirrel is cooked and eaten as an antidote for poison, and the bile is boiled as a remedy for asthma.

GENERAL REFERENCES: Kakati et al. 2006; Lambert 1990; Lee 1998; Lee, Lin, et al. 1993; Lee, Progulske, et al. 1986, 1993; Oshida, Lin, et al. 2000; Oshida, Satoh, et al. 1991; Oshida, Shafique, et al. 2004; Shafique et al. 2006; A. T. Smith and Xie 2008; R. W. Thorington and Heaney 1981; Umapathy and Kumar 2000; Walston, Duckworth, Sarker, et al. 2008.

Petaurista philippensis (Elliot, 1839)
Indian Giant Flying Squirrel

DESCRIPTION: Among members of the genus, the Indian giant flying squirrel is second in size only to *P. nobilis*. Its dorsal pelage is uniformly dark gray or black; the tail is long, and the pelage of the ventral surface is brownish to buff colored. The front of the ears is reddish.

SIZE: Female—HB 457.2 mm; T 536.7 mm.
Male—HB 457.7 mm; T 474.0 mm.
Sex not stated—HB 463.2 mm; T 538.1 mm; Mass 2268.0 g.

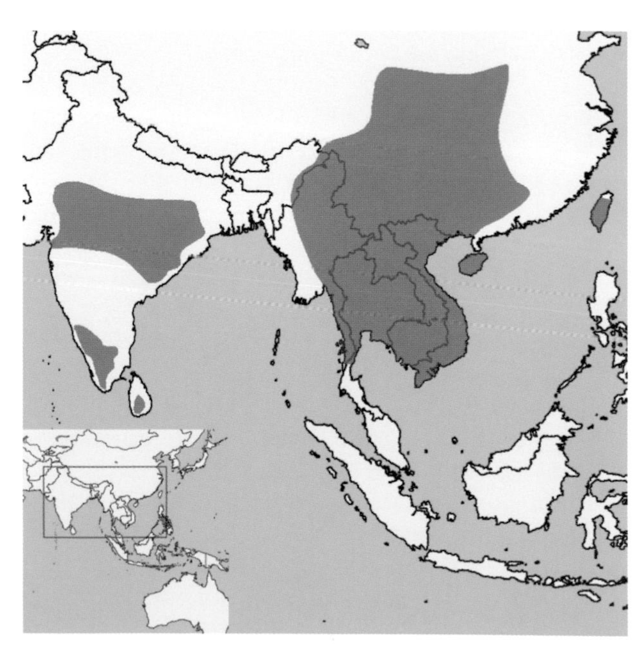

Petaurista philippensis grandis. Photo courtesy Shih-Wei Chang.

DISTRIBUTION: This species is found in India (from Mumbai and Rajastan in western India east to southern Bihar, and in the southern states of Karnataka, Kerala, and Tamil Nadu); Sri Lanka; several provinces in southern China (including Shaanxi, Sichuan, Yunnan, Guangxi, and Hainan); Taiwan; and through Myanmar, Thailand, Laos, Cambodia, and Vietnam. This species' distribution is extremely patchy in India.

GEOGRAPHIC VARIATION: Subspecies distinctions are quite variable, and they are poorly understood.

P. p. philippensis—India and Sri Lanka. The general color is brownish grizzled white, with the white being very conspicuous. The underparts are white. The patagium ("wing") may be tinged with rufous.

P. p. annamensis—southern Vietnam, Laos, and Cambodia. This form is chestnut rufous to rufous, with a white muzzle, white speckles, and a blackish tail.

P. p. cineraceus—Arakan (= Rakhine, Myanmar). The backs of the forearms and the ears are bay colored. The tail is drab gray. The general color is blackish to grayish.

P. p. grandis—Taiwan. This may instead be a subspecies of *Petaurista petaurista*, as is suggested by molecular data. It has bright red chestnut upperparts, and its underparts are rufescent. The dorsal hairs may be tipped in black.

P. p. hainana—Hainan Island (China). This may be a distinct species. The dorsal surface is grizzled rusty and black. The head is shining black, and the borders of the ears and lips are white. The patagium is blackish brown with a ferruginous border. The tail is all black except for the ventral surface of the tail base, which is grizzled rusty and black.

P. p. lylei—northwestern Thailand. This form is similar to *P. p. cineraceus*, but *P. p. lylei* is darker and more richly colored. The lower back is gray. The distal 7.5 cm of the tail are black. The eye ring, nose, and chin are dark brown. The gray on the lower cheeks fades to white as it approaches the throat. The underparts are white (with a buffy wash) to red. The forearms are dark brown, turning darker toward the phalanges. The phalanges are black, and the black color extends under the arm and along the edges of the patagium.

P. p. mergulus—Mergui Archipelago (Myanmar). The dorsal surface is grizzled grayish brown with a light olivaceus to chestnut rufous suffusion. The ventral surface is grayish white. The chin and throat are browner. The hands and feet are black, and the tail is grizzled hoary gray with a black tip.

P. p. nigra—northwestern Yunnan (China). No description is available.

P. p. rubicundus—Shaanxi and Sichuan (China). The ventral surface is a light bright rufous. The face, chin, neck, and feet are darker rufous. The tail tip is black. The throat and edges of the patagium are an intense rufous. The ears are almost hairless.

P. p. yunanensis—Yunnan (China). This may be a distinct

Petaurista philippensis. Photo courtesy Tarique Sani.

Petaurista philippensis. Photo courtesy Tarique Sani.

species. The dorsal surface of the body is tawny or ferruginous. It has a dull brown spot on the chin and a black eye ring. The ventral surface is white with a tinge of rufous to pinkish rufous.

Further work is needed on the geographic variation of this species. Recent morphometric and molecular analyses suggest that *P. p. yunanensis* and *P. p. hainana* could be elevated to species status, and that *P. philippensis grandis* may instead be a subspecies of *P. petaurista*.

CONSERVATION: IUCN status—least concern. Population trend—decreasing.

HABITAT: This species is found in subtropical hardwood and conifer forests, but it is reported to prefer the former in Taiwan. In the Western Ghats (India), *P. philippensis* consistently forages more often on forest edges and in open forest patches, and it occurs in both large and small forest fragments. Hence this species appears resilient to forest disturbance and fragmentation. However, on the island of Hainan (China), only larger forest fragments are occupied. In mainland China it is reported to occupy 5 of 12 zoogeographic regions.

NATURAL HISTORY: *P. philippensis* is highly folivorous. Leaf parts, for example, are estimated to account for 74 percent of this squirrel's annual diet in Taiwan, where it is estimated that 30 species of plants are frequently eaten. Fruits and young leaves are consumed in spring, summer, and autumn in Taiwan; mature leaves are eaten when other foods are not available. Elsewhere, leaves form 34–61 percent of the diet. In the Western Ghats (India), *P. philippensis* consumed fruit (figs of *Ficus racemosa*) more than any other diet item (44% of observations), followed by leaves (34%) and then flowers, bark, and lichens. Little is known about the behavior and population biology of *P. philippensis*. This nocturnal species is nearly exclusively arboreal, and it frequently relies on tree cavities for nesting. The Indian giant flying squirrel breeds in the spring and the fall. Litter sizes are one to two young, most commonly just one. The black-crested gibbon (*Nomascus concolor jingdongensis*) is reported to prey on *P. philippensis*. In the Western Ghats (India), playback calls of owl species are suggested as a potential census method for *P. philippensis*. In response to such calls, these flying squirrels are often seen, as well as heard producing alarm calls. Larger-bodied owls are assumed to be important predators, but details on predation attempts are not available. Humans also hunt this species for food and medicinal purposes throughout its range. A new species of the nematode genus *Breinlia* (*B. petaurista*) is reported from *P. philippensis*.

GENERAL REFERENCES: Babu and Jayson 2009; Chowattukunnel and Esslinger 1979; Kuo and Lee 2003; Nameer et al. 2001; Nandini and Parthasarathy 2008; A. T. Smith and Xie 2008; Sridhar et al. 2008; Walston, Duckworth, and Molur 2008b; Xiang et al. 2004; Yu, Yu, Pang, et al. 2006.

Petaurista xanthotis (Milne-Edwards, 1872) Chinese Giant Flying Squirrel

DESCRIPTION: The soft pelage of *P. xanthotis* is grayish yellow. Overall, the dorsal color is dark. The guard hairs are black at the base and whitish at the distal end. The throat is white, and the venter is gray. The patagium ("wing") is marked with an orange margin. The long tail has black and orange hairs. The feet are black, and the legs are orange.

SIZE: Female—HB 353.0 mm; T 340.0 mm.
Male—HB 407.7 mm; T 341.7 mm.
Sex not stated—HB 440.5 mm; T 379.1 mm; Mass 965.0 g.

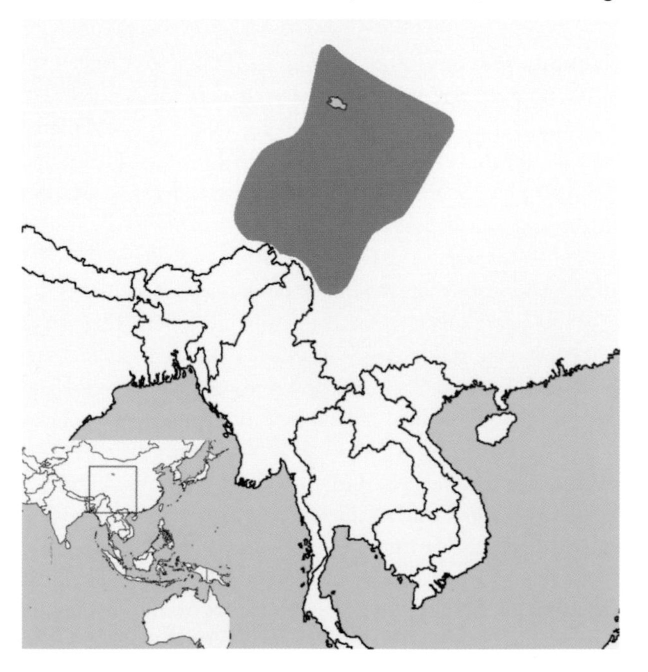

DISTRIBUTION: The Chinese giant flying squirrel is endemic to China, where it is restricted to western and central China (including eastern Tibet and Qinghai, Gansu, Sichuan, Yunnan, and Shanxi provinces).

GEOGRAPHIC VARIATION: Three subspecies are recognized.

P. x. xanthotis—Sichuan (China). The head and body are grizzled gray. The feet are blackish. There is a pale ochraceous spot at the inner base of the ear, and an orange spot behind the ear. The sides of the muzzle are white.

P. x. buechneri—Gansu to Sichuan (China). The general color of the dorsal surface is light grayish yellow to brownish yellow, blended with dark brown black.

P. x. filchnerinae—Gansu (China). The feet are lighter brown. The back and tail show more white, gray, and black. This subspecies is similar to *P. leucogenys*, except for the absence of white stripes on the head and throat.

CONSERVATION: IUCN status—least concern. Population trend—no information.

HABITAT: *P. xanthotis* is reported from coniferous (e.g., spruce [*Picea*]) montane forests, such as those in the Tibetan Plateau and the central provinces of China.

NATURAL HISTORY: Previously thought to be a subspecies of the *Petaurista leucogenys* complex, in 1992 *Petaurista xanthotis* was reclassified as a distinct species, based primarily on its more complex cheek teeth and the absence of a white streak below the ears. Molecular phylogenetic analyses (mitochondrial cytochrome *b* gene sequences) confirm that *P. xanthotis* is not included with other *Petaurista*, and that it is a distinct species from *P. leucogenys*. Similarly, more recent studies also confirm that *P xanthotis* is a distinct species, although they also suggest a close, but unresolved, relationship with *P. petaurista*. This nocturnal cavity nester is usually found at around 3000 m. Its diet includes shoots, leaves, and pine nuts. It does not hibernate, and it reproduces in the summer. The litter size is usually two.

GENERAL REFERENCES: Corbet and Hill 1992; McKenna 1962; Oshida 2006; Oshida, Shafique, et al. 2004; A. T. Smith and Johnston 2008b; A. T. Smith and Xie 2008; J. Wang 2000; Yu, Yu, Pang, et al. 2006.

Petinomys Thomas, 1908

Eight species are recognized in this genus.

Petinomys crinitus (Hollister, 1911)
Basilan Flying Squirrel

DESCRIPTION: Smaller than *P. mindanensis*, *P. crinitus* is similar to but slightly smaller than *Hylopetes nigripes*. It has small ears, each with three tufts of hair, one below and one on each side; grayish brown feet; and a brown tail, which is rather distichous.

SIZE: Male—HB 310.0 mm; T 260.0 mm.
Sex not stated—Mass 1130.0 g.

DISTRIBUTION: The Basilan flying squirrel is found on the islands of Basilan, Mindanao, Dinagat, and Siargao in the Philippines.

GEOGRAPHIC VARIATION: None.

CONSERVATION: IUCN status—least concern. Population trend—stable.

HABITAT: This species occurs in a variety of habitats, from lowlands to montane forests, although it seems to prefer oak (*Quercus*) woodlands at higher elevations. It also is reported to prefer primary stands and thus is likely to be at risk because of forest clearing for timber and agriculture.

NATURAL HISTORY: This rare species from four islands in the Philippines appears to be limited to the Mindanao Mammalian Province. Nothing is known about its ecology and behavior.

GENERAL REFERENCES: Dupont and Rabor 1973; Heaney and Rabor 1982; Heaney, Tabaranza, et al. 2006.

Petinomys fuscocapillus (Jerdon, 1847)
Travancore Flying Squirrel

DESCRIPTION: The upper parts are a rufescent fulvous or dark brownish hue. The cheeks, chin and underparts are rufous-white.

SIZE: Female—HB 319.7 mm; T 287.1 mm.
Sex not stated—HB 337.1 mm; T 250.0 mm; Mass 712.0 g.

Petinomys fuscocapillus. Photo courtesy Kalyan Varma.

DISTRIBUTION: This species is found in the Western Ghats of southern India and in the central and southern regions of the island of Sri Lanka. It is endemic to these areas, where its distribution is rather extensive but highly fragmented. Suitable habitat across the range is declining.

GEOGRAPHIC VARIATION: None.

CONSERVATION: IUCN status—near threatened, but close to vulnerable. Population trend—decreasing.

HABITAT: The Travancore flying squirrel occupies a variety of deciduous and evergreen forests and is often found in mountainous areas. It has also been observed feeding in plantations. This species is reasonably tolerant of disturbance, and in one study it showed higher densities in smaller forest fragments. It is sometimes found near villages.

NATURAL HISTORY: This solitary species nests in tree hollows and appears to spend most of its time in the canopy. Little information is available on its ecology, behavior, and population biology. Even where it exists, this species seems to occur in low numbers. Distribution maps may be misleading, because the areas it occupies are actually quite fragmented. Its primary threats come from habitat loss due to timbering, agriculture, and human development.

GENERAL REFERENCES: Molur et al. 2005; Rajamani et al. 2008b; R. W. Thorington, Musante, et al. 1996; R. W. Thorington, Pitassy, et al. 2002; R. W. Thorington, Schennum, et al. 2005; Umapathy and Kumar 2000.

Petinomys genibarbis (Horsfield, 1822)
Whiskered Flying Squirrel

DESCRIPTION: This is a medium-sized flying squirrel, chestnut colored on the shoulders, more orange on the rump, and more gray brown on the tail. The crown and shoulders are grayer on some animals. The gliding membrane is darker brown. The ventral surface varies from white or cream colored on the chin to more salmon colored on the abdomen and the hind legs, and particularly on the wing tips. *P. sagitta* should perhaps be included within this species.

SIZE: Female—HB 179.8 mm; T 192.5 mm; Mass 108.8 g.
Male—HB 157.3 mm; T 172.5 mm; Mass 69.5 g.
Sex not stated—HB 175.9 mm; T 171.4 mm.

DISTRIBUTION: The whiskered flying squirrel's range includes peninsular Malaysia, Sarawak, and Sabah (Malaysia); Brunei Darussalam; and Kalimantan, Java, and Sumatra (Indonesia).

GEOGRAPHIC VARIATION: None.

CONSERVATION: IUCN status—vulnerable. Population trend—decreasing.

HABITAT: The whiskered flying squirrel is found in older and secondary lowland forests, as well as adjacent plantations.

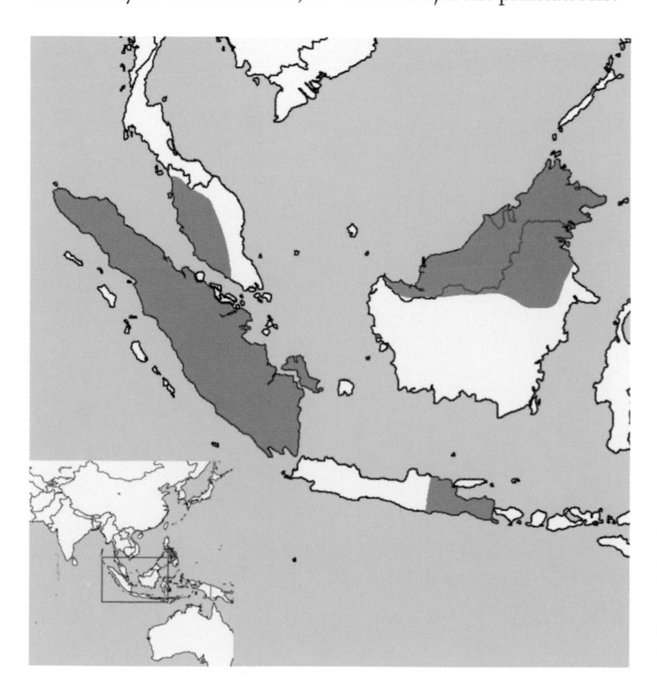

NATURAL HISTORY: Little is known about this species, because it is rarely detected in surveys. Two important surveys have turned up only two individuals.

GENERAL REFERENCES: Corbet and Hill 1992; Francis and Duckworth 2008b; Francis and Gumal 2008a; Hayssen 2008a; Medway 1969; Muul and Liat 1971; J. Payne and Francis 1985; R. W. Thorington and Darrow 2000; R. W. Thorington and Heaney 1981; R. W. Thorington, Pitassy, et al. 2002; R. W. Thorington, Schennum, et al. 2005; D. E. Wilson et al. 2006.

Petinomys hageni (Jentink, 1888) Hagen's Flying Squirrel

DESCRIPTION: The hairs on the dorsal surface of the head and back are slaty, with brown tips. A band of black-tipped hairs runs from the middle of the nose, around the eyes, to the ears. The hairs on the gliding membrane are black with reddish brown tips. The underparts of the squirrel are white with a reddish tint. The tail is brown.

SIZE: Female—HB 220.0 mm; T 215.3 mm; Mass 345.9 g.
Male—HB 260.8 mm; T 224.0 mm; Mass 358.6 g.
Sex not stated—HB 253.3 mm; T 238.3 mm; Mass 388.0 g.

DISTRIBUTION: Hagen's flying squirrel is reported from northern Sumatra (Indonesia) and Kalimantan (the Indonesian portion of the island of Borneo).

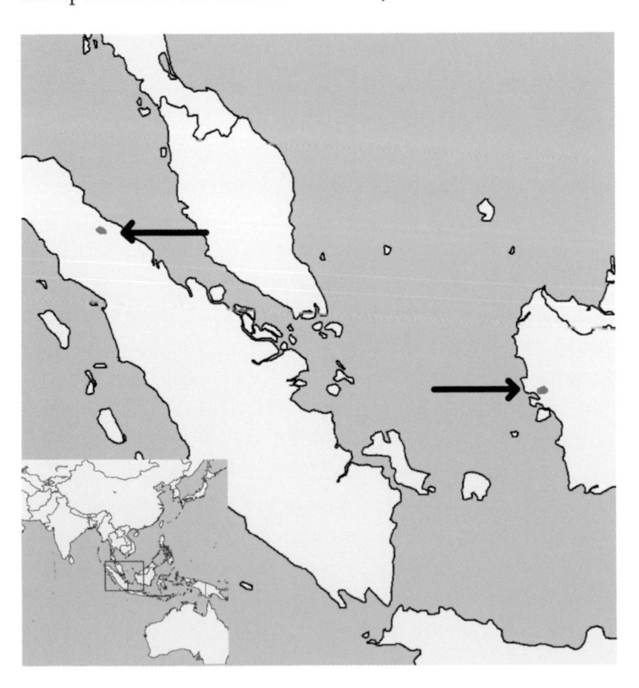

GEOGRAPHIC VARIATION: None.

CONSERVATION: IUCN status—data deficient. Population trend—no information.

HABITAT: *P. hageni* is reported from primary forests, but nothing else is known regarding this species' use of habitat.

NATURAL HISTORY: Nothing is known about the ecology, behavior, or general biology of this species.

GENERAL REFERENCES: Francis and Gumal 2008b; R. W. Thorington and Darrow 2000; R. W. Thorington, Pitassy, et al. 2002.

Petinomys lugens (Thomas, 1895) Siberut Flying Squirrel

DESCRIPTION: This large species is uniformly smoky brownish black. The tail is indistinctly distichous. Each cheek has a small wart, with three or four bristles on it.

SIZE: Female—HB 258.3 mm; T 225.8 mm.
Male—HB 250.0 mm; T 215.0 mm.
Sex not stated—HB 250.0 mm; T 222.5 mm; Mass 433.0 g.

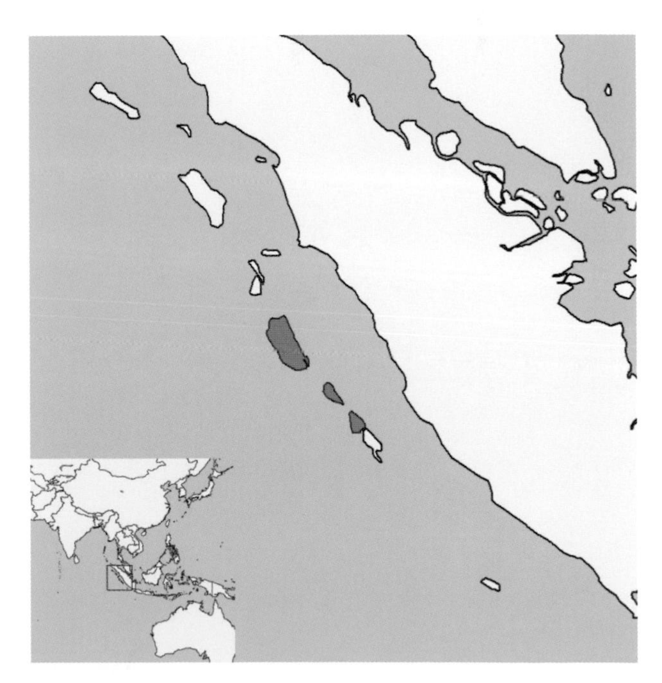

DISTRIBUTION: This species is found only on three islands in the Mentawai Archipelago: Siberut, Sipora and North Pagai islands, off the western coast of Sumatra (Indonesia).

GEOGRAPHIC VARIATION: None.

CONSERVATION: IUCN status—endangered. Population trend—decreasing.

HABITAT: The Siberut flying squirrel depends on tropical and subtropical forests that are rapidly being removed for agriculture and lumber. However, no details are available on its distribution and habitat use. This species is thought to occupy an area of less than 1200 km^2.

NATURAL HISTORY: Nothing is known about the biology of this species.

GENERAL REFERENCES: Chasen and Kloss 1927; Francis and Gumal 2008c; P. D. Jenkins and Hill 1982.

Petinomys mindanensis (Rabor, 1939) Mindanao Flying Squirrel

DESCRIPTION: The body and tail lengths of the Mindanao flying squirrel are greater than those of all other congeners, with the possible exception of *Petinomys fuscocapillus*. Compared with *P. crinitus*, also found in the Philippines, *P. mindanensis* has a more rounded rather than a flattened tail. During World War II, the holotype of this species was destroyed.

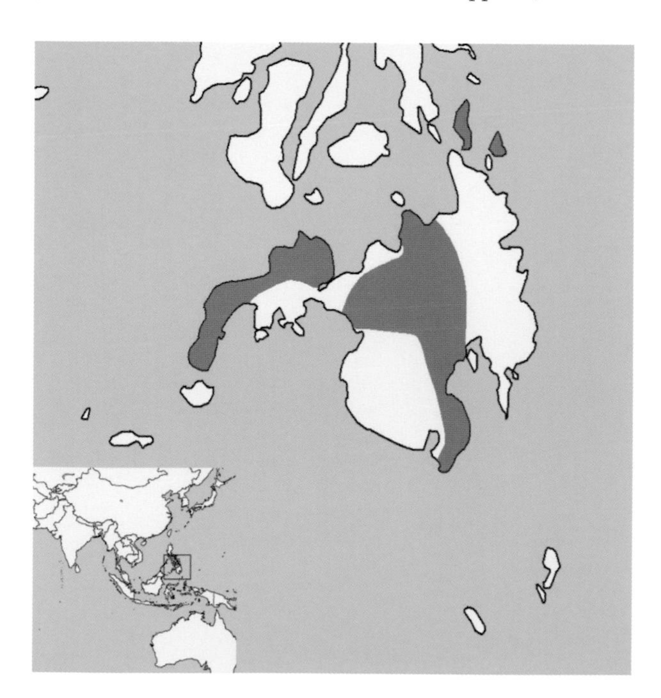

SIZE: Female—HB 341.5 mm; T 388.3 mm.
Male—HB 323.8 mm; T 348.4 mm.
Sex not stated—HB 289.3 mm; T 295.8 mm.

DISTRIBUTION: It is found on Dinagat, Siargao, and Mindanao islands (Philippines). On Mindanao Island it is reported from the provinces of Bukidnon, Davao del Sur, Lanao del Sur, Misamis Occidental, Misamis Oriental, and Zamboanga del Norte.

GEOGRAPHIC VARIATION: None.

CONSERVATION: IUCN status—least concern. Population trend—no information.

HABITAT: It is found in wet tropical and subtropical primary forests, from 500 to 1600 m, but seems to be most common at midelevation.

NATURAL HISTORY: More information is needed on the behavior, ecology, and reproduction of this species.

GENERAL REFERENCES: Chiozza 2008a; Hayssen 2008a; Heaney and Rabor 1982; Heaney, Balete, et al. 1998; Musser and Heaney 1992.

Petinomys setosus (Temminck, 1844) Temminck's Flying Squirrel

DESCRIPTION: This is a small flying squirrel. Specimens of *P. setosus* to the north of the Isthmus of Kra differ slightly from those to the south, and this may justify subspecies status. Specimens from the north are marginally larger in size, and they usually have a white-tipped tail. Northern specimens also have gray shoulders, a grizzled dorsum, white cheeks, and a black eye ring with a black line continuing to the nose. Those from the south have a blackish brown dorsum, a dark-colored patagium, and a blackish brown tail with whitish, pink, or buff-colored basal fur. The cheeks are gray, sometimes with a pinkish or yellowish wash. The throat and anterior ventral surface are whitish. Based on the sequence of the cytochrome *b* gene, this species is considered to be closely related to *Hylopetes phayrei* and *Glaucomys volans*; analyses of numerous morphological traits support the relationship with *G. volans*.

SIZE: Female—HB 116.7 mm; T 106.2 mm; Mass 38.2 g.
Male—HB 113.4 mm; T 94.1 mm; Mass 41.3 g.
Sex not stated—HB 118.9 mm; T 107.1 mm.

DISTRIBUTION: Temminck's flying squirrel has a highly fragmented distribution in northern Myanmar and northern Thailand, and in southern Thailand and southern peninsular Malaysia. Its range also includes Sumatra and the northern third of the island of Borneo, in Sabah and Sarawak (Malaysia), Brunei Darussalam, and extreme northeastern Kalimantan (Indonesia). This disjunct distribution, characterized by a separation of populations to the north and south at the Isthmus of Kra, is similar to that of 35 other mammal species and is attributed to repeated rises in sea level over the past 5 million years.

GEOGRAPHIC VARIATION: None.

CONSERVATION: IUCN status—vulnerable. Population trend—decreasing.

HABITAT: Temminck's flying squirrel has been collected or reported from both wet tropical primary forests and small village rubber plantations. In the southern part of its range, where this species is experiencing greater habitat loss, it is found in lowland forests, whereas in Thailand and probably Myanmar it is found in higher-elevation dry deciduous, tropical, or monsoon forests.

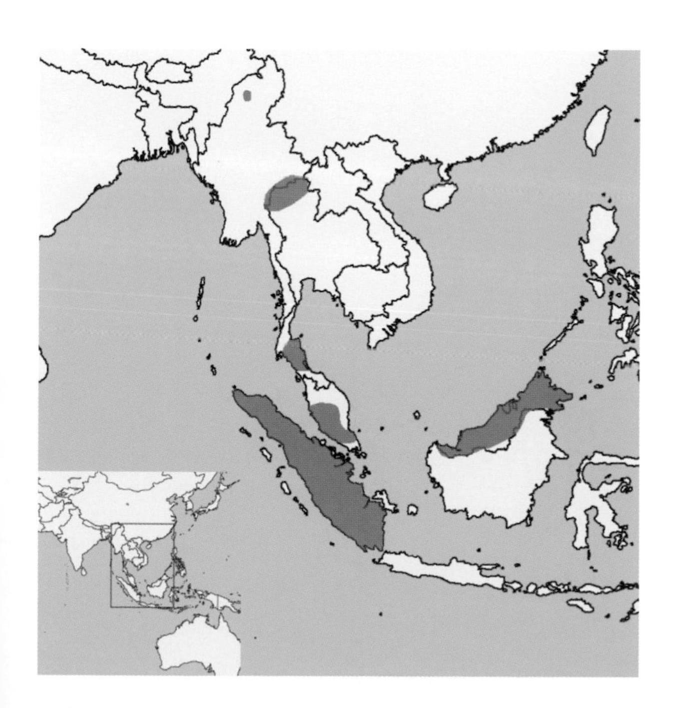

Petinomys setosus. Photo courtesy Jon Hall, www.mammalwatching.com.

NATURAL HISTORY: Little information is available on the behavior, ecology, and population biology of this species.

GENERAL REFERENCES: Corbet and Hill 1980; Dunn et al. 1968; Francis and Duckworth 2008c; McKenna 1962; Muul and Liat 1971; Muul and Thonglongya 1971; Oshida and Yoshida 1998; R. W. Thorington and Heaney 1981; R. W. Thorington, Musante, et al. 1996; R. W. Thorington, Pitassy, et al. 2002; Woodruff and Turner 2009.

Petinomys vordermanni (Jentink, 1890) Vordermann's Flying Squirrel

DESCRIPTION: This is one of the smallest flying squirrels. It has anterior-positioned eyes for binocular vision, an elongated coronoid process, and a highly developed condylar process. One specimen from Trengganu (Malaysia) had white spots on its back.

SIZE: Female—HB 118.8 mm; T 103.7 mm; Mass 37.8 g.
Male—HB 110.5 mm; T 98.6 mm; Mass 35.0 g.
Sex not stated—HB 105.4 mm; T 103.9 mm.

DISTRIBUTION: Vordermann's flying squirrel is found in southern Myanmar, possibly in Thailand on the Malay Peninsula, and on the island of Borneo, including Sabah and Sarawak (Malaysia), Brunei Darussalam, and Kalimantan

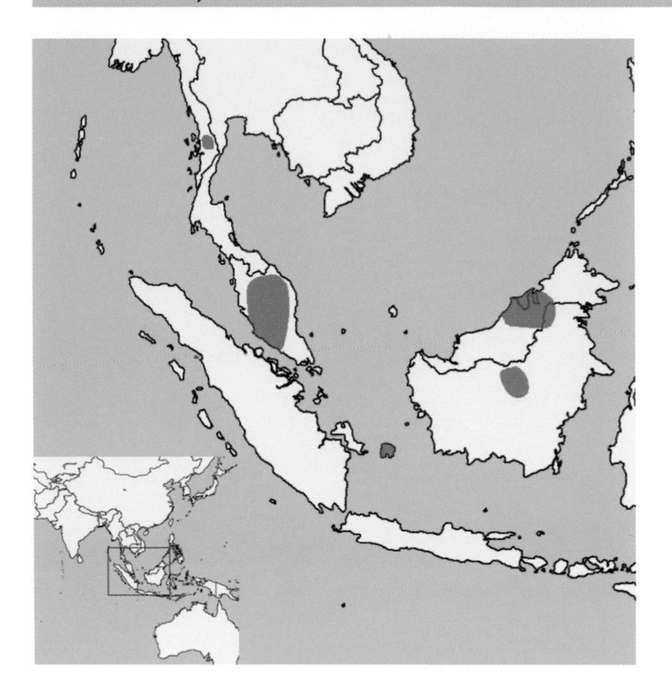

(Indonesia). Populations are reported from some of the eastern islands of Sumatra (Indonesia) as well. Although its distribution is considered highly fragmented, this may result from it being overlooked in surveys because of its arboreal and nocturnal behavior.

GEOGRAPHIC VARIATION: None.

CONSERVATION: IUCN status—vulnerable. Population trend—declining.

HABITAT: This species is reported to prefer lowland rainforest, and it is considered exclusively arboreal, restricting all of its activity to the canopy or subcanopy. It has been reported from rubber plantations, forests bordering swamps, and primary forests following selective timbering.

NATURAL HISTORY: Little is known about the behavior and population ecology of this species. It is considered threatened by deforestation for agriculture and timber. *P. vordermanni* is a definitive host for the ancanthocephalan *Moniliformis moniliformis*.

GENERAL REFERENCES: Deveaux et al. 1988; Francis and Gumal 2008d; W. P. Harris 1944; Hautier et al. 2009; Meijaard 2003b; Muul and Liat 1971; Muul and Thonglongya 1971; Nakagawa et al. 2007; R. W. Thorington, Pitassy et al. 2002.

Pteromys G. Cuvier, 1800

This genus has two species.

Pteromys momonga Temminck, 1844
Japanese Flying Squirrel

DESCRIPTION: The Japanese flying squirrel is about 15 percent of the size of the Japanese giant flying squirrel (*Petaurista leucogenys*) and 25 percent larger than *Glaucomys sabrinus*. Its dorsum is light brown or grayish brown; its belly and the ventral surface of the patagium ("wing") are white. The head is robust and rounded anteriorly.

SIZE: Female—HB 162.8 mm; T 140.5 mm.
Male—HB 154.0 mm; T 136.0 mm.
Sex not stated—HB 170.0 mm; T 120.0 mm; Mass 151.8 g.

DISTRIBUTION: *P. momonga* is endemic to 3 of the 12 islands in the Japanese Archipelago: Honshu, Shikoku, and Kyushu.

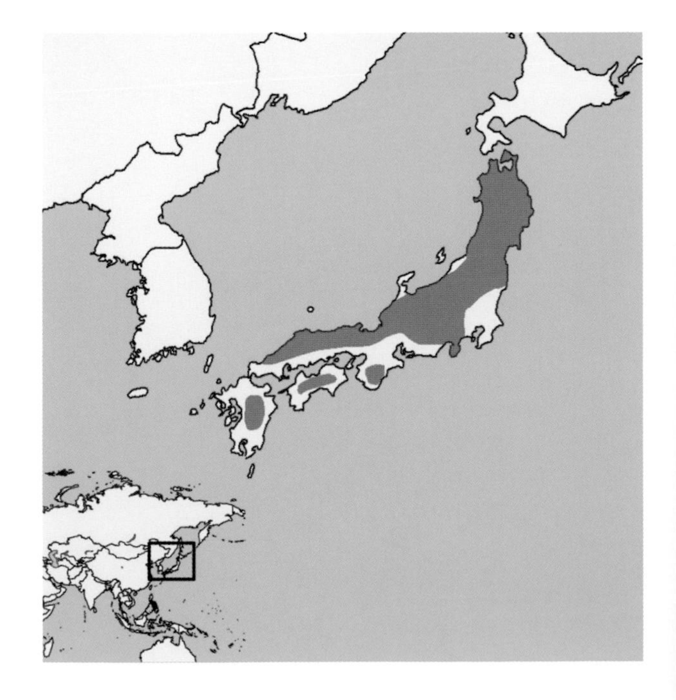

GEOGRAPHIC VARIATION: None.

Pteromys momonga. Photo courtesy Yusuke Iwasaki.

CONSERVATION: IUCN status—least concern. Population trend—no information.

HABITAT: This species typically occupies montane coniferous forests, just below tree line.

NATURAL HISTORY: *P. momonga* is omnivorous; it eats seeds, nuts, and fruits, but also leaves, buds, and bark. This species often occupies nest cavities and will inhabit artificial nest boxes readily. Tree bark is used for nest construction. The Japanese flying squirrel is locally common and generally widespread within its range. Although its range is limited, it is found in several protected areas and is not considered threatened by human activity. Its range overlaps less than expected by chance with its sister species, *P. volans*, although the latter is similar in its ecological tolerances and body size. Molecular analyses (based on the sequence of the mitochondrial cytochrome *b* gene) of several genera of flying squirrels suggest that the genus *Pteromys* is most closely related to *Petaurista*.

GENERAL REFERENCES: Andō 2005; Ishii and Kaneko 2008b; Lee and Liao 1998; Letcher et al. 1994; Millien-Parra and Jaeger 1999; Oshida, Hiraga, et al. 2000; Oshida and Yoshida 1999.

Pteromys volans (Linnaeus, 1758)
Siberian Flying Squirrel

DESCRIPTION: The Siberian flying squirrel is somewhat smaller (10%–20%) than its sister species, *P. momonga*. The dorsal pelage is gray or dark gray, and the ventral pelage is white to yellowish white. The tail is edged with dark-tipped hairs, and the feet are pale below and dark above. Based on sequences of the mitochondrial cytochrome *b* gene, the genus appears to be most closely related to *Petaurista*.

SIZE: Female—HB 160.8 mm; T 111.7 mm.
Male—HB 157.1 mm; T 112.9 mm; Mass 137.5 g.
Sex not stated—HB 168.7 mm; T 113.9 mm; Mass 131.3 g.

DISTRIBUTION: The Siberian flying squirrel is found in the boreal forests of the Palearctic region, from northern Finland south to the eastern shore of the Baltic Sea and eastward to Chukotka (Russia). It occurs in the southern Ural and Altai mountains of Russia; the Sakhalin Islands (Russia); Mongolia; northwestern, northeastern, and central China; the Korean Peninsula, and the island of Hokkaido (Japan).

GEOGRAPHIC VARIATION: Four subspecies are recognized.

P. v. volans—the type locality is Russia; this subspecies ranges from Finland to Chukotka, and into Mongolia. The upperparts are a uniform pale silvery gray. There is a narrow black eye ring, and the underparts are dull buffy white.

P. v. athene—Japan. The upperparts are drab gray. The underparts are dull whitish. The sides are washed with reddish brown.

P. v. buechneri—China and the Korean Peninsula. The overall

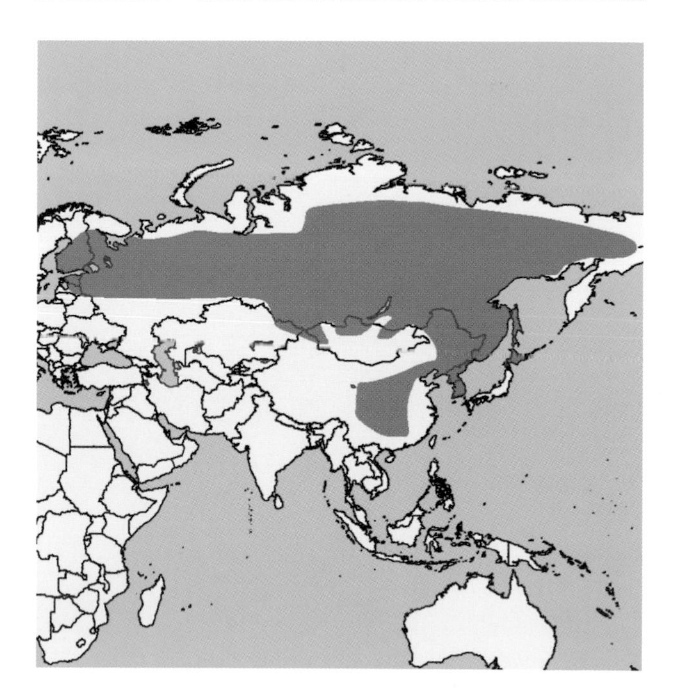

color, especially on the tail, is much darker than in the northern populations.

P. v. orii—island of Hokkaido (Japan). This form is similar to *P. v. volans*. The underparts are pure white. The cheeks are almost pure white. The tail is buffy on the margin and blackish on the median.

CONSERVATION: IUCN status—least concern. Population trend—declining.

HABITAT: This species is found in mature boreal closed-canopy forests and is reported to be dependent on primary-growth trees. Thus it is sensitive to logging and severe forest fragmentation. Various studies suggest, however, that the maintenance of well-connected habitat patches that meet this species' requirements may allow it to persist in a landscape managed for timber harvests. The Siberian flying squirrel shows a preference for mixed continuous forests with old conifers (e.g., Norway spruce [*Picea abies*]) and numerous deciduous trees, such as European aspen (*Populus tremula*); the deciduous trees provide critical nest cavities. A preference for spruce (*Picea*) over pine (*Pinus*) is also reported. In Finland it has been shown that this species is dependent on mixed forests that are dominated by spruce. As the spruce component of the habitat declines, the probability of local extinction of *P. volans* increases. Increased forest fragmentation has also contributed to its decline; however, *P. volans* can reside in fragmented systems and move reasonably well between forest fragments if adequate corridors of vegetation exist and suitable habitat components (e.g., spruce) are available in the matrix.

NATURAL HISTORY: This species feeds heavily on buds, leaves, fruit, catkins, twigs, lichen, and the mesophyll of conifer needles; it is less partial to seeds. In particular, large quantities of needles and leaves are reported to be consumed daily, and it appears that individual squirrels have preferred feeding trees. Occasional consumption of animal material (e.g., eggs, juvenile birds, and small mammals) is also reported. *P. volans* is reported to glide slowly (5-7 m/second), but to exhibit considerable maneuverability because of its low wing loading. This species is generally nocturnal, but activity peaks vary with the light/dark cycle. The animal may be active periodically, for up to 9 hours; it can also be active during the day when day length is long, despite lacking the color vision found in diurnal sciurids. Male home ranges vary with landscape structure (between 20 and 135 ha), and it appears that *P. volans* can move reasonably well between forest fragments, provided adequate corridors of vegetation exist and suitable habitat components (e.g., spruce) are available in the matrix. This species' use of habitat edges varies with the type of edges, overall landscape structure, and spatial scale.

Pteromys volans. Photo courtesy Marko Schrader.

Pteromys volans. Photo courtesy Yushin Asari.

Two peaks of reproduction are reported: one in March and another in April-May. It is estimated that a third of the animals in Finland breed twice annually, whereas those in other parts of the range (e.g., Russia) may only breed once annually, unless a female loses a litter. Gestation is 40-42 days; litter size is typically two to three, but can range from one to four. Details of their development are described in the literature. The young follow their mother during foraging bouts. This species is unique in that it shows female-biased juvenile dispersal, which is best explained by competition between mothers and daughters for limited resources. Mortality in juveniles appears to be highest prior to dispersal, rather than afterward. Relatively few predators, other than owls, are reported. At least two species of fleas are known from this host, one of which is host specific. A new trypanosome, *Trypanosoma kuseli*, is reported from *P. volans*. Phylogeographic studies suggest that unlike many terrestrial rodents, *P. volans* was forced into southern refugia during the Pleistocene.

GENERAL REFERENCES: Airapetyants and Fokin 2003; Desrochers et al. 2003; Hanski and Selonen 2008; Haukisalmi and Hanski 2007; Hurme, Kurttila, et al. 2007; Hurme, Mönkkönen, Nikula, et al. 2005; Hurme, Mönkkönen, Reunanen, et al. 2008; Johnson-Murray 1977; Mönkkönen et al. 1997; Oshida, Abramov, et al. 2005; Oshida, Hiraga, et al. 2000; Painter et al. 2004; Reunanen, Mönkkönen, and Nikula 2000; Reunanen, Mönkkönen, Nikula, et al. 2004; Reunanen, Nikula, and Mönkkönen 2002a, 2002b; Reunanen, Nikula, Mönkkönen, et al. 2002; Sato et al. 2007; Selonen and Hanski 2003, 2004a, 2004b; Selonen et al. 2001; Shar, Lkhagvasuren, Henttonen, et al. 2008; A. T. Smith and Xie 2008; R. W. Thorington and Darrow 2000; R. W. Thorington, Darrow, and Betts 1997.

Pteromyscus Thomas, 1908

This genus contains a single species.

Pteromyscus pulverulentus (Günther, 1873) Smoky Flying Squirrel

DESCRIPTION: This species is approximately twice the size of *Glaucomys sabrinus*. The smoky flying squirrel is brown to blackish on the dorsal surface and tail; the underparts are white to yellowish. The face is gray.

SIZE: Female—HB 230.8 mm; T 230.5 mm; Mass 235.0 g.
Male—HB 223.4 mm; T 222.6 mm; Mass 253.9 g.
Sex not stated—HB 250.3 mm; T 214.0 mm; Mass 268.5 g.

DISTRIBUTION: This species is found in southern Thailand, peninsular Malaysia (including Penang Island), Indonesia (Sumatra and Kalimantan), and in other parts of the island of Borneo (Sarawak and Sabah, and Brunei Darussalam).

GEOGRAPHIC VARIATION: Two subspecies are recognized.

P. p. pulverulentus—Malay Peninsula and Sumatra. The upperparts are brownish black, with a sprinkling of white. The underside of the patagium ("wing") is light grayish brown. The tail is brownish gray. The edge of the patagium is yellowish white.

P. p. borneanus—island of Borneo (divided among Malaysia, Brunei Darussalam, and Indonesia). The underparts are a clearer whitish.

CONSERVATION: IUCN status—endangered. Population trend—decreasing.

HABITAT: The smoky flying squirrel occurs primarily in lowland primary forests, below 3000 m. It is a relatively rare species that is even less common at higher elevations. Although a few secure populations are found on the island of Borneo, this species is considered endangered because of rapid habitat loss.

NATURAL HISTORY: This nocturnal species nests in tree cavities, but it also relies on exposed nests in undisturbed forests. It appears to be reproductive throughout the year, although only a small number of females have been observed to be reproductive at any one time. In western Malaysia, the average litter size for 14 females was 1.3 (range = 1-2), with an average pregnancy rate over a three-year period of only 9 percent.

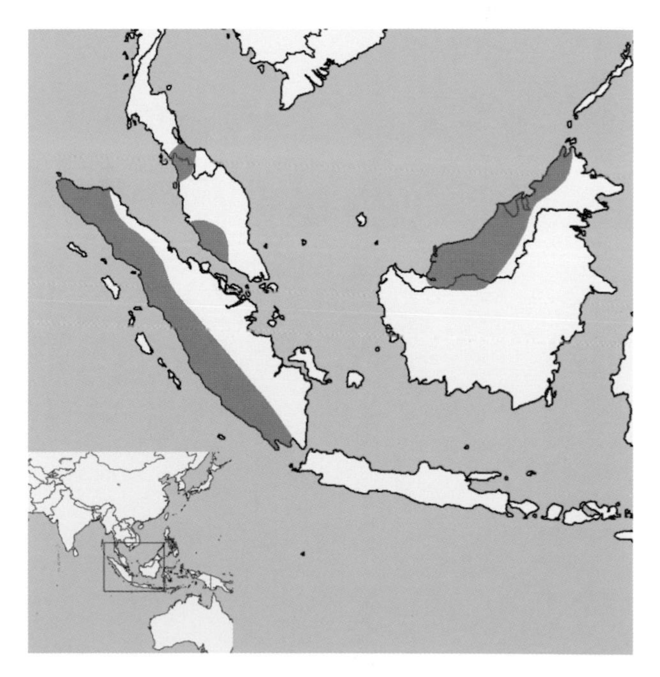

Similar wrist and carpal anatomy, important for maneuverability in flying squirrels, suggests a close phylogenetic relationship between *Pteromyscus* and *Aeretes, Belomys, Petaurista,* and *Trogopterus,* although a more thorough phylogenetic treatment using 80 morphological characters groups *Pteromyscus* only with *Aeromys, Belomys,* and *Trogopterus.*

GENERAL REFERENCES: Ahl 1987; Aplin and Duckworth 2008; Haslauer 2002; Muul and Liat 1971, 1974; R. W. Thorington and Darrow 2000; R. W. Thorington and Heaney 1981; R. W. Thorington, Pitassy, et al. 2002.

Trogopterus Heude, 1898

This genus contains a single species.

Trogopterus xanthipes (Milne-Edwards, 1867) Complex-Toothed Flying Squirrel

DESCRIPTION: The complex-toothed flying squirrel is a medium-sized flying squirrel. Its fur is gray at the base; brown to reddish at the tips; and, overall, appears a uniform reddish brown on the dorsum, and similar but lighter and browner on the venter. Long black hairs are found at the base of the ears and the tip of the tail. Its common name refers to the numerous ridges on the crown of the upper and lower cheek teeth. This species is considered most closely related to *Belomys.*

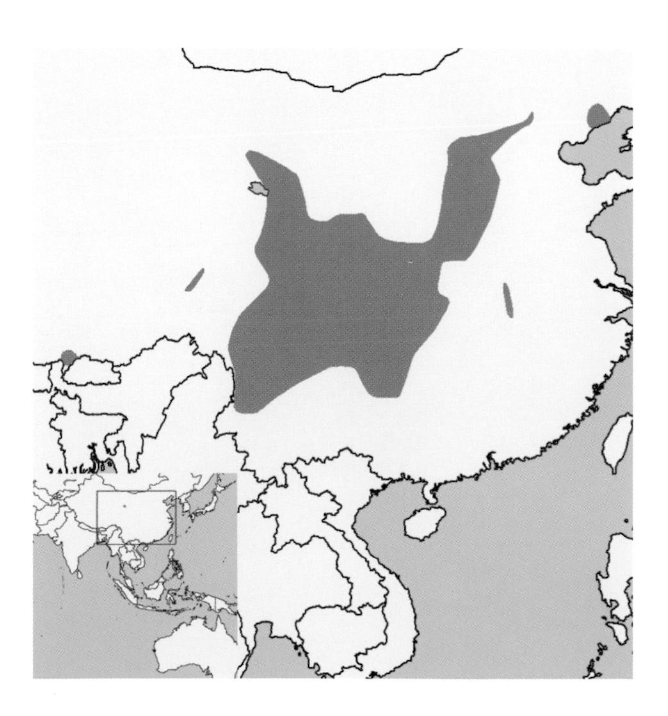

SIZE: Female—HB 309.9 mm; T 298.7 mm.
Male—HB 330.0 mm; T 270.0 mm.
Sex not stated—HB 300.0 mm.

Trogopterus xanthipes. Photo courtesy James Eaton/Birdtour Asia.

Trogopterus xanthipes. Photo courtesy Brad Schram, Arroyo Grande, California.

DISTRIBUTION: *T. xanthipes* is found in central China, from Yunnan in the south to Liaoning in the east. This species occupies a few small areas to the west, east, and northeast of its central contiguous range.

GEOGRAPHIC VARIATION: None.

CONSERVATION: IUCN status—near threatened. Population trend—declining.

HABITAT: The complex-toothed squirrel is found in montane temperate forests of oak (*Quercus*) and pine (*Pinus*), at elevations between 1360 and 2750 m. It is often associated with rocky cliffs and caves.

NATURAL HISTORY: This nocturnal species is reported to feed on oak leaves and nest in caves. Litter size varies from one to four, and gestation is reported to last between 78 and 89 days. Lactation is estimated to take 105 days. Nothing is known about its population biology, although this species is considered to be declining, due to habitat loss, logging, hunting, and removal for medicinal purposes. *T. xanthipes* is maintained in captivity because its dung is important in traditional Chinese medicine, the physiological basis of which has been explored.

GENERAL REFERENCES: Amori and Gippoliti 2003; Hayssen 2008a; Hayssen et al. 1993; Jeong et al. 2000; Kirk et al. 2008; Lee and Liao 1998; Scheibe et al. 2007; A. T. Smith and Johnston 2008f; R. W. Thorington and Darrow 2000; R. W. Thorington, Pitassy, et al. 2002; R. W. Thorington and Santana 2007; F. Wang 1985; Xiang et al. 2004.

Subfamily Callosciurinae Pocock, 1923

This subfamily includes 14 genera.

Callosciurus Gray, 1867

Callosciurus means "beautiful squirrel," and some of the 14 species in this genus match that epithet. They constitute the unstriped tree squirrels of Southeast Asia. The members of this genus are commonly omnivorous, feeding mainly on vegetable matter but probably eating animal material whenever it is available.

Callosciurus adamsi (Kloss, 1921)
Ear-Spot Squirrel

DESCRIPTION: The ear-spot squirrel has dull red underparts (the hairs are gray with reddish tips), a lateral white stripe with a dark stripe above it, and a buffy patch behind each ear that distinguishes it from the slightly larger plantain squirrel (*C. notatus*).

SIZE: Female—HB 170.3 mm (n = 4); T 148.7 mm (n = 3); Mass 150.0 g (n = 1).

Male—HB 157.0 mm (n = 1); T 158.0 mm (n = 1).

Sex not stated—HB 167.3 mm (n = 4); T 159.5 mm (n = 2); Mass 134.5 g (n = 2).

DISTRIBUTION: This species is present on the island of Borneo: in the lowlands and hills of Sabah and Sarawak (Malaysia) and Brunei Darussalam—below the altitudinal range of *C. orestes*—but it is also found at elevations of up to 900 m in the Kelabit Highlands of Sarawak (Malaysia).

GEOGRAPHIC VARIATION: None.

CONSERVATION: IUCN status—vulnerable. Population trend—decreasing.

HABITAT: This squirrel is found mostly in small trees in dipterocarp forests.

Callosciurus adamsi. Photo courtesy A. Coke Smith, www.cokesmithphototravel.com.

NATURAL HISTORY: No information is available.

GENERAL REFERENCES: J. Payne and Francis 1985.

Callosciurus baluensis (Bonhote, 1901)
Kinabalu Squirrel

DESCRIPTION: The venter of the Kinabalu squirrel is dark reddish, sometimes with a midventral dark line. The dorsum is gray black, with reddish speckling on the face and legs. It has short lateral stripes that are white over black.

SIZE: Female—HB 244.0 mm (*n* = 1); T 251.0 mm (*n* = 1).
Male—HB 242.5 mm (*n* = 2); T 247.5 mm (*n* = 2).
Sex not stated—HB 231.7 mm (*n* = 2); T 236.7 mm (*n* = 3); Mass 370.5 g (*n* = 2).

DISTRIBUTION: This squirrel is found above 300 m in elevation in Sabah and Sarawak (Malaysia) and Kalimantan (Indonesia).

GEOGRAPHIC VARIATION: We do not recognize any subspecies, but other authors recognize *C. baluensis baramensis* of Sarawak, which differs from *C. b. baluensis* of Sabah (Malaysia) by speckling on both the tail and the body.

CONSERVATION: IUCN status—least concern. Population trend—steady.

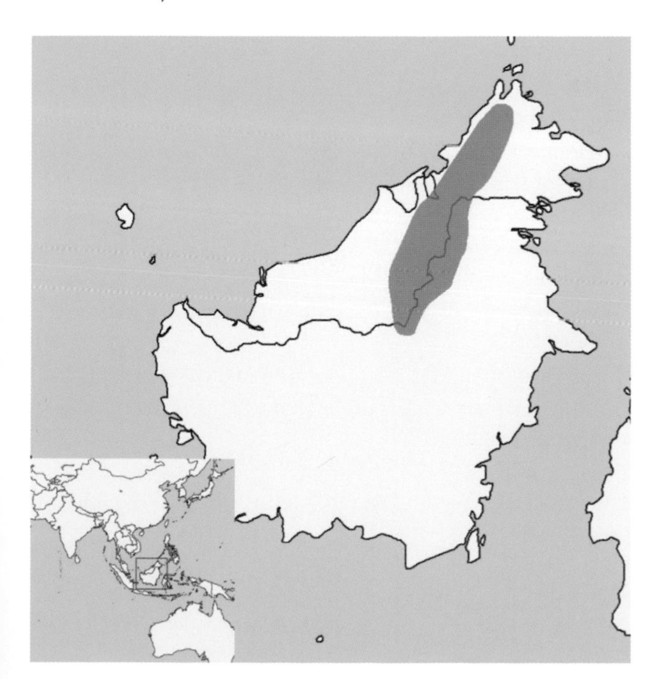

Callosciurus baluensis. Photo courtesy Amy B. Johnson, www.amysnaturephotos.com.

HABITAT: The Kinabalu squirrel occurs in the mountains of the northwestern part of the island of Borneo, such as at elevations of 300–1800 m on Mount Kinabalu (Sabah, Malaysia), in montane oak (*Quercus*) and lower cloud forests.

NATURAL HISTORY: This squirrel is mostly arboreal, but it is reported to descend to the ground occasionally.

GENERAL REFERENCES: J. Payne and Francis 1985.

Callosciurus caniceps (Gray, 1842)
Gray-Bellied Squirrel

DESCRIPTION: The gray-bellied squirrel has gray or silvery gray underparts, which may be washed with red; no stripes on the sides; and the upperparts and tail speckled olive brown, frequently with a black tail tip.

SIZE: Female—HB 222.3 mm (n = 11); T 219.4 mm (n = 9); Mass 312.9 g (n = 15).

Male—HB 215.5 mm (n = 11); T 242.5 mm (n = 15); Mass 316.4 g (n = 10).

Sex not stated—HB 215.9 mm (n = 74); T 230.4 mm (n = 72); Mass 266.9 g (n = 8).

DISTRIBUTION: It can be found in Thailand, peninsular Myanmar, peninsular Malaysia, and adjacent islands. It occurs at up to 1433 m on Mount Hijau (in the state of Perak) and at elevations of 3172 m in the Cameron Highlands of Malaysia.

GEOGRAPHIC VARIATION: Eight subspecies are recognized.

C. c. caniceps—mainland of Thailand and Myanmar. This form is bright orange on the back grading to agouti gray underneath, with an abruptly black tail tip. The bright orange is thought to be molted to agouti gray in the wet season.

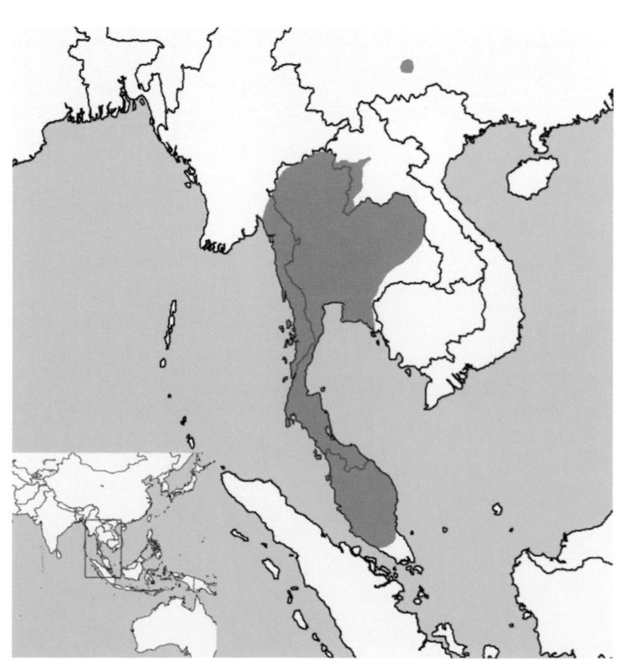

Callosciurus caniceps. Photo courtesy Nick Baker, www.ecologyasia.com.

C. c. adangensis—small islands off the coast of southwestern peninsular Thailand. This subspecies has a distinct black tail tip, and a buffy midline on the ventral surface of the tail.

C. c. altinsularis—High Island, Mergui Archipelago (Myanmar). This form is much paler than neighboring subspecies.

C. c. bimaculatus—upper peninsular Thailand and Myanmar and their offshore islands. This form lacks a reddish suffusion in the dorsal midline, but it shows this coloration on the sides. It has a distinct black tail tip and a silvery gray abdomen, sometimes suffused with red.

C. c. casensis—Chance Island, off the west coast of peninsular Thailand. This form is very pale gray on the underside of the tail.

C. c. concolor—peninsular Malaysia and lower peninsular Thailand. This form has a reddish suffusion in the midline of the back, and it usually lacks a black tail tip; it is silvery gray underneath.

C. c. domelicus—Domel, Bentinck, and Kisseraing islands, Mergui Archipelago (Myanmar). This form is dark gray ventrally, fairly dark dorsally, and intensely red on the sides.

C. c. fallax—Phangnan Island, off the east coast of Thailand. This form has a pale tail, but a darker pelage on the dorsum and venter.

CONSERVATION: IUCN status—least concern. Population trend—stable.

HABITAT: The gray-bellied squirrel is found in secondary and disturbed forests, and in plantations and gardens. In the Ulu Gombak Forest Reserve (Selangor, Malaysia), it occurs

Callosciurus caniceps bimaculatus. Photo courtesy Nick Baker, www.ecologyasia.com.

in lowland dipterocarp forests. In a lowland dipterocarp forest in the Krau Wildlife Reserve (Pahang, Malaysia), *C. caniceps* was captured only 8 percent as frequently as *C. notatus* (4 individuals versus 53 individuals live trapped, respectively). In China, it also occurs in dipterocarp forests with thick brushy vegetation. It may be found at elevations of up to 2500 m, but it is more common at lower elevations.

NATURAL HISTORY: In the Ulu Gombak Forest Reserve (Selangor, Malaysia), this squirrel is most active from 4 to 6 p.m., but it is also active in the morning. It is usually seen at heights between 0 and 10 m, less frequently at 10-15 m, and infrequently at greater heights or on the ground. One study notes that it feeds on 13 species of plants—on fruits (45%), leaves (21%), bark (17%), and flowers (10%)—with "others" comprising the rest of its diet. Elsewhere, insects are reported to be included in the diet. The home ranges of individuals overlap, both between and within sexes, with those of the males averaging 2.4-2.6 ha, and those of the females, 1.2-1.5 ha. The gray-bellied squirrel seems to respond differently to various kinds of predators. It reacts to terrestrial predators with a repeated staccato bark, which causes other conspecifics to climb upward and remain silent; it responds to an aerial predator with a low-frequency bark if the predator is distant and a "rattle" call if it is close; and it reacts to snakes with a "squeak" call that attracts other squirrels to join in mobbing the snake. Other gray-bellied squirrels respond to the "rattle" call by stopping any movement, and they join in the vocalizing if the aerial predator is near them. When a female comes into estrous, she attracts six to eight males and mates with four to six of them in a six-hour period. After mating with the female, a male gives a barking vocalization, like the antipredator call, for 12-35 minutes, during which time the female and the other males do not move. In the Ulu Gombak Forest Reserve, pregnant females have been trapped in most months of the year, and litter size is one to five, with a mean of 2.2 (n = 14). The nest is spherical, with a poorly defined lateral entrance. It has an outer wall of small twigs or coarse leaves, which is lined with shredded fibrous material. The nest is usually in the upper branches of a small tree or bush, but one was found on the ground at the foot of a bamboo clump, among leaf litter.

GENERAL REFERENCES: Abdullah et al. 2001; Lundahl and Olsson 2002; Moore and Tate 1965; Saiful et al. 2001; Tamura 1993; Tamura and Yong 1993.

Callosciurus erythraeus (Pallas, 1779)
Pallas's Squirrel

DESCRIPTION: This is an exceedingly variable squirrel that is generally recognizable by the combination of the reddish coloration of the venter and the olive brown coloration of the dorsum. *C. erythraeus* has been divided into several species by some taxonomists.

SIZE: Female—HB 217.4 mm (n = 142); T 216.6 mm (n = 10); Mass 375.1 g (n = 343).

Male—HB 227.0 mm (n = 15); T 205.3 mm (n = 15); Mass 359.2 g (n = 371).

Sex not stated—HB 209.3 mm (n = 52); T 176.9 mm (n = 46); Mass 286.5 g (n = 2).

DISTRIBUTION: This species is found west of the Irrawaddy River in Bhutan, India, Myanmar, and Tibet, as well as east of the Irrawaddy River in Myanmar, southern China, Laos, Cambodia, Vietnam, Thailand, peninsular Malaysia, and Taiwan. Pallas's squirrel has been introduced to Argentina, the Netherlands, Belgium, France, Italy, Hong Kong (China), and Japan. In 1935, 40 squirrels were introduced from Taiwan to

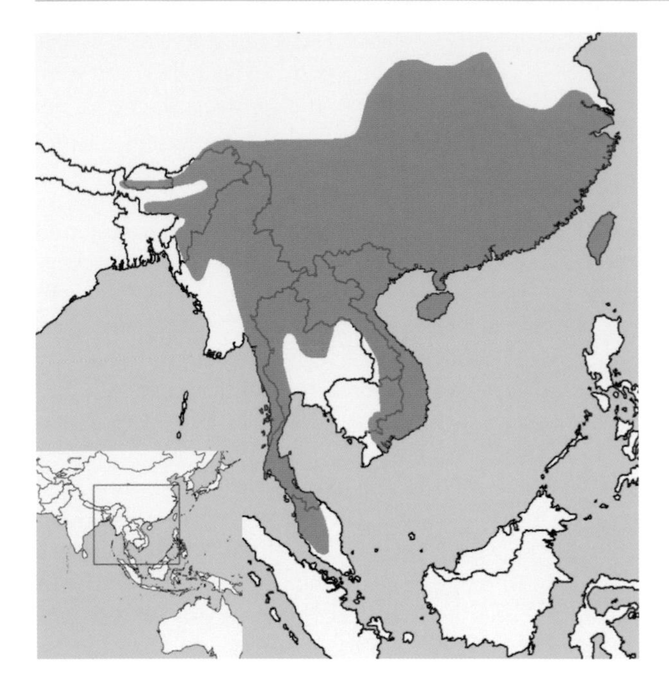

Japan, first to Izuoshima (= Izu Ōshima) Island (100 km south of Tokyo), then from Izu Ōshima to Eno-Shima Island, from which they then spread to neighboring Kamakura City. From Izu Ōshima Island, 100 animals were introduced to Tomogashima Island (at the mouth of Osaka Bay) in 1954. Numerous additional translocations have occurred since these initial introductions. This species was introduced into Argentina and became established in the province of Buenos Aires in 1973, where its geographic range expanded 1.6 km/year between 1999 and 2004. It has subsequently been introduced into Córdoba Province, Argentina.

GEOGRAPHIC VARIATION: Approximately 25 subspecies are generally recognized. In *The Mammals of the Indomalayan Region* (Corbet and Hill), these are divided into four groups of subspecies.

C. e. erythraeus group—from Sikkim to Assam (India) and through to Myanmar, western Thailand and peninsular Malaya, northern Vietnam, and Taiwan. The underparts are reddish brown or reddish (sometimes agouti), commonly with a midventral wedge or stripe of agouti dividing the reddish brown coloration into two parts.

C. e. flavimanus group—Laos, Cambodia, and southern Vietnam. The feet are paler than the back.

C. e. sladeni group—between the Chindwin and Irrawaddy rivers of Myanmar. The muzzle and feet are similar in color to the venter.

C. e. styani group—eastern China. The venter is pale cream to dull orange brown; the dorsum is pale and gray; and the feet are agouti but not dark.

CONSERVATION: IUCN status—least concern. Population trend—stable.

HABITAT: In Dampa Wildlife Reserve (Mizoram, eastern India), *C. erythraeus* has been observed with equal frequency per kilometer in primary and secondary forests (twice as pairs, nine times as individuals), but not in bamboo forests or near human habitations. It has been observed at elevations between 200 and 1000 m, but not at higher elevations. In Bhutan, it was found in heavily forested areas. In China, it most frequently occurs in tropical and subtropical forests at low elevations, but it is also found in subalpine coniferous forests, or at elevations above 3000 m in a mix of conifers and broadleaf trees. It will occupy pine (*Pinus*) plantations that are more than 16 years old if they are protected, but, for ones in unprotected areas, not until the plantations are 31 years old, apparently because human activities decrease the vegetation diversity on the forest floor. In peninsular Malaysia, *C. erythraeus* is described as being common in forests and woodlands in the hills. This species was introduced to Tomogashima Island in Japan, where it prefers both broadleaved and coniferous evergreen trees. In Argentina, these squirrels are reported in fruit plantations.

NATURAL HISTORY: Over their wide range, Pallas's squirrels seem to be very adaptable in their diets. On Tomogashima Island (Japan), they feed extensively on camellia flowers (*Camellia japonica*), bayberry fruits (*Myrica rubra*),

Callosciurus erythraeus. Photo courtesy Nick Baker, www.ecologyasia.com.

and Japanese pine seeds (*Pinus thunbergii*). In total, they eat 36 species of plants there: mainly camellia blossoms in January-March, leaves in April and May, and fruits (mostly seeds) in June-December. In addition, they consume the bark of 23 plant species, eat insects (mostly ants and cicadas), and drink water that accumulates in knotholes. In China, they have also been reported to feed on bird eggs and fledglings. In Malaysia, the stomach contents of three animals included a significant proportion of insects, as well as starchy vegetable matter. In Kamakura City (Japan), where they have been introduced and are now common, both males and females have overlapping home ranges, with fewer females overlapping with other females (2.3 in winter-spring, 1.6 in autumn-winter) than males overlapping with other males (12.6 in winter-spring, 8.7 in autumn-winter). Male home ranges were 2.6 ha in the winter-spring season and 1.2 ha in the autumn-winter season. Female home ranges were 0.7 ha and 0.5 ha in the same two seasons.

Nests in Malaysia were constructed of leafy green twigs among the outer branches of tall trees. In a more detailed study of nests on Tomogashima Island (Japan), the nests were described as being built in three layers—an outer layer of interwoven leafy twigs, a middle layer of twigs, and an inner layer of leaves and shredded material (e.g., the inner layer of bark). Nests were built predominantly by adult female squirrels. Subterranean nests were used in the winter on Tomogashima Island (Japan), and in Bhutan the squirrels were suspected of nesting in holes between the roots of trees, because they came down to the ground in the evening and when they were alarmed. In Kamakura City (Japan), the squirrels have antipredator calls for terrestrial predators, aerial predators, and snakes, each of which elicits appropriate responses from other squirrels. The males also have precopulatory calls and postcopulatory calls. The latter mimic the terrestrial predator call. Mating bouts start early in the morning and last for many hours. The most dominant male chases other males away and then mates with the female, after which he guards the female for an average of 33 minutes by giving the postcopulatory call, causing nearby males and the female to adopt their immobile antipredator response. The female subsequently mates with another male, mating 4-11 times in all during her estrus. Although not specified in the study, the assumption is that these other males also give a postcopulatory call. Mating bouts occur throughout the year, and the average litter size at weaning is 1.4 individuals.

GENERAL REFERENCES: Bertolino 2009; Corbet and Hill 1992; Guichón et al. 2005; Men et al. 2006; Oshida, Lee, et al. 2006; Setoguchi 1990, 1991; Shankar-Raman et al. 1995; Tamura 1995; Tamura et al. 1988.

Callosciurus finlaysonii (Horsfield, 1823) Finlayson's Squirrel

DESCRIPTION: This is also a quite variable species, differing both within and between subspecies; it ranges from all white, to all red, to all black. There may be more than one species included under this name.

SIZE: Female—HB 191.8 mm (n = 13); T 172.8 mm (n = 13).
Male—HB 190.1 mm (n = 14); T 173.4 mm (n = 14).
Sex not stated—HB 212.1 mm (n = 76); T 222.0 mm (n = 73); Mass 278.0 g (n = 2).

DISTRIBUTION: This species is found in southcentral Myanmar, Thailand, Laos, Cambodia, and Vietnam. It was introduced into Italy and Singapore.

GEOGRAPHIC VARIATION:

Insular forms include the following seven subspecies.

C. f. finlaysonii—Koh (= island) Si Chang (Thailand). This form is all white, with a yellow tinge on the back.

C. f. albivexilli—Koh Kut (Thailand). This form has a black tail with a white tip.

C. f. folletti—Koh Phai (Thailand). This form is grayish white.

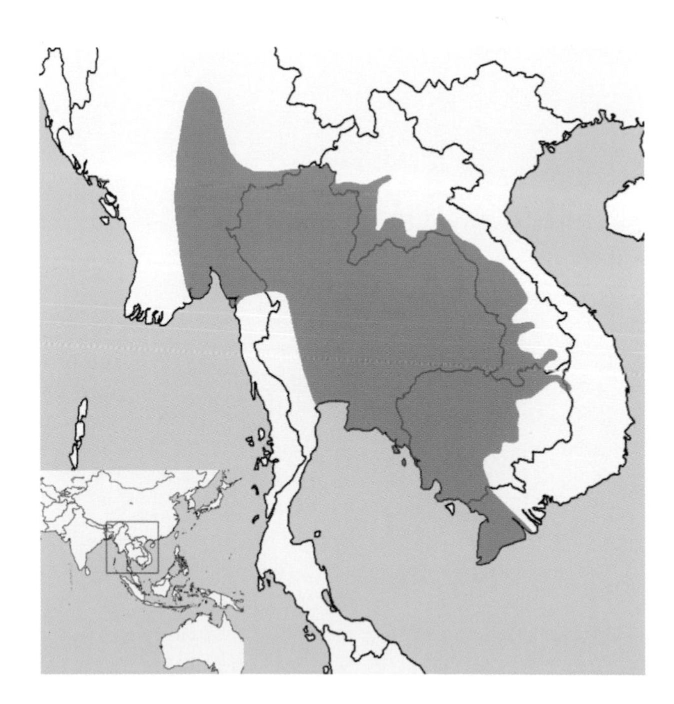

Callosciurus finlaysonii. Photo courtesy Pitchaya and Rattapon Kaichid.

C. f. frandseni—Koh Chang (Thailand). This form is red with gray flanks.

C. f. germaini—Condor (= Côn Son) Island (Vietnam). This form is all black.

C. f. harmandi—Phu Quoc Island (Vietnam). This form is reddish brown on the back, light orange red on the venter, and has a grayish white tail.

C. f. trotteri—Koh Lan (Thailand). This form is gray, with a white tail and black feet.

Mainland forms include the following nine subspecies.

C. f. annellatus—southern Laos and central and eastern Cambodia. This form is mainly reddish brown, with a cream band at the base of the tail.

C. f. bocourti—central Thailand. This form is quite variable, with a black back and a white to cream-colored head and venter.

C. f. boonsongi—northeastern Thailand. This form is mostly black above, gray on the venter, and with a white edge to the ears.

Callosciurus finlaysonii. Photo courtesy Pitchaya and Rattapon Kaichid.

Callosciurus finlaysonii bocourti. Photo courtesy A. Coke Smith, www.cokesmithphototravel.com.

C. f. cinnamomeus—southeastern Thailand and southwestern Cambodia. This form is usually all red, but it may be all olive agouti.

C. f. ferrugineus—Myanmar. This form is all red.

C. f. menamicus—northcentral Thailand. This form is all red except for the tip of the tail, which is buffy white.

C. f. nox—coastal Thailand, southeast of Bangkok. This form is completely black.

C. f. sinistralis—central to northwestern Thailand. This form is mostly reddish, with a reddish agouti back and a cream band at the base of the tail, similar to *C. f. annellatus*.

Callosciurus finlaysonii. Photo courtesy Pitchaya and Rattapon Kaichid.

Callosciurus finlaysonii annellatus. Photo courtesy Nick Baker, www.ecologyasia.com.

Callosciurus finlaysonii cinnamomeus. Photo courtesy Silvio Colaone.

C. f. williamsoni—Laos. This form has a chestnut coloration on the venter, abruptly separated from the red to orange coloration on the dorsum.

CONSERVATION: IUCN status—least concern. Population trend—stable.

HABITAT: A species of lowland forests, Finlayson 's squirrel is found in many habitats: dense forests, open forests, and coconut plantations. In 1915 it was common in the shrubby growth of Koh Si Chang.

NATURAL HISTORY: *Callosciurus finlaysonii* ecology has been studied in Italy, where this species was introduced in 1981. The animals are active most of the day, spend very little time on the ground, and forage most of the time. They feed predominantly on vegetable matter, including fruits, seeds, buds, and flowers in season; buds in the winter; and sap and bark in the autumn, winter, and spring. Hard nuts provided by visitors were cached in nests or in the trees, but never on the ground. To a small extent, these squirrels feed on insects, mostly ants.

Callosciurus finlaysonii. Photo courtesy Silvio Colaone.

GENERAL REFERENCES: Bertolino 2009; Bertolino, Mazzoglio, et al. 2004; Kloss 1915; Lekagul and McNeely 1977; Phuong et al. 2006.

Callosciurus inornatus (Gray, 1867) Inornate Squirrel

DESCRIPTION: The dorsal pelage and tail are deep olive agouti, with the tail frequently black tipped. The ventral pelage is bluish gray to light violet gray.

SIZE: Female—HB 203.0 mm (n = 8); T 204.8 mm (n = 6).
Male—HB 191.0 mm (n = 10); T 210.8 mm (n = 4).
Sex not stated—HB 225.0 mm (n = 2); Mass 325.0 g (n = 1).

DISTRIBUTION: This species is found in southern Yunnan (China), Laos, and northern Vietnam.

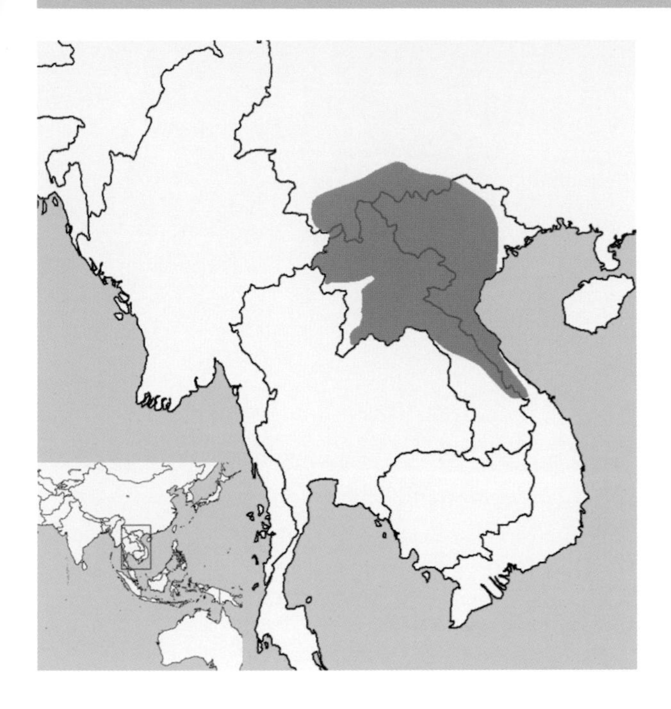

GEOGRAPHIC VARIATION: None.

CONSERVATION: IUCN status—least concern. Population trend—no information.

HABITAT: The inornate squirrel occurs in scrublands, degraded evergreen forests, and pristine evergreen forests.

NATURAL HISTORY: No information is available.

GENERAL REFERENCES: R. S. Hoffmann and Smith 2008.

Callosciurus melanogaster (Thomas, 1895) Mentawai Squirrel

DESCRIPTION: The Mentawai squirrel is a black-bellied *Callosciurus* species, with black feet and hands and a black coloration extending up the insides of the limbs. It is dark on the back as well, with a black head and obscure lateral side stripes, the dorsal one grizzled and the ventral one black, merging with the coloration of the abdomen.

SIZE: Female—HB 212.0 mm (n = 5); T 180.6 mm (n = 5).
Male—HB 209.3 mm (n = 9); T 179.6 mm (n = 28).
Sex not stated—HB 213.0 mm (n = 22); T 176.0 (n = 20); Mass 292.3 g (n = 3).

DISTRIBUTION: This species is found on the Mentawai Islands: Siberut, Sipora, and North and South Pagai (Indonesia).

GEOGRAPHIC VARIATION: Three subspecies are recognized.

C. m. melanogaster—Sipora Island (Indonesia). See description above.

C. m. atratus—North and South Pagai Islands (Indonesia). This form has a black tail, blackish brown underparts with a blackish gray belly, a black-and-russet dorsum, and obscure russet and black side stripes.

C. m. mentawi—Siberut Island (Indonesia). This form is similar to *C. m. melanogaster*, but with a reddish brown venter.

CONSERVATION: IUCN status—vulnerable. Population trend—decreasing.

HABITAT: On Siberut Island (Indonesia), *C. melanogaster* is a resident of primary mixed forests, but it is seen infrequently in primary mixed dipterocarp forests, swamp forests, and disturbed secondary growth.

NATURAL HISTORY: The Mentawai squirrel is most active in the early morning and late afternoon, and probably before sunrise and after sunset, to judge by its vocalizations. It is active at all heights in the forest—from 0 to 30+ m, although rarely seen on the ground—but it chooses heights above 25 m for resting. These squirrels have been observed feeding

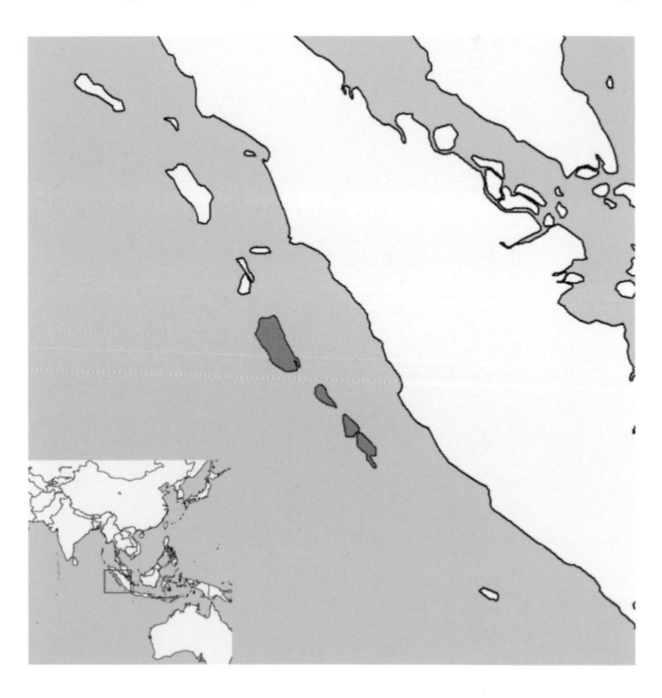

predominantly on fruits and young leaves, but 25 percent of the observations involved bark eating. The stomach contents of three animals contained 35 percent vegetable matter and 62 percent animal matter (mostly arthropods).

GENERAL REFERENCES: Chasen and Kloss 1927; G. S. Miller 1903; Thomas 1895; Whitten 1981.

Callosciurus nigrovittatus (Horsfield, 1823) Black-Striped Squirrel

DESCRIPTION: This species is a gray-bellied squirrel, with a buff stripe superimposed on a black stripe along each flank, a grizzled black and buffy dorsum, and more buffy shoulders.

SIZE: Female—HB 186.9 mm (*n* = 12); T 165.5 mm (*n* = 2); Mass 239.4 g (*n* = 7).

Male—HB 184.3 mm (*n* = 11); T 159.2 mm (*n* = 5); Mass 202.8 g (*n* = 7).

Sex not stated—HB 199.0 mm (*n* = 6); T 182.5 (*n* = 6); Mass 218.6 g (*n* = 7).

DISTRIBUTION: This species is found in peninsular Thailand; peninsular Malaysia; and Sumatra, Java, and adjacent small islands (Indonesia).

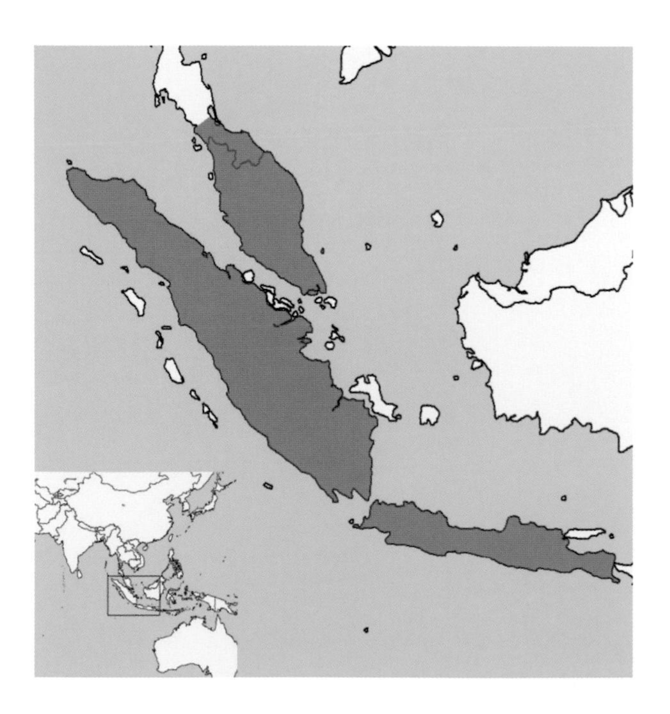

Callosciurus nigrovittatus. Photo courtesy Nick Baker, www.ecologyasia.com.

GEOGRAPHIC VARIATION: Four subspecies are recognized.

- *C. n. nigrovittatus*—Java (Indonesia). This form has an obscure buff lateral line.
- *C. n. bilimitatus*—Malay Peninsula (Thailand and Malaysia) and Tioman Island (Malaysia). This form has a sharply delineated buff line.
- *C. n. bocki*—highlands of Sumatra (Indonesia). This form has a bright and clearly delineated buff lateral line, and some animals have a pale patch behind the ear.
- *C. n. klossi*—Saddle Island, west of Kalimantan (Indonesia). This form is "blue-bellied," similar to *C. orestes*, but it is not as brightly colored and lacks a pale patch behind the ear.

CONSERVATION: IUCN status—near threatened. Population trend—decreasing.

HABITAT: In Malaysia, this species is an arboreal squirrel of the canopy and undercanopy of primary and secondary forests. In the Krau Wildlife Reserve (Pahang, Malaysia), *C. nigrovittatus* is 25 percent as common as *C. notatus* (13 individuals versus 53 individuals live trapped, respectively). It is approximately twice as common in disturbed areas as in places that are less disturbed by humans.

NATURAL HISTORY: At the Krau Wildlife Reserve (Pahang, Malaysia), the estimated population density of *C. nigrovittatus* is 5 individuals/100 ha. In the Ulu Gombak Forest Reserve (Selangor, Malaysia), a single female had a home range estimated, in two different ways, to be 1.4 or 2.0 ha. This species' diet includes fruits and a significant proportion of insects. The leaf nest is similar to that of the gray-bellied squirrel (*C. caniceps*), with an outer layer of twigs or coarse leaves and an inner layer of shredded materials. From 1948 to 1952 in the Ulu Gombak Forest Reserve, pregnant females were recorded in every month, most frequently in April-June (20%) and least often in October-December (8%). Litter size was one to four, with a mean of 2.2 young (*n* = 23 litters). In the Ulu Gombak Forest Reserve, the squirrels react to terrestrial predators with a repeated staccato bark and tail flicking, which causes conspecific squirrels to run up a tree and be silent. *C. nigrovittatus* responds to an aerial predator with a soft "chuckle" if not immediately threatened, and with a "rattle" sound if it is a close encounter. Nearby squirrels react by freezing in position, and perhaps by joining in the "rattle" call. In response to snakes, the squirrels give a "squeak" call, which causes other squirrels, sometimes of three different species, to join in mobbing the snake.

GENERAL REFERENCES: Lundahl and Olsson 2002; Medway 1969; G. S. Miller 1900, 1903, 1942; J. B. Payne 1980; H. C. Robinson and Wroughton 1911; Saiful et al. 2001; Tamura and Yong 1993.

Callosciurus notatus (Boddaert, 1785) Plantain Squirrel

DESCRIPTION: This is a red-bellied squirrel, with the ventral pelage always orange or reddish to deep chestnut, two side stripes of white over black, and the dorsal pelage and tail brown.

SIZE: Female—HB 237.8 mm (*n* = 25); T 175.0 mm (*n* = 25); Mass 227.9 g (*n* = 18).

Male—HB 233.6 mm (*n* = 26); T 186.5 mm (*n* = 26); Mass 233.9 g (*n* = 17).

Sex not stated—HB 201.7 mm (*n* = 42); T 183.4 (*n* = 28); Mass 219.4 g (*n* = 48).

DISTRIBUTION: This species is found in peninsular Thailand and peninsular Malaysia; Sumatra, Java, Bali, and Lombok (Indonesia); and at elevations up to 1700 m on the island of Borneo (divided among Malaysia, Brunei Darussalam, and Indonesia). It is widespread on smaller islands of the Sunda Shelf and on Salayer Island (south of Sulawesi [Indonesia]), where it was probably introduced.

GEOGRAPHIC VARIATION: Six subspecies are recognized.

C. n. notatus—Java (Indonesia). This form has a slight suffusion of buff on the abdomen.

C. n. albescens—restricted to North Sumatra (Indonesia). This subspecies is a paler form and is sometimes considered a distinct species.

C. n. diardii—southern coast of Java (Indonesia). This form is buff in the ventral pelage and around the eye; it is darker in the eastern part of the range.

C. n. miniatus—Malay Peninsula. This form has a gray venter with a slight brown to dull orange brown tinge; the pale flank stripe is narrow. The tip of the tail is reddish brown, particularly in the western part of the range.

C. n. suffusus—island of Borneo (divided among Malaysia, Brunei Darussalam, and Indonesia). This form resembles the Malayan forms, but the ventral color is a darker chestnut or maroon.

C. n. vittatus—Sumatra (Indonesia). This form resembles *C. n. miniatus*, but the venter is usually orange brown, with no red coloration in the tail tip.

CONSERVATION: IUCN status—least concern. Population trend—increasing.

HABITAT: *C. notatus* is the common arboreal squirrel of gardens, plantations, secondary forests, and fringe habitats. It is usually rare or absent in tall dipterocarp forests, but common in mangrove and swamp forests, and it can survive in monoculture plantations. The plantain squirrel is one of the generalist species dominating the mammalian fauna in a degraded peat swamp forest in Selangor (Malaysia), and it is one of the commonest species in areas disturbed by humans in the Krau Wildlife Reserve (Pahang, Malaysia). In four 1.1

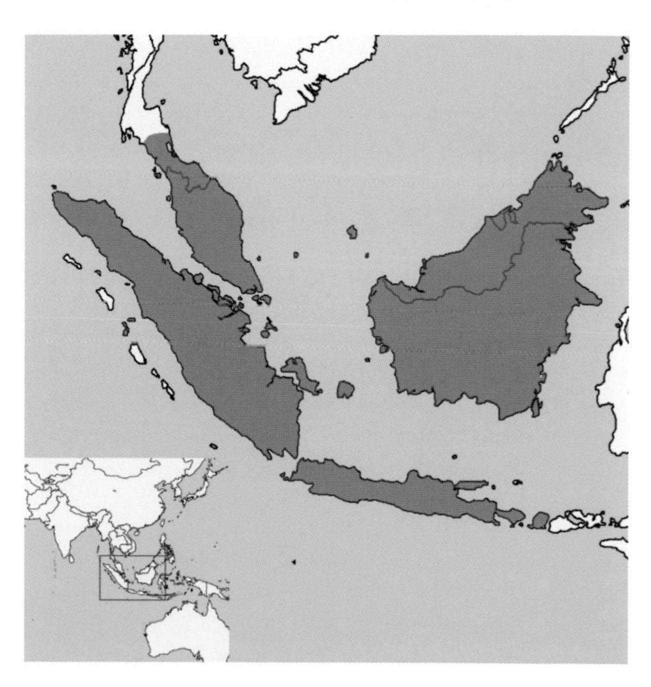

Callosciurus notatus. Photo courtesy Nick Baker, www.ecologyasia.com.

ha plots in 1999, it was trapped 10.7 and 7.8 times per thousand trap days in disturbed habitats, and 3.4 and 0.6 times per thousand trap days in areas less disturbed by human activity. It travels and feeds mainly in small trees.

NATURAL HISTORY: In the Krau Wildlife Reserve (Pahang, Malaysia), the plantain squirrel eats seeds (approximately 30%), fruits (10%, including some figs), leaves and shoots (25%), flowers (5%), bark and sap (25%), and insects and other arthropods (5%). In the Ulu Gombak Forest Reserve (Selangor, Malaysia), its foods include fruits (43%), bark (40%), and a variety of other plant parts, including flowers (4%) and leaves (1%) from 16 species of plants. On the island of Borneo, this species was recorded feeding on the fruit of *Rafflesia keithii*, a parasitic plant with colossal unisexual putrid-smelling flowers. The very small seeds of this plant appear to be well suited for dispersal in the feces of small mammals, such as this squirrel. Plantain squirrels commonly remove seeds and fruits from fruiting trees and take them elsewhere. They can be seen carrying fruit while traveling far from any fruit-bearing tree: perhaps to feed in a more protected area, perhaps to cache it. They usually eat the seeds after feeding on the fruit. They will gnaw the cambium of rubber trees and lick the latex. They are also known to eat insects—their stomach contents have been reported to be half insects by bulk.

Population density is 5.35 individuals/ km^2 in the Ulu Mudah Forest Reserve (Kedah, Malaysia) and 8.1 individuals/ km^2 in Kayan Mentarang National Park (East Kalimantan, Indonesia), but 244 individuals/ km^2 in the Krau Wildlife Reserve (Pahang, Malaysia). In lowland dipterocarp forests in the Ulu Gombak Forest Reserve (Selangor, Malaysia), the plantain squirrel is most active in the early morning, with its activity tapering off by 10 a.m., then increasing again from 4 to 6 p.m. It appears to avoid the ground, is most frequently observed at heights between 6 and 10 m, and is seldom seen above 20 m in the trees. In the Krau Wildlife Reserve, this species ranges from ground level to 24 m, but it is most commonly seen between 6 and 12 m. Its home range differs between males and females: 0.7 ha in males, with extensive overlaps between males and between males and females; and 1.8 ha in females, with no overlap between them, in the Ulu Gombak Forest Reserve.

C. notatus responds to terrestrial predators with a repeated staccato bark and tail flicking, to which other conspecific squirrels react by climbing upward and remaining quiet. The response to an aerial predator is a soft "chuckle," unless the predator is close, in which case the squirrel gives a "rattle" vocalization. Conspecifics react to the latter by freezing and, if they are nearby, by also giving a "rattle" call. This call is used more generally for a close encounter with a terrestrial predator. Snakes elicit "squeak" alarm calls, causing other squirrels to join in mobbing the snake. The nest is spherical in shape, with a poorly defined lateral entrance. It consists of a firm outer wall of twigs or leaves and is lined with fibrous material, such as shredded palm spathe. It is usually placed in the upper branches of a large bush or small tree. In the Ulu Gombak Forest Reserve, females mate with two to four males during a six- to seven-hour period, with five to seven males attending. After mating, a male gives a postcopulatory bark for 0.5–8 minutes. Between 1948 and 1952 in the Ulu Gombak Forest Reserve (Malaysia), females were pregnant in every month of the year, with the fewest in October–December (8%) and the most in April–June (29%). Litter size was one to four, with a mean of 2.2 young (n = 25 litters).

GENERAL REFERENCES: Abdullah et al. 2001; Becker et al. 1985; Bonhote 1901; Emmons et al. 1991; Lundahl and Olsson 2002; Norhayati et al. 2004; J. B. Payne 1980; Saiful et al. 2001; Tamura 1993; Tamura and Yong 1993.

Callosciurus orestes (Thomas, 1895)
Borneo Black-Banded Squirrel

DESCRIPTION: The dorsal pelage is brown and finely speckled. This species has a buffy spot behind the ear and side stripes that are buffy white over black. It is gray bellied, sometimes a dark gray, and sometimes paler with a reddish tinge.

SIZE: Female—HB 154.0 mm (*n* = 5); T 139.8 mm (*n* = 5).
Male—HB 148.0 mm (*n* = 4); T 157.5 mm (*n* = 4).
Sex not stated—HB 153.1 mm (*n* = 7); T 154.2 (*n* = 5); Mass 278.0 g (*n* = 1).

DISTRIBUTION: *C. orestes* is found on the island of Borneo, at middle elevations in Sabah and Sarawak (Malaysia) and in Kalimantan (Indonesia).

GEOGRAPHIC VARIATION: None.

CONSERVATION: IUCN status—least concern. Population trend—stable.

HABITAT: This species appears to be restricted to lower montane forests and possibly to upper dipterocarp forests. It is an uncommon squirrel on the island of Borneo, mostly found in the submontane primary forests of Sabah and Sarawak (Malaysia) in the northwestern part of the island, including in the Kelabit Highlands (at elevations above 1100 m), Mount Kinabalu (at 1100-1700 m), and Mount Dulit (at 1200 m).

NATURAL HISTORY: The Borneo black-banded squirrel is active in small and medium-sized trees. Fruits and black ants were found in the stomachs of two animals collected on Mount Kinabalu (Sabah and Sarawak, Malaysia).

GENERAL REFERENCES: J. Payne and Francis 1985.

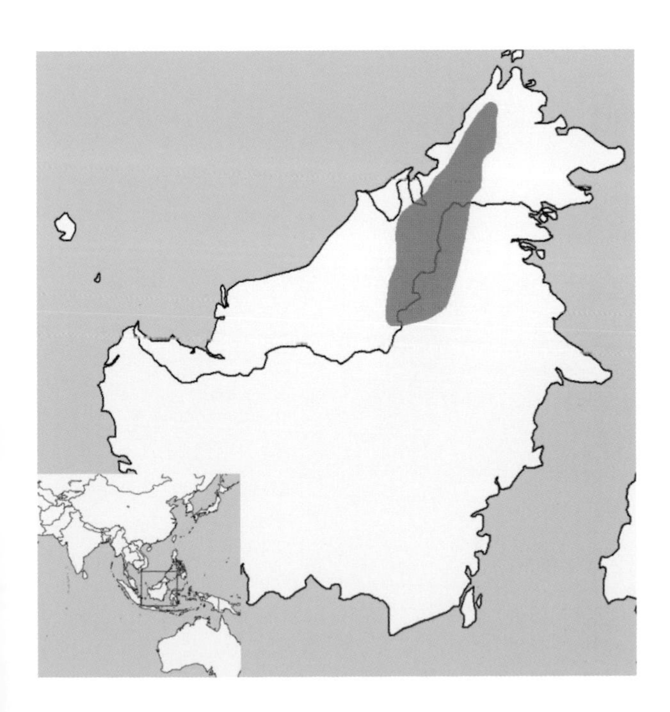

Callosciurus orestes. Photo courtesy Israel Didham.

Callosciurus phayrei (Blyth, 1856) Phayre's Squirrel

DESCRIPTION: The dorsal pelage is agouti gray, separated from the ventral pelage by a lateral black stripe between the forelimb and hind limb. The ventral pelage may be rich orange to very pale orange, but not red or gray. The feet are yellowish buff to pale orange; the tail has a midventral bright yellow stripe, 12-15 mm wide, and is black tipped.

SIZE: Female—HB 237.1 mm (*n* = 7); T 249.0 mm (*n* = 6); Mass 377.4 g (*n* = 6).

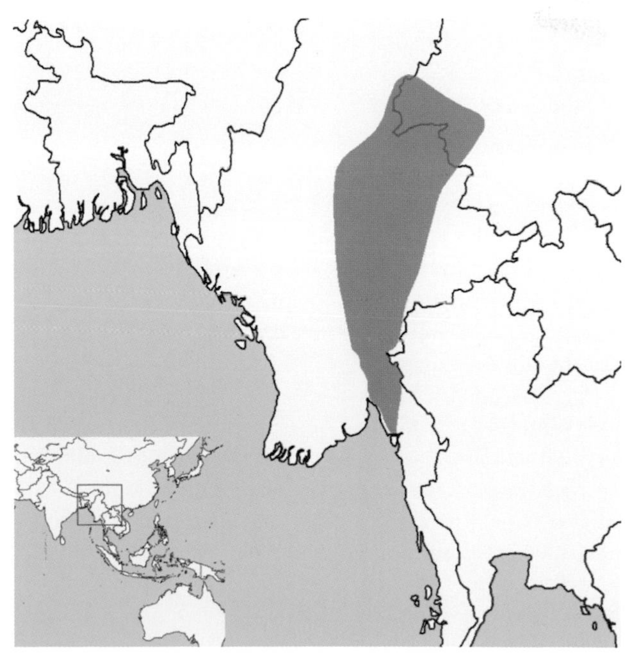

Male—HB 231.5 mm (n = 4); T 246.8 mm (n = 4); Mass 375.6 g (n = 2).

Sex not stated—HB 215.0 mm (n = 3); T 200.0 (n = 1); Mass 258.9 g (n = 2).

DISTRIBUTION: This species is found in southern China and, in Myanmar, from the upper Irrawaddy River and the Sittang River eastward to the Salween River and south to southern Myanmar.

GEOGRAPHIC VARIATION: None.

CONSERVATION: IUCN status—least concern. Population trend—no information.

HABITAT: This species occurs in tropical deciduous forests.

NATURAL HISTORY: *C. phayrei* is described as being a particularly active squirrel and an able leaper.

GENERAL REFERENCES: Moore and Tate 1965.

Callosciurus prevostii (Desmarest, 1822) Prevost's Squirrel

DESCRIPTION: Prevost's squirrel is an exceedingly variable species, sometimes tricolored and sometimes bicolored, commonly with a red belly separated from a black or dark back by a white stripe.

SIZE: Female—HB 237.7 mm (n = 45); T 233.3 mm (n = 44); Mass 361.9 g (n = 27).

Male—HB 238.9 mm (n = 43); T 233.0 mm (n = 42); Mass 353.7 g (n = 17).

Sex not stated—HB 241.4 mm (n = 32); T 234.7 (n = 21); Mass 403.2 g (n = 15).

This species varies greatly in size geographically. The length of the head and body differs by subspecies, ranging from 196 mm to 274 mm, and a plot of body length versus island size is curvilinear, with small and large islands having the smaller subspecies.

DISTRIBUTION: Prevost's squirrel is found from peninsular Thailand south to Sumatra, the island of Borneo (divided among Malaysia, Brunei Darussalam, and Indonesia), and adjacent small islands, as well as in northern Sulawesi (Indonesia), where it is probably an introduced species.

GEOGRAPHIC VARIATION: This is an extremely variable squirrel, with 47 named subspecies.

On peninsular Malaysia, two or three subspecies are recognized.

- *C. p. prevostii*—southern Malaysia, from southern Pahang and Negeri Sembilan south. This form is black on the back, with a lateral white stripe extending uninterrupted from the nose to the white thigh, and with a thin black line ventral to it between the limbs. The white cheeks are slightly grizzled.
- *C. p. humei*—westcentral Malaysia, in southern Perak and Selangor. This form has red on the venter that spreads over the whole forelimb and shoulder to the black of the back, interrupting the white side stripe, which extends from the shoulder onto the thigh and down to the heel.
- *C. p. wrayi*—northern Malaysia, from the Thai border into northern Pahang and northern Perak. This form is frequently considered to be the same as *C. p. humei*, because it differs only in having a less intense shoulder coloration.

On Sumatra (Indonesia), two or three subspecies are recognized:

- *C. p. melanops*—eastern Sumatra. This form has black cheeks; a black back; and a bright brownish red venter, extending as more rufous orange onto the forelimb and shoulder to the black back. The buffy white side stripe extends from the shoulder onto the thigh and down to the heel.
- *C. p. piceus*—northern Sumatra. This form is black on the dorsum and red on the venter, with no side stripe.

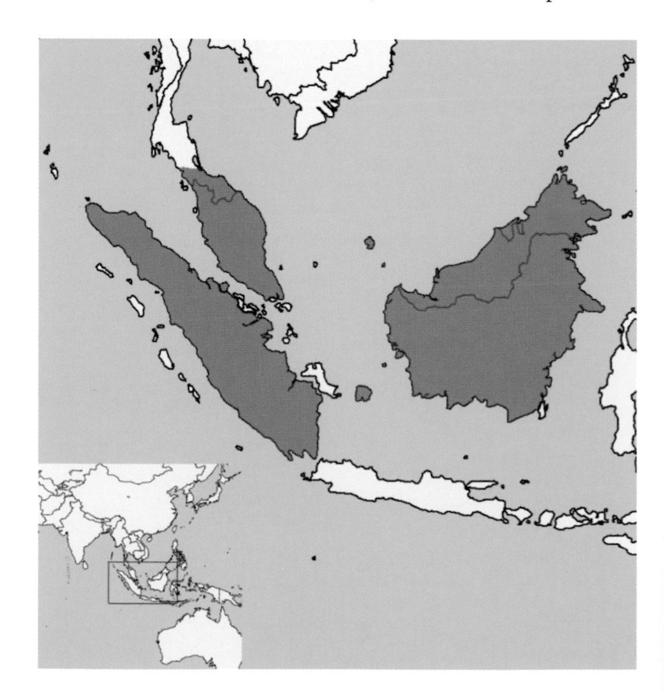

Callosciurus prevostii pluto. Photo courtesy A. Coke Smith, www.cokesmithphototravel.com.

C. p. rafflesii—southern Sumatra. This form is black on the back, and chestnut ventrally and on the forelimb and shoulder. It has a white lateral stripe extending from the shoulder and broadening posteriorly to the thigh. The cheeks are grizzled gray, with a white spot on the sides of the nose.

On the island of Borneo, eight subspecies have been recognized, all with dark red to orange underparts. Three other subspecies occur on small islands just off the Bornean coast.

C. p. atricapillus—eastern West Kalimantan, south of the Rajang River, and northern Central Kalimantan (Indonesia). This form has an olive buff dorsum, with a black head and rump.

C. p. atrox—West Kalimantan. This form has a black tail, with the rest of the dorsum grizzled olive buff.

C. p. borneoensis—southwestern Sarawak (Malaysia) and adjoining West Kalimantan (Indonesia). This form has a black dorsum, with the shoulders tinged with red. The thighs are gray, and both the thighs and the tail are grizzled. It has a distinct white side stripe.

C. p. caedis—Banggi and Balembangan islands (Malaysia, slightly north of the island of Borneo). This subspecies is all black on the back, without a lateral white stripe. It is similar to but geographically disjunct from *C. p. pluto* and *C. p. rufonigra*.

C. p. coomansi—an island in the Kapuas River delta (Indonesia). This form has a black dorsum, with the shoulders tinged bright rufus. The thighs are gray, and both the thighs and the tail are grizzled. It also has a distinct white side stripe.

C. p. palustris—just north of the Kapuas River (Indonesia). This subspecies resembles *C. p. borneoensis,* but it lacks the reddish tinge on the shoulders.

C. p. pluto—Sabah (Malaysia). This form is all black on the back, without a lateral white stripe. It is similar to but geographically disjunct from *C. p. caedis* and *C. p. rufonigra*.

C. p. rufonigra—Labuan Island (Malaysia). This form is all black on the back, without a lateral white stripe. It is similar to but geographically disjunct from *C. p. caedis* and *C. p. pluto*.

C. p. sanggaus—West Kalimantan, south of the Kapuas River (Indonesia). This form resembles *C. p. borneoensis,* but with paler shoulders and a completely black tail.

C. p. sarawakensis—Sarawak (Malaysia). This form's entire dorsum is grizzled gray, usually with a reddish tinge anteriorly.

C. p. waringensis—Central Kalimantan (Indonesia). This subspecies resembles *C. p. sanggaus,* but with paler thighs and less reddish shoulders.

There are also 15 smaller islands with named populations on them.

CONSERVATION: IUCN status—least concern. Population trend—decreasing.

HABITAT: *C. prevostii* occurs in tall and secondary forests. On the island of Borneo, it is widespread in the lowlands and hills, with the highest elevation records being slightly more than 1200 m. It will enter gardens and plantations in order to feed on fruits.

Callosciurus prevostii. Photo courtesy Jesse Cohen, Smithsonian Institution's National Zoo.

NATURAL HISTORY: At the Krau Wildlife Reserve (Pahang, Malaysia), the population density of Prevost's squirrels was 38 individuals/100 ha. The squirrels are most active in the early morning and late afternoon. They are usually arboreal, but they descend to the ground to cross gaps between trees. At the Krau Wildlife Reserve, the squirrels spend most of their time in the middle and upper parts of the canopy, at heights usually ranging from 6 m to more than 37 m above the forest floor. Their diet includes seeds (approximately 60%), fruit pulp (20%, with figs making up 10% of this), bark and sap (10%), leaves and sprouts (5%), and flowers (approximately 2%). They especially favor fruits with a sweet or oily flesh, and they will extract insects (including termites and beetle larvae) from dead wood. The squirrels frequently remove seeds and fruits from a fruiting tree and carry them elsewhere. The reasons for this behavior are not clear, and may be from several causes. Occasionally the squirrels cache fruits away from the source tree. They also transport fruits to feed juveniles, and they probably carry fruit to more protected areas before feeding on it themselves. Unlike some other species of *Callosciurus,* they usually drop the seeds after feeding on the fruits.

GENERAL REFERENCES: Becker et al. 1985; Heaney 1978; J. B. Payne 1980.

Callosciurus pygerythrus (I. Geoffroy Saint-Hilaire, 1831) Irrawaddy Squirrel

DESCRIPTION: These squirrels have a dark olive brown dorsal pelage. Their front legs, feet, and tail are grayer than the dorsal pelage; and the tail has a black tip. A cream or ochraceous buff hip patch appears seasonally. The ventral pelage is bluish gray, transitioning into a cream and orange buff coloration. The pelage varies seasonally, being brighter and with a hip patch during the wet season in the summer, and duller and without a hip patch during the dry season in the winter.

SIZE: Female—HB 189.2 mm (*n* = 17); T 175.1 mm (*n* = 16).
Male—HB 187.5 mm (*n* = 22); T 168.5 mm (*n* = 22).
Sex not stated—HB 203.2 mm (*n* = 5); T 150.0 (*n* = 1); Mass 252.0 g (*n* = 1).

DISTRIBUTION: This species occurs through Nepal, Bhutan, and northeast India to Myanmar, northern Vietnam, and Xizang (China).

GEOGRAPHIC VARIATION: Seven subspecies are recognized.

C. p. pygerythrus—between the Sittang and Irrawaddy rivers (Myanmar). This form is cinnamon rufous ventrally;

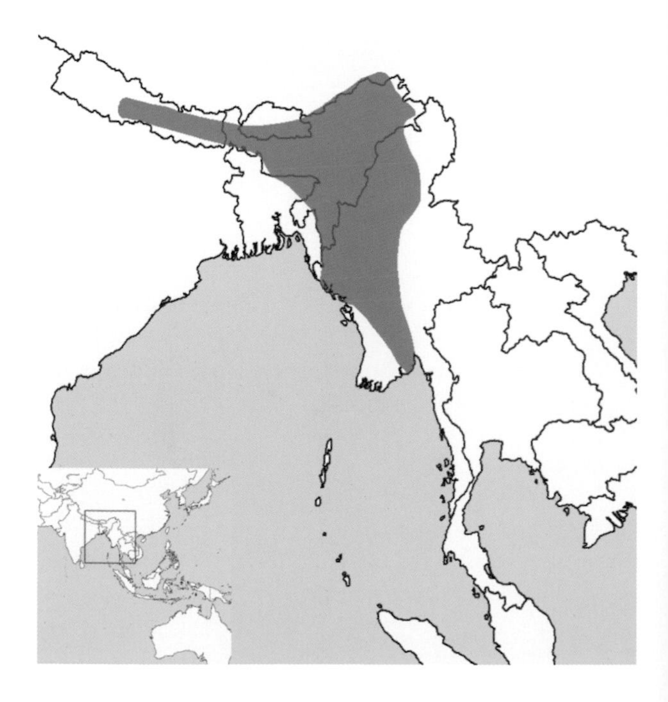

Callosciurus pygerythrus. Photo courtesy Raguib Uddin Ahmed.

paler on the chin, throat, and breast; olive brown dorsally; and has a black tail tip.

C. p. blythii—east of the Brahmaputra River, in Bangladesh and eastern India. The dorsal pelage varies from olive brown mid-dorsally to buffy brown laterally. This form has a grayish venter with a slightly buffy abdomen. It has seasonal white or creamy hip patches.

C. p. janetta—between the Irrawaddy River and the southern Chindwin River (Myanmar), and on the left bank of the Irrawaddy River (Myanmar). This form is pale ochraceous buff ventrally, and gray olive dorsally. The short black tail tip is inconspicuous. The cream flash mark on the hip is not seasonal.

C. p. lokroides—Nepal, Bhutan, and neighboring India. This form is agouti brown dorsally, more buffy laterally, grayish ventrally with a buffy abdomen, and has seasonal buffy orange hip patches.

C. p. mearsi—east of the Chindwin River (Myanmar). This form has an agouti dorsal pelage, ranging from buffy brown to light brownish olive; a buffy abdomen; a buffy eye ring; and seasonal white hip patches.

C. p. owensi—between the Brahmaputra and the upper Chindwin rivers (Myanmar). This form is more rufous dorsally than other subspecies in the summer, with the middle of the back cinnamon brown and the sides light brownish olive agouti. It has a seasonal ochraceous buff hip patch, separated from the ventral coloration by some of the agouti dorsal pelage.

C. p. stevensi—Xizang (China) and northeast Assam (India), west of the upper Chindwin River. This form has an agouti dorsal pelage, ranging from light brownish olive to cinnamon brown. The ventral pelage is gray, and it has seasonal buffy hip patches.

CONSERVATION: IUCN status—least concern. Population trend—no information.

HABITAT: In Dampa Wildlife Reserve (Mizoram, eastern India), *C. pygerythrus* was observed more frequently in secondary forests than in primary forests, and most often around settlements. In China, it occurs in rainforests, at elevations of 600–1300 m, often in cane shrubs at the edge of the forest or in banana plantations.

NATURAL HISTORY: This squirrel utilizes tree dens or makes grass-and-stick nests high in trees. It feeds on fruits, the flower buds of bananas, and insects. This species has been known to damage orange crops, and it has also been attracted to a bait of meat. It frequently forages on the ground. Females reproduce once each year, with a litter size of three to four young.

GENERAL REFERENCES: Corbet and Hill 1992; Ghosh 1981; Moore and Tate 1965; Shankar-Raman et al. 1995.

Callosciurus quinquestriatus (Anderson, 1871)
Anderson's Squirrel

DESCRIPTION: This dramatic squirrel has two white stripes that are separated by a midventral black stripe and flanked by two other longitudinal black stripes on its venter. The grizzled dorsal pelage is olive brown to olive yellow, with a rufous tint. The tail is like the back, but annulated with black and rufous coloration, and it has a black tip. *Callosciurus*

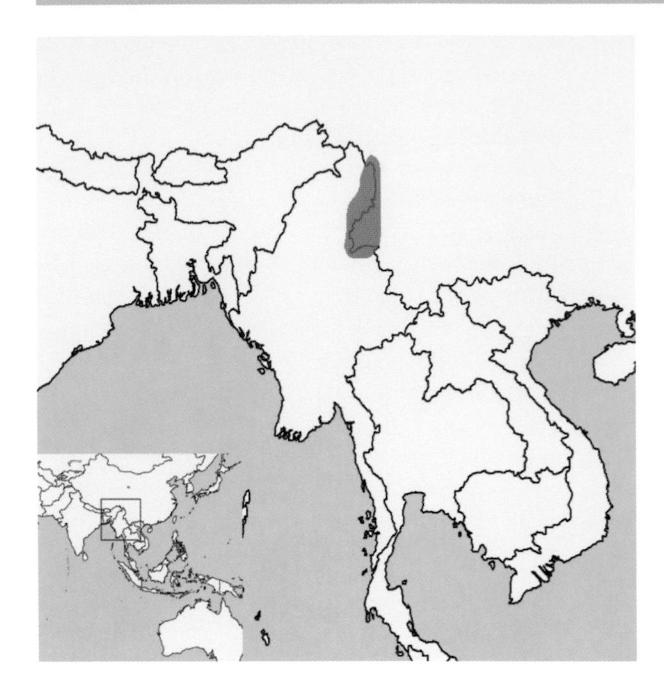

quinquestriatus has sometimes been considered to be conspecific with *C. erythraeus* (more precisely, with the *C. erythraeus* group of subspecies).

SIZE: Male—HB 226.5 mm (n = 2); T 202.0 mm (n = 1).
Sex not stated—HB 230.4 mm (n = 3); T 181.0 mm (n = 1).

DISTRIBUTION: This species is found in northeastern Myanmar and Yunnan (China).

GEOGRAPHIC VARIATION: Two subspecies are recognized.

C. q. quinquestriatus—southerly distribution. This form has shorter black stripes (with the midventral black stripe not extending to the inguinal region, and the lateral black stripes extending between the forelimb and hind limb) and longer white stripes extending to the inguinal region.

C. q. imarius—Kachin Province (northern Myanmar). This form has longer black stripes and shorter white stripes on the venter, with the midventral black stripe extending to the scrotum and the lateral ones extending onto the limbs.

***Callosciurus quinquestriatus*. Photo courtesy Fujie.**

CONSERVATION: IUCN status—near threatened. Population trend—decreasing.

HABITAT: Anderson's squirrel occurs in mountain forests, generally at elevations of 1000 m, and less commonly at lower elevations.

NATURAL HISTORY: These squirrels build nests in the outer branches of small trees. Their diet includes vegetable matter and insects.

GENERAL REFERENCES: R. S. Hoffmann and Smith 2008; Thomas 1926.

Dremomys Heude, 1898

This genus—the plain long-nosed squirrels of Asia—comprises six species, ranging from Nepal to southern China, Vietnam, and the island of Borneo. They are closely related to *Callosciurus*. *Dremomys* species are montane and forage in trees and on the ground.

Dremomys everetti (Thomas, 1890) Bornean Mountain Ground Squirrel

Dremomys everetti. Photo courtesy Peter Price.

DESCRIPTION: This is a dark squirrel, with the dorsal pelage brown speckled, the ventral pelage gray with buffy white-tipped hairs, and the tail black with red-tipped hairs.

SIZE: Female—HB 166.3 mm (n = 10); T 99.1 mm (n = 10).
Male—HB 155.9 mm (n = 8); T 98.3 mm (n = 8).
Sex not stated—HB 175.0 mm (n = 2); T 111.0 mm (n = 2); Mass 130.0 g (n = 2).

DISTRIBUTION: This species is found throughout the mountains in the northern and western parts of the island of Borneo: in Sabah and Sarawak (Malaysia), and in Kalimantan (Indonesia).

GEOGRAPHIC VARIATION: None.

CONSERVATION: IUCN status—least concern. Population trend—stable.

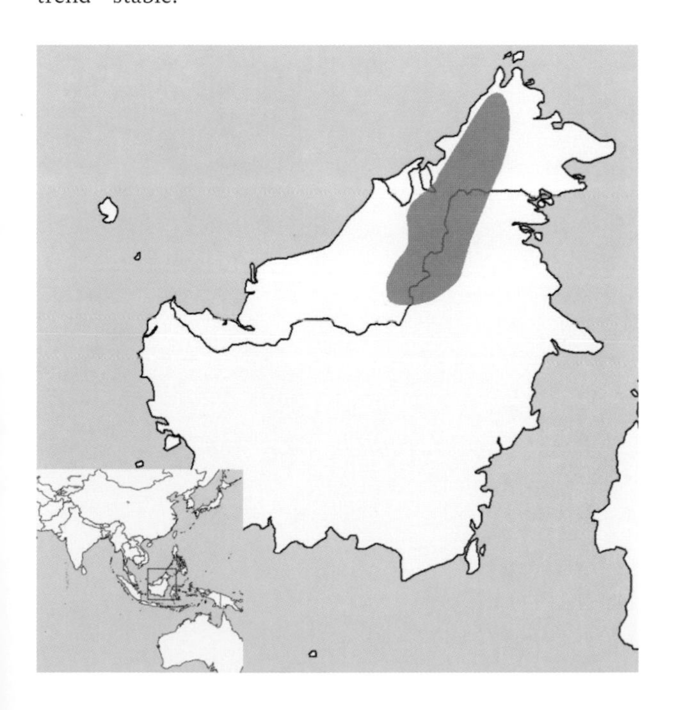

HABITAT: This is a terrestrial squirrel of the mountains in the northwestern part of the island of Borneo, occurring in forests at elevations above 975 m.

NATURAL HISTORY: *D. everetti* is reported to feed on insects (35% of the stomach contents of 19 animals), earthworms, fruits and nuts, and leaves and shoots.

GENERAL REFERENCES: D. D. Davis 1958, 1962; J. L. Harrison 1954; Moore 1958a; Moore and Tate 1965; J. Payne and Francis 1985.

Dremomys gularis Osgood, 1932 Red-Throated Squirrel

DESCRIPTION: The throat, chin, and neck are ochraceous tawny, contrasting with the dark blue gray of the rest of the ventral pelage. The ventral surface of the tail has a rich red coloration. *D. gularis* lacks the orange red flank patches of *D. pyrrhomerus*.

SIZE: Sex not stated—HB 216.0 mm (n = 10); T 168.0 mm (n = 10).

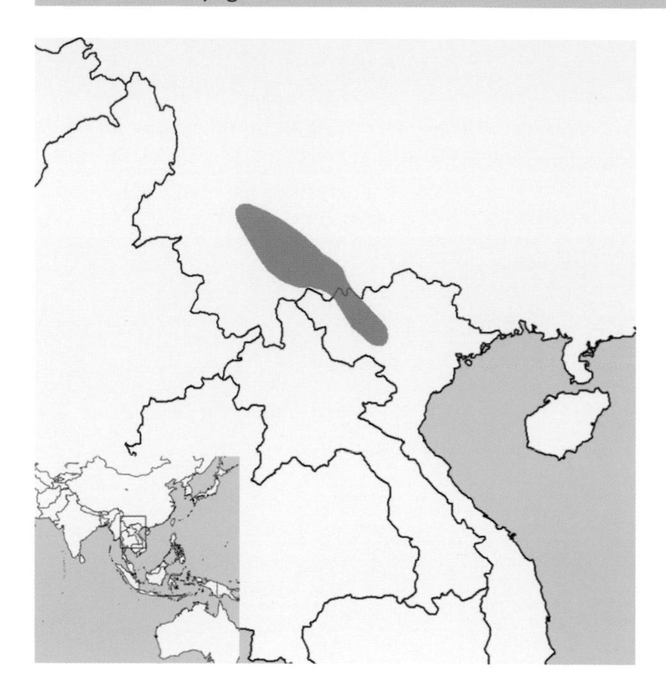

DISTRIBUTION: This species is found in the Red River Valley, from southcentral Yunnan (China) to northern Vietnam.

GEOGRAPHIC VARIATION: None. *D. gularis* was formerly considered to be a subspecies of *D. pyrrhomerus*.

CONSERVATION: IUCN status—least concern. Population trend—stable.

HABITAT: This species lives at high elevations, above the elevations at which *D. rufigenis* occurs.

NATURAL HISTORY: No information is available.

GENERAL REFERENCES: Osgood 1932; A. T. Smith and Xie 2008; Y. Zhang et al. 1997.

Dremomys lokriah (Hodgson, 1836) Orange-Bellied Himalayan Squirrel

DESCRIPTION: This species has a dull agouti dorsum and lacks a red coloration on the cheeks and hips. Ventrally, it is yellow to orangish and lacks a reddish brown perineal patch. The tail hairs are tipped with white; and the underside of the tail is black, mixed with a few orange hairs.

SIZE: Female—HB 194.1 mm (n = 2); T 125.0 mm (n = 1); Mass 172.5 g (n = 6).

Male—HB 180.4 mm (n = 27); T 131.9 mm (n = 28); Mass 180.2 g (n = 7).

Sex not stated—HB 194.1 mm (n = 2); T 125.0 mm (n = 1).

DISTRIBUTION: This species is found from central Nepal east through Bhutan to Tibet and Yunnan (China) as far as the Salween River, and south to northern Myanmar and the mountains of eastern India and extreme northeastern Bangladesh.

GEOGRAPHIC VARIATION: Six subspecies are recognized.

D. l. lokriah—Nepal, Bhutan, northeastern India, southern Tibet (in the vicinity of Mount Everest), and northern Myanmar. This form has a brown dorsal pelage, a bright orange ventral pelage with a pale throat, and no orange at the base of the tail.

D. l. garonum—Garo Hills and Khasi Hills of Assam (India) and extreme northeastern Bangladesh. This form is similar to *D. l. lokriah*, but the ventral pelage is paler yellow or ochraceous buff.

D. l. macmillani—Naga Hills and Chin Hills of Assam (India) to the west bank of the Chindwin River (Myanmar) and southeastern Tibet. This form has a coarsely grizzled olive gray dorsal pelage; a more tawny coloration on the nape and crown, with a thin black line extending from the nape to the lower back; deep ochraceous buffy

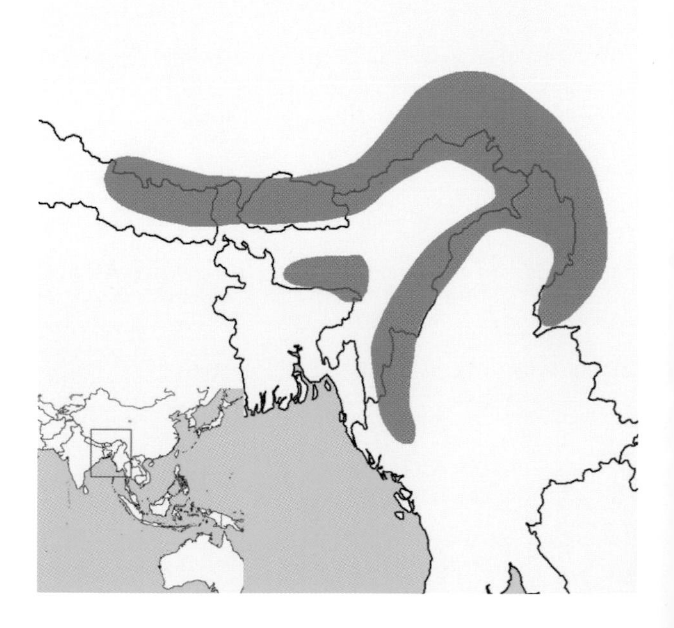

Dremomys lokriah. Photo courtesy Sudhir Shivaram.

patches behind the ears; and a bright buffy venter with a rufous patch in the inguinal area and at the base of the tail.

D. l. motuoensis—southeastern Tibet. This form lacks the black mid-dorsal stripe of *D. l. macmillani*, and has a much darker dorsal and ventral pelage—a darker yellowish brown dorsally—than the other subspecies.

D. l. nielamuensis—Tibet. This form is smaller than other subspecies, with a pale olive gray dorsal pelage and a pale yellow or orange ventral pelage.

D. l. pagus—Lushai Hills of Assam (India) and Chin Hills of western Myanmar. This form has a pale yellow ventral pelage with the gray bases of the hairs visible, lacks a rufous anal patch, and usually lacks a mid-dorsal black stripe.

CONSERVATION: IUCN status—least concern. Population trend—decreasing.

HABITAT: At lower elevations, this species occurs in subtropical forests and oak-rhododendron (*Quercus, Rhododendron*) forests. At higher elevations, *D. lokriah* is found in coniferous forests.

NATURAL HISTORY: The orange-bellied Himalayan squirrel nests in tree holes that are generally close to the ground. The nest is made of oak leaves and fern fronds, lined with grass. This species commonly feeds on insects, fruits (including those of mistletoe and *Pandanus furcatus*), and fallen nuts and fruits on the ground.

GENERAL REFERENCES: J. Li and Wang 1992.

Dremomys pernyi (Milne-Edwards, 1867) Perny's Long-Nosed Squirrel

DESCRIPTION: This species has an agouti gray dorsal pelage; it lacks a reddish cheek and a reddish hip patch. The ventral pelage is whitish, with a reddish brown patch in the perineal region. The ventral surface of the tail is buffy gray, not red.

SIZE: Female—HB 180.3 mm (n = 30); T 138.8 mm (n = 29); Mass 150.9 g (n = 7).

Male—HB 186.0 mm (n = 13); T 142.5 mm (n = 22); Mass 173.0 g (n = 10).

Sex not stated—HB 196.0 mm (n = 20); T 155.5 mm (n = 20); Mass 180.0 g (n = 2).

DISTRIBUTION: This species is found in northeastern India; northern Myanmar; Tibet and the Chinese provinces of Sichuan, Yunnan, Guizhou, Hubei, Anhui, Hunan, Jiangxi, Gansu, southern Shaanxi, Zhejiang, Guangxi, Guangdong, and Fujian; Taiwan; and northern Vietnam.

GEOGRAPHIC VARIATION: Eight subspecies are recognized.

D. p. pernyi—southern Gansu, southern Shaanxi, western Sichuan, and northwestern Yunnan (China) into western

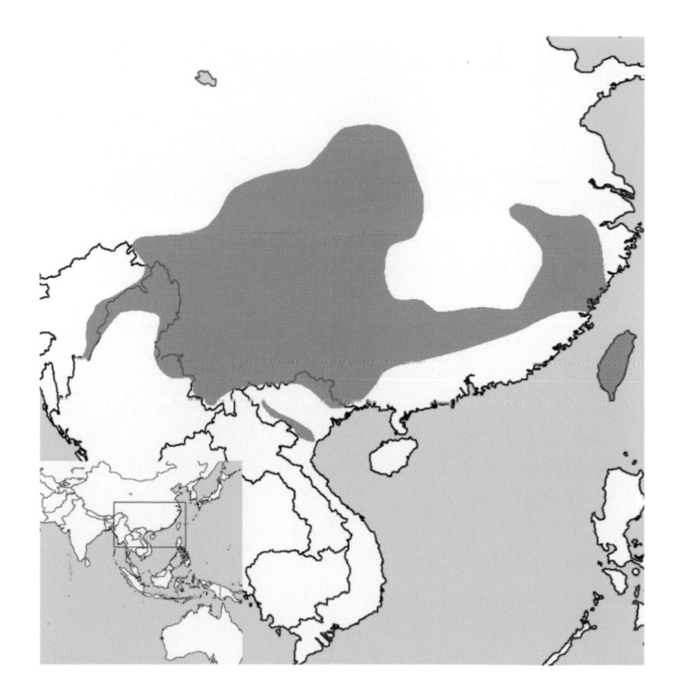

Dremomys pernyi. Photo courtesy Jeff B. Higgott, www.sequella .co.uk.

Tibet. This form is larger and lighter gray than the other subspecies.

D. p. calidior—Anhui, Zhejiang, Jiangxi, and Fujian (China). This subspecies is similar to *D. p. senex*, but its dorsal pelage is a warmer brown (almost olive brown), its ventral pelage is whiter, and it has buffy thigh patches.

D. p. flavior—Yunnan and Guangxi (China), extending into northern Vietnam. This form is smaller than *D. p. pernyi* and is darker olive on the dorsal pelage.

D. p. howelli—southwest Yunnan (China) and the Chin Hills of Myanmar and northeastern India. This form is generally darker above than the other species, with a short mid-dorsal stripe and a darker anal patch.

D. p. imus—northern Myanmar. This subspecies is larger than *D. p. howelli*, with the mid-dorsal stripe barely visible, and with a dull buff coloration on the front side of the limbs.

D. p. modestus—Guizhou, southern Hunan, Guangxi, and Guangdong (China). The dorsal pelage of this subspecies is dull brown, with inconspicuous ear patches; and the ventral pelage is tinged buff, especially posteriorly.

D. p. owstoni—Taiwan. The dorsal pelage of this form is brindled or grizzled buff and black, the ventral pelage is dull yellowish or orangish, and the throat is grayish white.

D. p. senex—Hubei and northern Guizhou (China). This subspecies has a whiter patch behind the ear than *D. p. pernyi*, the throat is white, and the underpart of the tail is more ochraceous than white.

CONSERVATION: IUCN status—least concern. Population trend—steady.

HABITAT: This species lives in coniferous and evergreen broadleaf forests, at elevations between 2000 and 3500 m.

NATURAL HISTORY: *D. pernyi* is mostly terrestrial. It has been known to prey on the eggs and chicks of Courtois's (= Blue-Crowned) Laughingthrush (*Garrulax [Dryonastes] courtoisi*) in Wuyuan (China). It inhabits protected planted pine (*Pinus*) forests in the Cangshan Mountains and the Erhai Lake National Reserve (Yunnan, China), but not until the pines are 6–10 years old or older. Squirrel abundance in this reserve is positively correlated with the amount of shrub growth, and these squirrels are less common in unprotected forests of the same ages outside the reserve, which have fewer shrubs. Human usage in unprotected forests causes a reduction in the diversity of the undergrowth, which delays the colonization and reduces the population levels of *D. pernyi*. These squirrels are considered important for the pines' seed dispersal.

GENERAL REFERENCES: G. M. Allen 1940; Corbet and Hill 1992; He and Lin 2006; Men et al. 2006.

Dremomys pyrrhomerus (Thomas, 1895) Red-Hipped Squirrel

DESCRIPTION: *D. pyrrhomerus* has a conspicuous red patch on each thigh, and the whole underside of the tail is vivid red.

SIZE: Female—HB 194.5 mm (n = 13); T 148.7 mm (n = 8); Mass 245.0 g (n = 1).

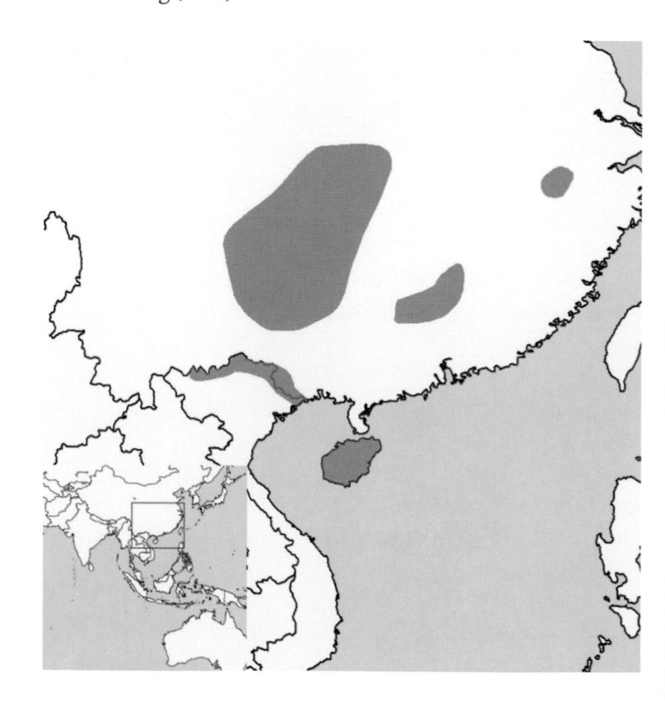

Dremomys pyrrhomerus. Photo courtesy Wang Tian Ye.

Male—HB 207.5 mm (*n* = 15); T 146.1 mm (*n* = 11).
Sex not stated—HB 198.8 mm (*n* = 11); T 151.5 mm (*n* = 11).

DISTRIBUTION: This species is found in central and southern China, extreme northern Vietnam, and Hainan Island (China).

GEOGRAPHIC VARIATION: Two subspecies are recognized.

D. p. pyrrhomerus—Sichuan, Guizhou, Hunan, Hubei, Anhui, Guangxi, Guangdong, and Fujian (south and central China), and extreme northern Vietnam. This form has conspicuous reddish patches on the thighs, but no red on the head.

D. p. riudonensis—Hainan Island (China). This subspecies also has conspicuous reddish hip patches, but its head is completely red.

CONSERVATION: IUCN status—least concern. Population trend—no information.

HABITAT: This squirrel acts like a rock squirrel, living in holes in rocky habitats and being almost completely terrestrial.

NATURAL HISTORY: *D. pyrrhomerus* is reported to be inactive aboveground during the winter months.

GENERAL REFERENCES: Moore and Tate 1965; A. T. Smith and Xie 2008.

Dremomys rufigenis (Blanford, 1878) Asian Red-Cheeked Squirrel

DESCRIPTION: This species is distinguished by its reddish cheeks, a rich red coloration in the anal area and on the underside of the tail, the lack of red thigh patches, and the lack of a reddish throat. Its dorsal pelage is dull olive gray agouti, and the ventral pelage is grayish white.

SIZE: Female—HB 187.1 mm (*n* = 37); T 149.4 mm (*n* = 36); Mass 198.3 g (*n* = 9).

Male—HB 191.3 mm (*n* = 42); T 144.9 mm (*n* = 31); Mass 190.4 g (*n* = 9).

Sex not stated—HB 199.6 mm (*n* = 4); T 159.3 mm (*n* = 3); Mass 240.0 g (*n* = 2).

DISTRIBUTION: This species is found in northeastern India; Yunnan, Guangxi, Hunan, and Anhui (China); northern and central Myanmar, Laos, Vietnam, and south through Thailand into peninsular Malaysia.

GEOGRAPHIC VARIATION: We tentatively recognize five subspecies.

D. r. rufigenis—southwestern Yunnan (China), central Myanmar, Thailand, Laos, and Vietnam. This form is dull olive gray agouti above, with buff or whitish ear patches.

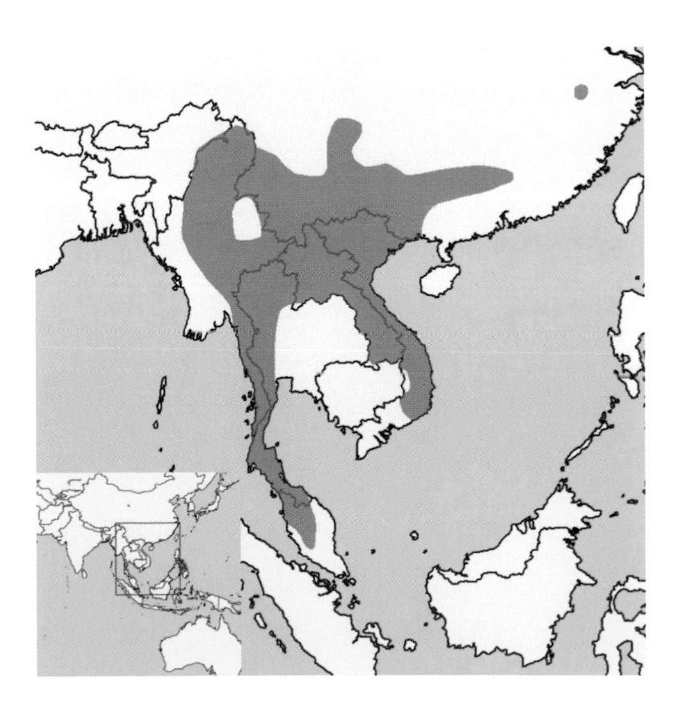

***Dremomys rufigenis*.** Photo courtesy Pitchaya and Rattapon Kaichid.

D. r. adamsoni—northeastern India, northern Myanmar, and western Yunnan. Compared with *D. r. rufigenis,* this form has a darker and more reddish dorsal pelage, and the patch behind the ear is larger and whiter.

D. r. belfieldi—peninsular Malaysia. Compared with *D. r. rufigenis,* this form has a paler and more grizzled dorsal pelage, redder cheeks, and a white (not buff) patch behind the ear.

D. r. fuscus—southern Vietnam. The dorsal pelage is very dark brown grizzled with buff, with chestnut-colored cheeks and deep chestnut on the thighs. The ventral pelage is creamy white, with deep chestnut on the underside of the tail.

D. r. ornatus—Yunnan, southwestern Guangxi, Hunan, and Anhui (China). This subspecies is larger than *D. r. rufigenis,* and it is more brightly colored than *D. r. belfieldi.*

CONSERVATION: IUCN status—least concern. Population trend—stable.

HABITAT: This is a species of the foothills, generally at elevations below 1500 m. In Malaysia, it occurs on the ridges and mountains between 800 and 1800 m, and as low as 400 m in tropical submontane forests.

NATURAL HISTORY: This species is partially terrestrial.

GENERAL REFERENCES: Endo et al. 2003; Moore and Tate 1965; A. T. Smith and Xie 2008.

Exilisciurus Moore, 1958

This genus of three species of pygmy squirrels occurs in the southern Philippine Islands, the major islands of Borneo (divided among Malaysia, Brunei Darussalam, and Indonesia) and Sumatra and Java (Indonesia), and some minor islands of the Sunda Shelf. *Exilisciurus* lacks facial stripes, in contrast to *Nannosciurus,* which has prominent facial stripes.

Exilisciurus concinnus (Thomas, 1888)
Philippine Pygmy Squirrel

DESCRIPTION: The Philippine pygmy squirrel is the largest of all species of pygmy squirrels. The dorsal pelage is dark brown with a slight reddish tint, and the ventral pelage is paler brown.

SIZE: Female—HB 87.0 mm (n = 2); T 68.0 mm (n = 2).
Male—HB 87.4 mm (n = 2); T 62.0 mm (n = 1).
Sex not stated—HB 85.9 mm (n = 8); T 65.5 mm (n = 6); Mass 28.1 g (n = 13).

DISTRIBUTION: This species is found on several of the Philippine Islands: Samar, Biliran, Leyte, Homonhon, Bohol, Dinagat, Siargo, Bucas Grande, Siquijor, Camiguin, Mindanao, and Basilan.

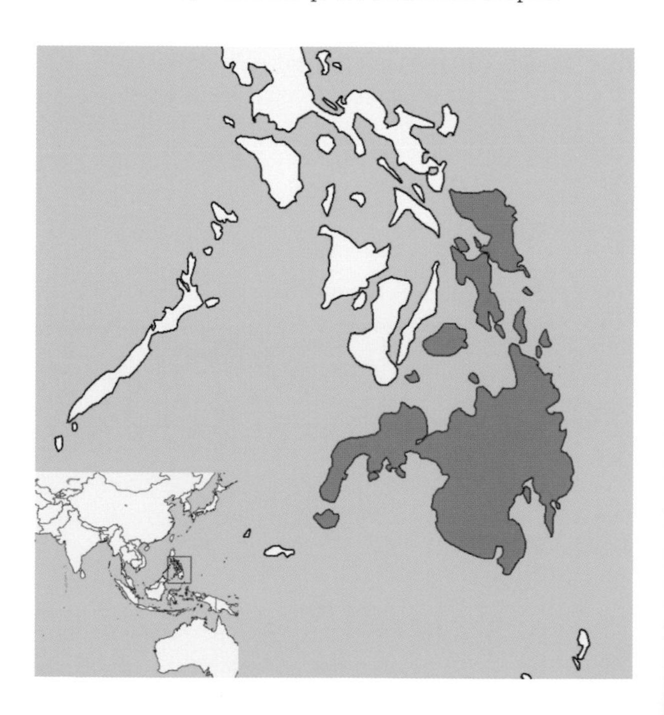

Exilisciurus concinnus. Photo courtesy Danilo S. Balete.

GEOGRAPHIC VARIATION: None.

CONSERVATION: IUCN status—least concern. Population trend—stable.

HABITAT: This species inhabits dense lowland and montane primary and secondary forests at middle elevations (from at least 700 to 6400 m). It also survives well in secondary habitats, such as small clearings where trees have been felled.

NATURAL HISTORY: Philippine pygmy squirrels are active from the early morning throughout the daylight hours. They have been observed in the low to midlevels of forests: on the trunks of large forest trees and the undersides of tree branches, in tree holes, and around stumps in forests with some human logging. They repeatedly produce brief high-pitched "chirping" vocalizations. When startled, they often run headfirst down tree trunks and hide among the root buttresses.

GENERAL REFERENCES: Heaney 1985; Rabor 1986, Rickart et al. 1993.

Exilisciurus exilis (Müller, 1838)
Least Pygmy Squirrel

DESCRIPTION: This is the smallest species of *Exilisciurus*. The dorsal pelage is an olive brown with rusty highlights, and the ventral pelage is pinkish buff. The eyes are encircled by narrow bands of darkly pigmented skin. The tail is thick but less bushy than in the other *Exilisciurus* species, and the ventral surface is redder than the dorsal surface.

SIZE: Female—HB 73.0 mm (n = 10); T 51.3 mm (n = 10); Mass 16.5 g (n = 5).

Male—HB 71.7 mm (n = 33); T 40.0 mm (n = 22); Mass 17.0 g (n = 3).

Sex not stated—HB 73.0 mm (n = 10); T 51.3 mm (n = 10); Mass 16.5 g (n = 5).

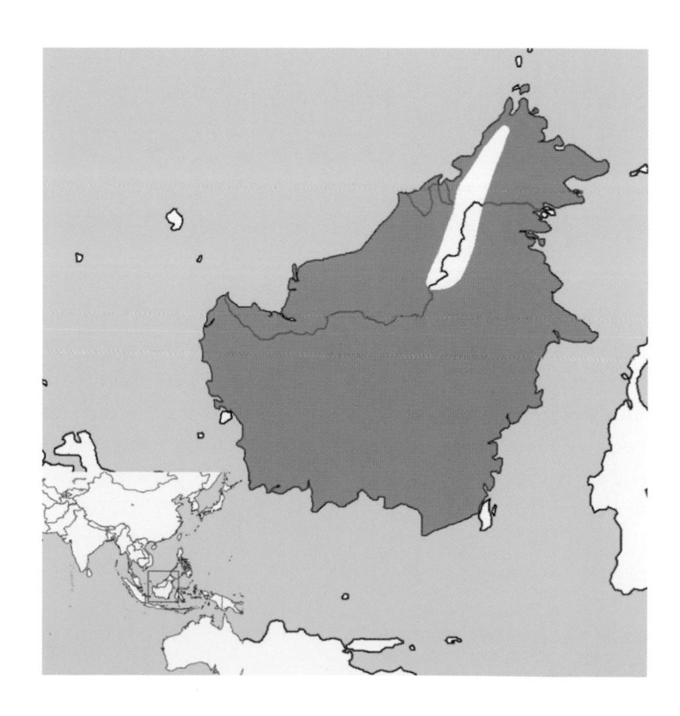

Exilisciurus exilis. Photo courtesy Pitchaya and Rattapon Kaichid.

DISTRIBUTION: This species is found on Banggi Island (Malaysia, slightly north of the island of Borneo) and on the island of Borneo (divided among Malaysia, Brunei Darussalam, and Indonesia).

GEOGRAPHIC VARIATION: Two subspecies are provisionally recognized.

E. e. exilis—island of Borneo (divided among Malaysia, Brunei Darussalam, and Indonesia). See description above.

E. e. retectus—Banggi Island (Malaysia). This form is less richly colored dorsally and more cream colored ventrally.

CONSERVATION: IUCN status—data deficient. Population trend—no information.

HABITAT: This species is found in dipterocarp forests, mostly at low elevations but as high as 1700 m in Sarawak. It is also known to survive well in old logged forests.

NATURAL HISTORY: These squirrels are frequently seen on the trunks of midstory trees, sometimes 12 m or more above the ground. They forage on tree trunks, and less frequently on vines or the buttresses of larger trees, where they feed on vegetable matter and small insects (small ants were found in the stomachs of three individuals). The nest was reported to be small and globular, 1 m above the ground, among the roots of a fallen tree. Females have a postpartum estrus and can be both pregnant and lactating at the same time.

GENERAL REFERENCES: D. D. Davis 1962; Heaney 1985; J. B. Payne 1980.

Exilisciurus whiteheadi (Thomas, 1887) Tufted Pygmy Squirrel

DESCRIPTION: This species is intermediate in size between the other two species of *Exilisciurus*. The dorsal pelage is dark brown, and the ventral pelage is paler. The ears are tufted with long white hairs.

SIZE: Female—HB 84.9 mm (n = 11); T 65.3 mm (n = 10); Mass 24.0 g (n = 2).

Male—HB 86.5 mm (n = 11); T 63.2 mm (n = 11); Mass 21.1 g (n = 4).

Sex not stated—HB 88.0 mm (n = 2); T 69.0 mm (n = 2); Mass 22.1 g (n = 6).

DISTRIBUTION: This species is found on the island of Borneo, distributed at elevations between 900 and 3000 m through the mountains of Sabah and Sarawak (Malaysia) and Brunei Darussalam, and in adjacent parts of West Kalimantan (Indonesia).

GEOGRAPHIC VARIATION: None.

CONSERVATION: IUCN status—least concern. Population trend—stable.

HABITAT: This squirrel occurs in dipterocarp and lower montane forests.

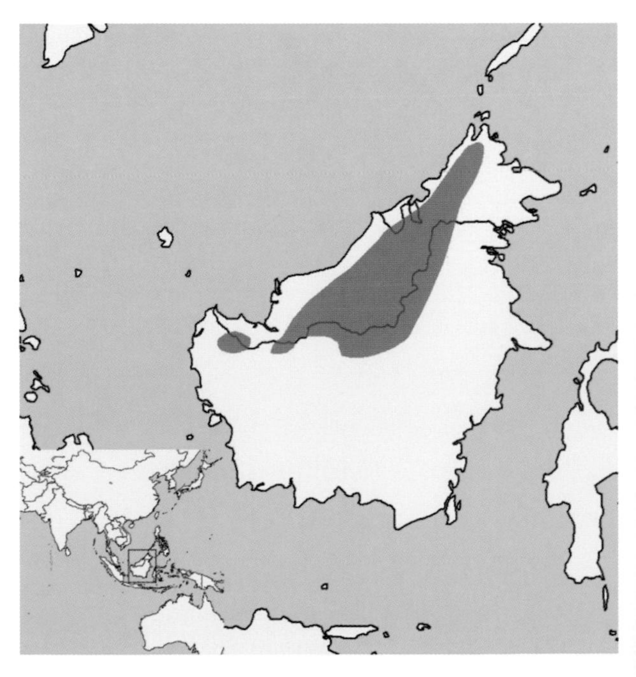

Exilisciurus whiteheadi. Photo courtesy John C. Mittermeier.

NATURAL HISTORY: *E. whiteheadi* is a bark gleaner, feeding on moss and lichens, and it is often seen on tree trunks (from the buttress to the canopy).

GENERAL REFERENCES: Conaway 1968; D. D. Davis 1958, 1962; Heaney 1985; Moore 1958a.

Funambulus Lesson, 1835

The genus *Funambulus* includes five species of striped squirrels that occur in India and Sri Lanka. Some are conspicuous and familiar inhabitants of urban and suburban regions. All of these species are omnivorous, feeding mostly on vegetable matter, but willing to feed on various kinds of animal material when they have the opportunity.

Funambulus layardi (Blyth, 1849)
Layard's Palm Squirrel

DESCRIPTION: *F. layardi* is the most colorful member of the genus. It has three longitudinal stripes on its back, with the mid-dorsal stripe, and sometimes the lateral stripes, brightly colored orange yellow. Ventrally it is a richly colored chestnut, yellow orange, or russet; and the tail has a red midventral line.

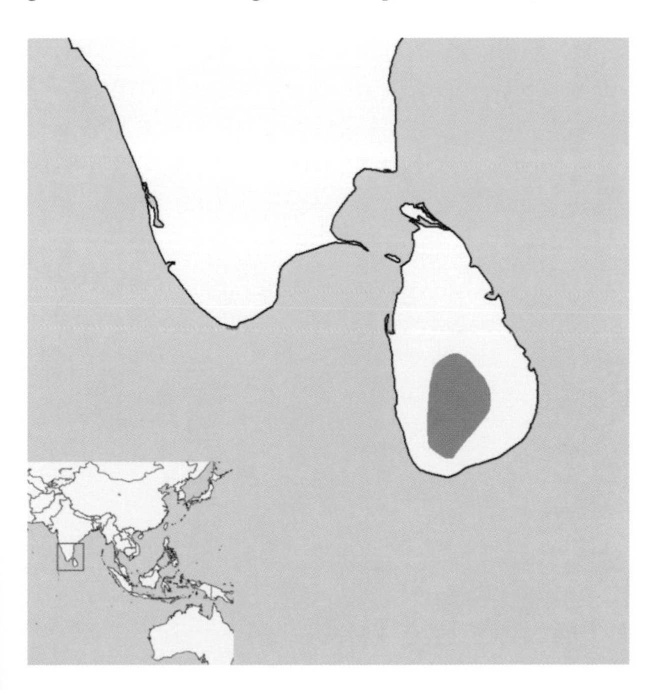

SIZE: Female—HB 154.0 mm (n = 12); Mass 168.0 g (n = 4).
Male—HB 144.0 mm (n = 2); T 158.0 mm (n = 2).
Sex not stated—HB 146.1 mm (n = 1); T 177.8 mm (n = 1).

DISTRIBUTION: This species is found in south and central Sri Lanka.

GEOGRAPHIC VARIATION: None.

CONSERVATION: IUCN status—vulnerable. Population trend—decreasing.

HABITAT: Layard's palm squirrel prefers evergreen rainforests.

NATURAL HISTORY: No information is available.

GENERAL REFERENCES: de A. Goonatilake et al. 2008.

Funambulus layardi. Photo courtesy Rajith Dissanayake.

Funambulus palmarum (Linnaeus, 1766) Indian Palm Squirrel

DESCRIPTION: This is the common three-striped squirrel of southern India. The mid-dorsal stripe is pale buffy, as is the ventral pelage. There is a red midventral line on the tail.

SIZE: Female—HB 146.6 mm (n = 79); T 157.7 mm (n = 24); Mass 99.2 g (n = 2).

Male—HB 149.3 mm (n = 81); T 147.7 mm (n = 30); Mass 117.5 g (n = 57).

Sex not stated—HB 146.1 mm (n = 1).

DISTRIBUTION: This species is found in India and in Sri Lanka. It has a limited area of geographic overlap with *F. pennantii* in the Central Provinces and along the west coast of India.

GEOGRAPHIC VARIATION: Three subspecies are recognized.

F. p. palmarum—western and southern India, to approximately 16° N latitude. This form is gray to brown on the dorsum, being more gray in the north and darker in the south.

F. p. brodiei—Sri Lanka. This is brightest-colored form, distinguished by lateral stripes that are more orange; it is darker in the Sri Lankan highlands.

F. p. robertsoni—eastern India, between 20° N latitude and 24° N latitude. This form is smaller than the other subspecies; it also differs from the more western animals, which are grayer on the head. This subspecies lacks any brown coloration.

CONSERVATION: IUCN status—least concern. Population trend—increasing.

HABITAT: This is more of a hill forest species than *F. pennantii*, and it inhabits deciduous rainforests, ascending to

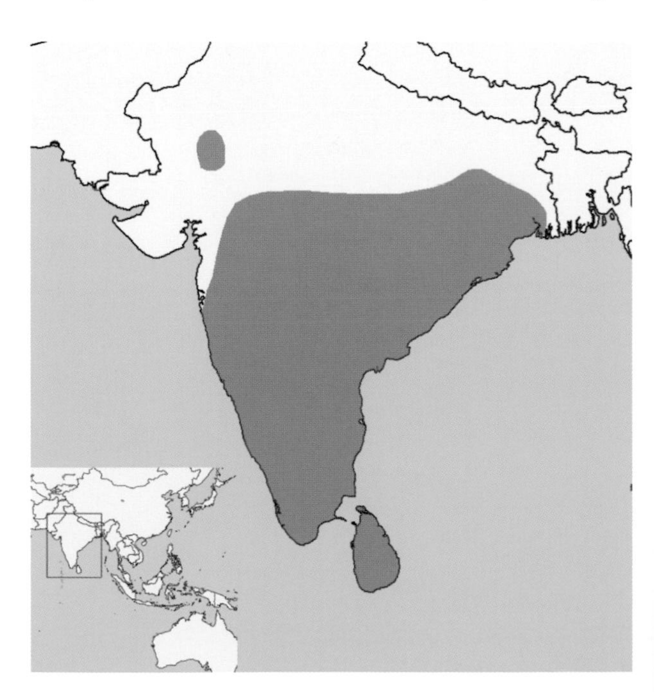

Funambulus palmarum. Photo courtesy Sudhir Shivaram.

elevations of about 1219-1372 m, where it will approach human environments, taking the place of *F. pennantii*. On the outskirts of the Western Ghats (India), *F. palmarum* occurs on the hills (at 1250 m), but it is replaced by *F. pennantii* on the Deccan plain (at 549 m). It occurs in deciduous rainforests throughout almost all of the Western Ghats, but it does not enter the wet evergreen forests. *F. palmarum* is more arboreal than *F. pennantii*, but it does descend to the ground.

NATURAL HISTORY: The Indian palm squirrel feeds on fruits, nuts, buds, young shoots, bark, nectar, and insects. At Point Calimere Wildlife Sanctuary (Tamil Nadu, India) it utilizes 50 species of plants. It feeds on the nectar of some plant species, including silk cotton trees (*Ceiba pentandra*), and probably helps with their pollination. Bark searching, probably for insects, is a frequent habit. This species builds a nest like that of a passerine bird—somewhat globular and placed on branches. The Indian palm squirrel's voice resembles that of *F. pennantii*, but it is more birdlike and slightly deeper in pitch. This species is an infrequent prey of the Indian Eagle Owl (*Bubo bengalensis*).

GENERAL REFERENCES: Nameer and Molur 2008a.

Funambulus pennantii Wroughton, 1905
Northern Palm Squirrel

DESCRIPTION: *F. pennantii* is the northern species of striped squirrel, with five pale longitudinal stripes on the darker dorsal surface. The mid-dorsal stripe reaches onto the tail, and the lateral stripes extend from the ears to the base of the tail. There are two lateral stripes on the head: a faint one extending from the ear to the eye, and a more distinctive one stretching from below the ear, passing below the eye, and onto the rostrum. It lacks the red midventral stripe on the tail seen in the other species.

SIZE: Female—HB 155.0 mm (n = 11); T 134.6 (n = 10); Mass 102.9 g (n = 41).

Male—HB 134.4 mm (n = 9); T 130.0 mm (n = 8); Mass 95.2 (n = 66).

DISTRIBUTION: This species occurs in southeastern Iran, east through Pakistan to Nepal, Bangladesh, northern and central India, and perhaps adjacent Afghanistan. It was introduced to the Andaman Islands (India), Perth (Western Australia), Israel, and perhaps to the area where it occurs in Assam (India).

GEOGRAPHIC VARIATION: Two subspecies are recognized.

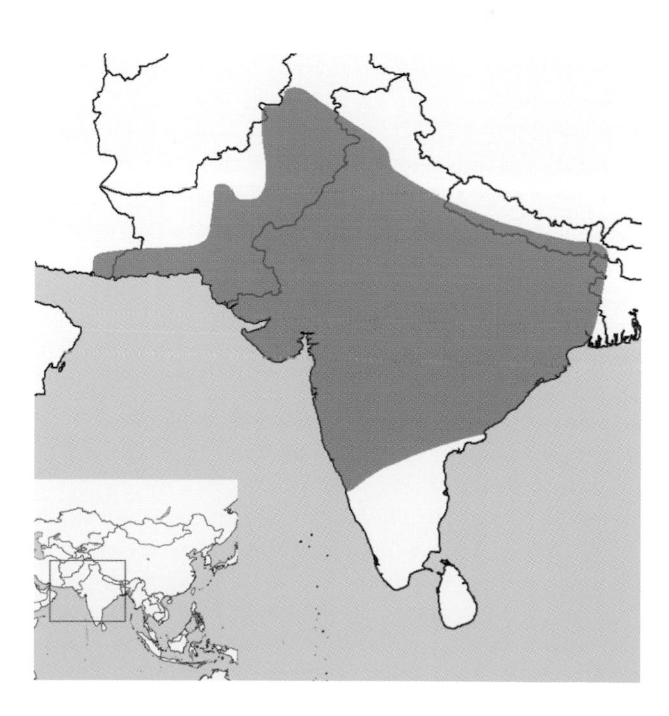

Funambulus pennantii. Photo courtesy Sudhir Shivaram.

F. p. pennantii—Nepal and India (except for northernmost India), and into Bangladesh. This form is darker.

F. p. argentescens—Pakistan and southeastern Iran. This form is paler on the dorsum, with a whiter tail; paler on the limbs, particularly the feet; and white on the venter.

CONSERVATION: IUCN status—least concern. Population trend—no information.

HABITAT: This species inhabits open plains and scrub country and is frequently found in the vicinity of villages.

NATURAL HISTORY: The northern palm squirrel builds globular nests around buildings, under tiles and eaves, in hollow trees, on tree branches, and among palm leaves. Adults are omnivorous, feeding on birds when they can catch them, honey from unprotected beehives, and insects (such as large quantities of termites and caterpillars) when they are available. Their home range in Jodhpur is reported to be 0.21 ha for males and 0.15 ha for females. Northern palm squirrels breed throughout the year, with peaks in March-April and July-August. Courtship involves mating chases, mating calls, and grooming. It lasts for a day, and the female may mate four to five times with one or more males. Litter size varies from two to four hairless young that have a 4-5 cm head and body length and a 1.5-2 cm tail length. Their eyes open at 10-15 days of age, and nursing continues until they are 25-30 days old.

GENERAL REFERENCES: Bertolino 2009; Nameer and Molur 2008b.

Funambulus sublineatus (Waterhouse, 1838) Dusky Palm Squirrel

DESCRIPTION: This is the smallest species of *Funambulus*. The three dorsal stripes tend to be obscured by the long fur. The venter is dull and drab, and this species lacks a red mid-ventral line on the tail.

SIZE: Female—HB 146.6 mm (n = 79); T 157.7 (n = 24); Mass 99.2 g (n = 2).

Male—HB 149.3 mm (n = 81); T 147.7 mm (n = 30); Mass 117.5 g (n = 57).

Sex not stated—HB 146.1 mm (n = 1).

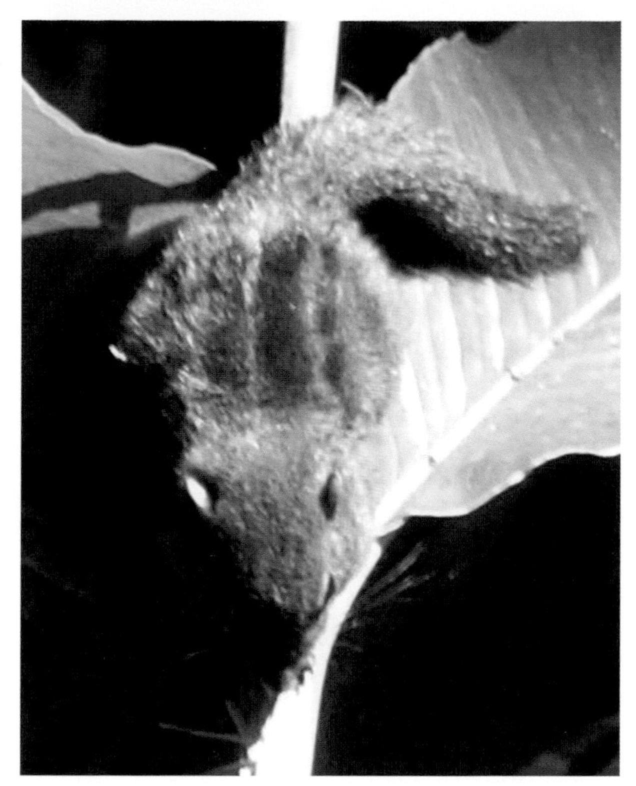

Funambulus sublineatus. Photo courtesy Rajith Dissanayake.

DISTRIBUTION: This species occurs in southwestern India and central Sri Lanka.

GEOGRAPHIC VARIATION: Two subspecies are recognized.

F. s. sublineatus—India. This form has narrower (3-5 mm) dark stripes between the pale midventral and lateral dorsal stripes.
F. s. obscurus—Sri Lanka. This subspecies has broader (6-8 mm) chestnut brown stripes.

CONSERVATION: IUCN status—vulnerable. Population trend—decreasing.

HABITAT: In southern India, near the Palni Mountains, this species occurs above the altitudinal range of *F. palmarum*, at elevations of 1372-2134 m. In Sri Lanka it is found in the jungles of the hills, from nearly the top of the mountains down to 610-762 m; in the southwestern part of the island it occurs in the jungles of the wet hilly country, almost down to the sea.

NATURAL HISTORY: In India, the dusky palm squirrel is an inhabitant of the scrub level of the forest, up to 2134 m in elevation, where it is preyed upon by the green viper (*Trimeresurus gramineus*).

GENERAL REFERENCES: Phillips 1980; Prater 1965; Rajamani et al. 2008a.

Funambulus tristriatus (Waterhouse, 1837) Jungle Palm Squirrel

DESCRIPTION: The jungle palm squirrel is the largest species of *Funambulus*. It is dark, with three pale buffy dorsal stripes, a pale venter, and a red midventral line on the tail.

SIZE: Female—HB 158.6 mm (n = 7); T 139.0 mm (n = 10).
Male—HB 159.8 mm (n = 10); T 143.4 mm (n = 10).
Sex not stated—HB 190.5 mm (n = 1); Mass 139.0 g (n = 1).

DISTRIBUTION: This species is found on the west coast of India, from below 20° N latitude to the southern tip.

GEOGRAPHIC VARIATION: Two subspecies are recognized.

F. t. tristriatus—south of 12° N latitude. This form has narrower and duller stripes.
F. t. numarius—north of 12° N latitude. This form has broader and brighter stripes.

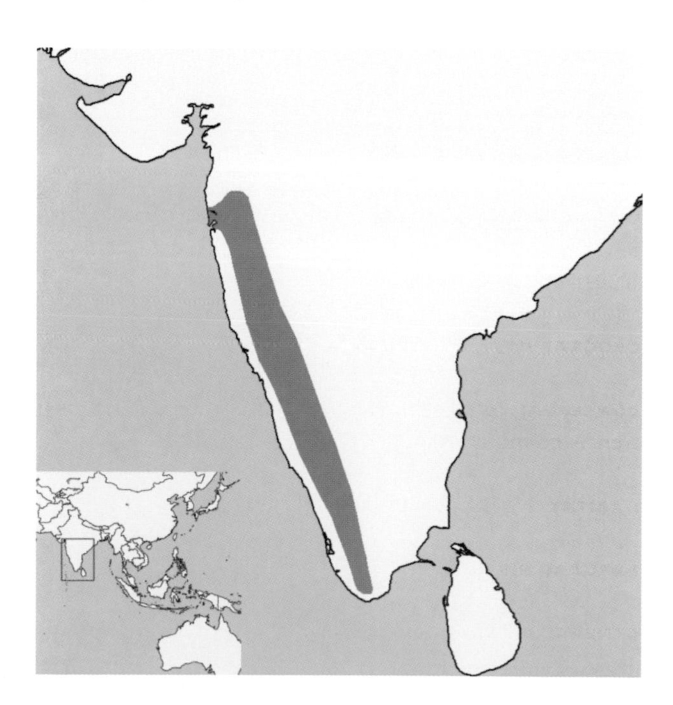

Funambulus tristriatus. Photo courtesy Rajith Dissanayake.

CONSERVATION: IUCN status—least concern. Population trend—decreasing.

HABITAT: *F. tristriatus* inhabits the evergreen rainforests of the Western Ghats (peninsular India). It is a resident of forests and cultivated fields in southwestern India.

NATURAL HISTORY: It feeds on crops, such as the male flowers of the coconut palm (*Cocos nucifera*) and grains of paddy (cultivated rice), but it also eats termites, caterpillars, beetles, scale insects, and other pests. Its nests are globular, averaging 22 cm in diameter; are placed between 2 and 29 m high in trees, although most were below 10 m and the median height was 5 m.

GENERAL REFERENCES: Molur and Nameer 2008.

Glyphotes Thomas, 1898

Glyphotes has one species, known only from the island of Borneo. It is sometimes treated as a species of the genus *Callosciurus*.

Glyphotes simus Thomas, 1898
Sculptor Squirrel

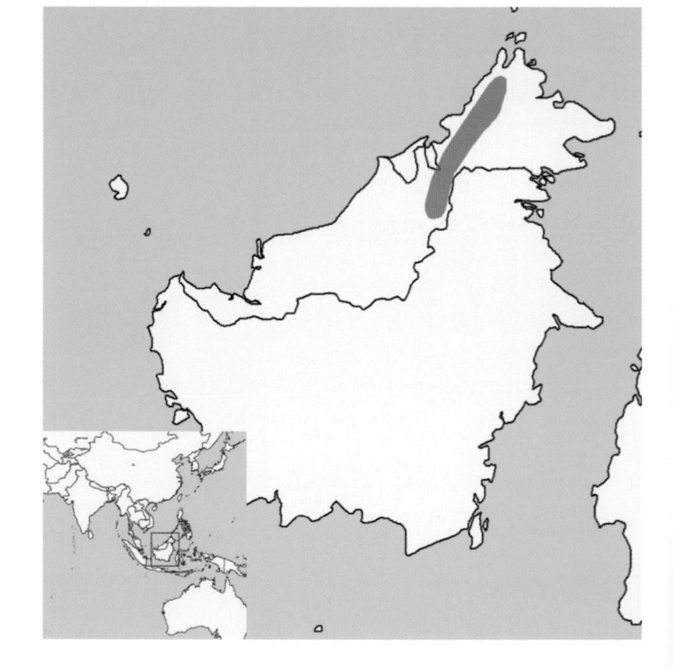

DESCRIPTION: This squirrel has a deep short blunt muzzle, a brownish finely speckled dorsal pelage, and a dull orange buff venter. Its lower incisors distinctively diverge.

SIZE: Sex not stated—HB 94-144 mm; T 95-106 mm.

DISTRIBUTION: It is found on the island of Borneo, in the mountains of Sabah and Sarawak (Malaysia), at elevations of 1000-1700 m.

GEOGRAPHIC VARIATION: None.

CONSERVATION: IUCN status—least concern. Population trend—no information.

HABITAT: No information is available.

NATURAL HISTORY: No information.

GENERAL REFERENCES: Hill 1959.

Hyosciurus Archbold and Tate, 1935

This genus contains two species of squirrels that are endemic to the island of Sulawesi (Indonesia). *Hyosciurus* are readily distinguished from other Sulawesi squirrels by their elongated muzzles and claws and their terrestrial lifestyle. Not much is known about these squirrels, and the information below is accurate but by no means comprehensive. The most current information available comes from G. Musser's recent publication on the Sulawesi squirrels and their associated sucking lice. More fieldwork on the natural history of the squirrels in this region of the world is needed.

Hyosciurus heinrichi Archbold and Tate, 1935
Montane Long-Nosed Squirrel

DESCRIPTION: The upperparts are dark brown flecked with buff and black. The underparts are dark brownish gray with a highly variable patterning of white fur midventrally. The white pelage can range from a swath of irregularly outlined white—running from the neck to the base of the tail—to an interrupted white stripe that is divided by the background brown pelage into patches of white on the chest and inguinal regions. The hair is reversed on the nape, with a median part of 30-40 mm. Each eye is encircled by a hairless dark buffy-colored ring. The tail color is similar to that of the upperparts. *H. heinrichi* can be distinguished from *H. ileile* by *H. heinrichi*'s white fur on the ventral surface; its darker dorsal color, larger ears, longer muzzle, and shorter body and tail; and its lighter weight; *H. heinrichi* tends to be more gracile and less stocky than *H. ileile*. The fur reversal on the nape is also longer in *H. heinrichi*.

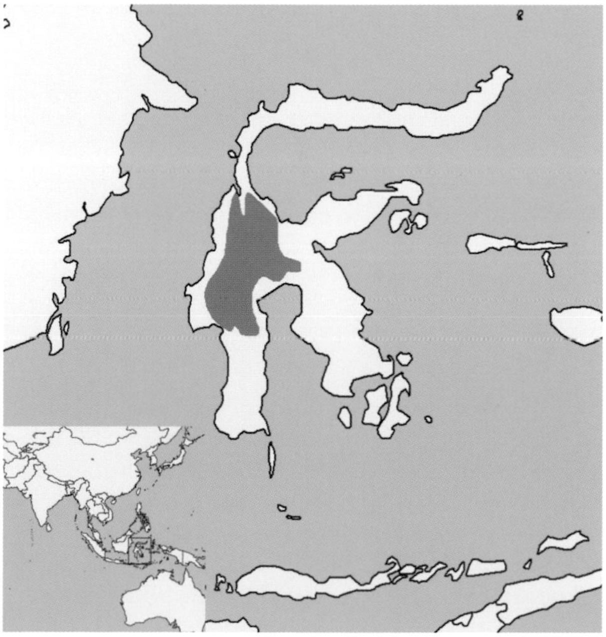

SIZE: Sex not stated—HB 195-240 mm; T 65-120 mm; Mass 228-370 g.

DISTRIBUTION: Its actual distribution is not fully known, but this squirrel is found in the mountains of the western central island of Sulawesi (Indonesia), at elevations of 1479 to 2287 m.

GEOGRAPHIC VARIATION: None.

CONSERVATION: IUCN status—least concern. Population trend—no information.

HABITAT: This squirrel lives in the primary forests of lower to upper montane tropical rainforests.

NATURAL HISTORY: *H. heinrichi* is a diurnal ground squirrel. It is secretive and not very vocal. Its described vocalizations are a series of single "chirps," sounding much like a bird and distinctively different from other Sulawesi squirrel vocalizations. As the degree of alarm increases, so does the frequency of the "chirps." When the animal is greatly alarmed or agitated, the "chirps" may carry tones of a "chuck" that may even change to a "chatter." This squirrel exposes its arthropod prey by rooting around in the leaf litter with its long muzzle and elongated claws. Its stomach contents indicate a diet of arthropods and fruits, mainly acorns. A litter consists of one or two young. Females have three pairs of teats: two inguinal pairs and one abdominal pair.

GENERAL REFERENCES: Musser et al. 2010.

Hyosciurus ileile Tate and Archbold, 1936
Lowland Long-Nosed Squirrel

DESCRIPTION: The upperparts are dark brown flecked with buff and black. The underparts are dark brownish gray with a highly variable pattern of cream fur midventrally. The cream-colored pelage can range from a swath of irregularly outlined white, running from the neck to the base of the tail, to patches of cream stripes separated by the background brown pelage on the chest and inguinal regions. The hair is reversed on the nape, with a median part of 10 mm. The tail color is similar to that of the upperparts. *H. ileile* can be distinguished from *H. heinrichi* by *H. ileile*'s stouter body; shorter muzzle, ears, and claws; and cream instead of white underparts. The upperparts of *H. ileile* tend to be brighter with more brownish tints, and the fur reversal part on the nape is also shorter in *H. ileile*.

SIZE: Sex not stated—HB 213–250 mm; T 70–125 mm; Mass 293–520 g.

DISTRIBUTION: The actual distribution of this squirrel is unknown, but collection sites show it occurring in the mountains in the northwestern area of the northern peninsula of Sulawesi (Indonesia), as well as in the northern area of the central core island.

GEOGRAPHIC VARIATION: None.

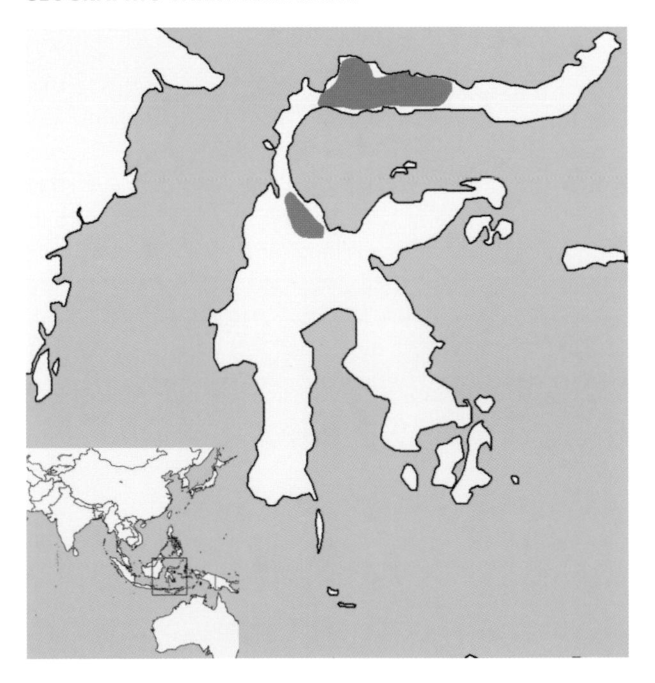

Hyosciurus ileile. Photo courtesy Guy G. Musser.

CONSERVATION: IUCN status—vulnerable. Population trend—decreasing.

HABITAT: This squirrel is documented as being found in lowland evergreen to lower montane tropical rainforests, at elevations between 168 and 1700 m.

NATURAL HISTORY: Diurnal and terrestrial, this ground squirrel is found in less dense distributions than *H. heinrichi*. Musser hypothesizes that the disparity in the two species' relative densities may be attributed to the lower density of oak (*Quercus*) trees—the source of the acorn food item—in habitats primarily occupied by *H. ileile*. More work needs to be done to tease out the relationships between the distributions of *Hyosciurus* and its resources. The vocalizations of the lowland long-nosed squirrel differ greatly from those of *H. heinrichi*. When disturbed, *H. ileile* emits growls from deep in its throat, making an "Errr, errr, grrr" in short bursts. The other documented vocalizations are emitted in the mornings and afternoons: a high-pitched "eee-eee" squeal lasting 3–5 seconds, or a single drawn-out "eee———e" squeal that starts high and drops to silence, or a quick double "eee, eee" squeal. Musser compares the squeal to the whine of an unoiled machine. A food test given to a live-trapped specimen of *H. ileile* showed that it would accept and eat large beetle larvae from rotting trunks and acorns from *Lithocarpus*, but it ignored adult forms of the beetle larvae, trap bait, carabid beetles, lizards, and earthworms. Both species of acorns (*Lithocarpus celebicus* and *Lithocarpus glutinosus*) were accepted, but *L. glutinosus* appeared to be favored. Its diet is assumed to be arthropods and fruits, with acorns and figs preferred. *H. ileile* forages in the leaf litter. Females have three pairs of teats: two inguinal pairs and one abdominal pair.

GENERAL REFERENCES: Musser et al. 2010.

Lariscus Thomas and Wroughton, 1909

This genus comprises four species of striped ground squirrels from Indonesia.

Lariscus hosei (Thomas, 1892)
Four-Striped Ground Squirrel

DESCRIPTION: This species is distinguished by a brown mid-dorsal stripe and an orange brown venter.

SIZE: Female—HB 189.0 mm (n = 2); T 86.0 mm (n = 2); Mass 215.0 g (n = 1).

Sex not stated—HB 181.0 mm (n = 2); T 126.0 mm (n = 2).

DISTRIBUTION: *L. hosei* is found in the northeastern portion of the island of Borneo (Sabah and Sarawak [Malaysia] and East Kalimantan [Indonesia]), at elevations as high as 1530 m on Mount Kinabalu (Sabah, Malaysia).

GEOGRAPHIC VARIATION: None.

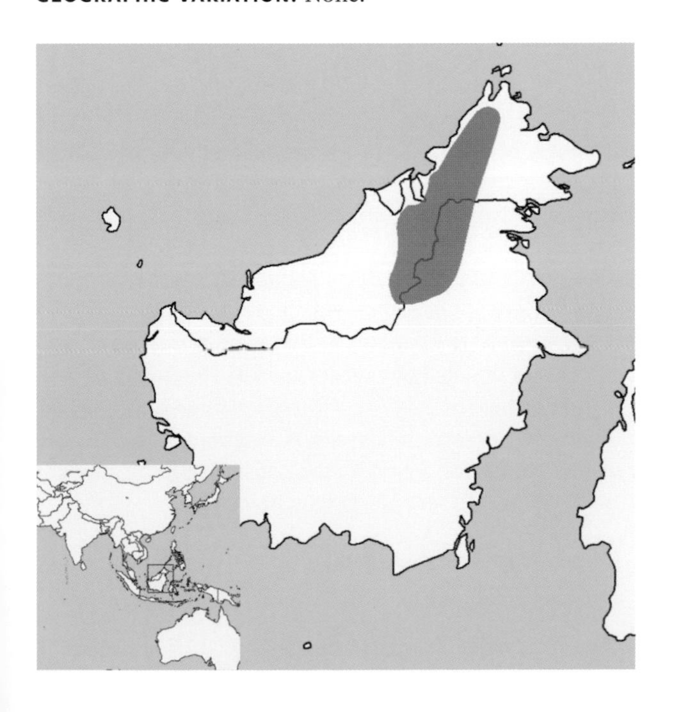

Lariscus hosei. Photo courtesy K. Wells.

CONSERVATION: IUCN status—near threatened. Population trend—decreasing.

HABITAT: *L. hosei* depends on primary forests in the island of Borneo; these are being deforested for timber at a rapid pace, which threatens its survival.

NATURAL HISTORY: This species is extremely rare, and no information has been published on its natural history.

GENERAL REFERENCES: Duckworth and Hedges 2008d.

Lariscus insignis (F. Cuvier, 1821)
Three-Striped Ground Squirrel

DESCRIPTION: The three-striped ground squirrel has a dark pelage with three dark stripes running longitudinally along its back. The venter is white or pale buff. The body fur is thin, and the tail is relatively flat.

SIZE: Female—HB 187.1 mm (n = 30); T 104.3 mm (n = 19); Mass 182.1 g (n = 12).

Male—HB 194.3 mm (n = 34); T 100.6 mm (n = 25); Mass 174.9 g (n = 11).

Sex not stated—HB 182.5 mm (n = 6); T 109.0 mm (n = 6); Mass 175.0 g (n = 2).

DISTRIBUTION: This species has been found in peninsular Thailand, Malaysia, Sumatra and Java (Indonesia), the island of Borneo (divided among Malaysia, Brunei Darussalam, and Indonesia), and on the small islands of Penang and Tioman (Malaysia), Singapore (where it is probably now extinct), and the small Indonesian islands of Bintang (Riau Islands), Tanabala (Batu Islands), and Sianan (Anambas Islands). It occurs at elevations of up to 1000 m.

GEOGRAPHIC VARIATION: Five subspecies are recognized.

L. i. insignis—Sumatra and eastern Java (Indonesia). This subspecies has a variable dorsal pelage, but it is generally dark. The ventral pelage is gray. The montane form on Sumatra is darker yet.

L. i. diversus—island of Borneo (divided among Malaysia, Brunei Darussalam, and Indonesia). This form resembles the Malayan forms, but it has a longer tail and orange brown flanks.

L. i. javanus—Buitenzorg (= Bogor) and western Java (Indonesia). This form has the dorsal and ventral pelage washed with rufous.

L. i. peninsulae—peninsular Malaysia, from Trang (Thailand) south. This form has a ventral pelage that is lighter and brighter yellow. The front feet are brown, and the hind feet are brown grizzled with yellow. The underparts are cream buff, fading to white on the chest and darkening to buff on the inner hind legs.

L. i. rostratus—Tanabala Island, Batu Island, and western Sumatra (Indonesia). This form has a darker pelage with broader black dorsal stripes, and the rostrum is more elongate.

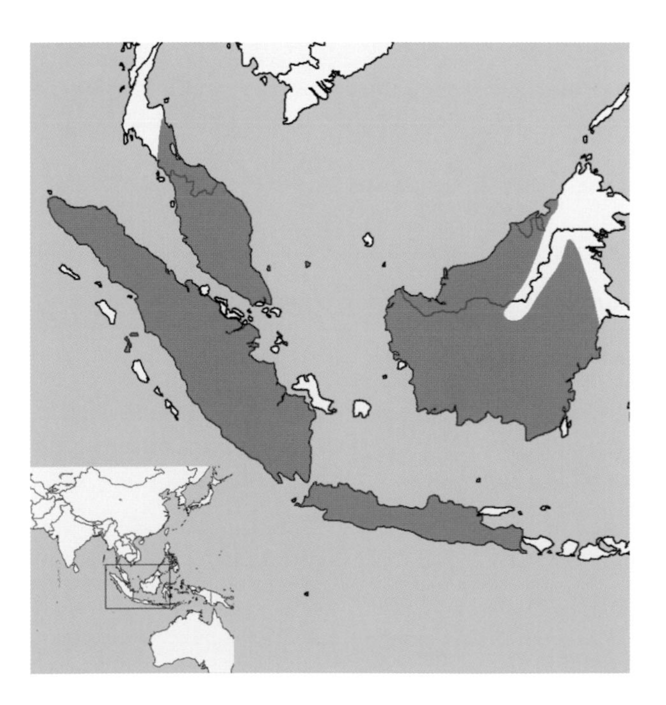

Lariscus insignis. Photo courtesy Dimar Adhi Perdana.

CONSERVATION: IUCN status—least concern. Population trend—decreasing.

HABITAT: *L. insignis* prefers lowland dipterocarp forests, but it can also survive well in deforested habitat. In a study in the Krau Wildlife Reserve (Pahang, Malaysia), three-striped ground squirrels were trapped slightly more often in areas disturbed by humans than in pristine areas.

NATURAL HISTORY: The three-striped ground squirrel is diurnal and terrestrial. It is found mostly on the ground and low on tree buttresses, fallen trees, and the like in rainforests, but it is also frequently seen in secondary forests. In addition, individuals have been observed in the limestone hills in Ranthan Ipoh in peninsular Malaysia. This species was previously believed to be relatively common, perhaps because of its ability to withstand human disturbance, but it is now believed to be less abundant. Research indicates that *L. insignis* may have a specialized diet, feeding on foods found in cool areas, which they appear to prefer. They are known to eat the buds of the *Rafflesia* flower in Gunung Gading National Park (Sarawak, Malaysia). A single adult female was recorded as having a home range of 0.1-0.2 ha in the Ulu Gombak Forest Reserve (Selangor, Malaysia).

GENERAL REFERENCES: Hedges et al. 2008; Lundahl and Olsson 2002; J. B. Payne 1980; Saiful and Nordin 2004; Saiful et al. 2001.

Lariscus niobe (Thomas, 1898)
Niobe Ground Squirrel

DESCRIPTION: The Niobe ground squirrel is the darkest of the *Lariscus* species. The dorsal pelage is so dark that the black lateral stripes are not visible. The ventral pelage is tinged dark. The tail is short, with a dark line of grizzled black and buff running along the center of the underside.

SIZE: Female—HB 194.6 mm (n = 13); T 85.6 mm (n = 13).
Male—HB 189.8 mm (n = 8); T 90.4 mm (n = 7).

DISTRIBUTION: This species is found in Indonesia, both in the mountains of western Sumatra and in the Idjen (= Ijen) Mountains of eastern Java.

GEOGRAPHIC VARIATION: Two subspecies are recognized.

L. n. niobe—Sumatra (Indonesia). See description above.
L. n. vulcanus—Java (Indonesia). The hairs on the tail are tipped with buff or tawny rather than white.

CONSERVATION: IUCN status—data deficient. Population trend—no information.

HABITAT: This species can be found in primary and secondary forests and in scrublands.

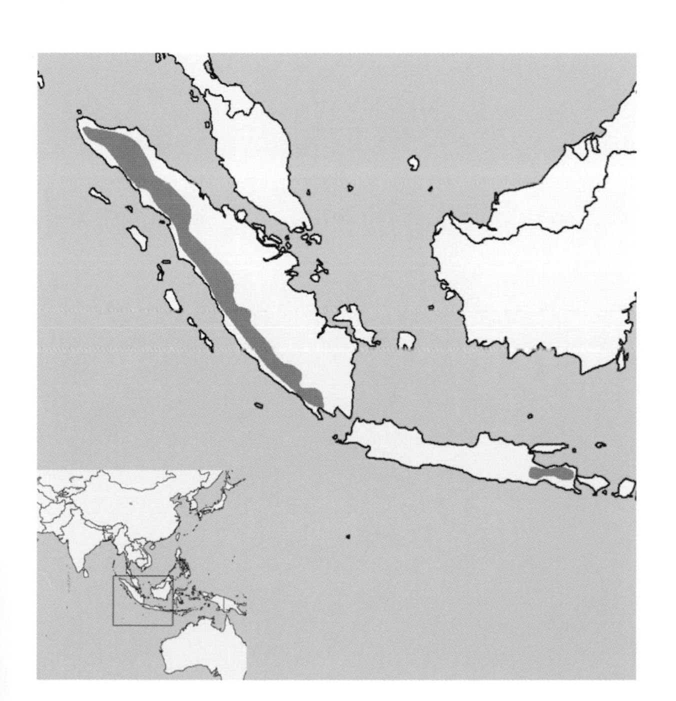

Lariscus niobe. Photo courtesy Israel Didham.

NATURAL HISTORY: No information is available.

GENERAL REFERENCES: H. C. Robinson and Kloss 1918.

Lariscus obscurus (Miller, 1903)
Mentawai Three-Striped Squirrel

DESCRIPTION: The Mentawai three-striped squirrel is distinguishable by its dark dorsum and faint longitudinal striping, as well as by its dark gray venter, which lacks any buff coloration and sometimes has a pale silvery central area. The animal also has a particularly long muzzle.

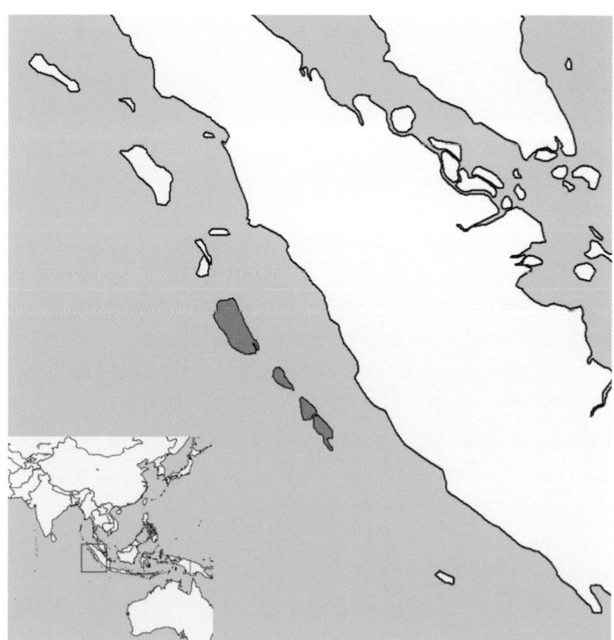

SIZE: Female—HB 202.0 mm (n = 20); T 86.4 mm (n = 17).
Male—HB 201.7 mm (n = 18); T 88.1 mm (n = 17).
Sex not stated—HB 187.0 mm (n = 19); T 86.0 mm (n = 17); Mass 242.7 g (n = 7).

DISTRIBUTION: This species is found in the Mentawai Islands, west of Sumatra: on Siberut, Sipora, and North and South Pagai islands (Indonesia).

GEOGRAPHIC VARIATION: Three subspecies are recognized. Records for Sipora Island are identified only at the species level.

L. o. obscurus—South Pagai Island (Indonesia). See description above.
L. o. auroreus—North Pagai Island (Indonesia). This form has a more reddish and less yellowish dorsal pelage.
L. o. siberu—Siberut Island (Indonesia). The ventral pelage is finely grizzled black and light ochraceous, and the muzzle and cheeks are slightly grayer. The tail is slightly blacker than the dorsal pelage.

CONSERVATION: IUCN status—near threatened. Population trend—decreasing.

HABITAT: This species probably prefers scrublands, but it tolerates human disturbance well, surviving in secondary forests and forest edges.

NATURAL HISTORY: The Mentawai three-striped squirrel is terrestrial, diurnal, and most active in the morning and midafternoon, judging by its vocalizations. Its diet consists primarily of fallen fruits. According to J. Whitten, the stomach contents of three individuals contained "79 percent" vegetable matter (mostly fruits and seeds) and "22 percent" animal matter (mostly arthropods). These squirrels nest in burrows and fallen trees.

GENERAL REFERENCES: Whitten 1981.

Menetes Thomas, 1908

This genus contains a single species, a ground squirrel of Indonesia, that is closely related to *Lariscus*.

Menetes berdmorei (Blyth, 1849)
Indochinese Ground Squirrel

DESCRIPTION: The dorsal coloration varies slightly from tawny agouti to brownish agouti, but there are always two whitish lateral lines on each side and a variable number of dark or black lines, ranging from none (in some dry forests) to three lateral lines and a mid-dorsal line (in the wettest forests). The venter is whitish or yellowish.

SIZE: Female—HB 182.7 mm; T 132.9 mm; Mass 172.0 g.
Male—HB 181.6 mm; T 138.2 mm; Mass 176.0 g.
Sex not stated—HB 180.9 mm; T 140.0 mm; Mass 190.0 g.

DISTRIBUTION: This species occurs in central Myanmar and southern Yunnan (China) south to Thailand, Laos, Cambodia, and southern Vietnam.

GEOGRAPHIC VARIATION: Seven subspecies are recognized. Records from northwestern Laos are identified only at the species level.

M. b. berdmorei—east of the Salween River (Myanmar) and south into peninsular Myanmar and Thailand. This form has lateral black stripes, but the mid-dorsal stripe is generally absent.

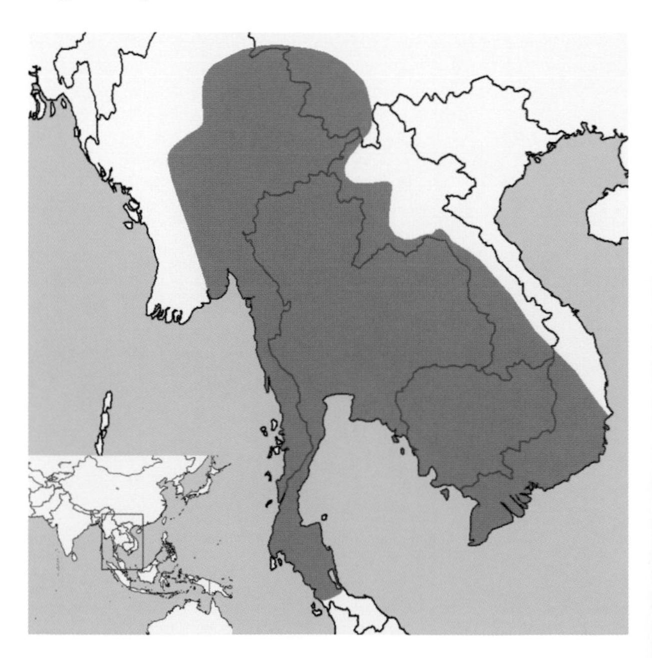

Menetes berdmorei consularis. Photo courtesy A. Coke Smith, www.cokesmithphototravel.com.

M. b. consularis—southern Yunnan (China) and western Thailand. It lacks all of the dark lines except the one lateral line between the two white stripes, and this may also be seasonally absent. The venter is yellowish white.

M. b. decoratus—between the Irrawaddy and Salween rivers (Myanmar). The general body color is olive, with the mid-dorsal black stripe and the lateral black stripes being prominent.

M. b. moerescens—southern Vietnam. The general body color is olive brown, with the mid-dorsal black stripe and the lateral dark stripes being prominent, but providing less of a contrast with the body color than in *M. b. decoratus*.

M. b. mouhotei—eastern Thailand, southern Laos, and Cambodia. The dorsum is grizzled gray brown, with one dark lateral stripe between the two whitish stripes. The venter is white.

Menetes berdmorei. Photo courtesy Phung my Trung, www.vncreatures.net.

Menetes berdmorei berdmorei. Photo courtesy Alex Vargas, www.pbase.com/alex_vargas.

M. b. peninsularis—peninsular Thailand, south of the Isthmus of Kra. It has very prominent black stripes.

M. b. pyrrocephalus—southern Vietnam. The general body color is brown tinted with black; the black midline stripe is not very prominent, and it fades out completely over the shoulders and toward the tail. A lateral dusky brown stripe is bordered above and below by yellowish white stripes.

CONSERVATION: IUCN status—least concern. Population trend—stable.

HABITAT: This is a ground squirrel of savanna and forest habitats. It is generally found in dry forests and scrublands, but it is also seen in dense rainforests.

NATURAL HISTORY: No information is available.

GENERAL REFERENCES: Cao et al. 1986; Mackenzie 1929; Murphy and Phan 2002; Van Peenen et al. 1970.

Nannosciurus Trouessart, 1880

This genus includes one species of pygmy squirrel.

Nannosciurus melanotis (Müller, 1840)
Black-Eared Squirrel

DESCRIPTION: This pygmy squirrel has a plain pale agouti pelage. However, it has a black line from the nose to the eye; a broad buffy stripe below it extending under the eye to a buffy patch behind the eye; and another thin black line below the stripe, extending to the mouth—all of which distinguish it from *Exilisciurus*. The back of the ear and a patch behind it are black.

SIZE: Female—HB 79 mm; T 70.5 mm; Mass 17 g.
Male—HB 79.6 mm; T 62.1 mm.
Sex not stated—HB 74 mm; T 65.5 mm.

DISTRIBUTION: This squirrel occurs in Sumatra and Java (Indonesia); the island of Borneo (divided among Malaysia, Brunei Darussalam, and Indonesia), except for the northern part; and adjacent small islands.

GEOGRAPHIC VARIATION: Four subspecies are recognized.

N. m. melanotis—Java (Indonesia). It has the darkest pelage of the subspecies, dorsally orange brown and ventrally pale orange brown. The nape of the neck is lighter brown than the crown.

N. m. bancanus—Bancan (= Bangka) Island (Indonesia). It has a yellow brown pelage and is most similar to *N. m. pulcher*; but *N. m. bancanus* is darker, has a more obscure nape patch, and the facial stripe is buffy.

N. m. borneanus—island of Borneo (divided among Malaysia, Brunei Darussalam, and Indonesia). It has a yellow brown pelage; a whitish facial stripe slightly tinged with buffy; and a distinct nape patch, which is not as pale and distinct as that of *N. m. pulcher*.

N. m. pulcher—Sumatra and Sinkep Island (Indonesia). It has a yellow brown pelage, a whitish facial stripe, and a distinct nape patch.

CONSERVATION: IUCN status—least concern. Population trend—decreasing.

HABITAT: This squirrel occurs in all dipterocarp forests in the lowlands, at elevations up to 1070 m.

NATURAL HISTORY: This species is a bark gleaner and is seen mostly on tree trunks and major branches.

GENERAL REFERENCES: Chasen and Kloss 1928; Heaney 1985; Hollister 1913; Taylor 1934; Thomas and Whitehead 1898.

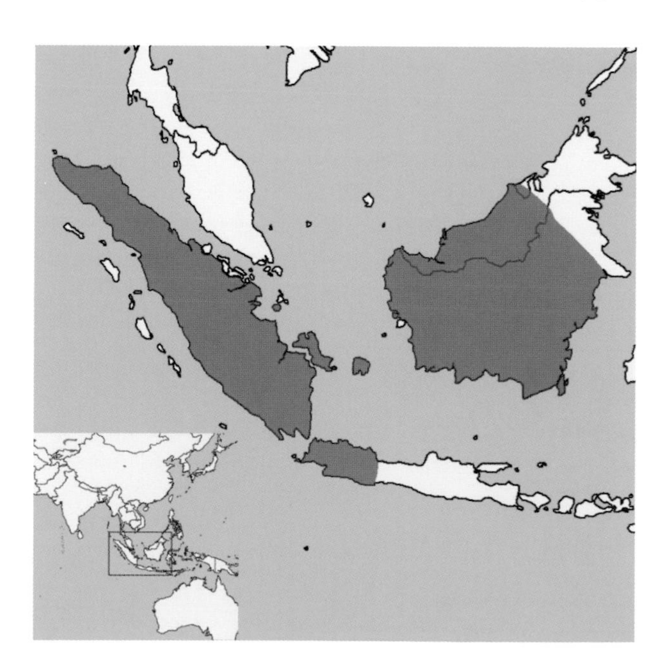

Nannosciurus melanotis. Photo courtesy David Bakewell.

Prosciurillus Ellerman, 1947

The genus *Prosciurillus* contains seven species that are endemic to the island of Sulawesi and to the smaller surrounding islands in Indonesia. J. R. Ellerman first proposed the new genus *Prosciurillus* for *Sciurus murinus,* after noticing that the Sulawesi dwarf squirrels and others in the Indomalayan region had a relatively short orbit. He compared *Prosciurillus* to the South American pygmy squirrel genus *Sciurillus,* even though the two groups are not phylogenetically close. Not much is known about *Prosciurillus.* The most current information available comes from G. Musser's recent publication on the Sulawesi squirrels and their associated sucking lice. More fieldwork on the natural history of the squirrels in this region of the world is needed.

Prosciurillus abstrusus Moore, 1958
Secretive Dwarf Squirrel

DESCRIPTION: This species is one of the smaller squirrels endemic to Sulawesi (Indonesia). The secretive dwarf squirrel's upperparts, tail, and head are dark brown flecked with pale buff. The underparts are dark grayish white; on some specimens, there is a pale buff wash on the chest region. The ears are conspicuously white on the dorsal surface. *P. abstrusus* is similar in size and coloration to *P. murinus,* but the coloration of *P. abstrusus* is duller, and the ears of *P. murinus* lack the white coloring on those of *P. abstrusus.*

SIZE: Sex not stated—HB 115-148 mm; T 72-130 mm.

DISTRIBUTION: This squirrel is known only from the type locality: on the Pegunungan (= mountain range) Mekongga in southeastern Sulawesi (Indonesia).

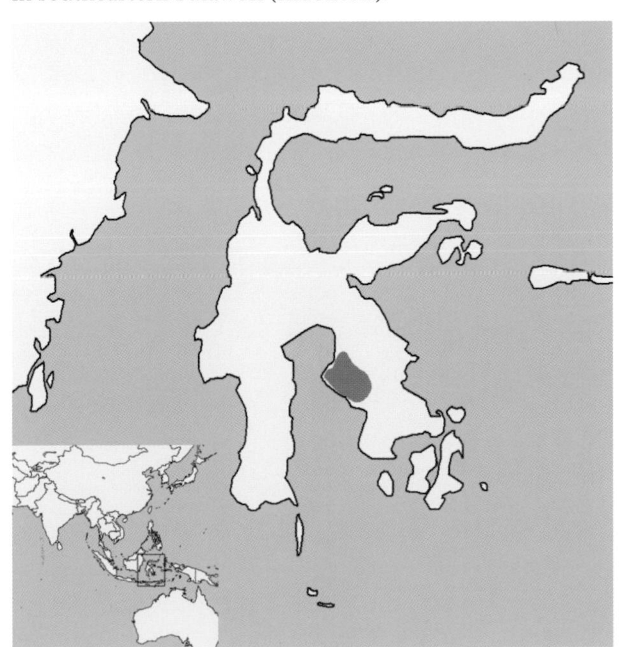

GEOGRAPHIC VARIATION: None. However, the information is insufficient.

CONSERVATION: IUCN status—data deficient. Population trend—no information.

HABITAT: *P. abstrusus* is found at elevations between 1500 and 2000 m, where the habitat is transitional from lower to upper tropical montane forests.

NATURAL HISTORY: Little is known about the secretive dwarf squirrel. Females have three pairs of teats: two inguinal pairs and one postaxillary pair.

GENERAL REFERENCES: Musser et al. 2010.

Prosciurillus alstoni (Anderson, 1879)
Alston's Squirrel

DESCRIPTION: The upperparts are dark brown with highlights of buff, orange, or black. The underparts are dark red to reddish brown. Most of these squirrels have white fur on the external parts of the ears, with prominent white ear tufts. The tail is patterned by alternating rings of black and buff.

SIZE: Sex not stated—HB 157-195 mm; T 135-180 mm; Mass 135-210 g.

DISTRIBUTION: On Sulawesi (Indonesia), it is found on the eastern central peninsula, the eastern part of the central core, and the southeastern peninsula. It also inhabits the outlying Indonesian islands of Pulau Buton and Pulau Kabaena.

GEOGRAPHIC VARIATION: None.

CONSERVATION: IUCN status—no information. Population trend—no information.

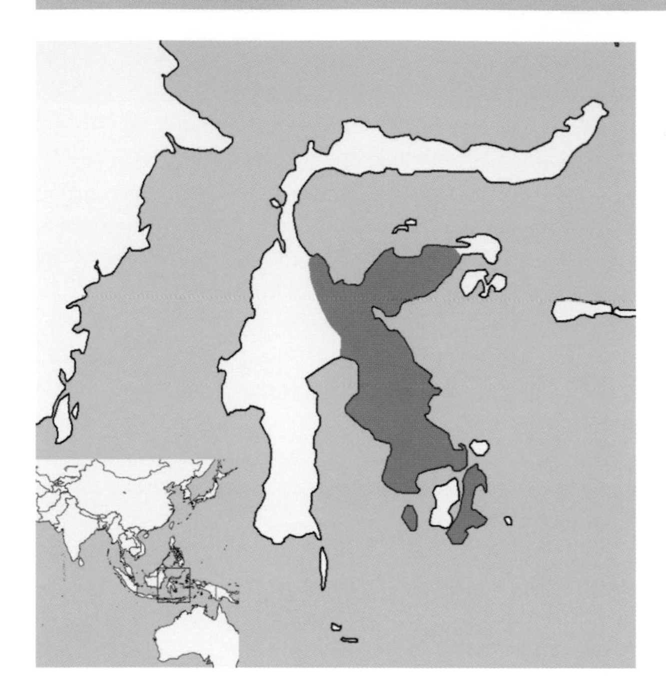

HABITAT: This squirrel occurs from the coastal plain to elevations up to 1200 m in the upper canopy of primary tropical lowland evergreen rainforests.

NATURAL HISTORY: *P. alstoni* travels through the upper canopy via pathways formed from branches and vines. To cross gaps in the canopy, Alston's squirrels descend to the forest floor to travel on fallen tree trunks. Their diet consists of soft fruits and insects; figs were most prevalent in the stomach contents. This species tends to be quiet and wary, giving alarm calls in response to approaching rain, human noises, and predators. Their regular calls and alarm calls are loud strident "chatters" and "chucks," similar to those of *P. topapuensis*. The squirrels vocalize individually, not in choruses. A single embryo was found in each of the pregnant females. Females have three pairs of teats: two inguinal pairs and one postaxillary pair.

GENERAL REFERENCES: Musser et al. 2010.

Prosciurillus leucomus (Müller and Schlegel, 1844)
Whitish Dwarf Squirrel

DESCRIPTION: The ears are bright ochraceous on the inside and black on the outside, with a prominent black ear tuft. This squirrel is distinguished by white or whitish buff or grayish white patches on the nape behind the ears. The underparts are reddish orange or ochraceous. The upperparts are olive brown and may be flecked with orange, buffy, or black. The tail is ringed with black and buff, terminating in a black tuft.

SIZE: Sex not stated—HB 165–188 mm; T 140–190 mm.

DISTRIBUTION: The whitish dwarf squirrel is found in the northern peninsula of Sulawesi and on the neighboring island of Pulau Lembeh (Indonesia).

GEOGRAPHIC VARIATION: Two subspecies are recognized.

P. l. leucomus—entire distributional range, except west of Gorontalo on the northern peninsula of Sulawesi (Indonesia). This form has prominent nape patches, and an ochraceous muzzle and cheeks.

P. l. occidentalis—west of Gorontalo, in the middle of the northern peninsula of Sulawesi (Indonesia). This form has a pale buff or gray muzzle and cheeks. The nape patches are diluted or even absent. The underparts are paler in color.

CONSERVATION: IUCN status—data deficient. Population trend—no information.

HABITAT: This squirrel can be found in coastal lowlands and at elevations up to 1700 m in montane forests.

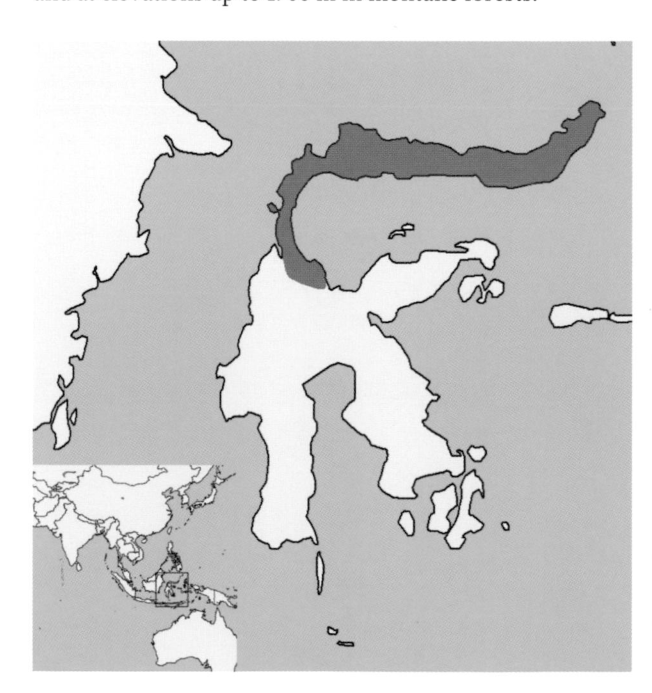

Prosciurillus leucomus. Photo courtesy Ingo Waschkies.

NATURAL HISTORY: Not much is known about *P. leucomus.* Based on the similarity of its body plan to *P. topapuensis* and *P. alstoni,* Musser conjectures that *P. leucomus* is a denizen of the canopy. Its diet is probably soft fruits and insects. Females have three pairs of teats: two inguinal pairs and one postaxillary pair.

GENERAL REFERENCES: Musser et al. 2010.

Prosciurillus murinus (Müller and Schlegel, 1844) Celebes Dwarf Squirrel

DESCRIPTION: This is the smallest of the endemic squirrels on Sulawesi (Indonesia). It has dark brown upperparts; and a tail flecked with buff, ochraceous, and black. It lacks ear tufts, nape patches, and dorsal stripes. The tail is shorter than the length of its body. The underparts are grayish. The eyes are circled by buffy rings. It is very similar in coloring and size to *P. abstrusus,* but *P. murinus* is smaller and the ears are uniformly dark.

SIZE: Sex not stated—HB 102-150 mm; T 55-120 mm; Mass 42-110 g.

DISTRIBUTION: The Celebes dwarf squirrel is ubiquitous all over Sulawesi (Indonesia), except perhaps for the eastcentral peninsula, where surveys for *P. murinus* have been poor. It is also found on the surrounding islands of Pulau Talise and Pulau Lembeh (Indonesia).

GEOGRAPHIC VARIATION: None.

CONSERVATION: IUCN status—data deficient. Population trend—no information.

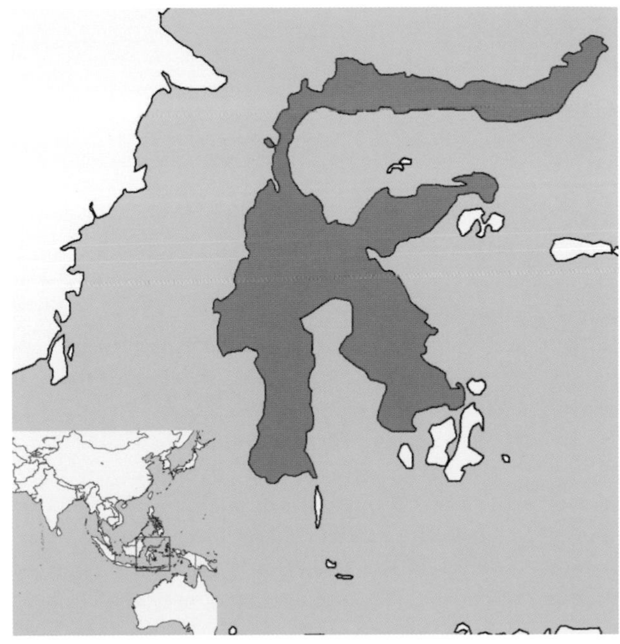

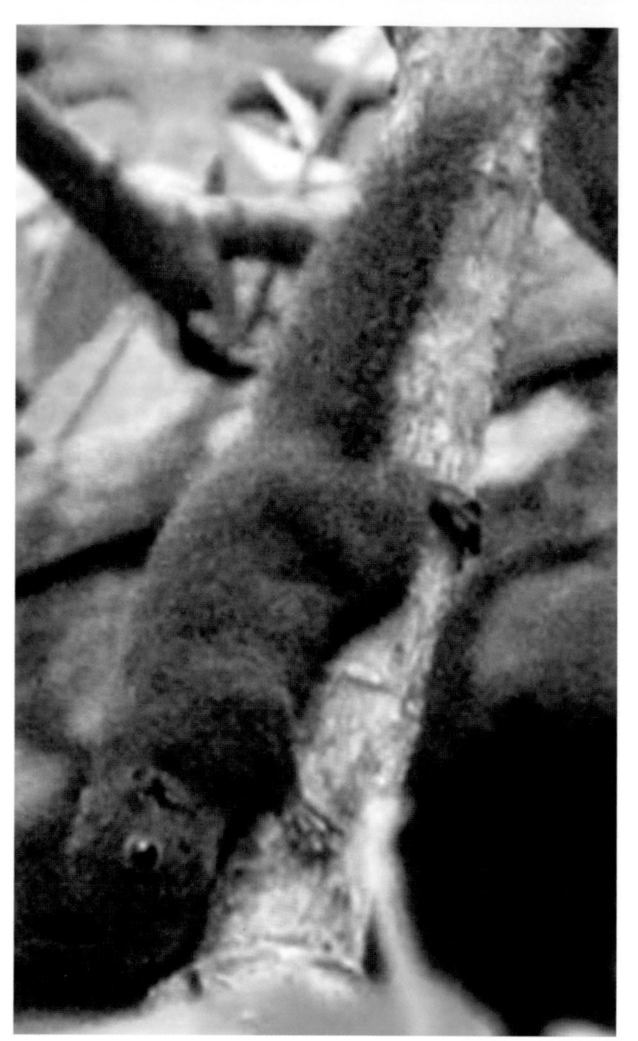

Prosciurillus murinus. Photo courtesy Iwan Hunowu.

HABITAT: This squirrel occurs along an altitudinal gradient from coastal lowlands to an elevation of 2200 m in the mountains. It lives in forest formations, ranging from tropical lowland evergreen rainforests to upper montane rainforests. On mountain ranges where *P. abstrusus* is found, it is thought that the altitudinal ranges of *P. murinus* and *P. abstrusus* are mutually exclusive, with *P. abstrusus* occurring at the higher elevations, where the forest changes from lower montane forest to lowland evergreen rainforest.

NATURAL HISTORY: *P. murinus* is diurnal and arboreal. This squirrel inhabits the primary forest understory and uses tree branches, vines, and fallen trunks to travel through the understory and along the ground. It may forage in the upper canopy and in the crowns of trees, and it has been observed to forage on tree trunks, where it gnaws into the bark. The Celebes dwarf squirrel has also been observed to forage with conspecifics, with no sign of aggression among them. Its stomach contents indicate a diet of soft fruits, seeds, and insects. This small squirrel is quite vocal and unwary of humans. Its vocalizations range from high-pitched "chucks" to whistles to birdlike staccato trills. Single embryos were found in each of the pregnant females examined. Females have three pairs of teats: two inguinal pairs and one postaxillary pair.

GENERAL REFERENCES: Musser et al. 2010.

Prosciurillus rosenbergii (Jentink, 1879)
Sanghir Squirrel

DESCRIPTION: The underparts are dark brownish gray to brownish buff. The upperparts are a rich dark chestnut brown. The tail is blackish. This squirrel can be distinguished from other similarly sized squirrels by its lack of ear tufts, nape patches, and a mid-dorsal line.

SIZE: Sex not stated—HB 190 mm; T 180 mm.

DISTRIBUTION: Historically, this squirrel was endemic to Kepulauan Sangihe (= Sangihe Islands, Indonesia), on the islands of Pulau Sangihe (the largest island), Pulau Siau, Pulau Tahulandang, and Pulau Ruang. Much of the land on these islands has been deforested or converted to planta-

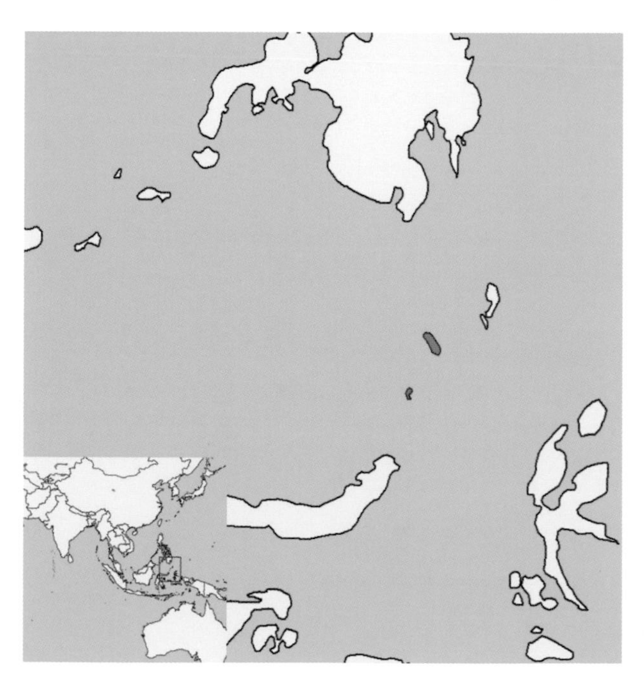

tions, and it is doubtful if the current range of *P. rosenbergii* is as extensive as it was when recorded in the late 1800s.

GEOGRAPHIC VARIATION: None.

CONSERVATION: IUCN status—least concern. Population trend—stable.

HABITAT: There is no information on *P. rosenbergii*, but considering that it is similar in morphology to *P. alstoni* and *P. topapuensis*, *P. rosenbergii* is suspected to dwell in the upper canopy.

NATURAL HISTORY: No information is available.

GENERAL REFERENCES: Musser et al. 2010; Riley 2002.

Prosciurillus topapuensis (Roux, 1910) Mount Topapu Squirrel

DESCRIPTION: This squirrel has black ear tufts that range from prominent to faint. The upperparts are brown with buff or black highlights. The underparts are dark gray washed lightly with pale buff or ochraceous, and some are tinted with silver. *P. topapuensis* lacks the mid-dorsal stripe and nape patches seen in other squirrels of similar coloring. It has a buffy ring around each eye, and a buffy-colored muzzle. The tail, which is shorter than the length of the head and body, is colored with rings of black and buff, with short buffy and black bands outlining the margins of the tail. The tail ends in a black terminal tuft.

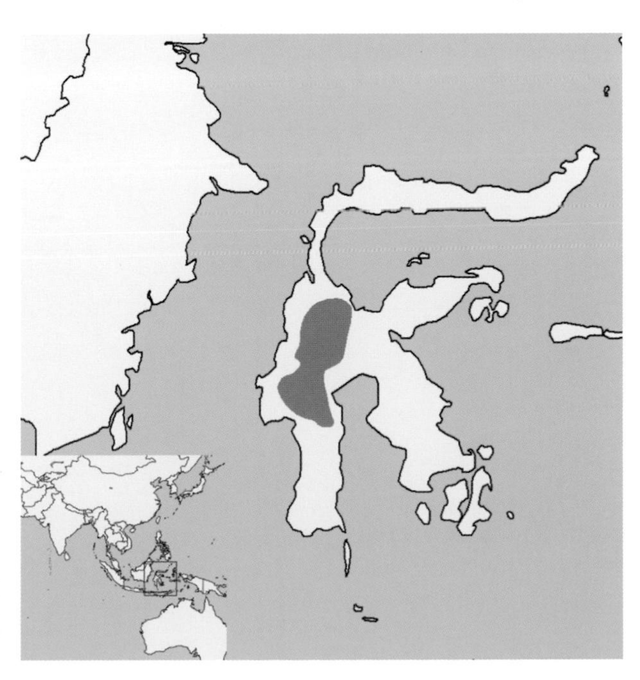

SIZE: Sex not stated—HB 155–190 mm; T 120–175 mm; Mass 130–210 g.

DISTRIBUTION: *P. topapuensis* can be found in the mountains in the western section of the central core island of Sulawesi (Indonesia).

GEOGRAPHIC VARIATION: None.

CONSERVATION: IUCN status—no information. Population trend—no information.

HABITAT: This squirrel is found at elevations from 350 to 2800 m, in habitats ranging from tropical lowland evergreen rainforests in the foothills to montane rainforest formations.

NATURAL HISTORY: *P. topapuensis* inhabits the upper canopy of primary forests, staying mostly in the tree crowns and on woody vines. When on the ground, these squirrels utilize living and dead vines and tree branches and trunks to transverse the ground and cross streams. They are diurnal, active mostly in the mornings and afternoons on sunny days, but they may be active all day during overcast weather. They are thought to place their leaf nests in tree cavities, especially those of strangler figs (*Ficus aurea*). Their stomach contents indicate that they have a diet of soft fruits, some seeds, and arboreal insects. Figs were prevalent in the stomach contents. This species' behavior ranges from wary and quiet to loud and aggressive. Their vocalizations are "chatters": loud, resonant, and scolding. The squirrels are sensitive to human voices and presence, the sounds of trees and limbs falling, approaching rain, and aerial predators; they will vocalize at all of these noises. The squirrels tend to call individually, not in a chorus. Females have three pairs of teats: two inguinal pairs and one postaxillary pair.

GENERAL REFERENCES: Musser et al. 2010.

Prosciurillus weberi (Jentink, 1890) Weber's Dwarf Squirrel

DESCRIPTION: Weber's dwarf squirrel is distinguished by its prominent black ear tufts and by the broad black mid-dorsal stripe extending from the neck to the base of the tail.

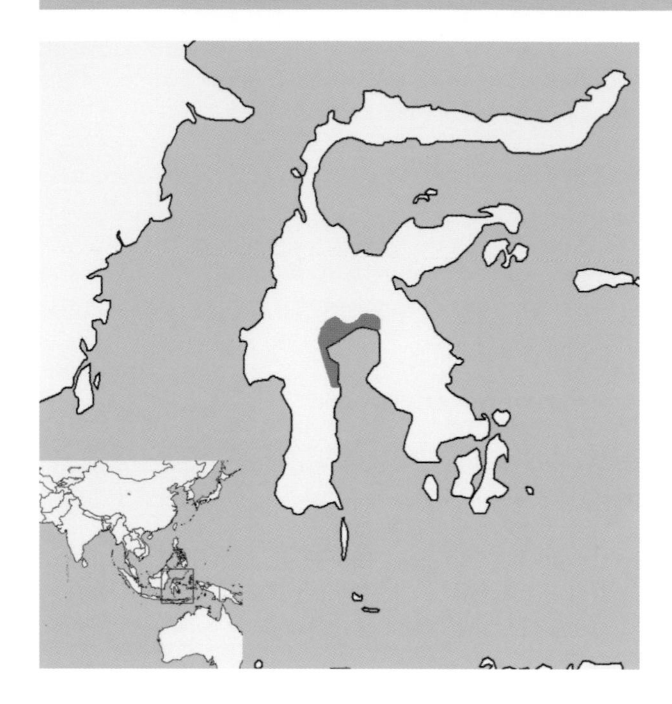

It lacks the nape patches seen in *P. leucomus*. The upperparts are dark brown with buffy, orange, and black highlights. The underparts are reddish orange.

SIZE: Sex not stated—HB 187 mm; T 142 mm.

DISTRIBUTION: Specimens are known from the central island of Sulawesi (Indonesia), in areas with elevations no higher than 100 m in the coastal lowlands around the central mountains and fringing the northwestern margin of the bay (Teluk Bone) in the southeastern part of the island. This species' total distribution is unknown.

GEOGRAPHIC VARIATION: None.

CONSERVATION: IUCN status—data deficient. Population trend—no information.

HABITAT: Little is known, but collection localities indicate that its habitat is tropical lowland evergreen rainforests.

NATURAL HISTORY: Nothing is known about the natural history of *P. weberi*. Due to its morphological similarity to *P. topapuensis* and *P. alstoni*, however, it is conjectured that *P. weberi* lives in the upper canopy and eats soft fruits, seeds, and insects.

GENERAL REFERENCES: Musser et al. 2010.

Rhinosciurus Blyth, 1856

This genus includes a single species from Southeast Asia.

Rhinosciurus laticaudatus (Müller, 1840) Shrew-Faced Squirrel

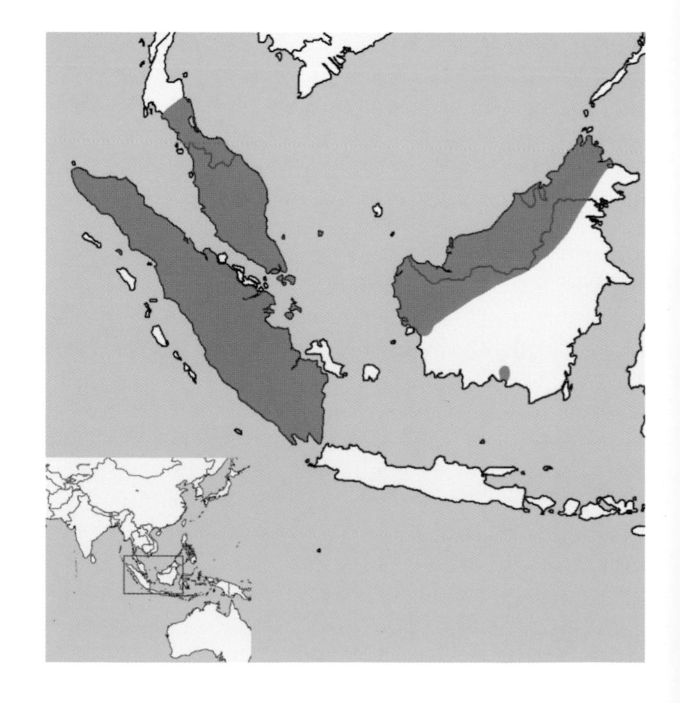

DESCRIPTION: This is a long-nosed ground squirrel with a dark brown dorsal pelage, and a white or buff ventral pelage. The short bushy tail is frequently held vertically.

SIZE: Sex not stated—HB 195-215 mm; T 115-140 mm.

DISTRIBUTION: This squirrel occurs on the Malay Peninsula, Sumatra (Indonesia), the lowlands and hills of the western part of the island of Borneo (divided among Malaysia, Brunei Darussalam, and Indonesia), and small adjacent islands.

GEOGRAPHIC VARIATION: Two subspecies are recognized.

R. l. laticaudatus—island of Borneo (divided among Malaysia, Brunei Darussalam, and Indonesia). See description above.

Rhinosciurus laticaudus. Photo courtesy Nick Baker, www.ecologyasia.com.

R. l. saturatus—Sumatra (Indonesia) and southern Malay Peninsula. This form is slightly darker.

CONSERVATION: IUCN status—near threatened. Population trend—decreasing.

HABITAT: *Rhinosciurus laticaudatus* is terrestrial and is found in tall and secondary forests.

NATURAL HISTORY: This squirrel feeds on insects.

GENERAL REFERENCES: Lekagul and McNeely 1977; J. Payne and Francis 1985; H. C. Robinson and Kloss 1919.

Rubrisciurus Ellerman, 1954

This genus contains a single species, endemic to the island of Sulawesi (Indonesia). Not much is known about this beautiful squirrel. The most current information available comes from G. Musser's recent publication on the Sulawesi squirrels and their associated sucking lice. More fieldwork on the natural history of the squirrels in this region of the world is needed.

Rubrisciurus rubriventer (Müller and Schlegel, 1844) Sulawesi Giant Squirrel

DESCRIPTION: This is the largest of the endemic squirrels in Sulawesi (Indonesia). The tail is reddish brown, and the reddish hue on the underparts extends to the inner and outer surfaces of the limbs. The bright reddish tone darkens to a reddish maroon on the shoulders, the sides of body, and the dorsal surfaces of the limbs. The head and back are brownish speckled with buff, orange, and black. The ears have prominent glossy black tufts. The eyes have a dark crescent above and are encircled by a wide buffy ring. The juvenile coat is similar in coloration to that of the adults, but muted.

SIZE: Sex not stated—HB 250-305 mm; T 180-255 mm; Mass 500-860 g.

DISTRIBUTION: This species is probably found all over the main island of Sulawesi (Indonesia), but there have not been adequate survey efforts to ascertain if it inhabits the east-central peninsula or the southwestern peninsula. Except for a few areas near the volcano Gunung Lompobatang, the southwestern peninsula has been deforested, and it is doubtful if this species occurs there. However, there are

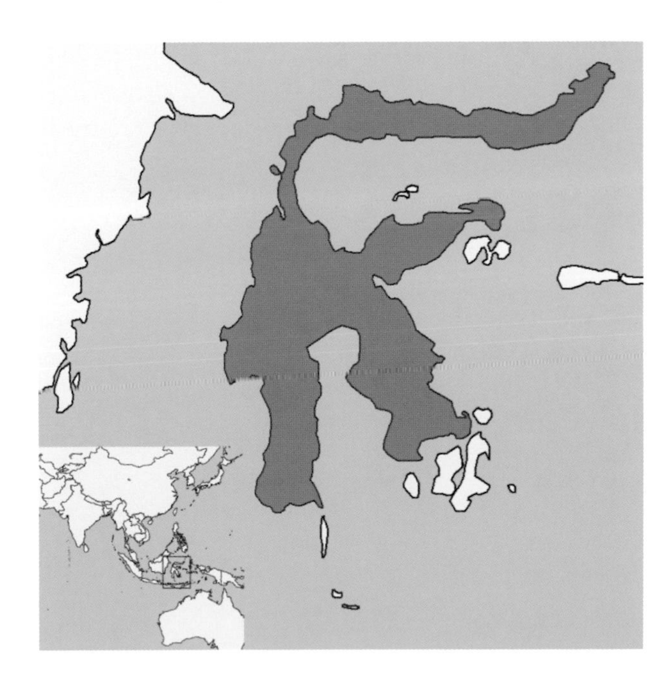

fossil records of its presence in this part of the island in prehistoric times.

GEOGRAPHIC VARIATION: None.

CONSERVATION: IUCN status—vulnerable. Population trend—decreasing.

HABITAT: The Sulawesi giant squirrel is found from the coast up to an elevation of 1512 m, in primary tropical lowland evergreen forests and in lower montane rainforests.

NATURAL HISTORY: This squirrel is diurnal, arboreal, and terrestrial. *R. rubriventer* forages and travels in the understory and on the ground, never in the canopy or higher. It nests in the forest understory. There is anecdotal evidence that the nests are large globular structures made from plant matter and built close to the ground in the cavities of large tree trunks. One nest, located by Musser and his team, was 30 cm in diameter, wedged in a hollow of a tree (*Pterospermum celebicum*), and constructed entirely of the long black fibers of the sugar palm (*Arenga pinnata*). Their diet is composed of fruits, seeds, and insects. These squirrels are known to be quiet, wary, and easily startled; sometimes the only evidence of their presence is the sound of loud gnawing on the fruit of the pohon pangi (*Pangium edule*). The skull morphology suggests that they are able to consume hard nuts and seeds. They also dig into rotting wood to expose insects. A single embryo has been observed in the few pregnant females caught. Females have four teats, arranged in pairs.

GENERAL REFERENCES: Musser et al. 2010.

Sundasciurus Moore, 1958

This is a genus of 17 species of small to medium-sized Southeast Asian squirrels.

Sundasciurus altitudinis (Robinson and Kloss, 1916)
Sumatran Mountain Squirrel

DESCRIPTION: *Sundasciurus altitudinis* is very similar to *S. tenuis*, but with much longer fur, more yellowish hands and feet, and a light gray hue on the underparts. There are trenchant differences in the skull anatomy of *S. altitudinis*, including the small size of the bullae and the longer rostrum.

SIZE: Sex not stated—HB 150 mm; T 115 mm.

DISTRIBUTION: This species occurs in the highlands of Sumatra (Indonesia), at elevations ranging from 488 to 2530 m or higher, in Atjeh (= Aceh) and Padang (on Mount Löser and on Kerinci Peak, respectively).

GEOGRAPHIC VARIATION: None.

CONSERVATION: No information is available.

HABITAT: The Sumatran mountain squirrel is only found "in heavy jungle . . . at any level from low in the bushes to high in the very tops of tall trees" (den Tex et al.).

NATURAL HISTORY: They feed on "curious acorn-like nuts that the natives call 'giseng'" (den Tex et al.).

GENERAL REFERENCES: den Tex et al. 2010.

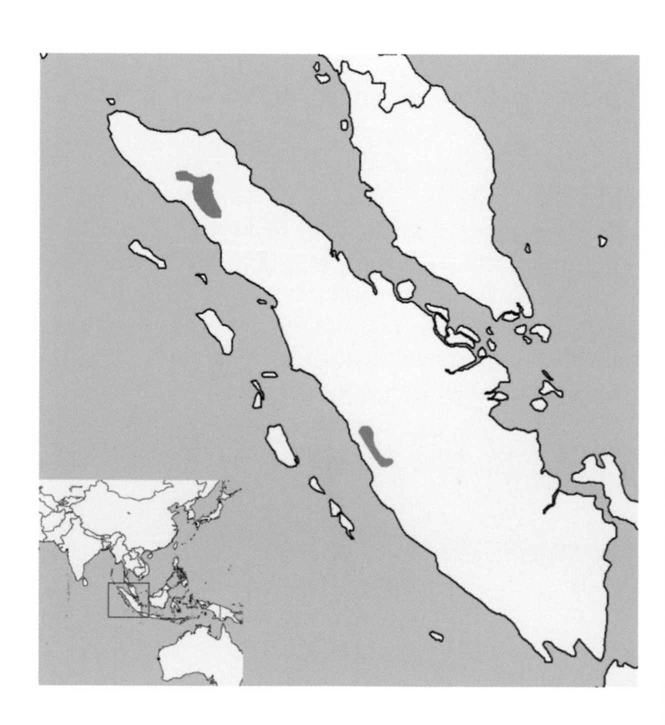

Sundasciurus brookei (Thomas, 1892)
Brooke's Squirrel

DESCRIPTION: The upperparts are speckled brown, and the underparts are gray with whitish tips on the hairs. It has an orange patch between the hind legs. The tail is barred.

SIZE: Female—HB 151.9 mm; T 122.1 mm; Mass 132.5 g.
Male—HB 153.9 mm; T 112.7 mm; Mass 124.0 g.
Sex not stated—HB 168.8 mm; T 136.0 mm; Mass 114.3 g.

DISTRIBUTION: This species is found in the northern, central, and western mountains of the island of Borneo (divided among Malaysia, Brunei Darussalam, and Indonesia), at elevations from 600 to 1500 m.

GEOGRAPHIC VARIATION: None.

CONSERVATION: IUCN status—least concern. Population trend—no information.

HABITAT: This squirrel is found in tall forests on hill ranges.

NATURAL HISTORY: *S. brookei* is diurnal and primarily arboreal.

GENERAL REFERENCES: Medway 1977; J. B. Payne 1980.

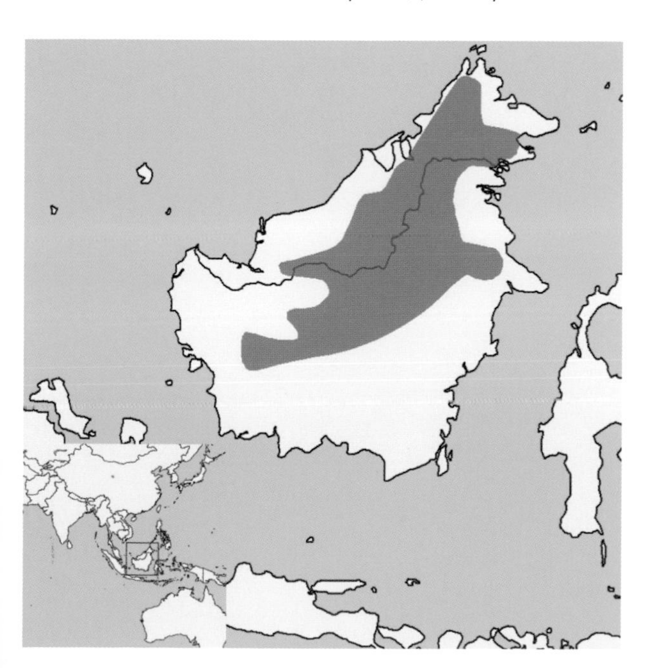

Sundasciurus brookei. Photo courtesy David Bakewell.

Sundasciurus davensis (Sanborn, 1952)
Davao Squirrel

DESCRIPTION: This species has a line that runs down the middle of the dorsal surface, from the nose to the base of the tail. This line has hairs with a black base and russet tips, and the line darkens as it progresses to the base of the tail. The cheeks are buffy colored. The underparts are apricot buff. The sides are clearly marked off from the back and the lighter underparts. The median of the tail is russet, bordered by black and an outer border of white. The ventral surface of the tail is less russet.

SIZE: Male—HB 198.0 mm; T 182.0 mm.

DISTRIBUTION: *S. davensis* is known only from the type locality: on Mindanao Island (Philippines).

GEOGRAPHIC VARIATION: None.

CONSERVATION: IUCN status—data deficient. Population trend—no information.

HABITAT: No information is available.

NATURAL HISTORY: No information is available.

GENERAL REFERENCES: Chiozza 2008b.

Sundasciurus fraterculus (Thomas, 1895)
Fraternal Squirrel

DESCRIPTION: This dull-colored species has a general coloration of fine grizzled brownish rufous on the upperparts, hands, feet, and the entire tail. There is more rufous on the cheeks and the thighs. The hairs of the underparts are slaty gray at the base and pale rufous white on the terminal ends. *S. fraterculus* most closely resembles *S. tenuis*, but the former may be distinguished by its smaller size and more uniform color.

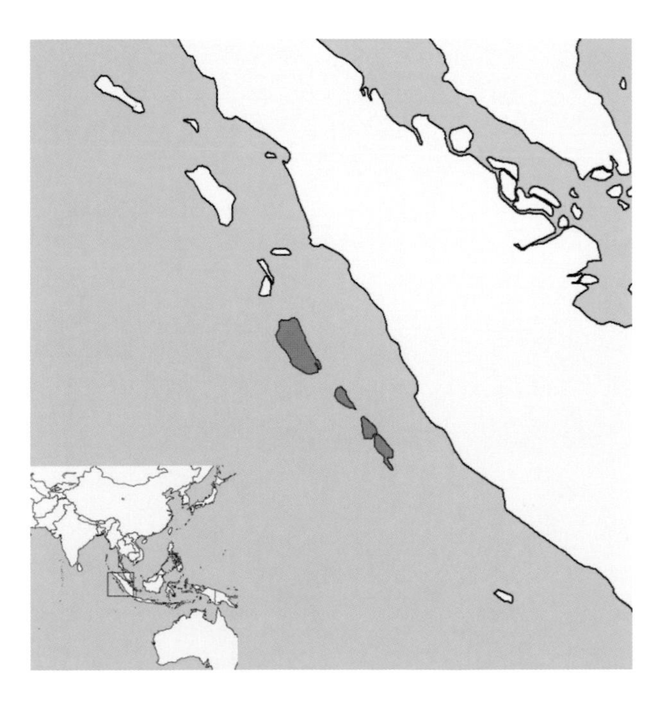

SIZE: Female—HB 116.3 mm; T 78.6 mm.
Male—HB 119.1 mm; T 75.1 mm.

DISTRIBUTION: This squirrel is found on the Indonesian islands of Siberut, Sipora, and North and South Pagai (west of Sumatra).

GEOGRAPHIC VARIATION: None.

CONSERVATION: IUCN status—endangered. Population trend—decreasing.

HABITAT: No information is available.

NATURAL HISTORY: No information is available.

GENERAL REFERENCES: Helgen and Aplin 2008.

Sundasciurus hippurus (I. Geoffroy, 1831)
Horse-Tailed Squirrel

DESCRIPTION: On the Malay Peninsula, the head, shoulders, and sides are speckled black and gray; the back is bright chestnut; the underparts are deep brick red; and the tail is very thick, bushy, and entirely glossy black. On the island of Borneo (divided among Malaysia, Brunei Darussalam, and Indonesia), the upperparts are reddish brown, with a gray head and shoulders.

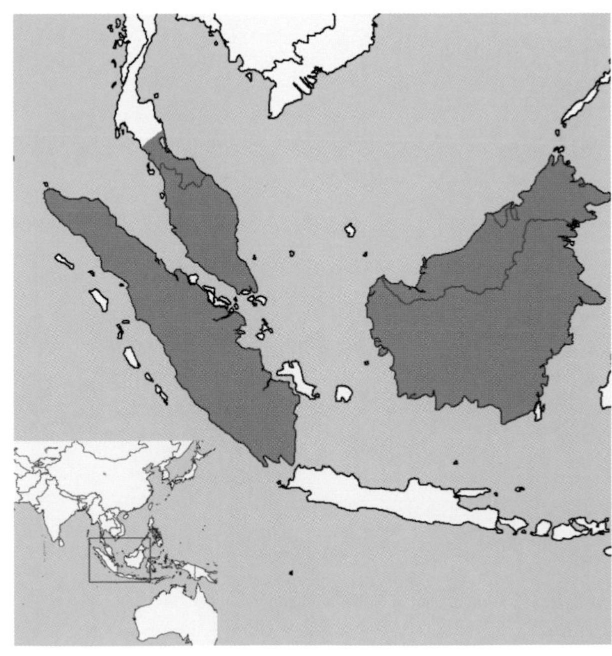

Sundasciurus hippurus. Photo courtesy Con Foley.

SIZE: Female—HB 234.9 mm; T 246.6 mm; Mass 432.7 g.
Male—HB 247.4 mm; T 234.9 mm; Mass 429.9 g.
Sex not stated—HB 235.2 mm; T 230.5 mm; Mass 363.0 g.

DISTRIBUTION: The horse-tailed squirrel is found in the southern Malay Peninsula (Thailand and peninsular Malaysia), Sumatra (Indonesia), and the island of Borneo (divided among Malaysia, Brunei Darussalam, and Indonesia).

GEOGRAPHIC VARIATION: Seven subspecies are recognized.

S. h. hippurus—range not known. See description above.
S. h. borneensis—Kalimantan (Indonesia), in ranges not occupied by *S. h. hippurellus*. This form has brick red underparts, and it is heavily grizzled gray on the shoulders and thighs. The tail is entirely dark, and thick and bushy.
S. h. hippurellus—coastal areas, in West Kalimantan (Indonesia) and at Sungai (= stream) Kapuas and Sungai Landak through Sarawak to the lower Sungai Rejang (Malaysia). The underparts are dark reddish; the tail is entirely dark and is thick and bushy, but not very similar to that of a horse. In some individuals the tail has a reddish tip.
S. h. hippurosus—west coast of Sumatra (Indonesia). This is a larger form.
S. h. inquinatus—ranges from Lawas in northern Sarawak (Malaysia) to Sungai Sebuku and Sungai Sembakung in northern East Kalimantan (Indonesia). The underparts are a dull orange.
S. h. ornatus—Minh Hai (= Cà Mau) Province (Vietnam). The sides of the head, the shoulders, the upper surfaces of the limbs, and the rump are grayish brown suffused with ochraceous on the flanks. The mid-dorsal area is washed with rufous in the anterior part, and the posterior part is glossy red brown bordered thinly with orange brown. The underparts are bright chestnut. The tail is dark brown on the basal third, and the tip is red.
S. h. pryeri—at low elevations throughout Sabah (Malaysia). This form has white underparts that are tinged with red in some individuals, and the tail is grizzled.

CONSERVATION: IUCN status—near threatened. Population trend—decreasing.

HABITAT: On the Malay Peninsula, it is confined to tall and secondary forests. It is found from the undercanopy to the ground, at elevations ranging from the lowlands to the hills of the main range, being recorded at 1006 m on Gunong (= Gunung, or Mount) Tahan. Usually *S. hippurus* is seen as a solitary individual or in pairs. On the island of Borneo it has been observed throughout the lowlands and the hills, except in South Kalimantan and the eastern parts of Central Kalimantan (Indonesia); there are also some records from higher elevations, such as at 1524 m on Mount Dulit in Sarawak (Malaysia).

NATURAL HISTORY: The horse-tailed squirrel is diurnal. Most often it is seen in small trees, but it sometimes travels on the ground. Its diet includes seeds, fruits, and insects. The most commonly occurring call is "chek . . . chek . . . chekchekchekchek . . ."

GENERAL REFERENCES: Medway 1969, 1977.

Sundasciurus hoogstraali (Sanborn, 1952) Busuanga Squirrel

DESCRIPTION: The sides are colored clay to tawny olive. The dorsum is darker than the sides, due to the dark-colored bases of the hairs and the presence of black hairs. The base of the tail is similar to the back. The face is grayish, and the hands and feet are dark brown. The underparts are paler, due to the presence of long grayish white hairs.

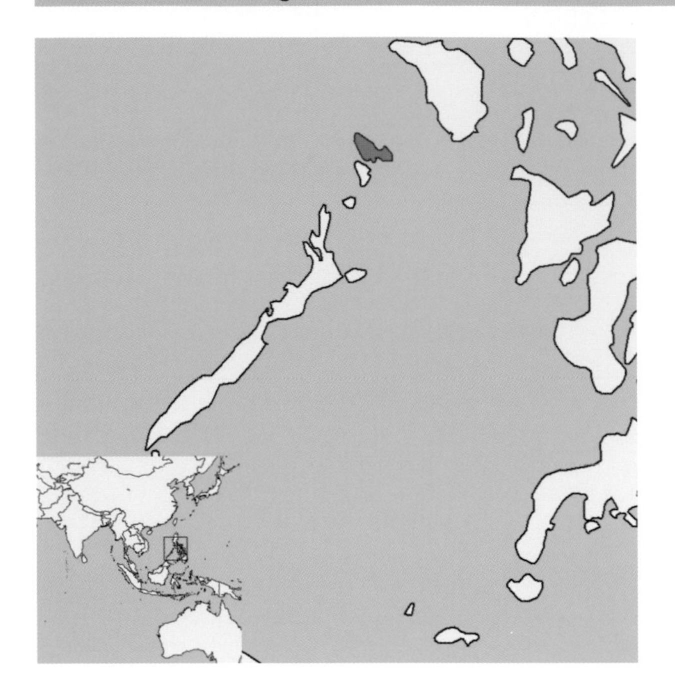

SIZE: Female—HB 202.0 mm; T 165.4 mm.
Male—HB 203.5 mm; T 160.5 mm.
Sex not stated—HB 204.8 mm; T 176.0 mm.

DISTRIBUTION: This squirrel is named for the island on which it is found, Busuanga Island (Philippines).

GEOGRAPHIC VARIATION: None.

CONSERVATION: IUCN status—least concern. Population trend—no information.

HABITAT: The Busuanga squirrel is found in primary and secondary lowland forests.

NATURAL HISTORY: No information is available.

GENERAL REFERENCES: Heaney, Balete, et al. 1998.

Sundasciurus jentinki (Thomas, 1887) Jentink's Squirrel

DESCRIPTION: The upperparts are rather pale, speckled brownish or creamy white, and have gray underfur. Jentink's squirrel has a distinctive creamy white "mustache," eye ring, and border to the ears. The tail is very thin, appearing banded dark and pale due to reddish, black, and white bands on the hairs.

SIZE: Female—HB 126.6 mm; T 112.6 mm; Mass 55.0 g.
Male—HB 130.6 mm; T 125.2 mm; Mass 60.0 g.
Sex not stated—HB 131.1 mm; T 118.1 mm.

DISTRIBUTION: This species is found in the northern part of the island of Borneo (in the mountains of Sabah and Sarawak [Malaysia] and Kalimantan [Indonesia]), at elevations above 900 m.

GEOGRAPHIC VARIATION: Two subspecies are recognized.

S. j. jentinki—Mount Kinabalu (900-3140 m), the Crocker Range, and Gunung (= Mount) Trus Madi in Sabah (Malaysia), as well as in the Sabah-Sarawak border hills, the

Sundasciurus hoogstraali. Photo courtesy Rey Sta. Ana.

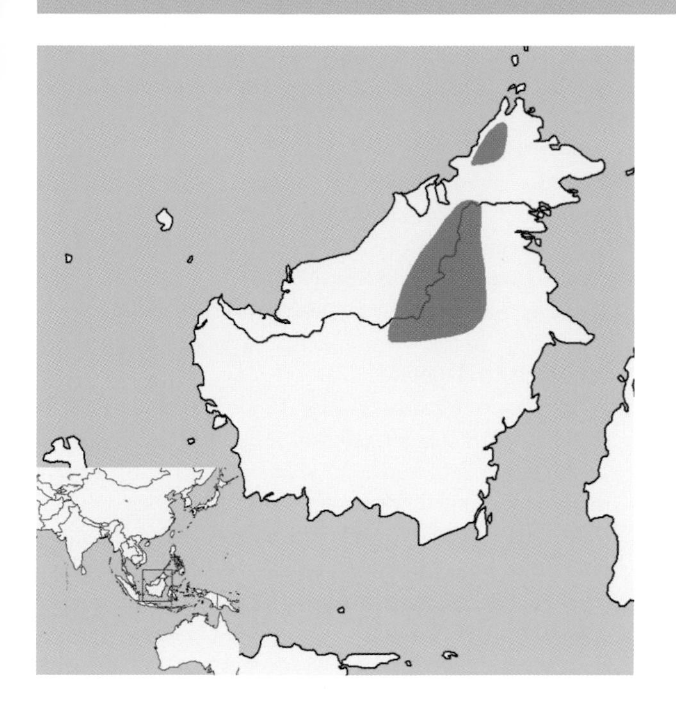

Kelabit Highlands, and other mountains in Sarawak (Malaysia). See description above.

S. j. subsignanus—Long Petak, Sungai (= stream) Telen (1170 m), Sungai Badang, and Sungai Kayan in East Kalimantan (Indonesia). *S. j. subsignanus* differs from *S. j. jentinki* in having buff rather than white tips on the tail hairs.

CONSERVATION: IUCN status—least concern. Population trend—no information.

HABITAT: It is confined to montane forests.

Sundasciurus jentinki. Photo courtesy Jon Hall, www.mammalwatching.com.

NATURAL HISTORY: This squirrel is diurnal. It is active mainly in the crowns of small trees, and it often follows flocks of mixed species of birds as they flush insects.

GENERAL REFERENCES: Medway 1977.

Sundasciurus juvencus (Thomas, 1908) Northern Palawan Tree Squirrel

DESCRIPTION: *S. juvencus* is similar to *S. steerii,* but the northern Palawan tree squirrel has a general brown color, with less rufous. The sides are gray, and the tail is tipped in black. The ventral surface is variable, ranging from all white to all rufous.

SIZE: Female—HB 195.5 mm; T 155.1 mm; Mass 245.3 g.
Male—HB 200.9 mm; T 171.8 mm; Mass 283.9 g.
Sex not stated—HB 204.6 mm; T 165.3 mm; Mass 259.0 g.

DISTRIBUTION: This squirrel is found in northern Palawan Island (Philippines).

GEOGRAPHIC VARIATION: None.

CONSERVATION: IUCN status—least concern. Population trend—stable.

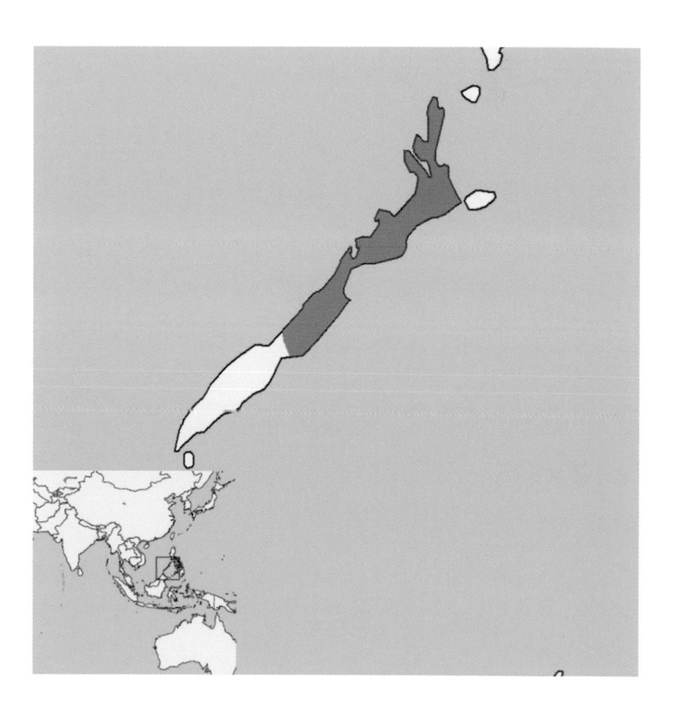

***Sundasciurus juvencus,* white morph.** Photo courtesy Pierre Fidenci, Endangered Species International.

HABITAT: *S. juvencus* is found in primary and secondary lowland forests.

NATURAL HISTORY: No information is available.

GENERAL REFERENCES: Heaney, Balete, et al. 1998.

Sundasciurus lowii (Thomas, 1892)
Low's Squirrel

DESCRIPTION: *S. lowii* is distinguished from *S. tenuis* by its creamy white underparts and proportionately shorter tail. The upperparts are speckled brown; the underparts are buffy white, sometimes with a reddish tinge. The pale reddish buff ring around each eye is rarely obvious from a distance, and the tail is short and bushy.

SIZE: Female—HB 137.5 mm; T 88.8 mm; Mass 76.4 g.
Male—HB 143.9 mm; T 92.9 mm; Mass 78.7 g.
Sex not stated—HB 130.2 mm; T 82.7 mm; Mass 83.5 g.

DISTRIBUTION: This species is found on the Malay Peninsula, Sumatra (Indonesia), the island of Borneo (divided among Malaysia, Brunei Darussalam, and Indonesia), and adjacent small islands.

GEOGRAPHIC VARIATION: Seven subspecies are recognized.

S. l. lowii—island of Borneo (divided among Malaysia, Brunei Darussalam, and Indonesia). See description above.

S. l. balae—Batu Islands (Indonesia). The underparts are nearly white.

S. l. bangueyae—Banggi Island (Malaysia, slightly north of the island of Borneo). The general color tends toward dark greenish olivaceous. The light rings on the hairs are buffy instead of ochraceous.

S. l. humilis—East Sumatra (Indonesia). This subspecies is similar in size and color to *S. l. seimundi,* but it has a darker general color.

S. l. natunensis—Natuna Islands (Indonesia). This form is smaller, with shorter feet and longer ears. The general color is grizzled rufous. The venter is white with a strong rufous wash. The anteorbital spots are rufous, not yellow; and the black patch behind the ear is absent. It should be noted that the type skin was prepared from a specimen preserved in alcohol, and O. Thomas cautioned that the color might have been reddened, as certain types of yellow turn red when introduced to an alcohol preparation.

S. l. robinsoni—Malay Peninsula (peninsular Thailand and peninsular Malaysia). This form is similar to *S. l. lowii,* but smaller.

S. l. seimundi—the type locality is Bliah, Kundur Island, Rhio [= Riau] Archipelago (Indonesia). This form is similar to *S. l. robinsoni,* but the dark dorsal color expands downward on the flanks, narrowing the white ventral area.

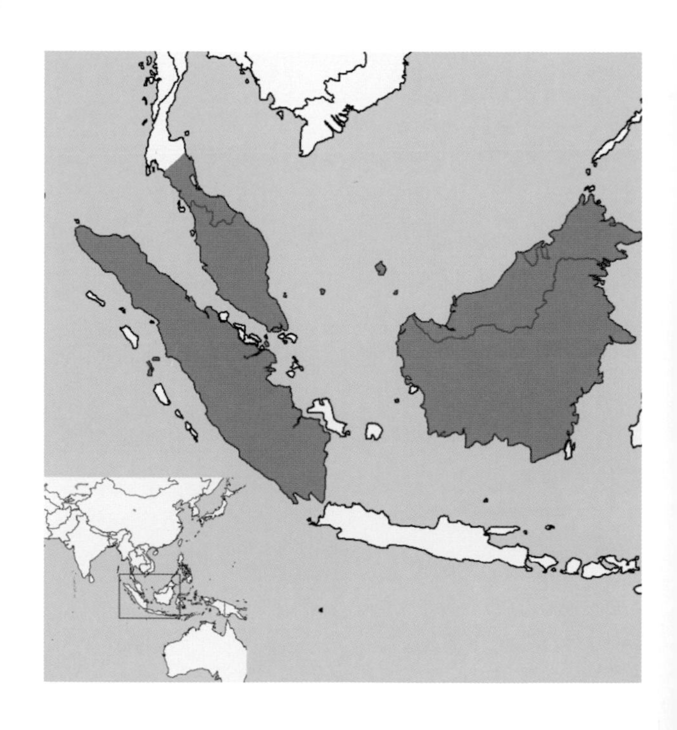

Sundasciurus lowii. Photo courtesy Morten Strange.

CONSERVATION: IUCN status—least concern. Population trend—stable.

HABITAT: This species is a scansorial inhabitant of lowland tall and secondary dipterocarp forests. A squirrel of the whole forest, *S. lowii* is frequently active in the undergrowth, brushwood, and fallen branches; only occasionally is it found high in the trees. On the Malay Peninsula it is found in the lowlands and foothills, at elevations up to at least 914 m, where it is widespread but nowhere common. On the island of Borneo (divided among Malaysia, Brunei Darussalam, and Indonesia), this squirrel occurs principally in lowland forests, usually at elevations below 900 m, but it is also recorded at up to 1400 m in the hills of the Kelabit Highlands (Sarawak, Malaysia).

NATURAL HISTORY: This squirrel is diurnal, most active in the early morning and late afternoon. It feeds predominantly on bark, but its diet also includes fruits, insects, and fungi. It travels and feeds in small standing trees, in fallen trees, and on the ground. It makes nests of plant fiber, similar to those of large treeshrews (*Tupaia tana*), and it also nests in tree hollows. Litter size is two to three, with a mean of 2.3 ($n = 4$). On the island of Borneo (divided among Malaysia, Brunei Darussalam, and Indonesia), five nests of *S. lowii* were observed. Of these, one was reported to be subterranean, with an entrance hole 6 cm in diameter. Another two were in tree cavities: one in an emergent tree at a height of 2 m, with nest material of leaves and plant fibers inside; the other in a stump at a height of 0.5 m. Two nests consisted of balls of woven plant fibers and leaves, 10–15 cm in diameter, at heights of 2 and 3.5 m between branches of palms.

GENERAL REFERENCES: J. B. Payne 1980; Medway 1977.

Sundasciurus mindanensis (Steere, 1890) Mindanao Squirrel

DESCRIPTION: The upperparts and tail are dark gray. The hair is black at the base, buffy in the middle, and then black at the tip. The mid-dorsal region is darker. The thighs and the dorsal surface of the feet are rufous gray. The nose and the eye region are rufous. The body and the lower surface of the legs are ashy.

SIZE: Female—HB 200.5 mm; T 191.1 mm.
Male—HB 203.3 mm; T 191.0 mm.
Sex not stated—HB 193.7 mm; T 171.6 mm; Mass 285.0 g.

DISTRIBUTION: This species is found on the island of Mindanao and on adjacent small islands (Philippines).

GEOGRAPHIC VARIATION: None.

CONSERVATION: IUCN status—least concern. Population trend—no information.

HABITAT: No information is available.

NATURAL HISTORY: No information is available.

GENERAL REFERENCES: Chiozza 2008c.

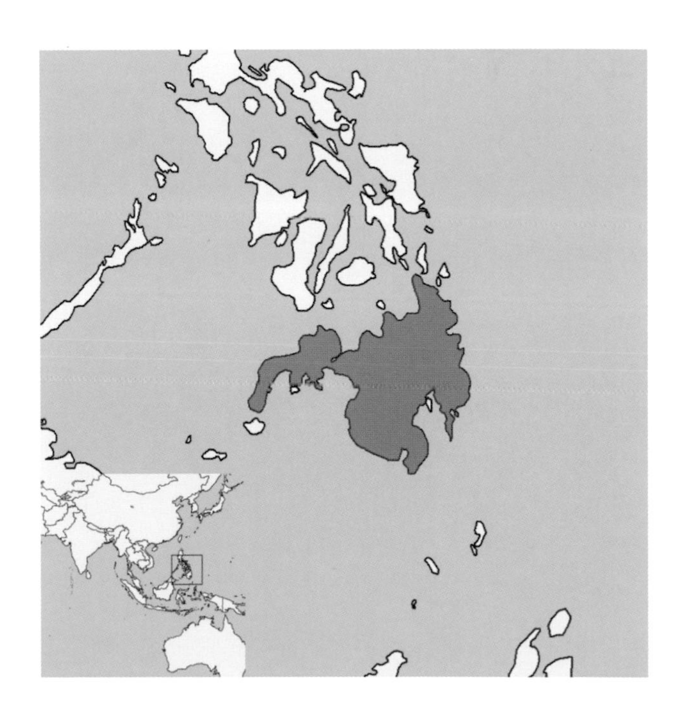

Sundasciurus moellendorffi (Matschie, 1898) Culion Tree Squirrel

DESCRIPTION: The ventral surface is cream colored. The tail is darker than the back, with more black hairs. The color of the head and upperparts is uniform. This description is based on the two specimens in the Smithsonian Institution's National Museum of Natural History. On one specimen the dorsal surface is light brown grizzled with white and darker brown, and there are white patches on the upper back. On the other specimen the dorsal surface is light brown russet grizzled with white, with the russet color strongest on the midline.

SIZE: Female—HB 200.0 mm; T 190.0 mm; Mass 190.0 g.
Male—T 218.0 mm.
Sex not stated—HB 207.0 mm; T 190.0 mm.

DISTRIBUTION: *S. moellendorffi* occurs on the Calamian Islands (Philippines), except for Busuanga Island.

GEOGRAPHIC VARIATION: None.

CONSERVATION: IUCN status—near threatened. Population trend—decreasing.

HABITAT: This species is found in primary and secondary lowland forests and coconut groves.

NATURAL HISTORY: No information is available.

GENERAL REFERENCES: Heaney, Balete, et al. 1998.

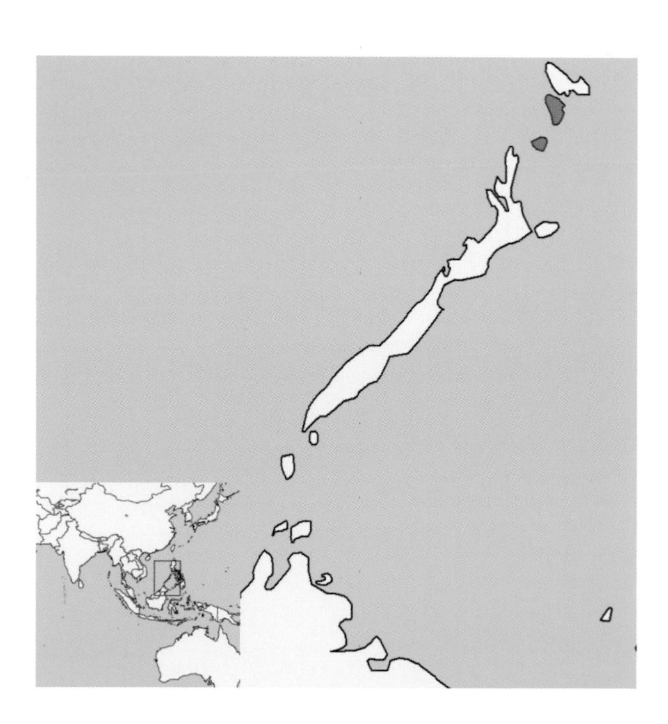

Sundasciurus philippinensis (Waterhouse, 1839) Philippine Tree Squirrel

DESCRIPTION: *S. philippinensis* is brown on the back. It has a reddish eye ring, a gray to dull orange brown venter, and an annulated tail.

SIZE: Female—HB 193.7 mm; T 192.1 mm.
Male—HB 193.0 mm; T 172.0 mm.
Sex not stated—HB 191.7 mm; T 219.0 mm; Mass 244.0 g.

DISTRIBUTION: This species is found in the Philippines, from the islands of Samar south through Bohol to Basilan, including Mindanao.

GEOGRAPHIC VARIATION: None.

CONSERVATION: IUCN status—least concern. Population trend—stable.

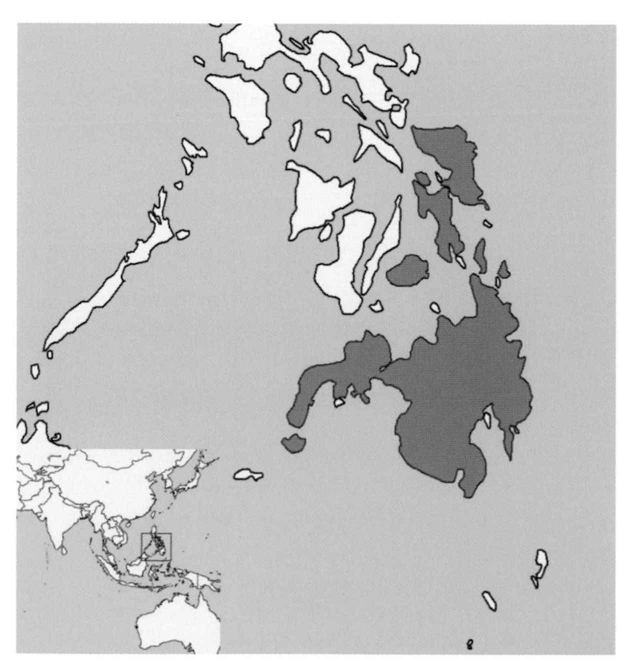

HABITAT: The Philippine tree squirrel is found at elevations up to 2100 m, in primary and secondary lowland and montane forests. It is also abundant near agricultural fields.

NATURAL HISTORY: No information is available.

GENERAL REFERENCES: Heaney, Balete, et al. 1998.

Sundasciurus rabori Heaney, 1979
Palawan Montane Squirrel

DESCRIPTION: The Palawan montane squirrel has a dark brown dorsal coloration, and a gray venter with light buff hair tips.

SIZE: Female—HB 163.0 mm; T 135.0 mm.
Male—HB 182.8 mm; T 139.8 mm.
Sex not stated—HB 174.5-185.5 mm; T 144.5 mm; Mass 163.0 g.

DISTRIBUTION: This squirrel is found in the mountains, at elevations above 800 m, on Palawan Island (Philippines). The type locality is Mount Matalingajan (Palawan).

GEOGRAPHIC VARIATION: None.

CONSERVATION: IUCN status—data deficient. Population trend—no information.

HABITAT: *S. rabori* is commonly found in upper-elevation forests.

NATURAL HISTORY: No information is available.

GENERAL REFERENCES: Heaney, Balete, et al. 1998.

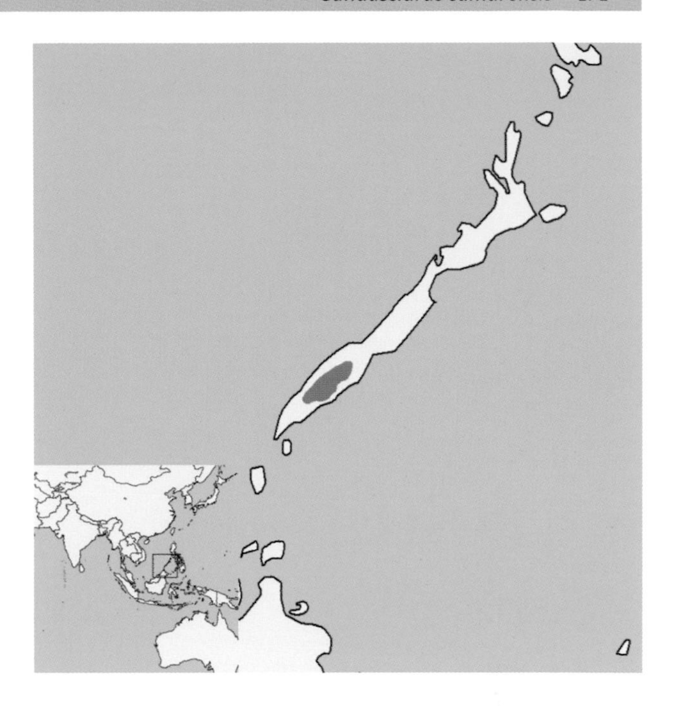

Sundasciurus samarensis (Steere, 1890)
Samar Squirrel

DESCRIPTION: The general color of *S. samarensis* is similar to *S. mindanensis*: the upperparts and tail are dark gray, and the body and the lower surface of the legs are ashy. *S. samarensis*, however, is a larger form, with the thighs, the nose, and the eye region the same color as the dorsal surface. The upper surface of the feet is dark gray.

SIZE: Female—HB 185.0 mm; T 158.5 mm; Mass 222.0 g.
Male—HB 190.6 mm; T 160.8 mm; Mass 243.0 g.
Sex not stated—HB 188.5 mm; T 165.5 mm.

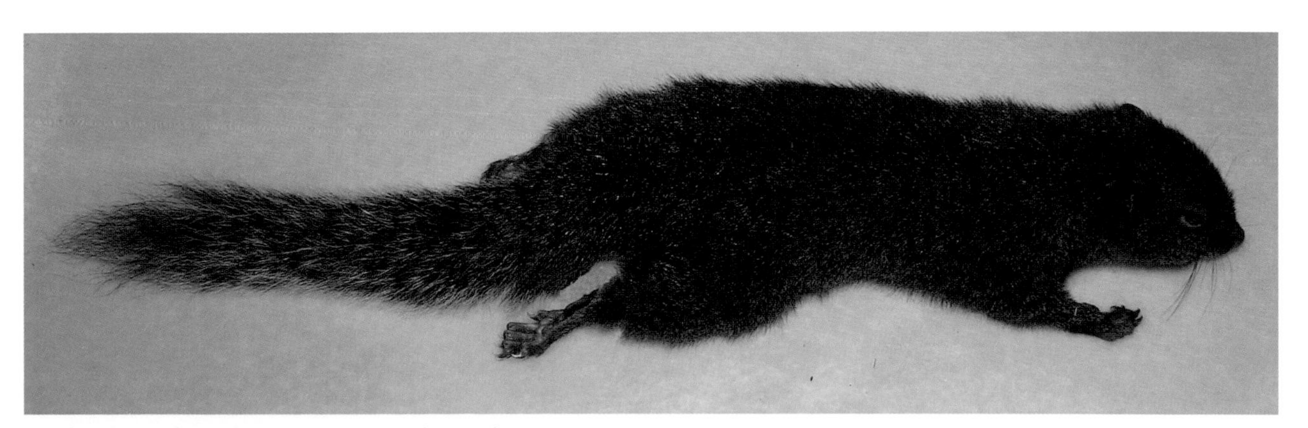

***Sundasciurus rabori*.** Photo courtesy Danilo S. Balete.

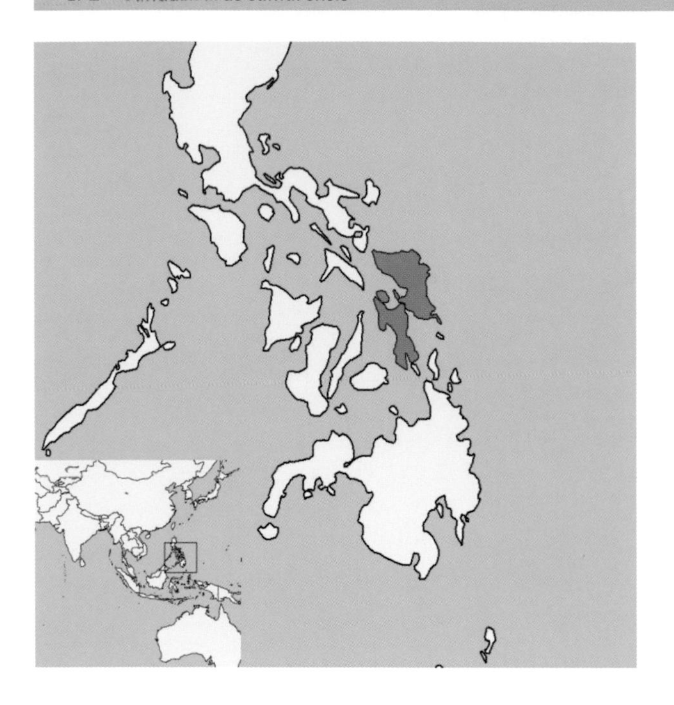

DISTRIBUTION: This species is found on the islands of Samar and Leyte (Philippines).

GEOGRAPHIC VARIATION: None.

CONSERVATION: IUCN status—least concern. Population trend—no information.

HABITAT: This squirrel may be found in primary and secondary lowland and montane forests, as well as in disturbed forests and upland agricultural areas.

NATURAL HISTORY: Samar squirrels may be hunted as pests, as they will consume corn (maize) and sweet potato crops.

GENERAL REFERENCES: Rickart et al. 1993.

Sundasciurus steerii (Günther, 1877) Southern Palawan Tree Squirrel

DESCRIPTION: The dorsal coloration is chestnut to light brown, and the venter is chestnut to dull brown.

SIZE: Female—HB 198.8 mm; T 155.8 mm; Mass 222.5 g.
Male—HB 200.5 mm; T 161.2 mm; Mass 257.1 g.
Sex not stated—HB 206.5 mm; T 160.3 mm.

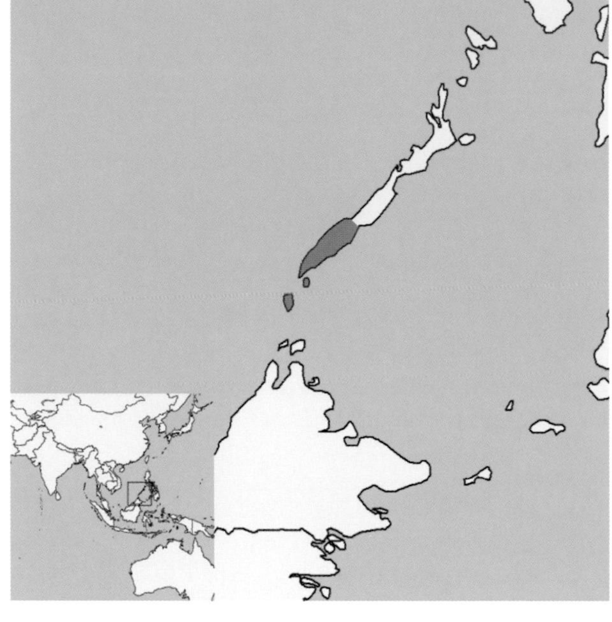

DISTRIBUTION: This squirrel is found in the Palawan Islands (Philippines).

Sundasciurus steerii. Photo courtesy Danilo S. Balete.

GEOGRAPHIC VARIATION: None.

CONSERVATION: IUCN status—least concern. Population trend—stable.

HABITAT: *S. steerii* can be found in coconut groves, plantations, and lowland forests.

NATURAL HISTORY: No information is available.

GENERAL REFERENCES: Heaney, Balete, et al. 1998.

Sundasciurus tahan (Bonhote, 1908) Upland Squirrel

DESCRIPTION: *S. tahan* is similar to *S. tenuis*, but *S. tahan* is larger and darker. The tail is annulated; and the tips of its guard hairs are buff, not white.

SIZE: Sex not stated—HB 155 mm; T 104 mm.

DISTRIBUTION: This species is found in the uplands of peninsular Malaysia, on Gunong (= Gunung, or Mount) Tahan and other peaks.

GEOGRAPHIC VARIATION: None.

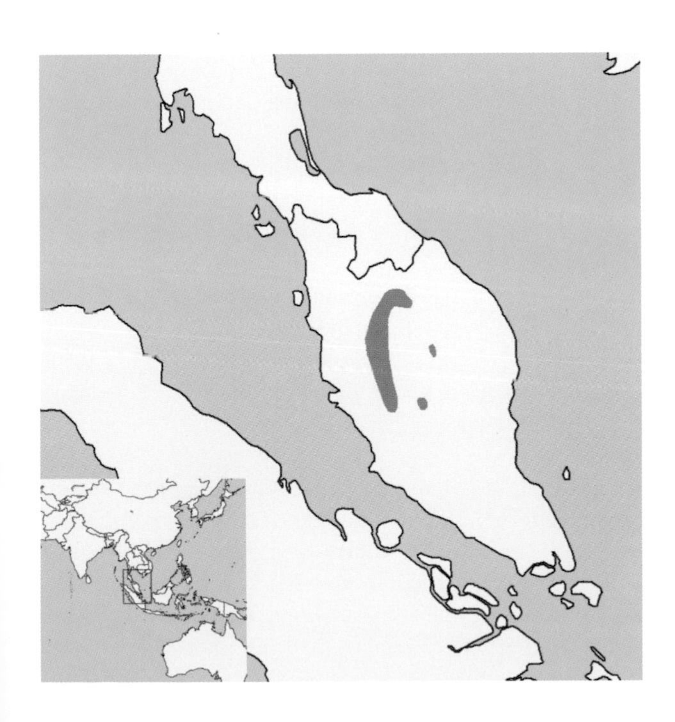

***Sundasciurus tahan*.** Photo courtesy Morten Strange.

CONSERVATION: No information is available.

HABITAT: No information is available.

NATURAL HISTORY: No information is available.

GENERAL REFERENCES: den Tex et al. 2010.

Sundasciurus tenuis (Horsfield, 1824) Slender Squirrel

DESCRIPTION: The upperparts are speckled brownish, and the hairs of the underparts are normally gray with white or buffy tips. *S. tenuis* is pale around the eyes and above the facial whiskers. It usually has an indistinct pale spot behind each ear. The tail is rather long and slender.

Sundasciurus tenuis. Photo courtesy Norman Lim.

SIZE: Female—HB 139.8 mm; T 114.8 mm; Mass 81.4 g.
Male—HB 140.7 mm; T 113.4 mm; Mass 85.2 g.
Sex not stated—HB 138.9 mm; T 114.1 mm; Mass 76.8 g.

DISTRIBUTION: This species can be found on the Malay Peninsula, Sumatra (Indonesia), the island of Borneo (divided among Malaysia, Brunei Darussalam, and Indonesia), and adjacent small islands.

GEOGRAPHIC VARIATION: Five subspecies are recognized.

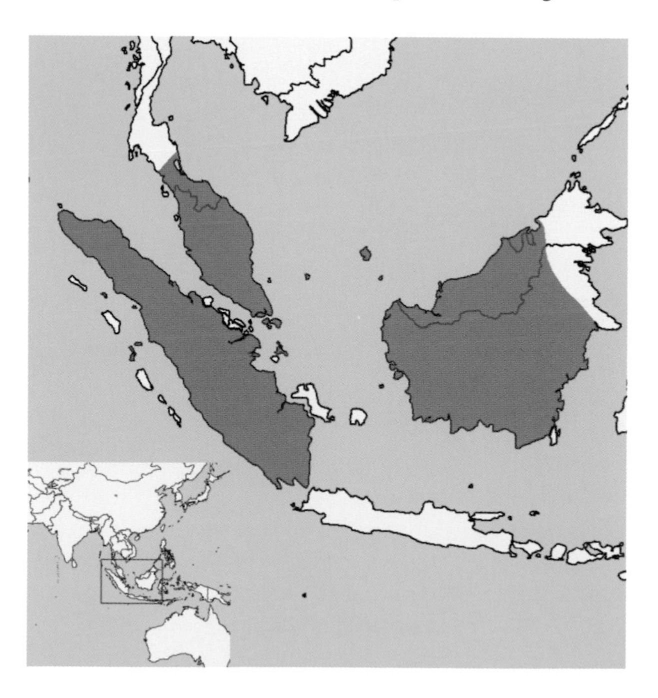

S. t. tenuis—southwest Thailand and south on the Malay Peninsula. It has a dull pelage. This form differs from *S. tahan* by its smaller size, paler upperparts, and the lack of a buffy suffusion on the inner thighs and inguinal region.

S. t. bancarus—Banjak (= Banyak) Islands (Indonesia). The upperparts are brown tinged with yellow. The gray underparts are washed with a whitish cream buff.

S. t. modestus—Sumatra (Indonesia). The underparts are cream gray. The dorsal surface is the typical speckled brown of this species, but with a rufous wash on the flanks. The tail is the same color as the dorsal surface. The tail tip is all black.

S. t. parvus—island of Borneo (divided among Malaysia, Brunei Darussalam, and Indonesia). The underparts are tinged buff. Otherwise the coloration is similar to *S. t. tenuis*.

S. t. procerus—Bunguran Island (Indonesia). This form is similar to *S. t. tenuis*, but smaller.

CONSERVATION: IUCN status—least concern. Population trend—decreasing.

HABITAT: *S. tenuis* occurs in tall and secondary forests throughout the lowlands and hills of the island of Borneo, except for most of Sabah (Malaysia) and northern and southeastern East Kalimantan (Indonesia). It was recorded at an elevation of 1650 m, but most records are from lower elevations. It appears to be replaced ecologically by *S. jentinki* in the mountain ranges in the northern and central parts of the island of Borneo.

NATURAL HISTORY: Slender squirrels are diurnal, and they are active mainly in small trees. Their diet includes the inner bark and insects from tree trunks, and fruits and seeds.

GENERAL REFERENCES: Medway 1977; J. B. Payne 1980.

Tamiops J. A. Allen, 1906

The four species in this genus are the small striped tree squirrels of Indochina, China, and peninsular Malaysia. They contrast with the species of *Funambulus* in having smaller rounded ears, frequently tipped with white, as well as a black longitudinal mid-dorsal stripe paralleled by two pairs of pale stripes that are separated by a darker stripe.

Tamiops maritimus (Bonhote, 1900)
Maritime Striped Squirrel

DESCRIPTION: This species is more olivaceous than other species of *Tamiops*, with the pale line on each side of the mid-dorsal stripe more similar to the color of the nape. The venter is buffy.

SIZE: Female—HB 121.2 mm; T 101.8 mm; Mass 56.5 g.
Male—HB 119.5 mm; T 101.1 mm; Mass 54.5 g.
Sex not stated—HB 122.9 mm; T 100.1 mm.

DISTRIBUTION: This squirrel is found in Hubei east to Anhui and Zhejiang, and south through Yunnan, Guangxi, Guangdong, and Hainan (China) to Laos, Vietnam, and Cambodia, as well as in Taiwan.

GEOGRAPHIC VARIATION: Four subspecies are recognized.

T. m. maritimus—Taiwan and coastal China. This form is grayer and of a more uniform coloration than *T. mcclellandii*; the mid-dorsal stripe is shorter and less distinct; and the dull white lateral stripes are narrow and short.

T. m. hainanus—Laos, Vietnam, and Hainan (China). This form is similar to *T. m. maritimus*, but slightly larger.

T. m. moi—southern Vietnam. This form is similar to *T. m. hainanus*, but it is claimed to be a bit larger, with slightly redder stripes lateral to the mid-dorsal stripe.

T. m. monticolus—montane areas west of the range of *T. m. maritimus*. This form has a distinct mid-dorsal stripe that extends to the base of the tail; and two lateral pale stripes that are broad, distinct, and also continue to the base the tail.

CONSERVATION: IUCN status—least concern. Population trend—stable.

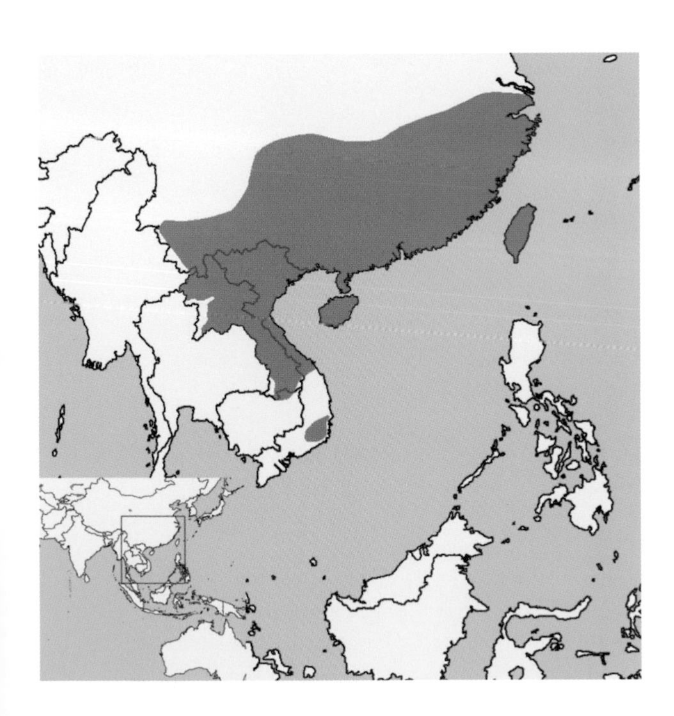

Tamiops maritimus. Photo courtesy James Eaton/Birdtour Asia.

HABITAT: On the mainland *T. maritimus* is a species of low-elevation broadleaved evergreen or mixed mesophytic forests. On Taiwan it occurs at higher elevations (from 2000 to 3000 m).

NATURAL HISTORY: The maritime striped squirrel is highly arboreal, but it descends to the herbaceous level to feed on the nectar of gingers (*Alpinia kwangsiensis*). It bites through the base of the flowers to get to the nectaries and thus does not pollinate them, so their seed set is reduced to less than 20 percent of the seed set of flowers that were not robbed by the squirrels. The diet of *T. maritimus* also includes insects, fruits, and seeds, and it is probably similar to that of *T. mcclellandii*. Its vocalization is described as a "cluck" or a short "chirrup."

GENERAL REFERENCES: J. A. Allen 1906; Robinson and Kloss 1922.

Tamiops mcclellandii (Horsfield, 1840)
Himalayan Striped Squirrel

DESCRIPTION: On this species, the mid-dorsal black stripe is not bisected by a longitudinal pale stripe (as seen in *T. rodolphii*). The pale stripe next to the mid-dorsal black stripe is fainter than the most lateral pale stripe. The facial stripe from the rostrum is continuous with the most lateral pale stripe of the back. The venter is ochraceous. *T. mcclellandii* is smaller than *T. swinhoei*, with a shorter and less dense pelage and more vivid stripes.

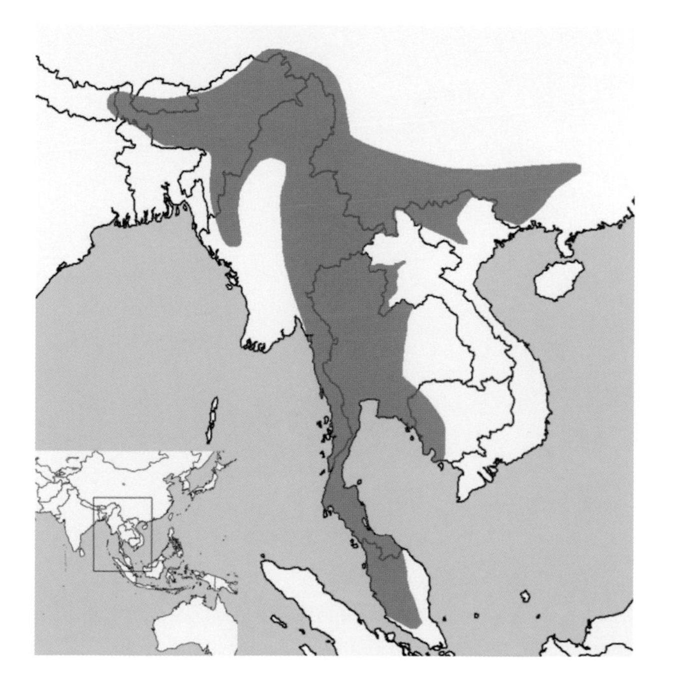

Tamiops mcclellandii. Photo courtesy Nick Baker, www.ecologyasia.com.

SIZE: Female—HB 112.9 mm; T 108.8 mm; Mass 51.8 g.
Male—HB 113.8 mm; T 108.3 mm; Mass 49.4 g.
Sex not stated—HB 107.6 mm; T 102.2 mm; Mass 39.0 g.

DISTRIBUTION: The Himalayan striped squirrel is found from eastern Nepal through Assam and Mizoram (India), northern and central Myanmar, Tibet and the Chinese province of Yunnan, northwestern Laos, northern Vietnam, Cambodia, and south through Thailand to peninsular Malaysia.

GEOGRAPHIC VARIATION: Six subspecies are recognized.

T. m. mcclellandii—northeastern India and northern Myanmar. It has a mid-dorsal black stripe (4 mm wide) extending from the nape to the rump; lateral dark lines extending from the shoulders to the rump; and, lateral to these, pale lines that are strongly colored yellow buff and pass from the rostrum to below the eyes and ears to the rump. The ventral coloration is buff, without any trace of red.

T. m. barbei—southern Myanmar. This form is reddish or reddish orange dorsally, and cinnamon rufous or ferruginous ventrally.

T. m. collinus—northeastern Thailand, neighboring Myanmar, Laos, Cambodia, and southern China. This form is more strongly striped than *T. m. mcclellandii* or *T. m. inconstans*, less red than *T. m. barbei*, darker dorsally than *T. m. kongensis*, and more orange and less yellow ventrally than *T. m. leucotis*.

T. m. inconstans—extreme northern Vietnam and adjoining China. This form is inconspicuously striped dorsally, and yellowish buffy or light orange ventrally.

T. m. kongensis—western and northern Thailand. This subspecies resembles a pallid form of *T. m. barbei*. It is generally light gray, rather than the reddish orange or yellowish coloration of *T. m. leucotis*, with the dorsal pale lines being pale buffy yellow and the venter being ochraceous buff.

T. m. leucotis—southern peninsular Thailand and peninsular Malaysia. This form lacks the general reddish coloration of *T. m. barbei* dorsally, and it is more yellowish ventrally.

CONSERVATION: IUCN status—least concern. Population trend—stable.

HABITAT: Himalayan striped squirrels live in tall trees in tropical or subtropical forests, but they may also be found in fruit trees in the vicinity of villages and in coconut plantations. On the Malay Peninsula this species tends to be a montane form. Farther north, it may be found at lower elevations. In China it is found at elevations above 700 m but below the altitudinal range of *T. swinhoei*.

NATURAL HISTORY: *T. mcclellandii* is a highly arboreal squirrel, seldom coming to the ground. It alternates quick dashes with motionless periods, usually with its tail straight out behind it. When frightened, it assumes a motionless position, with the head, body, and tail flattened out against a tree trunk. This species is commonly seen in pairs or in small groups. In some areas it is very common, although cryptic. It feeds on insects, fruits, and other vegetable matter. It shelters in tree holes. Its vocalization is a harsh "chick" call.

GENERAL REFERENCES: Blyth 1847; Bonhote and Skeat 1900; Moore 1958b; Thomas 1920.

Tamiops rodolphii (Milne-Edwards, 1867)
Cambodian Striped Squirrel

DESCRIPTION: The Cambodian striped squirrel has the mid-dorsal black stripe divided by a longitudinal thin pale brown stripe. The two more lateral pale lines are equally wide, and neither is paler than the other.

SIZE: Female—HB 117.7 mm; T 107.9 mm.
Male—HB 117.4 mm; T 111.1 mm.
Sex not stated: HB 119.5 mm; T 122.0 mm; Mass 56.0 g.

DISTRIBUTION: This squirrel can be found in eastern Thailand, Cambodia, southern Laos, and southern Vietnam.

GEOGRAPHIC VARIATION: Two subspecies are recognized:

T. r. rodolphii—southeastern Thailand, Cambodia, southern Laos, and southern Vietnam. This form has a dull brown

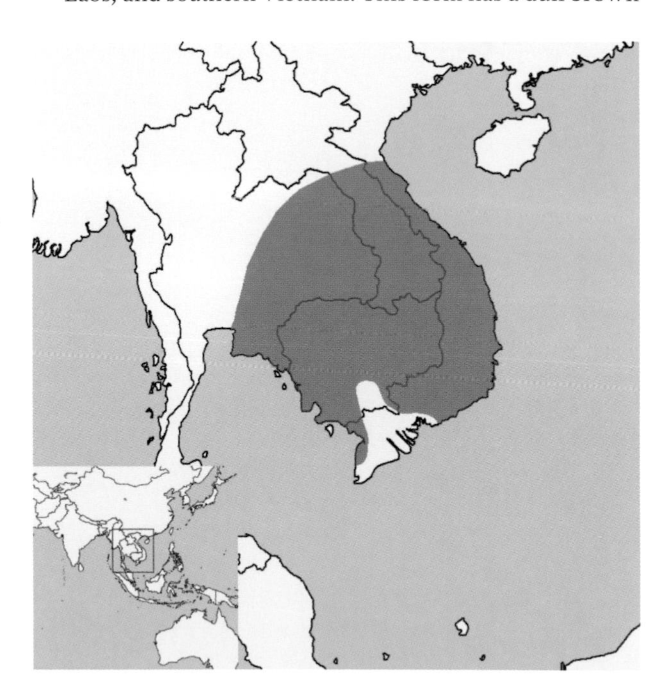

Tamiops rodolphii. Photo courtesy Phung my Trung, www.vncreatures.net.

head and nape, and a venter tending toward yellow or orange.

T. r. elbeli—Chaiyaphum and Khon Kaen provinces, Thailand. This form has a bright yellow brown head and nape, and a distinctive yellowish brown longitudinal line down the middle of the mid-dorsal black stripe.

CONSERVATION: IUCN status—least concern. Population trend—stable.

HABITAT: Little is known about the Cambodian striped squirrel, but its habitat is thought to be similar to *T. mcclellandii.*

NATURAL HISTORY: There is not much information about *T. rodolphii,* but its behavior is believed to be similar to *T. mcclellandii.*

GENERAL REFERENCES: Lekagul and McNeely 1977.

Tamiops swinhoei (Milne-Edwards, 1874) Swinhoe's Striped Squirrel

DESCRIPTION: *T. swinhoei* is quite variable, but it is generally larger than the other species in this genus, with a longer and denser pelage. It differs from the neighboring *T. mcclellandii* in that its most lateral pale line is broader, less brilliant, and usually stops at the shoulder rather than being continuous with the cheek stripe.

SIZE: Female—HB 130.7 mm; T 101.2 mm; Mass 87.9 g.
Male—HB 131.4 mm; T 99.7 mm.
Sex not stated—HB 132.6 mm; T 96.6 mm; Mass 78.0 g.

DISTRIBUTION: This squirrel is found in several Chinese provinces, from extreme southwestern Gansu, Shaanxi, and Henan south through Sichuan and Yunnan, to Tibet, far northeastern India, northern Myanmar, and northern Vietnam, with isolated populations in Hebei and Hunan (China).

GEOGRAPHIC VARIATION: Four subspecies are recognized.

T. s. swinhoei—southern China. This form has a yellow brown dorsal pelage, with a mid-dorsal stripe broader than in *T. mcclellandii* (9–10 mm) and a lateral pale line that is dull yellow brown and does not connect with the line on the cheek. The venter is buffy white.

T. s. olivaceus—Vietnam. This form has a much darker background color on the dorsum.

T. s. spencei—northern Myanmar. This subspecies has very obscure lateral pale stripes.

T. s. vestitus—northern China. This is a pale version of *T. s. swinhoei,* with the lateral lines being much paler than in the other subspecies.

CONSERVATION: IUCN status—least concern. Population trend—stable.

HABITAT: In the southwestern part of its range, Swinhoe's striped squirrel occurs at high elevations (from 2134 to 3048 m), above the altitudinal range of *T. mcclellandii.* It is found in broadleaved and coniferous forests, and in scrub jungles (bushes and tree rhododendrons) that are just below snowline.

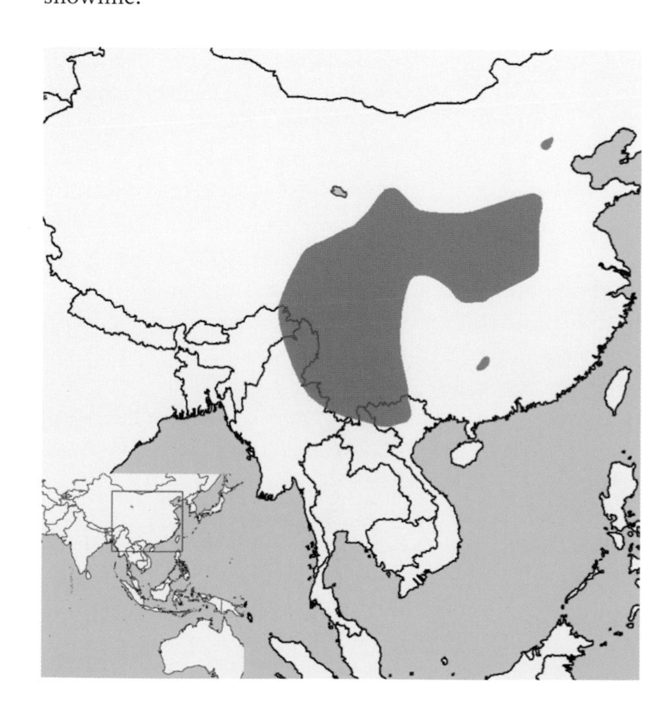

Tamiops swinhoei. Photo courtesy David Blank, www.animal diversity.ummz.umich.edu.

NATURAL HISTORY: These squirrels are usually arboreal and make nests in tree holes. They feed on insects, fruits, young shoots, and other vegetable matter. Their vocalizations are high-pitched and birdlike.

GENERAL REFERENCES: G. S. Miller 1915; Osgood 1932; Thomas 1921.

Subfamily Xerinae Osborn, 1910

This subfamily includes three tribes: the Xerini, the Protoxerini, and the Marmotini.

Tribe Xerini Osborn, 1910

This tribe contains three genera.

Atlantoxerus Forsyth Major, 1893

This genus of African ground squirrels includes a single species, found north of the Saharan Desert.

Atlantoxerus getulus (Linnaeus, 1758)
Barbary Ground Squirrel

DESCRIPTION: This is a striped ground squirrel, with a stripe on each flank and sometimes one down the middle of the back. The tail is bushy.

DISTRIBUTION: This species is endemic to the Middle and High Atlas Mountains in northwestern Africa, occurring through Western Sahara, Morocco, and northwestern Algeria. It was introduced in the Canary Islands of Spain: to Fuerteventura in 1965, Gran Canaria in 1996, and Lanzarote in 2006. The latter two populations appear to have been successfully removed.

SIZE: Sex not stated—HB 174 mm; T 133 mm (n = 37).

GEOGRAPHIC VARIATION: None.

CONSERVATION: IUCN status—least concern. Population trend—stable.

HABITAT: In its native African range, the Barbary ground squirrel lives at elevations between 150 and 4000 m in rocky areas of alpine meadows, forests, and deserts, such as the Atlas Mountains, argan (*Argania spinosa*) forests, and the Sahara Desert in Morocco and Algeria. On Fuerteventura (Canary Islands, Spain), this species thrives in rocky habitats as well as in semidesert scrublands and woodlands.

NATURAL HISTORY: Most observations of *A. getulus* have been conducted on Fuerteventura, where the animal is considered to be an invasive species, and these reports may not

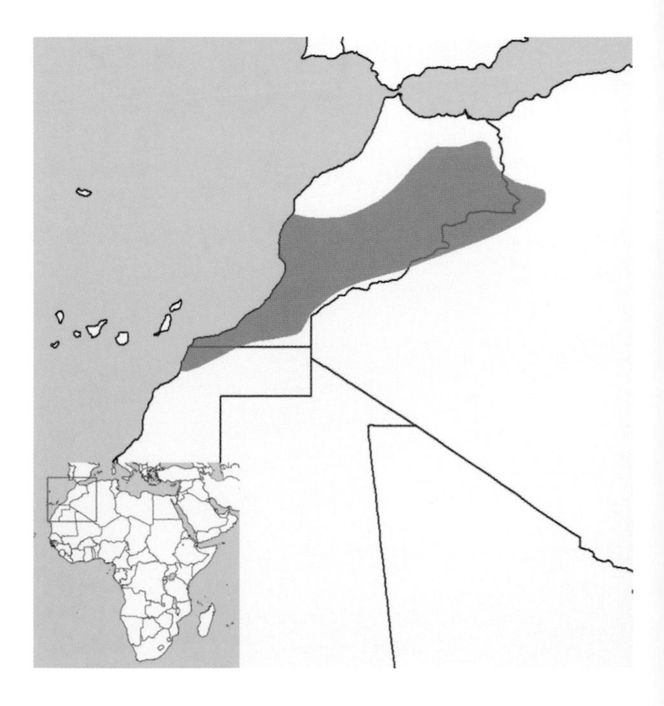

Atlantoxerus getulus. Photo courtesy Konstantin Simonov.

accurately reflect the behaviors of the squirrel in its native range. The Barbary ground squirrel is an omnivore, with 75 percent of its diet consisting of plant matter. The animal favors spurge (*Euphorbia*), *Atractylis*, *Mesembryanthemum*, mustard trees (*Nicotiana glauca*), *Salsola vermiculata*, lesser jack (*Emex spinosa*), *Scilla*, and *Asphodelus*. It also consumes several fruiting species—such as asparagus, *Rubia*, paddle cactus (*Opuntia*), boxthorn (*Lycium*), and *Prunus*—and disperses their seeds, perhaps even to the point of altering natural patterns of distribution. An opportunistic species, the Barbary ground squirrel feeds on crops when near a cultivated area. The animal occasionally consumes terrestrial mollusks, but its abundance is not dependent on snail density in Fuerteventura.

Barbary ground squirrels display predator vigilance behaviors common to many species of colonial ground squirrels. Throughout the day, adults rotate in shifts of up to an hour, lying on walls or other stone structures and monitoring the area surrounding the colony for threats. During the breeding season, males stand and call repeatedly from on top of the tallest pile of stones, attempting both to attract females and guard the region. These stone piles can be identified by the feces that collect around them. The Barbary ground squirrel's most common predators in Fuerteventura are housecats, Common Buzzards (*Buteo buteo*), Common Ravens (*Corvus corax*), and Common Kestrels (*Falco tinnunculus*). The squirrels also utilize walls and stone structures for thermoregulation, nesting, and protection against predators, entering their burrows when danger arises. Females spend a considerable amount of time in and around burrows as they care for their young, and they can often be seen in groups. Females come into estrus every four months and can produce up to three litters each year. Juveniles remain in the nest for as long as 6 weeks after birth. Some individuals within the Fuerteventuran and Moroccan populations of *A. getulus* have been identified as carrying *Acanthamoeba* and helminth parasites (*Brachylaima*, *Catenotaenia chabaudi*, *Dermatoxys getula*, *Protospirura muricola*, *Syphacia pallaryi*, and *Trichostrongylus*).

GENERAL REFERENCES: Bertolino 2009; Calabuig 1999; Gangoso et al. 2006; Gouat and Yahyaoui 2001; Linnaeus 1758; López-Darias 2006; López-Darias and Lobo 2008; López-Darias and Nogales 2008; Lorenzo-Morales et al. 2007; Machado and Domínguez 1982; Nogales, Nieves, et al. 2005; Nogales, Rodríguez-Luengo, et al. 2006; Petter and Saint-Girons 1965; Purroy and Varela 2003; Valverde 1957; Werner et al. 2005.

Spermophilopsis Blasius, 1884

This genus contains a single species of ground squirrel.

Spermophilopsis leptodactylus (Lichtenstein, 1823) Long-Clawed Ground Squirrel

DESCRIPTION: The long-clawed ground squirrel lacks external ears, and its upperparts are sand yellow to grayish yellow. The underparts are white. The terminal half of the tail is black underneath and light above, with a black fringe around the outside. There are two molts each year: the summer pelage is rough and bristly, and the winter pelage is long and silky. The several pairs of vibrissae on the ventral side of the body are also characteristic of this species. The feet are thickly furred all year long. As its name indicates, this species of squirrel has exceptionally long claws, more than 10 mm in length. Along with *Hyosciurus*, *Spermophilopsis* possesses the most well-developed claws.

SIZE: Sex not stated—HB 230-270 mm; T 27-81 mm; HF 59-62 mm.

DISTRIBUTION: *S. leptodactylus* is found on loess steppes, from the southeastern shore of the Caspian Sea east to Lake Balkhash (Kazakhstan), including Turkmenistan, Uzbekistan, southeastern Kazakhstan, northeastern Iran, northwestern Afghanistan, and western Tajikistan.

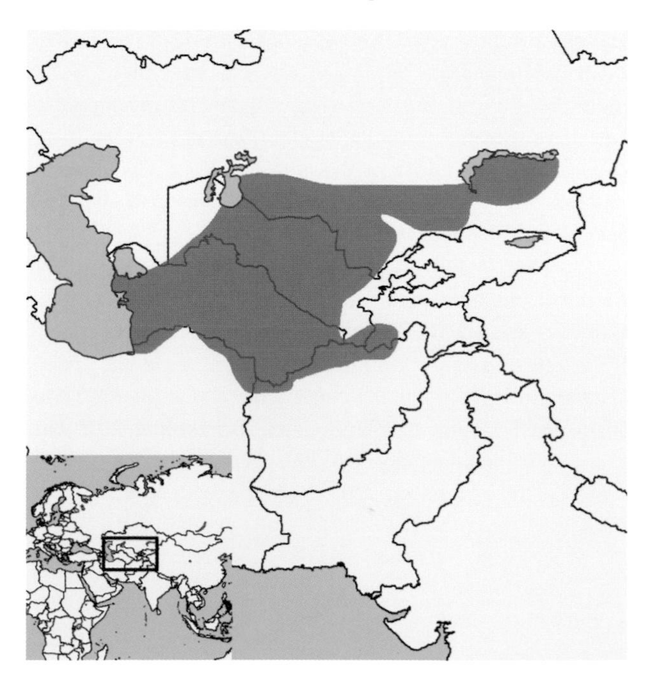

GEOGRAPHIC VARIATION: Three subspecies are recognized.

S. l. leptodactylus—the type locality is Karata, in "Russian Turkestan" (= Dagestan, Russia). See description above.

S. l. bactrianus—northern Afghanistan. The upperparts are pale fawn. The top of the head is slightly darker and browner. The rump is tinged with rufous. The tail is colored like the rump, and the tail also has a subterminal black ring and a pale fulvous tip. The outer surface of the limbs is bright fulvous.

S. l. heptopotamicus—"sandy deserts of Taukum [in Kazakhstan] on the left bank of the river Ili and near Bakanas;

Spermophilopsis leptodactylus. Photo courtesy K. A. Rogovin.

southern Prebalkans [in Bulgaria]" (Heptner and Ismagilov). No description is available.

CONSERVATION: IUCN status—least concern. Population trend—no information.

HABITAT: The long-clawed ground squirrel is found in sandy deserts, away from cultivated areas.

NATURAL HISTORY: This squirrel lives in small family groups, inhabiting burrows in brushy locales. It stays away from cultivated areas and thus has little interaction with farmers. But it has been known to uproot planted seed and destroy plants by consuming the underground roots. Although diurnal, it will adjust its activity to the season. During the summer it is in its burrow during the hottest part of midday, and during the winter it generally spends all its time out of its burrow, except when the weather gets very cold. It will venture up to 1000 m from the burrow to forage when resources are scarce. Its diet consists of fruits, seeds, bulbs, vegetation, and insects. Mating occurs in February and March. Three to six young are born in April or May. The skin of *S. leptodactylus* is marketed in substantial quantities in the fur trade.

GENERAL REFERENCES: Kullmann 1965; Lay 1967.

Xerus Hemprich and Ehrenberg, 1833

The genus *Xerus* contains four species. These are the ground squirrels of Africa south of the Sahara.

Xerus erythropus (É. Geoffroy, 1803) Striped Ground Squirrel

DESCRIPTION: The striped ground squirrel is a red sandy color, with pure white lateral stripes that stretch from its shoulders to its thighs. Its underbelly (where not bare) is thinly covered with white hair, and its face is highlighted with disconnected white around its eyes and behind its small ears. The hind feet are buff, and its other limbs are similarly lighter in color than the dorsal pelage. The red sandy color of the back extends to the end of the long tail, which fades gradually from black to white toward the tip. The tail is distichous rather than terete. *X. erythropus* can be distinguished from the sympatric species in its range by several features. In contrast to *X. rutilus*, which can also be found through Uganda, Sudan, and Kenya, the striped ground squirrel has a characteristic white or buffy lateral stripe running from the shoulders to the thighs, and it is also larger in size. This species differs from *X. inauris* and *X. princeps* by possessing three mammae rather than two.

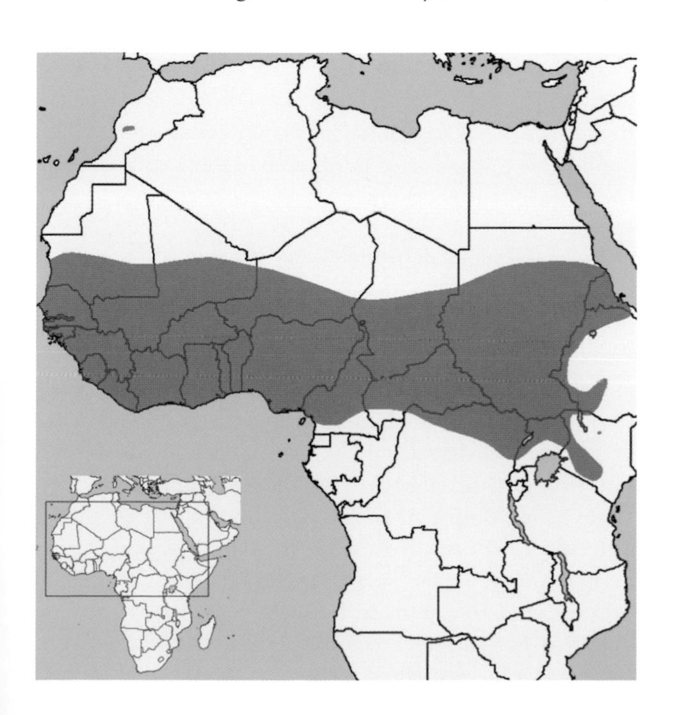

SIZE: Both sexes—HB 255 mm (223-290 mm); 219 mm (185-262 mm); Mass 500-950 g.

No sexual dimorphism is reported.

DISTRIBUTION: This squirrel's range stretches through three areas: (1) central Morocco; (2) northern Kenya; and (3) southern Mauritania, Senegal, Gambia, Guinea-Bissau, Guinea, Sierra Leone, Liberia, Côte d'Ivoire, southern Mali, Burkina Faso, Ghana, Togo, Benin, southern Niger, Nigeria, northern Cameroon, southern Chad, northern Central African Republic, northern Democratic Republic of the Congo, southern Sudan, northern Uganda, western Kenya, Ethiopia, and Eritrea. Abundance varies by region. For example, these squirrels are common in Kenya and Sierra Leone but rare in Morocco.

GEOGRAPHIC VARIATION: Six subspecies are recognized.

X. e. erythropus—West Africa, from southern Mauritania and Senegal to northeastern Nigeria, as well as in central Morocco. The animal is paler in color than the other subspecies, with a sandy-colored back and buff forearms.

X. e. chadensis—southeastern Niger through northeastern Nigeria, northeastern Cameroon, Chad, and Sudan. This subspecies is relatively small and pale. The dorsal pelage approaches a "pinkish buff" and the forearms are "cream-buff." The head and upper back are highlighted with a blacker coloration.

X. e. lacustris—northeastern Democratic Republic of the Congo and northwestern Uganda. Its dorsal pelage is darker than that of other subspecies, with more black, leading to an overall browner coat color. The region below the white lateral stripe is broader, and the forearms are a cinnamon brown.

X. e. leucoumbrinus—scattered distribution through Senegal, Cameroon, Sudan, and Ethiopia. Characteristic of species in the Sudan woodlands, this animal is darker, with brownish red tones, as opposed to yellowish or buffy; and it lacks the white patch posterior to the ears of *X. e. erythropus*.

X. e. limitaneus—southeastern Chad, northeastern Central African Republic, and Sudan. This subspecies is larger than *X. e. chadensis* and *X. e. leucoumbrinus*. The dorsal pelage is similar in color to *X. e. leucoumbrinus*, and it has two lateral lines: one pure white and the second darker than the dorsal coat.

X. e. microdon—Senegal, Gambia, Guinea, western Nigeria, and central Kenya. Common to the Guinea woodlands, this form has an overall pelage that is sandy buff, with black highlights on the crown of the head and stronger ochraceous coloring on the base of the tail. The forearms are a dull buff, moving into cream on the tops of the hands and on the feet.

CONSERVATION: IUCN status—least concern. Population trend—stable.

HABITAT: The striped ground squirrel is native to open woodlands or grasslands, mangrove and swamp forests, and coastal scrublands. It tolerates human disturbance remarkably well, often preferring secondary forests to primary ones, and it can be found in cultivated lands throughout much of its range. In eastern Africa, this squirrel lives at elevations between 600 and 1800 m.

NATURAL HISTORY: The striped ground squirrel is diurnal and terrestrial, although animals have been observed climbing low trees. Individuals emerge from their burrows well after dawn and return several hours before dusk. The squirrel's biphasic activity patterns are greatly influenced by the weather; morning activity is usually more intense than that in the afternoon, but it can be delayed by clouds or rain. When the midday sun becomes strong, the animals periodically retreat to the shade and press their bellies against the cooler ground. They are attracted to areas of human disturbance and can often be spotted foraging in cultivated fields or standing upright and alert in clearings and along roadsides. This species is primarily vegetarian, with a diet composed of soft fruits, leaves, flowers, acacia pods, dry seeds such as karangiya grass seeds (*Cenchrus biflorus*), insects, eggs, young birds, and small reptiles. They are adept at cracking the shells of nuts, which they usually carry to their burrows to open. In Kenya and many other dry regions of their range where food is scarce, these squirrels hoard food items in their burrows or bury them in pits scattered away from burrow entrances and camouflaged with a dried leaf or stone.

Xerus erythropus. Photo courtesy Edwin Schuller and Gisela J. van der Velden.

Striped ground squirrels forage alone and are not particularly social. When two meet aboveground, individuals briefly greet one another. A strict hierarchy is maintained during the meeting, with individuals assuming submissive and dominant roles. This behavior may be agonistic in same-sex encounters, although this species is not very aggressive or territorial. Underground, conspecifics share burrows with interchanging mixed-sex groups and even other smaller rodent species. Unlike other colony-forming species, their social lives do not revolve around the burrow system. They maintain large home ranges that cover a vast system of tunnels, and the animals frequently alternate between burrows. Females occupy home ranges varying between 1.3 and 12.4 ha in size and make occasional excursions up to 40 ha; males have larger home ranges than females and may travel farther in a single day (3 km). The squirrels scent-mark their home ranges by rubbing their cheek glands on rocks and trees. Although they bury their urine, the animals defecate aboveground near burrow entrances. Burrow structure is elementary and focused around a central den lined with grass and twigs. The shallow (less than 1 m deep) tunnel system contains one to three entrances, as well as several "pop holes" that end slightly below the soil surface and can be excavated quickly as emergency exits. At night the squirrels block entrances with loose soil. Although the animals frequent cultivated areas, burrows are rarely dug near human disturbances. Those that are appear crudely formed for temporary shelter while foraging; they consist of only one or two entrances and contain food debris but no bedding or other signs of nesting. In addition to burrows, the squirrels utilize rock and tree-root crevices as well as termite nests for refuge at night.

Striped ground squirrels typically move slowly, alternating short jumps with moments of stillness, during which they sniff the ground or stand erect to survey their surroundings. The tail is flicked constantly; and it is raised vertically while resting, arched behind the back when walking, and carried horizontally when moving quickly with long leaps. If mildly alarmed, the animal sits erect, extends its forelimbs, arches its tail over its back, and extends the hairs of its tail. In intense distress, the squirrel lowers its body to the ground and positions its tail close to its back. Juveniles often play by imitating escapelike behaviors, rapidly jumping and alternating the direction of movement. The striped ground squirrel emits indistinct "chatters" that resembles the vocalizations of other squirrel species. Multiple males may pursue a single female in a mating chase during the breeding season. Females give birth to litters of two to six kittens.

The animals are recognized as a major agricultural pest, damaging maize (corn), peanuts, yams, *Gmelina* seeds, young bolls of cotton, sweet potatoes, cassava, *Pinus caribaea* saplings, and the pods of legumes. Striped ground squirrels are easily tamed as pets, and they have survived in captivity for up to six years. They are bred for meat in Benin and have nearly been domesticated, but they are also hunted in the wild and eaten by humans. Natural predators include raptors, servals (*Felis serval*) and other wild cats, jackals (*Canis*), and snakes such as the puff adder (*Bitis arietans*). Wild striped ground squirrels can serve as hosts for organisms causing multiple zoonotic diseases, including ticks (*Hyalomma, Ornithodoros*), *Bunyaviridae* (Bhanja virus), nematodes (*Gongylonema*), and trypanosomes (*Trypanosoma xeri*). Their salivary glands can contain rabies or *Streptobacillus moniliformis* (Haverhill fever). Zoo populations have supported *Gongylonema macrogubernaculum*.

GENERAL REFERENCES: Angelici and Luiselli 2005; R. E. Buchanan and Gibbons 1975; Coe 1972; Craig et al. 1998; Delany 1975; Delany and Happold 1979; Dorst and Dandelot 1970; M. R. M. Ekué, pers. comm.; Ewer 1965, 1966, 1968; Herron and Waterman 2004; Herron, Waterman, et al. 2005; Hoogstraal 1955; Hopf et al. 1976; Hubálek 1987; G. E. Kemp et al. 1974; Key 1985, 1990a, 1990b; Kingdon 1974; Linn and Key 1996; Logan et al. 1993; Marinkelle and Abdalla 1978; O'Shea 1976; N. Robinson 1969; Rosevear 1969.

Xerus inauris (Zimmermann, 1780)
South African Ground Squirrel

DESCRIPTION: The South African ground squirrel is characterized by white stripes that extend laterally from the shoulders to the thighs, a faint white coloration around its large eyes, small exterior ear flaps, and a prominent white tail that is dorsoventrally flattened and marked by two black bands near the base. The squirrel's black skin is covered by sandy brown or cinnamon-colored dorsal fur and a white ventral pelage. The face, sides of the neck, and undersides of the limbs are also white. The body hair is short on the back and longer but sparser (sometimes even absent) on the belly. *X. inauris* strongly resembles *X. princeps* in appearance, but the latter is asocial and typically prefers mountainous terrain with more rocky areas. The two species are sympatric in some areas, and they sometimes burrow within close proximity of one another.

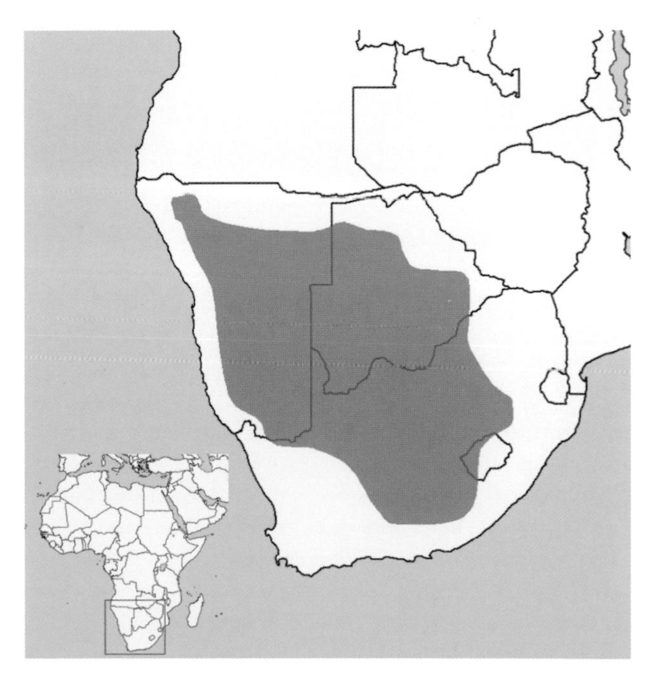

Xerus inauris. Photo courtesy Pitchaya and Rattapon Kaichid.

SIZE: Female—TL 435-446 mm; T 196-207 mm; Mass 444-600 g.

Male—TL 452-476 mm; T 194-211 mm; Mass 423-649 g.

Subtle sexual dimorphism exists, with males slightly larger.

DISTRIBUTION: The South African ground squirrel is distributed widely in Namibia, Botswana, South Africa, and Lesotho.

GEOGRAPHIC VARIATION: None.

CONSERVATION: IUCN status—least concern. Population trend—stable.

HABITAT: The South African ground squirrel lives in open terrain with sparse vegetation (sometimes dominated by the annual grass *Schmidtia kalahariensis*), such as karroid areas, grasslands, overgrazed land, floodplains, or dry watercourses. They prefer solid ground for making their burrows, but are also found in the dunes of the Kalahari Desert.

NATURAL HISTORY: This squirrel is a well-studied terrestrial and diurnal species, renowned for its ability to tolerate high temperatures and its sex-segregated social structure. It exhibits several behaviors that enable it to continue actively foraging when temperatures exceed 40°C. To decrease heat absorption, the animal turns its back to the sun and raises its dorsoventrally flattened tail over its back like a parasol, shielding its body from the sun. In an effort to minimize contact with the hot soil, an animal stands on its tiptoes while foraging. To further dissipate heat, the squirrel engages in "hearth rugging," in which it lies flat in a shady area with its limbs outstretched and its sparsely haired belly in close contact with the cool earth. The animal also utilizes "hearth rugging" to warm its body on cold mornings. These behaviors can raise or lower a squirrel's body temperature by over 5°C, and they allow an individual to forage for twice the amount of time. When temperatures become too extreme, the squirrels return to their burrows, which are built 60-70 cm below the ground and provide an insulated cool haven from the desolate heat above.

They spend their days feeding near burrow entrances. Their diet changes with the season, shifting from grasses and fruits during the wetter summer months to roots, stems, seeds, fruits, and insects during the drier season. The squirrels move over vast areas as they forage. For example, burrows spread over an area of 200-400 m^2 and have 22-30 access holes. Burrow clusters of individual female groups may be separated by several hundred meters, and males maintain home ranges as large as 12.5 ha. Neither males nor females are particularly aggressive, and colonies overlap their foraging grounds by an average of 26 percent, with individuals from different groups typically ignoring one another while feeding within close proximity. The social structure is segregated by sex and genetic relatedness into two kinds of colonies: matrilineal kin groups of adult females and mixed-sex groups of subadults that inhabit individual burrow clusters, and unrelated assemblies of males that rotate among their burrows. Both female and male colonies are composed of 6-12 animals. Only three reproductive females may exist at one time in a single female group, forcing subadult females to shift to another colony with fewer reproductive females once they are old enough to mate. Males live in vacant burrows except during mating, when they sleep in female burrow clusters. Bachelor groups occupy wide home ranges that span the home ranges of several female groups. Small groups of these males move among the females searching for receptive animals.

Females become sexually mature at approximately 8 months of age when only one adult female is present in a social group, and at 12 months when more than one reproducing female is present. Females spontaneously ovulate and remain in estrus for only four hours, mating with approximately four males during this time. The females give birth to one to three young in a nest burrow that they dig themselves, away from the group cluster. They remain there to raise the litter until the kittens are developed enough to emerge from the burrow, at which time they rejoin the social group. Lactation lasts an average of 52 days, and the young are weaned within seven days of joining the larger group. Once weaned, the young are cared for equally by all members of the female group. Females that successfully wean their young reenter estrus 56 days after the emergence of the litter, while females that lose their litters prematurely come into estrus 24 days after the loss. Females may produce up to three litters each year. South African ground squirrels are considered agricultural pests in some regions. They can serve as hosts to nematode (Ascaridae) parasites.

GENERAL REFERENCES: A. F. Bennett et al. 1984; Bouchie et al. 2006; Herzig-Straschil 1978; Lynch 1983; A. C. Marsh et al. 1978; Nel 1975; Skinner and Chimimba 2005; Skurski and Waterman 2005; Smithers 1971; Snyman 1940; Van Heerden and Dauth 1987; Waterman 1995, 1996, 1997, 1998, 2002; I. F. Zumpt 1968, 1970.

Xerus princeps (Thomas, 1929)
Damara Ground Squirrel

DESCRIPTION: *X. princeps* is sympatric and similar in coloration to *X. inauris*, but it is slightly larger in size and appears more grizzled than the latter species, due to the white-tipped hairs on the head and the lateral hip region. The tail is longer and features three black rings instead of two. The incisor teeth are yellow or orange-colored, rather than white as in *X. inauris*. The Damara ground squirrel's pale cinnamon brown dorsal coat and black skin contrast with the white pelage of the belly and the medial surfaces of the limbs, and the dorsal coat is marked by a white stripe running laterally from shoulder to thigh. The front of the face is white, the eyes are ringed by white lines, and the pinnae are small.

SIZE: Both sexes—HB 225-290 mm; T 210-282 mm; Mass 636 g.

Subtle sexual dimorphism exists, with males slightly larger.

DISTRIBUTION: The Damara ground squirrel is found from southwest Angola through the western side of Namibia to the northwestern edge of South Africa.

GEOGRAPHIC VARIATION: None.

CONSERVATION: IUCN status—least concern. Population trend—stable.

HABITAT: This squirrel is restricted to the Nama Karoo and Succulent Karoo biomes, where average annual rainfall ranges from 125 to 250 mm. This species lives in arid areas on rocky, hilly terrain with sparse vegetation.

NATURAL HISTORY: The Damara ground squirrel is diurnally active but sensitive to the extreme temperatures of its habitat. During the summer, the animals typically leave their burrows between 6:55 and 7:50 a.m. and return between 6:05 and 7:15 p.m., but they may leave later or return earlier to avoid high temperatures or heavy rain. After emerging from their burrows in the morning, individuals spend 10-15 minutes grooming and basking in the sun. They then depart from the burrow area to forage. They feed on grass stems and roots, mopane tree leaves (*Colophospermum mopane*, also referred to as *Copaifera mopane*), and plant lice. Individuals live alone or in small family groups consisting of two to four mothers with their young. Adult males are solitary but regularly visit two to seven female groups to mate. Gestation lasts 42-49 days and results in one to three young per litter. The young open their eyes within 21 days of birth. Social contact between individuals is generally rare, and no cohesive behaviors—such as allogrooming, playing, or affection—have been observed.

Damara ground squirrels build their burrows in rocky areas with few trees and little bush cover, such as gravel plains. Burrows are simple, composed of two to five openings and a single nest chamber with a floor located 67 cm below the surface. They tend to dig entrances under piles of stones or boulders, or beneath concrete slates near water pumps. Although they avoid sandy soils, they occasionally dig burrows on the plains and, in these instances, the en-

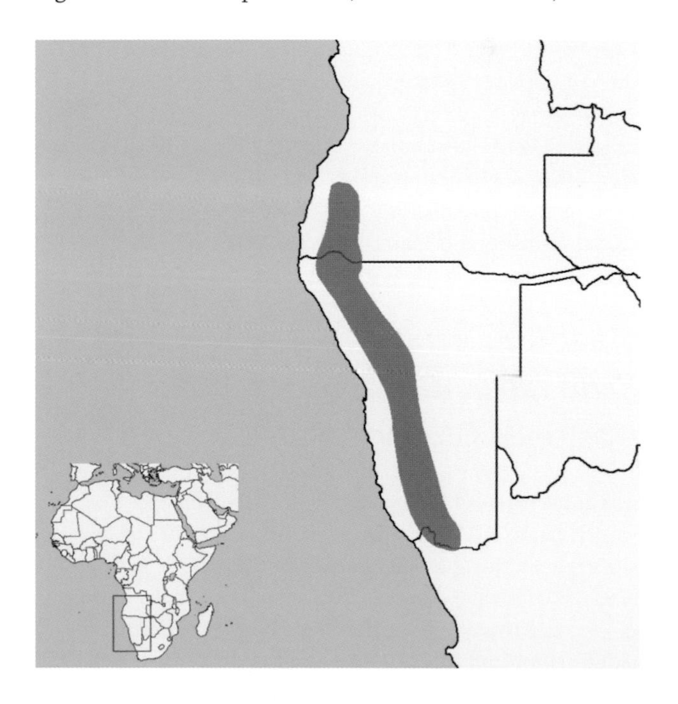

Xerus princeps. Photo courtesy Jane Waterman.

trances resemble those of *X. inauris*. *X. princeps* entrances are often accompanied by low mounds of earth and debris that were removed during excavation. Burrows are positioned 80 m or more away from other systems, and as near as 200 m to *X. inauris* colonies. Because of their insulation, burrows offer respite from the arid climate. During the summer, burrows remain at 25°C at night, in contrast to the outside temperature of 10°C, and at 32°C during the day, when the outside temperature is 38°C. The squirrels are highly resilient to the high temperatures of the area, however, and remain comfortably active at temperatures as high as 35°C. The animal utilizes its tail as a sunshade and moves to the shade of vegetation and rocks when the heat of the day becomes too strong. Damara ground squirrels do not tend to return to their burrows during the day; even when disturbed, they run to the nearest protective object, such as a rock or bush. They may stray up to 1 km away from their sleeping burrow. The Damara ground squirrel is a host for the pulicid flea (*Ctenocephalides connatus*).

GENERAL REFERENCES: De Graaff 1981; Haim et al. 1987; Herzig-Straschil and Herzig 1989; Herzig-Straschil et al. 1991; A. Roberts 1951; T. L. Robinson et al. 1986; Shortridge 1934a; Skinner and Chimimba 2005; Thomas 1929; Waterman and Herron 2004; F. Zumpt 1966.

Xerus rutilus (Cretzschmar, 1828)
Unstriped Ground Squirrel

DESCRIPTION: As its name suggests, the unstriped ground squirrel can be distinguished from other African ground squirrels by the absence of lateral stripes on its pelage. The dorsal coat is coarse in texture and ranges in color from pale tan to red brown, with individuals in drier regions having a paler coloration. The dorsum and the tail are grizzled with light and dark hairs, while the ventral pelage and the feet are uniformly paler than the dorsal coat. The hair on the belly is sparser than on the back of the animal. The eye ring is white or buff.

SIZE: Female—HB 229.1 mm (n = 16); T 182.6 mm (n = 16); Mass 252.0 g (n = 2).

Male—HB 268.1 mm (n = 13); T 187.8 mm (n = 13); Mass 306.7 g (n = 3).

Sex not stated—HB 226.2 mm (n = 8); T 202.5 mm (n = 2); Mass 368.8 g (n = 6).

DISTRIBUTION: The unstriped ground squirrel is an East African species common in the Somali-Masai arid zone. Its range stretches from northeastern Tanzania north through Kenya, eastern Uganda, Somalia, eastern Ethiopia, Djibouti, eastern Eritrea, and slightly into eastern and southern Sudan. An isolated population has been reported in the east-central part of Sudan, on the White Nile.

GEOGRAPHIC VARIATION: Eight subspecies are recognized. Records for Djibouti are identified only at the species level.

X. r. rutilus—Massawa (Eritrea) and slightly into eastern Sudan. See description above.

X. r. dabagala—northern Somalia. This form has a yellowish dorsal area.

X. r. dorsalis—western Kenya into eastern Uganda and southern Sudan. This form has a darker head, a fainter yellow tinge on the flanks, and white underparts and feet.

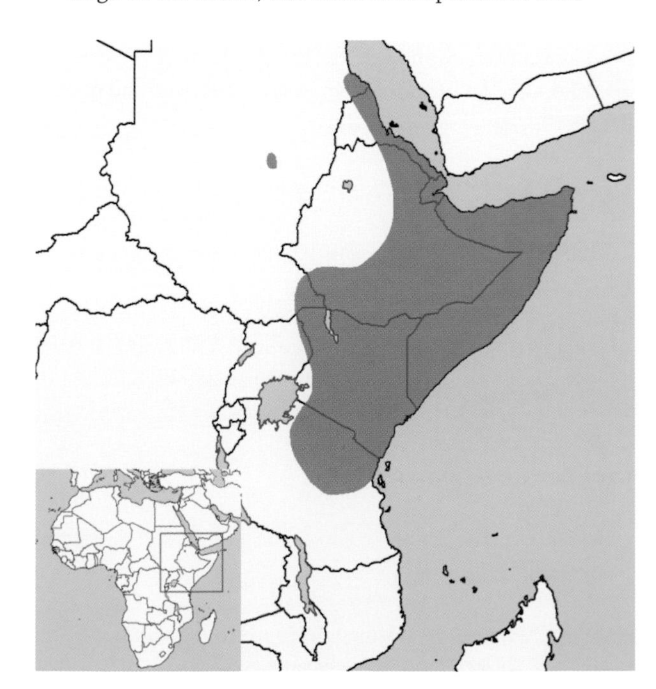

***Xerus rutilus*.** Photo courtesy B. D. Patterson.

X. r. intensus—Ethiopia. This form lacks the yellowish dorsal area of *X. r. dabagala*.

X. r. massaicus—Olorgesailie (north of Magadi, Kenya). This form has an overall white-yellowish pink color, speckled with black. The flanks are rufous pink without black speckling. The underparts are whitish yellow, with a brighter coloration along the sides of the chin, neck, and flanks. The red tinge is paler than in *X. r. rufifrons*.

X. r. rufifrons—northern Uaso Nyiro (Kenya). This form has a more extended and brighter rufous coloration on the front and head, and the dorsal pelage is more yellow.

X. r. saturatus—southeastern Kenya and northeastern Tanzania. The hands and feet are rufous, and there are indistinct red rings on the tail.

X. r. stephanicus—Ethiopia, Somalia, and northeastern Kenya. This form is pale yellow and rosy, with a reddish crown on the head. It has less black speckling.

CONSERVATION: IUCN status—least concern. Population trend—stable.

HABITAT: The unstriped ground squirrel occurs in dry savanna and scrubland habitats with precipitation levels of 800 mm or less. It inhabits thickets, gullies, and gravel flats; and it occupies elevational ranges from sea level to more than 2000 m. This species survives well in disturbed habitat and is often found in agricultural fields.

NATURAL HISTORY: A terrestrial and diurnal species, the behavior of the unstriped ground squirrel allows it to survive the extreme temperatures of its environment. Animals emerge from their burrows well after sunrise, and they proceed to bask in the sun and groom near their burrows for up to 30 minutes before beginning to forage. Thermoregulatory behaviors are displayed continuously while they feed—individuals alternate between foraging in hot open areas and resting in the nearby shade. They cool themselves by pressing their thinly haired bellies against the cool earth, similar to the behavior of *X. inauris*. The animals are omnivorous, and their diet consists of fruits (such as *Adansonia digitata*), seeds (such as *Acacia* and *Commiphora*), leaves, flowers, and insects. Burrow systems are spatially isolated and built with two to six entrances. Unstriped ground squirrels may situate the entrances of their burrows under thickets of vegetation, such as *Salvadora*, which reduces the temperature. In the shade of these thickets, temperatures typically remain around 35°C, in contrast to exposed soil-surface temperatures of over 67°C during midday in south Turkana (Kenya). When temperatures are high, individuals may limit their activity and remain in the thickets around their burrows for much of the day. This species is also known to occupy vacant holes excavated by other mammals, such as holes dug in termite mounds. The number of individuals in a colony varies from one to six individuals, ranging from a single-sexed group of animals, to solitary females with their young, to a single male with several females. Unstriped ground squirrels may also share a burrow with other species, such as *X. erythropus*.

Although linear social hierarchies exist within colonies, with males displaying dominance over females for food resources, this species is nonterritorial. Individuals move across large overlapping home ranges, varying in size from 7.0 ha for males to 1.4 ha for females. Home ranges encompass multiple burrow systems, and individuals may take refuge in another colony's burrow when endangered. Pregnant females occasionally move into a vacant burrow system toward the edge of their home range to give birth and raise their litters of one or two young. Once the young are weaned, the adult female may return to her original burrow, while the juveniles continue inhabiting their birth burrow. Unstriped ground squirrels interact frequently with one another through a variety of audible and visual displays. Aggressive behavior is exhibited through scolding calls, threatening lunges, chases, and occasional fights. When threatened by a superior member, subordinate individuals vocalize and piloerect their tails. Males also exhibit piloerection of the tail during sexual displays when approaching females. Although their life expectancy is not known in the wild, a wild-caught male survived 6 years and 56 days in captivity. *Xerus rutilus* is a host to the cestode *Catenotaenia geosciuri* and various ectoparasites, including the flea *Synosternus somalicus* and the specialist tick *Haemaphysalis calcarata*.

GENERAL REFERENCES: Coe 1972; O'Shea 1976, 1991.

Tribe Protoxerini Moore, 1959

This tribe contains six genera.

Epixerus Thomas, 1909

This genus includes a single species, a ground squirrel of West Africa.

Epixerus ebii (Temminck, 1853)
African Palm Squirrel

DESCRIPTION: These midsized lean squirrels are characterized by chevron-shaped bands of red, black, and gray on the underside of their thick gray-tipped tails. Their tails either droop or are carried parallel to the ground behind the body. The short pelage is a muted red color peppered with yellow and black fur on the dorsal side and some white on the chest. Their heads are broad, featuring prominent cheek muscles and large hairless ears that reach above the crown. The African palm squirrel is often misidentified as *Protoxerus stangeri*.

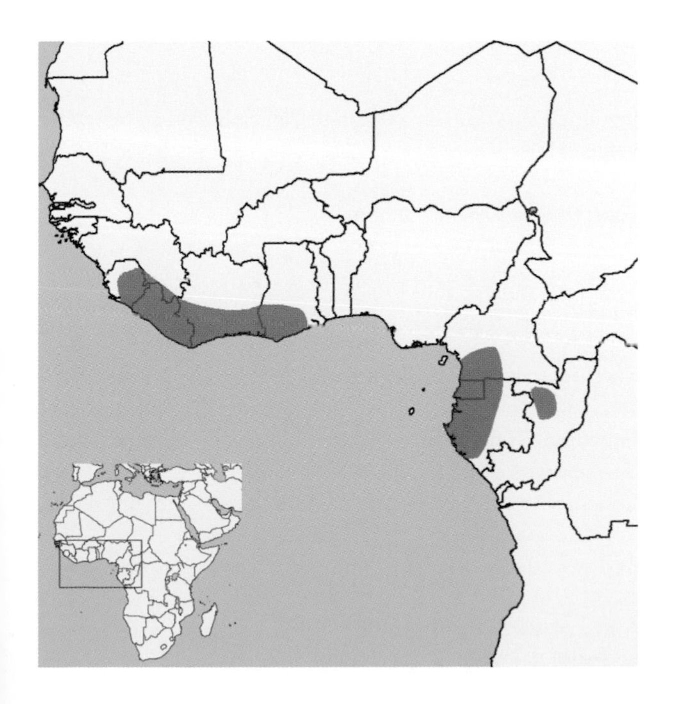

SIZE: Female—HB 278.3 mm (n = 6); T 284.3 (n = 4); Mass 388.0 g (n = 1).

Male—HB 288.3 mm (n = 7); T 277.0 mm (n = 5); Mass 652.0 g (n = 4).

Sex not stated—HB 280.7 mm (n = 3); T 290.0 mm (n = 2); Mass 577.4 g (n = 14).

DISTRIBUTION: This species is found in three pockets through West Africa: (1) from eastern Sierra Leone to southern Ghana, (2) from southern Cameroon to central Gabon, and (3) in the central part of the Republic of the Congo.

GEOGRAPHIC VARIATION: Three subspecies are recognized.

E. e. ebii—Côte d'Ivoire and Ghana. This subspecies has a rufous head and a brown dorsal region.

E. e. jonesi—Sierra Leone and Liberia. It has a red head and back.

E. e. wilsoni—Cameroon, Equatorial Guinea, Gabon, and the Republic of the Congo. It is distinguished by its muted red head and back.

CONSERVATION: IUCN status—least concern. Population trend—no information.

HABITAT: These squirrels favor evergreen rainforests, particularly undergrowth tree species such as raffia palms (*Raphia*). Individuals in Liberia were observed in wet and moist evergreen and semideciduous forests, and in Sierra Leone they were seen in montane forests, at elevations as high as 1020 m.

NATURAL HISTORY: The African palm squirrel's short period of daily activity, solitary social habits, wary personality, and skittish tendencies can make these animals difficult to observe in the forest, and this may contribute to the spe-

Epixerus ebii. Photo courtesy Julie Dewilde, Wildlife Conservation Society.

cies' reputation of being rare. However, the animal is well known and easily found by local people, who hunt this species with nets. Home ranges of adult males average 21.6 ha (*n* = 2), and a subadult female ranged over 13.9 ha (*n* = 1). The African palm squirrel nests in tree hollows with narrow entrances. They depart their dens at dawn and return in the afternoon, after only a few hours of activity. They maintain vigilance while feeding, often perching 0.5–1.5 m above the ground to enhance predator surveillance. The squirrels forage on the ground, eating select insects and fallen seeds and fruits. In Gabon they particularly favor thickly shelled nuts, such as panda nuts (*Panda oleosa*). The animals are unique in their ability to open these nuts: they either split the shells of old nuts or cut through the tough exterior shell of young nuts. Scattered larders (middens) of panda shells under low perches can be used to identify feeding sites of this species. In addition, African palm squirrels greatly facilitate the dispersion of panda nuts by hoarding seeds and burying them at distances of 20 m or more away from the parent tree. When alarmed, individuals climb to a low horizontal or vertical perch and scan for predators, their bushy tails hanging down. An animal will emit a series of staccato calls, either by quietly "chattering" their incisor teeth if mildly alarmed or announcing broad-frequency pulses if highly alarmed. The fecundity of this species has been poorly studied, and the only litter reported included two offspring in a nest. Females have been observed to have between eight and ten pairs of nipples.

GENERAL REFERENCES: Du Chaillu 1860; Emmons 1975, 1978, 1980; Grubb et al. 1998; Hoke et al. 2007; H. L. Kuhn 1964.

Funisciurus Trouessart, 1880

The 10 species of rope squirrels are inhabitants of the African forests, living at low to medium tree heights. Most are distinguished by lateral stripes. These species feed on a variety of foods, mostly consisting of vegetable matter, but they will eat insects when available.

Funisciurus anerythrus (Thomas, 1890)
Thomas's Rope Squirrel

DESCRIPTION: Thomas's rope squirrel is a medium-sized animal characterized by its long nose and by a bilateral pale stripe that runs from the shoulder to the hip. The head and dorsal pelage are brown with beige highlights and banded hairs, while the ventral coat ranges in color from white to gray to orange. The legs are a light caramel color and are tipped by well-developed claws. The relatively short tail is primarily red and is banded black to white toward the tip, but it appears dark with a white end when viewed from above. When at rest, *F. anerythrus* curls its tail vertically against its back. When moving, the squirrel keeps its tail up, but with the tip curling backward. Thomas's rope squirrel has small ears, and its eyes are framed above and below by beige-colored stripes.

SIZE: Female—HB 172.0 mm (*n* = 2); T 167.5 mm (*n* = 2).
Male—HB 176.7 mm (*n* = 3); T 166.3 mm (*n* = 3).
Sex not stated—HB 192.3 mm (*n* = 3); T 192.3 mm (*n* = 3); Mass 217.8 g (*n* = 18).

DISTRIBUTION: Endemic to Africa, this species is found from southern Benin to southern Nigeria and parts of Cameroon, Equatorial Guinea, northern Gabon, extreme southern Chad, the Central African Republic, the Democratic Republic of the Congo, western Uganda, and the tiny northern tip of Angola.

GEOGRAPHIC VARIATION: Four subspecies are recognized.

F. a. anerythrus—western Uganda south through the Democratic Republic of the Congo to Mount Kabobo and west

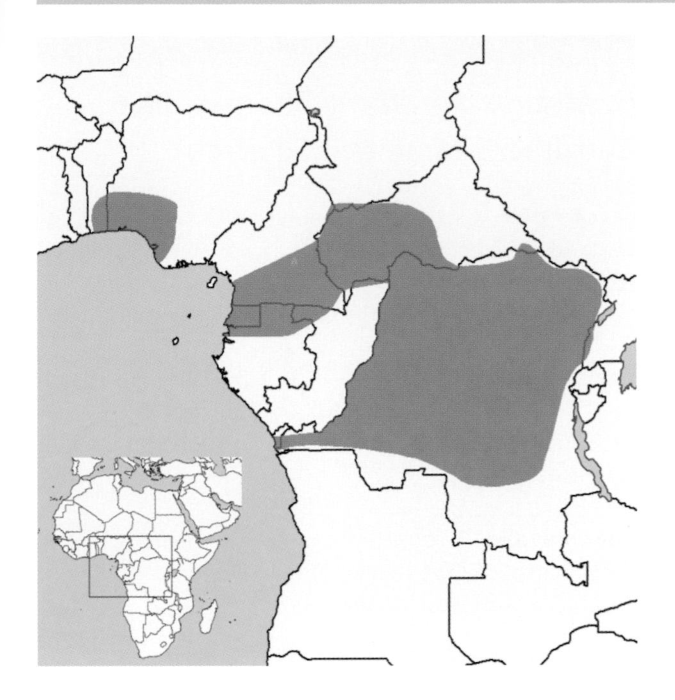

to the northern tip of Angola. The dorsal pelage is reddish brown, and the venter varies from yellow to red.

F. a. bandarum—Central African Republic into extreme southern Chad. The brown dorsum is tinged with beige hairs, and the underbelly is light gray to orange.

F. a. mystax—southern Cameroon to Gabon. This form has a reddish brown dorsum and an orange venter.

F. a. raptorum—Benin and Nigeria. This form has a grayer back and a whitish belly region.

CONSERVATION: IUCN status—least concern. Population trend—no information.

HABITAT: This squirrel favors habitats near permanent or seasonal water, including swamps (especially raffia palm [*Raphia*] swamps), gallery forests, river-based islands, and riversides. One known exception is the Gabon population of *F. a. mystax*, which avoids terra firma forests, even when the forest borders water.

NATURAL HISTORY: Thomas's rope squirrel is known to live in denser populations than any other African tree squirrel, a tendency complimented by its social nature. Although its community structure is unknown, this species' social behavior indicates that pairs may form monogamous bonds. In Gabon, individuals were seen with conspecifics 50 percent of the time (n = 431) and moved in pairs 28 percent of the time, grooming each other and lying on one another. Groups as large as six individuals have been observed feeding simultaneously in the same fruit tree. One heterosexual pair in captivity remained in contact while sharing a nesting box, yet competed aggressively for food. *F. anerythrus* is diurnal, scansorial, and omnivorous. It primarily consumes fruits and seeds (77% of the dry-mass stomach contents of individuals in Gabon, n = 15), but it also eats arthropods (20%), and mushrooms and green plants (3%). Squirrels that live in raffia palm (*Raphia*) swamps feed on the orange pericarp surrounding the seeds, often climbing in the crowns of the palms as they feed and temporarily caching the nuts in crevices above the ground. Commonly eaten arthropods include ants (92% occurrence) and termites (58% occurrence), among others. To locate these food items, the squirrels forage along the ground and as high as 13 m (average = 3.8 m).

The animal's vocalizations are frequent and distinct. Their low-intensity alarm call is a single "chuck" (45% of 166 call bouts) or a series of one to six rapid "chucks." As the squirrel calls, it rhythmically moves its tail vertically and stamps its feet: first the tail is quickly moved downward and it stamps its hindfeet, then the tail is slowly raised upward while it stamps its forefeet. Their high-intensity alarm call is particularly unique, consisting of two to four periodic pulses, followed by one to two long low-frequency whistles ("dada, dada, dadada . . . dadaweeeeeeou"), with only the final whistles audible from a distance. Thomas's rope squirrel has been recorded as averaging 1.2 embryos in eastern Zaire (n = 77 litters) and 1.5 embryos in Gabon (n = 6 litters). Females have two pairs of nipples. The young are raised in partially exposed nests situated among the vegetation of tree branches or the rachises of palm leaves that overhang the water. Nests are round, 20–24 cm in diameter, and composed of leaves lined with fibers or raffia palm (*Raphia*) leaflets. Individuals of this species have been identified as hosts for the monkeypox virus.

GENERAL REFERENCES: Emmons 1975, 1978, 1979, 1980; Jezek and Fenner 1988; Linnaeus 1758; Rahm 1970b; Rosevear 1969.

Funisciurus bayonii (Bocage, 1890) Lunda Rope Squirrel

DESCRIPTION: This medium-sized squirrel has an olive-tinged dorsal coat spotted with black, and a dull buff-colored stripe running along its side from the shoulder to the rump. The ventral pelage is a lighter gray grizzled with tan; and the tail, which is slightly longer than the body, is subtly ringed with dark black and gold. The eyes are ringed with buff. *F. bayonii* is often confused with the *F. congicus*, because both species occupy much of the same range (possibly being

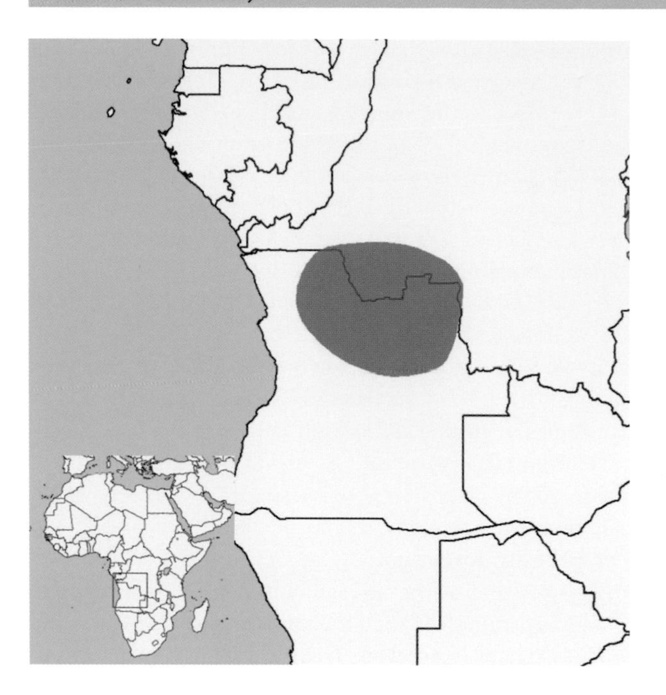

sympatric) and possess lateral stripes, but *F. bayonii* stands out as being the darker of the two species.

SIZE: Female—HB 250.5 mm (*n* = 3); T 199.0 mm (*n* = 3).
Male—HB 184.2 mm (*n* = 3); T 201.1 mm (*n* = 3).
Sex not stated—HB 177.5 mm (*n* = 2); T 150.0 mm (*n* = 2); Mass 135.0 g (*n* = 2).

DISTRIBUTION: The Lunda rope squirrel is found in Central Africa, from southern Democratic Republic of the Congo to northeastern Angola.

GEOGRAPHIC VARIATION: None.

CONSERVATION: IUCN status—data deficient. Population trend—no information. Sightings of the Lunda rope squirrel have not been reported since the mid-1900s, and the animal is probably uncommon or rare.

HABITAT: *Funisciurus bayonii* may prefer forest-savanna mosaic habitats and populate moist forests at low and medium elevations. Woodlands and savannas with Kalahari sand may also be favored.

NATURAL HISTORY: No information is available.

GENERAL REFERENCES: Amtmann 1966, 1975; Hayman 1951.

Funisciurus carruthersi Thomas, 1906
Carruther's Mountain Squirrel

DESCRIPTION: Carruther's mountain squirrel is a charismatic squirrel, characterized by its navy green dorsal pelage; light gray ventral pelage; and black-tipped, yellow-and-black striped tail. The eyes are ringed with pale-colored fur.

SIZE: Female—HB 224.0 mm (*n* = 3); T 191.7 mm (*n* = 3).
Male—HB 208.7 mm (*n* = 7); T 189.9 mm (*n* = 7).
Sex not stated—HB 229.0 mm (*n* = 2); T 192.5 mm (*n* = 2); Mass 268.0 g (*n* = 1).

DISTRIBUTION: *F. carruthersi* ranges from eastern Democratic Republic of the Congo through western Uganda and Rwanda to northwestern Burundi.

GEOGRAPHIC VARIATION: Four subspecies are recognized.

F. c. carruthersi—Ruwenzori (Burundi). See description above.
F. c. birungensis—Kalinzu and Kigezi forest reserves (Uganda). The sides are more yellowish green than the middle of the back. The tops of the ears are covered with grayish white hairs. The tail is mixed white, black, and ochraceous.
F. c. chrysippus—northwest of Lake Tanganyika (Democratic

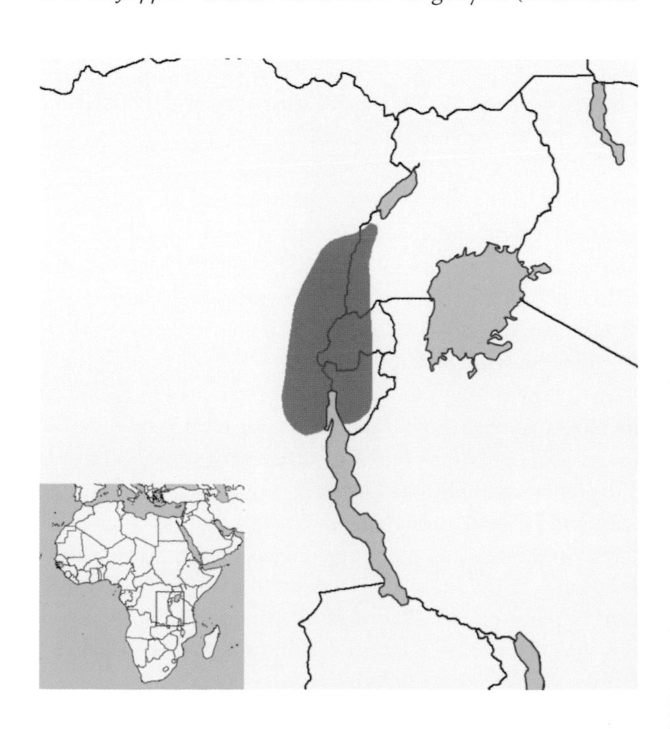

Funisciurus carruthersi. Photo courtesy Marcell Claassen.

Republic of the Congo). The ventral pelage is washed with yellow, and the sides of the face and ears are ochraceous.

F. c. tanganyikae—north of Lake Tanganyika, into Burundi. The body color is darker and the limbs are a brighter ochraceous color. The legs are a brighter ochraceous than the feet.

CONSERVATION: IUCN status—least concern. Population trend—no information.

HABITAT: Carruther's mountain squirrel thrives in dense moist montane forests, at elevations from 1500 to 2800 m. Individuals in the western part of the Albertine Rift in eastern Democratic Republic of the Congo prefer primary forests dominated by *Pygeum africanum*, as well as sougé (*Parinari excelsa*), *Conopharyngia durissima*, and *Albizia gummifera*; and secondary forests with *Macaranga kilimandscharica*, *Bridelia bridelifolia*, *Sapium ellipticum*, *Maesa lanceolata*, *Neoboutonia microcalyx*, and *Poliscias fulva*. The animal does not survive well in cultivated areas, but it can live in the high-altitude African redwood tree (*Hagenia*) that colonizes disturbed regions. *Funisciurus carruthersi* may be sympatric with *Heliosciurus ruwenzorii*, sharing both habitat and diet.

NATURAL HISTORY: The behavior and social structure of *F. carruthersi* are largely unknown, due to its skittish personality. The animals are active soon after dawn and throughout the day into the early evening, and they range from being on the ground to ascending into high tiers of vegetation. Individuals are most often observed alone, though they have been seen in pairs. They occasionally emit a hoarse-sounding "quack" that resembles the call of *P. lucifer*. Carruther's mountain squirrel is primarily vegetarian, preferring gourds and the fruits of *Bridelia*, *Alchornea*, *Carapa grandiflora*, and *Strombosia scheffleri*, but it may also consume insects. This species' most significant avian predators are Cassin's Hawk-Eagle (*Aquila africana*) and Ayres's Hawk-Eagle (*Hieraaetus ayresii*), as well as genets (*Genetta*). The squirrels construct large nests in woody vines and line them with small pieces of bark from a shrub known by the Bakiga people as *eminawa*. Females have six nipples, and they have been observed lactating in May.

GENERAL REFERENCES: Kaleme et al. 2007; Kingdon 1974; Linnaeus 1758; Schlitter 1989.

Funisciurus congicus (Kuhl, 1820)
Congo Rope Squirrel

DESCRIPTION: This small brown squirrel is characterized by the distinct pairs of stripes that run along its flank from the shoulder to the tail: a wide, cream-colored stripe on top, and a narrower dark brown stripe below. The dorsal pelage appears golden brown, with individual hairs that are black at the base and dark yellow or black at the tips. The ventral pelage, ear tips, cheeks, and throat are white; but the head is crested with brown fur. The eyes are framed above and below by white streaks. The outer surfaces of the limbs are cream colored, while the inner surfaces are white. The animal's tail is equal in length to its body, with bushy black-and-buff hair on top; the white underside of the tail is

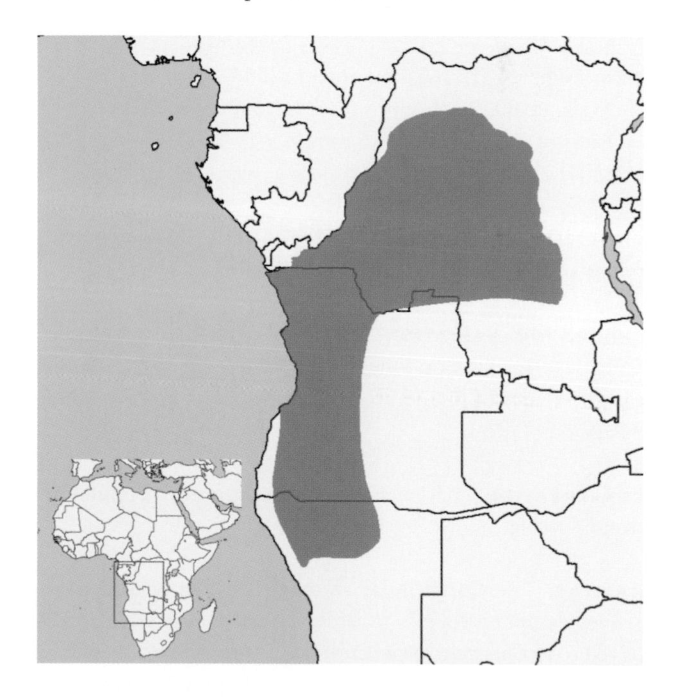

Funisciurus congicus. Photo courtesy Ian Merrill.

formed by individual hairs that are black at the base with alternating rings of buff and black, finished by a white-and-buff tip. The squirrel holds its tail against its back, with the last quarter curled. The pelage of the Congo rope squirrel varies with location, so that animals living in arid areas have lighter coats, and animals from more humid regions have darker coats with more golden highlights. The light and yellowish pelage of *F. congicus* makes it distinct from other *Funisciurus* species, which are characteristically darker and redder in coloration.

SIZE: Female—HB 160.5 mm (*n* = 8); T 162.4 mm (*n* = 7).
Male—HB 173.6 mm (*n* = 5); T 160.6 mm (*n* = 5).
Sex not stated—HB 150.5 mm (*n* = 2); T 165.0 mm (*n* = 2); Mass 111.2 g (*n* = 106).

DISTRIBUTION: This squirrel is found in the Democratic Republic of the Congo, Angola, and Namibia.

GEOGRAPHIC VARIATION: The subspecies that were originally described are no longer recognized, due to the significant influence of temperature and rainfall on body size and pelage coloration.

CONSERVATION: IUCN status—least concern. Population trend—stable.

HABITAT: The Congo rope squirrel inhabits coastal palm groves, mopane forests (*Colophospermum mopane,* also referred to as *Copaifera mopane*), and tall trees along waterways and projecting granite areas. This species prefers denser vegetation than the sympatric *Paraxerus cepapi,* and it can reach densities of 12-18 individuals/ha in optimal habitat.

NATURAL HISTORY: Diurnal animals, these squirrels are active from dawn (7:05-8:15 a.m., depending on whether the sky is clear or overcast) until several hours before sunset (6:00-6:50 p.m.). The squirrels are noticeably affected by the temperature fluctuations of the region. Their activity level is highest on cold mornings, probably in an effort to produce more body heat, whereas on hot days they shield their backs from the sun with their tails, with the white side of the tail face up to reflect the heat. At night Congo rope squirrels return to their nests, located in the junctions of tree branches or in holes in the trees. Nests are made of small pieces of vegetation, such as twigs, leaves, and grass. The animals spend their days foraging in the canopy and on the ground, in home ranges of 0.4-0.5 ha, although they prefer a low canopy of less than 2.5 m. These squirrels are seasoned acrobats of the trees, fleeing there if threatened; they then move quickly and efficiently through the branches to safety. When they do run along the ground, they position their tails above their backs, like question marks. Groups range across 0.4-0.5 ha per day.

Congo rope squirrels are omnivores, but while they occasionally consume mopane caterpillars (*Gonimbrasia belina*), they most commonly eat seeds, fruits, stems, and shoots. Favorites include mopane (*Colophospermum mopane,* also referred to as *Copaifera mopane*), velvet commiphora (*Commiphora mollis*), and *Grewia bicolor.* The squirrels are preyed upon by snakes, mammalian carnivores, and hawks, and they are parasitized by two species of ticks (*Haemaphysalis* and *Rhipicephalus*) and one species of flea (*Libyastus vates*). These squirrels live in groups of up to four individuals, and they remain in constant contact through visual observation and repeated high-pitched "chirps." They utilize scent to recognize individuals, and they strengthen social relationships by grooming one another. The animals emit specialized alarm calls to warn group members of dangers. For example, in response to a nearby raptor, an individual will give a series of high-pitched whistles, to which group members will respond by freezing. In the case of a ground predator, the squirrel will "chirp" as it flicks its tail. An entire group can jointly discourage a predator by "chattering" loudly from above while flicking their tails. Females have four nipples. Nothing is known about their reproductive behavior.

GENERAL REFERENCES: Amtmann 1966, 1975; Shortridge 1934b; Smithers 1983; Viljoen 1978, 1997a.

Funisciurus duchaillui Sanborn, 1953
Du Chaillu's Rope Squirrel

DESCRIPTION: This squirrel is gray along the limbs, feet, ventral side, and flanks, with the gray hair having white tips. The dorsal side and head are olive brown. Four dark brown dorsal stripes go from the base of the head to the tail; the hairs are annulated with greenish yellow, and their tips are black. The three stripes between the darker dorsal stripes are brownish yellow, and the tail is a bright red color with a black tip. The presence of gray pelage on the flanks and legs of *F. duchaillui,* and its larger size, distinguished this species from *F. isabella.*

SIZE: Female—Mass 205 g ± 21 g (n = 2).
Male—Mass 195 g ± 21 g (n = 2).
Sex not stated—HB 197 mm ± 1.3 mm (n = 4); T 210 mm ± 20 mm (n = 3); HF 45.0 mm ± 5.7 mm (n = 4).

DISTRIBUTION: This squirrel is found between the Ogooué River and the Ngounie River in central and south Gabon, and possibly in the Lékoumou region of southern Republic of the Congo.

GEOGRAPHIC VARIATION: None.

CONSERVATION: IUCN status—data deficient. Population trend—no information.

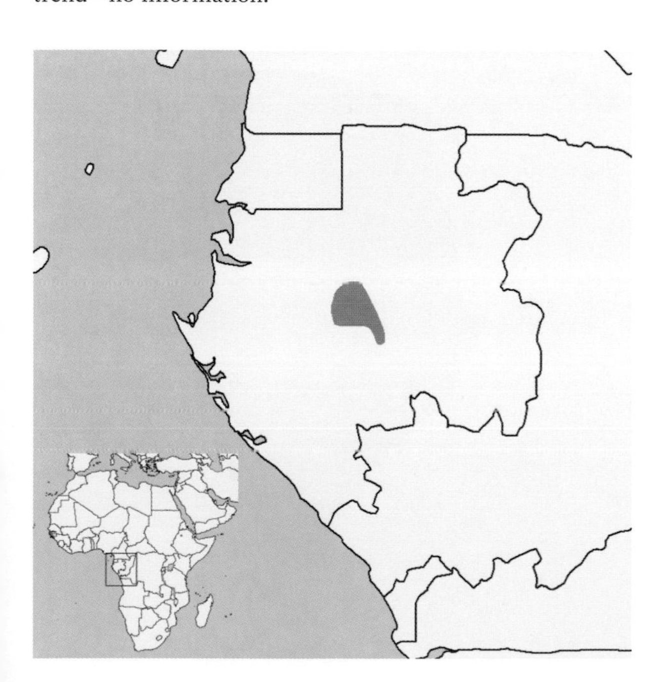

HABITAT: *F. duchaillui* lives in rainforests.

NATURAL HISTORY: Observed squirrels escape into the canopy.

GENERAL REFERENCES: Brugiere et al. 2005; Sanborn 1953b.

Funisciurus isabella (Gray, 1862)
Lady Burton's Rope Squirrel

DESCRIPTION: Lady Burton's rope squirrel is a small animal, with four black stripes down its golden brown dorsal pelage. The two median stripes extend from the crown to the tail, while the lateral stripes begin at the base of the neck. The four stripes are separated by brown stripes that are paler in color than the squirrel's shoulders. The brown back is grizzled with black hairs that have lighter tips. The ventral pelage is pale gray, with hairs that are gray at the bases and white toward the tips. The tail is thin and as long as the body, with a gradient of color from cream at the base to black toward the end, and is tipped with buff. The pelage pattern is similar to *F. lemniscatus,* except that in *F. isabella* the buff bands between the center stripes are darker than the fur between the lateral stripes in *F. lemniscatus.* Lady Burton's rope squirrel was named after the wife of Sir Richard Burton, who was the British Consul in Fernando Po (= Bioko Island, Equatorial Guinea) at the time the squirrel was collected.

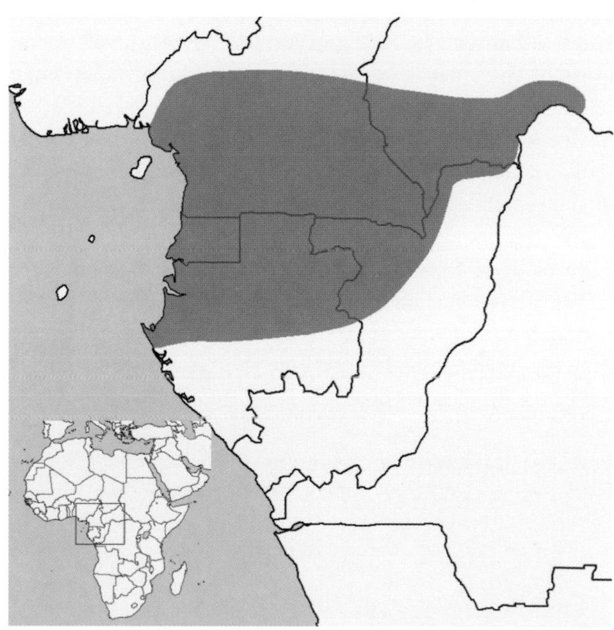

SIZE: Female—HB 165.3 mm (n = 6); T 161.3 mm (n = 6).
Male—HB 161.3 mm (n = 4); T 148.8 mm (n = 4).
Sex not stated—HB 162.8 mm (n = 4); T 155.0 mm (n = 2); Mass 107.1 g (n = 26).
There is reported sexual dimorphism.

DISTRIBUTION: Endemic to westcentral Africa, this species is present in patchy distributions. *F. isabella* is found from the western Cameroon Highlands as far east as western Central African Republic, and south through western Republic of the Congo, central Gabon, and Equatorial Guinea. A specimen was also identified from Brazzaville in southern Republic of the Congo, but this range extension is questionable.

GEOGRAPHIC VARIATION: None.

CONSERVATION: IUCN status—least concern. Population trend—no information.

HABITAT: Lady Burton's rope squirrel lives along a broad elevational range (up to 2100 m) in rainforests. This species prefers thick brush or vine tangles lower than 10 m in height, mature rainforests, and the dense secondary growth beside roads and in gardens and cultivated areas, habitats that it shares with *F. lemniscatus* in some parts of their range.

NATURAL HISTORY: These squirrels are diurnal and scansorial, foraging alone or in pairs (69% alone, 21% in pairs, and 10% in threes; n = 29 sightings). As omnivores, a vast majority of their diet consists of fruits and seeds (81% of the dry mass from stomach contents), with smaller amounts of green plants (9%), arthropods (6%, made up of ants, termites, and lepidopteran larvae), and fungi. These squirrels spend most of their day in vegetation above the ground. They make nests that resemble those of *F. lemniscatus,* consisting of dry leaves and fibers. Their frequent and distinct calls serve as the primary way to detect the animal's presence in its dense habitat. When mildly alarmed, Lady Burton's rope squirrel emits a progression of "chucks," singly (45% of 71 calls) or up to four in number. The animal's high-intensity alarm call is a sustained warbling sound, consisting of 2-10 connected frequency-modulated pulses of sound. The long tone is preceded by 1-14 repetitive shorter warbles. Females have four nipples but produce one kitten per litter.

GENERAL REFERENCES: Amtmann 1966; Bates 1905; Dubost 1968; Emmons 1975, 1978, 1979, 1980; Rosevear 1969.

Funisciurus lemniscatus (Le Conte, 1857) Ribboned Rope Squirrel

DESCRIPTION: *F. lemniscatus* is a small animal, with four black stripes stretching along the back from the base of the neck to the rump. The two middle stripes are separated by a muted brown band, while the outer stripes are separated from the inner stripes by light yellow lines. The dorsal pelage is brown, with the hairs banded black and buff; the ventral pelage is either white or cream colored, depending on the animal's geographical location. The dorsal surface of the long bushy tail is mixed black and buff, and the ventral surface is a pale yellow color. When at rest, the squirrel curls its tail over its back. The feet are narrow and elongated, which probably enhance its terrestrial mobility. This species resembles *F. isabella,* but the latter can be distinguished by its smaller size, distinct alarm calls, and different striping pattern, in which the two middle black stripes of *F. isabella* begin between the ears and all four black stripes are separated by pale yellow bands.

SIZE: Female—HB 167.6 mm (n = 10); T 135.8 mm (n = 10).
Male—HB 170.7 mm (n = 3); T 135.4 mm (n = 3).
Sex not stated—HB 169.3 mm (n = 4); T 160.0 mm (n = 2); Mass 140.9 g (n = 50).

DISTRIBUTION: This squirrel is endemic to West Africa, from central Cameroon south through Equatorial Guinea,

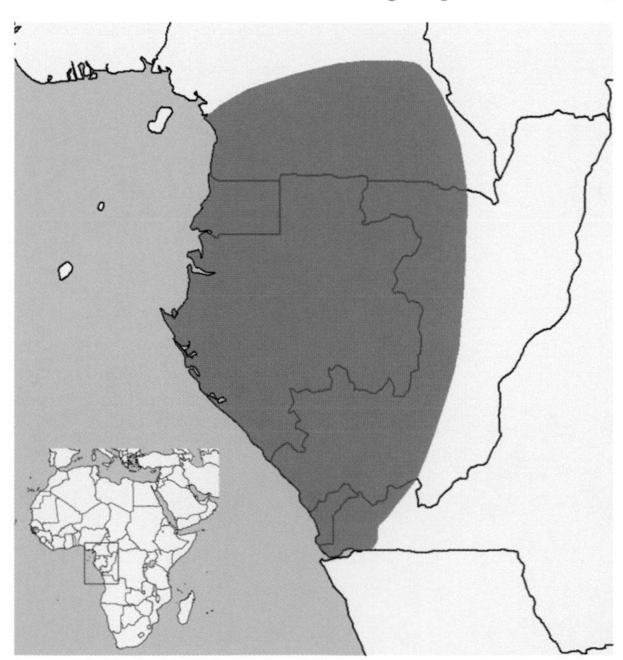

Funisciurus lemniscatus. Photo courtesy Julie Dewilde, Wildlife Conservation Society.

Gabon, western Republic of the Congo, western Democratic Republic of the Congo, and northern Angola.

GEOGRAPHIC VARIATION: Two subspecies are recognized.

F. l. lemniscatus—north of Gabon's Ogooué River. This form has a white ventral pelage.
F. l. mayumbicus—south of Gabon's Ogooué River. This form has a cream-colored ventral pelage.

CONSERVATION: IUCN status—least concern. Population trend—no information.

HABITAT: *F. lemniscatus* is rarely observed in disturbed vegetation, but it is abundant in lowland evergreen humid rainforests.

NATURAL HISTORY: Ribboned rope squirrels are diurnal and scansorial, spending most of the daylight hours away from their nests. Individuals leave their nests at dawn and return by late afternoon or nightfall. The squirrels forage primarily along the ground, and they are rarely seen in vegetation above 5 m. Ribboned rope squirrels possess a highly extendable tongue that they use to access the narrow living spaces of arthropods. This specialized feeding behavior enables the animals to consume mostly termites and ants. They are omnivorous, however, and they also feed on a large quantity of fruits and seeds. These squirrels spend a great deal of their time alone, although they occasionally socialize in groups of up to four individuals. Two adult males had home ranges of 1.0 and 1.24 ha and traveled at a mean rate of 51 m/hour, while a female stayed within a 1.6 ha area, moving at 43 m/hour; subadults have slightly smaller home ranges. This species maintains a structured social hierarchy and can aggressively defend resources such as food and space. This leads some individuals to keep their distance; for example, a male and a female with an overlapping home range of 0.5 ha remained 15–40 m apart from each other for two days. When observed in the wild, ribboned rope squirrels are typically giving alarm calls or foraging in pairs or small groups, with individuals spaced 5–20 m apart.

These squirrels frequently emit vocalizations. Their low-intensity alarm call consists of one to eight (most often three) repeated "chucks," and this is accompanied by a downward tail flick. An animal's high-intensity alarm call begins with single "chucks," followed by a more extended series of "chucks" or "chips" that are sounded in bursts of one to seven calls (most commonly two to three) that drop in frequency between groups of calls. Ribboned rope squirrels construct a system of leaf and den nests to which they escape in times of distress. For example, five individuals were identified as using 17 nests: 11 exposed leaf nests in small trees, 3 in tree hollows or lianas, and 3 in hollow logs lying on the ground. Leaf nests are 21 cm in diameter, with two to three entrances. They are built of large dried leaves and lined with fine plant fiber. These nests are located anywhere from on the ground to as high as 10 m, and they are placed either in an exposed area (on saplings at branch junctions, at the top of stumps, in hanging lianas, and against tree trunks) or tucked inside tree hollows, where they are often smaller, with openings that are slightly larger than the squirrel's body. If disturbed at night while resting in an exposed nest, the squirrels quickly relocate to another nearby nest. Females produce an average of 1.7 embryos and have four nipples. In Monte Mitra (Equatorial Guinea), hunters commonly snare ribboned rope squirrels and sell the skin and meat in local markets. Some wild populations of this species have been identified as hosts of the monkeypox virus.

GENERAL REFERENCES: Amtmann 1966; Emmons 1975, 1978, 1980; Fa and García Yuste 2001; Fenner 1994.

Funisciurus leucogenys (Waterhouse, 1842) Red-Cheeked Rope Squirrel

DESCRIPTION: The red-cheeked rope squirrel can be distinguished by its partial or pure white belly, which is tinged with orange in some subspecies. Its chest, chin, throat, and inner limbs are also white. The crown of the head is red with black highlights, and the backs of the ears are black. It is yellow across the shoulders; the back is red, grizzled with black hairs and tipped with orange toward the tail. The sides of the animal feature light orange brown stripes, which are faintly disrupted into spots toward the tail. Individuals vary highly in color between localities.

SIZE: Averages for this species are as follows:

Female—HB 206.7 mm (n = 10); T 147.2 mm (n = 10); Mass 251.9 g (n = 7).

Male—HB 204.4 mm (n = 8); T 148.8 mm (n = 11); Mass 271.4 g (n = 5).

Sex not stated—HB 192.5 mm (n = 2); T 165.0 mm (n = 2); Mass 250.0 g (n = 2).

Subspecies have also been noted as ranging in size:

F. l. leucogenys—HB 195 mm (185-202 mm); T 174 mm (160-183 mm).

F. l. auriculatus—HB 224 mm (213-240 mm); T 174 mm (151-200 mm).

F. l. oliviae—HB 194 mm (170-215 mm); T 152 mm (133-200 mm).

DISTRIBUTION: This species is found in southeastern Ghana, southern Togo, Benin, Nigeria, central Cameroon, southwestern Central African Republic, and Equatorial Guinea.

GEOGRAPHIC VARIATION: Three subspecies are recognized.

F. l. leucogenys—island of Fernando Po (= Bioko Island, Equatorial Guinea). This species is characterized by a pure white belly and pale lateral stripes.

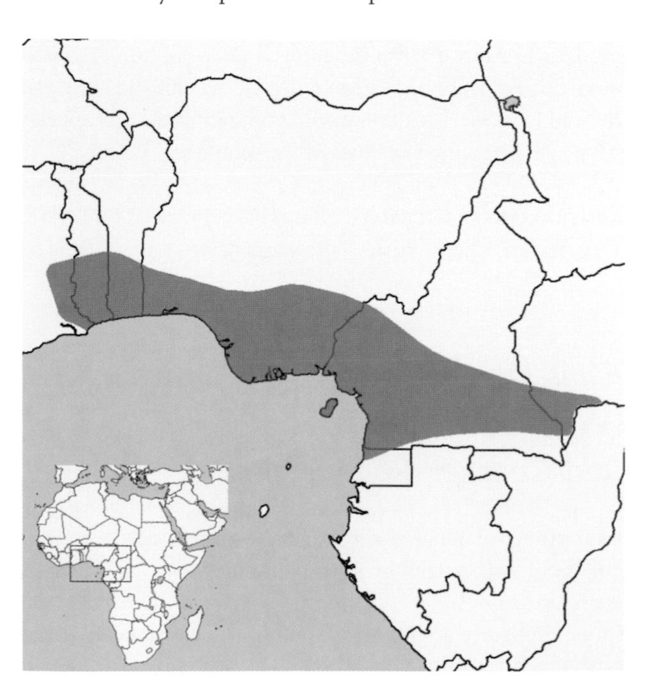

Funisciurus leucogenys. Photo courtesy Nik Borrow.

F. l. auriculatus—reported from the Benito River in Equatorial Guinea (15 mi from the river mouth), and also from Mussaka (on the lower Mungo River, east of Mount Cameroon) in Cameroon. This form has a unique whitish gray coloration on the neck and shoulders. The venter is partially or entirely orange, and the lateral stripes clearly change into seven rows of light spots toward the neck. The black ear hairs are denser and longer than on the other subspecies.

F. l. oliviae—patches from southeast Nigeria into Cameroon. This is the most widely distributed subspecies. These squirrels have a red crown and neck pelage, a dark patch of hair behind each ear, and an orange belly.

CONSERVATION: IUCN status—least concern. Population trend—no information.

HABITAT: The red-cheeked rope squirrel favors the valleys of westward-draining rivers among primary and secondary lowland and montane tropical rainforests. It tolerates mild disturbance within its habitat and lives within plantations greater than 2 ha, as well as in dense forest fringes.

NATURAL HISTORY: The red-cheeked rope squirrel is a terrestrial animal that forages on fruits and grains. It moves among a broad vertical assortment of microhabitats as it forages, ranging from shorelines and rocky ground to palms and forest trees. The squirrels nest in thick bush and in hollows within tree trunks and roots. Nests are lined with grass and dry leaves, and they have multiple entrances. Females appear to be in estrus continuously and can give birth every two to three months. Only one pregnant female has been documented in the wild, and she was carrying a single embryo in December. Red-cheeked rope squirrels are snared regularly on Bioko Island by local hunters.

GENERAL REFERENCES: Angelici and Luiselli 2005; Fa 2000; Grubb 2001; Rosevear 1969.

Funisciurus pyrropus (F. Cuvier, 1833)
Fire-Footed Rope Squirrel

***Funisciurus pyrropus*.** Photo courtesy Robert Barnes.

DESCRIPTION: This medium-sized squirrel is easily distinguished by its brilliant red limbs and face and white lateral stripes. It has a characteristic long nose and long narrow feet. The dorsal pelage is grizzled gray or black with hairs ringed black toward a buff tip, and its ventral pelage is pure white or ivory. The face is red below the crown and features pale hairs around the eyes and behind the ears. The long bushy tail is grizzled black and white with red highlights; the hairs are banded black at the base, red toward the middle, and white at the tip. When at rest, the squirrel curls its tail over its back; while moving, it holds the base of the tail vertically, with the tip curled backward or parallel to the ground behind its body.

SIZE: Female—HB 191.6 mm (*n* = 22); T 150.9 mm (*n* = 17); Mass 240.3 g (*n* = 7).

Male—HB 193.1 mm (*n* = 28); T 145.3 mm (*n* = 1); Mass 225.0 g (*n* = 33).

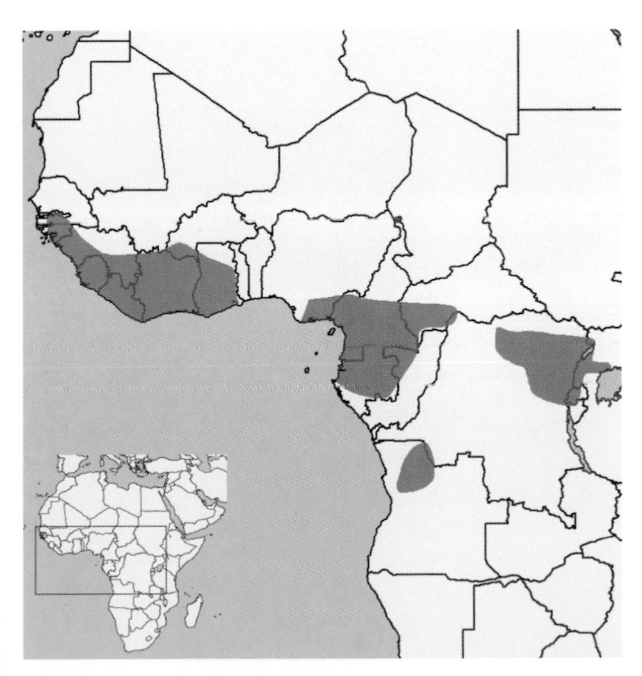

Sex not stated—HB 204.4 mm (*n* = 14); T 150.0 mm (*n* = 11); Mass 265.6 g (*n* = 6).

DISTRIBUTION: *F. pyrropus* is endemic to West Africa and is found in four disjunct distributions: (1) from southern Senegal southward through Gambia, Guinea-Bissau, Guinea, Sierra Leone, Liberia, Côte d'Ivoire, and western Ghana; (2) southeastern Nigeria, Cameroon, southern Central African Republic, Equatorial Guinea, western Republic of the Congo, northwestern Democratic Republic of the Congo, and Gabon; (3) northeastern Democratic Republic of the Congo through western Uganda and Rwanda and possibly Burundi; and (4) western Democratic Republic of the Congo to northern Angola.

GEOGRAPHIC VARIATION: Nine subspecies are currently recognized.

F. p. pyrropus—Gabon. See description above.

F. p. akka—eastern Republic of the Congo and Uganda. This subspecies lacks any red on the face, and the limbs and muzzle are tinged with orange.

F. p. leonis—Liberia. It has a deep rufous coloration, with red lateral regions.

F. p. leucostigma—southern Ghana. This form has a duller sienna red shading and red lines below the lateral stripes; it lacks any red on the crown.

F. p. mandingo—Gambia. The back is tan and black, and the limbs and ears are orange.

F. p. nigrensis—Nigeria, between the Cross River and the Niger River. The head is browner.

F. p. niveatus—northern Angola. This form has a grayish pelage.

F. p. pembertoni—from southern Cameroon to the Mayumbe Forest in the Republic of the Congo. This subspecies is brightly colored.

F. p. talboti—Mount Cameroon (Cameroon) and southeast Nigeria. The flanks are grizzled red and olive brown.

CONSERVATION: IUCN status—least concern. Population trend—stable.

HABITAT: The fire-footed rope squirrel is widespread within intact rainforests, such as tall evergreen forests and older secondary forests. It exhibits a moderate tolerance to disturbance and has been seen in some forest patches. This squirrel has been recorded at elevations from sea level to 1650 m on Mount Bintamane (= Mount Bintumane) in Sierra Leone.

NATURAL HISTORY: Fire-footed rope squirrels are diurnal and terrestrial. They are active soon after dawn and return to their nests in the late afternoon, between 3:30 and 6:00 p.m. Individuals are largely solitary, but occasionally they forage in pairs. Home ranges may differ by sex; one study tracked a male covering 5.2 ha, a female ranging over 1.0 ha, and a subadult female ranging over 2.3 ha. The animals forage slowly, with males moving approximately 61 m/hour and females moving 35 m/hour. Fire-footed rope squirrels forage on the ground, on fallen logs, and in vegetation below 1.5 m. These squirrels are omnivorous, and their long pointed noses allow them to raid nests and search small crevices for ants and termites. Despite this specialization, 80 percent of their diet consists of fruit and seeds. For their solitary nature, these squirrels are surprisingly vocal. When mildly alarmed, an animal emits a single or double "chuck" that resonates loudly through the forest. As it calls, it rhythmically moves its tail up and down. Their high-intensity alarm call is a quick-paced series of staccato machine gun-like "chucks" that lasts 20-40 seconds and sounds like "tatatata . . ." These calls are slightly similar to those of *Epixerus ebii* and are distinct from any other sympatric *Funisciurus* species. Another defining trait that separates this species from other *Funisciurus* is its tendency to nest on and under the ground. Although the animals occasionally build nests in the hollows of fallen logs, they prefer burrows and will sometimes utilize those excavated by other mammals, such as the forest giant pouched rat (*Cricetomys emini*) and the African brush-tailed porcupine (*Atherurus africanus*). Their own burrows are often situated near termite nests and built as simple runs with a central nest chamber and entrances on either side. When breeding, several males will pursue a single female in a mating chase. Females have four nipples but seem to raise only one to two kittens per litter.

GENERAL REFERENCES: Angelici et al. 2001; Emmons 1975, 1978, 1980; Rosevear 1969.

Funisciurus substriatus De Winton, 1899
Kintampo Rope Squirrel

DESCRIPTION: The Kintampo rope squirrel is a medium-sized squirrel with no characteristic markings other than one faint pale stripe running along each side. The greenish yellow or ochre dorsal pelage and limbs are speckled with black hairs; the ventral pelage, cheeks, and bases of the ears are paler ochre or brown, with hairs that are dark gray at the base and buffy brown at the tip. The lateral stripes are whitish, bordered ventrally by bands of darker hairs. The long tail is darker ochre than the dorsal pelage, with conspicuous rings of black and buff hairs. The dorsal tail hairs are black with short pale tips, and the lateral hairs are ochre with long cream-colored tips. This species strongly resembles *F. anerythrus* in form and color.

SIZE: Female—HB 161.0 mm (n = 5); T 155.0 mm (n = 5).
Male—HB 165.0 mm (n = 4); T 155.0 mm (n = 4).
Sex not stated—HB 180.8 mm (n = 13); T 170.0 mm (n = 2); Mass 186.1 g (n = 12).

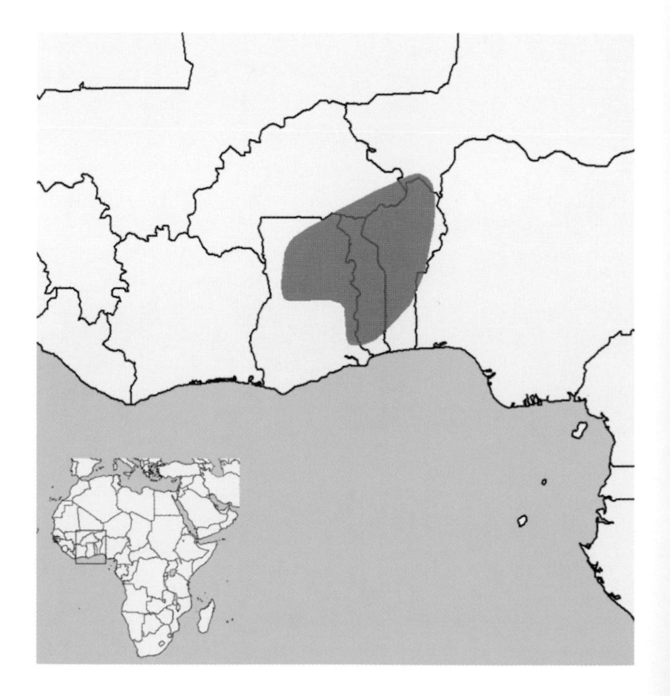

Funisciurus substriatus. Photo courtesy S. G. Davis.

DISTRIBUTION: *F. substriatus* ranges from southern Niger to southeastern Burkina Faso, Ghana, Togo, Benin, and extreme western Nigeria.

GEOGRAPHIC VARIATION: None.

CONSERVATION: IUCN status—data deficient. Population trend—no information.

HABITAT: The Kintampo rope squirrel is found in Guinea savanna habitat and drier patches, such as gallery forests. Individuals in Burkina Faso have been seen in rocky habitats and alongside rivers.

NATURAL HISTORY: A terrestrial species, the Kintampo rope squirrel is known to drop its tail when stressed, possibly as a defense mechanism against predators. The longest-living captive animal died at 6.9 years old.

GENERAL REFERENCES: Amtmann 1975; Refisch 1998; N. Robinson 1967; R. Weigl 2005.

Heliosciurus Trouessart, 1880

Sun squirrels comprise six species of medium to large African squirrels with banded tails. These species are omnivorous and will feed on animal matter (insects, birds, etc.) when they can obtain it.

Heliosciurus gambianus (Ogilby, 1835)
Gambian Sun Squirrel

DESCRIPTION: The Gambian sun squirrel exhibits a great deal of color variation between geographic regions, but it is generally duller and paler in color than other African squirrels. The fur on the back is buff or honey-colored speckled with black, giving it a grizzled appearance. The hairs on the back and the sides are short and gray, gray brown, or tan with black highlights. The ventral pelage, chest, throat, and inner surface of the limbs are white or light gray, and the hind feet are white or buff. The tail is ringed black and tan from the base to a black tip, and the hairs are tipped with white. Subspecies with more arboreal tendencies have longer and softer hairs, particularly in the dorsal pelage. *H. gambianus* can be distinguished from *H. rufobrachium* by its smaller size and white or gray thighs.

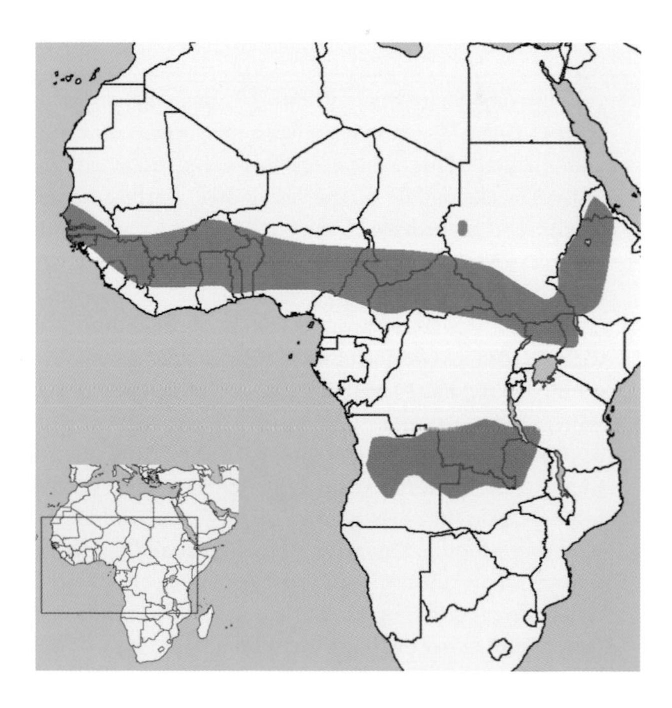

SIZE: Female—HB 204.0 mm (n = 13); T 230.3 mm (n = 13); Mass 328.6 g (n = 6).

Male—HB 217.7 mm (n = 12); T 239.9 mm (n = 11); Mass 245.0 g (n = 2).

Sex not stated—HB 200.6 mm (n = 51); T 201.5 mm (n = 2); Mass 212.9 g (n = 55).

DISTRIBUTION: This species is split into three geographic ranges through sub-Saharan Africa: (1) its northern range stretches through Senegal, Gambia, Guinea-Bissau, Guinea, Sierra Leone, Liberia, Mali, Côte D'Ivoire, Burkina Faso, Ghana, Togo, Benin, Nigeria, Cameroon, Chad, the Central African Republic, the Democratic Republic of the Congo, Sudan, Eritrea, Ethiopia, Uganda, and Kenya; (2) a small population is limited to a small patch of western Sudan; and (3) its southern range extends from Angola through the Democratic Republic of the Congo, Zambia, and Tanzania.

GEOGRAPHIC VARIATION: Sixteen subspecies are currently recognized.

H. g. gambianus—Gambia to western Nigeria. This squirrel prefers Guinea woodlands. It has a brown back and a light gray belly.

H. g. abassensis—Senegal to Sudan, south through central Cameroon and northeastern Congo, northern Uganda and eastern Kenya, and through southern Republic of the Congo, Angola and northern Zambia. It resembles *H. g. kaffensis*, but *H. g. abassensis* is paler, with shorter hair. The tops of the feet are golden, and the bottoms are reddish white. The hairs on the head and nose are slightly darker.

H. g. bongensis—Shari River (Central African Republic) to Bahr-el-Ghazal (Sudan). It prefers a moist riverine woodland habitat. This subspecies has a red tinge on the crown, neck, middle of the back, inner surface of the thighs, and at the base of the tail. The tail bands are indistinct, and the dorsal surfaces of the feet and arms are paler in color than in *H. g. gambianus*.

H. g. canaster—elevations around 1200 m in the foothills of the Jebel Marra (western Sudan). It is a paler gray color, with light red hairs around the anus and the base of the tail.

H. g. dysoni—west of Lake Rudolph (= Lake Turkana) in Kenya. This form has wide white eye rings and yellow highlights on the cheeks, crown, arms, hands, and feet.

H. g. elegans—Mount Elgon (western Kenya and northern Uganda). This subspecies has buffy-colored pelage on the back, the ears, the forearms, the back of the hands and feet, and the anal region. The rump and the base of the tail are tinged orange.

H. g. hoogstraali—Torit, Ikoto, and Obbo (Sudan). The back has a grizzled mixture of dark red, black, and white hairs; unconnected white stripes are above and below each eye. The cheeks, feet, hands, and lateral surfaces of the forearms are buff colored; the belly, throat, chin, edges of the hind feet, and the base of the tail are dark red.

Heliosciurus gambianus. Photo courtesy Sue Robinson.

H. g. kaffensis—Anderatscha (=Andracha), in Kaffa (Ethiopia). This subspecies has reddish dorsal pelage and a pure white belly. The limbs are red, with the tops of the feet a brighter rust color and the medial surfaces of the limbs a fainter reddish white. The tail is long and thick, with a black and white appearance.

H. g. lateris—White Nile (Sudan). It has a black and grayish white tail and gray feet.

H. g. limbatus—streamside forests in the Central African Republic and Cameroon. The thighs and the ventral and lateral sides are white; the back is faintly red. The tail bands are indistinct.

H. g. loandicus—northern Angola and the upper Lomani River (northern Katanga, Republic of the Congo). It has a yellow coloration on the hands, feet, and edges of the forearms.

H. g. madogae—Equatoria Province (Sudan) into Uganda. This subspecies is buffy, with a yellow ventral pelage and feet. The eyes are ringed with a wide white band; the anal region and the base of the tail are rufous.

H. g. multicolor—west of Lake Rudolph (= Lake Turkana) in Kenya, western Ethiopia, and Eritrea. This is a pale subspecies.

H. g. omensis—lower Omo River (Kenya). It has a pale white belly and a prominent white patch. The hands and feet are gray; and the back, sides, hips, and hind limbs are grayer than the dorsal pelage. The hairs of the buffy and black-ringed tail are distinctly tipped with pure white.

H. g. rhodesiae—west of the Mchinga Escarpment, in the High Plateau region (Zambia), northeastern Angola, and Tanzania. This subspecies has conspicuous white cheeks and silvery white discontinuous stripes above and below the eyes.

H. g. senescens—coastal woodland habitat in Senegal and a narrow belt of Guinea. It has distinct black and tan bands on the tail.

CONSERVATION: IUCN—least concern. Population trend—no information.

HABITAT: These arboreal squirrels prefer a dense habitat with tall trees, woodland savannas, seasonally flooded grasslands with an adjacent woodland, and rainforests. They are also found in low thickets and the woody vegetation near water sources. Fairly adaptable, the animals can live in high-forest farmlands and oil palm plantations, as well as in woodland areas burned by annual grass fires.

NATURAL HISTORY: *H. gambianus* is diurnal, emerging from its nest after dawn and returning before dusk. The animals spend most of their time in the highest branches of trees and are quite agile in their arboreal habitat. When they descend to the ground to feed, they move by bounding. These squirrels are omnivorous and opportunistic feeders, consuming everything from fruits, seeds, and acacia pods to insects, beetles, eggs, young birds, geckos, and lizards. They particularly like *Terminalia laxiflora* fruit, *Butyrospermum paradoxum*, and the husks of oil palm (*Elaeis*) nuts. In the wild, Gambian sun squirrels are solitary creatures, although parents remain with their kittens until the young disperse. This behavior changes in captivity, where individuals prefer to sleep with other animals, even with other species. The communication sounds of *H. gambianus* also vary between captivity and the wild. Captive Gambian sun squirrels repeatedly emit high-pitched squeaks and flick their tails when threatened or disturbed. In the wild, they use a variety of different calls, including a series of long "ker ker" sounds, a half-second trill, and a "chatter." The calls of young squirrels have been described as sounding particularly "musical." Whereas captive Gambian sun squirrels do not build nests, wild individuals form beds in the holes of woodland tree species. The squirrels select holes with entrances that are just large enough to admit the animals, and they line the cavities with fresh leaves and fibers from *Triplochiton* trees. Breeding probably takes place during two seasons, from July to August and December to January. The lifespan of *H. gambianus* has not been recorded from the wild, but one captive individual lived for 8 years and 11 months.

GENERAL REFERENCES: Delany 1975; Happold 1987; Kingdon 1974; Nowak 1999; Rosevear 1969; Setzer 1954; Watson 1975.

Heliosciurus mutabilis (Peters, 1852) Mutable Sun Squirrel

DESCRIPTION: The mutable sun squirrel is unique for the dramatic differences in its pelage coloration between molts, which appear to take place from October to May. This variation, in combination with its close resemblance to *H. rufobrachium* and *H. gambianus*, has caused a great deal of confusion over this species' identification. However, *H. mutabilis* generally has a paler pelage, distinct tail rings, and a clear demarcation between the dorsal and the ventral pelage, with the venter sparser and shorter haired than the dorsum.

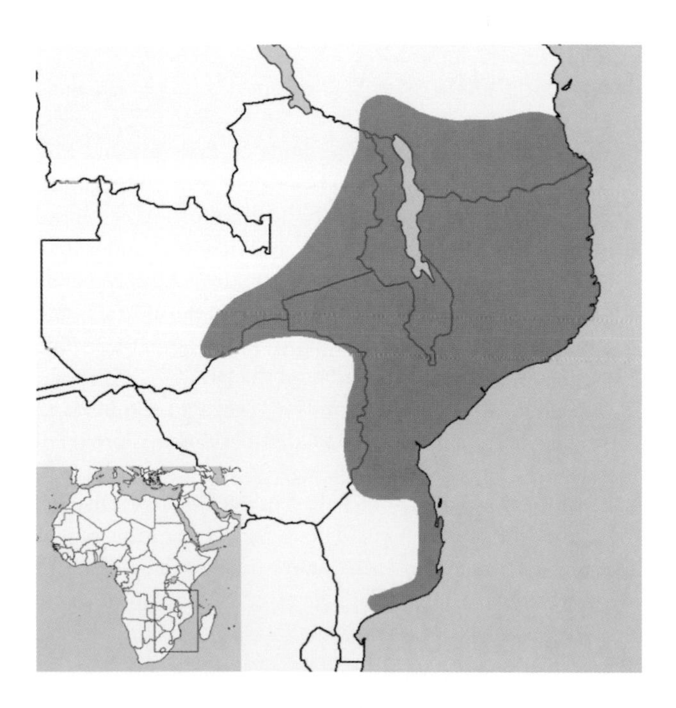

Heliosciurus mutabilis. Photo courtesy Bruce G. Marcot, used by permission.

SIZE: Female—HB 231.0 mm (*n* = 11); T 274.0 mm (*n* = 22); Mass 382.5 g (*n* = 2).

Male—HB 225.7 mm (*n* = 7); T 235.2 mm (*n* = 9); Mass 332.9 g (*n* = 2).

Sex not stated—HB 240.0 mm (*n* = 2); T 185.0 mm (*n* = 2); Mass 290.0 g (*n* = 2).

DISTRIBUTION: The range of this East African species stretches from eastern Zambia and northern Zimbabwe westward to central Tanzania and south through Malawi to southern Mozambique.

GEOGRAPHIC VARIATION: Five subspecies are recognized.

H. m. mutabilis—southern highlands of Tanzania and Mozambique. This subspecies has a dark brown dorsal pelage and a whitish, yellowish, or brownish gray ventral pelage. The tail rings are light, and they are subtler than in the other subspecies. Some individuals display a wide dark stripe along the back from the forehead to the tail, occasionally broadening on the flanks and the rump. Some also show a dark patch on the back.

H. m. beirae—southern highlands of Tanzania and Mozambique. This subspecies has distinct light-colored tail rings and a white ventral pelage.

H. m. chirindensis—Chirinda Forest in the Melsetter District, "Rhodesia" (Zimbabwe). The ventral coloration ranges from buffy white with an ochre tinge to deep ochre or gray brown. It has a paler pelage on the belly than on the flanks, and distinct tail rings.

H. m. shirensis—Zambia, western Malawi, far southwestern Tanzania, regions adjacent to Mozambique, and southeast Zimbabwe. The pelage is paler and lacks the dorsal black patch of *H. m. mutabilis*. The limbs and belly are white to buffy, and the tail exhibits 16 light tail rings.

H. m. vumbae—Zimbabwe-Mozambique highlands, spatially separated from *H. m. mutabilis* by 300 km. This subspecies has a dark pelage, with a uniform ventral and dorsal coloration. The dorsal band is dark and subtler than in the other subspecies.

CONSERVATION: IUCN status—least concern. Population trend—no information.

HABITAT: The mutable sun squirrel favors lowland or montane evergreen forests in Zimbabwe and Mozambique, as well as riparian forests and thickets in *Brachystegia* and *Julbernardia* woodlands.

NATURAL HISTORY: The mutable sun squirrel is a diurnal and relatively asocial species, living alone or with one other individual. These animals make nests in tree holes or thick tangles of vines on high branches. Individuals are known to consume *Kigelia* fruit and ivy leaves, but they may also eat insects and other vegetation. Little is known about their reproductive behavior; a female from eastern Zimbabwe was found pregnant with four young in August.

GENERAL REFERENCES: Ansell 1960; Skinner and Chimimba 2005; Smithers and Wilson 1979.

Heliosciurus punctatus (Temminck, 1853) Small Sun Squirrel

DESCRIPTION: The small sun squirrel is a dark-colored animal, with a brown coat grizzled with buff on its back and limbs. The ventral pelage is gray, and the thighs are dark gray. The long slender tail is ringed with alternating dark and pale-colored bands, and is lighter in color on the underside. It is sympatric with *H. rufobrachium* in parts of its range, but *H. punctatus* can be distinguished by its smaller size and a lack of red on the limbs. This species differs from *H. gambianus* in its distribution and its darker ventral pelage.

SIZE: Female—HB 184.8 mm (n = 28); T 207.0 mm (n = 28); Mass 168.6 g (n = 28).

Male—HB 182.3 mm (n = 30); T 201.4 mm (n = 30); Mass 165.9 g (n = 28).

Sex not stated—HB 190.0 mm (n = 20); Mass: 174.3 g (n = 19).

DISTRIBUTION: The small sun squirrel is found in the coastal countries of West Africa, from Sierra Leone through Liberia, Côte d'Ivoire, and Ghana. Its range may also extend to Guinea, but this has not been confirmed.

GEOGRAPHIC VARIATION: Two subspecies are recognized.

H. p. punctatus—coastal forested areas stretching from Liberia to the Volta River in Ghana. The form is darker in color.

H. p. savannius—inland savanna regions along Côte d'Ivoire. This form is paler in color.

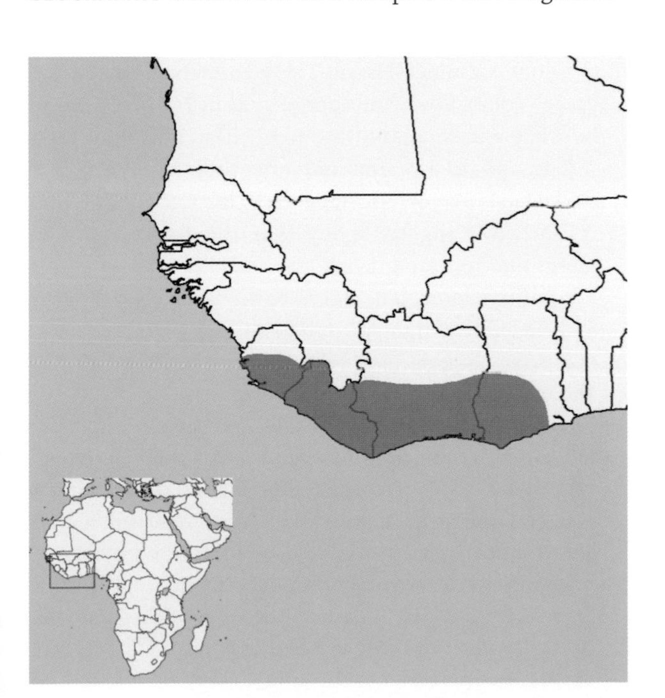

CONSERVATION: IUCN status—data deficient. Population trend—no information.

HABITAT: The small sun squirrel prefers lowland primary and secondary rainforests, as well as forest patches, savannas, and open areas. *H. punctatus* may be able to live in heavily disturbed agricultural areas, and it has been seen in farm bush.

NATURAL HISTORY: No information is available.

GENERAL REFERENCES: Rosevear 1969.

Heliosciurus rufobrachium (Waterhouse, 1842) Red-Legged Sun Squirrel

DESCRIPTION: The red-legged sun squirrel is a medium-sized animal, characterized by its dark brown or grayish coat, red-tinged legs, and thin tail banded with yellow and

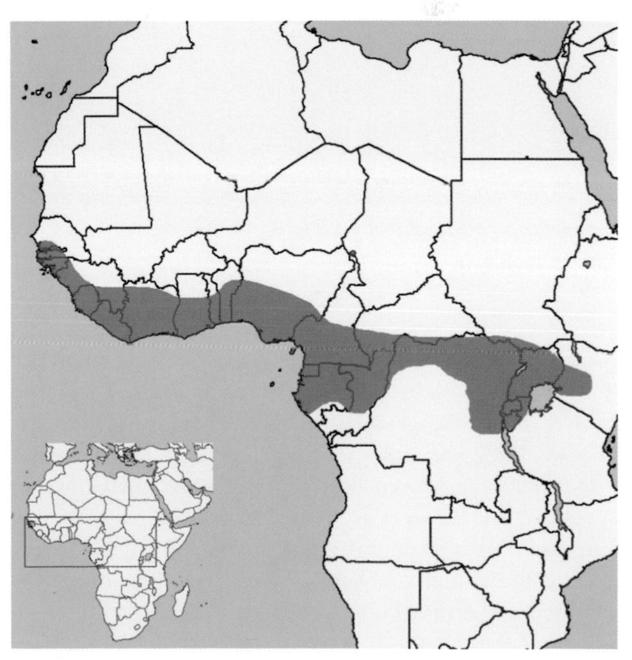

Heliosciurus rufobrachium maculatus. Photo courtesy Lars H. Holbech.

black rings. The body pelage varies regionally in color, but it is generally dark brown, gray, or red on the back and pale brown, red, or orange on the belly, with the entire body coat grizzled buff. The inner and outer sides of the limbs are bright rusty red, brown, or gray. The head is small, the ears are short, and the eyes are ringed by pale hairs. The tail is long and thin with faint rings of yellow and black, and it is usually held straight out behind the animal or drooped over a branch rather than curled against the back.

SIZE: Female—HB 274.0 mm (*n* = 3); T 249.3 mm (*n* = 7).

Male—HB 224.5 mm (*n* = 16); T 252.6 mm (*n* = 14); Mass 305.0 g (*n* = 1).

Sex not stated—HB 227.3 mm (*n* = 3); T 274.0 mm (*n* = 3); Mass 291.0 g (*n* = 1).

DISTRIBUTION: This squirrel is found in the coastal countries of West Africa and the countries of Central Africa, from southern Senegal through Gambia, Guinea-Bissau, Guinea, Sierra Leone, Liberia, Cote D'Ivoire, Ghana, Togo, Benin, Nigeria, Cameroon, Equatorial Guinea, Gabon, southern Central African Republic, the Republic of the Congo, northern Democratic Republic of the Congo, Uganda, Rwanda, Burundi, far northwestern Tanzania, and eastern Kenya.

GEOGRAPHIC VARIATION: Twenty-one subspecies are recognized by region.

H. r. rufobrachium—Bioko Island (Equatorial Guinea). The ventral pelage is light rust or gray.

H. r. arrhenii—Democratic Republic of the Congo. This subspecies is darker, with a rufous coloration on the nose and limbs.

H. r. aubryi—Gabon, southwestern Central African Republic, and the Democratic Republic of the Congo, between the Tshuapa and Kasai rivers. The dorsal pelage is very dark and features a reddish coloration on the flanks.

H. r. benga—Equatorial Guinea. This form is small, with the reddish coloration extending through the tail rather than ending at its base.

H. r. caurinus—Guinea-Bissau. These animals show the "average" coloration of subspecies within *H. rufobrachium*, but they exhibit a tannish brown chest, a white ventral pelage, and red forearms and inner thighs.

H. r. coenosus—Democratic Republic of the Congo. The ventral pelage is slightly lighter than the dorsal coat. The hands and feet are grayish tan, and the outer forearms are buffy colored. The tail is dark, speckled with gray and ringed with light-colored bands.

H. r. emissus—Nigeria. It is distinguished by its small size and brown underfur.

H. r. hardyi—Côte d'Ivoire. This subspecies is paler than the others, with a dully whitish ventral pelage and buff feet.

H. r. isabellinus—Togo to eastern Nigeria. The red coloration is very faint in these animals, and is replaced with a darker coloration; the tail is heavily ringed with black.

H. r. keniae—western slope of Mount Kenya (Kenya). This form has yellow highlights on the sides of the head; the tail is black with yellow and black bands.

H. r. leakyi—eastern Kenya, around Garissa. We found no available description.

H. r. leonensis—Sierra Leone. This subspecies is similar to *H. r. caurinus*, but bolder in color. It has a darker dorsal coat, redder limbs, and grizzled ochre and black feet.

H. r. lualabae—Democratic Republic of the Congo. This form is small, dark, and finely speckled, with white or buffy tail rings.

H. r. maculatus—eastern Sierra Leone to Ghana. This subspecies is dark, with a deep red coloration on the forearms and the inner surfaces of the thighs.

H. r. medjianus—Democratic Republic of the Congo. The ventral pelage is uniformly dark, with a conspicuous white chest area. The throat and the medial surfaces of the limbs are pale or dull red.

H. r. nyansae—Nyando River Valley (western Kenya), Rwanda, Burundi, and western Tanzania. The chin, the sides of the head, and the medial and lateral sides of the feet are a rusty red color. The tail is black with gray rings.

H. r. obfuscatus—southeastern Nigeria and Mount Cameroon (Cameroon). This subspecies is very dark, with deep brown and ochraceous feet and a black and ochraceous tail.

H. r. occidentalis—Cutia (Guinea-Bissau). The head, back, and limbs are uniform in color. The ears, throat, neck, and lower part of the face are russet red, as are the base of the tail and the inner sides of the thighs. The tail is circled by approximately 25 black rings set against a russet red base.

H. r. pasha—Democratic Republic of the Congo. The back and the base of the tail are more rufous, and the feet and belly are light rufous. The inner surfaces of the thighs lack a rufous suffusion. The belly is thinly covered with whitish gray hairs and is clearly demarcated from the flanks.

H. r. rubricatus—Democratic Republic of the Congo. This subspecies is very red, with a pale rufous coloration on the ventral pelage, throat, sides of the head, and medial surfaces of the limbs; it has a darker chestnut rufous coloration on the distal and lateral surfaces of the limbs and the underside of the tail.

H. r. semlikii—Semliki River (Democratic Republic of the Congo). This subspecies resembles *H. r. nyansae*, but *H. r. semlikii* is distinguished by a grayer overall appearance and finer speckling on the dorsal pelage. The midline on the back is faintly yellow.

CONSERVATION: IUCN status—least concern. Population trend—no information.

HABITAT: The red-legged sun squirrel is found in a range of natural habitats, including lowland evergreen tropical forests, mangrove swamps, and drier savanna and gallery forest areas. This species tolerates human disturbances well and can be found in secondary forests, gardens, and agricultural lands, especially cocoa and oil palm plantations.

NATURAL HISTORY: Red-legged sun squirrels are diurnal and arboreal, spending most of their time foraging in the middle and upper parts of the canopy. They leave their nests after dawn and return well before sunset. Nests are built by placing fresh leaves and twigs inside narrow hollows in trees or large branches, 1-20 m above the ground. Although 95 percent of their diet is composed of fruits, seeds, and green vegetation, individuals spend most of their foraging time hunting for arthropods. Red-legged sun squirrels, resembling mongooses with their sleek appearance and slinking gait, stalk meticulously through crevices among tree branches and lianas for ants, termites, and the larvae of moths and butterflies; they will even break apart dead wood to locate their prey. They capture insects with the forepaws and/or mouth and kill them by biting the head end, which they consume first. Captive squirrels adeptly caught and consumed flying birds inside their cages, and they also ate the birds' eggs. *H. rufobrachium* is not known to hoard food, except for prized food items such as birds. Red-legged sun squirrels may follow mixed flocks of insectivorous birds to increase their foraging efficiency as well as to receive increased protection against predators.

Individuals move as singles, pairs, or (very rarely) in threes. When partnered with company, both wild and captive red-legged sun squirrels are sociable; they nest, groom, and play with conspecifics. Although wild squirrels forage freely with one another, captive squirrels maintain a strict dominance hierarchy for food. *H. rufobrachium* females give birth twice a year to a litter of one or two young. These squirrels are less vocal than other species of *Heliosciurus* and are not often heard calling. Their low-intensity alarm call sounds like a single or double bark and is repeated one to three times for several minutes. Their high-intensity alarm call is composed of a low-amplitude descending frequency whine followed by a short quick trill of low-frequency notes; all these vocalizations are repeated several times at long intervals, distinctively sounding like the cooing of a dove. Each alarm call is accompanied by a stiff circular sweep of the tail while the feet are stamped. When sitting with a hanging tail, the animal moves it stiffly in random directions.

GENERAL REFERENCES: Emmons 1975, 1980; Morse 1970; Rahm 1970a; Rosevear 1969.

Heliosciurus ruwenzorii (Schwann, 1904)
Ruwenzori Sun Squirrel

DESCRIPTION: The Ruwenzori sun squirrel is a medium- to large-sized squirrel, distinguished from other *Heliosciurus* species by the white stripe running down the ventral pelage of some subspecies. The dorsal pelage and limbs are medium gray; the chin, throat, chest, and ventral pelage are pure white. The long slender tail is conspicuously ringed with gray and white bands.

SIZE: Female—HB 209.0 mm (n = 10); T 249.3 mm (n = 7).

Male—HB 224.5 mm (n = 16); T 252.6 mm (n = 14); Mass 305.0 g (n = 1).

Sex not stated—HB 227.3 mm (n = 3); T 274.0 mm (n = 3); Mass 291.0 g (n = 1).

DISTRIBUTION: This species is found in eastern Democratic Republic of the Congo, western Uganda, Rwanda, and Burundi.

GEOGRAPHIC VARIATION: Four subspecies are recognized.

H. r. ruwenzorii—Ruwenzori Mountains, at elevations from 1900 to 2000 m (Democratic Republic of the Congo). See description above.

H. r. ituriensis—mountains west of Lake Albert (on the border between the Democratic Republic of the Congo and Uganda), near Djalasinda (= Djalasiga) and Djugu. It has a darker ventral pelage and tail than *H. r. ruwenzorii*, with paler brown hair on the hind feet.

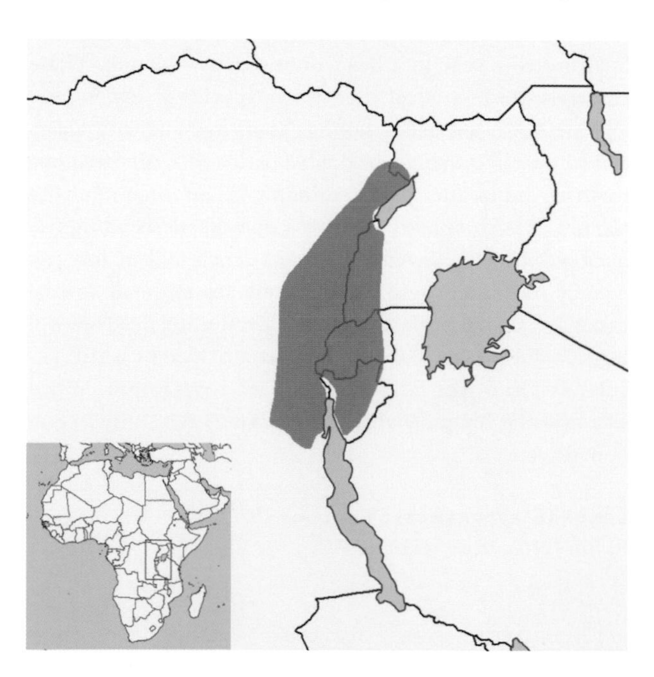

Heliosciurus ruwenzorii. Photo courtesy Jon Hall, www.mammalwatching.com.

H. r. schoutedeni—Democratic Republic of the Congo, Uganda, and Rwanda, in the mountains to the west and northwest of Lake Edward and Lake Kivu, as well as the Virunga Mountains. The lower face and feet are brown. A tan coloration surrounds the white strip on the ventral pelage.

H. r. vulcanius—alpine forests, at elevations above 2000 m, in the Nyungwe Forest National Park (Rwanda), through northwest Burundi to the northwest of Lake Tanganyika (Democratic Republic of the Congo). This form is characterized by a brown black dorsal pelage speckled with buff. It has a red coloration on the feet and alongside the white stripe on the ventral pelage.

CONSERVATION: IUCN—least concern. Population trend—no information.

HABITAT: The Ruwenzori sun squirrel is generally found in alpine habitats ranging from 1600 to 2700 m in elevation, and it survives in patches of secondary forest and cultivated land. Its subspecies tolerate a spectrum of disturbances, with *H. r. schoutedeni* preferring forest edges and lightly wooded areas, *H. r. ituriensis* selecting transitional and gallery forests, *H. r. vulcanius* preferring sparsely forested and cultivated areas, and *H. r. ruwenzorii* generalizing among all montane forest types.

NATURAL HISTORY: Ruwenzori sun squirrels are diurnal and arboreal, living near the ground and in lower vegetation rather than in the higher canopy. They build open nests of grass and leaves in the crooks of branches or vegetation. These squirrels are primarily vegetarian but occasionally consume insects, and they may hoard food items. They forage on the stems and leaves of several species, most notably various trees (*Parinari holstii, Syzygium cordatum, Conopharyngia holstii, Carapa,* and *Urera hypselodendron*), and the lichen *Usnea*. Ruwenzori sun squirrels are quite opportunistic in farmlands and plantations, feeding on guavas, papayas, bananas, and palm nuts when available. Little information is available about the social or reproductive structure of this species. Individuals usually remain solitary or socialize in pairs. One pregnant female was found in March with three large young, indicating that females may come into estrous and mate again before the previous litter is weaned and independent. They most commonly vocalize to others with a loud "chattering" call. As an individual moves through the trees, it holds its tail horizontally out behind it, parallel to the substrate.

GENERAL REFERENCES: Kingdon 1974; Rahm and Christiaensen 1963; Thomas 1909.

Heliosciurus undulatus (True, 1892) Zanj Sun Squirrel

DESCRIPTION: The Zanj sun squirrel is a large animal, characterized by its brown gray dorsal pelage and long thin tail

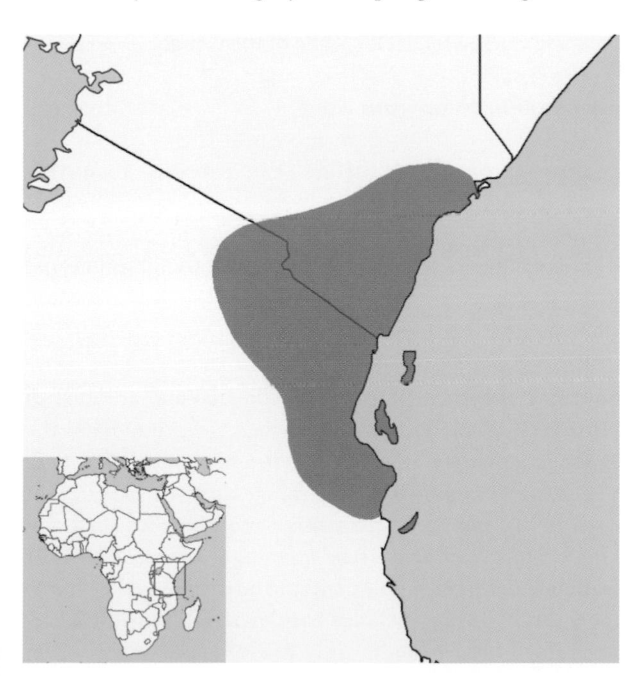

***Heliosciurus undulatus*.** Photo courtesy William T. Stanley.

ringed with 10–14 alternating black and pale bands. The ventral coat ranges from light gray to ochre. The color of the face, nose, and feet is similar to that of the dorsal pelage, with highlights of light gray to orange ochre. The intensity of coloration on the dorsal pelage varies geographically, with squirrels from higher elevations being darker and richer in color, animals to the north paler, and individuals from the south duller and grayer.

SIZE: Female—HB 235.3 mm (n = 4); T 270.0 mm (n = 4).
Male—HB 241.1 mm (n = 4); T 261.9 mm (n = 7).
Sex not stated—HB 232.2 mm (n = 20); T 277.4 mm (n = 20); Mass 315.0 g (n = 2).

DISTRIBUTION: This species is found in southeastern Kenya and northeastern Tanzania, as well as on the Tanzanian islands of Pemba, Zanzibar, and Mafia.

GEOGRAPHIC VARIATION: None.

CONSERVATION: IUCN—data deficient. Population trend—no information.

HABITAT: The Zanj sun squirrel lives at elevations below 2000 m in lowland and montane forests and thickets, as well as in secondary growth. This species is typically found in coastal or riverine forest areas, but records from elevations as high as 2000 m on Mount Kilimanjaro indicate that this species is present through a wide altitudinal range.

NATURAL HISTORY: Little information is available about this species. *H. undulatus* is omnivorous, foraging for most of the year on fruits, seeds, palm dates, leaves, and buds and consuming insects during some seasons. Individuals nest in the hollows of tree trunks or thick branches.

GENERAL REFERENCES: Grimshaw et al. 1995; Grubb 1982; Kingdon 1974, 1997.

Myosciurus Thomas, 1909

This genus includes a single species of pygmy squirrel, the only one found in Africa.

Myosciurus pumilio (Le Conte, 1857) African Pygmy Squirrel

DESCRIPTION: The African pygmy squirrel is distinguished most obviously by its small mouselike size. The dorsal pelage and head are a warm brown speckled with buff; the eyes are encircled by pale buff rings. The ventral pelage is light brown. The ears are brown, lined with bright pale buff on the inner and outer fringes. The feet are long and narrow.

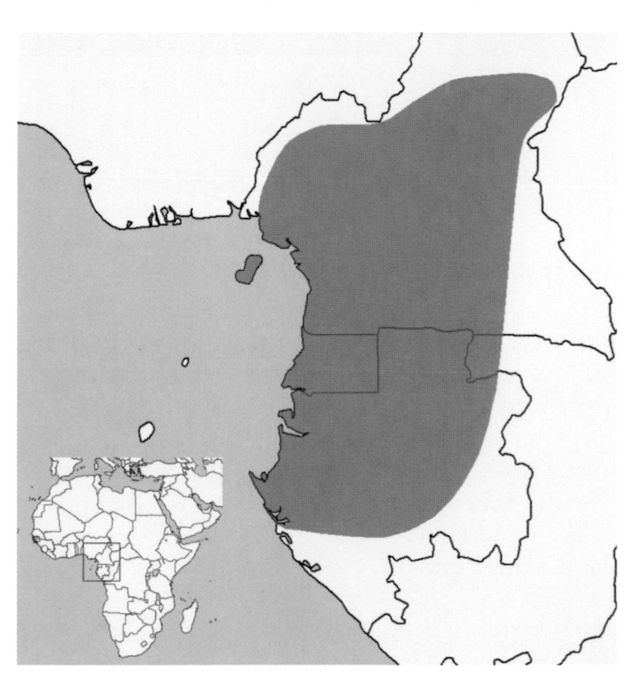

The thumbs on the forefeet are not visible externally and are reduced to a vestigial structure among the wrist bones. The tail is slightly shorter than the body length, slender, and reddish black. *M. pumilio* holds its tail straight out behind its body rather than curled over the back.

SIZE: Female—HB 74.8 mm (n = 5); T 50.0 mm (n = 1).
Male—HB 72.1 mm (n = 12); T 55.3 mm (n = 12).
Sex not stated—HB 71.3 mm (n = 2); T 55.0 mm (n = 2); Mass 16.1 g (n = 10).

DISTRIBUTION: The African pygmy squirrel is distributed from Cameroon southward through Equatorial Guinea, Gabon, and northwestern Republic of the Congo.

GEOGRAPHIC VARIATION: None.

CONSERVATION: IUCN status—least concern. Population trend—no information.

HABITAT: The African pygmy squirrel is found in lowland tropical primary forests, and it may be able to live in secondary growth.

NATURAL HISTORY: *M. pumilio* is diurnal and arboreal. It utilizes all vegetation levels between the ground and the canopy, but it primarily remains 0-5 m above the ground. The African pygmy squirrel forages on the surfaces of large living and dead tree trunks and branches, first ripping off pieces of bark, then scraping the surface with its teeth to extract the edible content and dropping the chip to the ground. Its diet includes bark material (possibly a bac-

***Myosciurus pumilio*.** Photo courtesy Nik Borrow.

terial or fungal film), fruits, ants, and termites. *M. pumilio* moves quickly around a tree, using a lizardlike locomotion in which it flattens its small body against the substrate, splays its elongate limbs around the trunk or branch, and secures itself with its hooked claws. Its small size and the elongate toes and missing thumbs on the forefeet enable this specialized feeding technique. These squirrels do not cache their food. Individuals spend most of their time alone, but occasionally they socialize in pairs. They communicate with a simple alarm call composed of repeated sequences of slow low-amplitude pulses, forming an overall "piping" sound. A calling individual carries its tail stiffly, straight out behind its body, and twitches the base sideways while keeping the tip centered. The tail serves as an effective alarm display, since the squirrels spend most of their time on tree trunks, with only their backs visible. Foraging continues while they emit calls. African pygmy squirrels probably nest within tree holes. Females may produce two young, although more research is needed to confirm this.

GENERAL REFERENCES: Emmons 1979, 1980.

Paraxerus Forsyth Major, 1893

These 11 species, the bush squirrels of Africa, are generally medium-sized animals. Bush squirrels are omnivorous, but principally vegetarian. They inhabit savanna woodlands and low brush. Some of these species are able to live under extreme environmental conditions, using behavioral adaptations.

Paraxerus alexandri (Thomas and Wroughton, 1907) Alexander's Bush Squirrel

DESCRIPTION: A small squirrel, this species has two black stripes on the back, separated by a mid-dorsal tawny yellowish stripe, and bordered by a dorsolateral white stripe on an olive or yellowish background. It has white on the back and the edges of the ears.

SIZE: Female—HB 102.9 mm; T 110.4 mm; Mass 40.2 g.
Male—HB 105 mm; T 127 mm; Mass 54.5 g.
Sex not stated—HB 106.4 mm; T 103.2 mm; Mass 45 g.

DISTRIBUTION: Alexander's bush squirrel is found in northeastern Democratic Republic of the Congo and Uganda.

GEOGRAPHIC VARIATION: None.

CONSERVATION: IUCN status—least concern. Population trend—no information.

HABITAT: *P. alexandri* is found in lowland tropical moist forests, at elevations below 1500 m, in tall relatively mature forests and occasionally in fallow plantations.

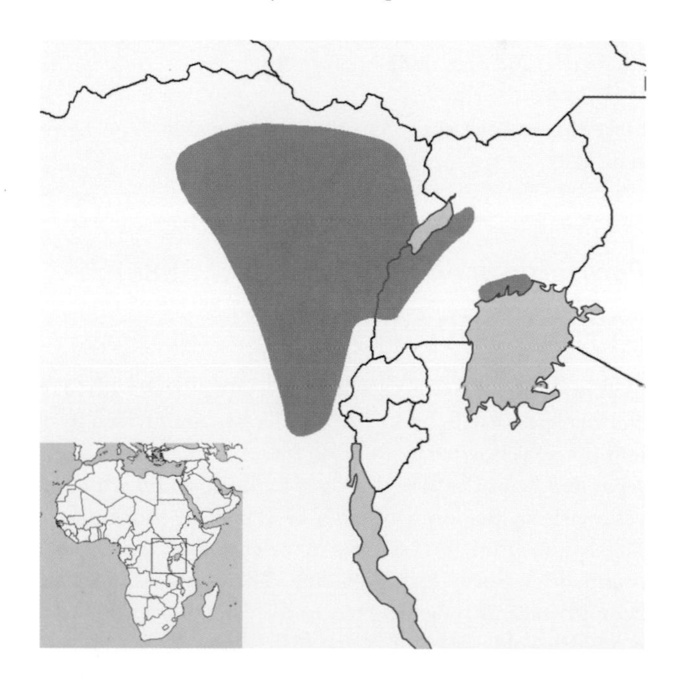

Paraxerus alexandri. Photo courtesy Nik Borrow.

NATURAL HISTORY: This small squirrel forages singly or in pairs, principally on tree trunks. It feeds on insects, vegetable matter, and occasionally on resin.

GENERAL REFERENCES: Grubb 2008; Kingdon 1974; Leirs et al. 1999.

Paraxerus boehmi (Reichenow, 1886)
Boehm's Bush Squirrel

DESCRIPTION: *P. boehmi* is a small striped squirrel, with four black stripes on the back. The middle two are divided by a mid-dorsal yellowish to greenish russet stripe, and they are separated from the lateral stripes by dorsolateral white to yellowish stripes on a reddish to yellowish background. There is no white on the backs of the ears. The eyes are surrounded by a brownish gold ring. There is a black stripe from the nose through the eye to the ear, bordered by pale brownish gold. The tail is greenish yellow brown mixed with black. The long hairs of the tail tip are black with small golden brown rings. *P. boehmi* is similar in size and color to *Funisciurus congicus*.

SIZE: Female—HB 129.6 mm; T 146.0 mm; Mass 74.8 g.
Male—HB 124.6 mm; T 135.6 mm; Mass 70.5 g.
Sex not stated—HB 129.8 mm; T 138.1 mm; Mass 60 g.

DISTRIBUTION: This squirrel is found in Sudan, northern and eastern Democratic Republic of the Congo, Uganda, Rwanda, Burundi, northwestern Tanzania, and northern Zambia.

GEOGRAPHIC VARIATION: Four subspecies are recognized.

P. b. boehmi—Marungu (southern Democratic Republic of the Congo). The upper side is olive intermixed with black. Single hairs are grayish black at the base and have olive gold brownish bands at their tips. Of the three light stripes along the length of the back, the two outer ones are whitish, and the midstripe is olive gold brown; all three are bordered by black stripes. The underbody and the insides of the arms and thighs are yellowish white. The throat region is a pale yellow, while the chin area is a rust yellow.

P. b. antoniae—small area above Stanley Falls (Democratic Republic of the Congo). The general ground color is pale greenish yellow. The median pale stripe is more yellowish; the outer two pale stripes are more yellowish white and run only halfway along the body, from the loins. The underparts are washed with a bright yellowish buff.

P. b. emini—northern and eastern Democratic Republic of

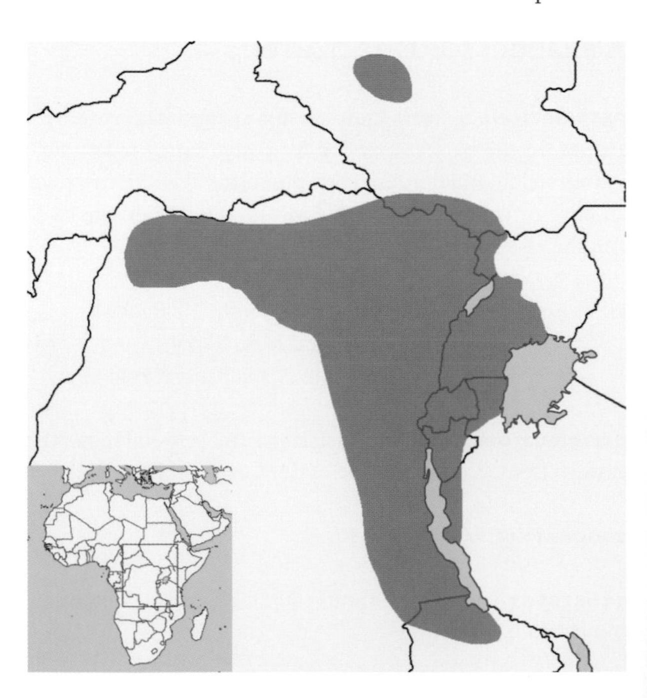

Paraxerus boehmi. Photo courtesy Marcell Claassen.

the Congo, Uganda, northwestern Tanzania, Rwanda, Burundi, and northern Zambia. The space between the dorsal stripes is more olive green white. The ground color is a more olive green.

P. b. gazellae—eastern Equatoria (Sudan). The body color is more pale and gray, with shorter dorsal stripes and with the outer stripes about 2.5 cm long. The underparts are grayish with little olive infusion.

CONSERVATION: IUCN status—least concern. Population trend—no information.

HABITAT: Boehm's bush squirrel is common in forests dominated by *Cynometra* trees, as well as in lowland, swamp, and montane forests, at elevations up to 2300 m.

NATURAL HISTORY: *P. boehmi* is a tree squirrel, occasionally found on the ground, but usually seen low down on tree trunks and branches. It feeds predominantly on insects (e.g., ants and caterpillars), as well as on mushrooms, fruits, and resin. The nest, found in tangles of vines, is built of twigs and leaves and lined with shredded bark. Courtship involves a mating chase, and breeding may occur throughout the year.

GENERAL REFERENCES: Kingdon 1974; Rahm and Christiaensen 1963.

Paraxerus cepapi (A. Smith, 1836) Smith's Bush Squirrel

DESCRIPTION: Smith's bush squirrel is a medium-sized yellowish brown squirrel that lacks distinguishing colors or markings. The dorsal pelage and short limbs are brown, yellow brown, or gray; the ventral pelage is gray white with a yellow or buff coloration on the chest. The pelage color varies geographically. The cheeks are light yellow brown, and the face features white stripes above and below the eyes. The long bushy tail is grizzled black and yellowish brown.

SIZE: Female—HB 226.6 mm (n = 9); T 183.7 mm (n = 7).

Male—HB 228.2 mm (n = 9); T 180.5 mm (n = 8); Mass 243.0 g (n = 1).

Sex not stated—HB 238.5 mm (n = 2); T 200.0 mm (n = 2).

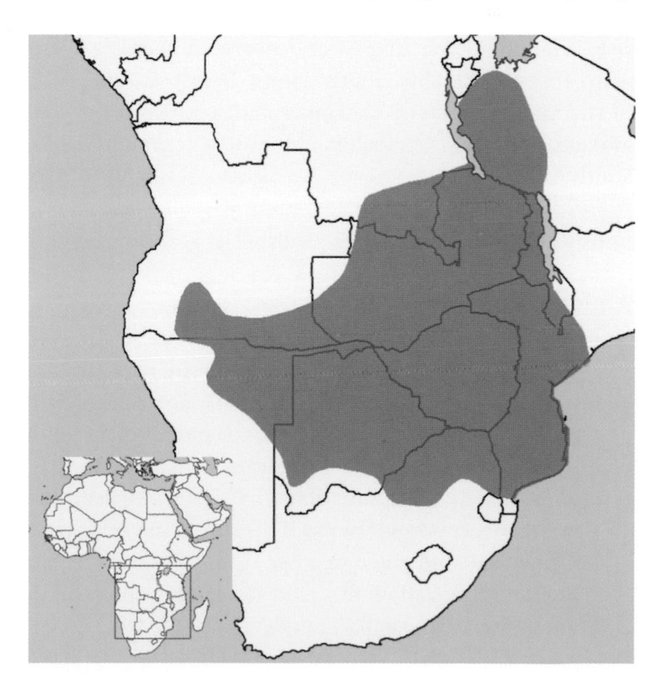

Paraxerus cepapi. Photo courtesy B. D. Patterson.

DISTRIBUTION: *P. cepapi* is present in East and Central Africa, from southern Angola and northern Namibia westward through Zambia, southeastern Democratic Republic of the Congo, western Tanzania, and southward through Malawi, Botswana, Zimbabwe, Mozambique, and northern South Africa.

GEOGRAPHIC VARIATION: Ten subspecies are recognized.

P. c. cepapi—Transvaal (South Africa), southern Botswana, and Zimbabwe. See description above.

P. c. bororensis—north of the Zambezi River, near "Namabieda," Boror (Mozambique). It is a darker warmer brown than *P. c. cepapoides*, with a grayer coloration on the sides of the body and the lower portion of the hind limbs.

P. c. carpi—around the junction of the Messenguezi and Zambezi rivers (Mozambique). This form is smaller (nearly half the size) and paler in color than *P. c. cepapi.* It has white or white yellow feet, and an orange yellow coloration on the thighs and the underside of the tail.

P. c. cepapoides—Beira (Mozambique). This subspecies has a more rusty coloration than *P. c. cepapi*, with tawny highlights on the dorsal and thigh pelage.

P. c. chobiensis—northern Namibia and northern Botswana, into southern Angola. The ventral pelage and toes are whiter than in *P. c. cepapi.*

P. c. phalaena—central and northwestern Ovamboland (Namibia). It is distinguished by gray pelage on the back, the crown of the head, the shoulders, the hips, and the legs. The feet are paler buffy white than in *P. c. cepapi.*

P. c. quotus—southeastern Katanga (Democratic Republic of the Congo). This subspecies has a darker coloration.

P. c. sindi—southern Malawi and the Tete District near the Zambezi River (Mozambique). The midline of the underside of the tail is ochraceous, and the ventral pelage is white.

P. c. soccatus—north of the Zambesi River, in Malawi around Vwaza, near the Hewe River. This subspecies has grayish white feet and lacks the yellow highlights on *P. c. cepapi*, particularly on the limbs and flanks.

P. c. yulei—northeastern Zambia, western Tanzania, and northern Malawi. This subspecies is larger, and it has grayer shoulders with light tan gray sides. The ventral color is white with gray highlights on the belly; and the dorsal pelage is paler, grizzled with tan. The feet are gray white to white yellow.

CONSERVATION: IUCN status—least concern. Population trend—stable.

HABITAT: Smith's bush squirrel is present in savanna woodlands where the trees offer suitable nesting holes—especially in mopane (*Colophospermum mopane*, also referred to as *Copaifera mopane*) and acacia woodlands—and in mixed associations such as *Acacia/Terminalia* and *Acacia/Combretum*. This squirrel is less common in combined *Brachystegia/Julbernardia* woodlands and *Baikiaea* woodlands, possibly because the trees do not provide ideal holes for nesting.

NATURAL HISTORY: *P. cepapi* is diurnal and arboreal. Its activity period is influenced by the extreme temperatures of its environment, which leads the squirrel to bask in the sun on cold winter mornings and rest in the shade during the heat of summer afternoons. Smith's bush squirrel spends most of its time in the trees, running and jumping quickly and adeptly from branch to branch, but it also descends to the ground occasionally to feed on fallen fruits. This species is primarily vegetarian and consumes a diversity of plant structures, as well as some arthropods. In South Africa, individuals forage opportunistically in each season on more than 30 species of plants, eating seeds, berries, flowers, stems, and leaves, but preferring the seeds and gums of acacia and the seeds and flowers of aloes. It also feeds on termites. In East Africa the animals eat bulbs, nuts, seeds, insects, bird eggs, aloe and euphorbia leaves, and the fruits of *Sclerocarya, Pterocarpus*, and *Kigelia*. Smith's bush squirrels are tolerant of disturbed areas and generally prefer to forage on vegetation that grows quickly and produces many fruits and seeds, such as mangoes and other cultivated fruits found in gardens and plantations. The animals store food items at the bases of tree trunks and thus play a central role in the dispersal of savanna species. Foods are eaten while held with the forefeet. The squirrels actively drink water, which they obtain from tree holes in the wild.

Smith's bush squirrels are social and live in groups composed of one or two adults and several juveniles. The groups nest in trees, rocky crevices, house roofs, or on the ground in holes lined with grass and leaves. The animals clean out and reline their nests often, which may reduce the number of parasites. The social hierarchy is obvious within groups, particularly during feeding and interactions with conspecifics, such as allogrooming, chasing, and fighting. Scent-marking play is utilized in establishing social organization. These squirrels are highly territorial, and they mark areas 0.3–1.26 ha in size with scent by mouth-wiping, urinating, and anal-dragging. They express aggression frequently with vocalizations, chases, and fights, which occasionally result in the death of one individual. Territories are defended, except during mating season, when foreign squirrels are permitted to enter. Smith's bush squirrels utilize numerous vocalizations to communicate. Most calls consist of a "click," "rattle," or whistle of decreasing intensity and descending pitch, and many types may be combined in different situations. When disturbed in their nests, the squirrels emit grunts and growls. Courting individuals and females communicate with "clicks," and males give a low-pitched nasal murmur during mating. They have a low-intensity alarm call composed of a series of three "chir" or "click" sounds spaced over several seconds, which indicates alertness, warning, or aggression for territorial defense. Their high-intensity alarm call consists of six to seven high-pitched notes emitted at one-second intervals, and it resembles a bird call or whistle.

Mating occurs only in the morning and is initiated by the female, which emits rattlelike calls. The male responds with his own mating call and chases the female, during which time both sexes make clicking noises and flick their tails. The male allogrooms the female during the mating process, and both individuals autogroom once copulation is complete. Females gestate for 56 days. They produce one litter each year in the wild, but up to three litters in captivity, with the captive interbirth interval lasting 60–63 days. The timing of breeding differs regionally through this species' range. Females in Botswana tend to give birth during the warm wet months rather than the cold dry months, and they do not produce any young during May and September. In South Africa the squirrels reproduce seasonally, with most females giving birth from October to January. The young are born precocious and develop at a typical rate, opening their eyes by day 8, climbing beyond the nest by day 19, consuming solid foods by day 21, weaning between days 29 and 42, and reaching sexual maturity within 6–10 months. At this time, the subadults are evicted from the group. Both parents care for and groom their young, although males sometimes practice infanticide in order to mate again with the females more quickly. Smith's bush squirrels are preyed upon by raptors, snakes, and carnivorous mammals. They are also vulnerable to ectoparasites, including one species each of chigger and louse, two species of mites, four species of fleas, and seven species of ticks, as well as the parasitic nematode *Syphacia paraxeri* and bacteria transmitted by ticks.

GENERAL REFERENCES: Ansell and Dowsett 1988; de Graaff 1981; de Villiers 1986; Kingdon 1974; Smithers 1971, 1983; Viljoen 1975, 1977a, 1977b, 1977c, 1983a, 1989, 1997b.

Paraxerus cooperi (Hayman, 1950) Cooper's Mountain Squirrel

DESCRIPTION: Cooper's mountain squirrel is a medium-sized dark brown animal, distinguished by its golden rufous thighs and feet. The dorsal pelage is dark brown grizzled

they usually rest through the hottest period of the day. These squirrels move quickly and adeptly along branches, utilizing all elevations from the ground to the treetops. They are primarily vegetarian and have a broad diet of fruits, seeds, buds, flowers, roots, bulbs, acacia gum, and animal matter. Their predators probably include buzzards, snakes, and genets (*Genetta*). Ochre bush squirrels are social and live either in pairs or small groups. Vocal communication is used; when threatened, an individual emits a high-pitched "burr" while flicking its tail. Before mating, males and females engage in extensive courtship activities, including chasing and grooming one another and arching the tail over the body as a signal of reproductive interest. Breeding probably occurs during most months of the year. Females produce two to three young, and they may nest in pairs with other females to raise their litters jointly. Females occasionally transport their young by carrying them in their mouths. The young emerge from the nest at 2-3 weeks of age.

GENERAL REFERENCES: Amtmann 1975; Kingdon 1974, 1997.

Paraxerus palliatus (Peters, 1852)
Red Bush Squirrel

DESCRIPTION: The red bush squirrel varies geographically in color and size. The dorsal coat and cheeks are brown grizzled with buff. The ventral pelage, limbs, and feet are bright red or yellow. The base of the long bushy tail is grizzled brown, and the remaining two-thirds are rufous. Body size and color difference are associated with climate and habitat, with larger darker individuals living in humid forests, and smaller paler squirrels living in dry forests.

SIZE: Female—HB 206.9 mm (*n* = 19); T 201.4 mm (*n* = 16); Mass 307.4 g (*n* = 73).

Male—HB 212.3 mm (*n* = 14); T 206.8 mm (*n* = 14); Mass 312.0 g (*n* = 91).

Sex not stated—HB 205.9 mm (*n* = 95); T 196.7 mm (*n* = 168); Mass 308.5 g (*n* = 361).

DISTRIBUTION: The red bush squirrel is distributed along the coast of East Africa, with some populations extending inland along riverine forests and others inhabiting the islands of Mafia and Zanzibar (Tanzania). This species ranges through southeastern Somalia, eastern Kenya, eastern Tanzania, Malawi, southeastern Zimbabwe, Mozambique, and northeastern South Africa. Habitat requirements create a patchy distribution, limited primarily to coastal areas, riverine forests, or mountains (at elevations up to 2000 m). Populations are recorded from south of the Jube (= Jubba) River (Somalia), the Tana River (Kenya), the Ruaha River (Tanzania), Mount Mulanje (Malawi), Mount Selinda (Zimbabwe), and Lake Saint Lucia and the Ongoye Forest (South Africa).

GEOGRAPHIC VARIATION: Six subspecies are recognized. Records for Malawi are identified only at the species level.

P. p. palliatus—coast of eastern Tanzania and northern Mozambique. It has rufous feet.

P. p. frerei—Mafia and Zanzibar islands (Tanzania). This subspecies has black feet.

P. p. ornatus—Ngoye Forest of the Eshowe District, Zululand (= KwaZulu-Natal), in South Africa. It has a large body, and its brownish black dorsal pelage is grizzled with buff. Its ventral pelage is rufous orange. The tail is dark brownish black tinged with rufous throughout.

P. p. sponsus—coast of eastern Mozambique, south of the Save River. *P. p. sponsus* resembles *P. p. palliatus* in appearance, but it is spatially separated.

P. p. swynnertoni—Chirinda Forest (southeastern Zimbabwe). This species has a cinnamon rufous belly and cheeks, and a grizzled black and buff dorsal pelage and face. It resembles *P. p. palliatus* and *P. p. ornatus* in color, but it is smaller in size.

P. p. tanae—southern Somalia through eastern Kenya south to the Pangani River (northern Tanzania). The tail is completely rufous orange.

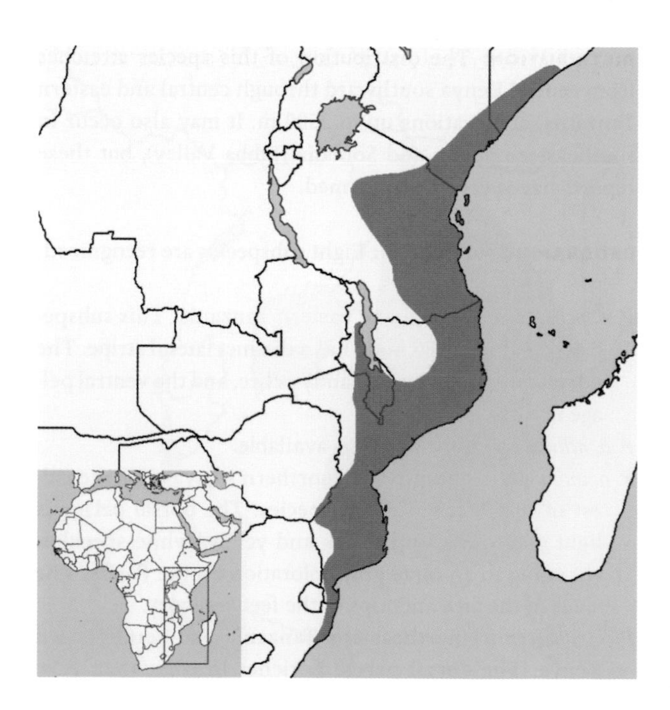

Paraxerus palliatus. Photo courtesy Lee. E. Harding.

CONSERVATION: IUCN status—least concern. Population trend—no information.

HABITAT: The red bush squirrel is found in a wide range of habitats, including dry and moist evergreen forests, woodlands, dune forests, riverine forests, and thickets. It lives primarily along the coast or by rivers, but it is also found on islands and in mountainous regions, at elevations as high as 2000 m.

NATURAL HISTORY: The red bush squirrel is diurnal and arboreal. It is fairly abundant through its range. but this species is difficult to observe in the wild, due to its shy nature. Individuals spend much of their time foraging on the ground, consuming seeds, fleshy fruits, nuts, and invertebrates. They scratch underneath tree bark to find insects, and they have been observed stalking prey and displaying other hunting behaviors in captivity. Individuals hoard small amounts of large seeds. These squirrels most commonly live alone or in pairs, but they also nest in groups of two to four individuals. Nests are built in the holes of baobabs (*Adansonia*) and *Kigelia* trees. Home ranges vary in size by habitat. In evergreen moist forests, *P. p. palliatus* maintains a mean range of 3.18 ha for males and 2.19 ha for females; whereas in coastal forests and thickets, the mean home range size is 4.17 ha for males and 0.73 ha for females. A female may give birth to one to two litters of one to two young each year. In southern Africa gestation lasts 60-65 days, and the young are born during the wet season, from August to March. The female separates from other squirrels before giving birth and builds a new nest, lined with leaves,

Paraxerus palliatus. Photo courtesy Frédéric Salein, France, www.flickr.com/photos/fredericsalein/.

in a tree hole. She cleans the nest, perhaps to reduce the number of parasites that are present. While the young are small, the female prevents other adults from entering and aggressively responds to other squirrels, including the male. When the young are old enough, the female permits the male to join the family group. After reaching subadulthood, the young are forced to leave the nest. Red bush squirrels communicate with vocalizations, visual cues, and olfactory signals. They commonly flick their tails and fluff their fur, particularly when in thick vegetation. Individuals emit a number of sounds, including murmurs, hisses, growls, clicks, twitters, and barks. Males express sexual interest in females by murmuring and by chasing them, which may help stimulate the onset of estrus. Red bush squirrels also engage in urine dribbling and anal dragging to scent-mark areas.

GENERAL REFERENCES: Amtmann 1975; Kingdon 1974; Smithers 1983; Viljoen 1980, 1983a, 1983b, 1986, 1989.

Paraxerus poensis (A. Smith, 1830) Green Bush Squirrel

DESCRIPTION: *P. poensis* is a small squirrel, with an olive green dorsal coat and yellow underparts. The body hair is thick and soft. The eyes are ringed by yellow fur. The tail is thin and dark olive in color. This species has short limbs, small broad feet, and thick curved claws that enable it to climb with ease.

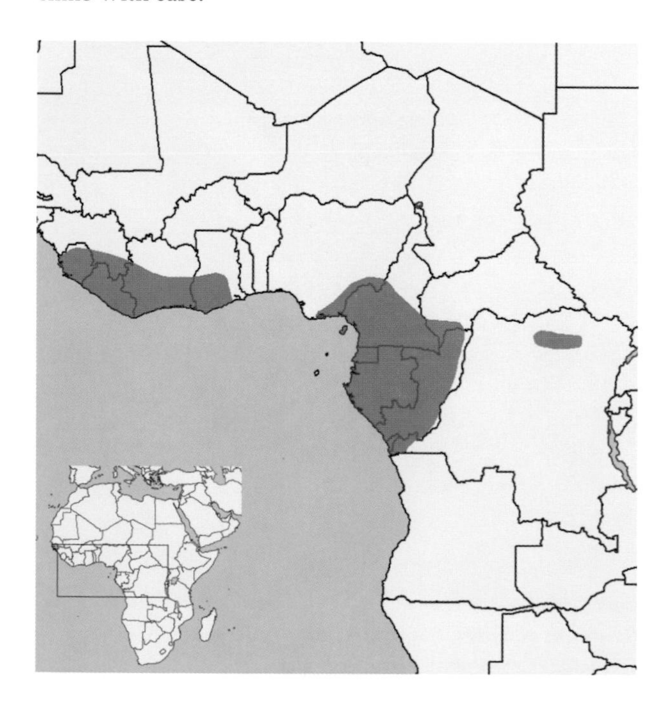

Paraxerus poensis. Photo courtesy Alain Devez.

SIZE: Female—HB 154.2 mm (n = 17); T 159.1 mm (n = 17); Mass 125.0 g (n = 2).

Male—HB 155.0 mm (n = 22); T 166.1 mm (n = 26); Mass 144.5 g (n = 2).

Sex not stated—HB 156.8 mm (n = 4); T 160.0 mm (n = 3); Mass 130.6 g (n = 11).

DISTRIBUTION: The green bush squirrel is distributed through sub-Saharan Africa, in several discontinuous ranges. It is found from Sierra Leone through the coastal states to west of the Volta Basin (Ghana), as well as from east of the Niger River (southeastern Nigeria) southward to eastern Democratic Republic of the Congo. Its presence has been reported in Bata in the south and Medje and Niapa (= Niapu) in the north of the Democratic Republic of the Congo. It is also present on Bioko Island (Equatorial Guinea) and the Democratic Republic of São Tomé and Príncipe. This species has been recorded from elevations as high as 1600 m on Mount Bintamane (= Mount Bintumane) in Sierra Leone.

GEOGRAPHIC VARIATION: None.

CONSERVATION: IUCN status—least concern. Population trend—no information.

HABITAT: The green bush squirrel is present in lowland tropical moist forests, brush, and secondary forests around villages and agricultural areas. This species is considered a pest in cocoa plantations.

NATURAL HISTORY: *P. poensis* is arboreal, foraging in tree branches for fruits, seeds, and arthropods, as well as consuming flying insects and bird eggs in captivity. These squirrels locate arthropods by meticulously searching stems, bark, and the crannies of branches. Green bush squirrels live alone or in pairs that are possibly monogamous. They nest in open leaf nests or tree hollows lined with leaves or fibers. Females give birth to one to two kittens in each litter. Males and females reside together in the nest while raising their litter, and both appear to care for the young. This species has a single alarm call that is accompanied by a complex visual display. While emitting a single "buzz" sound, the animal freezes motionless while holding its tail stiffly behind its body with the tip curved upward. Between repeats of the sound, it quickly jerks its tail upward to a nearly vertical position, so that the tail resembles a C shape. While moving its tail, the animal hops or stamps its feet. The "buzz" and its subsequent movements are then repeated 100 or more times.

GENERAL REFERENCES: Amtmann 1966; Ellerman 1940; Emmons 1975, 1978, 1979, 1980; Hollister 1919; Moore 1959; Rosevear 1969; Thomas 1916.

Paraxerus vexillarius (Kershaw, 1923) Swynnerton's Bush Squirrel

DESCRIPTION: Swynnerton's bush squirrel is a large brown animal, distinguished in some individuals by the orange rufous color on the tip of the tail. The dorsal pelage is olive green to brown, and the ventral pelage is gray. The limbs and feet are orange rufous. The face is rufous on the nose and mouth, and in a wide stripe from the ear to the eye. The long tail is brown with pale rings at the base, buff in the middle, and orange at the distal tip. *P. vexillarius* can be distinguished from other *Paraxerus* species by having a rufous coloration only on the limbs and tail, rather than on the belly (*P. palliatus*) or the entire body (*P. lucifer*).

***Paraxerus vexillarius*.** Photo courtesy William T. Stanley.

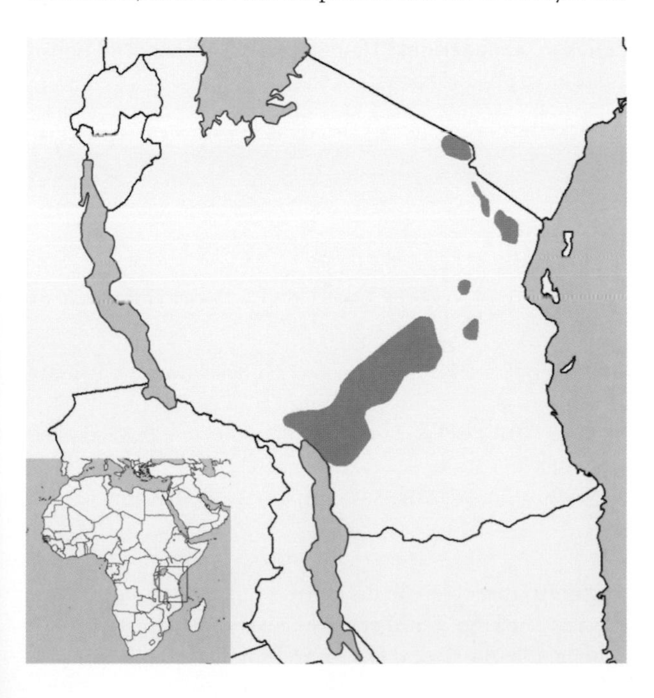

SIZE: Female—HB 226.6 mm (n = 9); T 183.7 mm (n = 7).

Male—HB 228.1 mm (n = 9); T 180.5 mm (n = 8); Mass 243.0 (n = 1).

Sex not stated—HB 238.5 mm (n = 2); T 200.0 mm (n = 2).

DISTRIBUTION: Swynnerton's bush squirrel is distributed through Tanzania in an area no greater than 20,000 km^2. It is primarily limited to the Eastern Arc Mountains, having been recorded from Mount Kilimanjaro (between 1900 m and the heath zone), the Usambara range (Amani, Lushoto, Magamba, and Mazumbai), the Ulugurus (Bagiro), and the Udzungwa range (Kigogo). It has also been recorded from the Ukinga Mountains and the Katessa Forest. The small patchy distribution and restricted habitat have generated concern over the preservation of this species.

GEOGRAPHIC VARIATION: Two subspecies are recognized. Records for the Ukinga Mountains and the Katessa Forest are identified only at the species level.

P. v. vexillarius—Usambara, Uluguru, and Udzungwa mountains (eastern Tanzania). See description above.

P. v. byatti—Mount Kilimanjaro (Tanzania). This subspecies has no orange on the tip of the tail, a more ochraceous buff dorsal pelage, and a darker belly.

CONSERVATION: IUCN status—near threatened. Population trend—no information.

HABITAT: Swynnerton's bush squirrel is limited to montane forests.

NATURAL HISTORY: Little is known about this species. *P. vexillarius* is arboreal, and it forages on fruits and seeds.

GENERAL REFERENCES: G. M. Allen 1939; G. M. Allen and Loveridge 1933; Amtmann 1975; Ellerman 1940; Kingdon 1974, 1997.

Paraxerus vincenti Hayman, 1950
Vincent's Bush Squirrel

DESCRIPTION: Vincent's bush squirrel is a medium-sized to large animal with a black and red body. The dorsal pelage, limbs, and the tops of the feet are grizzled black. The ventral coat, nose, and eye region are deep rufous. The crown of the head and the cheeks are dark brown. The long tail is black brown with rufous on the distal tip. *P. vincenti* strongly resembles *P. palliatus,* but the former has a darker rufous coloration on the ventral pelage.

SIZE: Female—HB 208.0 mm (n = 2); T 213.5 mm (n = 2).
Male—HB 214.7 mm (n = 3); T 206.0 mm (n = 3).
Sex not stated—HB 212.0 mm (n = 5); T 209.0 mm (n = 5).

DISTRIBUTION: Vincent's bush squirrel is found only on Mount Namuli in central Mozambique (15.21° S, 37.04° E), at elevations from 1200 to 1850 m. Few surveys have been carried out in this region, and it is not known whether this species is present in the surrounding mountains.

GEOGRAPHIC VARIATION: None.

CONSERVATION: IUCN status—endangered. Population trend—decreasing.

HABITAT: Vincent's bush squirrel inhabits montane moist evergreen forests.

NATURAL HISTORY: No information is available.

GENERAL REFERENCES: Hayman 1950; Smithers and Lobão Tello 1976; Viljoen 1989.

Protoxerus Forsyth Major, 1893

This genus of African ground squirrels includes two species of large animals that are prominent in West and Central Africa.

Protoxerus aubinnii (Gray, 1873)
Slender-Tailed Squirrel

DESCRIPTION: *P. aubinnii* is a large brown squirrel. The dorsal pelage, head, and limbs are dark brown grizzled with yellow or buff. The ventral pelage is a paler brown. The ears are small, round, and covered in thick hair. The long tail is black and grizzled with yellow or buff toward the tip.

SIZE: Female—HB 240.0 mm (n = 7); T 304.4 mm (n = 7); Mass 454.3 g (n = 4).
Male—HB 239.7 mm (n = 9); T 300.7 mm (n = 9); Mass 415.0 g (n = 3).
Sex not stated—HB 255.0 mm (n = 3); T 300.0 mm (n = 2); Mass 525.0 g (n = 1).

DISTRIBUTION: The slender-tailed squirrel is found in West Africa, through southeastern Sierra Leone, southern Guinea, Liberia, Côte d'Ivoire, and southwestern Ghana.

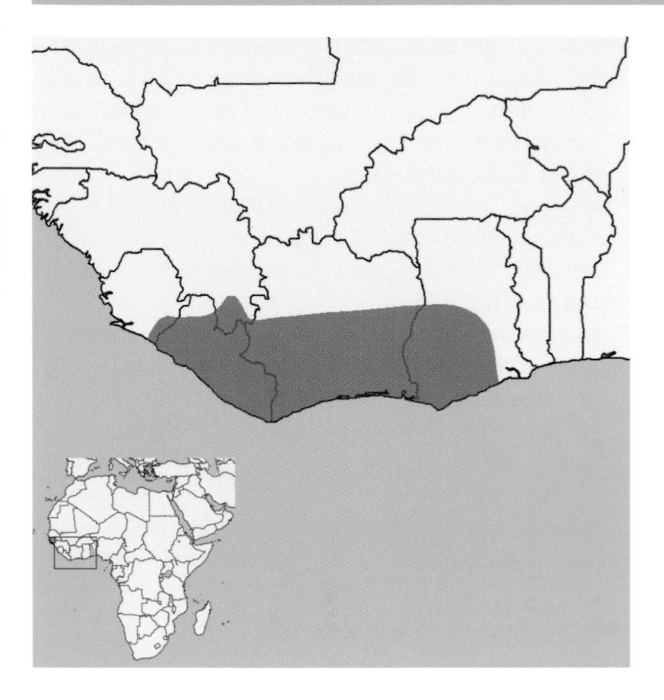

GEOGRAPHIC VARIATION: Two subspecies are recognized.

P. a. aubinnii—Côte d'Ivoire and Ghana. See description above.

P. a. salae—Sierra Leone, Liberia, and southern Guinea. This is a darker subspecies, with bands on the tip of the tail.

CONSERVATION: IUCN—data deficient. Population trend—no information.

HABITAT: The slender-tailed squirrel inhabits the undergrowth of lowland tropical moist forests and is often found in raffia palm (*Raphia*) swamps.

NATURAL HISTORY: *P. aubinnii* forages on the husks of raffia palms. No other biological information is known.

GENERAL REFERENCES: Booth 1960; Grubb et al. 1998; Rosevear 1969.

Protoxerus stangeri (Waterhouse, 1842)
Forest Giant Squirrel

DESCRIPTION: The forest giant squirrel is one of the largest and most abundant squirrels in the areas where it occurs. The dorsal pelage and limbs are brown grizzled with black and yellow. The belly is sparsely haired and white, yellow, or dark; the head is gray. The long bushy tail is gray with subtle dark-colored bands. While moving, the animal holds its tail straight out behind its body. At rest, it hangs the tail below the body.

SIZE: Female—HB 276.7 mm (n = 18); T 298.6 mm (n = 16); Mass 760.8 g (n = 5).

Male—HB 279.5 mm (n = 19); T 300.1 mm (n = 19); Mass 538.3 g (n = 3).

Sex not stated—HB 304.6 mm (n = 17); T 307.5 mm (n = 2); Mass 658.35 g (n = 35).

DISTRIBUTION: The forest giant squirrel is distributed through West and Central Africa, in Sierra Leone, southern Guinea, Liberia, Cote d'Ivoire, Ghana, Togo, Benin, Nigeria, Cameroon, Central African Republic, Equatorial Guinea (including Bioko Island), Gabon, the Republic of the Congo, the Democratic Republic of the Congo, central Uganda, western Kenya, far western Rwanda, and far western Burundi. Two isolated records indicate that populations may also exist in Tanzania and the Angola Escarpment (in Angola and from Angola into southern Democratic Republic of the Congo). This species lives at elevations up to 2000 m.

GEOGRAPHIC VARIATION: Twelve subspecies are recognized.

P. s. stangeri—Bioko Island (Equatorial Guinea). See description above.

P. s. bea—Kakamega Forest (western Kenya). It has a light

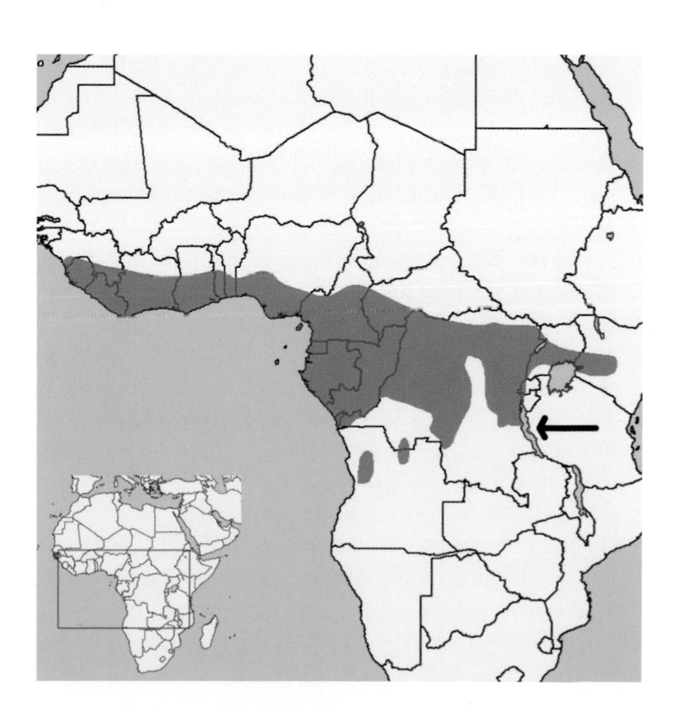

Protoxerus stangeri. Photo courtesy Nik Borrow.

rust-colored rump and limbs, a light gray head, and a white throat.

P. s. centricola—Mount Nkungwe, Tanzania. It resembles *P. s. bea*, but with richer rufous on the rump and hind legs and a grizzled russet back. There is a gray mask on the face, and a single black mark beneath each ear.

P. s. cooperi—Sango Bay Forest (Uganda). This subspecies has a grizzled yellow back and yellowish orange hind legs, with a pale yellow ventral pelage. The head is paler gray, with orange and black patches behind the ears. There is no red on the forearms.

P. s. eborivorus—"Ovenga River" (Gabon), Cameroon, Equatorial Guinea, and Central African Republic. It is characterized by a black and red dorsal pelage, a rufous and whitish head, a gray throat and chest, a brown abdomen, and red feet. The tail is black at the base, changing to rufous and then black and white rings in the middle sections, with a black distal tip. This subspecies is recorded as eating the animal matter on freshly fallen elephant tusks.

P. s. kabobo—Mount Kabobo (Democratic Republic of the Congo). This form has a white head, yellow nape of the neck, and ochraceous tawny back limbs.

P. s. kwango—Kwango (Democratic Republic of the Congo). The crown of the head is light buff, the dorsal pelage is warm buff, and the back and limbs are yellow ochre or burnt sienna. The sparse ventral pelage is dark, becoming a russet red color on the chest.

P. s. loandae—high forest in northern Angola.

P. s. nigeriae—Togo, Benin, and Nigeria. The throat and chest are white, with a distinct white line separating the dorsal and ventral pelages.

P. s. personatus—mouth of the Congo River (Democratic Republic of the Congo). This subspecies has deep yellow highlights in the dorsal pelage, a black head grizzled with white, an off-white throat and chest, and rufous on the top of the forefeet. The tail is mostly red and black with a white and black tip.

P. s. signatus—Lodja, upper Lukenye River (southern Democratic Republic of the Congo). A uniform black stripe separates the dorsal and ventral pelages; the hands and feet are a rich rufous.

P. s. temminckii—Sierra Leone and Liberia eastward to Ghana. This form has a white throat and chest, a white line separating the flank from the belly, and white hair on the insides of the limbs. Some individuals display a dark or black spot above the sides of the neck.

CONSERVATION: IUCN status—least concern. Population trend—no information.

HABITAT: The forest giant squirrel is found in lowland evergreen rainforests, open grass woodlands, swamp forests, and montane forests. It occurs among a variety of forest qualities, including mature, secondary, and edge forests; in plantations; and in gardens with trees.

NATURAL HISTORY: The forest giant squirrel is an arboreal species that lives almost exclusively in the upper forest canopy. Individuals are active from dawn until late afternoon. Their diet consists primarily of fruits and seeds from tree and liana species such as *Panda oleosa, Coula edulis,* and various species of *Klaineodoxa, Irvingia, Elaeis* (oil palm), *Pseudospondias, Musanga, Parinari, Chrysophyllum, Carapa, Caloncoba, Cordia, Urera,* and *Phoenix*. They also consume small amounts of vegetation and arthropods, and they have even been reported to kill nesting hornbills. This species' major predators are eagles and other large raptors, probably because of the squirrels' arboreal location. They nest in tree hollows with small entrances or in the forks of branches. They line their beds with fresh leaves and twigs, and they may use the same nests for many seasons. Individuals live alone, and they avoid conspecifics except to mate or raise their young. They prefer to forage in solitude, to the point of aggressively chasing other squirrels out of fruit trees, and they do not engage in social interactions (e.g., play or allogrooming). Subadult females maintain home ranges of 3.2-5.0 ha. Individuals communicate vocally with two main calls. Their mild alarm call is composed of repeated sniffs or sneezes, occasionally alternated with "clicks" of the incisor teeth. Their high-intensity alarm call is a series of pulses of descending frequency, which resembles a short whinny; this is repeated every 5-20 seconds. The calls are not accompanied by visual displays, but this squirrel does respond to foreign objects by fluffing its tail and moving it from side to side above the back. Males chase estrous females before mating. Females appear to reproduce one to two times annually, and they produce one to two young in each litter.

GENERAL REFERENCES: Amtmann 1966; Emmons 1975, 1978, 1979, 1980; Gautier-Hion et al. 1980; Kingdon 1971; Rosevear 1969.

Tribe Marmotini Pocock, 1923

This tribe includes 13 genera.

Ammospermophilus Merriam, 1892

This genus contains five species.

Ammospermophilus harrisii (Audubon and Bachman, 1854)
Harris's Antelope Squirrel

DESCRIPTION: The dorsum is gray, suffused with reddish to yellowish brown near the head and along the front and hind limbs, grading to gray through the tail. A single thin white line on each side parallels the spine. The venter is white, extending slightly up the sides. The dorsal surface of the tail is charcoal, and the underside is gray, resulting from a mix of black and white hairs.

SIZE: Both sexes—HB 229-245.8 mm (range = 216-267 mm); T 70.9-84.6 mm (range = 67-92 mm); Mass 113-150 g.

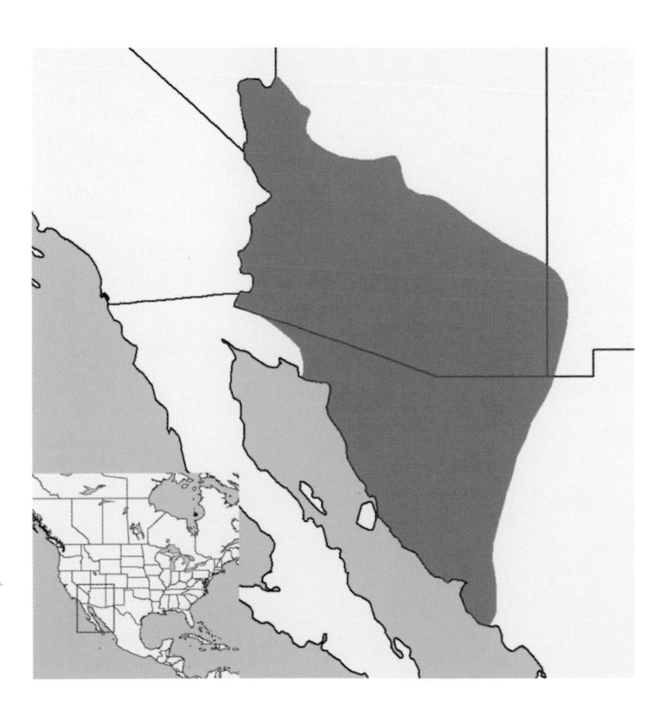

DISTRIBUTION: Harris's antelope squirrel occurs from Arizona to southwestern New Mexico (USA) and adjoining Sonora (México).

GEOGRAPHIC VARIATION: Two subspecies are recognized.

A. h. harrisii—along the east bank of the Colorado River in Arizona (USA) and ending along the Sea of Cortez's Sonoran coast. This form tends to be slightly darker, with more contrast, than *A. h. saxicolus*.

A. h. saxicolus—eastern 60 percent of the range. This is a paler subspecies.

CONSERVATION: IUCN status—least concern. Population trend—no information.

HABITAT: Harris's antelope squirrels are found in a variety of desert habitats, and they thrive in areas with cacti and desert shrubs.

NATURAL HISTORY: This squirrel is diurnal, active year-round, and does not store fat. Harris's antelope squirrels are well adapted for climbing cacti and boulders. *A. harrisii* lives in shallow inconspicuous burrows often associated with structures such as rocks, cacti, or shrubs; the squirrels sometimes store large seeds within their burrows. A nest chamber provisioned with bedding is found in each burrow system, especially for pregnant females. Breeding occurs each year, starting in December, and it can continue well into spring. After a 30-day gestation, a single litter averaging 6.5 is most common, but two litters may occur in mild climates. Litters emerge at 4-5 weeks, wean by 7 weeks, and the young are fully grown by about 217 days. Harris's antelope squirrels are omnivorous. They feed heavily on cactus fruits and flesh, consume seeds that they often store in their cheek pouches, and will eat insects. Although these squirrels are quite conspicuous, densities are low and individuals are scattered throughout most of the range. They travel a maximum of

Ammospermophilus harrisii. Photo courtesy Randall D. Babb.

274 m (on average) and thus have modest home ranges. Harris's antelope squirrels are extremely energetic and rarely remain motionless, except when sitting on their haunches atop a tall boulder or cactus. Their alarm vocalization is a long pure-noted trill. These squirrels hold their tails straight upward when scurrying about the ground, stopping to forage for seeds on the desert floor. Due to their small body size and diurnal habits, *A. harrisii* is a likely prey of foxes, coyotes (*Canis latrans*), procyonids, small felids, raptors, and snakes. In urban areas Harris's antelope squirrels occasionally become pests in gardens, but never at more than a localized scale.

GENERAL REFERENCES: Best, Titus, Caesar, et al. 1990; Bolles 1988.

Ammospermophilus insularis (Nelson and Goldman, 1909)
Espiritu Santo Island Antelope Squirrel

DESCRIPTION: *A. insularis* is gray to slate gray on the dorsum and head, with a white stripe running along each side. The venter is white to cream. The snout and legs are suffused with light orange to rust. The tail is grizzled gray with a whitish underside and a band of black near the tip of the tail.

SIZE: Both sexes—TL 229 mm (210-240 mm); T 78 mm (71-83 mm).

DISTRIBUTION: This squirrel is endemic to Espiritu Santo Island, Baja California Sur (México).

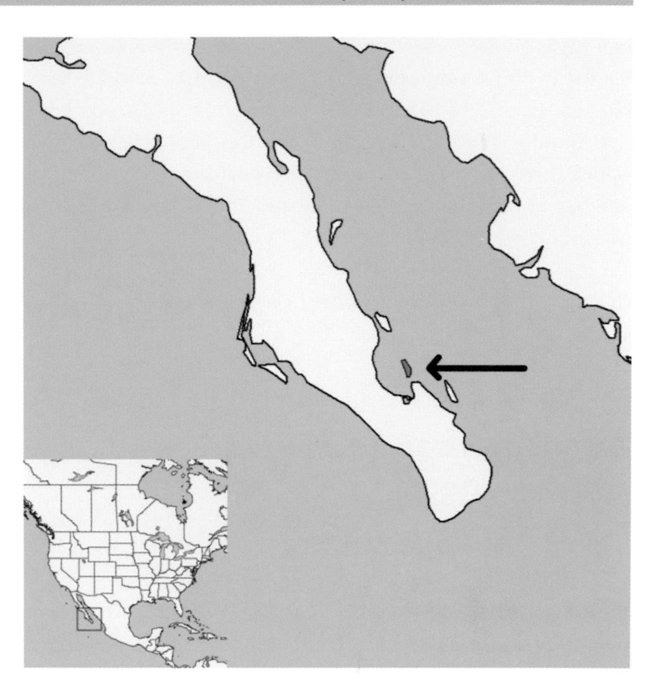

GEOGRAPHIC VARIATION: None. *A. insularis* was proposed as a subspecies of *A. leucurus* by Álvarez-Castañeda, and Helgen et al. followed this classification.

CONSERVATION: IUCN status—not listed. Population trend—no information. *A. insularis* is considered threatened in México.

HABITAT: Espiritu Santo Island is characterized by rocky volcanic slopes with sparse vegetation, in which this species is found.

NATURAL HISTORY: *A. insularis* is diurnal, but little else is known about this small island endemic. This squirrel's call

Ammospermophilus insularis. Photo courtesy Troy L. Best.

is a high-pitched short trill, which apparently is especially useful in the rocky open habitat on Espiritu Santo Island. No genetic differentiation exists between *A. insularis* and the mainland *A. leucurus*. However, the frequent absence of an upper third premolar on the island species suggests that this species should be at least a subspecies of Baja California peninsular *A. leucurus*.

GENERAL REFERENCES: Álvarez-Castañeda 2007; Best, Caesar, et al. 1990; Helgen et al. 2009.

Ammospermophilus interpres (Merriam, 1890) Texas Antelope Squirrel

DESCRIPTION: The dorsum is gray suffused with brown near the head and along the front and hind limbs, grading to gray through the tail. A single thin white line on each side parallels the spine. The venter is white. The dorsal surface of the tail darkens to black at the distal end; the ventral surface is white with two black bands.

SIZE: Both sexes—HB 226 mm (220-235 mm); T 74.2 mm (68-84 mm); Mass 99-122 g.

DISTRIBUTION: *A. interpres* is found from New Mexico and west Texas (USA) to Chihuahua, Coahuila, and Durango (México).

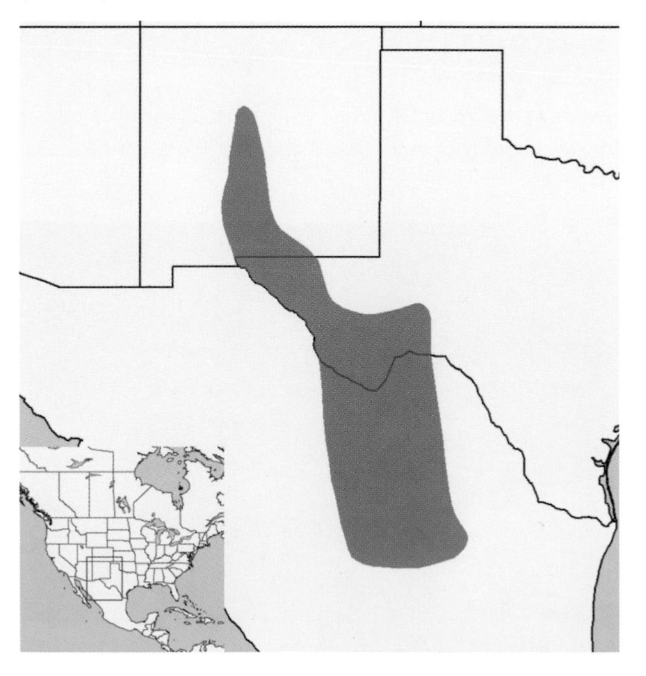

Ammospermophilus interpres. Photo courtesy Maryann Eastman.

GEOGRAPHIC VARIATION: None.

CONSERVATION: IUCN status—least concern. Population trend—no information.

HABITAT: Texas antelope squirrels live in rocky habitats on and around desert mountain ranges; they often can be found in deserts, grasslands, and woodlands where boulders occur in proximity to juniper (*Juniperus*) and large shrubs.

NATURAL HISTORY: *A. interpres* is diurnal. It is active year-round but may hibernate during periods of extreme cold, as individuals accumulate fat deposits in the fall of each year. Texas antelope ground squirrels are well adapted for climbing trees and rocks. *A. interpres* lives in shallow (< 1 m) inconspicuous burrows that lack a mound and are often associated with structures such as boulders, trees, or shrubs; these squirrels may also use rocky crevices. A nest chamber provisioned with bedding exists in each burrow system. Breeding occurs each year in February; a single litter of 5-14 young is most common, but there may be two litters in the mild climates inhabited by these squirrels. Litters emerge in April. Texas antelope squirrels feed heavily on seeds, fruit, and cactus, but they will also include insects in their diet. Densities are low and patchy in all habitats. These squirrels are extremely energetic and rarely remain motionless, except when sitting on their haunches atop a tall boulder, cactus, or shrub. Their alarm vocalization is a mellow rolling-but-harsh trill that appears to be an adaptation to structurally complex environments. When running, Texas antelope squirrels hold their tails curled over their backs. Due to their small body size and diurnal habits, *A. interpres* is a likely prey of foxes, coyotes (*Canis latrans*), procyonids, small felids, raptors, and snakes. This species may be eaten in México; nowhere is the Texas antelope squirrel considered a pest.

GENERAL REFERENCES: Best, Lewis, et al. 1990; Bolles 1988.

Ammospermophilus leucurus (Merriam, 1889) White-Tailed Antelope Squirrel

DESCRIPTION: White-tailed antelope squirrels are gray to slate gray on the dorsum and head, with a white stripe running along each side. The venter is white to cream. The snout and legs are suffused with light orange to rust. The tail is grizzled gray with a whitish underside and a band of black near the tip of the tail. An orbital eye ring is present.

SIZE: Female—HB 210 mm (202–216 mm); T 61 mm (55–66 mm); Mass 100.6 g (96.2–104.5 g).

Male—HB 212 mm (188–220 mm); T 56 mm (42–71 mm); Mass 111.1 g (103.7–116.8 g).

DISTRIBUTION: *A. leucurus* is found in southeastern Oregon and eastern California east to Colorado and New Mexico (USA) and south to Baja California Sur (México).

GEOGRAPHIC VARIATION: Nine subspecies are recognized.

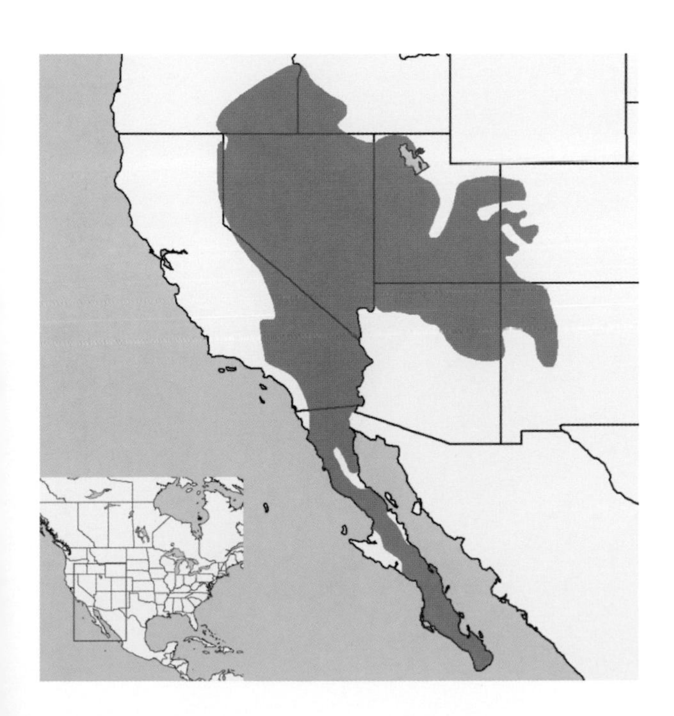

Ammospermophilus leucurus. Photo courtesy Doris Potter/ Focus on Nature Tours, Inc.

A. l. leucurus—the vast majority of the northern and western range in the USA and extreme northern Baja California (México). See description above.

A. l. canfieldiae—central Baja California Sur (México). The pelage is darker over all colored areas than in *A. l. leucurus*, and the orbital ring is more pronounced.

A. l. cinamomeus—primarily northern Arizona, southeastern Utah, and western Colorado (USA). The pelage tends toward a vinaceous cinnamon suffusion.

A. l. escalante—primarily southwestern Utah (USA). This form is redder in color than *A. l. leucurus*. The feet are whitish, a distinguishing characteristic shared only with *A. l. leucurus*.

A. l. extimus—southern portion of Baja California Sur (México). Closely resembling *A. l. canfieldiae*, but in comparison *A. l. extimus* is larger and darker colored, tending toward mikado brown (from Ridgway color standards); in addition, the winter pelage is shorter, thinner, and more hispid. The tail is longer than in *A. l. leucurus*.

A. l. notom—northeastern Utah (USA). This form is distinguished by the combination of its redder color, the lack of all white and black bands on the hairs of the head and dorsum, and the pale pinkish cinnamon feet.

A. l. peninsulae—northern Baja California (México). The pelage is darker, with a more orange cinnamon suffusion over the colored areas than in *A. l. canfieldiae*, and the orbital ring is less pronounced.

A. l. pennipes—western Colorado and New Mexico (USA). In summer, the pelage color is grayer and more brownish than in *A. l. cinamomeus*, but more vinaceous than in *A. l. leucurus*. In winter, the underparts are white; the pelage color is more grayish than in *A. l. cinamomeus*, and more vinaceous than in *A. l. leucurus*. The lateral stripe is creamy white.

A. l. tersus—extreme northwestern Arizona (USA). This form is similar in size to *A. l. leucurus*, but the color is darker.

CONSERVATION: IUCN status—least concern. Population trend—stable.

HABITAT: White-tailed antelope ground squirrels are found in sandy and rocky desert habitats, ranging from low-elevation valley floors up to pinyon-juniper (*Pinus, Juniperus*) forests, often among shrubs and cacti.

NATURAL HISTORY: This squirrel is diurnal, active year-round, and does not estivate or store fat, even in the high temperatures that accompany desert life. White-tailed antelope ground squirrels require free water in their arid habitats. They exhibit behavioral thermoregulation by reducing their activity in the heat of the day, and their sparsely haired tail appears to be effective for dissipating heat. *A. leucurus* lives in shallow (< 1 m) burrows spread throughout a 3-8 ha home range, probably this widespread both for escape and for heat avoidance. Burrows contain a nest chamber that is provisioned with bedding. Nearly all adults more than 1 year old breed each year, from February-June in the north and March-September in the south; usually a single litter of 5-14 young (average = 8) is produced each year after a gestation of 30-35 days. There is a latitudinal gradient in litter size, with northern populations in Oregon averaging 9.3 young/litter and southern populations in Baja California averaging 5.8 young/litter. The young emerge at 50-60 days and are weaned soon afterward. Juveniles disperse from their natal area. Adults live a solitary life, with dominance rank well established between neighbors. Naso-oral sniffing and contact appear to be a major means of maintaining a rigid dominance hierarchy among neighbors. Aggression can lead to lunging and side-to-side displays, chases, tooth chattering, and boxing and other physical combat. White-tailed antelope ground squirrels feed heavily on green vegetation, cactus, yucca, and assorted seeds, but they will also eat insects or scavenge vertebrate carcasses. As one of the few diurnal rodents in the desert, *A. leucurus* appears to minimize its interaction with other species. Densities are rather low, but its relative body size means that *A. leucurus* accounts for the majority of rodent biomass in many desert ecosystems. Two alarm vocalizations are given—shrill rapid "chitters" and high-pitched trills—that appear to be adaptations to open environments. Their predators include foxes, coyotes (*Canis latrans*), procyonids, small felids, raptors, and snakes. White-tailed antelope squirrels are not hunted or trapped, due to their small size. However, in some areas it will visit cultivated crops, orchards, and gardens.

GENERAL REFERENCES: Belk and Smith 1991; Karasov 1983; O'Farrell and Clark 1984; Whorley and Kenagy 2007.

Ammospermophilus nelsoni (Merriam, 1893) Nelson's Antelope Squirrel

DESCRIPTION: Nelson's antelope squirrels have a buff to tan dorsum with a suffusion of yellow; and a single thin white stripe on each side, parallel to the spine. The venter is white to cream. The tail is buff gray on the dorsal surface and creamy white underneath.

SIZE: Female—HB 238 mm (230-256 mm); T 72 mm (67-78 mm); Mass 154.5 g (141.8-179 g).

Male—HB 249 mm (234-267 mm); T 73.1 mm (66-78 mm).

DISTRIBUTION: *A. nelsoni* is found in the San Joaquin Valley of southern California (USA).

GEOGRAPHIC VARIATION: None.

CONSERVATION: IUCN status—endangered. Population trend—decreasing.

HABITAT: Nelson's antelope squirrels are found in dry flat or rolling terrain with slopes less than 10°-14°—sparse shrublands and sparse grasslands—on alluvial and loamy soils,

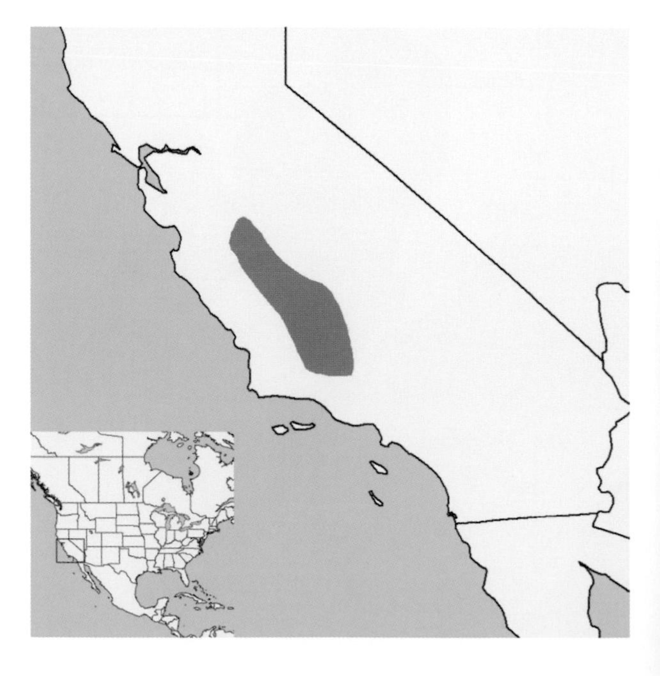

Ammospermophilus nelsoni. Photo courtesy Maggie Smith.

soils with sandy or gravelly texture, or fine-grained soils that are nearly brick hard when dry.

NATURAL HISTORY: This squirrel is diurnal and active year-round, but it may estivate during periods of extreme heat. Nelson's antelope squirrels are well adapted for desert life, due to their pale color, which highly reflects solar radiation; and they are agile at climbing shrubs. *A. nelsoni* lives in shallow yet often complex burrows (usually excavated by kangaroo rats) that can have numerous openings and are often associated with small gullies or shrubs. Nest chambers have never been observed within the burrows. Breeding occurs in late winter and spring, and a single litter of 6–11 (average = 8.9) is born 26 days later (in March) within the burrows, where females nurse their young. Litters emerge at weaning in April, when spring rains result in the major season for plant growth.

Nelson's antelope squirrels feed heavily on green vegetation and on seeds, which they collect in their cheek pouches; scavenged vertebrate flesh and insects can also make up significant portions of their diet. Densities are patchy in the poor quality habitat often inhabited by *A. nelsoni*; however, small colonies often form in suitable soil. Home ranges are 4.4 ha for each sex. Nelson's antelope squirrels are less energetic than other antelope squirrels, especially in extreme heat and cold; they often take refuge in the shade of shrubs or in burrows. *A. nelsoni*'s alarm vocalization is a short deep-pitched trill (relative to other *Ammospermophilus*); this species is extremely quiet, with soft calls. When running, Nelson's antelope squirrels dart back and forth irregularly, often with their tails curled over their backs, but these animals are more inclined to remain motionless unless startled by sound. Mortality rates are high, with greater than 80 percent unable to survive past 1 year of age. *A. nelsoni* is a prey of badgers (*Taxidea taxus*), foxes, coyotes (*Canis latrans*), and raptors. The major threats to conservation of this species are conversion of their habitat to agriculture and the spread of dense foliage from exotic species. The barren habitats in which Nelson's antelope squirrels live ensure that this species is not an economic pest, nor is it harvested.

GENERAL REFERENCES: Best, Titus, Lewis, et al. 1990; Cypher 2001; Germano et al. 2001; U. S. Fish and Wildlife Service 1998.

Callospermophilus (Merriam, 1897)

This genus has three species.

Callospermophilus lateralis (Say, 1823) Golden-Mantled Ground Squirrel

DESCRIPTION: *C. lateralis* is recognized by the golden brown mantle on the back of the head; two white stripes down the back, each framed by two black stripes; and pale whitish or yellow sides. Compared with *Tamias*, the golden-mantled ground squirrel does not have a median black line, and its tail is smaller and less bushy. Compared with its sister species, *C. madrensis*, found in México, *C. lateralis* has this distinctive mantle, more pronounced and more prominent dorsal stripes, and shorter white stripes. Compared with *C. saturatus*, found in the northern and northwestern portion of its range, *C. lateralis* has a distinctive golden mantle and a larger head.

SIZE: Female—HB 176.5 mm; T 83.4 mm; Mass 159.7 g.
Male—HB 180.0 mm; T 87.2 mm; Mass 178.0 g.
Sex not stated—HB 174.0 mm; T 88.0 mm; Mass 245.7 g.

DISTRIBUTION: Golden-mantled ground squirrels are found in mountainous areas of western North America, from central British Columbia (Canada) to southern New Mexico (in the Rocky Mountains), and from the Columbia River south

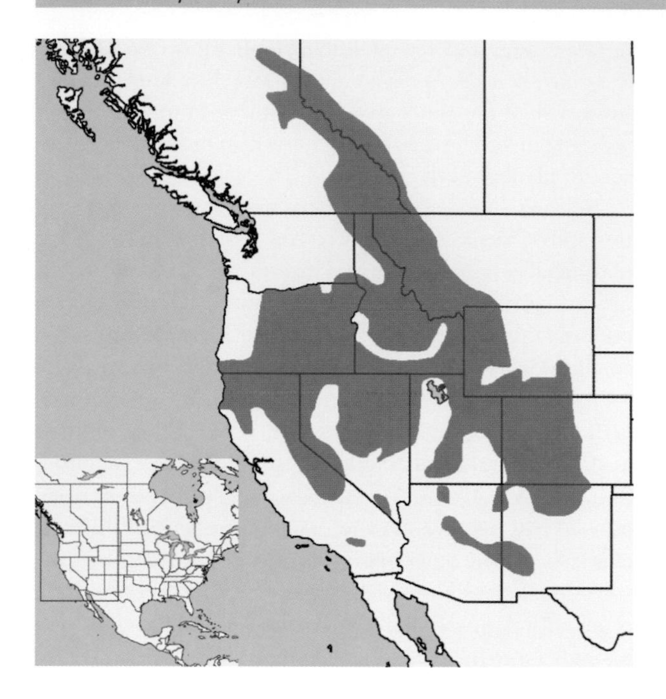

to southern California and Nevada (USA). Their range is somewhat fragmented, especially in the south.

GEOGRAPHIC VARIATION: Thirteen subspecies are recognized.

C. l. lateralis—mountain slopes of western states, from Wyoming south to Arizona and New Mexico (USA). The mantle is rusty yellow to bright chestnut.

C. l. arizonensis—San Francisco Mountains, Arizona (USA). This is a larger and richer-colored form. The ventral surface of the tail is dull gray.

C. l. bernardinus—San Bernardino Mountains, California (USA). The mantle is dull fulvous, and the head is dark.

C. l. castanurus—Wasatch Mountains (Utah) north to Wyoming (USA). The mantle is bright rusty chestnut. The back is grizzled red brown, black, and yellow.

C. l. certus—Charleston Mountains, southern Nevada (USA). It is "distinguished by pale general coloration in combination with dark russet under side of tail . . . a dark, rich russet, instead of ochraceous-buff or ochraceous-tawny tone" (Goldman). The mantle is tawny to tawny ochraceous.

C. l. chrysodeirus—mountains of southern and eastern Oregon south to Tulare County, California (USA). The inner black stripes are well defined. The mantle is rusty chestnut, while the rest of the back is grizzled gray, light brown, and black.

C. l. cinerascens—Alberta (Canada) south through Idaho and Montana to Yellowstone National Park (USA). The inner black stripes are relatively short. The mantle is dark chestnut red, and the rest of the back is grayish. The dorsal surface of the tail is mixed black and pale buff.

C. l. connectens—Homestead, Oregon (USA). The mantle is cacao brown. The ventral surface of the tail is cinnamon, and the dorsal surface is fuscous black.

C. l. mitratus—South Yolla Bolly Mountain, California (USA). The back is paler gray and brown. The mantle is russet to dark rusty chestnut.

C. l. tescorum—Alberta and British Columbia (Canada), near the Moose Pass branch of the Smoky River. The mantle is extensive and dark reddish brown. This form is larger, and the back is brownish gray.

C. l. trepidus—Pine Forest Mountains, Humboldt County, Nevada (USA). The mantle is light ochraceous. The rest of the back is grizzled ochraceous, black, and white.

C. l. trinitatus—Trinity Mountains, east of Hoopa Valley, California (USA). This is a larger form, with a darker general color. The ventral side of the tail is dark chestnut.

C. l. wortmani—Sweetwater County (Wyoming) south into the northwestern corner of Colorado (USA). The inner pair of black stripes is usually missing. The mantle is a paler reddish brown. The hands and feet are almost white. The general color is paler.

CONSERVATION: IUCN status—least concern. Population trend—stable.

Callospermophilus lateralis. Photo courtesy Phil Myers.

HABITAT: *C. lateralis* occupies a wide range of habitats (usually at elevations between 1220 and 3960 m), ranging from sagebrush, chaparral, and pinyon communities to above the alpine tundra. It also occupies open mountain meadows and rocky slopes, and it will frequently colonize burned and timbered areas; most areas of occupation have a significant rock component. The golden-mantled ground squirrel is often associated with stands of pine (*Pinus contorta, P. flexilis,* and *P. ponderosa*), spruce (*Picea engelmannii*), Douglas fir (*Pseudotsuga menziesii*), and quaking aspen (*Populus tremuloides*).

NATURAL HISTORY: Classified as an omnivore, *C. lateralis* is opportunistic in its feeding habits, but it is highly granivorous in the late summer and autumn, when the seeds of conifers and oak acorns are available. It is known to cache seeds, but details on this behavior are not available. Its consumption of seeds may deter forest regeneration in some areas. These squirrels also feed on a variety of shrub and herb species, and they consume several kinds of animal material, including insects, bird eggs, young birds, lizards, mammals, and roadkilled conspecifics. In an experimental situation, *C. lateralis* has been shown to forage selectively on artificial diets of polyunsaturated fats, which are readily found in plants, are not produced by mammals, and are critical for hibernation.

Golden-mantled ground squirrels excavate burrows that are used for escaping predators, nesting, and storing food. Burrows, which are up to 90 cm deep and 183 cm long, are found near rocks, stumps, and logs; sometimes they are in logs, stumps, and rock crevices. Burrows can be simple or complex, but frequently they are more complex than those of *Tamias minimus*; the burrows often have two or more entrances, usually oriented to the southwest. The length of hibernation varies with temperature, altitude, and snow depth, but it usually begins between late August and early November and ceases between late March and early May. Experimental restriction of caloric intake prior to hibernation delays the onset of hibernation, but apparently this has no effect on the length of the torpor bouts; it results in a loss of fat-free mass but not lipid mass. In the autumn, this species undergoes five to six periodic bouts of torpor, in which body temperature steadily decreases and the length of each torpor event increases, until the animal enters winter hibernation. *C. lateralis* is able to prevent the pulmonary surfactant secretion that typically occurs in most mammals when exposed to fluctuations in body temperature. The secretory and regulatory pathways involved in this squirrel's secretion of surfactants are insensitive to thermal changes, and its cells are able to survive at 0°C–5°C, suggesting an adaptation that facilitates hibernation. During hibernation the diaphragm muscle becomes hypertrophied and shows greater oxidative capacity; both changes are probably adaptations that enhance gas exchange during torpor. Hibernation is also aided by a suite of other physiological traits: for example, the ability to control breathing, ventilation, and respiratory chemosensitivity; an unusually high level of plasticity in the skeletal muscles; and the profiles, expression, and activity of specific muscle enzymes.

Breeding begins soon after emergence, although some females may not become reproductively active until a few weeks after emergence. Gestation is estimated to range between 26 and 33 days, and litter sizes average five (but are lower at higher elevations). The young weigh about 6 g at birth, are weaned after 35 days, and are the last to enter hibernation. The golden-mantled ground squirrel has been reported to live for up to 7 years in the wild. The animals are generally silent, but they do rely on high-pitched alarm calls and tail waving when threatened; in experimental situations they have been shown to learn to associate novel sounds with the presence of a potential predator. This species is prey for at least four species of carnivorous mammals and two raptor species. Ectoparasites associated with *C. lateralis* include 11 species of fleas, 2 ticks, and 3 sandflies. Its internal parasites consist of at least 7 species of 5 genera of protozoans and 1 nematode species. The golden-mantled ground squirrel also is a frequent host of *Yersinia pestis*, the bacterium responsible for sylvatic (bubonic) plague. *C. lateralis* fossils are reported from the late Pleistocene; some deposits suggest that in the past, this species' range may have extended 260 km farther south, and to much lower elevations, than it does today.

GENERAL REFERENCES: Anthony 1928; Bartels and Thompson 1993; Bihr and Smith 1998; Frank 1994; A. W. Linzey and Hammerson 2008d; J. A. MacDonald and Storey 2002, 2004; Ormond et al. 2003; Pulawa and Florant 2000; Rourke et al. 2004; Shriner 1999; Staples and Hochachka 1998; Zimmer and Milsom 2001, 2002.

Callospermophilus madrensis (Merriam, 1901) Sierra Madre Ground Squirrel

DESCRIPTION: Although similar in color to its sister species *C. lateralis*, *C. madrensis* is duller in its coloration, with shorter and more faded dark stripes, and longer white stripes. It also has a smaller body size and a shorter tail. The skull is narrower, and the brain case is more arched than that of *C. lateralis*.

SIZE: Female—HB 168.0 mm; T 65.0 mm; Mass 152.0 g.
Male—HB 169.0 mm; T 64.0 mm; Mass 151.0 g.
Sex not stated—HB 175.0 mm; T 58.0 mm.

DISTRIBUTION: *C. madrensis* is found in northwestern México. Historically, specimens were only reported from southwestern Chihuahua (México), but this species has now been documented in the state of Durango as well.

GEOGRAPHIC VARIATION: None.

CONSERVATION: IUCN status—near threatened. Population trend—no information.

HABITAT: *C. madrensis* occurs in conifer forests of the Sierra Madre Occidental biotic province above the pinyon pine belt, typically at elevations between 3000 and 3750 m. This species is associated exclusively with conifer forests of fir (*Pseudotsuga*), pine (*Pinus*), juniper (*Juniperus*), and sometimes quaking aspen (*Populus tremuloides*).

NATURAL HISTORY: Between two and five embryos have been reported from four individuals, and lactation has been observed in July. Nothing else is known about the population biology, ecology, behavior, or genetics of this ground squirrel. One species of sucking lice (*Enderleinellus suturalis*) is reported from *C. madrensis*. The Sierra Madre ground squirrel is significantly threatened by habitat loss due to deforestation.

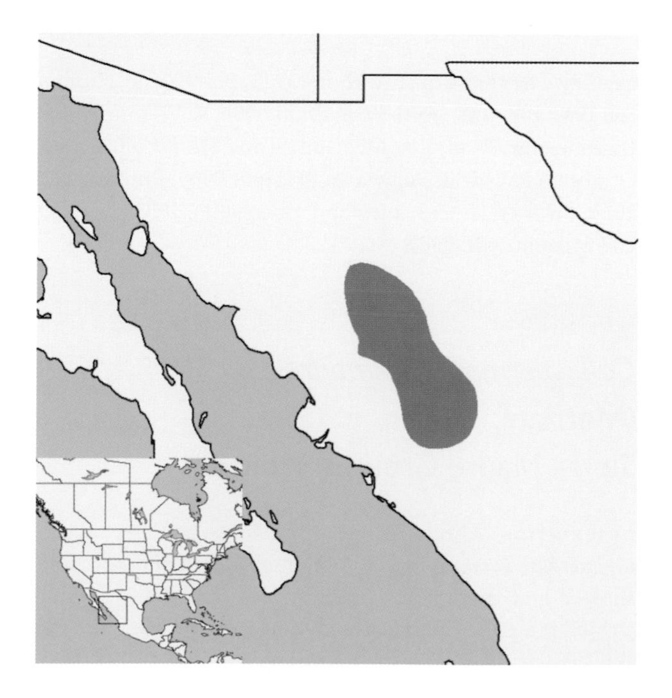

Callospermophilus madrensis. Photo courtesy Juan Cruzado.

GENERAL REFERENCES: Álvarez-Castañeda, Castro-Arellano, Lacher, et al. 2008b; Arita and Ceballos 1997; Best and Thomas 1991; Durden and Musser 1994a, 1994b; R. G. Harrison et al. 2003; Herron, Castoe, et al. 2004; Sánchez-Cordero et. al. 2005; Servin, Alonso-Perez, et al. 2003; Servin, Sánchez-Cordero, et al. 1996.

Callospermophilus saturatus (Rhoads, 1895)
Cascade Golden-Mantled Ground Squirrel

DESCRIPTION: Cascade golden-mantled ground squirrels are grizzled gray brown to charcoal on the dorsum, suffused with buff to ochre. The top and sides of the head and the shoulders have a variable and sometimes indistinct russet mantle. The most distinguishing characteristic for this species is the single white dorsal stripe that extends from the shoulder to the hip on each side; a black line offsets each white line on both the top and the bottom. The eye ring is pinkish buff. The chin is often white to buff; the cheeks are straw yellow to russet; the facial stripes that are characteristic of chipmunks (*Tamias*) are absent. The feet and venter are buff to straw yellow. The tail has a dark grizzled upper side and a buff to straw yellow to ochraceous underside.

SIZE: Female—HB 300 mm (286-312 mm); T 106.5 mm (92-116 mm).

Male—HB 305 mm (287-315 mm); T 110.9 mm (100-118 mm).

Both sexes—Mass 200-350 g.

DISTRIBUTION: *C. saturatus* is found in the Cascade Mountains of southwestern British Columbia (Canada) and western Washington (USA).

GEOGRAPHIC VARIATION: None.

CONSERVATION: IUCN status—least concern. Population trend—stable.

HABITAT: Cascade golden-mantled ground squirrels can be found in talus, montane meadows, forest clearings, sagebrush steppes, krummholz, open pine forests, and closed subalpine and alpine forests.

NATURAL HISTORY: This species is diurnal. Golden-mantled ground squirrels hibernate in burrows for eight to nine months, with adults emerging as early as April or May, followed by yearlings one to two weeks later. Mating occurs soon after emergence and extends for two weeks; yearling females breed, but typically not yearling males, which delay breeding until their third year. After a gestation of 28 days, the year's only litter, averaging four (range = 1-5) young, is born in a burrow. The young emerge in about 36 days from their natal burrows, appearing through July and early August. Natal dispersal is not sex biased; juvenile males move an average of 182 m, and females average 158 m, after leaving their natal areas. Adults and yearlings live solitarily, as this is a relatively asocial ground squirrel. Adults and yearlings immerge to hibernate in mid-August to late September; juveniles remain aboveground into November and early December. Burrows are used for escape, nesting, and hibernation. Burrow entrances typically are placed near objects such as rocks, boulders, shrubs, or trees; multiple tunnels descend to a plant-lined nest chamber. Their diet is rich in fungi (especially during the fall), as well as the shoots, leaves, fruits, and the seeds of a variety of grasses, forbs, shrubs, and trees; these squirrels will also eat insects and scavenge carrion. They generally forage on the ground, but they will (rarely) climb into bushes and conifers for food or cover. Their major predators include mustelids, canids, felids, and raptors. The alarm call of *C. saturatus* is a multinote trill call, with a high note followed by a rapid series of lower notes, but it also has components in the ultrasonic range. Cascade golden-mantled ground squirrels are not hunted or trapped, but they are often welcome (or unwelcome) guests in picnic areas and by cabins.

Callospermophilus saturatus. Photo courtesy Takeo Kawamichi.

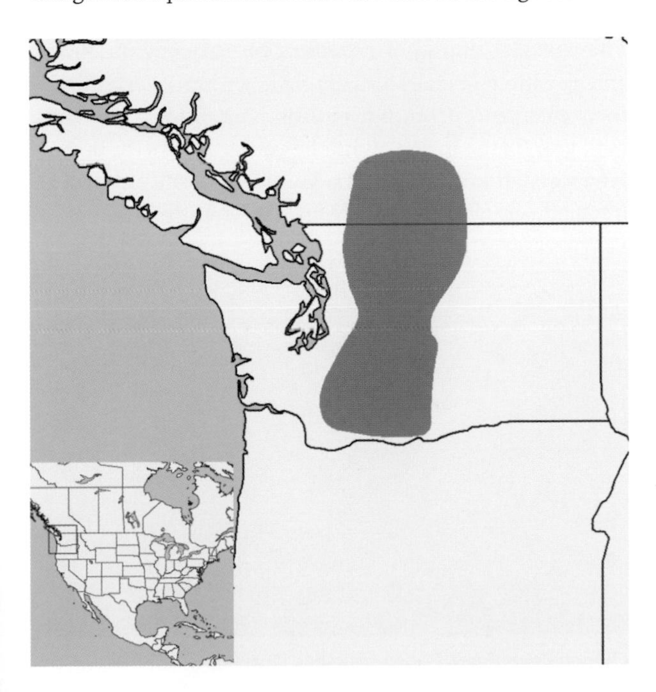

GENERAL REFERENCES: Cheng and Leung 1997; Eiler and Banack 2004; Trombulak 1987, 1988.

Cynomys Rafinesque, 1817

This genus contains five species of prairie dog.

Cynomys gunnisoni (Baird, 1858) Gunnison's Prairie Dog

DESCRIPTION: Gunnison's prairie dog is buff to pale yellow on the dorsum, grizzled with small amounts of black. The underside is white to cream, with a gradual transition from the darker dorsum. The head is colored similarly to the dorsum, but it is often cream to white and lighter than the rest of dorsum; there is often a faint black patch between the eyes and the sides of the snout. The tail fades to a pale buff or white tip. *C. gunnisoni* is the smallest of the prairie dogs.

SIZE: Female—TL 325 mm (309-338 mm); T 54 mm (46-61 mm); Mass 644 g (465-750 g).

Male—TL 335 mm (317-390 mm); T 51 mm (40-60 mm); Mass 816 g (460-1300 g).

Males are about 136 percent of female mass at hibernation.

DISTRIBUTION: Gunnison's prairie dog is found in southeastern Utah, southwestern Colorado, northeastern Arizona, and northwestern New Mexico (USA).

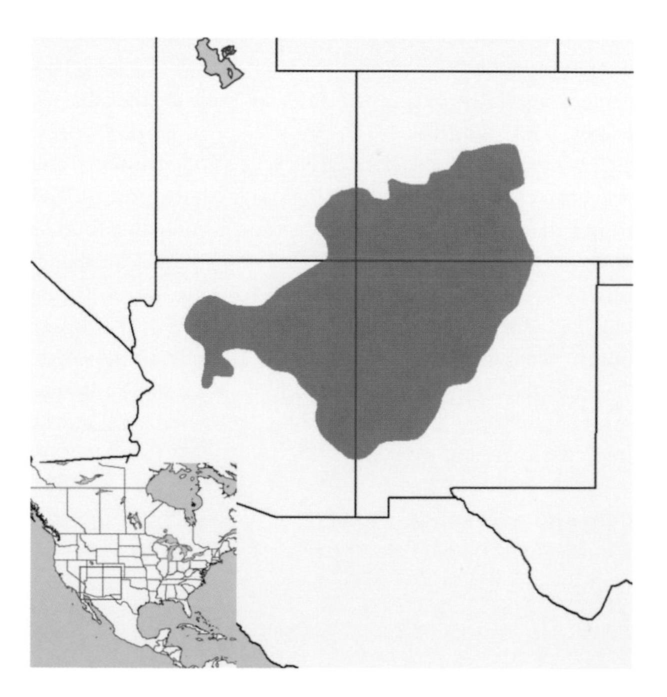

GEOGRAPHIC VARIATION: Two subspecies are recognized.

C. g. gunnisoni—eastern portion of the range. This form is a slightly smaller, paler, and buff-colored montane form.

C. g. zuniensis—western portion of the range. It is a slightly larger, darker, and more cinnamon-colored grassland form.

CONSERVATION: IUCN status—least concern. Population trend—decreasing.

HABITAT: Gunnison's prairie dogs inhabit high-elevation open mountain valleys and plateaus, but they can thrive in modest shrub and open pinyon-juniper (*Pinus, Juniperus*) savannas.

NATURAL HISTORY: This species is diurnal. It hibernates in burrows for up to five months in many parts of its range, typically from late October to February. The burrows are generally 2-3 m deep and average as much as 13 m long, with multiple entrances (generally three to four entrances per burrow, but sometimes up to six). Females attain sexual maturity and mate as yearlings; each female can wean only one litter per year. Males sometimes mate as yearlings, but they commonly delay their first mating until they are 2 years old. The seasonal timing of reproduction varies with latitude and elevation. Females usually mate within several days of their emergence from hibernation. One litter is produced each year (in the spring), with the young born in the burrow after a gestation of 28-30 days. Multiple paternity of litters is frequent, approaching 80 percent. The young remain below-

Cynomys gunnisoni. Photo courtesy Doris Potter / Focus on Nature Tours, Inc.

ground for about 5-6.5 weeks, when four to five (range = 1-7) emerge from the nursery burrow. Weaning usually is not complete until one to three weeks after the juveniles first emerge from the nursery burrow, and communal nursing is common following their first emergence. *C. gunnisoni* is highly social. Social groups (clans) are organized into territories that generally contain one adult male, one or more adult females, yearlings, and young-of-the-year; neighboring groups have a minimal overlap, whereas the overlap among group members is nearly complete. In most populations females are philopatric, and clans are composed of close female kin. Amicable interactions among adults and juveniles include kissing, sniffing anal and oral glands, and playing; hostile interactions among adults of all species include fights, chases, and territorial disputes.

The burrows of *C. gunnisoni* increase habitat heterogeneity, and the plant communities associated with these burrows create significant increases in that region's biodiversity. Gunnison's prairie dogs feed on grasses, forbs, sedges, and shrubs; insects and other animal foods are rarely eaten. They do not cache food. Gunnison's prairie dogs can distinguish between potential threats and even identify specific characteristics of the threats, and they code this information into their elaborate alarm calls. Survivorship in the first year after emergence from the nursery burrow is usually only about 50 percent for both sexes, and less than 15 percent survive through their second year. Males sometimes live as long as 5 years, and females occasionally can be as much as 6 years old. The principal predators of *C. gunnisoni* are raptors, canids, felids, and mustelids, including endangered black-footed ferrets (*Mustela nigripes*). Colonies suffer drastic population declines and are often extirpated during outbreaks of flea-borne sylvatic (bubonic) plague. Gunnison's prairie dogs are hunted for recreation in much of their range. They are also sometimes considered as competitors for forage in cattle country or as a threat to livestock, due to their open burrows that grazing animals might step into. Although both of these concerns are exaggerated, *C. gunnisoni* have been poisoned and shot for pest removal in many areas of their range. On the other hand, some researchers consider *Cynomys* to be a keystone species for ecosystem health.

GENERAL REFERENCES: Bartz et al. 2007; Davidson and Lightfoot 2008; Haynie et al. 2003; Hoogland 1998, 1999, 2003a, 2003b; Pizzimenti and Hoffmann 1973; Rayor 1988; Slobodchikoff et al. 2009.

Cynomys leucurus (Merriam, 1890)
White-Tailed Prairie Dog

DESCRIPTION: The dorsum is yellowish buff color frosted with black, with a paler venter. There is a dark brown or black stripe above each eye, with some dark brown or black below each eye. The tail is buff to white at the base, with a suffusion of cinnamon and white at the tip.

SIZE: Female—TL 353 mm ± 4 mm (315-375 mm); T 40-65 mm; Mass 925 g (675-1200 g).

Male—TL 371 mm ± 4 mm (342-399 mm); T 40-65 mm; Mass 1139 g (750-1700 g).

DISTRIBUTION: *C. leucurus* is found in extreme southcentral Montana, western and central Wyoming, northeast Utah, and northwest Colorado (USA).

GEOGRAPHIC VARIATION: None.

CONSERVATION: IUCN status—least concern. Population trend—decreasing.

HABITAT: White-tailed prairie dogs inhabit xeric and relatively high-biomass mixed stands of grasses, forbs, and shrubs in high-elevation meadows, flats, and gently rolling hillsides.

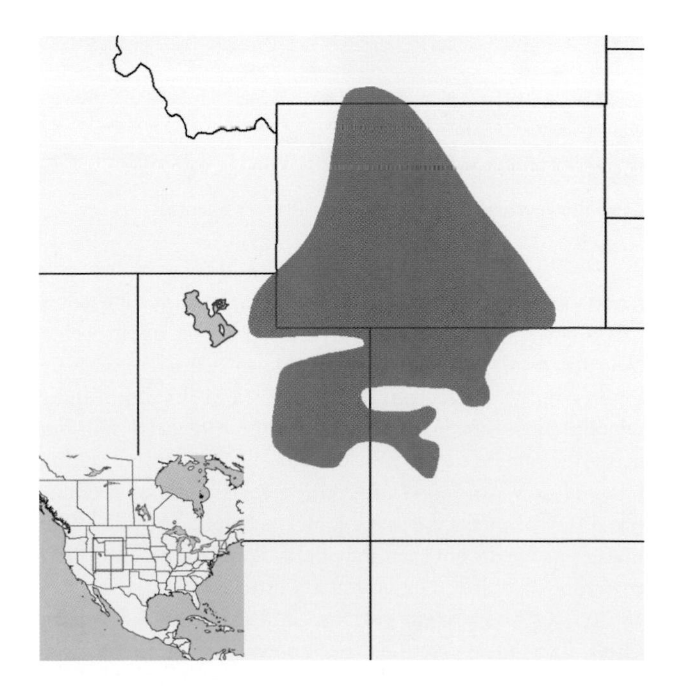

Cynomys leucurus. Photo courtesy Robert Shantz.

NATURAL HISTORY: *C. leucurus* is diurnal, and it hibernates in burrows during lengthy winters that last up to seven months. Adult males usually emerge in February, with females coming aboveground two to three weeks later. Adults usually enter hibernation in late summer or early fall, but juveniles remain active into October or November. Mating takes place within days after emergence. Females produce one litter per year, beginning at 1 year of age; males also mature at 1 year, but they may delay copulations until their second year of life. The young are born after a gestation of 28-30 days and remain underground for about 5.5 weeks. Litter size when juveniles first appear aboveground from the nursery burrow is most commonly four or five (range = 1-8). Weaning is nearly complete when the juveniles first emerge from the nursery burrow, and communal nursing following their first emergence is rare. Annual survivorship of adult males ranges from 12 to 56 percent; for adult females, from 21 to 62 percent; and for juveniles, from 5 to 39 percent. Both males and females live as long as 5 years and possibly longer, but no study has followed marked individuals beyond 5 years. Survivorship in the first year after emergence from the nursery burrow is usually only about 50 percent for both sexes. Juvenile females are philopatric and remain on or near their natal area.

White-tailed prairie dogs are less social than black-tailed prairie dogs (*C. ludovicianus*), but their levels of coloniality and sociality are equivalent to those in Utah prairie dogs (*C. parvidens*) and Gunnison's prairie dogs (*C. gunnisoni*). Burrows can be recognized by a ring of excavated soil around the entrance; however, these rings are less conspicuous in *C. leucurus* burrows than in those of the other species. Burrows are often complex—up to 10 entrances are common, and occasionally approach 30—and those that are 2-3 m deep may contain a nest. Colonies vary considerably in size and density. *C. leucurus* clans are small, and individuals forage in overlapping ranges within the colony. Females within clans are usually close relatives. Larger colonies do detect predators more quickly than smaller colonies, which suggests an important benefit to aggregations. Adult home ranges are about 1-2 ha in size. Amicable interactions among adults and juveniles include kissing, sniffing anal and oral glands, and playing; hostile interactions among adults of all species include fights, chases, and territorial disputes.

C. leucurus feeds on grasses, forbs, flowers, and seeds. Individuals sit on their haunches or stand upright on two feet to scan for predators. Alarm calls are given in response to mustelids, felids, canids, and raptors, which are this species' principal predators. Colonies of white-tailed prairie dogs appear to increase biodiversity. The rediscovery of the black-footed ferret (*Mustela nigripes*)—believed to be extinct—near Meteetsee, Wyoming (USA), occurred on a white-tailed prairie dog town. Historically, white-tailed prairie dogs were hunted for food; they were often (and still are) also used for target practice or sport. As a potential pest, white-tailed prairie dogs were frequently poisoned in the 1900s in an attempt to reduce their presumed competition for forage with for-profit grazing of livestock. Like the other species of prairie dogs, white-tailed prairie dogs are highly susceptible to sylvatic (bubonic) plague, which is transmitted via fleas.

GENERAL REFERENCES: Bakko and Brown 1967; T. W. Clark 1977; T. W. Clark et al. 1971; Hoogland 1981, 2003b; B. Miller et al. 1996.

Cynomys ludovicianus (Ord, 1815)
Black-Tailed Prairie Dog

DESCRIPTION: Black-tailed prairie dogs are buff to brown to cinnamon on the dorsum, with a pale buff to white venter. The region above the eyes and around the snout can be paler than other parts of the dorsum. The tail is dark brown to black on the distal third.

SIZE: Female—HB 373.5 mm (354.5-397.8 mm); Mass 819 g (n = 430) in autumn, 689 g (n = 276) in winter, and 696 g (n = 613) in spring.

Male—HB 373.5 mm (354.5-397.8 mm); Mass 905 g (n = 217) in autumn, 750 g (n = 149) in winter, and 801 g (n = 281) in spring.

Males are 10-15 percent heavier than females.

DISTRIBUTION: The black-tailed prairie dog is found in a few localities across a broad area: extreme southern Saskatchewan (Canada); and Montana to western North Dakota and south to New Mexico and western Texas. It was recently reintroduced to southeastern Arizona (USA) and northeastern Sonora and northern Chihuahua (México).

GEOGRAPHIC VARIATION: Two morphologically similar subspecies are recognized.

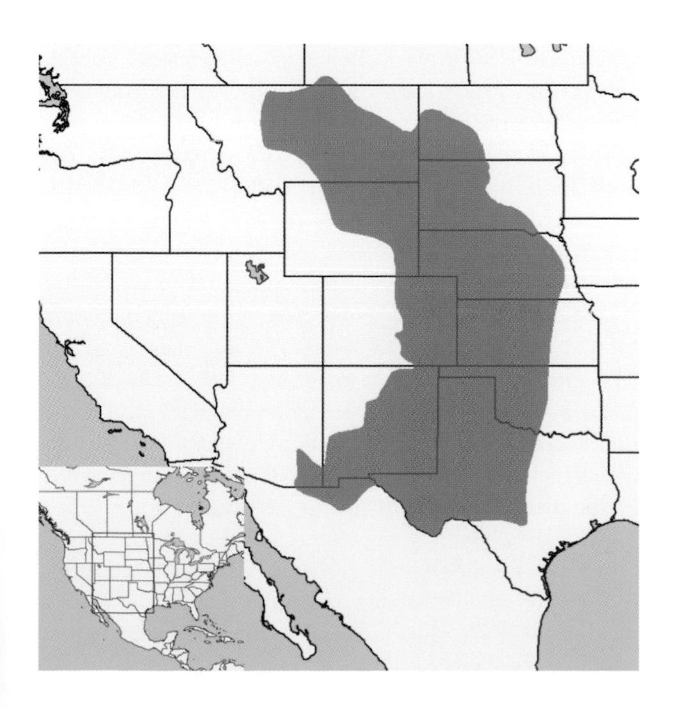

C. l. ludovicianus—covers the vast majority of the range. There are no external differences between the subspecies.

C. l. arizonensis—southwestern portion of the range.

CONSERVATION: IUCN status—least concern. Population trend—decreasing. *C. ludovicianus* is a recent candidate for threatened status in the USA, and this species is considered threatened in México.

HABITAT: Black-tailed prairie dogs occupy open flat or gently sloping grasslands with low and relatively sparse vegetation, in areas with fine- to medium-textured soils. They are commonly found in shortgrass and mixed-grass prairies in the western portion of their range; and in tallgrass prairies in the eastern portion, primarily where grass height is reduced by grazing, mowing, or agriculture. This species can occur in overgrazed livestock range and in open vacant lots at the edges of urban areas.

NATURAL HISTORY: *C. ludovicianus* is diurnal and active throughout the year, except during periods of prolonged and exaggerated cold, when individuals may enter temporary torpor and remain underground. Adults emerge in March or early April and enter hibernation in July or August, whereas juveniles remain active into October or November. Mating occurs during a two- to three-week period, when females enter a single day of estrus and copulate underground. The breeding season varies with latitude. This species is polygynous; multiple paternity occurs occasionally (less than 25%), and a first-male advantage exists for sires. Most males and females copulate for the first time when they are almost 2 years old; however, a few females and males (less than 10%) produce young when they are yearlings. Males are more likely to delay their first reproduction than females. After a gestation of 34-35 days, the young are born in the burrow. Females can wean one litter per year; although most adult females mate, only about 89 percent successfully wean a litter, and just 54 percent of yearling females are successful. Litters of three or four (range = 1-6) young appear aboveground at 6 weeks and are weaned soon afterward. Natal dispersal is male biased, with most males, when they are yearlings, leaving their natal area during the late spring or early summer; breeding dispersal by older adult males is also common. Mean lifetime reproductive success is higher in males (7.1 weaned offspring) than females (3.9 weaned offspring), but it is more variable for males. Female reproductive success is positively impacted by the previous year's rainfall and by breeding early in the season. Heavy long-lived animals have the highest lifetime reproductive success.

Black-tailed prairie dogs are social and live in large complexes called towns (or colonies). Most burrows can be recognized by a ring of excavated soil around the entrance and a small mound; some peripheral burrows lack any mound. Burrows are modest, with two entrances (but occasionally as many as six). There is a nest within lengthy burrows (typically exceeding 10-15 m long and penetrating 2-3 m deep). Black-tailed prairie dog colonies vary considerably in size and density, but some are known to extend for many kilometers; a prairie-dog town in Oklahoma (USA) was estimated to be 35 km in length, and such descriptions are common in the early literature. *C. ludovicianus* is highly social. Colonies of *C. ludovicianus* consist of territorial family groups called coteries. Coterie size (i.e., the number of adults and yearlings living in the same territory) can be as small as 2 or as large as 26, but the typical coterie contains 1 adult male, 2-3 adult females, and 1-2 yearlings of both sexes. Coterie territories average about 0.3 ha in size, and their boundaries remain remarkably stable over time. A typical coterie has approximately 70 burrows, and all members of a coterie can use any of these entrances, with one exception: pregnant and lactating females vigorously defend nursery burrows. Females within the same coterie are almost always close kin (mothers, daughters, sisters, nieces, etc.). The coterie structure reduces this species' effective population size, due to matrilineal relationships, and fine-scale genetic differentiation is detectable. Playing, greeting by sniffing anal and oral glands, kissing, and allogrooming are affiliative behaviors shared between group members (except when females are nursing their young), but aggression—including physical combat—is directed at individuals from outside the coterie. The territorial two-syllable "jump-yip" call is given as an individual stands on hind feet and throws its front legs and head upward, and this call commonly elicits similar responses from members of its own and adjacent coteries. Infanticide is the primary source of juvenile mortality, accounting for the loss of 39 percent of the juveniles, especially when in the burrow or immediately after emergence. Sources of infanticide (in order of importance) are marauding females that actively seek and kill the offspring of other coterie members, female abandonment of their own offspring, immigrant male takeovers of coteries, and immigrant females. Weaning usually is not complete until one to three weeks after the juveniles first emerge from the nursery burrow, and communal nursing is common following this first emergence. Survivorship is about 50 percent in the initial year of life, with a maximum longevity of 8 years for females and 5 years for males.

***Cynomys ludovicianus*.** Photo courtesy Elaine Miller Bond.

C. ludovicianus can often be seen feeding on grasses, forbs, thistles, flowers, cactus, and seeds; individuals also clip vegetation without ingestion, in an effort to maintain lines of sight. Individuals sit on their haunches or stand upright on two feet to scan for predators. Vigilance functions mostly to improve the probability of detecting a predator, but it also monitors the activities of nearby conspecifics (many of whom might be infanticidal). Alarm calls are given in response to the large-bodied mustelids (badgers [*Taxidea taxus*], weasels [*Mustela*], and black-footed ferrets [*Mustela nigripes*]), bobcats (*Lynx rufus*), coyotes (*Canis latrans*), snakes, and raptors that function as this species' principal predators. The vocal repertoire of *C. ludovicianus* is robust, with at least 12 known sounds, but only "jump-yips" and alarm calls are common. Black-tailed prairie dogs benefit from a large colony size by their ability to detect predators early and to spend more time feeding and less time being vigilant. Their area of occupancy has been reduced from about 40 million ha historically to about 766,400 ha, a decline of 98 percent. Habitat loss is considered a major threat as significant amounts of grassland (perhaps 40%) are being converted to agriculture. Urbanization also continues to remove sizeable amounts of habitat. Large and recently isolated fragments near other prairie dog colonies, flat areas, and those with substantial grass cover are most likely to support prairie dog populations in urban areas. Historically, black-tailed prairie dogs were hunted occasionally for food, often for target practice or sport, and for pest removal; shooting remains popular in some portions of their range. They were often (and sometimes still are) viewed as a significant competitor for forage or as a threat to livestock, due to their open burrows that grazing animals might step into; both of these concerns are exaggerated. As a potential pest, black-tailed prairie dogs were often poisoned in the 1900s in an attempt to extirpate them within their small isolated range, resulting in massive declines. Poisoning continues in some portions of their range. Some researchers, however, consider *Cynomys* to be important for soil quality, as a vital contributor to other wildlife habitats, and as an indicator of ecosystem health. Black-tailed prairie dogs are highly susceptible to outbreaks of sylvatic (bubonic) plague, a bacterial disease introduced to North America in the early 1900s. These outbreaks rapidly eliminate local populations, due in part to the high densities of this highly social species.

GENERAL REFERENCES: Ceballos, Mellink, et al. 1993; Hoogland 1995, 1996, 2003b, 2006; J. A. King 1955; S. D. Miller and Cully 2001; U.S. Fish and Wildlife Service 2004; Winterrowd et al. 2009.

Cynomys mexicanus (Merriam, 1892) Mexican Prairie Dog

DESCRIPTION: Mexican prairie dogs are a grizzled pinkish buff on the dorsum; the head is often slightly darker in coloration, with the lower muzzle showing patches of buff. The venter is yellow to buff suffused with dark brown to black. The tail is long relative to other prairie dogs, and the distal half is tipped in black. The ears are small and pressed against the head.

SIZE: Female—T 100 mm ± 8.9 mm; Mass 931 g ± 35 g.

Male—T 105 mm ± 6.6 mm; Mass 1065 g ± 60 g in August.

Both Sexes—TL 389 mm (380–440 mm).

DISTRIBUTION: *C. mexicanus* is found in an area of less than 800 km² in Coahuila and San Luis Potosí (and perhaps in Zacatecas and Nuevo León) in northcentral México, at elevations of 1600–2000 m.

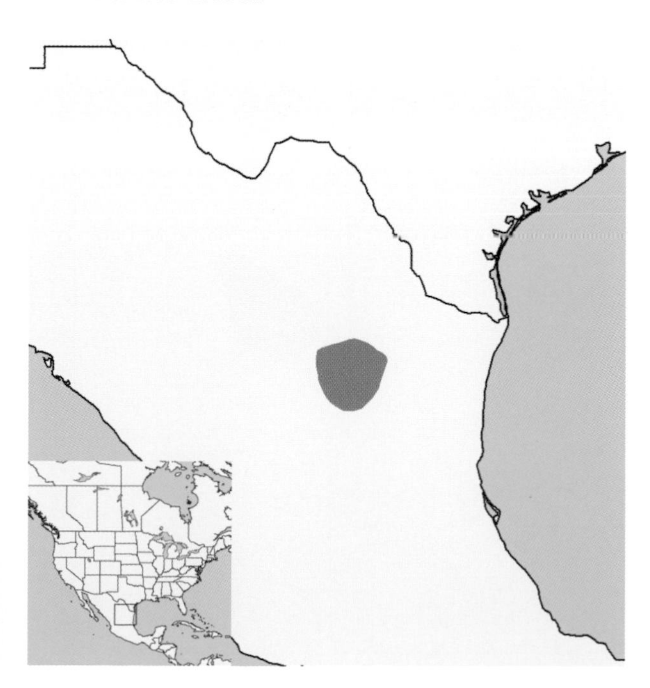

Cynomys mexicanus. Photo courtesy Rene Ortega Arguello, arguellorene@hotmail.com.

GEOGRAPHIC VARIATION: None.

CONSERVATION: IUCN status—endangered. Population trend—a slow decline. *C. mexicanus* is also federally listed as endangered in México.

HABITAT: Mexican prairie dogs inhabit open plains and plateaus that contain well-drained soil supporting grasses and herbs; these areas are often fragmented within desert and agricultural landscapes. *C. mexicanus* can also be found in cattle pastures and in fields that are degraded from grazing by goats.

NATURAL HISTORY: *C. mexicanus* is diurnal. Mexican prairie dogs are the least studied species in the genus. They do not hibernate, probably due to the low elevations and low latitude of the sites they inhabit. Nights are spent in burrows that are often provisioned with clipped vegetation for bedding; burrows have small mounds of excavated soil, in a circle with a radius of 1–2 m, at their entrance. Mexican prairie dogs are generalist herbivores with a flexible dietary strategy. Individuals forage on the ground for a variety of grasses, forbs, flowers, and seeds. Their diets differ in areas grazed by cattle and goats, with only a modest dietary overlap. In fact, Mexican prairie dogs are able to persist in pastures degraded by goats. The breeding season of *C. mexicanus* is protracted, and it may extend from January to late April; many males have descended testes (indicative of reproductive condition) from December until July. Estrus peaks in February and March. After a gestation that probably is between 30 and 35 days, females produce a single litter (averaging four juveniles) in burrows. The young nurse for 21-23 days, until they are weaned by late April or early May. Infanticide is not known; however, females vigorously defend nesting burrows from males, suggesting this possibility. Mexican prairie dogs are highly social and form coteries composed of one to two adult males, one to four adult females, yearlings, and young-of-the-year of each sex. Natal dispersal appears to occur during the second year. Group members are amicable and greet each other by investigating one another's cheeks. Amicable interactions among adults and juveniles include kissing, sniffing anal and oral glands, playing, and allogrooming; hostile interactions among adults of all species include fights, chases, and territorial disputes. No information is available about communal nursing.

Adults of both sexes can be seen in an alert posture, sitting on their haunches and being vigilant. *C. mexicanus* has two calls: an alarm vocalization, given in response to intruders and predators; and an elation call, emitted as an apparent affiliative behavior. Their principal predators are raptors, kit foxes (*Vulpes macrotis*), badgers (*Taxidea taxus*), and coyotes (*Canis latrans*), all of which frequently stalk and attack Mexican prairie dogs. Large rattlesnakes may also prey on these prairie dogs, especially the young. In addition, sylvatic (bubonic) plague is a known cause of local extinctions. Loss of habitat, due to its conversion to agriculture and livestock grazing, is a major threat to their persistence. Mexican prairie dogs are considered (erroneously) as competitors with livestock for forage, and they have been the target of intensive and widespread poisoning historically, which may have reduced their range by 65 percent over the last 150 years.

GENERAL REFERENCES: Ceballos, Mellink, et al. 1993; Ceballos and Wilson 1985; Hoogland 2003a, 2003b; Mellado and Olvera 2008; Trevino-Villarreal 1990.

Cynomys parvidens (Allen, 1905)
Utah Prairie Dog

DESCRIPTION: Utah prairie dogs are buff to cinnamon to clay on the dorsum, with a light cinnamon to pale buff venter. The upper lip and the chin are a pale buff. A dark brown or black stripe is found above each eye, with some dark brown or black below the eye. The tail grades from a proximal buff to a distal white.

SIZE: Female—TL 320 mm (290-368 mm); T 52 mm (47-56 mm); Mass 785 g ± 15 g in June-July.

Male—TL 341 mm (299-370 mm); T 54 mm (49-62 mm); Mass 1080 g ± 15 g in June-July.

DISTRIBUTION: This species is found in southwestern and southcentral Utah (USA).

GEOGRAPHIC VARIATION: None.

CONSERVATION: IUCN status—endangered. Population trend—decreasing, with fewer than 8000 individuals. *C. parvidens* is listed as threatened under the U.S. Endangered Species Act.

HABITAT: This species is found in open grasslands and sagebrush flats in mountain valleys with well-drained soils. *C. parvidens* appears to avoid slopes and areas where vegetation obstructs its ability to view its surroundings.

NATURAL HISTORY: The Utah prairie dog is diurnal. It hibernates in burrows for lengthy winters that last up to seven months. Adults typically emerge in February and enter hibernation in late summer or early fall, whereas juveniles remain active into October or November. Females usually

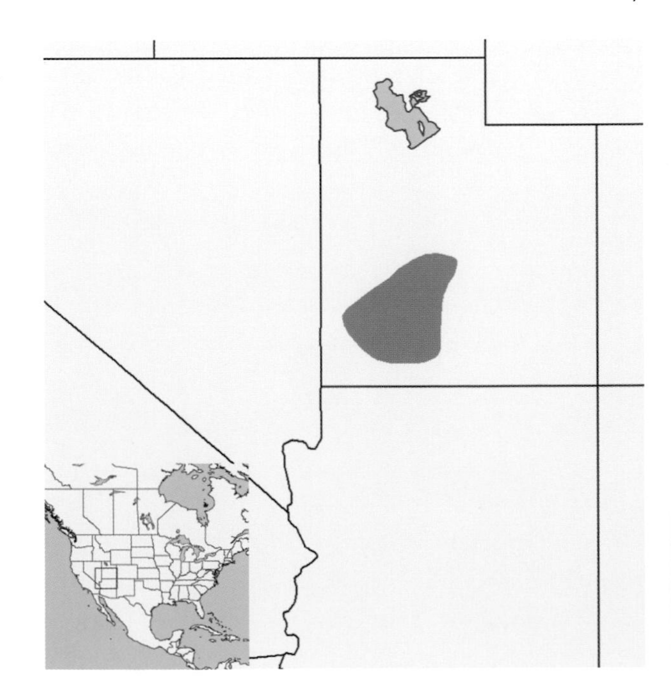

Cynomys parvidens. Photo courtesy Elaine Miller Bond.

mate within several days of their emergence from hibernation; multiple paternity is common (more than 70%). Females attain sexual maturity and mate as yearlings. Males sometimes mate as yearlings, but they commonly delay their first mating until they are 2 years old. The young are born after a gestation of 28–30 days, and they remain underground for about 5.5 weeks. Litter size when juveniles first appear aboveground from the nursery burrow is most commonly four or five (range = 1–7). Females produce one litter per year, and although most females mate, only about 67 percent successfully wean a litter. The young appear aboveground when they are 6 weeks old and are weaned soon afterward. Survivorship in the first year after emergence from the nursery burrow is usually only about 50 percent for both sexes, and fewer than 30 percent remain alive at 2 years of age. Males sometimes live as long as 7 years, and females occasionally can be as much as 8 years old. Utah prairie dogs are social, and they live in colonies (often called towns). Their burrows can be recognized by the ring of excavated soil around the entrance. Burrows can be complex, with up to 10 entrances and 1–3 nests. Burrows typically do not exceed 6 m in length and perhaps 3 m in depth. Colonies vary considerably in their size and density. *C. parvidens* is highly social, and the typical social units (known as clans) are composed of an adult male, several adult females, and immature young less than 2 years of age. Amicable interactions among adults and juveniles include kissing, sniffing anal and oral glands, and playing; hostile interactions among adults of all species include fights, chases, and territorial disputes. Adult females within a social group are always close kin. Following the first emergence of juveniles from their nursery burrows, females commonly will nurse the offspring of other females until weaning is complete (in another one to three weeks); beneficiaries of such communal nursing are usually close kin, such as grandoffspring, nieces, and nephews.

C. parvidens can often be seen eating grasses, forbs, flowers, seeds, and the fruits and seed of shrubs. Alfalfa (*Medicago sativa*) planted for hay is a favorite food. In addition, these prairie dogs are known to consume insects and feed on cattle feces. Individuals sit on their haunches or stand upright on two feet to scan for predators. Vigilance functions mostly to improve the probability of detecting a predator, but it is also used to monitor the activities of nearby conspecifics. Infanticide by adult males is known to occur in *C. parvidens*. Alarm calls are given in response to the large-bodied mustelids, felids, canids, and raptors that function as this species' principal predators. Historically, Utah prairie dogs were hunted occasionally for food and frequently for target practice or sport. They were often (and sometimes still are) falsely viewed as a significant competitor for forage or as a threat to livestock, due to open burrows that can be stepped into by grazing animals. However, some researchers consider *Cynomys* to be a keystone species for ecosystem health. As a potential pest, Utah prairie dogs were often poisoned in the 1900s in an attempt to extirpate them within their small isolated range, resulting in massive declines from populations estimated to have been perhaps 95,000 individuals over 1800 km^2. *C. parvidens* is also susceptible to outbreaks of sylvatic (bubonic) plague that rapidly eliminate local populations.

GENERAL REFERENCES: Haynie et al. 2003; Hoogland 2001, 2003b, 2006, 2009; Manno 2007; Pizzimenti and Collier 1978; U.S. Fish and Wildlife Service 1991.

Ictidomys J. A. Allen, 1877

This genus comprises three ground squirrel species.

Ictidomys mexicanus (Erxleben, 1777)
Mexican Ground Squirrel

DESCRIPTION: Mexican ground squirrels have an olivaceous gray to sepia brown back, with nine rows of white to buff spots from the back of the head to the rump. The snout and chin are yellowish to ochre to cinnamon. The white to buff eye ring is conspicuous. The sides and venter are buff to yellow. The tail is grizzled black, suffused and frosted with the coloration of the venter.

SIZE: Both sexes—TL 322-380 mm; T 124-166 mm; Mass 217-398 g.

DISTRIBUTION: This species is found in central México, in Aguascalientes, Jalisco, Guanajuato, Querétaro, Hidalgo, México D.F., Tlaxcala, and Puebla.

GEOGRAPHIC VARIATION: None.

CONSERVATION: IUCN status—least concern. Population trend—stable.

HABITAT: Mexican ground squirrels inhabit grassy pastures and meadows, grassy shrublands and savannas, and arid regions.

NATURAL HISTORY: This species is diurnal. Mexican ground squirrels hibernate in burrows for 5-6 months, beginning in September (males) or October (females) through March or April. Males emerge first in late March, and females two to four weeks later. Mating occurs in the spring. A litter averaging four (range = 3-5) young is born in a burrow. Nursing females are found in the population from June to September; males with scrotal testes occur from March to July. Juveniles enter hibernation in late October and November. Numerous short burrows are used for temporary or escape cover, and they often appear to be the burrows of pocket gophers (*Thomomys*). Burrows for residence are less than 1.5 m long and can have one or more nest chambers within the labyrinth of tunnels. Colonies of many individuals occur in favorable habitat; however, levels of sociality are low. *I. mexicanus* is an omnivore that feeds heavily on the leaves, shoots, flowers, and seeds of grasses, forbs, shrubs, and cacti, but it will also consume significant amounts of insects and other animal matter when available, especially beetles in the spring. Most small carnivores—including felids, canids and mustelids, raptors, and snakes—prey on *I. mexicanus*. Their alarm call is a shrill whistle. Mexican ground squirrels are not hunted for food or pelts; however, they are shot or poisoned as a local nuisance.

***Ictidomys mexicanus*.** Photo courtesy Juan Cruzado.

GENERAL REFERENCES: Cothran 1983; Cothran and Honeycutt 1984; Martinez et al. 1999; Millán-Peña 1998; Schwanz 2006; Valdez and Ceballos 1991, 2003; Young and Jones 1982.

Ictidomys parvidens (Erxleben, 1777) Rio Grande Ground Squirrel

DESCRIPTION: Rio Grande ground squirrels have an olivaceous gray to light brown dorsum, with nine rows of white to buff spots from the back of the head to the rump. The snout and the chin are yellowish to ochre to cinnamon. The white to pale buff eye ring is conspicuous. The sides and venter are white to pale buff. The tail is grizzled black, suffused and frosted with the coloration of the venter.

SIZE: Female—TL 216-303 mm; T 80-124 mm; Mass 90-152 g.

Male—TL 243-325 mm; T 80-130 mm; Mass 100-210 g.

DISTRIBUTION: The Rio Grande ground squirrel is found in southern New Mexico and western Texas (USA) to Coahuila, Nuevo León, Tamaulipas, Durango, and Zacatecas (México).

GEOGRAPHIC VARIATION: None.

CONSERVATION: IUCN status—not listed. Population trend—stable. *I. parvidens* is not ranked by IUCN, because it has only recently been recognized as a distinct species, different from its sister taxon, *I. mexicanus*. Mexican ground squirrels are listed as a species of least concern.

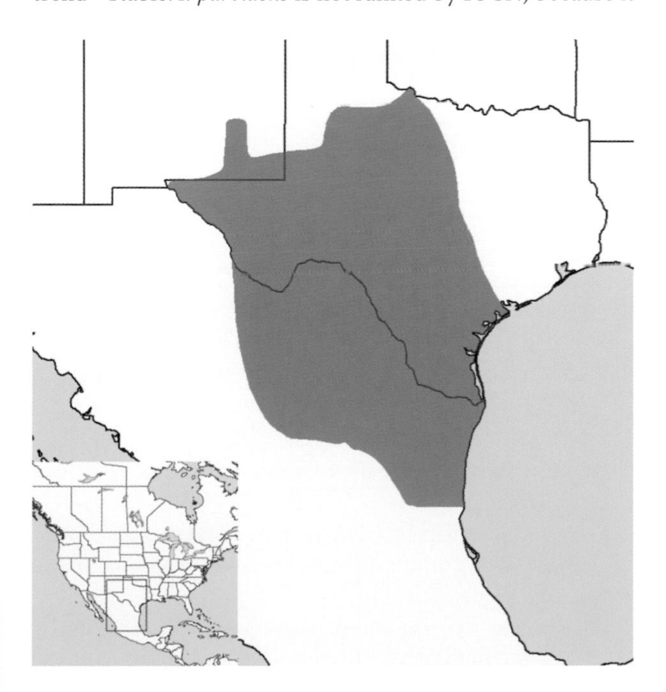

Ictidomys parvidens. Photo courtesy Troy L. Best.

HABITAT: Rio Grande ground squirrels inhabit grassy habitats and grassy shrublands in arid regions. They can also be found in human-maintained grassy habitats, such as cemeteries and golf courses.

NATURAL HISTORY: This species is diurnal. In mild climates, Rio Grande ground squirrels may be active for much of the winter; however, most hibernate in burrows for about seven months, beginning in late July through September. All males immerge by late August, whereas females may stay aboveground until the fall. Juveniles enter hibernation in September and October. Males emerge first in March and feed ravenously to increase their body mass, in apparent preparation for mating opportunities with females that emerge two to four weeks later. The mating season can be delayed upon emergence and is prolonged (30-71 days) during March and April. After a gestation of 28-30 days, a litter of 2-10 young (average = 5-6) is born in the burrow; they emerge 30-54 days later, in late May-August. The young are weaned by August. Numerous short burrows are used for temporary or escape cover. Burrows for maternity and hibernation are as deep as 1.5 m and are more convoluted, with a nest chamber and sometimes multiple openings. This species is colonial in good habitat but relatively asocial, although tolerant of home range overlap. The Rio Grande ground squirrels traverse relatively small areas, generally less than 90 m in diameter.

I. parvidens is an omnivore that feeds heavily on the leaves, shoots, flowers, and seeds of grasses, forbs, shrubs, and cacti, but it will consume significant amounts of insects and other animal matter when available. Adult survivorship is high, with 53-63 percent of the adults remaining alive during the year, whereas juvenile survival is only about 30 percent. Most small carnivores—including felids, canids and mustelids, raptors, and snakes—prey on *I. parvidens*. Their alarm call is a shrill whistle. Rio Grande ground squirrels are not hunted for food or pelts; however, they are shot or poisoned as a local nuisance.

GENERAL REFERENCES: Baudoin et al. 2004; Mandier and Gouat 1996; Schwanz 2006; Yancey et al. 1993; Young and Jones 1982.

Ictidomys tridecemlineatus. Photo courtesy Phil Myers, Animal Diversity Web, animaldiversity.org.

Ictidomys tridecemlineatus (Mitchill, 1821) Thirteen-Lined Ground Squirrel

DESCRIPTION: The back of thirteen-lined ground squirrels is marked with alternating light and dark longitudinal stripes. Five cinnamon to sepia to black dark stripes are usually evident, with a series of white to buff spots or squares along the middle of each stripe. Alternating between these dark stripes are six light cream to buff stripes; sometimes these light stripes are broken into spots. Several additional lines or spots can be seen lower on the sides. The chin, cheeks, eye rings, and sides of the nose are buff to cinnamon buff. The venter is buff to cinnamon buff to pinkish buff. The tail is fuscous black on the dorsal surface, suffused with brown and frosted with buff; the ventral side of the tail is fuscous black throughout, with russet brown at the base and cinnamon buff near the tip.

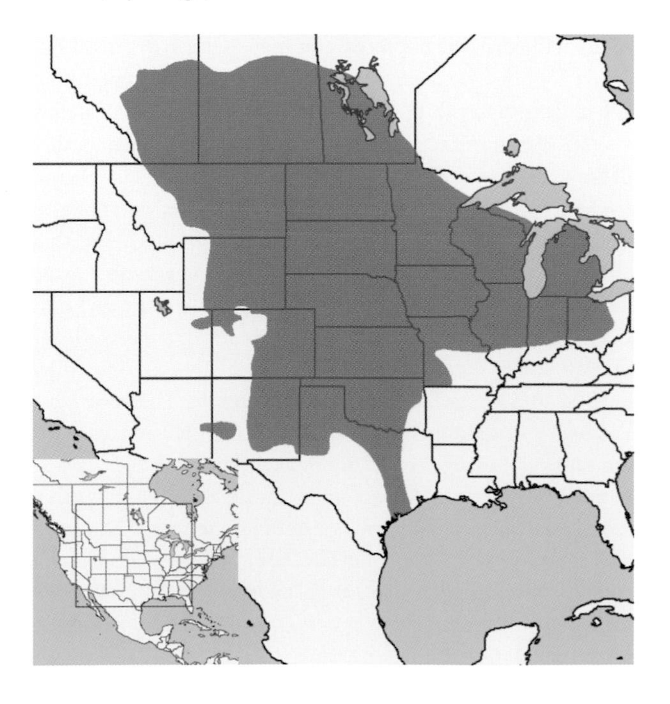

SIZE: Both sexes—HB 170-310 mm; T 60-132 mm; Mass 110-270 g.

DISTRIBUTION: This species is found in the Great Plains, from southcentral Canada south through eastern Utah and eastern Arizona to central Texas and east to Michigan and Ohio (USA).

GEOGRAPHIC VARIATION: Ten subspecies are recognized.

I. t. tridecemlineatus—central USA and southcentral Canada. See description above.

I. t. alleni—northwestern Wyoming (USA). This is a diminutive and pale subspecies.

I. t. arenicola—eastern New Mexico, southwestern Colorado, western Kansas, western Oklahoma, and northwestern Texas (USA). This is a small and pale form, with light brownish dark stripes.

I. t. blanca—southcentral Colorado (USA). This is a small pale subspecies.

I. t. hollisteri—central New Mexico (USA). This form has a reddish tinge to the dark stripes and a strong buff tinge to the light stripes.

I. t. monticola—eastcentral Arizona and westcentral New Mexico (USA). It has a reddish tinge to the dorsum and the underside of the tail.

I. t. olivaceus—northeastern Wyoming and westcentral South Dakota (USA). This form is much darker on the dorsum, with the light stripes and spots a pale yellowish olivaceous.

I. t. pallidus—northern Great Plains (USA). This is another small and pale form.

I. t. parvus—southcentral Wyoming, northeastern Utah, and northwestern Colorado (USA). This diminutive form has two of the medial light dorsal stripes broken into spots.

I. t. texensis—central Texas, Oklahoma, southeastern Kansas, and southwestern Missouri (USA). This subspecies is smallish, with a pale back and underside of the tail, and a reddish tinge.

CONSERVATION: IUCN status—least concern. Population trend—stable. *I. t. monticola* is a species of special concern in Arizona (USA).

HABITAT: Thirteen-lined ground squirrels are restricted to dry and sandy soils of open and very short grasslands, meadows, shrublands, roadsides, cultivated fields, airfields, golf courses, and suburban lawns.

NATURAL HISTORY: This species is diurnal. Thirteen-lined ground squirrels hibernate for five to eight months in burrows. Emergence begins in mid-March, with adult males emerging through April or even May, which is when females appear aboveground. Mating occurs soon afterward. Males range widely to find females, although they do defend breeding territories. Male-male aggression is reduced at this time, as males queue while waiting to mate with a female. Copulation most frequently takes place aboveground, and females mate with multiple males. After a gestation of 27-28 days, a litter of eight (range = 6-13) young is born within the burrow; only a single litter is produced, except in the southern portion of the range, where a second litter may be born. Juveniles emerge at about 5 weeks and are weaned by 6-7 weeks of age; all juveniles disperse from their natal area and reach adult size by 11 weeks. Adults begin to enter hibernation as early as late July, and young-of-the-year immerge by October, after accumulating considerable fat stores that will enable them to reach sexual maturity by their first spring. Burrows of at least two types are constructed: escape burrows, which are short, dispersed, and inconspicuous temporary shelters; and nesting burrows, which are up to 8 m long and 0.5 m deep, with a grass-lined nest chamber. Burrow entrances are plugged each night. Although relatively asocial, individuals do aggregate in areas of good-quality habitat. They scent-mark with glands near the cheek that are rubbed against objects. Territories are defended in at least some populations, whereas in others interindividual overlap is considerable. Male home ranges (5.0 ha) are larger than those of females (0.5-0.8 ha). *I. tridecemlineatus* is more omnivorous than many ground squirrels; this species eats the shoots, leaves, flowers, seeds, and fruits of grasses, forbs, and trees, as well as considerable numbers of insects (and even small vertebrates). Seeds are sometimes stored in burrows over the winter. Their main predators include canids, mustelids, felids, raptors, and snakes. Their alarm call is a soft whistled trill, often given in close proximity to the burrow entrance. Calls are voiced primarily in response to terrestrial predators, not aerial ones. Furthermore, females with their young call most frequently, suggesting that the function of the alarms is to warn their progeny. Thirteen-lined ground squirrels are not hunted, but they are occasionally trapped or poisoned for removal. *I. tridecemlineatus* is a common species in the manicured grass landscapes throughout much of its range, and it is tolerated by humans, due to its low densities, small burrow structures, and intriguing appearance.

GENERAL REFERENCES: Arenz and Leger 1999; Cothran 1983; Cothran and Honeycutt 1984; Devenport et al. 2000; Luna and Baird 2004; Schwagmeyer 1980, 1985.

Marmota Blumenbach, 1779

The genus, comprising the marmots, consists of 15 species.

Marmota baibacina (Kastschenko, 1899) Gray Marmot

DESCRIPTION: Gray marmots are lightly grizzled brown to charcoal, with a prominent suffusion of tan to buff to yellow suggesting a gray color. The venter is tan to bright rust. The snout is often dark brown with white to buff patches around the nose or chin. The tail is a lighter color at the base, with a dark brown to black tip.

SIZE: Both sexes—HB 460-650 mm; T 130-154 mm; Mass 4250-6500 g.

DISTRIBUTION: This squirrel is found in the Altai and Tien Shan Mountains in southwestern Siberia (Russia), southeastern Kazakhstan, Kyrgyzstan, Mongolia, and Xinjiang

(China). It was introduced into the Caucasus Mountains (Dagestan, Russia).

GEOGRAPHIC VARIATION: Two subspecies are recognized.

M. b. baibacina—eastern range of the Altai and Tien Shan Mountains. It is characterized by its more cinnamon dark back, with a lighter-colored venter.

M. b. centralis—western and central ranges of the Altai Mountains. It has a rust to orange venter.

CONSERVATION: IUCN status—least concern. Population trend—no information.

HABITAT: Gray marmots are found in gently sloping open (or, less frequently, boulder-strewn) mountain steppes, alpine meadows, and ridgetops, at elevations up to 4000 m.

NATURAL HISTORY: This squirrel is diurnal. Gray marmots hibernate for seven to eight months and overwinter in communal groups, with as many as 10 individuals to a burrow. They form multiburrow colonies. After mating in May and early June, gray marmots produce a single litter of two to six young after approximately 40 days of gestation. Maturation is slow, and it may not occur for three years. *M. baibacina* nest in burrows and appear to form communal groups. They forage primarily on grasses and herbaceous material growing in open grasslands and valleys. Their major predators are wolves (*Canis lupus*), foxes, domestic dogs, polecats, felids, and large raptors. Gray marmots are hunted for meat, pelts, and components of indigenous medicines; hunting pressure appears to be the major threat facing this species, which seems to be declining within its range.

***Marmota baibacina*.** Photo courtesy Kenneth B. Armitage.

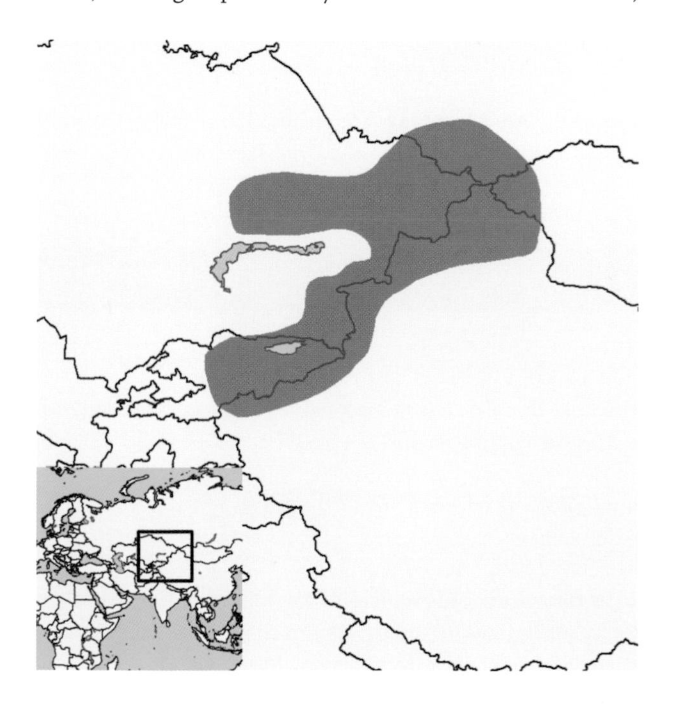

GENERAL REFERENCES: Armitage 2009; Demberel and Batbold 1991; Kolesnikov et al. 2009; Rogovin 1992.

Marmota bobak (Müller, 1776) Bobak Marmot

DESCRIPTION: Bobak marmots have a relatively uniform coloration, with a straw yellow to rust-colored dorsum and only a slightly lighter-colored venter. A variable portion of the head is dark brown to black. The tail darkens near the tip to brown or black.

SIZE: Both sexes—HB 490–575 mm; T 106–130 mm; Mass 5700 g.

DISTRIBUTION: *M. bobak* is found on the steppes of eastern Europe, from Ukraine and Russia to northern and central Kazakhstan. It was introduced to the Caucasus Mountains (Dagestan, Russia).

GEOGRAPHIC VARIATION: Three subspecies are recognized.

M. b. bobak—western portion of the range, through the Volga River. This form has a generally darker coloration on the body and head.

M. b. kozlovi—fragmented central portion of the range, near Ulyanovsk (Russia). A description in English of this subspecies is not available to the authors. It was described by Fokanov.

M. b. tschaganensis—eastern portion of the range. This form has a generally light straw yellow coloration.

CONSERVATION: IUCN status—least concern. Population trend—stable.

HABITAT: Bobak marmots are found in a variety of steppe habitats, including lowland, mixed grass, arid, and wormwood (*Artemisia*) steppes. Relative to other marmots, *M. bobak* thrive in rolling plains, grasslands, and on the periphery of cultivated fields.

NATURAL HISTORY: This species is diurnal. Bobak marmots hibernate in social groups in burrows for six to seven months. Burrows are complex, and they may be up to 4-5 m deep; they are recognizable by the pronounced mounds of excavated soil surrounding the entrance. About 60 percent of females produce litters of four to seven young in late spring or early summer; the young delay dispersal until their third year, when sexual maturity occurs. *M. bobak* forms colonies of several families, with a dominant male and adult female, 2-year-old and 1-year-old offspring, and young-of-the-year as the fundamental unit; however, social groups may be unstable, especially in areas with considerable human influence. Bobak marmots feed on green forbs, bulbs, flowers, and grass shoots. Much like prairie dogs (*Cynomys*) in North America, bobak marmots sit on their haunches in an alert posture and produce a single alarm call in response to threats. A variety of oral and cheek secretions are used by adults to scent-mark burrow entrances, mounds, and other structures. *M. bobak* may live to be 7.4 years old in captivity. Bobak marmots are hunted for food and pelts, and they have been a staple food for humans during periods of famine.

Marmota bobak. Photo courtesy Sergey Levykin and Grigoriy Kazachkov.

GENERAL REFERENCES: Nikol'skii 2009; Nikol'skii and Savchenko 1999.

Marmota broweri (Hall and Gilmore, 1934) Alaska Marmot

DESCRIPTION: The Alaska marmot is dark colored, with a gray to charcoal dorsum suffused with cinnamon, especially near the haunches. The venter is gray to charcoal. The tail is a uniform gray to brown. The head is dark brown to charcoal and relatively uniform in color.

SIZE: Female—HB 540-600 mm; T 130-160 mm; Mass 2500-3500 g.

Male—HB 580-650 mm; T 150-180 mm; Mass 3000-4000 g.

DISTRIBUTION: This marmot is found in the Brooks Range of northern Alaska (USA), from near the coast of the Chukchi Sea to the Alaska-Yukon border and south through the Ray Mountains to the Yukon River. Its extension into the Yukon or the Northwest Territories (Canada) is not yet confirmed.

GEOGRAPHIC VARIATION: None. *M. broweri* was once considered a subspecies of *M. caligata*. However, chromosomal number and molecular analyses confirm its status as a separate species.

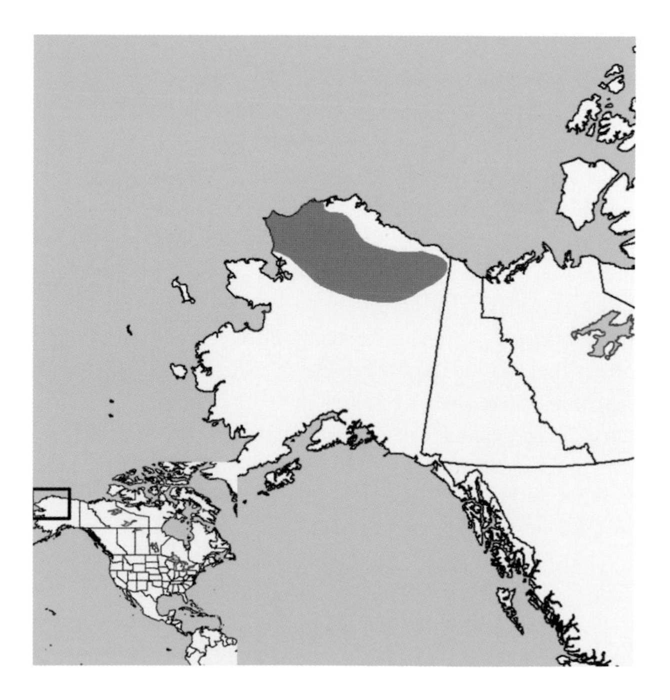

***Marmota broweri*.** Photo courtesy Dave Robichaud.

CONSERVATION: IUCN status—least concern. Population trend—stable.

HABITAT: *M. broweri* inhabits open alpine tundra habitats—such as boulder fields, talus slopes, and rock outcrops—in the alpine tundra of northern Alaska.

NATURAL HISTORY: Alaska marmots hibernate for six to eight months in winter in burrows that also provide protection throughout the year. Litters of four to five are born in late spring or early summer, when this species' major foods—such as grasses, sedges, and herbaceous plants—become readily available. Their main predators appear to be grizzly bears (*Ursus arctos horribilis*), which dig the animals from their burrows. *M. broweri* is rarely hunted or trapped for food or pelts.

GENERAL REFERENCES: Bee and Hall 1956; Gunderson et al. 2009.

Marmota caligata (Eschscholtz, 1829) Hoary Marmot

DESCRIPTION: Hoary marmots have a pale cream to white anterior dorsum and venter; and a yellow to tan posterior dorsum, rump, and tail. The feet are often black. The head is also cream to white, but with dark brown to black patches around the snout, crown, and chin.

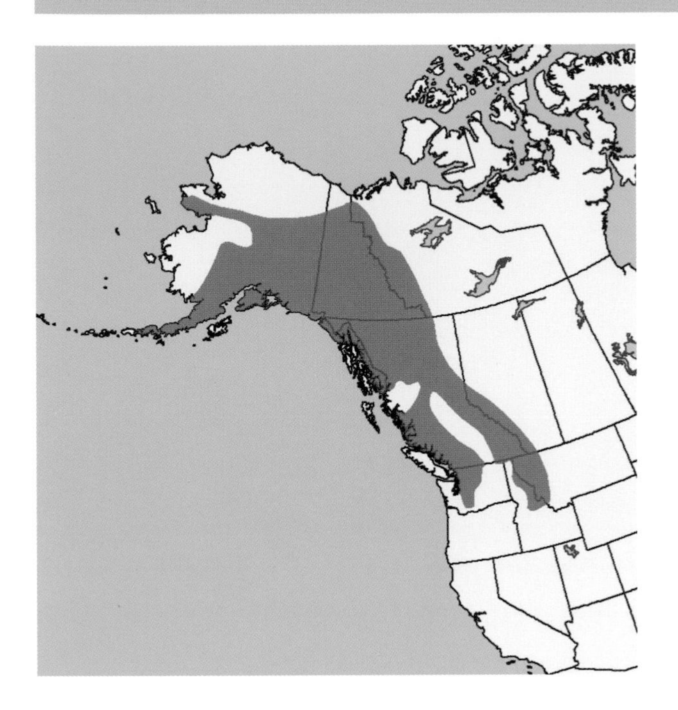

SIZE: Both sexes—HB 450-600 mm; T 170-250 mm; Mass 3600-9000 g.

Males are larger than females.

DISTRIBUTION: This marmot ranges from the Yukon River through central and southern Alaska (USA) and the Yukon and Northwest Territories (Canada) south to Washington, Idaho, and western Montana (USA).

GEOGRAPHIC VARIATION: Three subspecies are recognized.

M. c. caligata—western Yukon (Canada) and most of Alaska south of the Yukon River, including Montague Island and Glacier Bay (USA). See description above.

M. c. cascadensis—southern and western British Columbia (Canada) and Washington (USA). The head and feet are browner, and the underparts are darker.

M. c. okanagana—southern Yukon and the Northwest Territories into eastern British Columbia and western Alberta (Canada) to Idaho and Montana (USA). This is a lighter-colored form.

CONSERVATION: IUCN status—least concern. Population trend—stable.

HABITAT: Hoary marmots inhabit open rocky talus slopes and alpine tundra throughout their range. Such habitats are often disjunct, so the distribution of *M. caligata* at local and landscape scales is often fragmented.

NATURAL HISTORY: Hoary marmots hibernate in burrows, often as family groups, during the many months of the long winters throughout their distribution. They emerge in late spring, when females give birth to a litter of four to five young; litters are produced every two years. They feed in open meadows on grasses and herbaceous species during the growing season; their large gut capacity permits them to exist on even low-quality forage. They communicate by scent-marks and vocalizations. These vocalizations primarily convey information on predation risks, and their calls differ for aerial and terrestrial predators. Hoary marmots are highly social; they typically exist in family groups composed of a dominant male, adult female(s), yearlings, young-of-the-year, and subordinate males on the periphery of the dominant male's home range. Dominant females appear to suppress the reproductive activity of other adult female colony members. The mating system appears to be facultative, with monogamy and polygyny observed within a colony, although polygyny appears to be more common in the

Marmota caligata. Photo courtesy U.S. Fish and Wildlife Service.

southern portion of their distribution. *M. caligata* is rarely hunted or used for food or pelts.

GENERAL REFERENCES: Barash 1989; Blumstein 1999; Kyle et al. 2007.

Marmota camtschatica (Pallas, 1811) Black-Capped Marmot

DESCRIPTION: Black-capped marmots have a grizzled buff to yellow to gray dorsum. The venter is cinnamon to rust to russet. The front legs and shoulders are often paler and tend toward olive or buff. The tail is brown to charcoal to black. The most conspicuous feature is a black "cap" on the head, which extends from beneath the eye and from the nose to the back of the skull, then narrows into a mid-dorsal line extending to the midback, often with smaller extensions from the eye beneath the ear to the shoulder.

SIZE: Both sexes—HB 495 mm (460-530 mm); Mass 4000 g (2900-5000 g).

DISTRIBUTION: The black-capped marmot is found in eastern Siberia, in isolated populations from Transbaikal to Kamchatka (Russia).

GEOGRAPHIC VARIATION: Three subspecies are recognized.

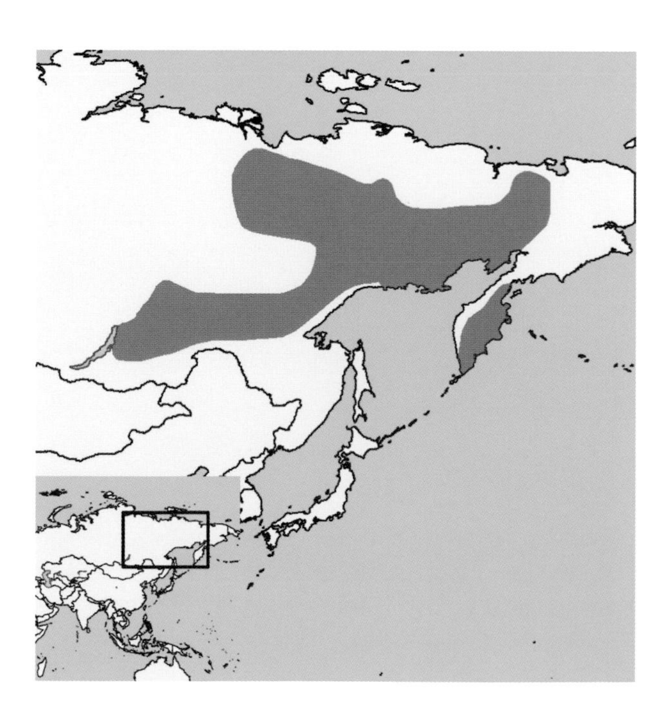

Marmota camtschatica. Photo courtesy Kenneth B. Armitage.

M. c. camtschatica—eastern portions of the range. This is the largest of the subspecies, and the most boldly marked.
M. c. bungei—central portion of the range. This form is intermediate in size, with a fainter coloration than *M. c. camtschatica*.
M. c. doppelmayri—southern and western portions of the range. This form is the smallest of the subspecies, with less prominent black markings and more suffusion of brown.

CONSERVATION: IUCN status—least concern. Population trend—no information. The Baikal population is considered rare.

HABITAT: Black-capped marmots inhabit open alpine meadows within low mountain forests, and dry well-drained sites with silt and boulders within mountain dryad-green moss tundra.

NATURAL HISTORY: *M. camtschatica* hibernates during winter months (often for a period of eight months) and emerges in May. The winter burrows have depths down to the permafrost. These marmots hibernate in large groups, perhaps even in multifamily groups. They overwinter, plugging the burrow entrances and using a great quantity of nest bedding. They plaster the nest walls and thicken the ceiling of the burrows with crushed stone to maximize the indoor winter temperatures. In winter these marmots arouse periodically, independent of their hibernation conditions. There are age differences in the duration of the hibernation bouts, the arousal times, and the quantity of energy expended. Black-capped marmots live in colonies with many burrows, and groups consist of a reproductive pair and their young from previous litters. They forage on a variety of grasses, herbs, and moss. They maintain daily rhythms of activity, even above the Arctic Circle under the constant daylight in summer. Their home ranges are 10–15 ha in size. Mating occurs before spring emergence, and litters of five to six young remain in their natal group for at least three years. Black-capped marmots appear to increase the vegetation diversity on the periphery of their home ranges, suggesting that they may be a keystone species. Low-density populations of *M. camtschatica* are often hunted for food and pelts at unsustainable rates; overgrazing and mining have contributed to localized declines.

GENERAL REFERENCES: Nikolsky et al. 1991; Semenov, Ramousse, and Le Berre 2000; Semenov, Ramousse, Le Berre, and Tutukarov 2001; Semenov, Ramousse, Le Berre, Vassiliev, et al. 2001; Tokarsky and Valentsev 1994.

Marmota caudata (I. Geoffroy, 1844) Long-Tailed Marmot

DESCRIPTION: Long-tailed marmots have long coarse dorsal fur that is golden orange suffused with black; a black dorsal midline or saddle is often apparent. The venter is rusty orange. The head typically has a dark brown to black cap that extends from the snout to the back of the skull and includes the eyes and the cheeks. The tail is relatively long and bushy, with a solid black tip. Bare eye patches are larger on males than on females.

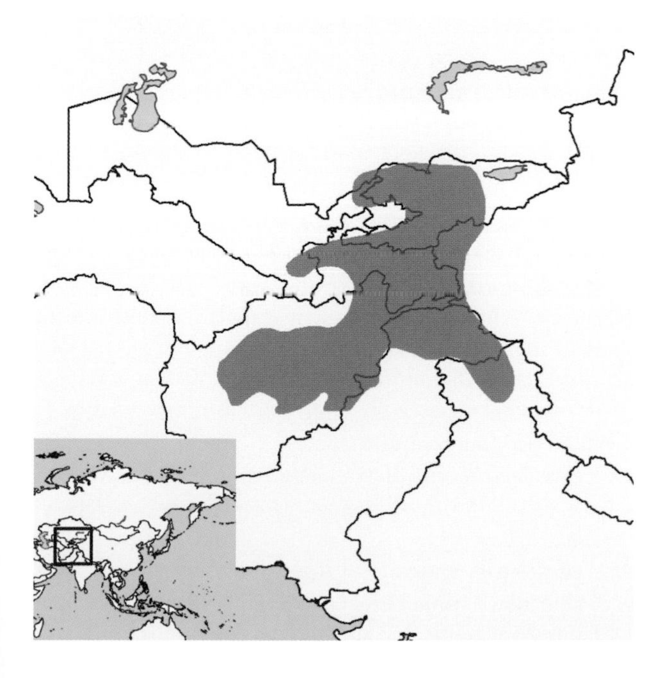

Marmota caudata. Photo courtesy Oleg Brandler.

SIZE: Sex not stated—HB 500 mm; Mass 3000–5000 g.

Their weight peaks prior to their entrance into hibernation.

DISTRIBUTION: Long-tailed marmots are found from the western Tien Shan Mountains through the Pamirs (Kyrgyzstan, Tajikistan) to Hindu Kush (Afghanistan), Pakistan, Kashmir (India), and the mountains of extreme western Tibet and the Chinese province of Xinjiang.

GEOGRAPHIC VARIATION: Three subspecies are recognized.

M. c. caudata—Kashmir region. This form often has a very dark brown to black dorsum.

M. c. aurea—Turkistan. This form is often called the golden marmot, due to the brilliant yellow to orange venter and the suffusion of this coloration over some of the dorsum.

M. c. dichrous—Afghanistan region. This form is grayish, grizzled on the sides of the neck and the shoulders. The belly has more brown on it.

CONSERVATION: IUCN status—least concern (near threatened). Population trend—no information.

HABITAT: Long-tailed marmots inhabit the high alpine flats and meadows often associated with glacial and terminal moraines; they may occur in alpine scrub with some dwarf juniper (*Juniperus*). Their habitats often have very low levels of plant productivity.

NATURAL HISTORY: *M. caudata* is diurnal. Long-tailed marmots hibernate for periods of seven to eight months in the winter and emerge in late spring or early summer from their burrows. Only 17 percent of females breed during this species' short active season. Gestation is about 30-33 days, after which a litter of two to six young is born. Weaning is earlier when food resources are abundant. More than 60 percent of juveniles die during the first year; 22 percent of this mortality is due to infanticide by new resident males, which are apparently trying to avoid providing resources for another male's progeny. Juveniles delay dispersal and remain in their natal group for at least three years. Most of the dispersal appears to occur after sexual maturity, when males move to groups with fewer individuals than their natal group. Nonjuvenile survival rises in years when food is abundant. *M. caudata* is a gregarious species; these marmots live in large colonies frequently consisting of monogamous pairs, but there can be groups of seven adult animals plus their young and yearlings. The home ranges of the social groups are large, averaging 3.0 ha. Group members will sleep together and generally behave amicably toward each other. *M. caudata* forage for long hours on a variety of legumes, herbs, grasses, and small shrubs to accumulate subcutaneous fat reserves for the winter. Individuals are often seen basking on rocks or at burrow entrances, or standing on their haunches surveying their surroundings for potential threats. Individuals emit complex alarm calls that warn conspecifics about the relative risk of predation, and they use a variety of less apparent vocalizations in other social contexts. An assortment of large raptors, canids, and felids are frequent predators. Males and females each scent-mark from orbital glands, doing so primarily in the vicinity of their burrows; males are more responsive to novel scents than females. In some portions of their range, long-tailed marmots are hunted and trapped for food, pelts, and medicinal purposes; they are not considered pests, except where they are believed to compete with livestock for forage.

GENERAL REFERENCES: Blumstein and Arnold 1998; Habibi 2003; T. J. Roberts 1977.

Marmota flaviventris (Audubon and Bachman, 1841) Yellow-Bellied Marmot

DESCRIPTION: The dorsal surface of the head is brown to black, usually with white patches in between the eyes and the snout, on the sides of the snout, and on the apex of the lower jaw. The side of the neck is buff to yellow to orange. The venter is buff to yellow. The feet are buff to brown.

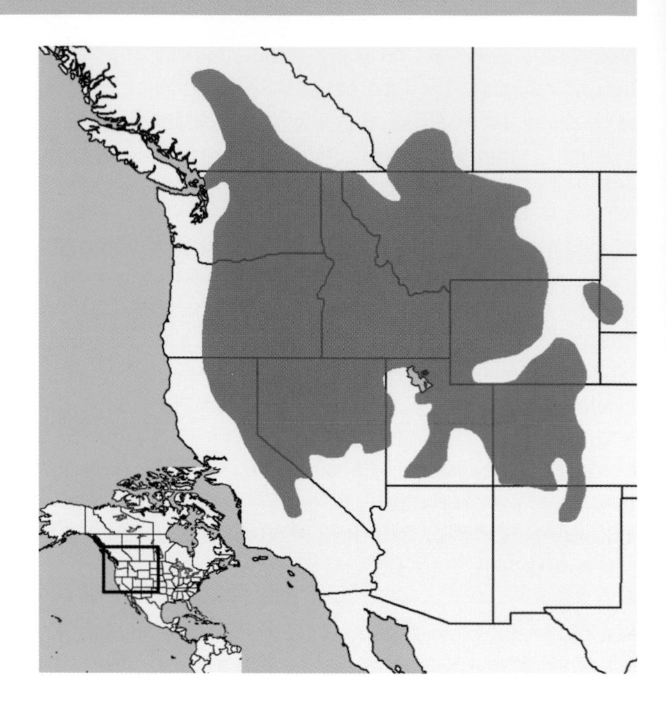

SIZE: Females—Mass 2800 g (1590-3970 g) after hibernation.
Males—Mass 3900 g (2950-5220 g) after hibernation.
Both sexes—HB 470-700 mm; T 130-220 mm.
Males tend to be longer than females.

DISTRIBUTION: Yellow-bellied marmots are found from southern British Columbia and Alberta (Canada) south to California, Nevada, southern Utah, and northern New Mexico (USA). There is also an isolated population from northeastern Wyoming to western South Dakota (USA).

GEOGRAPHIC VARIATION: Eleven subspecies are recognized.

M. f. flaviventris—Cascade and northern and central Sierra Nevada ranges of Oregon and California (USA). The dorsal surface is grizzled russet and white. The upper back has an indistinct buffy mantle. The underparts and feet are ochraceous.

M. f. avara—throughout the northwestern range. This form is smaller and paler.

M. f. dacota—endemic to the Black Hills of South Dakota (USA). This subspecies, while similar to *M. f. nosophora*, has redder fur ventrally.

M. f. engelhardti—central to southwestern Utah (USA). This is a smaller form, with dark red on the upperparts and feet.

M. f. fortirostris—endemic to the White Mountains on the California-Nevada border (USA). This subspecies is similar to *M. f. sierrae*, but smaller and paler.

M. f. luteola—southcentral Rocky Mountains, from southern Wyoming to New Mexico (USA). This form resembles *M. f. nosophora*, but *M. f. luteola* has a whitish mantle on the upper back and yellowish underparts.

M. f. nosophora—throughout the northeastern portion of the range and south to northern Utah. This form has a golden buff mantle on the upper back; the general color is more ochraceous dorsally and redder ventrally.

M. f. notioros—known only from a single specimen in southcentral Colorado (USA). This form is larger and darker than the others. While similar to *M. f. obscura*, *M. f. notioros* is a darker reddish brown dorsally and a more chestnut brown ventrally.

M. f. obscura—northcentral New Mexico (USA). Dorsally, this form is dark brown with whitish grizzling. The face does not show the typical white markings of the other subspecies.

M. f. parvula—central Nevada (USA). This form is smaller and darker than *M. f. avara*.

M. f. sierrae—southern Sierra Nevada, California (USA). The general color is similar to *M. f. flaviventris*, but *M. f. sierrae* is redder, and the mantle is reduced or absent.

CONSERVATION: IUCN status—least concern. Population trend—stable.

HABITAT: *M. flaviventris* is found in mountains on grass- and forb-dominated meadows with loose talus, in flat grassy inland valleys, and on the edges of irrigated hayfields.

NATURAL HISTORY: Yellow-bellied marmots are diurnal. They hibernate as individuals in separate burrows for extended periods (often for seven to eight months, depending on the elevation.) Burrow systems are modestly complex; frequently there are several entrances leading to a terminal nest chamber 1-2 m belowground. Conspicuous trails often connect the burrow entrances. Escape burrows that provide only short-term cover are less complex, with just a single entrance and a short length. Males emerge before females and will tunnel down through the snow to mate with females ending hibernation; male testes regress in size just two weeks after emergence. Gestation is about 30 days, after which a litter of four to five young is produced. Adult females rarely reproduce in successive years.

The social system consists of one or more female kin groups within the territory of a male; the mating system is polygynous. Juveniles overwinter in the natal area; males and many females disperse as yearlings. Mothers and daughters often form matrilines when some females remain in their natal area, and these lineages can persist over several generations. Behaviors are predominantly amicable between social-group members, whereas nongroup members are aggressively chased. The survival of juveniles can be highly variable, ranging from 15 to 90 percent; whereas annual adult survival typically ranges from 65 to 80 percent. Female survival is greater than that of males; maximum longevity in the wild exceeds 14 years for females and 8 years for males. The majority of the losses occur during hibernation. Their principal predators are canids, mustelids, bears, and large raptors. Individuals can often be seen in open meadows foraging for grasses, forbs, flowers, and sedges during early- to midmorning and midafternoon. Communication is most often visual and auditory, but yellow-bellied marmots also make scent-marks on structures and burrows. Moderately complex calls are produced by animals surveying their surroundings while standing on their haunches. "Chirps" are by far the most common calls, and they become more frequent both as the perceived risk increases and with some ground predators (such as canids). Yellow-bellied marmots are rarely hunted for food or pelts; they can be viewed as pests when found near orchards or crops.

Marmota flaviventris. Photo courtesy Nichole Cudworth.

GENERAL REFERENCES: Anthony 1928; Armitage 1991; Armitage and Schwartz 2000; Frase and Hoffmann 1980; O. A. Schwartz et al. 1998.

Marmota himalayana (Hodgson, 1841) Himalayan Marmot

DESCRIPTION: Himalayan marmots are generally light in coloration, with a cream to buff to yellow dorsum that often has varying patches of charcoal to black. The venter is buff to tan to light brown. The snout is often dark brown to black, and sometimes the coloration extends to the forehead. The tail is buff to light brown at the base with a charcoal to black tip.

SIZE: Both sexes—HB 475-670 mm; T 125-150 mm; Mass 4000-9215 g.

DISTRIBUTION: *M. himalayana* is found in the mountains of western China, extending south into Ladakh (Pakistan) and northern India, and from Nepal, Bhutan, and extreme northeastern India to Tibet.

GEOGRAPHIC VARIATION: Two subspecies are recognized.

M. h. himalayana—western portion of range. These forms are lighter in color and without rust to orange colors.

M. h. robusta—eastern portion of the range. Chinese forms have a buff and black mix on the dorsum, with buff to orange buff on the venter. Their subspecific status is questionable.

CONSERVATION: IUCN status—least concern. Population trend—no information.

HABITAT: *M. himalayana* is found in alpine meadows, upland grasslands, and desert conditions, often on slopes with small shrubs, where soil can be readily excavated. One of the highest-living mammals in the world, the Himalayan marmots are found at elevations of 5670 m. *M. himalayana* is also the southernmost-occurring species of marmot.

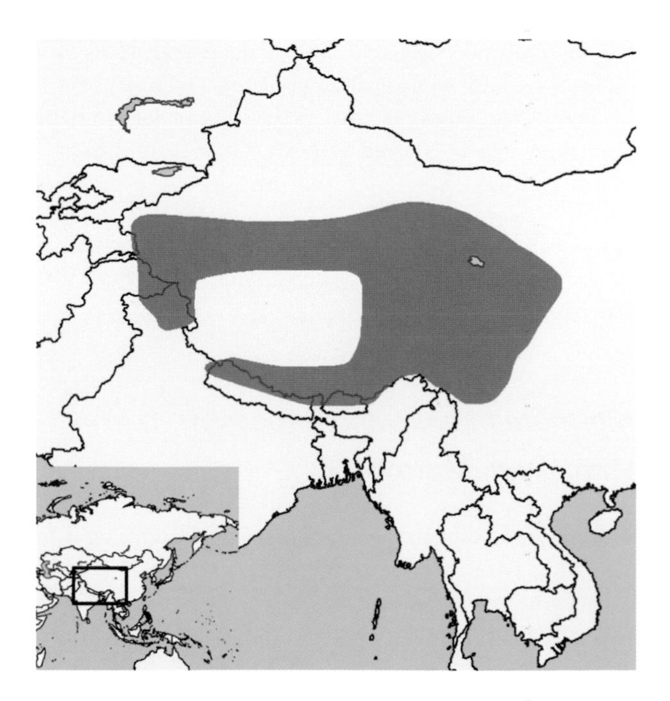

Marmota himalayana. Photo courtesy David Blank, www.animaldiversity.ummz.umich.edu.

NATURAL HISTORY: This species is diurnal. Himalayan marmots hibernate for six to seven months in burrows, and they emerge for a short intensive season of foraging on the grasses, forbs, roots, flowers, and bulbs found in cold dry meadows. *M. himalayana* can persist in overgrazed pastures throughout their range and still find forage, although these conditions may not be ideal. Burrows are often extensive, with large mounds, and they reach deep belowground (more than 10 m in some cases); summer burrows may be shallower. Colonies are highly variable in size, ranging from isolated family groups to expanses of more than 100 burrows. The young are born in the burrow after a gestation of about 30 days; litter size averages 4.8 young in high-density populations and 7 at low densities. Natal dispersal is delayed, and social groups typically consist of a single dominant adult male, an adult female, some 2-year-olds, plus 1-year-olds and young-of-the-year. Females can reproduce in their second year. Interactions among group members are highly amicable; however, intruders are greeted with displays, tooth chatters, and chases. A single relatively complex alarm call—consisting of two distinct sounds—is produced in response to threats or to the perception of predation risk. Snow leopards (*Panthera* [*Uncia*] *uncia*), raptors, wolves (*Canis lupus*), and the dogs of herdsmen appear to be this species' principal predators. Their conservation status reflects the rapid loss of alpine and subalpine meadows, due to an intensive conversion to agriculture. *M. himalayana* is hunted and trapped locally for food, pelts, and medicinal purposes.

GENERAL REFERENCES: Bibikow 1996; Molur et al. 2005; Nikol'skii and Ulak 2006; Oli 1994; T. J. Roberts 1977.

Marmota kastschenkoi (Stronganov and Judin, 1956)
Forest Steppe Marmot

DESCRIPTION: Forest steppe marmots are dark brown on the dorsum; however, they may be slightly paler on the underside, varying from a dark yellowish brown to tawny or rufous. The snout is brown, sometimes with buff to tawny patches around the nose and a white patch on the chin.

SIZE: Both sexes—HB 495-640 mm; T 124-200 mm; Mass 3000-6300 g in spring, 4600-9000 g in autumn.

DISTRIBUTION: Apparently a relict population was found in the forest steppe region south of Novosibirsk, Kemerovo, and Tomsk in southwestern Siberia (Russia).

GEOGRAPHIC VARIATION: None.

CONSERVATION: IUCN status—not listed. Population trend—no information. *M. kastschenkoi* is listed as declining rapidly in the *Red Books* of Tomsk and Kuzbass (Russia).

HABITAT: Forest steppe marmots are found on slopes in open forest steppes and meadows. They make use of abandoned buildings and cemeteries.

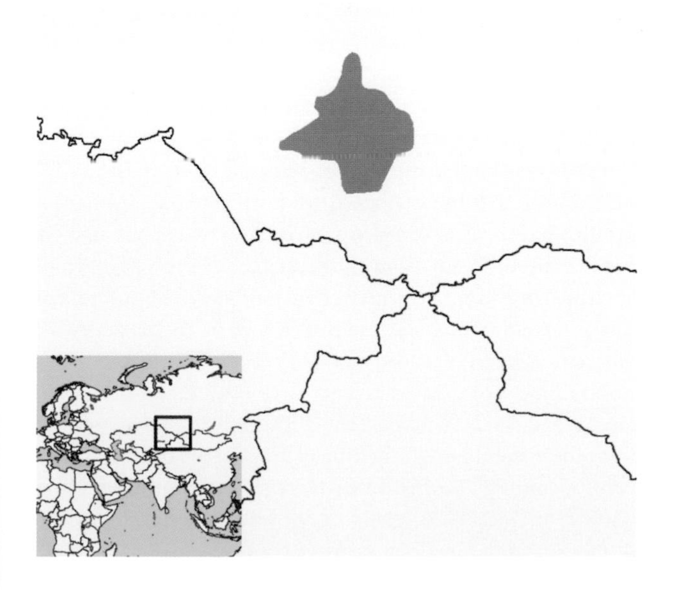

Marmota kastschenkoi. Photo courtesy Viktor Mashkov.

NATURAL HISTORY: This species is diurnal. Forest steppe marmots hibernate within burrows for seven to eight months during the winter. Females produce a single litter of about two young once each summer. Maturation is slow, and it may not occur for three years. They form multiburrow colonies with other individuals and typically persist in family groups consisting of a dominant male, two to three adult females, and four to eight immature individuals (from newborn to 3 years old). The territories of family groups are about 1 ha. Forest steppe marmots nest in burrows with multiple openings; these burrows serve as winter hibernacula. The animals also use less developed burrows on the periphery of the colony for escape. Burrows typically are found on hillsides and have a raised platform of soil that facilitates vigilance. Dispersing males (3-year-olds) travel great distances (up to 15 km/day) and they eventually often settle on the periphery of a colony. Their alarm calls are high-pitched short whistles. In areas of heavy human influence, these marmots may shift to nocturnal foraging. Forest steppe marmots forage primarily on young grasses, shoots, and herbaceous material growing in open forest steppes and valleys. *M. kastschenkoi* eats garden crops such as beans and peas, but this behavior is extremely rare and localized; this species is not considered to be a pest on any significant scale. Human hunting of *M. kastschenkoi* for food appears to be a significant threat in some localities, as this species is declining within its range. Forest steppe marmots may serve as a flagship species for preservation of the forest steppe within western Siberia.

GENERAL REFERENCES: Armitage 2009; Brandler 2003; Brandler et al. 2008; Galkina et al. 2005; Polyakov 2005; Polyakov and Baranova 2007; Skalon and Gagina 2006.

Marmota marmota (Linnaeus, 1758) Alpine Marmot

DESCRIPTION: Alpine marmots have a highly variable coloration. The dorsum ranges from a grizzled tan to brown to russet; the venter is not grizzled and is paler in color, ranging from white to yellow to orange. The ventral coloration can extend up the sides to form a light-colored saddle on the back. The head is tan to dark brown, with the coloration extending to the shoulders. There is a white to cream patch around the snout and the chin. The tail is typically paler at the base and darkens to a charcoal tip. Albinism has also been documented.

SIZE: HB 500-600 mm; T 150 mm; Mass 2500-5000 g.

DISTRIBUTION: This marmot is found in the Alps of France, Switzerland, Italy, Germany, Austria, and Slovenia; the Tatra Mountains (Slovakia and Poland), and the Carpathian Mountains (Romania). It was introduced into the French Pyrenees, eastern Austria, northern Serbia, and Montenegro.

GEOGRAPHIC VARIATION: Two subspecies are recognized.

M. m. marmota—the most widespread form, found throughout the western Alps. It is the source for the vast majority of translocations and reintroductions, such as into Romania.

M. m. latirostris—isolated form in the High Tatra Mountains of Slovakia and Poland. There is minimal external distinction between the subspecies.

CONSERVATION: IUCN status—least concern. Population trend—stable. *M. m. latirostris* is found in a restricted range and is considered rare and threatened.

HABITAT: Alpine marmots typically inhabit alpine meadows and high-elevation pastures, most frequently on south-facing slopes at elevations of 1200-3000 m.

NATURAL HISTORY: Alpine marmots are diurnal. They hibernate in social family groups for six to seven months in burrows, immerging and plugging the burrows by late September. Burrow systems are extensive and complex, 3-10 m in length, often with several entrances leading to a terminal nest chamber 1-2 m belowground. Conspicuous trails often connect the burrow entrances, due to their frequent use during the growing season. Escape burrows that provide only short-term cover are short and less complex, with just a single entrance. Mating occurs around the time of emergence, in late April or early May. These marmots are socially monogamous; however 30-33 percent of the litters are the product of extra-pair matings, resulting in the young from those litters being of mixed parentage. Most often the extra-pair young can be attributed to a sire outside the social group. Extra-pair matings are most likely when the male is genetically different from the female, and the young from such matings have a greater chance of survival and reproduction than young sired within the pair. Dispersal by *M. marmota* is costly, because mortality increases and the likelihood of locating a vacancy decreases with distance. By delaying dispersal, 12-22 percent of the subordinates are able to become dominant in their natal area, with another 50 percent becoming dominant nearby; this leads, however, to a pattern of increased levels of inbreeding. Moreover, the dominant adult female suppresses the reproduction of other

Marmota marmota. Photo courtesy Kenneth B. Armitage.

females in the family group, again reducing the genetic diversity of the group.

Gestation is 33-34 days, after which a litter of two to four young is produced. Adult females rarely reproduce in successive years, so only about 50 percent of them produce a litter in a given year. Female reproductive success is positively correlated with body condition and experience. The social system of *M. marmota* is one based on family groups consisting of an adult male, an adult female, and young less than 3 years of age within a territory. Juveniles overwinter in the natal area, and older male offspring assist their younger siblings with social thermoregulation; this benefit from male offspring may be the reason for a male-biased sex ratio in some populations. Males and many females disperse in their third year; however, inbreeding is known to occur, with sons mating with their mothers when dispersal is delayed. Behaviors such as greeting, allogrooming, and other physical contacts are predominantly amicable between social-group members, and nongroup members are aggressively chased. Immigrant adult males that assume control of a social group kill juveniles. The survival rate of juveniles is high, averaging 50-65 percent, whereas annual adult survival typically ranges from 50-90 percent; no sex differences are evident in survivorship. The majority of the losses occur during hibernation. This species' principal predators are canids and large raptors. Individuals are frequently observed foraging in open subalpine and alpine meadows for grasses, forbs, flowers, and bulbs. Communication is most often visual and auditory. Olfactory communication through scent-marking on structures and burrows is apparently used to mark territories. The animals survey their surroundings when standing on their haunches, and they produce two alarm calls: a common whistle and a descending whistle. The common whistle becomes more frequent as the perceived risk increases. In some countries Alpine marmots are game animals and are occasionally hunted for food; they are also harvested for their pelts and their fats/oil (for medicinal purposes).

GENERAL REFERENCES: Arnold 1988; Cohas et al. 2008; Hacklander et al. 2003; W. J. King and Allaine 2002; Mann et al. 1993.

Marmota menzbieri (Kashkarov, 1925) Menzbier's Marmot

DESCRIPTION: Menzbier's marmot has a straw- to tan-colored dorsum with a frosting of dark brown to black. The venter is cream to buff to tan; this coloration is prominent on the front limbs and extends to the cheek. The head is light to dark brown, sometimes with a buff patch on the snout and the chin. The tail is straw to light brown at the base with a dark brown to black tip.

SIZE: Both sexes—HB 490 mm; T 120 mm; Mass 3000-4000 g.

DISTRIBUTION: This marmot is found in the western Tien Shan Mountains in southern Kazakhstan, western Kyrgyzstan, and northeastern Uzbekistan. Its range is extremely restricted (200 km^2).

GEOGRAPHIC VARIATION: Two subspecies are recognized.

Marmota menzbieri. Photo courtesy Alexander Esipov.

M. m. menzbieri—western Tien Shan Mountains. This form has two seasonal pelages: in autumn the coat is more black on the dorsal surface; in summer the coat is more brownish yellow on the dorsal surface. The ventral surface is grayish brown; the throat, chest, and inner surfaces of the forelimbs are more rufous in color. There is a yellowish gray patch extending from beneath the eye to beneath the anterior edge of the ear.

M. m. zachidovi—northern Tien Shan Mountains, separated by more than 100 km from *M. m. menzbieri*. We were unable to find an available description.

CONSERVATION: IUCN status—vulnerable. Population trend—decreasing.

HABITAT: Menzbier's marmots are found in open high-elevation alpine and subalpine meadows. They also tolerate areas with open juniper (*Juniperus*) meadows.

NATURAL HISTORY: This species is diurnal. Menzbier's marmots hibernate in social groups for seven to eight months in burrows; mate in burrows, or soon upon exiting them; and emerge for a short intensive season of foraging on forbs, roots, rhizomes, grasses, flowers, and bulbs. Burrows are relatively complex, with an average of 3.5 entrances per burrow, and they extend 2-3 m belowground. A small litter (averaging 2.5 young in the south, and 4-5 in the north) is produced. Some 2-year-olds in this species may be able to breed. Natal dispersal is delayed, and social groups typically consist of a monogamous pair (a single dominant adult male and an adult female), 2-year-olds, 1-year-olds, and young-of-the-year. *M. menzbieri* is a highly social species. Individuals in the wild are known to live to be more than 10 years old. Interactions among group members are highly amicable; however, intruders are not welcome and are greeted with displays, tooth chatters, and chases. They frequently scent-mark territories with their oral glands. A single relatively simple alarm call is produced in response to threats or to the perception of predation risk. Raptors, wolves (*Canis lupus*), and the dogs of herdsmen appear to be their principal predators. Their conservation is threatened by the rapid loss of alpine and subalpine meadows, due to an intensive conversion to agriculture; individuals may be forced to move seasonally to avoid such influences.

GENERAL REFERENCES: Allaine 2000; Bibikow 1996; Nikol'skii 2007b; Ognev 1963.

Marmota monax (Linnaeus, 1758) Woodchuck, Groundhog

DESCRIPTION: Woodchucks are relatively small North American marmots. Their tail is relatively short and cinnamon to dark brown to black. The dorsum is grizzled gray to cinnamon to dark brown, often appearing frosted because of the light-colored tips of the guard hairs. The venter is gray to tan to rufous and dark brown. White to tan patches often surround the nose and the apex of the lower jaw. The front legs can be suffused with rufous to cinnamon. The feet are dark brown to black.

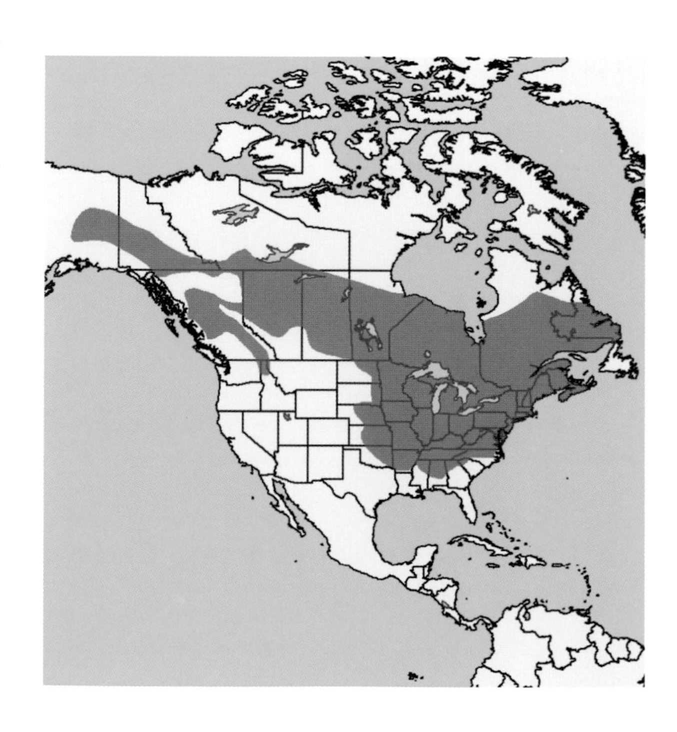

***Marmota monax*.** Photo courtesy Lisa H. Ware.

SIZE: Female—Mass 3526 g (n = 410) year-round.
Male—Mass 3826 g (n = 433) year-round.
Both sexes—HB 418-665 mm; T 100-155 mm.

DISTRIBUTION: Woodchucks range widely from central Alaska (USA) eastward through central and eastern Canada and southward into the southeastern USA.

GEOGRAPHIC VARIATION: Nine subspecies are recognized.

M. m. monax—throughout the southern portion of the range in the USA. This form tends to be relatively pale gray brown, suffused with buff on the dorsum, with a whitish buff to pale venter.

M. m. bunkeri—extreme southwestern portion of the range, in Kansas and Nebraska (USA). This is a large form that tends to be relatively pale gray brown suffused with buff on the dorsum; the venter is whitish buff to pale ochre.

M. m. canadensis—throughout most of Canada (from the Northwest Territories to Québec) and into the northeastern USA. This is a small-bodied form, with a strong reddish cast on the dorsum and venter.

M. m. ignava—northern Québec and Labrador (Canada). This is a large form that is a dark orange cinnamon color frosted with buff.

M. m. johnsoni—endemic to extreme southeastern Québec, on the south bank of the St. Lawrence Seaway (Canada). The hairs on the back are vinaceous cinnamon, turning to orange cinnamon on the shoulders and flanks. The tail and ventral surface of body are cinnamon rufous. The throat is tawny colored.

M. m. ochracea—Alaska (USA) and northwestern Canada. It is a strong reddish brown on the dorsum, with a lighter venter and a pinkish cinnamon tail.

M. m. preblorum—throughout the New England region (northeastern USA). This form is a medium-sized woodchuck with a pale reddish hue on the dorsum; the venter is pinkish cinnamon.

M. m. pretensis—British Columbia (Canada) and the extreme northwestern USA. It has a strong reddish color above and below.

M. m. rufescens—eastern prairies of the Dakotas east through the northern Great Lakes (USA) and southeastern Ontario (Canada) into New England (USA). This subspecies tends to be more reddish to pinkish buff on the dorsum and venter.

CONSERVATION: IUCN status—least concern. Population trend—stable.

HABITAT: *M. monax* inhabit old fields, meadows, meadow-forest ecotones, hedgerows, orchards, the edges of croplands, highway roadsides, and urban parklands.

NATURAL HISTORY: This species is diurnal, although some crepuscular and even nocturnal aboveground activity has been reported in urban areas. Woodchucks hibernate alone in burrows, maintaining body temperatures that are slightly above those of the burrow and cycling through bouts of torpor and arousal. They emerge after two to six months, in midspring; the length of hibernation is positively correlated with latitude. Burrows are relatively simple, and they can be located in open fields or in adjacent deciduous or mixed forests. In urban settings, burrows can be located under human structures, such as buildings or bridges. Groundhog (= woodchuck) Day in the USA and Canada is celebrated on February 2; the myth suggests that if a woodchuck emerges on that day and sees its shadow, this predicts a delayed spring. The roots of this holiday are found in several European countries, where Candlemas Day dates from 300 to 500 BC and is the day when Christians celebrated the purification of the Virgin Mary; the celebration traditionally occurred halfway between the winter solstice and the vernal equinox. This time period is also of significance for spring planting in Europe: the emergence of hibernating animals such as hedgehogs and badgers (*Meles meles*) was believed to be a signal for the time when spring crops might be planted. When Europeans immigrated to North America, the woodchuck was selected as a proxy in the absence of the commonly used species in their homeland.

Males often emerge earlier than females; mating occurs soon after females arouse and emerge from hibernation. The mating system is polygynous. Each year 50-80 percent of adult females may reproduce; yearling females can reproduce, but usually in lower proportions than adults. As many as nine—but typically three to five—young are born in the burrow after a gestation of 31-32 days. The young emerge from the burrow after 5-6 weeks, in early summer to midsummer. Most juveniles disperse from their natal area at the end of their first growing season; however some (especially females) remain near their natal area. Although independent after dispersal, woodchucks continue to grow for two to three years before achieving adult body size. Annual cycles of fat deposition in preparation for hibernation changed after two years when individuals were transported to the Southern Hemisphere. Woodchucks are generalist herbivores and eat an incredible diversity of forbs, grasses, sedges, flowers, fruits, and buds. They can climb trees to consume leaves and (occasionally) fruits or soft seeds. They will feed on orchard fruits and the vegetation in hayfields. They sometimes are known to eat insects and to scavenge small vertebrates.

Adults are solitary and territorial. Male home ranges vary from 0.5 to 4.0 ha, with female home ranges being about half that size. Females, in particular, tend to overlap only a little in their home ranges, whereas the overlap between males and females is more significant. Because individuals are relatively asocial, interactions among adults are typically agonistic. Kinship does influence interactions, however, and close relatives demonstrate slightly more amicable and slightly less aggressive interactions toward each other. Olfactory communication appears important to woodchucks; they scent-mark on structures and plants using well-developed glands in the upper lip and the corner of the mouth. Vocalizations are less varied in woodchucks than in marmot species that are more social. Calls are rarely given, and the woodchucks' high-pitched alarms are probably to warn offspring of potential dangers. Their predators include most raptors, felids, canids, and large mustelids that occur within the distribution area of *M. monax*. Woodchucks are hunted for food and sport in much of their range. They are considered a nuisance on golf courses and in urban areas, where they dig under structures. In orchards and croplands, *M. monax* can cause considerable crop losses in local settings.

GENERAL REFERENCES: Anonymous 1990; Caire and Sloan 1996; Jordheim 1990; Kaufman and Kaufman 2002; Kwiecinski 1998; Maher 2009; Roehrs and Genoways 2004; S. S. Robinson and Lee 1980; Stevenson 1990; Tumlinson et al. 2001; Zervanos et al. 2010.

Marmota olympus (Merriam, 1898) Olympic Marmot

DESCRIPTION: The Olympic marmot has a yellow brown to light brown dorsum, with a long and densely haired tail of similar color. The snout is white, and a white band often extends back to between the eyes. The venter is similar in coloration to the dorsum but can be more rufous or gray. The dorsal pelage appears to fade to yellow over the summer.

SIZE: Male—HB 740 mm (720-750 mm); T 219 mm (210-237 mm); Mass 4100 g (4100-4300 g; n = 6) after hibernation, 9300 g (8500-11000 g; n = 5) before hibernation.

Female—HB 680 mm (670-690 mm); T 186 mm (180-192 mm); Mass 3100 g (2700-3500 g; n = 4) after hibernation, 7100 g (6700-7500 g; n = 6) before hibernation.

Male-biased sexual dimorphism is apparent.

DISTRIBUTION: This species is endemic to the Olympic Mountains of western Washington (USA).

GEOGRAPHIC VARIATION: None.

Marmota olympus. Photo courtesy Duke Coonrad, www.flickr.com/photos/skagitman/.

CONSERVATION: IUCN status—least concern. Population trend—declining.

HABITAT: *M. olympus* inhabits open grass, sedge, and forb subalpine and alpine meadows and talus slopes near and above timberline; most of its habitat is within the borders of Olympic National Park (USA).

NATURAL HISTORY: This species is diurnal. It hibernates in burrows in communal social groups for seven to eight months, emerging in late spring. Polygynous colonial groups are the rule, commonly consisting of a single dominant adult male; fewer than three adult females; and numerous 1-year-olds, 2-year-olds, and young-of-the-year. Olympic marmots are among the most social of squirrels. Colonial territories typically average 5.4 (2-8.7) ha. Subordinate males can live on the periphery of colonies, but typically they are not part of the social group unless the dominant male dies. Burrows are numerous within colonies and are employed for maternal nests, sleeping dens, and escape. Larger burrows are most often used for sleeping, and they usually have small mounds of soil and rocks at their entrance from frequent excavations. Sleeping chambers are often lined with dry plant material. *M. olympus* forages on the ground in meadows, eating sedges, herbs, and grasses; it will excavate the roots of some forbs. Olympic marmots may remove up to 30 percent of the biomass; however, this grazing promotes a high level of plant diversity. The bark and cambium of conifers are sometimes gnawed. This species will occasionally eat animal material; cannibalism has been observed. Dry grasses may be collected underground for either food or bedding. Under the best conditions, females typically reproduce in alternating years. Litter size is about four, and the young remain in their natal area until at least their third year, when dispersal occurs. About 53 percent of the young-of-the-year and 85 percent of the yearlings survive, with no difference between the sexes. Mortality occurs mostly during the winter, especially in years with little insulating snowpack. Their major predators are raptors, as well as larger mustelids, canids, and felids. Scent-marking on rocks and plants, using their oral and cheek glands, is common. Four distinct alarm calls are produced in response to potential threats; and a number of whistles, tooth chatters, growls, "yips," and "chucks" are used in a variety of contexts. Due to this species' localized distribution, high-elevation habitat, and protected status, *M. olympus* is not harvested for food, pelts, or sport. Olympic marmots are not considered to be pests.

GENERAL REFERENCES: Barash 1973, 1989; Blumstein et al. 2001; Edelman 2003; Griffin et al. 2009.

Marmota sibirica (Radde, 1862) Tarbagan Marmot

DESCRIPTION: Tarbagan marmots are often a light grizzled buff on the dorsum but vary in color, with some very pale

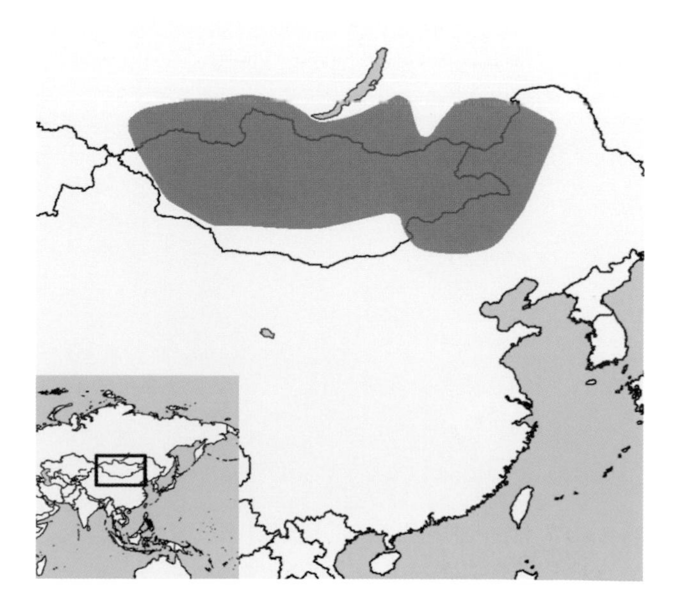

cream individuals reported. The venter, throat, front legs, and hind flanks are often white or pale cream. The feet can be tan to rust to dark brown. The head often is brown to black from the snout to behind the eyes. The tail increases in darkness toward the tip, which is often dark brown to black.

SIZE: Both sexes—HB 360-495 mm; T 112-121 mm; Mass 5000-8000 g.

DISTRIBUTION: This species is found in southwestern Siberia, Tuva, and Transbaikal (Russia); northern and western Mongolia; and Inner Mongolia and Heilongjiang (China).

GEOGRAPHIC VARIATION: Two subspecies are recognized.

M. s. sibirica—eastern portions of the range, in steppe and grassland habitats. The dorsum in the fall is straw yellow sprinkled with brown, and the venter is rust colored.
M. s. caliginosus—northern and western part of the range, in rugged mountainous habitats. No description is available.

CONSERVATION: IUCN status—endangered. Population trend—decreasing.

HABITAT: *M. sibirica* inhabits open steppes, semideserts, forest-steppes, and grassy mountain slopes and valleys.

NATURAL HISTORY: This species is diurnal. Tarbagan marmots hibernate in burrows for extended winters of six to seven months, emerging in April-May. These marmots hibernate in large groups of up to 20 individuals. Mating begins in March and April, after their arousal from hibernation. After a gestation of 40-42 days, litters ranging from three to eight young emerge in June and remain with their natal group for at least three years. First-year survivorship is estimated at 70-80 percent, if females are not removed by trapping and hunting. Tarbagan marmots live in large colonies with many burrows; the burrows have conspicuous circular mounds of soil that initially are bare, but eventually are colonized by herbaceous vegetation. Family groups of *M. sibirica* consist of a reproductive pair and their young from previous litters, up to 3 years of age. Densities are often more than 1 marmot/ha and in protected areas can approach 7 marmots/ha. Tarbagan marmots of all ages forage on grasses, a diversity of forbs, and a few woody plants. Where *M. sibirica* overlaps with *M. baibacina*, Tarbagan marmots are found more in valley bottoms, while the latter (grey marmots) inhabit ridgetops and more alpine environments. Predators of Tarbagan marmots include wolves (*Canis lupus*), foxes, domestic dogs, bears, snow leopards (*Panthera [Uncia] uncia*), and several large eagles and hawks. These marmots have a relatively simple single alarm call. *M. sibirica* is a keystone species in the steppe communities that it inhabits, because the large colonies harbor high levels of diversity and receive more attention by predators than might otherwise be expected. Their low-density populations are often hunted for food, oil, and pelts at unsustainable rates; overgrazing and mining have contributed to localized declines. Recent restrictions on hunting seem to have had a positive impact on Tarbagan marmots; however, rangewide declines appear to be continuing. Furthermore, sylvatic (bubonic) plague causes local extinctions, although resistant populations are known.

GENERAL REFERENCES: Clark et al. 2006; Reading et al. 1998; Rogovin 1992; Townsend 2006; Townsend and Zahler 2006; Van Staalduinen and Werger 2007.

Marmota sibirica. Photo courtesy Kevin Tierney, www.flickr.com/photos/kevintierney/.

Marmota vancouverensis (Swarth, 1911)
Vancouver Island Marmot

DESCRIPTION: Vancouver Island marmots have a dark brown to sepia dorsum and venter. The color of old pelage fades to tawny or olive brown. A white patch surrounds the snout and the lower jaw. Flecks or small patches of white can also be found on the forehead and the venter; sometimes white patches are aggregated as a line along the midline of the venter.

SIZE: Female—HB 661 mm (n = 12); T 193 mm (n = 12).
Male—HB 695 mm (n = 6); T 220 mm (n = 6); Mass 3000 g in May, 6500 g in September.

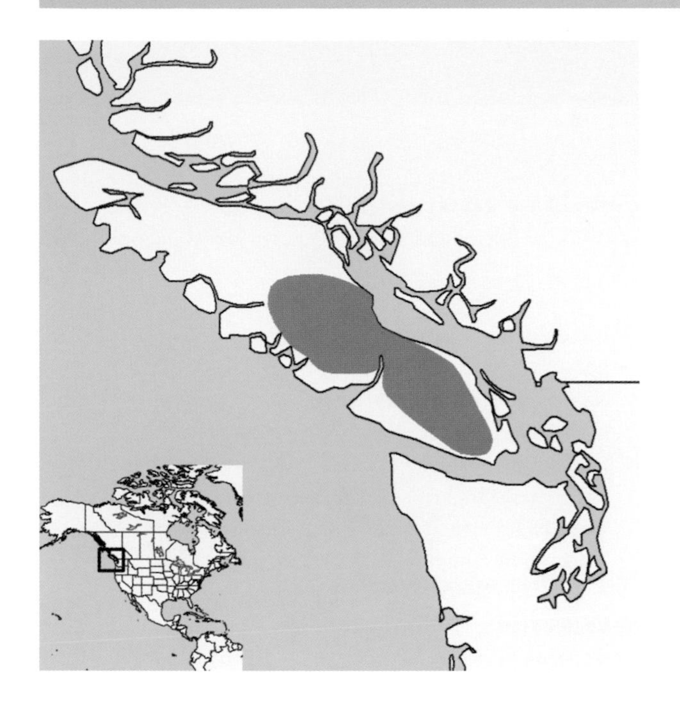

Marmota vancouverensis. Photo courtesy Oli Gardner, www.marmots.org.

DISTRIBUTION: This species is only found on Vancouver Island (British Columbia, Canada).

GEOGRAPHIC VARIATION: None.

CONSERVATION: IUCN status—critically endangered. Population trend—decreasing. There are fewer than 100 individuals persisting in the wild. A captive breeding program has been initiated.

HABITAT: *M. vancouverensis* is most commonly found in open alpine meadows, avalanche bowls, or clearcuts, at elevations of 1000-1500 m.

NATURAL HISTORY: This species is diurnal. *M. vancouverensis* hibernates through the long winters, which last up to eight months, in family groups within burrows. Litter size ranges from three to five young; females often produce litters in consecutive years, but they can also skip one or two years. They live in restricted family groups that include an adult male, one to three adult females, yearlings, and a single litter produced by one of the female group members. Two-year-olds disperse from their natal area. This species' daily activity is bimodal, with the animals foraging in open habitats for grasses, forbs, succulent fruits, and fern fronds. The vocal repertoire of Vancouver Island marmots is rich; it consists of a variety of descending, ascending, and flat single-note calls that increase in length with the level of alarm. Trills are given when these marmots are highly alarmed. A fifth call type, a "kee-aw," seems to maintain vigilance levels among other *M. vancouverensis*. Predation levels are high and may be responsible for this species' decline in recent years. Golden Eagles (*Aquila chrysaetos*), wolves (*Canis lupus*), and mountain lions (*Puma concolor*) are the most significant predators of Vancouver Island marmots.

GENERAL REFERENCES: Aaltonen et al. 2009; Bryant 2005; Nagorsen 1987.

Notocitellus A. H. Howell, 1938

This genus has two species.

Notocitellus adocetus (Merriam, 1903) Tropical Ground Squirrel

DESCRIPTION: *N. adocetus* is smaller and paler in color than *N. annulatus*, the most similar congener found within its range. In addition, the ears are small, and the rostrum is shorter but broader than that of *N. annulatus*; unlike the latter species, there are no tail rings. Other species found within or near its range include *Ictidomys mexicanus* and *Otospermophilus variegatus*; the latter can be distinguished by its nine dorsal rows of white spots. Unlike *O. variegatus*, *N. adocetus* has tawny or buffy pelage on the sides of the head, and a closed supraorbital foramen.

SIZE: Female—HB 168.3 mm; T 131.9 mm.
Male—HB 175.5 mm; T 150.2 mm.
Sex not stated—HB 194.0 mm; T 156.0 mm.

DISTRIBUTION: The tropical ground squirrel is found in westcentral México, in four Mexican states (Jalisco, Michoacán, México D.F., and northern Guerrero) within the Trans-Mexican Volcanic Belt. Disjunct populations may result from introductions, due to the pet trade.

GEOGRAPHIC VARIATION: Three subspecies are recognized. Records for Jalisco and México D.F. are identified only at the species level.

N. a. adocetus—Michoácan (México). The general color is grayish or buffy.

N. a. arceliae—Guerrero (México). There are tail marks in a V shape. The color of the incisors is raw sienna.

N. a. infernatus—southwestern portion of the species' range, in Michoacán and northern Guerrero (México). This form is smaller and more marked with a dark blackish coloring on the muzzle than *N. a. adocetus*.

CONSERVATION: IUCN Status—least concern. Population trend—stable.

HABITAT: *N. adocetus* is typically found in xeric vegetation and rocky habitats, often among cliffs and canyons; it is also associated with agricultural areas, where it can cause significant crop damage. The native vegetation in its habitat is often mesquite (*Prosopis*) and barrel cactus (*Cephalocereus*).

Notocitellus adocetus. Photo courtesy Troy L. Best.

NATURAL HISTORY: These squirrels typically dig burrows that are branched and up to 60 cm deep. The tropical ground squirrel is technically omnivorous, but it is strongly granivorous whenever seeds are available. These animals have been reported to feed on the seeds and fruits of *Acacia, Prosopis, Crescentia,* and *Prunus.* This species' reliance on agricultural crops—corn (maize), sorghum, and beans—may contribute to occasional range expansions. The tropical ground squirrel is diurnal, social, and generally active year-round, although estivation may occur in the warmest months of the year, when food is less abundant. These squirrels frequently vocalize, giving high-pitched "chirps" in the presence of humans and potential predators. Only one louse (*Neohaematopinus traubi*) is reported from *N. adocetus,* but ectoparasites are likely to be common. Although not listed as threatened in México, it is one of 122 vertebrate and vascular plant species, and one of 11 mammal species, considered key species for conservation within the Transvolcanic Belt of central México.

GENERAL REFERENCES: Best 1995g; de Grammont and Cuarón 2008; Hayssen 2008a; Herron, Castoe, et al. 2004; Sánchez-Cordero et al. 2005; Velázquez et al. 2003.

Notocitellus annulatus Audubon and Bachman, 1842
Ring-Tailed Ground Squirrel

DESCRIPTION: This squirrel is similar to *N. adocetus,* but *N. annulatus* is larger, with a longer narrower rostrum. The tail of *N. annulatus* has obvious pale-colored rings; its pelage is darker and more reddish than that of *N. adocetus.*

SIZE: Female—HB 219.3 mm; T 207.6 mm; Mass 386.3 g.
Male—HB 221.2 mm; T 215.2 mm; Mass 260.0 g.
Sex not stated—HB 209.5 mm; T 208.0 mm.

DISTRIBUTION: The ring-tailed ground squirrel occupies a narrow range along the Pacific coast of western México, from Nayarit through Jalisco, Colima, Michoacán, and northern Guerrero.

GEOGRAPHIC VARIATION: Two subspecies are recognized.

N. a. annulatus—northern third of the species' range, from Nayarit to Jalisco (México). See description above.

N. a. goldmani—southern two-thirds of the species' range, from Jalisco to Guerrero (México). The reddish tinge on the face, neck, thighs, and tail is less extensive and intense.

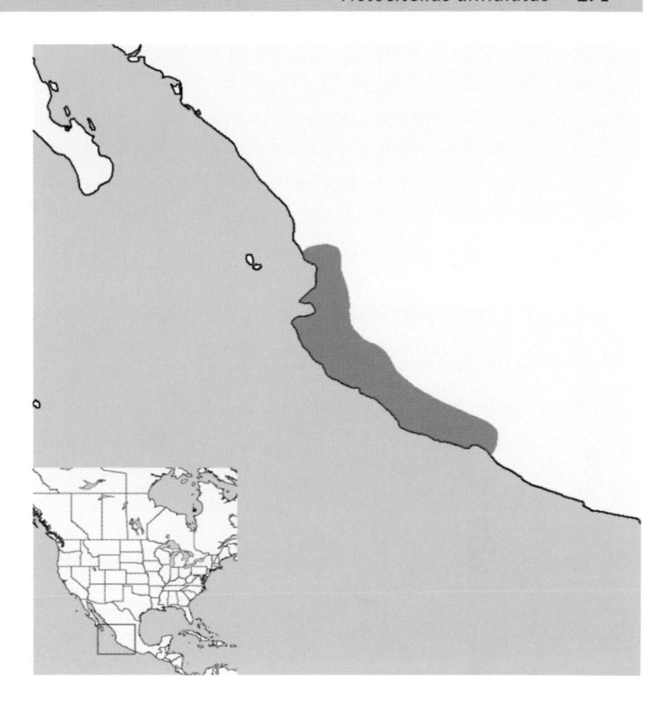

CONSERVATION: IUCN status—least concern. Population trend—no information.

HABITAT: The ring-tailed ground squirrel is found at elevations from sea level to 1200 m. It typically occurs in dry tropical deciduous forests, sometimes in association with figs (*Ficus*), but also oil palm (*Elaeis*), mesquite (*Prosopis*), or catclaw (*Acacia*). Although ground burrows in sand or rocky cliffs and outcrops are common, *N. annulatus* is also reported to nest in tree cavities.

NATURAL HISTORY: This species is primarily granivorous, eating a variety of seeds, fruits, and nuts from the plants

Notocitellus annulatus. Photo courtesy Troy L. Best.

within its habitats, including various agricultural crops. It is agile and can often be sighted in trees, usually a few meters off the ground. *N. annulatus* vocalizes and quickly seeks shelter in the presence of intruders. Breeding occurs during the dry season (between December and June), but little is known about the reproductive behavior of the ring-tailed ground squirrel. The baculum is rather unique among this group of ground squirrels, with a short broad spoon-shaped distal end, lined with a single row of toothlike projections. Predators are not reported for this species, and the only parasite observed from *N. annulatus* is the botfly (*Cuterebra*).

GENERAL REFERENCES: Álvarez-Castañeda, Castro-Arellano, Lacher, et al. 2008a; Best 1995h.

Otospermophilus Brandt, 1844

This genus contains three species.

Otospermophilus atricapillus W. Bryant, 1889
Baja California Rock Squirrel

DESCRIPTION: The Baja California rock squirrel is recognized by a dark triangular patch on the back of the head, the neck, and the anterior portion of the back. Hairs on the posterior end of the body and the tail are dark at the base and a pinkish buff color at the tip. This species has a whitish eye ring. It is distinguished from *O. variegatus* by the dark triangular patch on top of the head and between the shoulders, and from *O. beecheyi* by the longer tail and the marginally smaller head/skull.

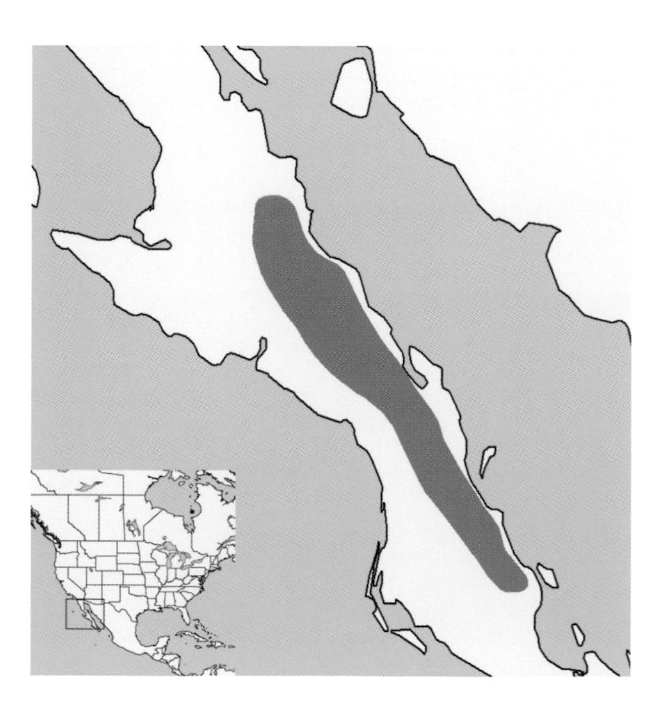

SIZE: Female—HB 231.2 mm; T 196.5 mm; Mass 350.0 g.
Male—HB 237.8 mm; T 192.6 mm; Mass 505.0 g.
Sex not stated—HB 300.0 mm; T 235.0 mm.

DISTRIBUTION: This species' range extends along the eastern edge of southern Baja California and central Baja California Sur (México). The entire area of its geographic range is less than 5000 km^2, but it is highly fragmented.

GEOGRAPHIC VARIATION: None.

CONSERVATION: IUCN status—endangered. Population trend—declining.

HABITAT: This species is found in volcanic mountains with desert shrub vegetation. Dominant plant species typically found within its habitat includes several cacti—cardon (*Pachycereus pringlei*), choya (*Opuntia cholla*), and pitaya (*Lemaireocereus thurberi* and *Machaerocereus gummosus*)—and other desert plants such as mesquite (*Prosopis juliflora*), palo blanco (*Lysiloma candidum*), torote (*Bursera cerasifolia*), and ashy limberbush or lomboy (*Jatropha cinerea*). The Baja Cali-

***Otospermophilus atricapillus*.** Photo courtesy Eric Mellink.

fornia rock squirrel is considered a habitat specialist that is subject to considerable habitat fragmentation.

NATURAL HISTORY: This species has received very little attention. No information is available on its behavior, population biology, or general ecology. *O. atricapillus* has been collected during winter months and is thus considered either active throughout the year or as having a limited period of torpor. This species may be preyed on by coyotes (*Canis latrans*), and it is host to at least one species of sucking louse (*Enderleinellus orborni*).

GENERAL REFERENCES: Álvarez-Castañeda, Arnaud, et al. 1996; Álvarez-Castañeda, Castro-Arellano, and Lacher 2008; Castro Arellano and Ceballos 2005; Durden and Musser 1994a, 1994b; Goodwin 2009; Grajales et al. 2003; R. G. Harrison et al. 2003; Helgen et al. 2009.

Otospermophilus beecheyi (Richardson, 1829) California Ground Squirrel

DESCRIPTION: *O. beecheyi* has a grayish brown back speckled with white spots, and sometimes a grayish white collar on the back, the neck, and the sides. The venter is yellowish white to light brown; the tail is bushy; and white eye rings are usually present. This species shows sexual dimorphism (larger males) and temporal, spatial, and latitudinal variation in body size.

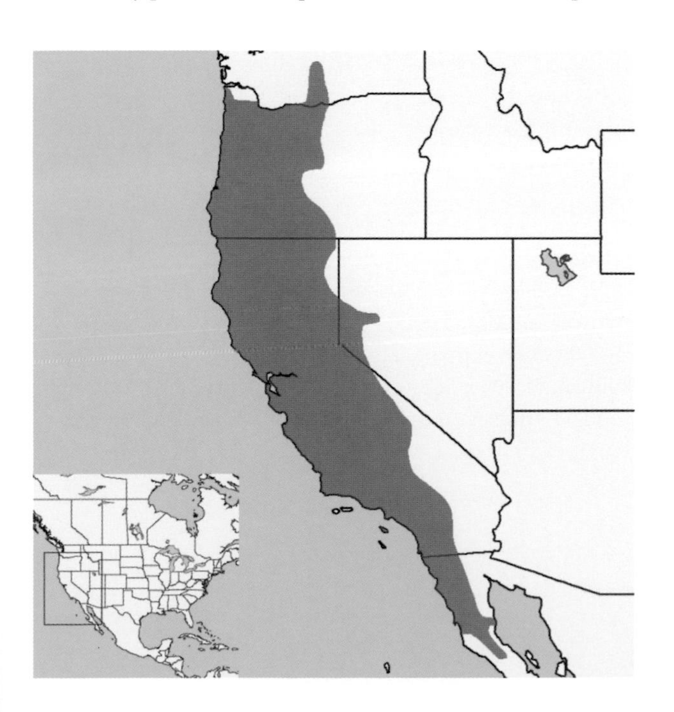

SIZE: Female—HB 254.3 mm; T 171.0 mm; Mass 508.5 g.
Male—HB 273.2 mm; T 181.7 mm; Mass 621.3 g.
Sex not stated—HB 246.9 mm; T 179.0 mm; Mass 599.5 g.

DISTRIBUTION: The California ground squirrel ranges from western Washington through most of California (USA) to Baja California, México.

GEOGRAPHIC VARIATION: Eight subspecies are recognized.

O. b. beecheyi—westcentral California (USA). See description above.
O. b. douglasii—Washington and the Columbia River south to San Francisco Bay, California (USA). The shoulder patch is dark, ranging from brown to black.
O. b. fisheri—California, from Madera County south to the southern San Joaquin Valley and east to the Panamint Mountains (USA), and extending farther south to Baja California (México). The general color is paler, with more silvery gray on the sides of the neck and the shoulders.
O. b. nesioticus—Santa Catalina Island and the Santa Barbara Islands, California (USA). The general color is darker, with less yellowish brown. The top of the head and the ears are black. The sides of the neck are gray, extending into a stripe of gray along the shoulders. A patch between the shoulders is mixed black and tawny ochraceous.
O. b. nudipes—Sierra Juarez, Baja California (México). Its general color is darker than *O. b. rupinarum*, and it has a silvery white mantle.
O. b. parvulus—Argus Mountains, California (USA). This form is similar to *O. b. fisheri*, but *O. b. parvulus* is smaller. It is also paler than the typical *O. b. beecheyi*.
O. b. rupinarum—Catavina, Baja California (México). It is similar to *O. b. nudipes*, but *O. b. rupinarum* is grayer on the dorsum, with many light spots. The venter is very pale whitish.
O. b. sierrae—Lake Tahoe, in California and Nevada (USA). This subspecies is similar to *O. b. beecheyi*, but *O. b. sierrae* is less buffy in general, with the sides of the head grayer; the venter is paler whitish.

CONSERVATION: IUCN status—least concern. Population trend—stable.

HABITAT: *O. beecheyi* occurs in a variety of habitats, including meadows, grasslands, chaparral, successional stands, and agricultural lands. It is often found along roadsides and in disturbed areas.

Otospermophilus beecheyi. Photo courtesy A. Coke Smith, www.cokesmithphototravel.com.

NATURAL HISTORY: The California ground squirrel is herbivorous; its diet consists of a broad range of plant material—more than 20 species of grasses, legumes, and forbs—but it frequently includes seeds and nuts (e.g., the acorns of *Quercus*) in the diet when available. Both the overall energy content and the squirrel's assimilation efficiency are higher for seeds than for foliage, but the reduced availability and lower water content of seeds probably explains the high percentage of foliage in this species' diet. Other items consumed occasionally include mosses, lichens, and arthropods. *O. beecheyi* exhibits sexual dimorphism; males are slightly but significantly larger than females. Variation in body mass in California ground squirrels shows a Bergmann's cline, with the smaller animals found in the southern part of the range, in more xeric environments. Such variation corresponds primarily with precipitation and probably results from the seasonality of resources, caused by differences in rainfall. Reproduction occurs soon after emergence from the burrow, the timing of which varies with latitude. Both males and females mate several times within a breeding season. The young are able to reproduce in their first year. Gestation is estimated to be 30 days, and litter sizes range from 5 to 11, depending on elevation and latitude. Infanticide is common in *O. beecheyi* and, where systematically studied, appears to be performed almost exclusively by neighboring mothers, who frequently cannibalize the victims.

This species is abundant throughout its range; population densities are reported to vary between 8 and 92 animals/ha. Home ranges overlap considerably; the average size varies from 270 to 375 m^2 for males, and 616 to 902 m^2 for females. Females typically maintain larger home ranges than males, and home ranges do not change appreciably in size during and after the breeding season. Numerous carnivorous mammals, raptors, and snakes prey on *O. beecheyi*. The young of California ground squirrels are particularly susceptible to predation by the northern Pacific rattlesnake (*Crotalus oreganos oreganos*). *O. beecheyi* has shared a long evolutionary history with this snake and, in response, has apparently evolved a suite of behavioral and physiological adaptations for thwarting successful predation, including the ability to detoxify the venom with specific blood proteins, the creation of distraction displays by tail flicking, and the ability to assess predation risks by the sounds of the rattle. Squirrels will induce the snake to rattle by kicking sand, so that they can assess specific risk characteristics such as body temperature, the probability of a strike, and the body size of the snake, all of which are correlated with the rate at which the rattle is shaken. Moreover, a California ground squirrel flags its tail in the presence of both rattlesnakes and gopher snakes (*Pituophis melanoleucus*), but it increases the blood flow (and thus heat) to the tail only in the presence of a rattlesnake, which hunts mammalian prey with infrared detectors. Such a reaction by the squirrel decreases the probability of a successful attack by rattlesnakes. Differences in alarm calls between adults and juveniles, as well as variations in the responses of adults to the two call types, are linked to differences in the level of predation risk between juveniles and adults. The California ground squirrel is an important reservoir host of *Crypotosporidium parvum* and the human bacterial pathogen *Bartonella washoensis*. In California this species is considered one of the most significant vertebrate pests for agriculture; although it is also thought to be a competitor with grazing cattle, detailed studies on its diet and energy consumption suggest otherwise. While nonlethal methods of control (habitat modification and translocations) have been attempted, large-scale attempts at controlling crop damage will most likely require an integrated method of pest management.

GENERAL REFERENCES: Anthony 1928; Atwill et al. 2004; Biardi, Chien, et al. 2006; Biardi, Coss, et al. 2006; Blois et al. 2008; Boellstorff and Owings 1995; Coss and Biardi 1997; Hanson and Coss 2001a, 2001b; Helgen et al. 2009; Kosoy et al. 2003; A. W. Linzey, Timm, et al. 2008a; R. E. Marsh 1994; Rabin et al. 2006; Schitoskey and Woodmansee 1978; Swaisgood, Owings, et al. 1999; Swaisgood, Rowe, et al. 1999; Van Vuren et al. 1997.

Otospermophilus variegatus (Erxleben, 1777)
Rock Squirrel

Otospermophilus variegatus. Photo courtesy Randall D. Babb.

DESCRIPTION: This species has a variable pelage, as its Latin epithet suggests. The dorsum can range from grayish mixed with cinnamon buff to light brown to bone brown to dark blackish brown, sometimes with the head and the shoulders, or nearly the entire dorsum, being black. The head is pinkish buff to cinnamon buff to seal brown to fuscous black. The prominent eye ring is white to buff to tawny. The venter is grayish white to cinnamon buff. The tail is large and bushy, resembling that of a tree squirrel at first, with a mixed black or brown and buffy white coloration.

SIZE: Female—Mass 450-796 g.
Male—Mass 470-875g.
Both sexes—HB 430-540 mm; T 174-263 mm.

DISTRIBUTION: Rock squirrels are found in Nevada east to Colorado and southwestern Texas (USA), and south to Puebla (México).

GEOGRAPHIC VARIATION: Eight subspecies are recognized.

O. v. variegatus—central Nayarit and northern San Luis Potosí to Puebla and Oaxaca (México). This large form has a blackish head, a gray dorsum, and a grizzled black and white tail.

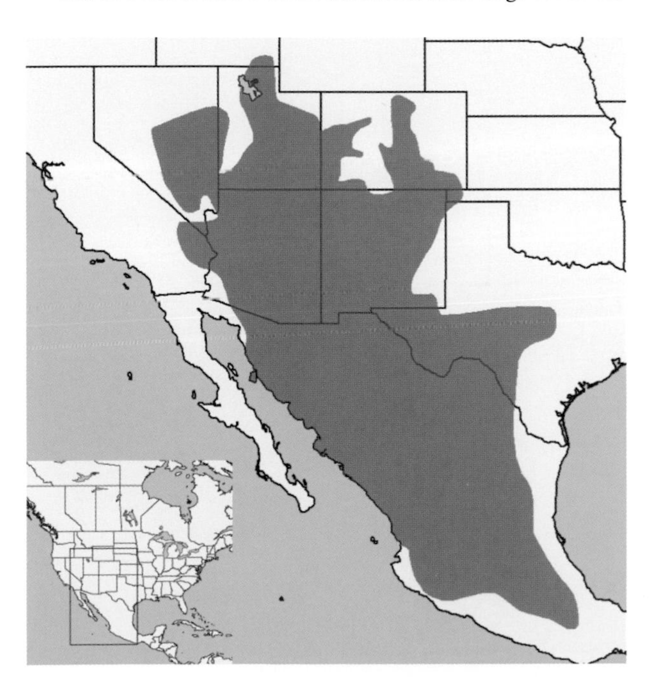

O. v. buckleyi—central Texas to the Rio Grande (USA). This is a small form, with a dark dorsal pelage.

O. v. couchii—Coahuila northwest into Chihuahua, south into San Luis Potosí, and to western Tamaulipas (México). It is a small pale form, dominated by brown, including a dark brown cap on the head and the shoulders.

O. v. grammurus—northern Colorado and New Mexico (USA) south to Sonora and central Chihuahua (México), and from southeastern California to westcentral Texas (USA). This form is also small and pale, but it lacks the dark coloration on the head.

O. v. robustus—eastcentral Nevada (USA). This is a larger form. The head is brownish black, the upper back is white, and the lower back is cinnamon buff. The eye ring is creamy white.

O. v. rupestris—eastern Sonora and central Chihuahua to central Nayarit and central Zacatecas (México). This subspecies is pale; and the dorsum is dominated by brown or buff, including a dark brown head with buff underparts.

O. v. tularosae—southcentral New Mexico (USA). This form lacks a black head or dorsum, and it has a tawny posterior.

O. v. utah—northern Utah to southern Nevada and northwestern Arizona (USA). This is also a diminutive form, but with a dark dorsum, due to an increased infusion of a tawny color.

CONSERVATION: IUCN status—least concern. Population trend—stable.

HABITAT: Rock squirrels inhabit semiarid regions with rocky canyons, cliffs, and hillsides. They can also be found in urban areas.

NATURAL HISTORY: *O. variegatus* is diurnal. Rock squirrels hibernate in burrows for one to six months, and they may remain active throughout the winter in mild weather. Males emerge first and, unlike many other ground squirrels, are not yet ready to breed. After two to three weeks, females emerge and are capable of breeding within about one week. The breeding season lasts up to two months between March and July, a prolonged period for ground squirrels. After a gestation of 30 days, a litter averaging 4.8 (range = 3-8) young is born in the burrow; females may be able to have two litters a year in mild climates, but evidence for this is scarce. Lactating females are known from May until October. The young emerge at 8 weeks of age and are weaned after two to four more weeks. Most juveniles grow slowly and do not reach adult weight until midway through their second year of life.

Burrows are almost always found among rocks, boulders, and trees that can serve as excellent safe locations; however, rock squirrels have been reported to use arboreal nests occasionally. Burrows are moderately complex, with a few openings. The burrows are up to 6 m long and have a nest chamber at a depth of more than 1.0 m. Rock squirrels tend to be colonial, with several females in a colony. They have overlapping home ranges and well-defined dominance relations. Scent-marking is a significant means of communication. Sociality is moderate (at best), with little interaction between individuals despite considerable spatial and temporal overlap between adults. Natal philopatry is female biased, and all males leave their birthplace. The overlapping home ranges of females (3.8-4.5 ha) are smaller than those of males (7.9 ha).

Rock squirrels are primarily herbivorous, eating such foods as buds, flowers, nuts, fruits, cacti, tree seeds (acorns, juniper berries) and cultivated crops (grapes and corn/maize). However, they also often eat animal matter, such as insects, earthworms, and small vertebrates. Rock squirrels are very capable climbers and will forage in trees that are laden with fruit. Their primary predators are raptors, canids, felids, and mustelids. *O. variegatus* has a stereotyped response to snakes: the squirrels make a rapid approach, flag the tail from side-to-side, and kick soil at the potential predator. Their alarm calls consist primarily of long and short whistles that are difficult to locate. Rock squirrels typically are not hunted; however, they are shot and poisoned for removal in situations where they become pests. No major conservation threats are known for this common and adaptable species.

GENERAL REFERENCES: Botello et al. 2007; Groves et al. 1988; Ortega 1987, 1990a, 1990b; Owings et al. 2001; Shriner and Stacey 1991.

Poliocitellus A. H. Howell, 1938

This genus has only one species.

Poliocitellus franklinii (Sabine, 1822)
Franklin's Ground Squirrel

DESCRIPTION: Franklin's ground squirrel has a grizzled gray coat with a yellowish wash on the rump and venter. Its tail is long and bushy, but less so than the similar-sized Richardson's ground squirrel (*Urocitellus richardsonii*); the Columbian ground squirrel (*U. columbianus*) is more rusty in color. With the exception of the yellowish wash in the pelage and more rounded shoulders, *P. franklinii* resembles the eastern gray squirrel (*Sciurus carolinensis*).

SIZE: Female—HB 237.9 mm; T 128.1 mm; Mass 424.9 g.
Male—HB 232.4 mm; T 127.7 mm; Mass 461.2 g.
Sex not stated—HB 235.0 mm; T 136.0 mm; Mass 607.0 g.

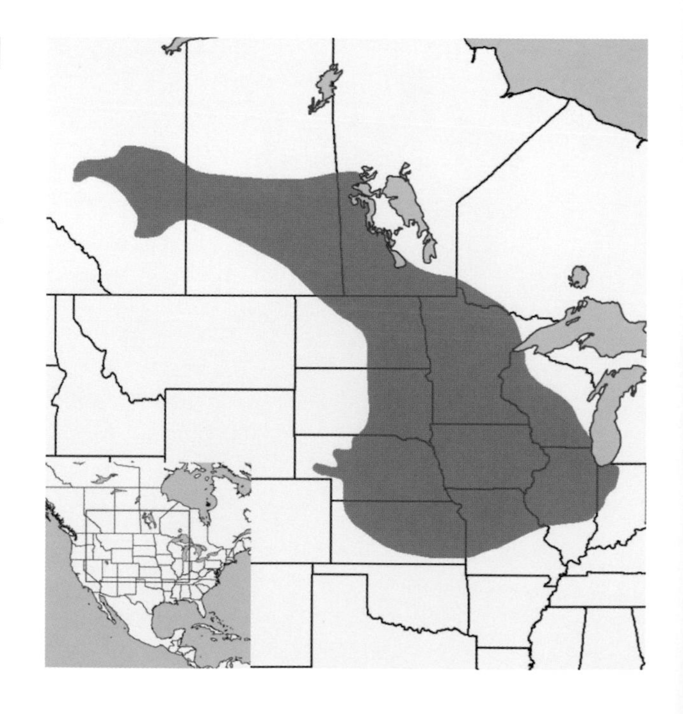

DISTRIBUTION: Franklin's ground squirrel is found in the northern Great Plains: in parts of Canada and the northcentral USA. Its range extends from central Alberta to southern Ontario (Canada), south through North Dakota and Minnesota to Kansas, Missouri, Illinois, and Indiana (USA). In much of the USA, the distribution of *P. franklinii* is highly fragmented, and populations are rapidly disappearing because of loss of habitat.

GEOGRAPHIC VARIATION: None.

CONSERVATION: IUCN status—least concern. Population trend—declining.

HABITAT: The preferred habitats of Franklin's ground squirrel are thicker grasslands or grassland-prairie ecotones. Tall dense vegetation is the primary variable that seems to best characterize suitable habitat of this species.

NATURAL HISTORY: This omnivorous ground squirrel shows seasonal shifts in its diet: from primarily vegetation in the spring (which includes roots, shoots, and grasses) to seeds and fruits later in the summer, when the animal is preparing for hibernation. *P. franklinii* is reported to consume a great variety of animal material, including bird eggs, adult birds, frogs, mice, rabbits, and a variety of insects. This species may be a particularly important predator of ground-nesting birds. Its daily energy requirements are estimated at 3.08 calories $g^{-1} h^{-1}$. *P. franklinii* is intolerant of conspecifics; males are often solitary outside the breeding season. It is considered the most asocial of all the ground squirrel species. These squirrels scent-mark with their oral, dorsal, and anterioposterior glands; juveniles are capable of kin recognition, but whether this influences nepotism is not yet known. *P. franklinii* is diurnal, with the majority of its activity concentrated in the middle of the day. Two studies noted densities ranging between 1.25 and 2.5 animals/ha. Home ranges over the course of a year average 24.6 ha for males and 8.7 ha for females. Franklin's ground squirrels build complex branched burrows with two to three openings. Different branches of the burrow are used for specific activities (i.e., nesting, storing food, defecation). Adults enter hibernation any time between early July and late August, but young animals remain above ground and continue feeding until late August-October. Males emerge one to two weeks before females and establish dominance hierarchies, as opposed to the territories of other species of ground squirrels. Males and females sometimes reside together in the burrow, but males will always disperse after mating. Gestation is 28 days. Litter sizes, based on embryo counts from two studies and actual births from a third, averaged 9.4, 7.5, and 7.9, respectively. Juvenile Franklin's ground squirrels disperse at 9-11 weeks of age. In dispersing, both males and females can travel more than 1 km (up to 3.6 km for one male) across open fields, but they typically avoid crossing major roadways; males usually disperse farther than females.

Poliocitellus franklinii. Photo taken by Paul Huber.

Badgers (*Taxidea taxus*) are reported to be their primary predator, although at least five other species of carnivorous mammals, and several species of raptors and snakes, are also reported to prey on Franklin's ground squirrels. This species is host to a rather rich community of ectoparasites, at least one of which shows high prevalence and intensity levels. These ectoparasites include 2 ticks, 1 mite, and 11 species of fleas, 9 of which are not reported for other ground squirrels. Not surprisingly, given the animal material in its diet, *P. franklinii* is also host to more helminths than are typically observed in other species of ground squirrels. These parasites include 2 cestodes, 4 trematodes, and 5 nematode species. Three species of coccidians (*Eimeria*) also reside in the digestive tract of this squirrel. The pulmonary disease adiaspiromycosis, caused by the fungus *Emmonsia crescens* (= *Chrysosporium parvum* var. *crescens*), is reported in Franklin's ground squirrels. *P. franklinii*, along with several other rodent species, may serve as a reservoir host for *Bartonella* infections. The earliest fossils for *P. franklinii* are reported from the Middle to Late Pleistocene (Irvingtonian period).

Although this species is relatively common in most of the northern parts of its range in Canada, its southern and eastern populations are rapidly declining and experiencing local extinctions, due to habitat loss and fragmentation. Continued fragmentation may prevent successful dispersal between populations and, in turn, lead to a loss of genetic variability.

However, other factors—such as the extermination of ground squirrels to reduce both crop damage and waterfowl depredation—have lowered population numbers; mowing and herbicide use along railroad beds and other potential corridors have increased habitat loss and fragmentation. Poisoning of pocket gophers (*Geomys bursarius*) also causes significant mortality for *P. franklinii*. Franklin's rock squirrel is classified as Imperiled (S2) or Rare (S3) in 6 of the 14 states or provinces where they were once found, although it may be expanding its range northward in Minnesota. This species' secretive behavior probably contributes to lower estimates of its presence and abundance; track tubes may be useful in assessing the presence of *P. franklinii* in many parts of its range.

GENERAL REFERENCES: Conover et al. 2005; Hare 2004; Huebschman 2003; Jannett et al. 2007; Jardine et al. 2006; Johnson and Choromanski-Norris 1992; T. L. Lewis and Rongstad 1992; J. M. Martin and Heske 2005; J. M. Martin et al. 2003; Ostroff and Finck 2003; Pergams et al. 2008; Wiewel et al. 2007.

Sciurotamias Miller, 1901

This genus contains only two species of rock squirrels. Among the ground squirrels, *Sciurotamias* shows relatively primitive dental morphology, similar to that of *Ammospermophilus*, *Callospermophilus*, and *Otospermophilus*.

Sciurotamias davidianus (Milne-Edwards, 1867)
Père David's Rock Squirrel

DESCRIPTION: Père David's rock squirrel is paler in coloration than its congener *S. forresti*, and it has no stripe along its sides. Overall its pelage is olive gray with a yellowish wash; it is darker above and light below. A dark line across the face is also reported for some individuals.

SIZE: Female—HB 204.0 mm; T 142.9 mm.
Male—HB 212.3 mm; T 140.7 mm.
Sex not stated—HB 261.8 mm; T 140.2 mm; Mass 260.0 g.

DISTRIBUTION: This species is found in southern Gansu east to Ningxia, Hebei, southern Liaoning, and Shandong, and south to southwestern Sichuan, Guizhou, Guangxi, Hubei, and Anhui (China). Père David's rock squirrel was originally believed to be introduced to Belgium; however, recent genetic analyses demonstrate that the introduced animals instead were *Callosciurus erythraeus*.

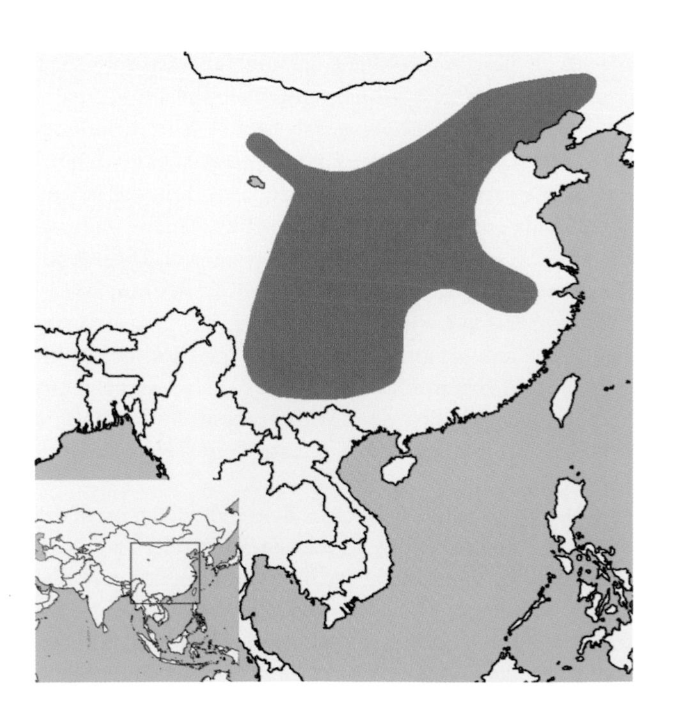

***Sciurotamias davidianus*.** Photo courtesy James Eaton/Birdtour Asia.

GEOGRAPHIC VARIATION: Three subspecies are now reported.

- *S. d. davidianus*—Sichuan, Gansu, Ningxia, Shaanxi, Shanxi, Henan, Shandong, Hebei, Tianjin, Beijing, and Liaoning (China). This form is lighter gray on the dorsum; its venter is more gray and less buff brown.
- *S. d. consobrinus*—northwestern Gansu, western Sichuan, northeastern Yunnan, and Guizhou (China). The fur is richer and darker in color, and the feet are black. Postauricular patches may be absent.
- *S. d. saltitans*—southcentral provinces of Sichuan, Chongqing, Guizhou, Guangxi, Hubei, Henan, and Anhui (China). The dorsum is brown.

CONSERVATION: IUCN status—least concern. Population trend—no information.

HABITAT: *S. davidianus* is typically associated with rocky habitats, where it nests in underground cavities.

NATURAL HISTORY: The diet of Père David's rock squirrel includes a variety of seeds and nuts, which it carries with the aid of cheek pouches and then stores. It both scatter- and larder-hoards seeds and, in the process, significantly influences the dispersal and establishment of several tree species, being both a seed predator for some (e.g., wild apricot [*Prunus armeniaca*]) and an important dispersal agent for others (e.g., Liaodong oak [*Quercus liaotungensis*] and cultivated walnut [*Juglans regia*]). Recent studies demonstrate similar ecological and evolutionary interactions between *S. davidianus* and oaks (*Quercus*), as is shown for members of *Sciurus* and oaks in North America. For example, Père David's rock squirrel performs embryo excision on early-germinating acorns of white oak species to arrest early germination and the loss of seed stores, in the same way that is reported for the eastern gray squirrel (*Sciurus carolinensis*). This suggests a convergence of behavioral strategies in these two distantly related sciurids. Other comparisons between *Sciurotamias* and *Sciurus*, however, show that the former may be more dependent on olfactory sense than spatial memory for retrieving and managing scatter-hoards. Two species of sucking lice are reported from *S. davidianus*, but little else is known about the ecology and population biology of this species.

GENERAL REFERENCES: Bertolino 2009; Durden and Musser 1994a; Goodwin 2009; Lu and Zhang 2004, 2008; A. T. Smith and Johnston 2008c; A. T. Smith and Xie 2008; Tate 1947; R. W. Thorington, Miller, et al. 1998; W. Wang, Ma, et al. 1999; W. Wang, Zhang, et al. 2007; Xiao et al. 2010; H. M. Zhang and Zhang 2008; H. M. Zhang et al. 2008; Y. Zhang et al. 1997; Z. B. Zhang et al. 2005

Sciurotamias forresti (Thomas, 1922) Forrest's Rock Squirrel

DESCRIPTION: The dorsal pelage is grayish to dark brown; the sides are lighter, sometimes with an ochraceous wash. The back and sides are separated by a pale thin stripe that runs along both sides of the body. The ventral surface is pale, but it is similar in color to the sides.

SIZE: Female—HB 224.0 mm; T 160.0 mm.
Sex not stated—HB 233.1 mm; T 154.0 mm.

DISTRIBUTION: This species is endemic to Yunnan and southern Sichuan (China). Although *S. forresti* may occur farther south outside of China, no data exist to verify this assertion.

GEOGRAPHIC VARIATION: None.

CONSERVATION: IUCN status—least concern. Population trend—no information.

HABITAT: Forrest's rock squirrel is usually associated with rocks, talus slopes, and cliffs, and it is reported to occupy only 2 of 12 zoogeographic regions of China, usually at ele-

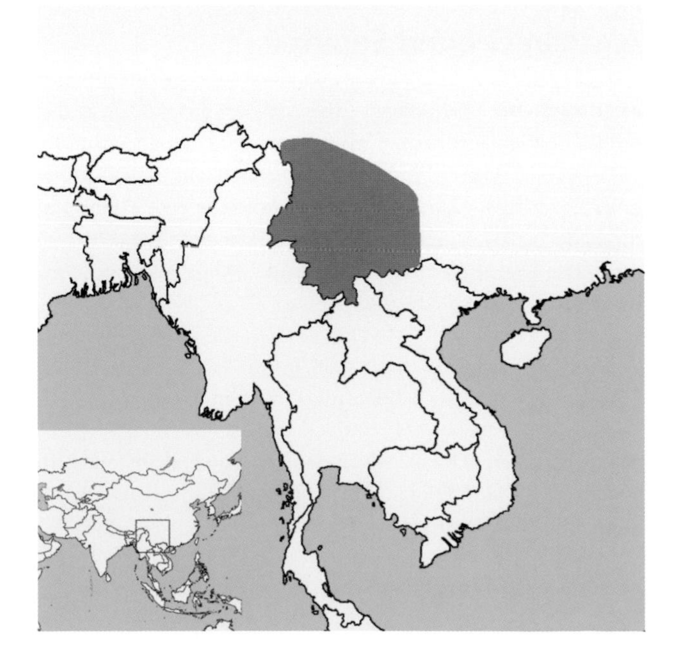

Sciurotamias forresti. Photo courtesy John and Jemi Holmes.

vations of about 3000 m. Vegetation in its habitat is described as scrubby.

NATURAL HISTORY: There is very little data available on the behavior, ecology, and reproductive biology of *S. forresti*. Morphological comparisons suggest that the two species of *Sciurotamias* are related to chipmunks (*Tamias*), and *S. forresti* may have a natural history similar to that of chipmunks. Forrest's rock squirrel is reported to host one species of sucking louse. Fossils of this species from the late Pleistocene are reported from Huanglong Cave (Hubei Province).

GENERAL REFERENCES: Durden and Musser 1994b; Hayssen 2008c; A. T. Smith and Johnston 2008d; A. T. Smith and Xie 2008; R. W. Thorington, Miller, et al. 1998; Wu et al. 2006; Xiang et al. 2004.

Spermophilus F. Cuvier, 1825

This genus comprises 15 species, commonly known as ground squirrels.

Spermophilus alashanicus Büchner, 1888
Alashan Ground Squirrel

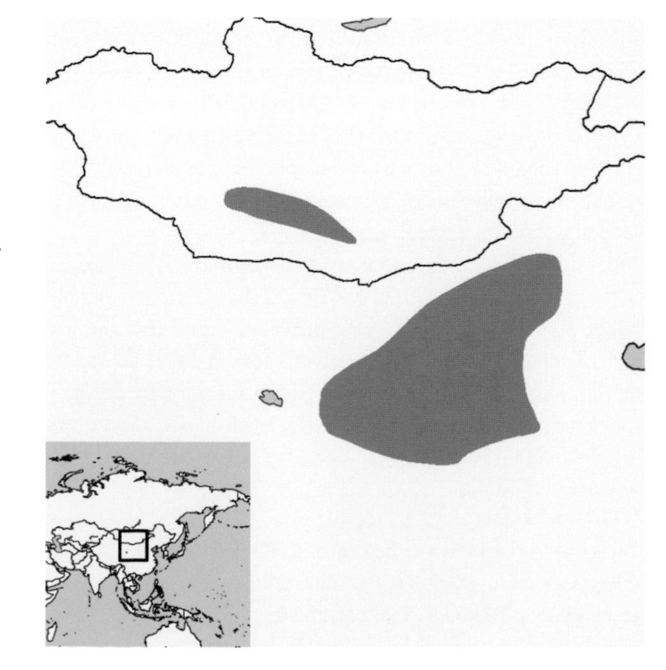

DESCRIPTION: The pelage of the Alashan ground squirrel is reddish or pinkish brown in summer, and lighter and more yellowish in winter. A white stripe runs from the ear to the nose, forming a distinctive ring around the eye. The top of the head is darker, and the ventral surface of the tail is reddish. The legs and the tail are relatively short. The auditory bullae are long and narrow.

SIZE: Male—HB 222.7 mm; T 68.3 mm.
Sex not stated—HB 199.4 mm; T 71.6 mm.

DISTRIBUTION: The Alashan ground squirrel is found in southcentral Mongolia and northcentral China (Inner Mongolia, Qinghai, Ningxia, and Gansu).

GEOGRAPHIC VARIATION: None.

Spermophilus alashanicus. Photo courtesy Danny Yee.

CONSERVATION: IUCN status—least concern. Population trend—decreasing.

HABITAT: This species is found in open meadows of steppe, montane, and alpine terrain, up to an elevation of 3200 m. It usually is found in areas with grasses, shrubs, and forbs, but it also occurs in dry sandy areas of deserts in central China and in grasslands near the Gobi Desert.

NATURAL HISTORY: *S. alashanicus* is herbivorous, feeding on herbs, plants, and grains, including crops. Between three and six young per litter are usually produced, although litter sizes as high as nine and as few as one have been reported. The young are born in June. *S. alashanicus* hibernates. High-pitched vocalizations are frequently produced by Alashan ground squirrels. Little else is known about its natural history.

GENERAL REFERENCES: R. G. Harrison et al. 2003; Hayssen 2008a, 2008b; Helgen et al. 2009; Shar, Lkhagvasuren, and Smith 2008; A. T. Smith and Xi 2008.

Spermophilus brevicauda Brandt, 1843
Brandt's Ground Squirrel

DESCRIPTION: Brandt's ground squirrel is small relative to other species of *Spermophilus*. It has a yellow brown dorsal pelage with light spots on the dorsum. The tail is short, varying in color from yellow to orange. A light eye ring is usually visible.

SIZE: Female—HB 197.0 mm; T 42.0 mm.
Sex not stated—HB 284.5 mm; T 50.8 mm.

DISTRIBUTION: Brandt's ground squirrel is found along the border of Kazakhstan through northwestern China to the border of Mongolia; and between Kazakhstan and Kyrgyzstan, in the "Zaisan depression south and westward along the Tien Shan Mountains to the vicinity of Almaty, on both sides of the Kazakhstan-Chinese border" (R. W. Thorington and Hoffmann).

GEOGRAPHIC VARIATION: None.

CONSERVATION: IUCN Status—least concern. Population trend—no information.

HABITAT: This species is usually associated with semidesert vegetation in dry steppes; however, limited information is available on its habitat use.

NATURAL HISTORY: Brandt's ground squirrel lives in underground burrows near vegetation. It hibernates in winter months and estivates when summer temperatures are high. Omnivorous in its diet, this species relies on shoots and bulbs from a variety of vegetation; it sometimes climbs or digs for food. *S. brevicauda* has limited vocalizations. Nothing is known about its population or reproductive biology. No predators or parasites are reported.

GENERAL REFERENCES: R. G. Harrison et al. 2003; A. T. Smith and Johnston 2008e; A. T. Smith and Xie 2008.

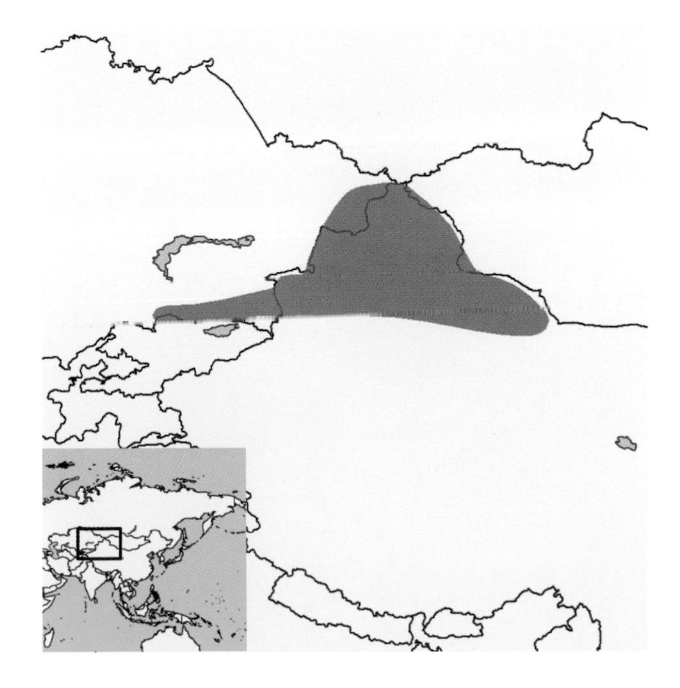

Spermophilus citellus (Linnaeus, 1766) European Ground Squirrel, European Souslik

DESCRIPTION: The back is light grayish brown, often with a yellow wash; the back and the sides are speckled with small light-colored spots. The limbs are short, and the forelimbs are especially well developed for digging.

SIZE: Female—HB 167.6 mm; T 50.6 mm; Mass 202.3 g.
Male—HB 146.3 mm; T 59.5 mm; Mass 255.7 g.
Sex not stated—HB 205.0 mm; T 65.0 mm; Mass 290.0 g.

DISTRIBUTION: The European ground squirrel's range now extends from the Czech Republic through Hungary, Slovakia, and southeastern Europe, and from southwestern Ukraine southward through Moldova, Romania, and Bulgaria to European Turkey. Its distribution in Macedonia and northern Greece is highly fragmented. *S. citellus* is now extinct in Germany, possibly in Croatia, and in Poland (although it has been reintroduced here).

GEOGRAPHIC VARIATION: Four subspecies are recognized. Records for southwestern Ukraine, Moldova, northeastern Greece, and European Turkey are identified only at the species level.

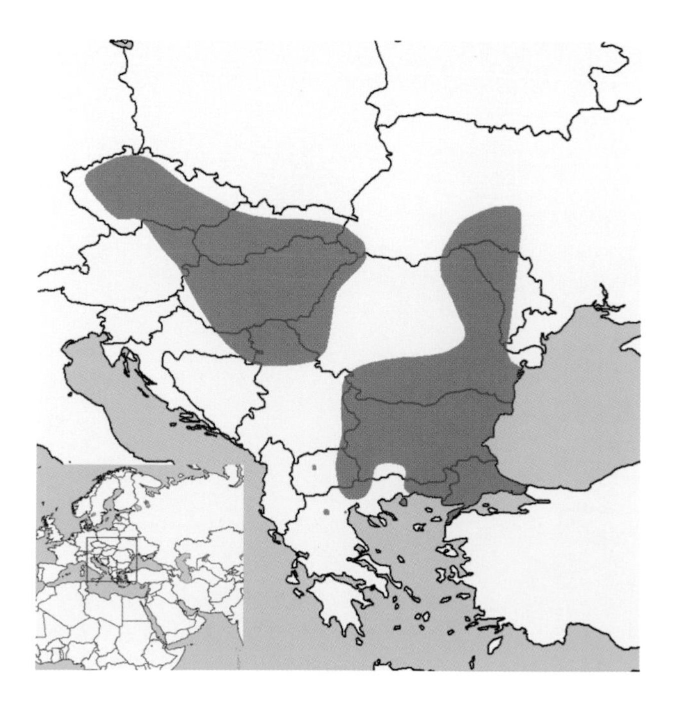

Spermophilus citellus. Photo courtesy Fabrice Schmitt.

S. c. citellus—Austria, Czech Republic, Slovakia, and Hungary. See description above.
S. c. gradojevici—Macedonia. The color is more uniform pale yellow on the upperparts and buff yellow on the underparts. The dark subterminal band of the tail tends more toward brown than black.
S. c. istricus—plains of the lower Danube River (Romania). The dorsum is conspicuously sprinkled with white spots.
S. c. martinoi—Bulgaria. No description is available.

CONSERVATION: IUCN status—vulnerable. Population trend—decreasing.

HABITAT: The natural habitats of the European ground squirrel are shortgrass steppe communities and open forest stands. It avoids exceptionally wet soils and thicker ground vegetation. It is often found in human-created habitats (e.g., on lawns and other mowed grasslands, such as pastures and golf courses).

NATURAL HISTORY: *S. citellus* is omnivorous; it feeds on shoots, flowers, roots, and bark, but it also consumes seeds when available. Like other sciurids, it is opportunistic and includes animal material (when available) in its diet. The European ground squirrel is diurnal, concentrating most of its activity in the middle of the day, but it will also show a midday drop in activity in response to high ambient temperatures. This obligate hibernator constructs well-developed and complex burrow systems, excavated with forelimbs that are highly specialized, even for a ground squirrel. This species is somewhat social and polygynous; males appear to assist in the excavation of burrows later occupied by their mates and their young. However, males that exhibit this parental investment are often individuals with lower mating success. As has been observed in many ground squirrels, earlier litters are larger than those produced later in the season. In *S. citellus*, this is apparently due to differences in maternal investment. Females show signs of two periods of estrus: one soon after emergence, and another later in summer that does not result in reproduction, because males are sexually inactive at that time. It is suggested that the hormonal changes associated with the second estrus may increase fat storage for hibernation and thus allow females to enter estrus soon after their spring emergence. Populations of *S. citellus* undergo significant cycling, and they sometimes crash; densities vary from a low of 5 to as many as 60 animals/ha. During periods of low population density, male aggression is reduced, juveniles are able to reproduce, home ranges increase in size, and litter sizes are larger. In contrast, at high densities male aggression is increased, reproductive success among males is more variable, the animals lose mass, and emigration rates increase. Early reproduction appears to correspond with lower growth rates, higher levels of aggression, and possibly reduced health. Other than three species of *Eimeria*, few parasites are reported from this host. Evidence of fragmentation, genetic isolation, and inbreeding suggest that many populations are in peril, especially on the periphery of this species' range.

GENERAL REFERENCES: Aschauer et al. 2006; Çolak and Özkurt 2002; Coroiu et al. 2008; Everts et al. 2004; Golemansky and Koshev 2007; I. E. Hoffman et al. 2003; Huber, Hoffmann, et al. 2001; Huber, Millesi, and Dittami 2002; Huber, Millesi, Walzl, et al. 1999; Hulová and Sedlácek 2008; Koshev 2008; Lagaria and Youlatos 2006; Millesi, Hoffmann, and Huber 2004; Millesi, Hoffmann, Steurer et al. 2002; Millesi, Huber, Dittami et al. 1998; Ruediger et al. 2007; Millesi, Huber, Everts, et al. 1999; Millesi, Huber, Pieta, et al. 2000; Millesi Strijkstra, et al. 1999; Ozkurt et al. 2007; Strauss, Mascher, et al. 2007; Váczi et al. 2006.

Spermophilus dauricus Brandt, 1843
Daurian Ground Squirrel

DESCRIPTION: *S. dauricus* is a small ground squirrel with a short tail. The pelage is buffy or grayish brown, with no dorsal spots or lines. There is a black ring near the distal end of the tail, and the tip is light yellow.

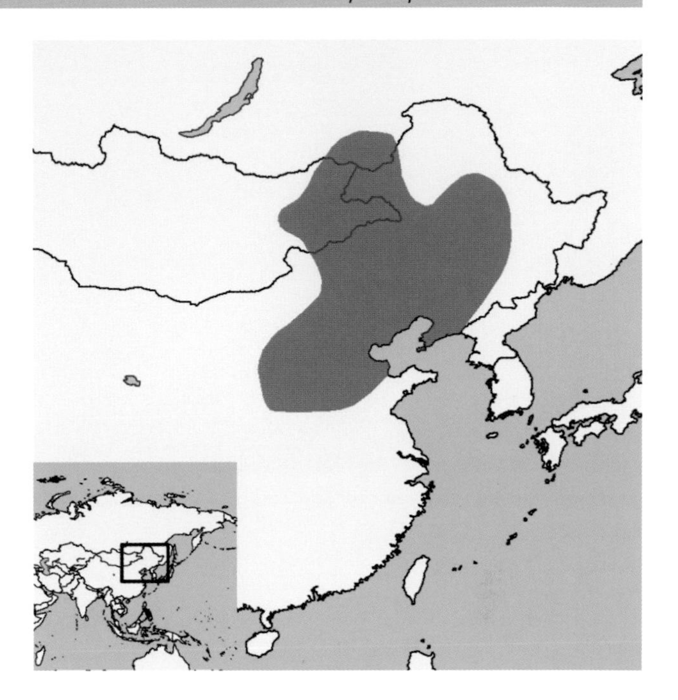

SIZE: Female—HB 190.2 mm; T 55.3 mm.
Male—HB 191.4 mm; T 63.2 mm.
Sex not stated—HB 196.1 mm; T 59.1 mm; Mass 223.8 g.

DISTRIBUTION: This species' range extends from Transbaikal (Russia) into Mongolia and northeastern China.

GEOGRAPHIC VARIATION: We consider this species to be monotypic, but three subspecies are sometimes recognized: *S. d. dauricus*, *S. d. mongolicus*, and *S. d. ramosus*.

CONSERVATION: IUCN status—least concern. Population trend—no information.

HABITAT: *S. dauricus* is found in open plains, steppes, and deserts. It occurs along the northern edge of the Gobi Desert into northern Mongolia and Siberia, and along the eastern edge of the Gobi into China, where it is found in nine provinces.

NATURAL HISTORY: The diet of the Daurian ground squirrel consists mostly of plant material and seeds, including some grains and other crops. The squirrels hibernate in simple underground burrows that are usually just a few meters in length, with nests that are less than 1 m belowground. One litter is produced annually (in the spring), ranging in size from two to nine young. Theoretical life-table analyses of *S. dauricus* from northeastern China and the Columbian ground squirrel (*Urocitellus columbianus*) show that the two species have similar life-history traits; further, these analy-

Spermophilus dauricus. Photo courtesy Susan Fox, www.fox studio.biz.

ses suggest a tradeoff between fecundity and survival in both species. The Daurian ground squirrel is an efficient hibernator, dropping its body temperature to 4.8°C and employing nonshivering thermogenesis. There is strong genetic differentiation between *S. dauricus* and three other species of eastern palaearctic ground squirrels (*S. relictus, Urocitellus parryi,* and *U. undulatus*).

GENERAL REFERENCES: Cai et al. 1992; Gündüz et al. 2007a; Q. Li et al. 2001; Luo and Fox 1990; A. T. Smith and Xie 2008; Tsvirka, Spiridonova, et al. 2008; S. Q. Wang et al. 2002.

Spermophilus erythrogenys Brandt, 1841 Red-Cheeked Ground Squirrel

DESCRIPTION: The color of the back ranges from pale yellow straw gray with faint whitish speckling to a more saturated straw gray tinged with rust and patterned with straw yellow spots. The back of the head is gray brown or gray straw, and there is a broad chestnut brown spot beneath the eye.

SIZE: Female—HB 192.8 mm; T 46.1 mm.
Male—HB 187.8 mm; T 41.3 mm; Mass 335.0 g.
Sex not stated—HB 215.1 mm; T 46.5 mm; Mass 355.0 g.

DISTRIBUTION: This species is found in eastern Kazakhstan and southwestern Siberia (Russia). *S. erythrogenys* formerly included an isolated population from Mongolia and Inner Mongolia (China). In this publication, that isolated population, *S. pallidicauda,* is given full species status.

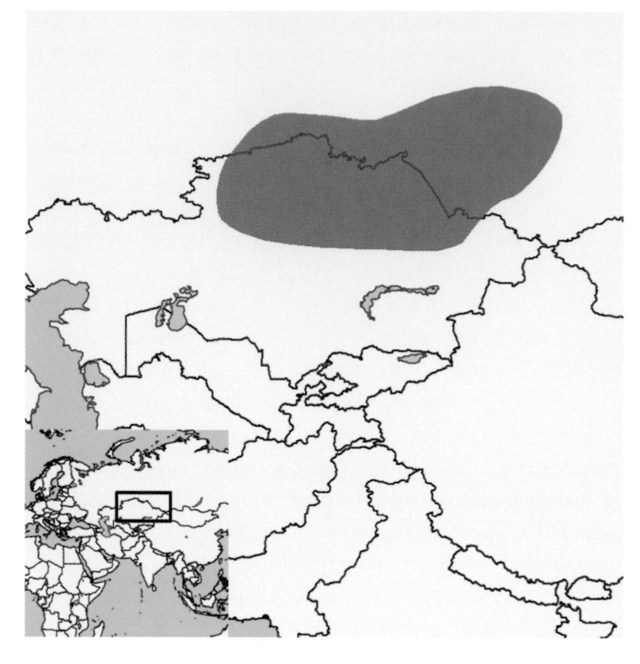

GEOGRAPHIC VARIATION: None.

CONSERVATION: IUCN status—least concern. Population trend—stable.

HABITAT: This squirrel is found in xeric plains, typically with dry steppe and semidesert vegetation, but in the north it lives on the edge of birch (*Betula*) and European aspen (*Populus tremula*) forests, while in the south it is found in mountains, at elevations up to 2100 m.

NATURAL HISTORY: The red-cheeked ground squirrel's diet includes cereals and other grains, but this species may also consume animal material, like most other sciurids. It hibernates from late August until late March-April in simple but deep burrows. Litter sizes are reported to vary from seven to nine. This species is typically common throughout its range, and it has large populations. This squirrel is considered susceptible to habitat loss as a result of activities related to agricultural, grazing, and nomadic livestock management practices, and it is also subject to hunting pressure. Results from an RAPD-PCR genetic analysis demonstrate that *S. erythrogenys* hybridizes with *S. major* in the region between the Tobol and Ishim rivers in northern Kazakhstan and southern Russia. Red-cheeked ground squirrels can host a number of diseases, including encephalitis and tularemia, a disease caused by the bacterium *Francisella tularensis,* which is transmitted to humans either by direct contact or by ticks, flies, and mosquito vectors. *S. erythrogenys* is host to at least

Spermophilus erythrogenys. Photo courtesy Yurii Danilov.

two species of coccidian parasites (*Eimeria berkinbaevi* and *E. callospermophili*).

GENERAL REFERENCES: Dawaa 1972; Rogovin and Shenbrot 1995; Shar and Lkhagvasuren 2008; Spirodonova, Chelomina, Starikov, et al. 2005; Spirodonova, Chelomina, Tsuda, et al. 2006; Wilber et al. 1998.

Spermophilus fulvus (Lichtenstein, 1823) Yellow Ground Squirrel

DESCRIPTION: This ground squirrel has brownish gold guard hairs and ash gray wool hairs.

SIZE: Female—HB 224.0 mm; T 71.4 mm.
Male—HB 284.3 mm; T 85.4 mm; Mass 290.0 g.
Sex not stated—HB 323.0 mm; T 102.0 mm; Mass 596.0 g.

DISTRIBUTION: The range of the yellow ground squirrel extends east from the Caspian Sea and the Volga River (Russia), across central Kazakhstan, to Lake Balkhash; and south through Uzbekistan, western Tajikistan, and Turkmenistan. Other populations are found in northeastern Iran and northern Afghanistan. A disjunct population occurs in western Xinjiang (China). At least four introductions outside this species' native range have been successful.

GEOGRAPHIC VARIATION: Three subspecies are recognized.

S. f. fulvus—region between the Caspian Sea and the Aral Sea. This form is as described above, with a buff coloration on the abdomen.
S. f. hypoleucos—northeastern Iran. This form has a white belly, with a slight buffy appearance.
S. f. oxianus—southern portion of the range. This is a smaller subspecies, similar in appearance to *S. f. fulvus*, but with the head slightly grayer.

CONSERVATION: IUCN status—least concern. Population trend—no information.

HABITAT: *S. fulvus* is found in desert and semidesert ecosystems.

NATURAL HISTORY: The yellow ground squirrel is primarily herbivorous, feeding largely on the rhizomes and tubers of desert plants; however, few details are available on its dietary habits. This species is considered territorial, defending a relatively large area in which it constructs a single burrow, where it hibernates from approximately mid-September to mid-May. *S. fulvus* is a relatively abundant and common species, although hunting pressure has contributed to its

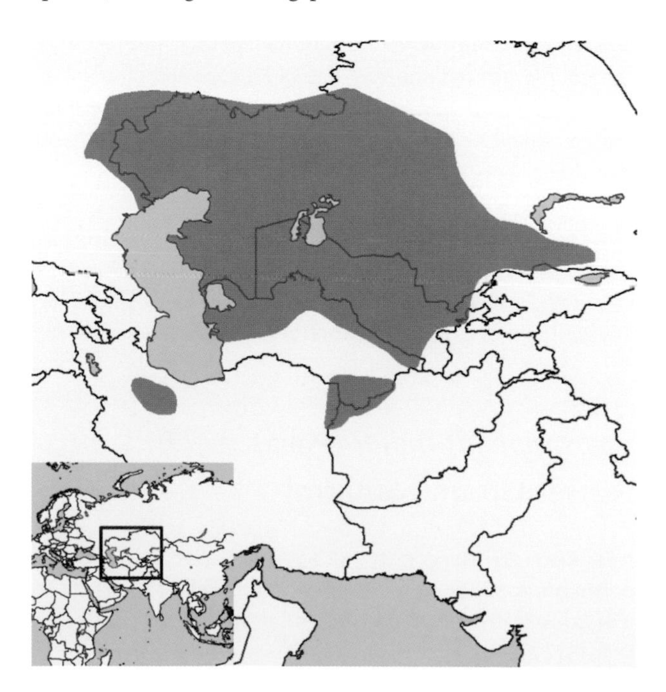

Spermophilus fulvus. Photo courtesy Sergey Cherenkov.

decline in some areas. The yellow ground squirrel is one of several rodents in the region—including the great gerbil (*Rhombomys opimus*), the midday gerbil (*Meriones meridianus*), and the little ground squirrel (*S. pygmaeus*)—that may maintain and transmit sylvatic (bubonic) plague. Interestingly, the yellow ground squirrel has been reported to use burrows of the great gerbil when available. *S. fulvus* hybridizes with *S. major*. At least seven species of coccidians (*Eimeria*) have been reported from *S. fulvus*.

GENERAL REFERENCES: Bertolino 2009; Ermakov, Surin, Titov, Tagiev, et al. 2002; Gage and Kosoy 2005; Herron, Castoe, et al. 2004; Kashkarov and Lein 1927; Özkurt et al. 2007; Tsytsulina, Formozov, and Sheftel 2008; Wilber et al. 1998.

Spermophilus major (Pallas, 1779) Russet Ground Squirrel

DESCRIPTION: Russet ground squirrels have a rather dark ochre brown dorsum mottled with light whitish rust spots. The tops of the head and the snout are often gray. A pronounced yellow rust spot occurs below each eye. The sides are ochre to rust. The venter is whitish to yellow to ochre. The tail is ochre brown above and rust below.

SIZE: Sex not stated—HB: 275 mm (253-320 mm); T 83.1 mm (73-105 mm); Mass 500-570 g.

DISTRIBUTION: *S. major* occurs on the steppes between the Volga and Irtysh rivers (from Russia south into northern Kazakhstan), and it has been introduced to the northern Caucasus Mountains (Russia).

GEOGRAPHIC VARIATION: None.

CONSERVATION: IUCN status—least concern. Population trend—no information.

HABITAT: Russet ground squirrels are found on mixed grassy plains, including steppes and agricultural plains. They also occur in the forest-steppe zone and in more arid habitats. They are expanding their range along roads and rivers—using dams, grassy berms, and banks—and they may also be enlarging their distribution by anthropogenic introductions.

NATURAL HISTORY: This species is diurnal. Russet ground squirrels hibernate in burrows from mid-June (males) or August (females and juveniles) until spring emergence some 6.5-8.5 months later. Mating occurs soon after emergence.

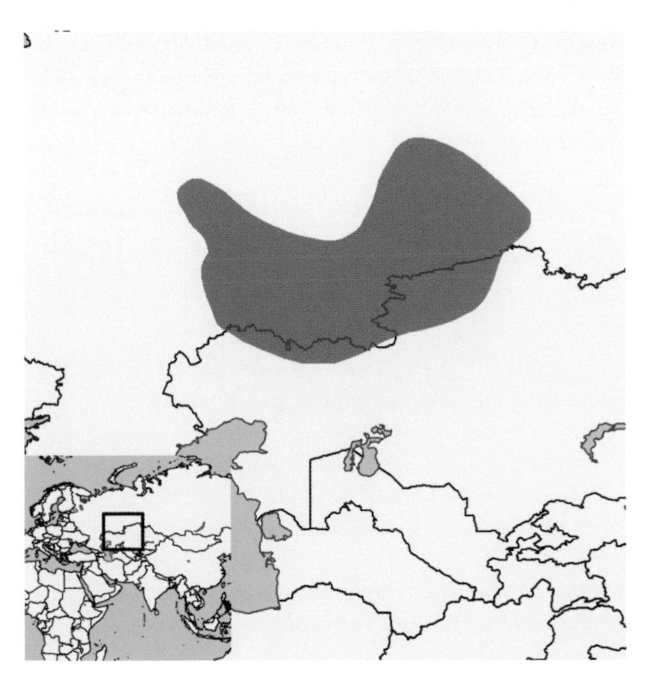

Spermophilus major. Photo courtesy Sergey Titov.

The young are born in spring within the burrow, and they emerge in late spring. Litter size is large, averaging from seven to eight. Dominance-subordinate relationships are well defined, and the most common interaction is avoidance behavior. Aggression occurs in about 25 percent of the interactions. Their social and mating system is one of territorial polygyny, with weakly expressed intermale competition. Females are less mobile and are found in small territories within the home range of males; as a result, females are behaviorally monogamous. Females tend to settle in or near their natal area and adjacent to their mothers, whereas males tend to disperse farther. Russet grounds squirrels feed on the leaves and seeds of forbs and grasses, some bulbs and roots, and cereal grains (millet, oats, wheat) where available; they opportunistically eat ground-nesting bird eggs and nestlings. Nest burrows are typically 1-2 m in length and depth, with a single entrance and nest chamber being most common; however, short shallow burrows are used for escape. Maternity burrows may be more circuitous and have a second entrance. Alarm calls are relatively short and high pitched. Raptors appear to be their primary predators. *S. major* is hunted and trapped in some areas, primarily to reduce crop damage, but also for pelts.

GENERAL REFERENCES: Ermakov and Titov 2000; Titov 2003a, 2004; Titov et al. 2008.

Spermophilus musicus (Ménétries, 1832) Caucasian Mountain Ground Squirrel

DESCRIPTION: Caucasian Mountain ground squirrels are brownish gray on the dorsum, with small faint rust gray spots. The flanks are light brownish gray with a suffusion of yellow. The head is dark brownish gray, with straw gray to tan cheeks. The eye rings are straw yellow. The venter is whitish gray to straw yellow. The tail is a grayish brown on the dorsal surface; the underside of the tail is pale, sometimes with a suffusion of rust; the tip is dark brown to black, sometimes frosted in white. Similar chromosome morphology, molecular and genetic features, and vocalizations suggest that *S. musicus* may be conspecific with *S. pygmaeus*.

SIZE: Sex not stated—HB 220 mm (205-240 mm); T 34-50 mm.

DISTRIBUTION: This species is endemic to the northern Caucasus Mountains (Russia, just north of Georgia).

GEOGRAPHIC VARIATION: None. Helgen et al. propose this species as a subspecies of *S. pygmaeus*.

Spermophilus musicus. Photo courtesy Mikhail Golubev.

CONSERVATION: IUCN status—near threatened. Population trend—no information.

HABITAT: Caucasian Mountain ground squirrels occupy moist or xerophytic alpine meadows, pastures, grass steppes, and cereal-grain fields.

NATURAL HISTORY: This species is diurnal. Caucasian Mountain ground squirrels hibernate in burrows, beginning in late August and September for adults, and September or October for juveniles; emergence occurs from March to May depending on the elevation. Individuals mate soon after emergence; after a 22-day gestation, a litter of two to four young is born in the burrow. Burrows typically have short main passages with a single nest chamber, less than 0.5 m deep, under a rock or shrub. Complex burrows with long bifurcating tunnels, several nest chambers, and multiple exits may be used for maternity or hibernation. This species feeds on shoots, flowers, seeds, leaves, and the bulbs of forbs and grasses, rarely consuming animal matter. Overwinter mortality averages about 40 percent. Foxes, domestic dogs, mustelids, and raptors are their principal predators. The major threat to their conservation is the conversion of alpine meadows for grazing. *S. musicus* is occasionally hunted for food or pelts, and it can be a localized pest in grain fields; however, the greatest impact of this species is as a carrier of sylvatic (bubonic) plague.

GENERAL REFERENCES: Emelianov 1983; Emelianov et al. 1982; Ermakov, Titov, et al. 2006; Helgen et al. 2009; Nikol'skii et al. 2007; Tkachenko et al. 1985; Trufanov and Golubev 1982.

Spermophilus pallidicauda (Satunin, 1903) Pallid Ground Squirrel

DESCRIPTION: Pallid ground squirrels have a pale dorsum that is pinkish buff to straw-colored. The eyelids are white. A reddish brown spot below the eye is connected to the russet snout and the back of the neck by a faint white to buff line. The feet and limbs are white to buff. The venter is white to buff. The tail is white to straw yellow, with a rusty core through the distal portion.

SIZE: Sex not stated—HB 198-233 mm; T 35-53 mm.

DISTRIBUTION: This species is endemic to Mongolia and adjacent eastern Xinjiang and Inner Mongolia (northern China).

GEOGRAPHIC VARIATION: None.

CONSERVATION: IUCN status—least concern. Population trend—no information.

HABITAT: Pallid ground squirrels inhabit the steppes and grasslands of the Gobi Desert.

NATURAL HISTORY: *S. pallidicauda* is diurnal. Pallid ground squirrels live in large colonies of varying densities. They are extremely shy and quickly return to their burrows when

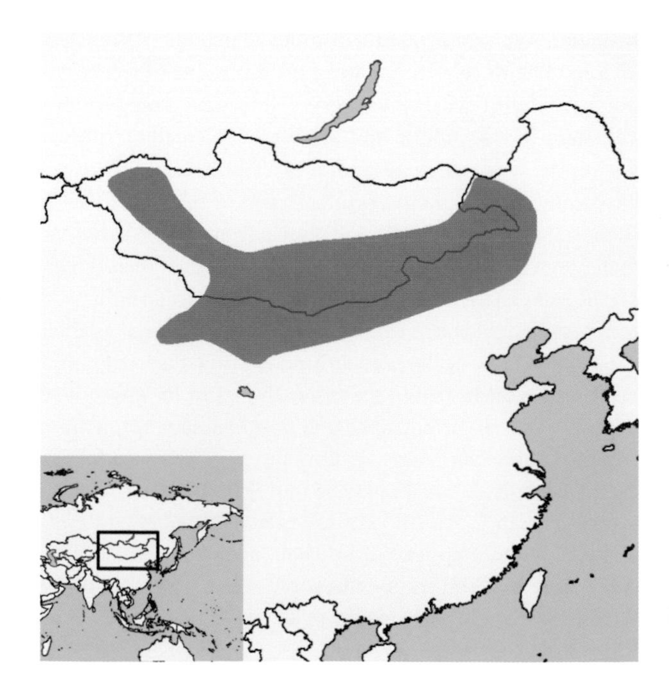

Spermophilus pallidicauda. Photo courtesy Oleg Brandler.

threatened. The greatest challenge to this species' conservation is likely to be overgrazing in the semiarid to arid grasslands they inhabit. They are also occasionally hunted for food and pelts. Little is known about *S. pallidicauda*, which has recently been elevated from its former status as a subspecies of *S. erythrogenys*.

GENERAL REFERENCES: G. M. Allen 1940; A. T. Smith and Xie 2008; Tsvirka, Chelomina et al. 2006.

Spermophilus pygmaeus (Pallas, 1778) Little Ground Squirrel

DESCRIPTION: The dorsum varies from a pale gray, suffused with straw yellow, to a brownish gray. The faint dorsal spots may be greatly reduced or even absent. The head is often brighter than the dorsum, with a reddish spot above the eye. The sides are a pale straw yellow suffused with gray. The feet are white to straw yellow. The tail varies greatly, from white to buff to brown to charcoal.

SIZE: Sex not stated—HB 175-260 mm; T 25-50 mm.

DISTRIBUTION: This species is found in southeastern Ukraine (including the Crimean peninsula) through the southern Ural Mountains (Russia) and Kazakhstan.

GEOGRAPHIC VARIATION: Four subspecies are recognized.

S. p. pygmaeus—lower Volga River and the Ural Mountains (Russia). See description above.

S. p. brauneri—Crimean peninsula (Ukraine). This subspecies is pale in color; the dark tail has a rufous cast.

S. p. herbicolus—central portion of the range. This form tends to have rust on the head and tail.

S. p. mugosaricus—eastern portion of the range, including Kazakhstan. This subspecies is paler, with cinnamon to ochraceous spotting.

CONSERVATION: IUCN status—least concern. Population trend—decreasing.

HABITAT: Little ground squirrels are found in sparsely vegetated arid and semiarid grasslands.

NATURAL HISTORY: This species is diurnal. *S. pygmaeus* hibernates in burrows for five to eight months a year. Males emerge first in the spring, often in March or April, followed by the females. Males actively pursue females and attempt to repel other interested males. Gestation is 25 to 26 days. Litter sizes typically average six to eight offspring, which are born underground. Due to their hot arid environments, individuals may enter torpor in the heat of June and July, and then reemerge or transition from torpor to hibernation. Males are the first to immerge, in August. Short shallow burrows without nest chambers are used for escape; longer and more convoluted burrows, which may reach 1 m in depth and have one or more nest chambers, are used for maternity and hibernation. These squirrels live in colonies with intrasexual territoriality, where females have small territories that are defended from other females, and males have large territories that overlap those of several females. *S. pygmaeus* feeds on a variety of plant tissues, including leaves, shoots, and seeds, as

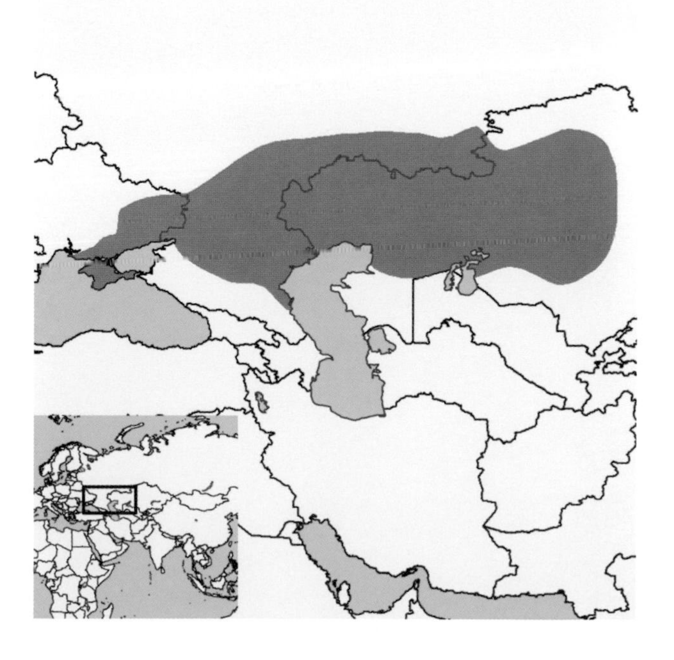

Spermophilus pygmaeus. Photo courtesy Sergey Levykin and Grigoriy Kazachkov.

well as underground stems, roots, and bulbs. Raptors and foxes appear to be their major predators. Their high-pitched alarm calls are modestly complex. Little ground squirrels are known to feed heavily on pasture grasses and vegetable crops, and individuals are shot and poisoned to remove localized damage.

GENERAL REFERENCES: Gladkina and Skalinov 1987; Nikol'skii 2007a; Ognev 1963.

Spermophilus ralli (Kuznetsov, 1948) Tien Shan Ground Squirrel

DESCRIPTION: Tien Shan ground squirrels have a grayish brown to grayish yellow dorsum. The sides are lighter in color, grading to a straw gray venter. The faint spots on the dorsum are often hard to detect. The tail is yellow to light rust, with a dark brown to black band and a white to yellow tip.

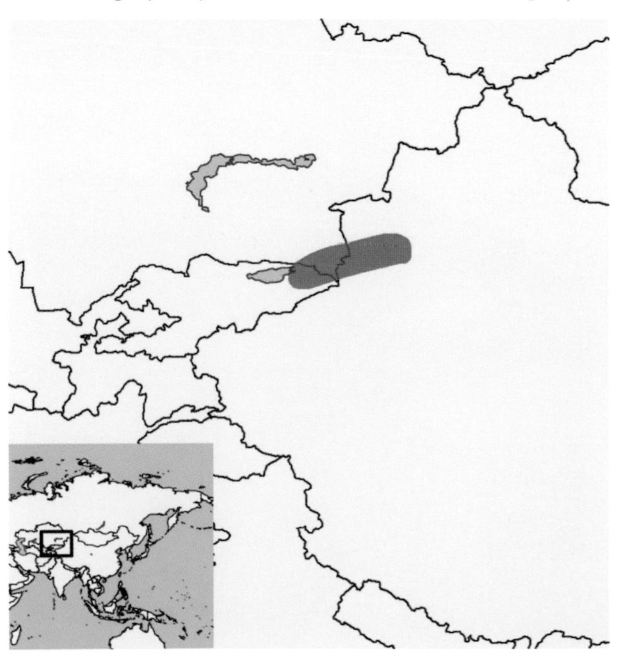

SIZE: Sex not stated—HB 200-240 mm; T 60-75 mm; Mass 290-405 g.

DISTRIBUTION: This species is endemic to the eastern Tien Shan Mountains of Kyrgyzstan, Kazakhstan, and China.

GEOGRAPHIC VARIATION: None.

CONSERVATION: IUCN status—least concern. Population trend—no information.

HABITAT: Tien Shan ground squirrels inhabit meadows within their restricted range.

NATURAL HISTORY: This species is diurnal. Tien Shan ground squirrels hibernate within burrows from late summer (August-September) until late winter (February-March). Mating occurs after emergence in later winter or spring. A litter of three to seven is born in the burrow after a gestation of 25-27 days. *S. ralli* forages on grasses, forbs, and insects. Many burrow openings are clustered together, with a hibernation chamber 1-2 m belowground. These animals may be seen calling at the burrow entrance; their call is a quiet "squeak." Tien Shan ground squirrels are harvested for food. Long considered a subspecies of *S. relictus*, recent genetic evidence caused *S. ralli* to be elevated to a species instead.

GENERAL REFERENCES: R. G. Harrison et al. 2003; Ognev 1963; A. T. Smith and Xie 2008.

Spermophilus relictus (Kashkarov, 1923) Relict Ground Squirrel

DESCRIPTION: Relict ground squirrels have a dorsum that is a mix of gray, cinnamon, and straw yellow, with indistinct speckles. The head is slightly paler, a gray suffused with straw. The sides and the venter are light gray to straw yellow. The tail is cinnamon grayish yellow at the base and darkens near the tip to a grizzled dorsal appearance, with the underside more cinnamon to rust; the tail is frosted with white to cream.

SIZE: Both sexes—HB 242 mm (230-270 mm); T 69.8 mm (58-76 mm).

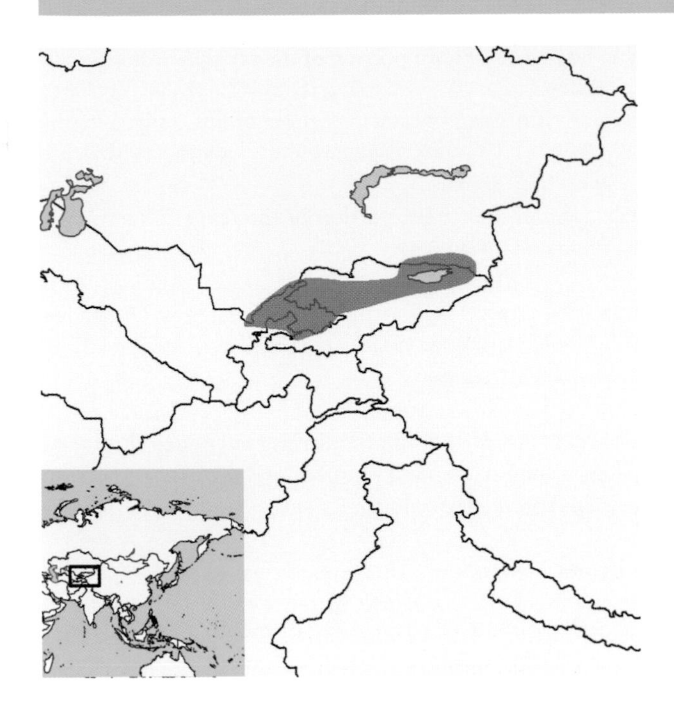

DISTRIBUTION: This species is found in the Tien Shan Mountains in eastern Uzbekistan, Kyrgyzstan, and southeastern Kazakhstan.

GEOGRAPHIC VARIATION: None.

CONSERVATION: IUCN status—least concern. Population trend—no information.

HABITAT: Relict ground squirrels are found in mountain steppes, piedmont steppes, and meadows of mixed grasses with good drainage.

NATURAL HISTORY: This species is diurnal. Relict ground squirrels hibernate within burrows from late summer (August-September) until late winter or spring (February-early May). A litter of three to six is born in the burrow, with the number of young greater at lower elevations. *S. relictus* forages on roots, bulbs, grasses, forbs, and flowers; insects can also be consumed in substantial numbers when available. Their burrow structure is relatively simple, most often a single tunnel with a nest chamber. Burrow openings are clustered together, with a hibernation chamber 1-2 m belowground; summer burrows, used primarily for escape, are shallower (< 1.0 m). *S. relictus* is relatively shy and quickly seeks the cover of burrows after detecting a potential threat; these squirrels can also climb trees. Their call is a quiet low "squeak." Relict ground squirrels are harvested for food and pelts.

***Spermophilus relictus*.** Photo courtesy Cyril Ruoso.

GENERAL REFERENCES: R. G. Harrison et al. 2003; Ognev 1963; Tsvirka, Spiridonova, et al. 2008; Volzheninov et al. 1986.

Spermophilus suslicus (Güldenstaedt, 1770) Speckled Ground Squirrel, Suslik

DESCRIPTION: Speckled ground squirrels have rich auburn chestnut brown to light grayish brown coats, often suffused with ginger. The coats have small light spots that range from white to buff; often these spots are very pronounced, but

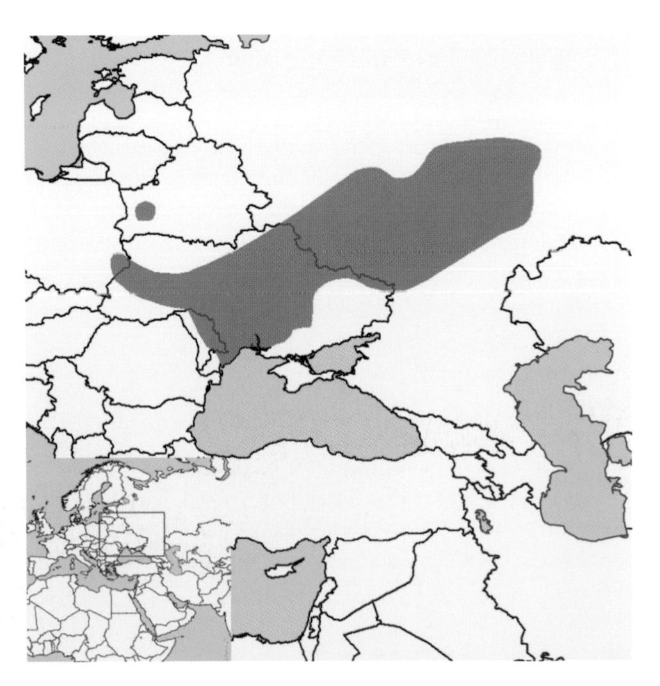

***Spermophilus suslicus*.** Photo courtesy Ilya Volodin.

they can be faint in some forms. The head has small spots; the chin and the eye rings are white to buff. There is a chestnut brown spot below each eye. The chest and front legs are cinnamon to ochre, with the venter being a lighter shade. The tail is grizzled dark brown to black near the base, sometimes suffused with reddish; the underside is paler. The tail is frosted with white to straw yellow at the tip.

SIZE: Both sexes—HB 180-260 mm; T 36-57 mm; Mass 180-220 g.

DISTRIBUTION: This species is found on the steppes of eastern and southern Europe, including Poland, Ukraine, Russia north to the Oka River and east to the Volga River, and Moldova. There is also an isolated population in western Belarus.

GEOGRAPHIC VARIATION: Three subspecies are recognized.

S. s. suslicus—southern portion of the range. See description above.

S. s. boristhenicus—western portion of the range, in the Ukraine. This is a rather brightly colored subspecies, with faint spots.

S. s. guttatus—northern portion of the range. This subspecies is relatively dark.

CONSERVATION: IUCN status—near threatened. Population trend—decreasing. *S. suslicus* is considered endangered in Poland and Moldova.

HABITAT: Speckled ground squirrels prefer open shortgrass habitats, such as steppes, pastures, and roadsides. They can be found on or near cultivated fields.

NATURAL HISTORY: This species is diurnal. Speckled ground squirrels hibernate in burrows from August and September to February and March. Mating occurs during a brief period in spring, just after their emergence from hibernation; significant aggression and well-defined dominant-subordinate relations are evident at this time. After a gestation of 22-27 days, a litter of three to eight young is born within the burrow; it will be the only one produced for the year by the female. The young emerge in June and July. Speckled ground squirrels may share territories, with the males often covering larger areas than the females. The social and mating system of this species may be monogamous at low densities, but it appears to be polygynous when females aggregate their territories. Territories range from 0.013 to 0.055 ha, with those of the females slightly smaller than those of the males, especially when females are nursing their young, and in early spring, when males are searching for females. Juveniles disperse before their first hibernation, and males in particular move far from their natal area. Short and shallow escape burrows are common; deep (1-2 m) and more complex maternity and hibernation burrows are less so. Their diet consists primarily of grasses and cereal grains, with insects and other animal matter occasionally eaten. Overwinter mortality can be as high as 70 percent, but it is variable, depending on winter temperatures and snowpack. More than half of the juveniles perish before the end of their first year, and mortality is highest for late-born young. Infanticide is known to occur in *S. suslicus*. Adult survival is higher, and some individuals can live for up to 6 years. Many populations appear to show evidence of genetic bottlenecks in the past, perhaps due to the highly variable demography of this species. Significant predators of speckled ground squirrels include mustelids, foxes, and raptors. In response to potential predators, individuals give alarm calls that are highly variable, weakly modulated, high-pitched, and powerful. Juvenile calls cannot be easily distinguished from those of adults. Major threats to persistence of this species are the conversion of land to agriculture, urbanization, and removal by shooting and poisoning. *S. suslicus* is often considered a pest when in the vicinity of crops (especially grain). Western populations have a different karyotype ($2n = 36$), and this chromosomal difference may lead in the future to the separation of *S. suslicus* into two distinct species, in order to reflect this.

GENERAL REFERENCES: Biedrzycka and Konopinski 2008; Matrosova et al. 2009; Shekarova et al. 2008; Titov 2003a, 2003b; Trunova and Lobkov 1997.

Spermophilus taurensis (Gündüz et al., 2007) Taurus Ground Squirrel

DESCRIPTION: Taurus ground squirrels have a straw yellow dorsum with a slightly grizzled appearance, due to a fine flecking of dark brown to black with a light suffusion of red. The chin and cheeks are buff; there are small buff postauricular patches. The feet are pale buff. The venter is white to gray. The tail is relatively broad. The dorsal coloration of the tail is the same as the body, and the underside is straw to reddish; the black tail tip is distinctive.

SIZE: Female—HB 201 mm; T 61 mm; Mass 201 g.

Spermophilus taurensis. Photo courtesy Mustafa Sozen.

DISTRIBUTION: This species is restricted to the Taurus Mountains of southern Turkey.

GEOGRAPHIC VARIATION: None.

CONSERVATION: IUCN status—least concern. Population trend—no information.

HABITAT: This species is found in montane habitats, at elevations above 1000 m.

NATURAL HISTORY: *S. taurensis* is diurnal. Ecological information is lacking on Taurus ground squirrels, due to their recent description. This species diverged from *S. citellus,* with which it shares a similar karyotype ($2n = 40$), some 2.5 MYA. The range of *S. taurensis* is small (17,255 km^2), and the Taurus ground squirrel has low genetic diversity. This montane endemic may need to be conserved, because ground squirrels in the Taurus Mountains have no legal protection and are viewed as pests.

GENERAL REFERENCES: Gündüz et al. 2007a, 2007b.

Spermophilus xanthoprymnus (Bennett, 1835) Asia Minor Ground Squirrel

DESCRIPTION: Asia Minor ground squirrels have a brownish straw-colored dorsum that lacks spots but can appear a bit grizzled, because of black bands on individual hairs. The snout and the cheeks are grayish and faint yellow. The eye rings are pale buff to cream colored. The sides, the venter,

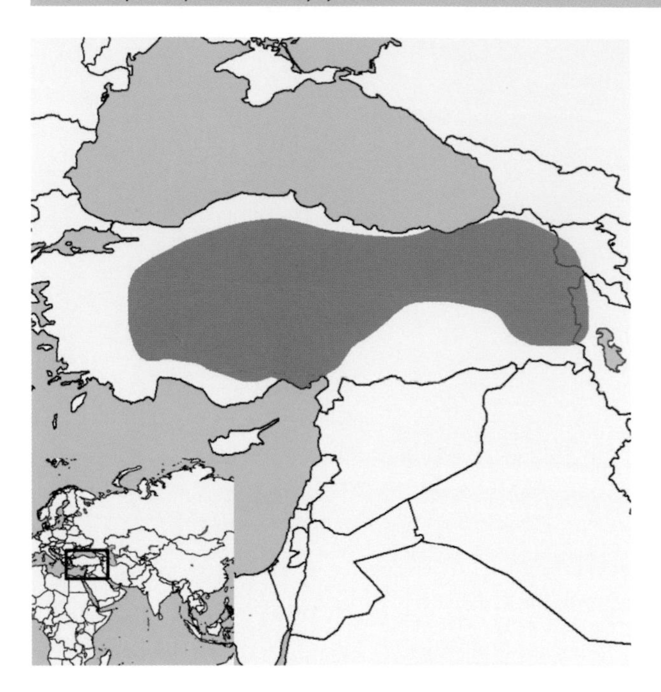

Spermophilus xanthoprymnus. Photo courtesy Ebru Diker.

and the limbs are dull yellow with pinkish buff. The tail is a dull straw rust on the upper surface and lighter, with a suffusion of rust, below.

SIZE: Female—Mass 172-174 g at emergence.
Male—Mass 272-312 g at emergence.
Both sexes—HB 145-233 mm; T 33-35 mm.

DISTRIBUTION: This species is found in Turkey, extreme western Armenia, and northwestern Iran.

GEOGRAPHIC VARIATION: None.

CONSERVATION: IUCN status—near threatened. Population trend—decreasing.

HABITAT: Asia Minor ground squirrels inhabit areas with shortgrass steppe vegetation, but they can also be found on rolling rocky mountain slopes and in association with agricultural grain fields.

NATURAL HISTORY: This species is diurnal. Asia Minor ground squirrels hibernate in underground burrows from mid-August for adults, and a month later for juveniles. Males emerge about 10 days earlier than females, in February or March, and in much better body condition. Mating occurs three to six days after the females' emergence, and the season lasts for three to four weeks. Mating occurs belowground, and a copulatory plug forms in the reproductive tract of the female, due to the coagulation of semen. Physical combat among males occurs during the breeding season. One litter is produced each year, beginning with the second year of life. The young are born in the burrow after a gestation of 25-26 days and emerge from April to July, about 30 days after birth. The litter size of *S. xanthoprymnus* averages 5.8 (range = 4-9) in captivity and 4.7 (range = 3-9) in the wild. Juveniles appear to disperse from their natal area, and individuals live alone in a burrow within a larger colony. The sex ratios of adults are female-biased, suggesting a high male mortality rate due to male-biased natal dispersal and intermale aggression. Raptors are this species' principal predators. Asia Minor ground squirrels have a loud whistle, given when they are startled, and they seek burrows for refuge. Burrows often have a rock near the entrance. Burrows may drop precipitously and then wind to a terminal nest chamber, which is often lined with grass and forbs. Short shallow burrows may be used for escape cover. Asia Minor ground squirrels are herbivores and feed on bulbs, roots, forbs, seeds, and grain crops. They are not hunted or trapped, probably due to their small size, but they can be a localized pest of grains and other crops.

GENERAL REFERENCES: Çolak and Özkurt 2002; Demirsoy et al. 2006; Gür and Barlas 2006; Gür and Kart Gür 2005; Kart Gür et al. 2009; Özkurt et al. 2007.

Tamias Illiger, 1811

This genus contains 25 species. Chipmunks are commonly considered to be either a single genus, *Tamias*, or three genera—*Tamias*, *Eutamias*, and *Neotamias*. In the latter case, *Tamias* would contain the eastern chipmunk, *Eutamias* would consist of the Siberian chipmunk, and *Neotamias* would include the remaining 23, mostly western, species. Here we treat the chipmunks as a single genus, *Tamias*.

Tamias alpinus Merriam, 1893
Alpine Chipmunk

DESCRIPTION: This small chipmunk is pale in coloration. The venter is orangish, and the back has four whitish stripes. The lateral stripes are wide and white, but often pale. *T. alpinus* is smaller than all sympatric (or parapatric) species (e.g., *T. amoenus*, *T. quadrimaculatus*, *T. senex*, *T. speciosus*, and *T. umbrinus*), except *T. minimus*. *T. alpinus* is distinguished from *T. minimus* by the former's shorter tail and larger ears and skull.

SIZE: Female—HB 106.4 mm; T 75.5 mm.
Male—HB 104.6 mm; T 69.5 mm.
Sex not stated—HB 105.3 mm; T 78.5 mm; Mass 35.8 g.

DISTRIBUTION: This species' range is limited to a narrow portion (less than 20,000 km^2) of the higher elevations of the Sierra Nevada in central California, from Tuolumne to Tulare counties (USA).

GEOGRAPHIC VARIATION: None.

CONSERVATION: IUCN status—least concern. Population trend—stable.

HABITAT: *T. alpinus* is found in the Sierra Nevada of California (USA), at elevations between 2300 and 3900 m, primarily in the alpine life zone (more than 3000 m in altitude), but also in the Hudsonian and Canadian life zones. It is a highly insular species, whose prevalence and occurrence probably depend on the physical structure of its habitat. This consists primarily of meadows, talus slopes, boulder fields, open stands of lodgepole pine (*Pinus contorta*), and small patches of whitebark pine (*P. albicaulis*). *T. alpinus* spends considerable time in rocky habitats, which may allow a quick and efficient escape from predators. Their habitat is generally xeric, as it is exposed to full sun, although maximum daily temperatures do not exceed 20°C.

NATURAL HISTORY: Alpine chipmunks will consume some pine seeds, but they depend more heavily on the smaller seeds of sedges, grasses, and forbs found throughout their habitat. Most information on their diet is based on details recorded from the cheek pouches of a few individuals; the cheek pouches of one animal contained approximately 5000 seeds. It is assumed that this species larder-hoards, like other species of *Tamias*; however, early natural-history observations suggest scatter-hoarding behavior as well. *T. alpinus* is also considered a predator of bird eggs and nestlings, and this species may limit the breeding range of the Asian rosy finch (*Leucosticte arctoa*). No nests have been reported for alpine chipmunks. The range of *T. alpinus* overlaps with those of five other species of *Tamias*, but alpine chipmunks are fully sympatric with only one, *T. speciosus*. Two species of lice (*Neohaematopinus pacificus* and *Hoplopleura arboricola*) and one species of mite (*Ornithonyssus sylviarum*) parasitize it; no predators are reported, although the short-tailed weasel (*Mustela erminea*) is known from the habitat of *T. alpinus*. Alpine chipmunks are territorial; aggressive displays are common toward conspecifics and often include a range of vocalizations (e.g., high-pitched calls, and lower-pitched "chucks" and "chips") as well as vertical tail displays. They are reported to enter hibernacula in late October and emerge

Tamias alpinus. Photo courtesy Oleksandr Holovachov.

in June, while their habitat is still covered in snow. This allows time for territorial establishment and subsequent foraging during the limited growing season. Alpine chipmunks produce young in early summer; nestlings are active by late July-early August, and young-of-the-year are reported to reach adult size by October. Litter size is between four and five young. The distribution of *T. alpinus* at lower elevations is limited by territorial exclusion by its larger congener, *T. speciosus*, which is reported to be more aggressive, more successful in territorial disputes in the wild, and more efficient at foraging in the dense vegetation where the two species are sympatric. In captivity, however, *T. alpinus* appears more aggressive and dominant over *T. speciosus*, which suggests that the outcome of aggressive encounters in the wild are in part influenced by habitat structure.

GENERAL REFERENCES: Clawson et al. 1994a; R. M. Davis et al. 2008; Hayssen 2008a, 2008b, 2008c; A. W. Linzey and Hammerson 2008e; Piaggio and Spicer 2000.

Tamias amoenus J. A. Allen, 1890
Yellow-Pine Chipmunk

DESCRIPTION: *T. amoenus* is a small member of the genus. The dorsal surface is reddish brown and dark in appearance. The five dark stripes are usually black. The lateral dark stripes are each bordered by a white stripe. The more median light stripes are grayish. Females are larger than males, and *T. amoenus* in general is smaller in body mass than *T. speciosus*. It is suggested that genital bones may be necessary to distinguish *T. amoenus* from several other species.

SIZE: Female—HB 123.2 mm; T 90.3 mm; Mass 50.6 g.
Male—HB 119.3 mm; T 86.5 mm; Mass 58.3 g.
Sex not stated—HB 119.3 mm; T 95.5 mm; Mass 43.0 g.

DISTRIBUTION: The yellow-pine chipmunk ranges from central British Columbia and the southwestern edge of Alberta (Canada) south to central California and east to central Montana and northwestern Wyoming (USA). It also occurs in a limited number of locations in western and northern Nevada.

GEOGRAPHIC VARIATION: Fourteen subspecies are recognized. The subspecies exhibit subtle differences in coloration. Genetic studies reveal considerable variation within this species and show that two subspecies (*T. a. canicaudus* and *T. a. cratericus*) are not grouped with the 12 other clades, some of which are not consistent with current subspecific designations.

T. a. amoenus—USA, in the "transition and boreal zones from northwestern California north through central and eastern Oregon and Washington" (Anthony). The five dark lines on the back are black, well defined, and

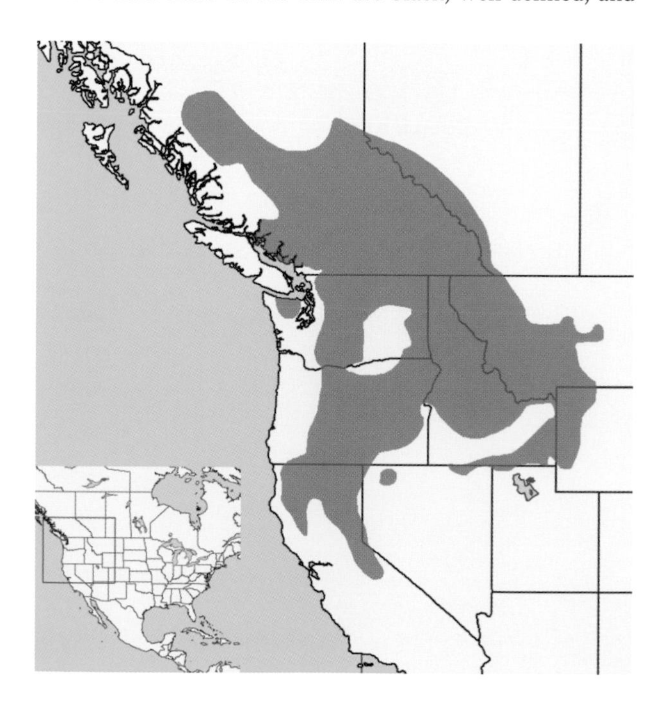

***Tamias amoenus*.** Photo courtesy A. Coke Smith, www.cokesmithphototravel.com.

sprinkled with rufous. The two inner light stripes are grizzled gray, and the outer pair of light stripes are white. The head is mixed gray, rufous, and black; the ear patch is dark.

T. a. affinis—Canada, in the "transition and Canadian zones of southern British Columbia" (Anthony). This form is indistinguishable in appearance from *T. a. amoenus*.

T. a. albiventris—type locality is on the boundary between Asotin and Garfield counties, Washington (USA). The upperparts are grayish, and the underparts are whitish or cream. The inner pair of light stripes are smoke gray and the outer pair of light stripes are whitish.

T. a. canicaudus—USA, in the "transition zone of eastern Washington" (Anthony). The tail is edged with gray. The general color on the dorsum is vinaceous gray, and the dorsal stripes are broad.

T. a. caurinus—Olympic Mountains, Washington (USA), up to the timberline. This form is similar to *T. a. amoenus*, but darker. The ear patch is reduced in size, and the dark stripes are black.

T. a. celeris—Pine Forest Mountains, Humboldt County, Nevada (USA). This is a smaller and paler form, similar to *T. a. monoensis*.

T. a. cratericus—Butte County, Idaho (USA). This is a dull gray form. The light stripes are all smoke gray, with the outer pair lighter.

T. a. felix—Mount Baker Range, British Columbia (Canada). There is a heavy suffusion of ochraceous on the sides, the cheeks, and the underside of the tail. The stripes are broad. The upperparts have a rusty brown tone.

T. a. ludibundus—Canada, "along the boundary line region between Alberta and British Columbia" (Hollister). This is the largest of the subspecies. The sides are dark and tawny. The underparts are yellowish.

T. a. luteiventris—"transition and Canadian zones from southern Alberta [Canada] south into Montana and Wyoming [USA]" (Anthony). This form is similar to *T. a. amoenus*, with a strong suffusion of ochraceous buff on the underparts.

T. a. monoensis—USA, "on the arid crest and east wall of the central Sierra Nevada (California), where it is characteristic of the Canadian zone" (Grinnell and Storer). This form is paler and grayer, with whiter light stripes.

T. a. ochraceus—USA, "only in the Siskiyou Mountain region of northern California and southern Oregon" (Howell). This form is larger than *T. a. amoenus*, and the head and the dorsum are more ochraceous. The dark dorsal stripes are less black.

T. a. septentrionalis—Canada, on the "north shore of Ootsa Lake, British Columbia" (Cowan). This is a large form. The inner pair of light stripes are reddish brown anteriorly, becoming paler posteriorly. The outer pair of light stripes are white with a faint reddish brown wash.

T. a. vallicola—USA, "confined to the Bitterroot Valley (Montana) and the adjacent foothills, but the exact limits of its range are not known" (Howell). This form is similar to *T. a. amoenus*, but the overall color is paler.

CONSERVATION: IUCN status—least concern. Population trend—stable.

HABITAT: The yellow-pine chipmunk primarily inhabits drier forests of the transition zone, but it may occur in the lower Canadian zone. In the far southern parts of its range, it is found over a broad span of elevations, up to about 3000 m. *T. amoenus* prefers meadows of forbs, grasses, and sedges or brushy habitats that are often associated with open stands of Jeffrey pine (*Pinus jeffreyi*) or lodgepole pine (*P. contorta*). Overall, *T. amoenus* is a habitat generalist and may coexist with other mammal species because of this flexibility. Multiple studies report significantly higher capture rates of *T. amoenus* in logged forests, regenerating stands, and clearcuts than in primary or mature stands, underscoring this species' preference for open habitat.

NATURAL HISTORY: Like many of its congeners, *T. amoenus* eats the seeds and fruits of many of the plant species in its habitat, as well as flowers, tubers, fungi, vegetation, and animal material (e.g., bird eggs and insects). It also consumes large quantities of hypogeous (underground) fungi, for which it probably serves as a significant spore dispersal agent. Although the yellow-pine chipmunk larder-hoards large quantities of seeds and other plant material for hibernation, numerous studies show that it regularly scatter-hoards seeds and may be important for the dispersal of several plant species (e.g., antelope bitterbrush [*Purshia tridentata*] and chinquapin [*Castanopsis sempervirens*]) and otherwise wind-dispersed pines (e.g. Jeffrey pine [*Pinus jeffreyi*] and sugar pine [*Pinus lambertiana*]). As suggested for other species of *Tamias*, a low availability of mast results in more time spent foraging and less time devoted to vigilance and other activities; this, in turn, may influence winter survival. Detailed field and laboratory studies by Vander Wall and associates also demonstrate how numerous factors (e.g., masting patterns, seed quality, cache-site selection, cache depth, risk of pilferage, olfactory capabilities, spatial memory) influence the scatter-hoarding decisions of *T. amoenus*, and, in turn, the potential for seed dispersal and establishment by this rodent.

Population densities average 1.25-7 animals/ha, depending on location and habitat quality. Like other members of the genus, *T. amoenus* is diurnal, but inactive during the winter months when hibernating. Although grassy nests are reported in aboveground vegetation, this species usually burrows belowground, often under stumps, logs, and rocks. Emergence from hibernation occurs by April, and mating takes place by early May. Paternity studies suggest a promiscuous mating system, with low variability in male mating success and multiple mates for both sexes. There is also strong evidence for sperm competition and little evidence for inbreeding. One litter is produced per year; litter sizes average four to five. *T. amoenus* is less able to deal with temperature extremes than other congeners. In *T. amoenus*, testosterone levels in plasma appear to reduce glucocorticoid levels, and thus fat reserves and overwinter survival. This species is aggressive toward other species of *Tamias*, but it is subordinate to *T. townsendii* and occasionally dominant over *T. minimus*. Such aggression may influence the elevational distribution of these species relative to that of *T. amoenus* in some parts of its range. A number of mammals and birds are reported to prey on yellow-pine chipmunks. Ectoparasites include the botfly (*Cuterebra*), at least seven species of fleas, one tick, and one phoretic mite; high flea burdens are reported. Surveys of endoparasites are not available, although this species' highly granivorous diet probably predisposes it to few internal parasites.

GENERAL REFERENCES: Anthony 1928; Briggs and Vander Wall 2004; Demboski and Sullivan 2003; Good, Demboski, et al. 2003; K. M. Kuhn and Vander Wall 2008; A. W. Linzey and Hammerson 2008f; Morris 1996; Place 2000; Roth and Vander Wall 2005; Schulte-Hostedde, Gibbs, et al. 2000, 2001; Schulte-Hostedde and Millar 2000, 2004; Schulte-Hostedde, Millar, and Gibbs 2004; Schulte-Hostedde, Millar, Gibbs, et al. 2002; T. P. Sullivan and Klenner 2000; T. P. Sullivan, Lautenschlager, et al. 1999; T. P. Sullivan, Sullivan, and Lindgren 2000; T. P. Sullivan, Sullivan, Lindgren, et al. 2005; Sutton 1992, 1995; Thayer and Vander Wall 2005; Vander Wall 1993a, 1993b, 1993c, 1993d. 1993e, 1995a, 1995b, 1995c, 1995d, 1998, 2000, 2003; Vander Wall, Beck, et al. 2003; Vander Wall and Peterson 1996; Waters and Zabel 1998.

Tamias bulleri J. A. Allen, 1889
Buller's Chipmunk

DESCRIPTION: *T. bulleri* has nine dorsal stripes: five black and four pale (gray or grayish white) stripes. The more median three black stripes are dark in color, and the outer two are lighter, browner, and shorter in length. The sides are brownish yellow; the inside of the ears is rust colored.

SIZE: Female—HB 134.0 mm; T 91.5 mm; Mass 74.9 g.
Male—HB 131.9 mm; T 84.6 mm; Mass 66.1 g.
Sex not stated—HB 130.4 m; T 104.8 mm.

DISTRIBUTION: This is the southernmost species of the genus, found in the Sierra Madre in southern Durango, western Zacatecas, and northern Jalisco (México).

GEOGRAPHIC VARIATION: None.

CONSERVATION: IUCN status—vulnerable. Population trend—decreasing.

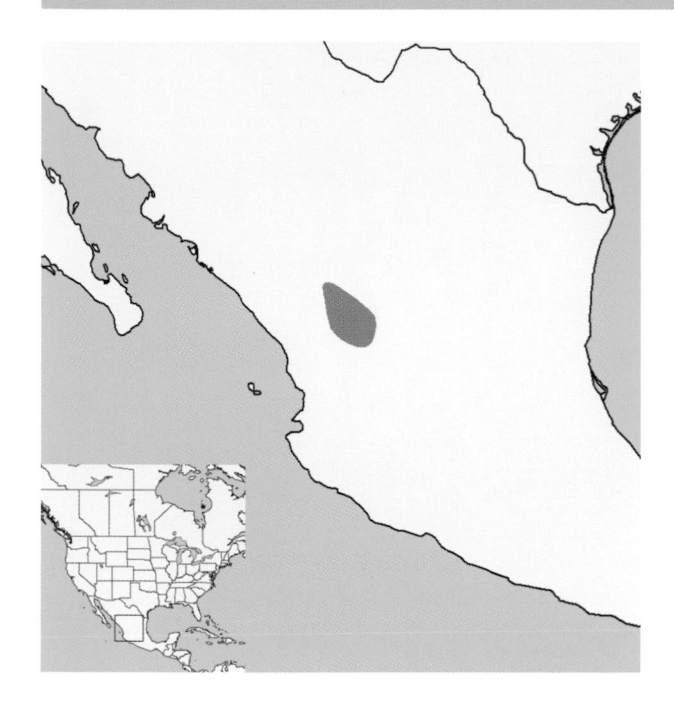

HABITAT: *T. bulleri* occupies an extremely small range at the juncture of three western states in México, in the Sierra Madre Occidental biotic province. The region is characterized by rolling mountainsides, with sharply cut canyons, at elevations of 2100-2400 m. At higher altitudes the vegetation consists mostly of pine (*Pinus*), a few other species of conifer, oak (*Quercus*), and quaking aspen (*Populus tremuloides*). At lower elevations the habitat is typical of the scrub oak vegetation of the upper Sonoran zone, such as manzanita (*Arctostaphylos pungens*), mountain mahogany (*Cercocarpus*), and *Ceanothus*. *T. bulleri* is replaced by *T. durangae* on the western slopes of the Sierra Madre (between Río Nazas and Río Mezquital) and by *T. dorsalis* on the eastern slopes (north of Río Nazas). *T. bulleri* prefers more heavily wooded forests, whereas *T. dorsalis* is found in more open xeric habitats.

NATURAL HISTORY: *T. bulleri* has been observed feeding on the catkins of oak (*Quercus*), juniper seeds (*Juniperus*), and the shoots of pine stems (*Pinus*), but no other information is available on its diet. Predators are not known, and only one flea (possibly *Monopsyllus polumus*) and one unidentified nematode have been reported from *T. bulleri*. The Buller's chipmunk appears to prefer forests with a significant component of rocks or downed woody debris. In Jalisco it appears to be sympatric with *Otospermophilus variegatus* and *Sciurus nayaritensis*. *T. bulleri* is reported to nest in tree cavities and ground dens. The little that is known about its reproduction includes reports of two females with two and three embryos each (on 18 July and 26 June, respectively), lactating females (between 27 June and 20 July), and a subadult on 13 May. Although one study reports sexual dimorphism in this species, another argues against it. In addition to the characteristic vocalizations of *Tamias* ("chucks," "chips," and trills), this species produces a sound unique to the genus.

Tamias bulleri. Photo courtesy Celia López-González, Instituto Politécnico Nacional, México.

GENERAL REFERENCES: Álvarez-Castañeda, Castro-Arellano, Lacher, et al. 2008c; Bartig et al. 1993; Escalante, Espinosa, et al. 2003; Escalante, Rodríguez, et al. 2005.

Tamias canipes (V. Bailey, 1902) Gray-Footed Chipmunk

DESCRIPTION: This species is recognized by the gray coloration of the dorsal surface of its hind feet. *T. canipes* is small, but it is larger than *T. minimus*.

SIZE: Female—HB 131.6 mm; T 107.8 mm.
Male—HB 128.7 mm; T 98.4 mm.
Sex not stated—HB 128.9 mm; T 100.0 mm; Mass 70.0 g.

DISTRIBUTION: The gray-footed chipmunk is found in the Capitan, Jicarilla, Gallinas, and Sacramento mountains of southeastern New Mexico, as well as in Texas, where it is found only in the Sierra Diablo and the Guadalupe Mountains of western Texas (USA).

GEOGRAPHIC VARIATION: Two subspecies are recognized.

T. c. canipes—occupies most of the species' range. It is not dimorphic.

T. c. sacramentoensis—only in the Sacramento Mountains of New Mexico (USA). It is dimorphic; males are smaller than females.

CONSERVATION: IUCN status—least concern. Population trend—stable.

HABITAT: This is a woodland species that appears to prefer forest edges with a high percentage of downed woody debris. *T. canipes* occurs in both the transition and Canadian life zones, at elevations between 1600 and 3600 m, where it is found in a variety of forest types (even including scrub forests) and in lava flows and rocky outcrops. However, this species is most common and abundant in stands of Douglas fir (*Pseudotsuga menziesii*) and ponderosa pine (*Pinus ponderosa*). The gray-footed chipmunk is reported to be significantly more abundant in commercially thinned stands than in more mature unthinned stands. *T. canipes* feeds on the seeds of conifers; various fruits, including those of juniper (*Juniperus*); and the acorns of oak (*Quercus*), which are reported to be the mainstay of their diet from late summer through winter.

Tamias canipes. Photo courtesy Bob Barber.

NATURAL HISTORY: *T. canipes* nests in fallen logs, stumps, and underground dens, but it is also a skillful climber, even ascending scrub oaks to harvest acorns. Diurnal activity peaks occur in the early morning, when this species is feeding. Their alarm calls, produced frequently in response to disturbances and potential predators, are characterized as high-pitched trills, similar to those reported for *T. cinereicollis* and *T. quadrivittatus*. Such vocalizations are usually followed by a rapid retreat. The young are born between late May and June and mature to adult size by early autumn. Only one litter is produced per year, and little information is available on litter size. *T. canipes* is reported to hibernate by October, though there is limited fat deposition prior to hibernation. The animals remain in the nest through the winter, where they are assumed to depend on food stores. No specific reports of predation are available, and *T. canipes* is not sympatric with any other conspecifics. Two species of *Eimeria* (*E. cochisensis* and *E. dorsalis*) are the only parasites reported from *T. canipes.*

GENERAL REFERENCES: Best, Bartig, et al. 1992; Cameron and Scheel 2001; A. W. Linzey, Clausen, et al. 2008; Piaggio and Spicer 2001; Wampler et al. 2008.

Tamias cinereicollis (J. A. Allen, 1890) Gray-Collared Chipmunk

DESCRIPTION: Gray-collared chipmunks have five brown to black stripes running along their dorsum. Two prominent white stripes on each side of the face are offset by brown. The venter is white to buff. The sides below the stripes are rufescent. The cheeks, the neck, the shoulders, and the rump are pale gray. The tail is gray suffused with orange brown.

SIZE: Both sexes—HB 223 mm (208-243 mm); T 80-113 mm; Mass 55-70 g.

DISTRIBUTION: This species is found in Arizona and New Mexico (USA).

CONSERVATION: IUCN status—least concern. Population trend—stable.

GEOGRAPHIC VARIATION: Two subspecies are recognized.

T. c. cinereicollis—eastcentral Arizona and western New Mexico (USA). See description above.

T. c. cinereus—westcentral New Mexico (USA). It is paler and grayer than *T. c. cinereicollis,* and has small reddish patches.

HABITAT: Gray-collared chipmunks favor clearings and the edges of high-elevation pine (*Pinus*), spruce (*Picea*), and fir (*Abies*) forests; these forests are often the most mesic in the region. This species can also be found in lower-elevation pine, Douglas fir (*Pseudotsuga menziesii*), and oak-juniper (*Quercus, Juniperus*) habitats.

NATURAL HISTORY: *T. cinereicollis* is diurnal. It may hibernate, but it probably enters torpor for only moderate periods of time. It is active from March to November, coming aboveground to forage during warm periods in winter. Nests are placed under logs, stumps, and roots, or in the hollows of trees or woodpecker cavities. A single litter averaging five (range = 4-6) young is born in the first half of June, after a gestation of about 30 days. Nursing lasts 41-45 days; the young appear aboveground by late July. *T. cinereicollis* is often seen perching on logs and stumps, sitting upright on its haunches while vigilant. When not alarmed, it gives a mellow call of short "chucks," but in response to a threat it gives a shrill rapid "chipper." Its diet includes the seeds of herbaceous plants, conifer seeds, acorns, fruits, fungi, green vegetation, and insects. *T. cinereicollis* may store food in logs, cavities, or crevices. Gray-collared chipmunks are excellent climbers and commonly forage in trees. This species responds positively to modest forest-thinning efforts; however, heavier thinning and fire treatments have resulted in the local decline of gray-collared chipmunks.

Tamias cinereicollis. Photo courtesy Robert Shantz.

GENERAL REFERENCES: Converse, Block, et al. 2006; Converse, White, et al. 2006; Hilton and Best 1993.

Tamias dorsalis (Baird, 1855)
Cliff Chipmunk

DESCRIPTION: The dorsum is pale gray with faint or indistinct stripes. The center stripe of faint brown to charcoal is sometimes all that is visible; an additional two longitudinal stripes may be discernable, but no others. There is a white patch behind each ear, and two white stripes on the face. The sides are gray to faint reddish brown. The venter is white to cream. The tail is grizzled gray above with reddish brown below.

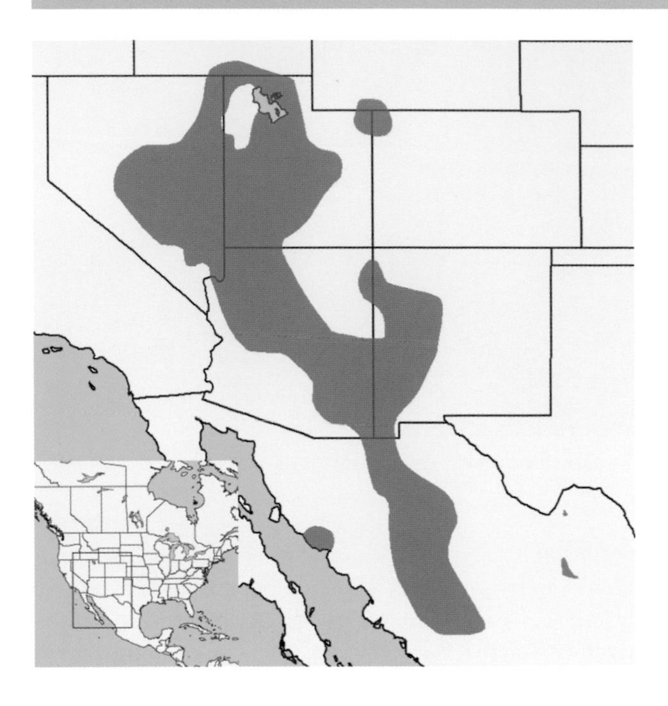

SIZE: Female—Mass 70-74 g.
Male—Mass 61-64.5 g.
Both sexes—TL 230 mm (217-249 mm); T 100 mm (85-115 mm).

DISTRIBUTION: This species is found in southern Idaho, eastern Nevada, Utah, southwest Wyoming, and northwest Colorado south through Arizona and western New Mexico (USA) to western Coahuila and northwest Durango, and in coastal Sonora (México).

GEOGRAPHIC VARIATION: Six subspecies are recognized.

T. d. dorsalis—Arizona and New Mexico (USA) southward along the Sierra Madre to the southern extent of the range in México. It is a large subspecies, with relatively dark and distinct colors.

T. d. carminis—two isolated and disjunct localities in Coahuila (México). This subspecies is dark, with more distinct stripes.

T. d. grinnelli—most of the species' range in Nevada (USA). It is a pale form, with less distinctive and narrower stripes on the face and the body.

T. d. nidoensis—central Chihuahua (México). No description is available.

T. d. sonoriensis—an isolated disjunct population in coastal Sonora (México). This form is distinctly smaller than *T. d. dorsalis*, found to the east and north of *T. d. sonoriensis*.

T. d. utahensis—northeastern portion of the range, including Utah, Idaho, Wyoming, and Colorado (USA). This form is smaller and paler in nearly all characters.

CONSERVATION: IUCN status—least concern. Population trend—stable.

HABITAT: Cliff chipmunks are found in a variety of habitats in different portions of their range; however, a common structural component is usually rocks, boulders, or cliffs. Major habitats include montane forests of pine (*Pinus*) and spruce (*Picea*); pinyon-juniper (*Pinus, Juniperus*) and oak (*Quercus*) woodlands; riparian areas; or shrublands.

NATURAL HISTORY: This squirrel is diurnal. Cliff chipmunks do not hibernate and can be active year-round in favorable weather; individuals will remain in nests during poor weather and may enter short bouts of torpor, emerging rapidly to forage when the weather permits. *T. dorsalis* has a prolonged breeding season. Males are reproductively active from January to July, whereas most females breed in April and June. Litters of four to six young are born in nests from late April to August. Nests occur in rocky crevices, piles, or cliffs; underground burrows; and trees. The young emerge from May to September. A few females are able to produce two litters per year, but most females seem to be unable to reproduce even once every year, suggesting slow population growth. Cliff chipmunks are primarily terrestrial but frequently climb rocks and cliffs, and even shrubs and trees. *T. dorsalis* is a generalist herbivore, eating many parts of forbs, grasses, shrubs, cacti, and trees; however, it is opportunistic and will occasionally consume insects and other animal matter. Seeds are the principal component of cliff chipmunk diets; these are gathered in cheek pouches and transported to dispersed temporary cache sites or to nests for storage and use over winter. Individuals are shy and secretive as they scurry about their home ranges, which average from 0.9 to 1.3 ha. Groups of females

Tamias dorsalis. Photo courtesy Melissa J. Merrick.

foraging together have been reported, with short quiet "bark chirps" given by group members. Due to the modest size of these chipmunks, they are prey for a variety of raptors, mustelids, felids, canids, and snakes. When threatened, *T. dorsalis* runs for the protective cover of rocks and logs rather than climbing on vegetation. They give short "chirps" when excited; their alarm calls are sharp short but loud bursts uttered while they are standing upright on their hind limbs. Cliff chipmunks are not trapped or hunted; occasionally they are a nuisance near homes or in picnic areas.

GENERAL REFERENCES: Dobson, Pritchett, et al. 1987; Hart 1992; Rompola and Anderson 2004.

Tamias durangae (J. A. Allen, 1903) Durango Chipmunk

DESCRIPTION: *T. durangae* is distinguished by the nine alternating light and dark stripes on the dorsum. It is similar in appearance to *T. bulleri*, but it is distinguished from that species by its cinnamon dorsal pelage and the reddish brown color on the ventral surface of the tail.

SIZE: Female—HB 135.0 mm; T 98.4 mm; Mass 83.8 g.
Male—HB 135.0 mm.
Sex not stated—HB 227.7 mm; T 102.2 mm.

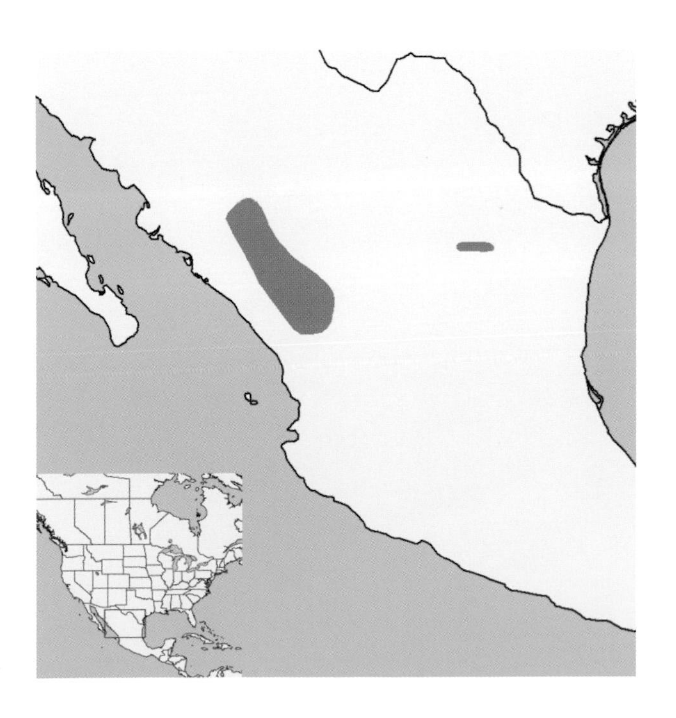

Tamias durangae. Photo courtesy Troy L. Best.

DISTRIBUTION: This species has a fragmented distribution in México, with one subspecies occurring in southwestern Chihuahua southward through western Durango; the other is found in southeastern Coahuila, possibly extending into Nuevo León.

GEOGRAPHIC VARIATION: Two subspecies are recognized.

T. d. durangae—Sierra Madre, from the southwestern corner of Chihuahua southward through the extreme western third of Durango (México). See description above.

T. d. solivagus—isolated in southeastern Coahuila, along the border of Nuevo León (México). No description is available.

CONSERVATION: IUCN status—least concern. Population trend—no information.

HABITAT: *T. d. durangae* appears to prefer mesic forests of mixed pine (*Pinus*) and oak (*Quercus*) in the Sierra Madre Occidental. *T. d. solivagus* occurs in similar forests of pine, fir (*Abies*), and quaking aspen (*Populus tremuloides*), at elevations above 2700 m in the Sierra Madre Oriental biotic province. Other than its apparent preference for more mesic conditions, nothing else is known about its habitat use.

NATURAL HISTORY: The Durango chipmunk is particularly secretive; little is known about its natural history. Its diet is assumed to include mostly acorns and conifer seeds, but few studies are available. Too little information on reproduction is known to report estimates of behavior, breeding season, litter sizes, and gestation period.

GENERAL REFERENCES: Álvarez-Castañeda, Castro-Arellano, Lacher, et al. 2008d; Best, Burt, et al. 1993; Callahan 1980.

Tamias merriami J. A. Allen, 1889
Merriam's Chipmunk

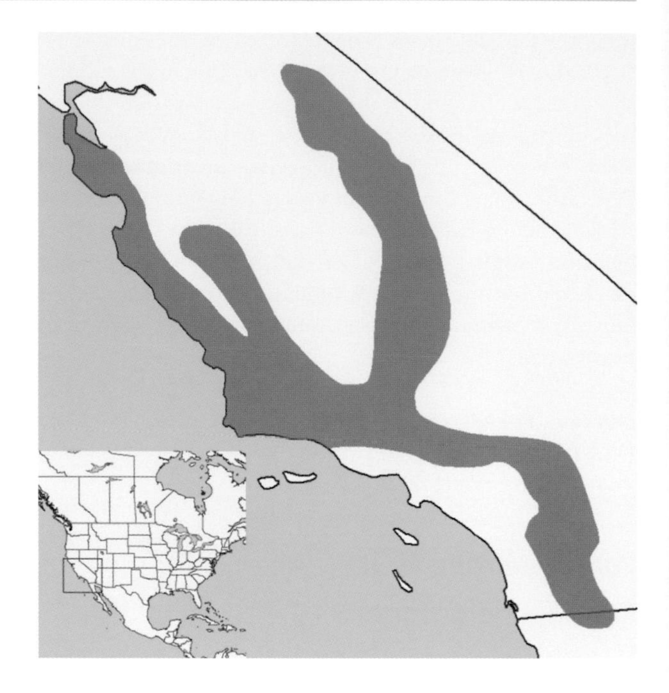

DESCRIPTION: Distinguished by its larger size and long bushy tail (up to 97% of its head and body length), Merriam's chipmunk also has long narrow ears, a white venter, and gray and brown dorsal stripes that are usually equal in width.

SIZE: Female—HB 135.3 mm; T 112.4 mm.
Male—HB 131.6 mm; T 106.4 mm; Mass 68.0 g.
Sex not stated—HB 133.4 mm; T 116.1 mm; Mass 71.3 g.

DISTRIBUTION: Merriam's chipmunk is found along the coast in the southern Sierra Nevada, from south of San Francisco to Columbia, California (USA), and then southward farther inland to extreme northern Baja California (México).

GEOGRAPHIC VARIATION: Three subspecies are recognized.

T. m. merriami—in California (USA), its range extends southward of that of *T. m. pricei* along the coast and northward through the southern Sierra Nevada, as well as inland and south to northern Baja California (México). This subspecies is intermediate in pelage darkness.

T. m. kernensis—its narrow range in California is limited to xeric pine forests of the Kern Basin and eastern slopes of the southernmost portion of the Sierra Nevada (USA). This is the palest of the subspecies, showing little distinction of the dorsal stripes in the winter pelage.

T. m. pricei—coastal areas of California, from San Francisco to 125 km southward (USA). This is the largest and darkest of the subspecies, and it also has the longest tail.

Tamias merriami. Photo courtesy Don Roberson.

CONSERVATION: IUCN status—least concern. Population trend—stable.

HABITAT: Merriam's chipmunk is found only in the upper Sonoran and transition life zones, from sea level to 2940 m in elevation. Although *T. merriami* occurs in a wide range of habitats, it usually occupies shrubby chaparral vegetation often associated with pine (*Pinus*) and oak (*Quercus*). Downed woody debris and rocks are also important habitat components.

NATURAL HISTORY: Merriam's chipmunk has a broad diet, including the seeds, fruits, and nuts of numerous plant species; insects; and other animal material when available. Acorns and conifer seeds appear to be the mainstay of its diet, and in some years acorns may occur in the diet year-round. This species may be important for forest regeneration, although detailed studies on its caching behavior are not available. Drought conditions reduce this chipmunk's densities, probably as a result of decreased food supply. Merriam's chipmunk is usually active throughout the year; it hibernates only at the highest elevations within its range. *T. merriami* is diurnal, showing peaks of activity in the morning and afternoon, separated by a midday rest. Both underground nests and tree cavities are used for shelter, although the latter appear to be preferred for rearing young. Shelters used by *T. merriami* are often prepared by other species, such as woodpeckers, pocket gophers (*Thomomys*), and ground squirrels. The tail of *T. merriami*, the longest and bushiest within the *Tamias* species, appears to be used for thermoregulation as well as in balance. Tail length is negatively correlated with elevation. When water is limited, *T. merriami* appears to show behavioral and physiological adjustments to reduce water loss.

GENERAL REFERENCES: Best and Granai 1994a; R. M. Davis et al. 2008; A. W. Linzey, Timm, et al. 2008b.

Tamias minimus Bachman, 1839
Least Chipmunk

DESCRIPTION: As one of the smaller members of the genus, *T. minimus* is nearly indistinguishable from several similar-sized congeners within its range. However, subtle differences in cranial morphology or pelage coloration allow it to be separated out from similar species in each part of its range. The ventral surface of the tail varies from reddish yellow to brown.

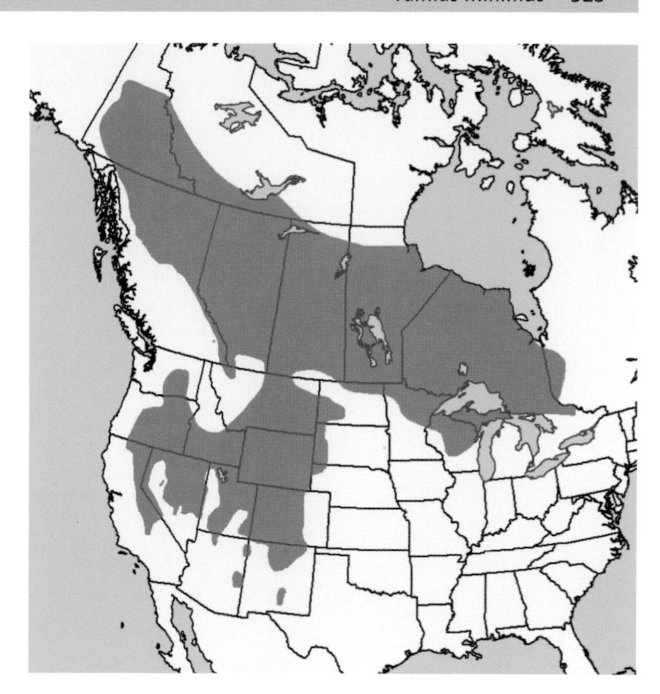

SIZE: Female—HB 114.2 mm; T 82.4 mm; Mass 46.4 g.
Male—HB 109.3 mm; T 80.3 mm; Mass 43.7 g.
Sex not stated—HB 106.0 mm; T 84.9 mm; Mass 50.5 g.

DISTRIBUTION: *T. minimus* ranges from central Yukon east to western Québec (Canada) and Michigan (USA) and south through the Cascade Range, the Sierra Nevada, and the Rocky Mountains in the western states, with isolated populations in Arizona and southern New Mexico (USA).

GEOGRAPHIC VARIATION: Twenty-one subspecies are recognized in this geographically diverse species.

T. m. minimus—"Found on plains and Sonoran plateaus in Wyoming, northeastern Utah, and Northwestern Colorado" (Anthony). The general color is dull gray, with the upperparts grizzled gray and buff, and the underparts white. The hairs on the ventral surface of the tail are rufous, banded with black and edged with buffy.

T. m. arizonensis—type locality is the Prieto Plateau at the south end of the Blue Range, Greenlee County, Arizona (USA). This form is similar in color to *T. m. consobrinus*, but *T. m. arizonensis* is grayer, with less tawny. The shoulders are washed with pale smoke gray. The tail is bright tawny on the ventral surface.

T. m. atristriatus—Peñasco, New Mexico (USA). The sides are grayish fulvous, and the underparts are yellowish. The dorsal light stripes are rusty gray and buffy white. Although similar to *T. m. operarius*, this form is slightly larger, darker, and duller in color.

T. m. borealis—Canada and upper regions of Montana (USA). The autumn pelage is pale yellowish gray on the upperparts, and the sides are washed with yellowish brown. The ventral surface is grayish white tinged with buff. The dark dorsal stripes are black, and the light dorsal stripes are gray. The outer pair of light stripes are white tinged with yellow.

T. m. cacodemus—"Sheep Mountain, Big Bad Lands" (= Scruton Mountain, White River Badlands, Pennington County), South Dakota (USA). This form is similar to *T. m. pallidus*, but *T. m. cacodemus* is paler, with a very long tail.

T. m. caniceps—Yukon south to north British Columbia (Canada). This form is similar to *T. m. borealis*, but *T. m. caniceps* is grayer, especially on the head, the feet, and the tail. The postauricular patches are more prominent, and the underparts are pure white.

T. m. caryi—San Luis Valley, Colorado (USA). This form is similar to *T. m. minimus*, but *T. m. caryi* is paler and grayer. The face stripes are white.

T. m. chuskaensis—Chuska Mountains, in Arizona and New Mexico (USA). The coloring of the dorsal stripes in *T. m. chuskaensis* is similar to *T. m. minimus*, but the general coloring is perhaps paler. The dorsal surfaces of the feet are gray. The underside of the tail is yellowish to reddish brown, outlined with pale buff.

T. m. confinis—Bighorn Mountains, northern Wyoming (USA). This form is similar to *T. m. consobrinus*, but *T. m. confinis* is larger, and the upperparts are more buffy in the winter. The blackish dorsal stripes are mixed with ochraceous tawny.

T. m. consobrinus—Utah and Colorado, north along the boundary between Idaho and Wyoming (USA). The ventral surface of the tail is fulvous.

T. m. grisescens—central Washington (USA). This form is similar to *T. m. pictus*, but *T. m. grisescens* is smaller, the coloration is grayer and less buffy, and the dark dorsal stripes are narrower. The tail is paler and grayish.

T. m. hudsonius—northern Manitoba (Canada). This form is similar to *T. m. borealis*, but *T. m. hudsonius* is a darker gray form. The ventral surface of the tail is paler. There is less reddish brown on the shoulders, the back, and the rump. The hind legs are dark gray. The dark dorsal stripes are relatively broader.

T. m. jacksoni—Oneida County, Wisconsin (USA). This form is similar to *T. m. borealis*, but the upperparts and the tail of *T. m. jacksoni* are more intensely tawny, with a darker face and head. The tail is also darker.

T. m. neglectus—southern Manitoba and southern Ontario (Canada) south to the northern parts of Minnesota and Wisconsin (USA). This form is similar to *T. m. borealis*, but *T. m. neglectus* is larger, and the colors are brighter.

***Tamias minimus*.** Photo courtesy James N. Stuart.

T. m. operarius—western Colorado, expanding into northern New Mexico, the eastern edge of Utah, and southern Wyoming (USA). This form is similar to *T. amoenus*, but *T. m. operarius* lacks the distinct black color on the posterior aspect of the ear. The tail is longer. The upper surface of the tail is strongly fulvous.

T. m. oreocetes—southernmost boundary between Alberta and British Columbia (Canada), south into Montana (USA). The upperparts are gray, tinged with yellow on the sides. The top of the head is grizzled gray. The outer pair of light dorsal stripes are white. The undersurface of the tail is pale fulvous, bordered with black and fringed with ochraceous; the upper surface of the tail is grizzled yellow.

T. m. pallidus—Montana, eastern North Dakota, and northeastern Wyoming, expanding west over the boundary into South Dakota and Nebraska (USA). This form is similar to *T. m. minimus*, but *T. m. pallidus* is larger, and the sides are more washed with ochraceous.

T. m. pictus—southern Idaho and northern Utah (USA). The upperparts are more pallid than *T. m. minimus*. The upperparts are slate gray, and the sides are pale yellow buff. The median dark dorsal stripes are black, with the outer dark stripes seal brown edged with rufous. The outer pair of dorsal stripes are white, and the median light stripes are slate gray. The upper surface of the tail is black and yellowish gray, and the under surface is dark yellowish buff bordered with black and fringed with yellowish gray.

T. m. scrutator—southcentral part of Washington, central

Oregon, and southwestern Idaho, south to the eastern side of the Sierra Nevada in California, and west into Nevada (USA). This form is smaller, with cinnamon sides. This form is similar to *T. m. consobrinus,* but *T. m. scrutator* is lighter colored and less rufescent. In comparison with *T. m. pictus, T. m. scrutator* has a shorter tail, and the top of the head is darker.

T. m. selkirki—known only from the type locality at Paradise Mine, near Toby Creek, 19 miles west of Invermere, British Columbia (Canada). The outer dark stripes are brown, and the inner dark stripes get blacker. The inner light stripes are heavily colored with gray and brown. The postauricular patch is grayish white. The sides are cinnamon buff. The undersurface of the tail is between pinkish cinnamon and cinnamon buff.

T. m. silvaticus—region around the boundary between Wyoming and South Dakota (USA). This form is large, and the general color is drab. The sides are ochraceous buff. The ventral surface of the tail is ochraceous orange fringed with black. In comparison with *T. m. pallidus, T. m. silvaticus* has darker and redder upperparts. The dorsal stripes and the top of head are darker.

CONSERVATION: IUCN status—least concern. Population trend—stable. *T. m. atristriatus* is critically endangered, and *T. m. selkirki* is vulnerable.

HABITAT: *T. minimus* lives in open areas relatively free of downed woody debris, and in a riparian system along the Green River (upland riparian habitat in Utah, Wyoming, and Colorado [USA]). Microhabitat use by the least chipmunk is often determined by the presence of conspecifics. For example, in western Colorado in the pinyon-juniper-sagebrush community (*Pinus edulis, Juniperus scopulorum, Artemisia*), *T. minimus* occurs close to trees, except when *T. rufus* is abundant; *T. minimus* then shifts to the surrounding sagebrush habitat.

NATURAL HISTORY: This species is largely granivorous, but its diet depends on the seasonal availability of other foods and frequently includes fruits and flowers, arthropods (primarily insects), and leaves. The least chipmunk is reported to construct leaf nests or occupy tree cavities, although it most commonly relies on underground burrows for hibernation and nesting. Such burrows are relatively shallow (0.5-1 m deep), short (0.4-3.5 m long), and usually have one to three entrances and one or two chambers. Extensive larder-hoards of food are not reported from the nests of *T. minimus.* Females enter estrus within seven days of emergence. Gestation is 28-30 days. Average litter sizes across this species' range, estimated from embryo counts and placental scars, average 4.0-6.4 and 4.0-5.5, respectively. This species typically produces only one litter per year. Lactating females are reported as early as May and as late as August, depending on the location. Home range in one study in Alberta (Canada) averaged 1.22 ha and 0.66 ha for males and females, respectively. Most other estimates are based on trapping data, and they are substantially smaller and possibly less accurate. Least chipmunks are considered relatively trap shy, less prone to capture, and thus probably underrepresented in trap surveys.

This species appears to be subordinate to congeners and other ground squirrels when aggressive interactions are involved, although its ability to exploit resources efficiently may allow it to successfully compete in some circumstances. Both *T. amoenus* and *T. minimus* occur in the Rocky Mountains, though *T. minimus* appears to be restricted to the alpine zone. Experimental removal of *T. amoenus* results in replacement by *T. minimus,* indicating that the former is more dominant. However, it is also apparent that *T. minimus* can survive in both habitats, whereas *T. amoenus* cannot. Least chipmunks vocalize frequently and exhibit a diversity of calls; several types of alarm call are used (five to six), but detailed analyses of the structure and context of their vocalizations are not available. Predators are not reported, possibly because of this species' small size, its agility, and its secretive behavior. Ectoparasites reported from *T. minimus* are 2 lice, 1 mite, 4 ticks, 11 fleas, and 1 fly (the botfly [*Cuterebra*]). The least chipmunk can harbor several potential diseases of concern to humans. Although refractory to Colorado tick fever, *T. minimus* is known to carry the virus that causes this disease. It also appears to have the potential to serve as a reservoir host for the bacterium that causes Lyme disease, although it is unlikely that it hosts the tick that transmits this bacterium. It is considered highly susceptible to the bacterium (*Yersinia pestis*) that causes sylvatic (bubonic) plague. It is also known to serve as a reservoir host for the tick *Ixodes spinipalpis,* which is the vector of *Borrelia bissettii,* the cause of human granulocytic ehrlichiosis, a recently discovered tick-borne disease similar to Rocky Mountain spotted fever. The least chipmunk is known to carry strains of the bacteria *Bartonella,* which are known human pathogens. Reports of *T. minimus* from the fossil record are numerous.

GENERAL REFERENCES: Anthony 1928; DeNatale et al. 2002; Ditto and Frey 2007; Guralnick 2007; Hadley and Wilson 2004a, 2004b; Kosoy et al. 2003; A. W. Linzey and Hammerson 2008g; Morris 2005; Root et al. 2001; Vander Haegen et al. 2002; Verts and Carraway 2001.

Tamias obscurus J. A. Allen, 1890
California Chipmunk, Dusky Chipmunk

DESCRIPTION: The pelage is marked by five dark chestnut-colored dorsal stripes, which are distinct in the winter pelage but drab and inconspicuous by spring. Compared with *T. merriami*, *T. obscurus* is slightly smaller in body size and cranial dimensions, and it has a grayer venter; they are sympatric in parts of their ranges in southern California.

SIZE: Female—HB 128.3 mm; T 117.0 mm.
Male—HB 125.3 mm.
Sex not stated—HB 124.1 mm; T 103.3 mm; Mass 69.0 g.

DISTRIBUTION: This species' range is fragmented; it extends from isolated populations in the San Bernardino and San Jacinto mountains in San Bernardino County, California (USA) to separate populations in Sierra de San Francisco and central Baja California, México.

GEOGRAPHIC VARIATION: Three subspecies are recognized.

T. o. obscurus—central fragment of the species' range that extends from extreme southern California (USA) southward for about 300 km into Baja California (México). The upperparts are gray suffused with brown. The dorsal stripes are inconspicuous. The tail is shorter.

T. o. davisi—occupies the two northernmost fragments of the species' distribution in southern California (USA). The throat and upper chest are gray.

T. o. meridionalis—isolated in a small area of central Baja California (México). This is a smaller and grayer form.

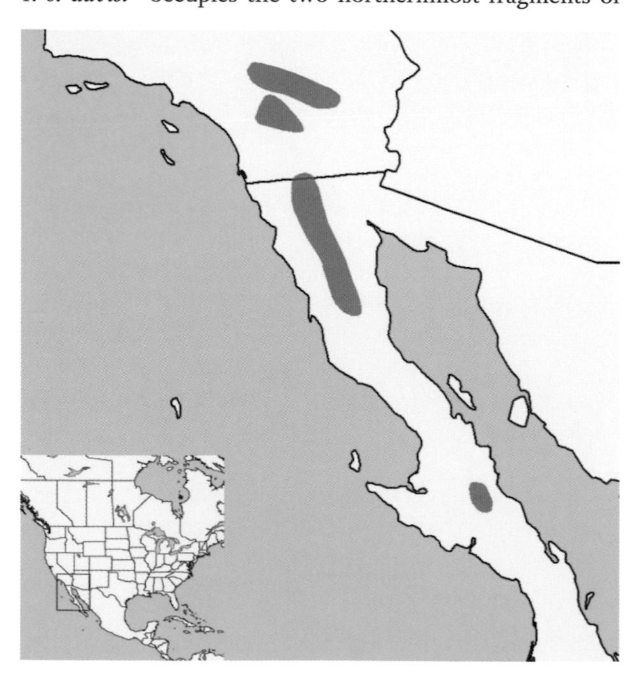

Tamias obscurus. Photo courtesy Nicolas Ramos-Lara.

CONSERVATION: IUCN status—least concern. Population trend—stable.

HABITAT: This species' range consists of at least four distinct fragments; within each of these, populations appear to be even further fragmented by low-lying deserts. *T. obscurus* is found primarily in arid pinyon-juniper (*Pinus, Juniperus*) and pine-oak (*Pinus, Quercus*) stands in the desert areas of southern California to central Baja California, usually at elevations between 1200 and 3000 m. However, specific patterns of habitat use vary with location and elevation. In southern California, for example, isolated populations of *T. o. davisi* occupy pine-fir-oak (*Pinus, Abies, Quercus*) forests of the transition zone; pinyon-juniper (*Pinus, Juniperus*) stands of the upper Sonoran zone; lodgepole pine-chinquapin (*Pinus contorta, Castanopsis sempervirens*) communities of the Canadian life zone; and even live oak-chaparral (*Quercus chrysolepis, Adenostoma* [chamise]) habitats at elevations less than 1500 m. *T. obscurus* also often occurs on rock outcrops or talus slopes throughout its range, especially in the San Pedro Mártir Mountains (Baja California, México). At the extreme southern edge of the species' range, *T. o. meridionalis* occupies palm-cactus stands of the lower Sonoran life zone.

NATURAL HISTORY: *T. obscurus* is active year-round and, like other chipmunks, is diurnal. Its daily activity is bimodal, with peaks in the early morning and late in the day. Details of burrows and nests, which vary widely, are reported from boulder fields, fallen logs, and cacti (for *T. o. meridionalis*). Like its congeners, *T. obscurus* feeds on seeds and fruits, especially acorns, which it frequently larder-hoards in burrows. Predators are not reported for this species; parasites include four species of *Eimeria*, one flea, and one mite. Sympatric populations of *T. merriami* and *T. speciosus* occur within the range of *T. obscurus*. The breeding season appears to be particularly long for this species; it begins as early as January and continues to as late as early June. The southernmost populations of this species (*T. o. meridionalis*) appear to reproduce earliest and are reported to have young in the nest by February. *T. obscurus* produces litters of three to four individuals and, although speculation exists that two litters may be produced, one is more likely.

GENERAL REFERENCES: Best and Granai 1994b; A. W. Linzey, Timm, et al. 2008c.

Tamias ochrogenys (Merriam, 1897) Yellow-Cheeked Chipmunk, Redwood Chipmunk

DESCRIPTION: *T. ochrogenys* is distinguished from other members of the *townsendii* complex (*T. senex*, *T. siskiyou*, and *T. townsendii*) by its larger body size, dark dorsal pelage, thinner tail, the structure of its genital bones, and its vocalizations. In addition, its body size is larger, its pelage is darker, and its tail is shorter than those of *T. sonomae* and *T. merriami*, which can be found in the southern part of the range of *T. ochrogenys*.

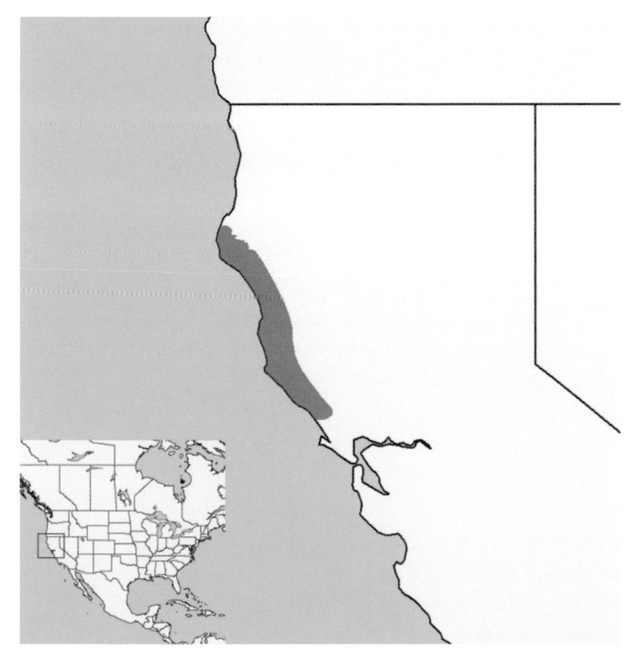

***Tamias ochrogenys*.** Photo courtesy Lisa D. Walker, www.sunriseseaglass.com.

SIZE: Female—HB 152.5 mm; T 115.1 mm; Mass 94.1 g.
Male—HB 147.8 mm; T 109.0 mm; Mass 72.7 g.
Sex not stated—HB 148.7 mm; T 113.8 mm; Mass 94.1 g.

DISTRIBUTION: The yellow-cheeked chipmunk is found along the coast of northern California, from the Van Duzen River south to southern Sonoma County (USA).

GEOGRAPHIC VARIATION: None.

CONSERVATION: IUCN status—least concern. Population trend—stable.

HABITAT: This species is found in the coastal redwoods (*Sequoia sempervirens*) of the northwestern coast of California (USA). Its preferred habitat appears to be moist conifer forests with thicker underbrush. This species is limited to forests within 40 km of the coast, at elevations from sea level to 1280 m.

NATURAL HISTORY: Although detailed studies on the diet of *T. ochrogenys* are not available, anecdotal reports suggest that it feeds on a broad selection of seeds (from grasses and shrubs); fruits, such as raspberries (*Rubus*); acorns (*Quercus*); fungi (especially in the late winter and spring); and even insects. No predators are reported for yellow-cheeked chipmunks, and the few parasites definitively recorded from this species include the botfly (*Cuterebra*) and one louse (*Hoplopleura arboricola*). Little information is available about reproduction and development in this species. Males are known to be reproductively active between March and June. Four fetuses (43 mm) were reported from one female in mid-March. Visual detection of this species is difficult because of the thick underbrush of its habitat, but *T. ochrogenys* can be recognized by its distinctive low-frequency two-part "chip" call. This "chip" call, most similar to that produced by *T. townsendii*, is of a lower frequency and has a distinctive rhythm compared with that of other species in the *townsendii* species-group. Hybridization between *T. ochrogenys* and *T. senex* is hypothesized, based on the intermediate "chip" vocalizations in Humboldt County, California (USA), south of the Eel River.

GENERAL REFERENCES: Gannon et al 1993; A. W. Linzey and Hammerson 2008h.

Tamias palmeri (Merriam, 1897) Palmer's Chipmunk

DESCRIPTION: This is a relatively small chipmunk, but it is slightly larger than its only sympatric congener, *T. panamintinus*. The overall pelage is dark to brown. The shoulders are grayer, the body is browner, and the ventral surface is paler than those characteristics of *T. panamintinus*.

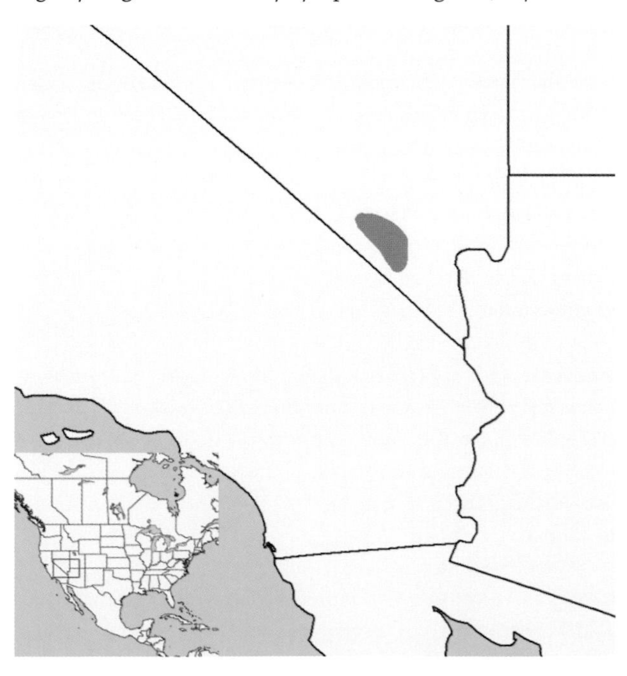

***Tamias palmeri*.** Photo courtesy Chris Lowrey.

SIZE: Female—HB 126.1 mm; T 80.3 mm; Mass 55.2 g.
Male—HB 126.6 mm; T 98.0 mm; Mass 52.4 g.
Sex not stated—HB 125.5 mm, T 94.0 mm, Mass 59.7 g.

DISTRIBUTION: *Tamias palmeri*'s entire range is restricted to the Spring Mountains of Clark County in southern Nevada (USA), where it is surrounded and isolated by desert.

GEOGRAPHIC VARIATION: None.

CONSERVATION: IUCN status—endangered. Population trend—declining.

HABITAT: Palmer's chipmunk is found at elevations from 2100 m to timberline (3600 m), but population densities are highest between 2400 and 2550 m. This species is frequently associated with rock crevices, rock outcrops, caves, and fallen timber, and it is reported to occur in several forest communities: bristlecone pine (*Pinus aristata*); singleleaf pinyon pine-Utah juniper (*Pinus monophylla, Juniperus osteosperma*); white fir-ponderosa pine (*Abies concolor, Pinus ponderosa*); and mountain mahogany-manzanita (*Cercocarpus ledifolius, Arctostaphylos pungens*).

NATURAL HISTORY: Palmer's chipmunk is especially granivorous and feeds heavily on conifer seeds and fruits; but, like other congeners, it is opportunistic and includes fungi, as well as arthropods and other animal material, in its diet. Bark, lichens, vegetation, and flowers are also reported from this species' stomach contents. Significant differences in activity and aggression have been observed, with males showing more aggression at certain times of the year. This species' vocalizations are described as "chips," "chippering," "chucks," and growls, with the first being the most common. "Chip" calls vary from 5-17 calls/5 seconds. Females call at a higher maximum frequency than males. Reproduction is thought to occur in April and May, and parturition in late May and June, although young animals have been observed throughout the summer. Gestation is estimated at more than 33 days. Litter sizes average approximately four young, but as many as seven embryos have been reported. Detailed studies of this species' reproduction, growth, and development are found in the literature. *T. palmeri* is sympatric with *T. panamintinus*, although the latter generally occurs at lower elevations. No predators are reported. Parasites include one acanthocephalan (*Moniliformis moniliformis*), three nematodes (*Heteroxynema cucullatum, Pterygodermatites coloradensis*, and *Syphacia eutamii*), and one flea (*Monopsyllus eumolpi*).

GENERAL REFERENCES: Best, Clawson, et al. 1994a; R. M. Davis et al. 2008; Gannon and Stanley 1991; Lowrey et al. 2008.

Tamias panamintinus Merriam, 1893
Panamint Chipmunk

DESCRIPTION: The Panamint chipmunk is similar in size to *T. amoenus*, but with smaller feet and ears, a paler pelage, lighter dorsal stripes, and a broader skull. *T. panamintinus* is more reddish in color than *T. minimus*. The baculum is distinct from but most similar to that of *T. speciosus*.

SIZE: Female—HB 119.6 mm; T 87.1 mm; Mass 54.1 g.
Male—HB 117.1 mm; T 89.1 mm.
Sex not stated—HB 107.2 mm; T 91.7 mm; Mass 53.2 g.

DISTRIBUTION: This species is found in the mountains of southeastern California and southwestern Nevada (USA).

CONSERVATION: IUCN status—least concern. Population trend—stable.

GEOGRAPHIC VARIATION: Two subspecies are recognized.

- *T. p. panamintinus*—the entire species' range, except the Kingston Mountains of northeastern San Bernardino County, California (USA). See description above.
- *T. p. acrus*—isolated by desert to 100 km² of the Kingston Mountains, California (USA). This form is smaller and darker.

HABITAT: This species occurs in the upper Sonoran and transition zones (1230-3180 m) of the Great Basin and is often sympatric or parapatric with *T. amoenus, T. merriami, T. minimus, T. palmeri, T. speciosus*, and *T. umbrinus*. The Panamint chipmunk is found primarily in pinyon juniper (*Pinus, Juniperus*) forests, but it also occurs at higher elevations in stands of limber pine (*P. flexilis*) and bristlecone pine (*P. aristata*). Although the animals climb and occasionally nest in trees, they reside primarily in rocky cliffs and ledges and seek rocky crevices for refuge.

NATURAL HISTORY: The Panamint chipmunk feeds heavily on the seeds of juniper (*Juniperus*) and pinyon pine (*Pinus*), but is also opportunistic (like other members of the genus), feeding on vegetation, flowers, and arthropods when seeds

***Tamias panamintinus*.** Photo courtesy Don Roberson.

and fruits are not available. *T. panamintinus* appears to be a more facultative hibernator that many of its local congeners, awakening during milder winter days and emerging earlier than some species, such as *T. palmeri*. Details on this species' oxygen consumption, variations in body temperature, organ weights, and vocalizations are available in the literature.

GENERAL REFERENCES: Best, Clawson, et al. 1994a; Hayssen 2008a, 2008b, 2008c; Laudenslayer et al. 1995; A. W. Linzey and Hammerson 2008i.

Tamias quadrimaculatus Gray, 1867
Long-Eared Chipmunk

DESCRIPTION: This species is distinguished from most other chipmunks by its long narrow ears, darker reddish coloration, and postauricular patches of white pelage. Within the *townsendii* species-group it is distinguished by its larger size, long and narrow rostrum, and zygomatic width.

SIZE: Female—HB 138.2 mm; T 94.5 mm; Mass 87.4 g.
Male—HB 135.8 mm; T 95.4 mm; Mass 78.1 g.
Sex not stated—HB 135.6 mm; T 102.6 mm; Mass 82.8 g.

DISTRIBUTION: The long-eared chipmunk occurs in the Sierra Nevada of eastcentral California (Plumas to Mariposa counties) and extreme westcentral Nevada, USA.

GEOGRAPHIC VARIATION: None.

CONSERVATION: IUCN status—least concern. Population trend—stable.

HABITAT: The long-eared chipmunk is typically found in stands of mixed conifers: ponderosa pine (*Pinus ponderosa*), Douglas fir (*Pseudotsuga menziesii*), and sugar pine (*Pinus lambertiana*). It is often associated either with dense forest stands or open stands with thick ground cover or dense understory. Downed woody debris, brush, stumps, and snags are important habitat components.

NATURAL HISTORY: Although its diet varies with habitat and season, *T. quadrimaculatus* is reported to feed most heavily on hypogeous (underground) fungi, consuming lesser amounts of seeds, leaves, flowers, fruits, and arthropods. Caching behavior by *T. quadrimaculatus* is thought to result in the regeneration of conifer species and thick understory brush; the latter may prevent conifer reforestation when it occurs. Hibernation takes place from late November to late March, during which time long-eared chipmunks rely on large stores of seeds. This species is more active and less territorial than other species of *Tamias*, but it is still aggressive, especially during the breeding season. It vocalizes frequently, producing several types of calls: various "chucks" and "chips" associated with aggression, alarm responses, and breeding. Tail flicks are often associated with "chipping" calls. Breeding occurs in late April to early May; gestation is 31 days; and litter size ranges from two to six young. Only one litter is produced each year. The home ranges of males and females are reported to average 0.88 ha and 0.48 ha, respectively. Peak densities reach approximately 1.0/ha. Predators are probably common, but the only act of predation that was observed was by a rattlesnake (*Crotalus*). This

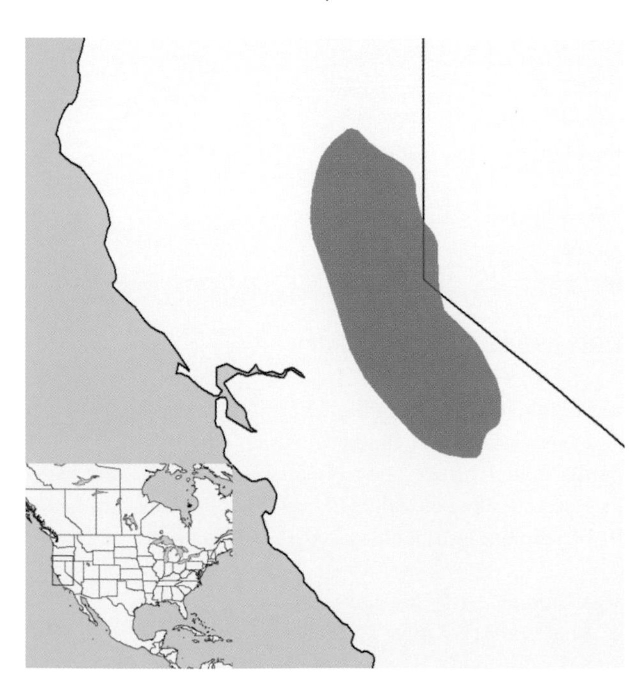

Tamias quadrimaculatus. Photo courtesy Mike Cong.

species has been observed to carry *Yersinia pestis*, the bacterium responsible for sylvatic (bubonic) plague, and the only parasites recorded from the long-eared chipmunk are fleas (*Catallagia sculleni, Diamanus montanus, Monopsyllus ciliatus, M. eumolpi,* and *Oropsylla idahoensis*) and ticks (*Demacentor andersoni* and *Haemaphysalis leporispalustris*).

GENERAL REFERENCES: Clawson et al. 1994b; A. W. Linzey and Hammerson 2008j; Piaggio and Spicer 2000; Sutton 1995.

Tamias quadrivittatus (Say, 1823)
Colorado Chipmunk

DESCRIPTION: Colorado chipmunks have a reddish to cinnamon head, sometimes mixed with gray; two white to cream stripes on the face; and a black stripe through the eye, offset with cinnamon. A gray postauricular patch is often present. The median dorsal stripe is dark gray to black, with two lateral dark stripes that are often slightly lighter in color. The dark stripes are offset by white to cream to buff light stripes. The sides, the hips, and the feet are ochraceous to cinnamon. The venter is cream to white.

SIZE: Female—HB 132.0 mm.
Male—HB 125.3 mm.
Both sexes—T 99.6 mm; Mass 59.6 g (45.2–69.8 g).

DISTRIBUTION: This species is found in eastern Utah, Colorado, northeast Arizona, northern New Mexico, and extreme western Oklahoma (USA).

GEOGRAPHIC VARIATION: Three subspecies are recognized.

T. q. quadrivittatus—throughout the entire range, except southern New Mexico (USA). See description above.

T. q. australis—southcentral New Mexico, within the Organ Mountains (USA). This form is grayer, and the dorsal surface of the feet is grayish brown.

T. q. oscuraensis—central New Mexico, within the Oscura Mountains (USA). This is the smallest of the subspecies. The sides are reddish brown, and there is more reddish brown on the dorsal surface of the feet.

CONSERVATION: IUCN status—least concern. Population trend—stable. *T. q. australis* and *T. q. oscuraensis* are threatened in New Mexico.

HABITAT: The Colorado chipmunk occurs in habitats that are mainly rocky, with conifers or shrubs. It is often found in pinyon-juniper (*Pinus, Juniperus*), scrub oak (*Quercus*), ponderosa pine (*Pinus ponderosa*), or spruce-fir (*Picea, Pseudotsuga, Abies*) forests.

NATURAL HISTORY: This species is diurnal. Colorado chipmunks probably do not hibernate and can be active yearround in favorable weather; individuals will remain in nests during poor weather and may enter short bouts of torpor,

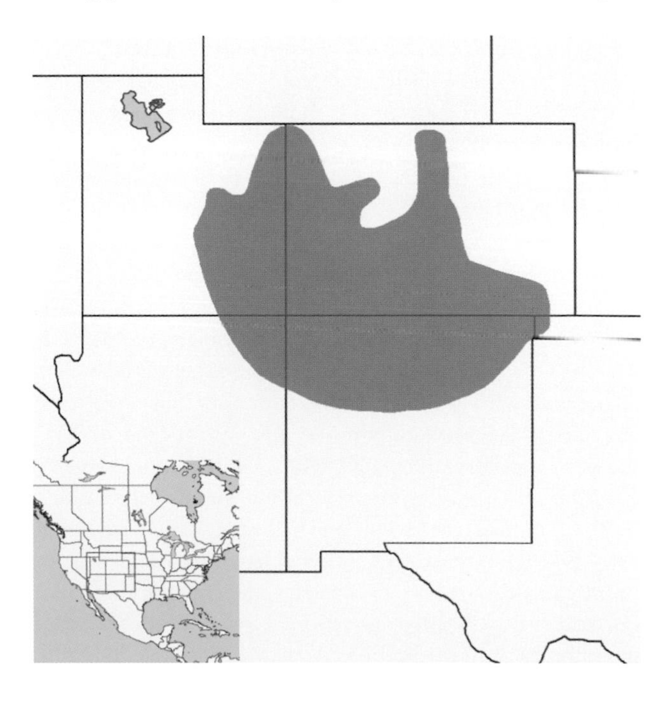

Tamias quadrivittatus. Photo courtesy James N. Stuart.

emerging rapidly to forage when weather permits. *T. quadrivittatus* begins breeding in spring (April or May), but it may have a bimodal breeding season in the south, during February and again during July. Litters of two to six young are born in nests from May to June. One litter is generally produced each year; however, the occurrence of pregnant and lactating females in late summer in some areas suggests that there is either a second litter or late breeding. Nests are found in crevices, under rocks, and in underground burrows. The young are weaned at 6-7 weeks and disperse by the fall. Colorado chipmunks are primarily terrestrial but frequently climb shrubs and trees in search of food. *T. quadrivittatus* is a generalist herbivore, eating the seeds and fruits of forbs, grasses, shrubs, and trees; however, this chipmunk is opportunistic and will frequently consume insects and other animal matter. As is the case for many other chipmunks, seeds are the principal component of Colorado chipmunk diets. Seeds are gathered in cheek pouches and placed in crevices or under rocks, or transported to an underground location near the nest chamber, for storage and use over the winter. Individual home ranges average 2.7-2.8 ha. This species' overwinter survival rate is less than 33 percent, with local extirpation common. Due to the modest size of Colorado chipmunks, they are preyed on by a variety of raptors, mustelids, felids, canids, and snakes. *T. quadrivittatus* runs for the protective cover of rocks or logs, or else it climbs trees when under threat. "Chips" and trills are loud high-pitched calls often given from a safe vantage point when the animal is alarmed. Colorado chipmunks are not trapped or hunted; occasionally they are a nuisance near homes or in picnic areas.

GENERAL REFERENCES: Bergstrom and Hoffmann 1991; Best 1994; Rivieccio et al. 2003; R. M. Sullivan 1996.

Tamias ruficaudus (A. H. Howell, 1920) Red-Tailed Chipmunk

DESCRIPTION: This species is a large chipmunk. It has an orangish dorsum, and five black to brownish dark dorsal stripes interspersed with four gray to yellowish brown stripes. The belly is white to cream colored, and the ventral surface of the tail is reddish with black and pink borders. Where *T. ruficaudus* is sympatric or parapatric with subspecies of *T. amoenus* and *T. minimus,* it can be distinguished by its larger body and/or larger skull and its distinctly reddish tail.

SIZE: Female—HB 127.2 mm; T 102.9 mm; Mass 63.2 g.
Male—HB 121.8 mm; T 98.3 mm; Mass 57.1 g.
Sex not stated—HB 125.3 mm; T 109.1 mm; Mass 67.3 g.

DISTRIBUTION: This species' range extends from southeastern British Columbia and southwestern Alberta (Canada) to northeastern Washington and east through Idaho to western Montana (USA).

GEOGRAPHIC VARIATION: Two subspecies are recognized.

T. r. ruficaudus—eastern half of the species' range. This subspecies has a longer baculum than *T. r. simulans.*

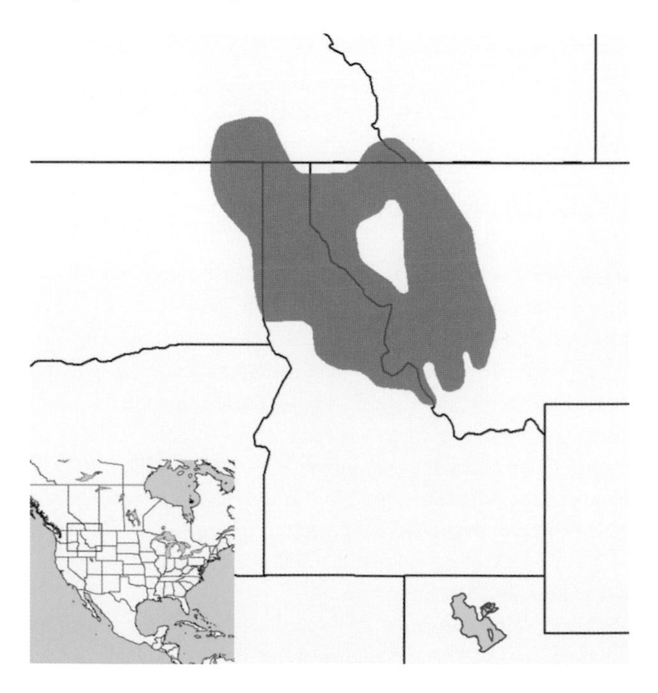

***Tamias ruficaudus*.** Photo courtesy Terry Gray, Moscow, Idaho.

T. r. simulans—western half of the species' range. *T. r. simulans* can be distinguished from *T. r. ruficaudus* by its paler sides and the paler ventral surface of its tail.

CONSERVATION: IUCN status—least concern. Population trend—stable. *T. ruficaudus* is considered at potential risk in Alberta and British Columbia (Canada), where available habitat and survey data are limited.

HABITAT: *T. ruficaudus* is found in a variety of coniferous forests, often occurring in open patches or edges where there is thick ground vegetation. It inhabits various tree assemblages in different areas, ranging from low- to high-elevation forests: from ponderosa pine (*Pinus ponderosa*) forests, to mesic riparian stands, to high-elevation subalpine stands. The red-tailed chipmunk is typically associated with significant ground cover, downed woody debris, and thick brush. Substantial snow cover occurs throughout much of its range.

NATURAL HISTORY: The red-tailed chipmunk is primarily granivorous-frugivorous, feeding on the seeds and fruits of a wide variety of trees and shrubs, depending on the habitat. It mainly forages on the ground and spends a considerable amount of its time storing food in a larder. Scatter-hoarding may occur, but this has not been well studied. Like many other species of *Tamias*, *T. ruficaudus* usually nests underground, beneath rocks, fallen logs, and stumps. Details on burrow construction are not available, but anecdotal observations suggest a burrow system similar to that of *T. amoenus*. Aboveground nests in trees and shrubs have been reported. Nests are usually constructed of grasses and lichens. The red-tailed chipmunk is diurnal, but it is generally only active aboveground between April and October. Vocalizations are reported between mother and offspring but are not well studied. Mating occurs in late April to mid-May in lower elevations and latitudes, and from early to late May in locales where conditions are more severe. Average litter size varies from 2.5 to 5.2 but is usually between four and six; it is consistently lower at higher elevations. Gestation is estimated to be 31 days. The young are born as early as late May and as late as late June, depending on the elevation. Limited information is available on the demographics and population densities of *T. ruficaudus*. *T. ruficaudus* is typically found at elevations above those of *T. amoenus* and below those of *T. minimus*, although altitudinal overlap with the latter may occur. *T. amoenus* and *T. ruficaudus* show similar characteristics, but they remain segregated in most parts of their range. Limited information on predators and parasites is available. *T. ruficaudus* is only known to host one species of flea (*Ceratophyllus ciliatus*).

Early comparisons of the morphology (e.g., the cranium and baculum) of *T. ruficaudus* suggested that *T. r. simulans* and *T. r. ruficaudus* might be two distinct species. A more recent analysis of the cytochrome *b* sequence, in conjunction with morphological studies of the baculum, supports the concept that these are two genetically distinct entities, but it also identifies a zone of hybridization and therefore an absence of complete reproductive isolation. These cytochrome *b* data also reinforce the idea that this species experienced multiple range contractions, repeated isolation in numerous Pleistocene refugia, and a recent range extension to the north. Molecular analyses have documented hybridization with *T. amoenus canicaudis* during the Pleistocene but suggest no recent gene flow between the two. The fossil record suggests that *T. ruficaudus* may have intermittently bred with as many as 10 other species in the *quadrivittatus* group, with which it is now generally allopatric, with a different bacular morphology.

GENERAL REFERENCES: R. P. Bennett 1999b; Best 1993a; Good, Demboski, et al. 2003, Good, Hird, et al. 2008; Good and Sullivan 2001; A. W. Linzey and Hammerson 2008k; Nagorsen et al. 1999; Patterson and Heaney 1987; Pearson et al. 2003.

Tamias rufus (Hoffmeister and Ellis, 1980) Hopi Chipmunk

DESCRIPTION: Hopi chipmunks are pale orange to cinnamon in general tone. The medial dorsal stripe is chestnut; lateral longitudinal stripes are less apparent; the interspersed light strips are white to grayish. The hips, the flanks, and the head are pale gray, and there are large white postauricular patches. The sides and the feet are bright pale reddish orange. The venter is buff to pale orange. The tail is grizzled charcoal suffused with reddish on the dorsal surface, and a bright red orange with a black border on the ventral side.

SIZE: Both sexes—Body 120 mm (93-148 mm); Head 35.1 mm (34.5-35.7 mm); Mass 59.3 g in September, 47.9 g in March. Females are slightly larger than males.

DISTRIBUTION: This species is found in eastern Utah, western Colorado, and northeastern Arizona.

GEOGRAPHIC VARIATION: None.

CONSERVATION: IUCN status—least concern. Population trend—stable.

HABITAT: Hopi chipmunks are found in various rocky habitats, although most often in open pinyon-juniper (*Pinus, Juniperus*) woodlands.

NATURAL HISTORY: This species is diurnal. *T. rufus* likely does not hibernate; rather, it enters into short bouts of torpor during extreme weather and will emerge from burrows on warm winter days or feed on cached food stores near its nest. Hopi chipmunks do store more fat (10%-20% of their body mass) than most western North American chipmunks, suggesting the potential for prolonged torpor or perhaps hibernation. Males are reproductively active from February to late April; mating typically occurs in February and March. A single litter of about five young is produced in nests after a gestation of 30-33 days. The young emerge from nests in May and are weaned after 6-7 weeks. Growth is somewhat slow, and sexual maturity is not reached until 10-11 months of age; thus reproduction does not occur until individuals are yearlings. Nests are often in burrows beneath rocks or shrubs or in crevices. Hopi chipmunks are primarily herbivores, feeding heavily on the seeds of forbs, shrubs, and trees; but they are opportunistic and will eat other plant parts, fungi, and insects. Individuals have home ranges of 1.0-1.3 ha that include many trees and shrubs, as well as cliffs, which the squirrels climb in search of food. *T. rufus* collects seeds in its cheek pouches and retreats to the cover of trees, shrubs, or rocks to feed. It caches seeds under rocks or in crevices. This species' primary predators are probably coyotes (*Canis latrans*), foxes, weasels, raptors, and snakes. Hopi chipmunks are not hunted or trapped; however, they can become a nuisance in croplands, gardens, and campgrounds.

Tamias rufus. Photo courtesy Doris Potter/Focus on Nature Tours, Inc.

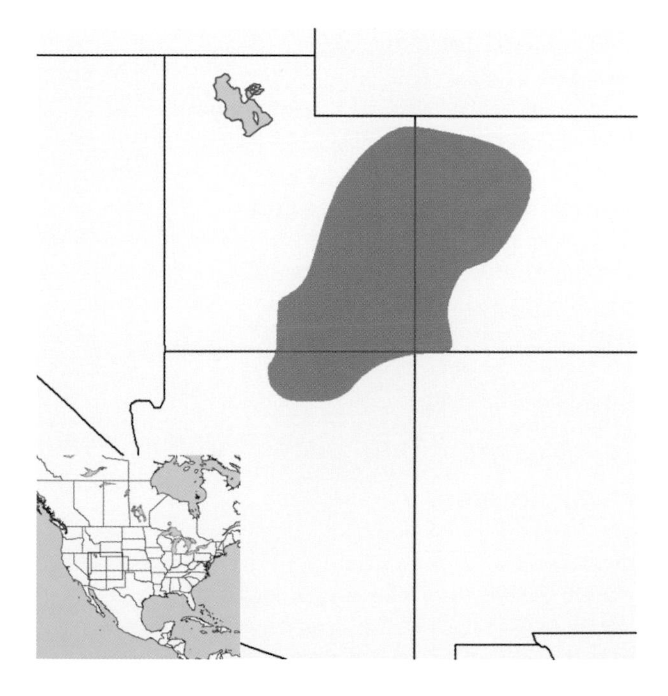

GENERAL REFERENCES: Burt and Best 1994; Root et al. 2001.

Tamias senex (J. A. Allen, 1890) Shadow Chipmunk

DESCRIPTION: Among *Tamias*, this species is noted for its large size, but the body size and pelage characteristics of *T. senex* vary throughout its range (see below). It is distinguished by a grayish wash on the head and the dorsal hindquarters, well-defined eye stripes, postauricular patches, and the distinctively shaped baculum. Coastal populations have a dark olive dorsum and a grayish pink venter, whereas inland populations are light olive with a pale gray wash on the dorsum and creamy white underparts. The pale gray wash is most apparent on the head, the hips, and the tail. Two white to buff lines on the side of the face are offset by ochraceous to red orange brown. It has white to buff eye rings, and white to cream postauricular patches. There is a dark cinnamon to sepia to black longitudinal stripe with two faint additional stripes on each side; white to grayish white stripes offset the dark stripes. The stripes generally fade before reaching the rump. The thin tail is grizzled gray suffused with reddish orange, with an ochraceous underside; the tail is often frosted with buff.

SIZE: Female—HB 147.5 mm; T 107.3 mm; Mass 94.0 g.
Male—HB 146.3 mm; T 103.7 mm; Mass 86.0 g.
Sex not stated—HB 142.6 mm; T 102.9 mm; Mass 89.7 g.

DISTRIBUTION: The range of *T. senex* extends from central Oregon southward through the Sierra Nevada in California and the northwestern tip of Nevada to the southern terminus of its range in California's Yosemite National Forest, as well as across northern California from the coast to the northern Nevada border.

GEOGRAPHIC VARIATION: This species was previously considered to be monotypic, but inland populations of *T. senex*, like those of *T. siskiyou*, are smaller and paler than those on the Pacific coast, and the interspecific convergence of these characters is so significant that the two species are nearly indistinguishable farther inland. Genital characters, however, indicate that the two are separate species, thus supporting the designation of one new subspecies for each of the two species.

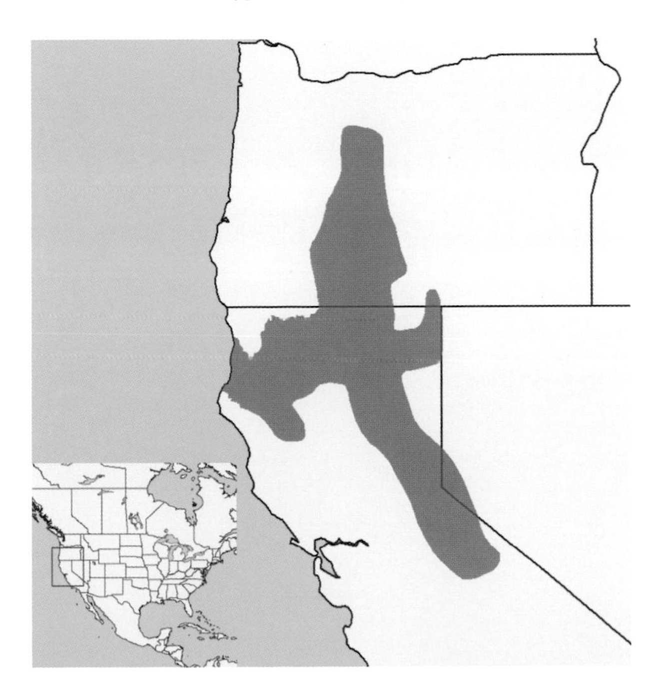

Tamias senex. Photo courtesy Chris Wemmer.

T. s. senex—found inland in the Sierra Nevada, Cascade, and other ranges east of coastal redwood (*Sequoia sempervirens*) forests (USA). This subspecies is generally ochraceous tinged with pale gray, and it is slightly larger than *T. s. pacifica*. While similar in size and color to *T. quadrimaculatus*, *T. s. senex* is grayer, with less white behind the ears.

T. s. pacifica—narrow belt of coastal redwood forests on the northwestern California coast, inland to 32 km (USA). The new subspecies is similar in coloration (dark olive) and size to *T. ochrogenys*.

CONSERVATION: IUCN status—least concern. Population trend—stable.

HABITAT: The shadow chipmunk is found in the Canadian and upper transition life zones, preferring dense brushy understory in closed canopy primary forests that are generally free of disturbances such as logging. *T. senex* is ecologically isolated from congeneric species (*T. alpinus*, *T. amoenus*, *T.*

merriami, T. quadrimaculatus, T. sonomae, T. speciosus, and *T. umbrinus*) found within its range. It is common in montane conifer forests of the Sierra Nevada, at approximately 1200–1800 m in elevation, and also in chaparral slopes. Hibernation takes place from late November to mid-March, and breeding occurs for one month following emergence. Litter sizes vary between three and five, and the young are weaned late, usually near the end of June.

NATURAL HISTORY: Various dietary studies suggest that *T. senex* is a truffle specialist, feeding heavily on hypogeous (underground) fungi, and almost exclusively so at certain times of the year. Also included in its diet are conifer seeds, flowers, and arthropods. An active arboreal species, *T. senex* often nests high in trees. Home range estimates are less than 3.5 ha but depend on the type and abundance of food. "Chip" vocalizations are intermediate in frequency between those of other *Tamias* species, such as *T. ochrogenys* and *T. siskiyou,* and consist of three to five syllables per "chip." No predators are reported. Four species of fleas are the only parasites noted, of which *Monopsyllus ciliatus* and *M. eumolpi* are the most common.

GENERAL REFERENCES: Gannon and Forbes 1995; Gannon and Lawlor 1989; E. R. Hall 1981; Kain 1985; A. W. Linzey and Hammerson 2008l; Sutton 1987; Sutton and Nadler 1974; Sutton and Patterson 2000; J. A. Wilson et al. 2008.

Tamias sibiricus (Laxmann, 1769)
Siberian Chipmunk

DESCRIPTION: This is a small chipmunk, with five black dorsal stripes. The pelage on the top of the head, the sides, and the spaces between the dark dorsal stripes is yellowish brown; the dorsal surface has a lateral dark-and-white stripe on both sides. The tail is shorter than the body length.

SIZE: Female—HB 150.5 mm; T 108.2 mm; Mass 96.2 g.
Male—HB 149.3 mm; T 106.5 mm; Mass 93.4 g.
Sex not stated—HB 147.5 mm; T 116.2 mm; Mass 99.5 g.

DISTRIBUTION: This species ranges across northern European and Siberian Russia to Sakhalin Island (Russia) and the southern Kurile Islands (claimed by Japan); it also occurs from extreme eastern Kazakhstan to northern Mongolia, China, Korea, and the island of Hokkaido (Japan). Its range appears to be expanding naturally in the north, and—by introductions—to England, France, Belgium, the Netherlands, Denmark, Germany, Switzerland, Austria, Italy, China (Hong Kong), and Japan (Honshu, Shikoku, and Kyushu islands), primarily as a result of the pet trade.

GEOGRAPHIC VARIATION: Nine subspecies are recognized.

T. s. sibiricus—Altai Mountains (Russia, Kazakhstan, Mongolia, and China) and Siberia (Russia). This is a darker and less brightly colored form.
T. s. asiaticus—region north of the Sea of Okhotsk, extending west into northcentral Siberia and east into the Kamchatka Peninsula (Russia). This form has a gray rump.
T. s. lineatus—Sakhalin Island (Russia) and the island of Hokkaido (Japan). No description is available.
T. s. okadae—South Kurile Islands (Japan). The head, the lower back, and the rump are a deep rusty red. The tail is nearly black with hoary hair tips. The underparts are nearly white.
T. s. ordinalis—China. This form is similar to *T. s. senescens,* but *T. s. ordinalis* is a much paler form.
T. s. orientalis—Siberia, northeast China, and Korea. This form is similar to *T. s. senescens,* but *T. s. orientalis* is brighter and ruddier. Also, the clearly defined white supraorbital stripe extends to the nose. The outer dorsal light stripes are ruddier than the inner stripes.
T. s. pallasi—no information is available.
T. s. senescens—China. The general color of *T. s. senescens* is similar to *T. s. ordinalis.*
T. s. umbrosus—Kansu (= Gansu) (China). This form differs from *T. s. senescens* by the lack of gray hairs on the head

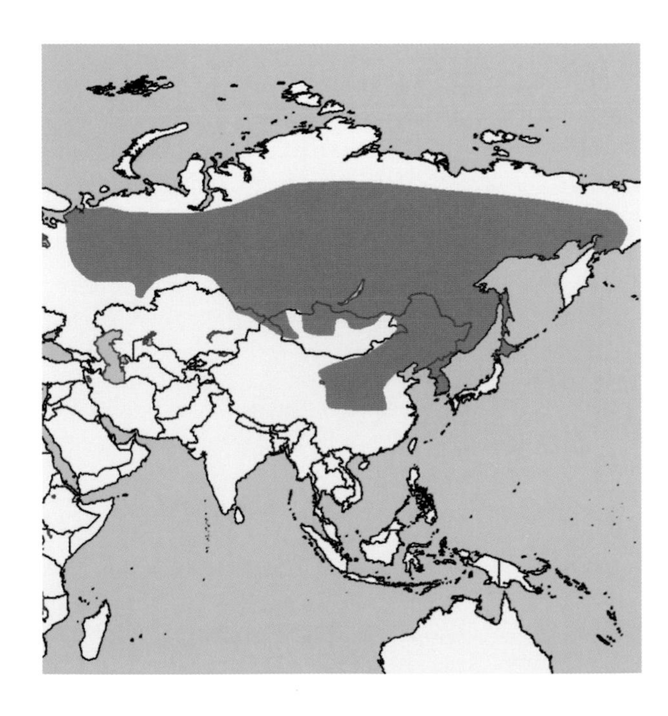

***Tamias sibiricus*.** Photo courtesy David Blank, www.animaldiversity.ummz.umich.edu.

and the shoulders of *T. s. umbrosus*. The head color is darker and duller.

CONSERVATION: IUCN status—least concern. Population trend—stable.

HABITAT: Found in boreal forests across its range, *T. sibiricus* is almost exclusively a resident of coniferous forest habitats (e.g., mixed stands, including Siberian pine [*Pinus sibirica*]). In Japan the species is reported from stands dominated by oak (*Quercus*).

NATURAL HISTORY: The Siberian chipmunk is primarily granivorous, feeding consistently on conifer seeds but also eating herbaceous plants and animal material, including insects, bird eggs, nestlings, and mollusks. *T. sibiricus* is apparently a significant predator of the dusky warbler (*Phylloscopus fuscatus*); under some circumstances it can inflict 80-90 percent mortality on these warblers, and this probably influences nest-site selection by the bird. The Siberian chipmunk is somewhat arboreal and will, on occasion, nest in stumps and logs. However it usually constructs relatively simple shallow burrows that are used for nesting and larder-hoarding food. The burrow structure differs during early and late hibernation. Early in hibernation, burrows usually consist of a single nest chamber and a plug that blocks the entrance. Later during hibernation, the tunnel is blocked further and one or two secondary chambers are constructed. In the spring these secondary chambers are then filled by the soil excavated to construct a new tunnel. Each burrow is occupied either by a single chipmunk or by a mother and her young. Siberian chipmunks are abundant throughout their range, but little demographic data is available. *T. sibiricus* is reported to migrate during years of food shortages. Home ranges reported for an introduced population in France averaged 1.86 ha for males and 0.76 ha for females; these chipmunks showed considerable site fidelity over a two-year period and home range overlap was significant (more than 80%) over one year, but male home ranges overlapped more with those of females than with those of other males. Adults enter hibernation before juveniles, and females do so before males. The onset of hibernation varies annually and is influenced in part by burrow and food availability. Females enter estrus for a single day, usually within a few days after emerging from hibernation. Distinctive vocalizations, consisting of three calls, are given during estrus; one is for advertisement, and all three are involved in courtship. Details from a few observations on copulation are reported in the literature. Gestation is 30-31 days. Litter sizes average 4.4 (range = 1-8) and 4.7 (range = 2-7) in two lab studies, and 4.6 (range = 3-6) in the wild in Japan. One litter per year is produced in the wild, but two are possible in captivity. This species is not territorial.

In some parts of its range (e.g., Italy), introduced populations of *T. sibiricus* may contribute to the decline of the endangered *Sciurus vulgaris*. In China *T. sibiricus* appears to be a reservoir host for *Cryptosporidium parva*, a protozoan pathogen of humans and cattle. A minimum of four species of nematodes are reported from *T. sibiricus*, but at least one is from nonnative populations. *T. sibiricus* can host the bacterium *Yersinia enterocolitica* in Japan, a potential pathogen that causes yersiniosis in humans. Vision, including the structure and function of the retina and the optic nerve, has been studied extensively in *T. sibiricus*. Probably because it is maintained easily in captivity, this species is frequently used for biomedical studies. Primary neoplasia of the liver has been reported in Siberian chipmunks.

GENERAL REFERENCES: Bertolino 2009; Bertolino, Currado, et al. 2000; Bertolino and Genovesi 2005; Blake 1992; Blake and Gillett 1988; Ellis and Maxson 1979; Forstmeier and Weiss 2002, 2004; Gromov et al. 1965; Kaneko and Hashimoto 1981; M. Kawamichi 1989, 1996; T. Kawamichi and Kawamichi 1993; Koh, Wang, et al. 2009; Koh, Zhang, et al. 2010; Levenson et al. 1985; T. Li et al. 2004; Marmet et al. 2009; Matsui et al. 2000; Niethammer and Krapp 1978; Oshida and Yoshida 1994; Pisanu et al. 2009; A. T. Smith and Xie 2008; Tsytsulina, Formozov, Shar, et al. 2008; Vourc'h et al. 2007.

Tamias siskiyou (A. H. Howell, 1922) Siskiyou Chipmunk

DESCRIPTION: This is a large chipmunk, similar in characteristics to *T. ochrogenys* and *T. senex*, but with a distinctive baculum and baubellum.

SIZE: Female—HB 146.7 mm; T 107.2 mm.
Male—HB 144.5 mm; T 105.0 mm.
Sex not stated—HB 144.4 mm; T 105.2 mm; Mass 75.0 g.

DISTRIBUTION: This species is found in the Siskiyou Mountains, from central Oregon to the coast of northern California (USA).

GEOGRAPHIC VARIATION: Like *T. senex*, *T. siskiyou* shows considerable variation between inland and coastal populations, with each of the two populations showing pelage characteristics similar to those of *T. senex*. Two subspecies are recognized.

T. s. siskiyou—populations within 32 km of the coast in Oregon and California (USA). Coastal populations have a darker tawny pelage and larger cranial features.
T. s. humboldti—all inland populations in Oregon and California (USA). This form shows a grayish wash with grayish white stripes.

Tamias siskiyou. Photo courtesy Phillip Colla / Oceanlight.com.

CONSERVATION: IUCN status—least concern. Population trend—stable.

HABITAT: On the coast, this species is associated with mature forests of redwood (*Sequoia sempervirens*), Douglas fir (*Pseudotsuga menziesii*), western hemlock (*Tsuga heterophylla*), and cedar (*Chamaecyparis lawsoniana*); it is also often associated with oaks (*Quercus*) and maples (*Acer*). Inland populations are typically found in mature stands of sugar pine (*Pinus lambertiana*), Jeffrey pine (*Pinus jeffreyi*), cedar, and Douglas fir. Siskiyou chipmunks are more abundant in areas with greater amounts of coarse woody debris, and they are less common in riparian forests than in adjacent upland stands.

NATURAL HISTORY: The diet of *T. siskiyou* includes mostly seeds and fruits of various trees and shrubs, hypogeous (underground) fungi, and insects. This species is highly arboreal, but it appears to do well in logged areas with recent regeneration. It is an important dispersal agent of hypogeous fungi in a variety of forest management regimes. The Siskiyou chipmunk breeds in early to mid-April and produces three to six young after a 28-day gestation period. Little is known about the predators and parasites of *T. siskiyou*.

GENERAL REFERENCES: Jacobs and Luoma 2008; A. N. Johnston and Anthony 2008; A. W. Linzey and Hammerson 2008m; Piaggio and Spicer 2001.

Tamias sonomae (Grinnell, 1915) Sonoma Chipmunk

DESCRIPTION: Most likely a member of the *townsendii* group, the Sonoma chipmunk is usually recognized by its reddish coloration, larger size, and the white trim on its tail.

SIZE: Female—HB 138.3 mm; T 106.6 mm.
Male—HB 134.3 mm; T 106.4 mm.
Sex not stated—HB 133.2 mm; T 111.6 mm; Mass 70.0 g.

DISTRIBUTION: *T. sonomae* occurs in northwestern California (USA), from Siskiyou County south to San Francisco Bay. It is found from the upper Sonoran to transition life zones, below an elevation of 1800 m.

GEOGRAPHIC VARIATION: Two subspecies are recognized.

T. s. sonomae—throughout the species' range. *T. s. sonomae* is larger than *T. s. alleni*.
T. s. alleni—only in the extreme southwestern portion of the species' range, near San Francisco Bay in California (USA). The head and the venter of *T. s. alleni* are darker in coloration.

CONSERVATION: IUCN status—least concern. Population trend—stable.

HABITAT: *T. sonomae* is found in open stands of redwood (*Sequoia sempervirens*), drier low-elevation stands of ponderosa pine (*Pinus ponderosa*), and scrub oak (*Quercus*) habitat, such as chaparral; it also occurs along streams or open brushy fields, and in stands of digger pine (*Pinus sabiniana*), white fir (*Abies concolor*), or red fir (*A. magnifica*).

NATURAL HISTORY: Little information is reported on the feeding habits of the Sonoma chipmunk, but it is assumed to consume conifer seeds, the acorns of scrub oaks (*Quercus*), and the seeds and leaves of other chaparral plants. The Sonoma chipmunk breeds once per year, usually in the spring, but often earlier at sea level. Litters range in size from three to five young. After weaning, females often remain at the nest while males disperse. This results in female kin clusters, and it has been hypothesized that this facilitates the evolution of alarm calling (via kin selection), as is reported for *T. striatus*. Females produce alarm calls more often than males. The only observed predator for *T. sonomae* is the Red-Tailed Hawk (*Buteo jamaicensis*). Parasites include two lice (*Hoplopleura arboricola* and *Neohaematopinus pacificus*) and the larval or nymph stages of three tick species (*Dermacentor occidentalis*, *Ixodes pacificus*, and *I. spinipalpis*).

***Tamias sonomae*.** Photo courtesy David Mclelland.

GENERAL REFERENCES: Best 1993b; Burke da Silva et al. 2002; Casher et al. 2002; R. M. Davis et al. 2008; A. W. Linzey and Hammerson 2008n.

Tamias speciosus Merriam, 1890
Lodgepole Chipmunk

DESCRIPTION: This is a midsized member of the genus. The dorsal stripes are distinct alternating bands of light and dark that extend through the eye and across the side of the face to the nose. The sides are cinnamon to yellowish orange or light brown in color. The skull is broad, and the rostrum is short and narrow at the anterior end. The tail is dark, especially a short subterminal portion of the ventral surface.

SIZE: Female—HB 127.1 mm; T 88.5 mm; Mass 62.7 g.
Male—HB 122.2 mm; T 86.8 mm; Mass 56.8 g.
Sex not stated—HB 124.4 mm; T 94.8 mm; Mass 59.2 g.

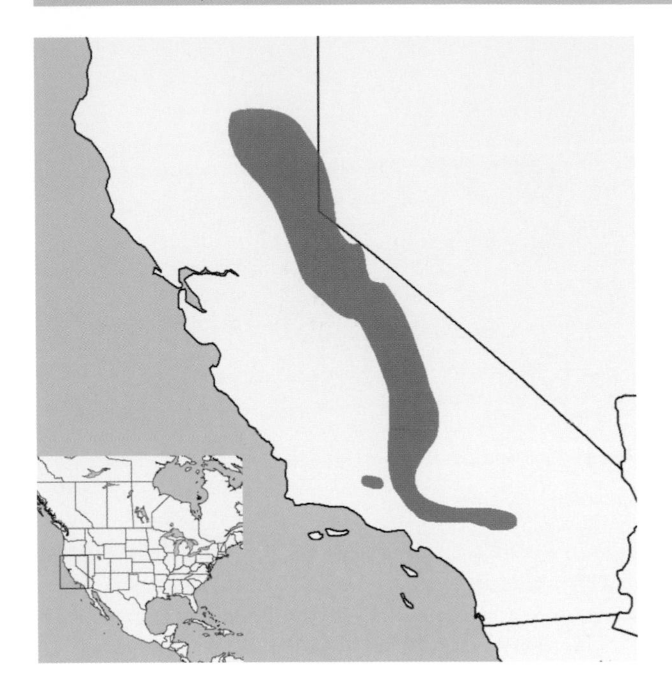

DISTRIBUTION: This species is found in California and extreme westcentral Nevada (USA), occurring in the Sierra Nevada from Mount Lassen to the San Bernardino Mountains and in neighboring mountains in southcentral California.

GEOGRAPHIC VARIATION: Four subspecies are recognized.

T. s. speciosus—southern third and terminus of the species' range. This form is isolated on several mountains south of the Sierra Nevada in southcentral California (USA).

T. s. callipeplus—restricted to Mount Pinos, California (USA) and isolated from the remainder of the species' range. The darker band on the underside of the tail is shorter than that of the other subspecies.

T. s. frater—northern half of the species' range, in the northern Sierra Nevada (USA). It is distinguished from the other subspecies by its broader skull, more reddish sides, and narrower lateral stripes.

T. s. sequoiensis—middle of the species' range, in the southern Sierra Nevada (USA). The underside of its tail is lighter than that of the other subspecies.

CONSERVATION: IUCN status—least concern. Population trend—stable.

HABITAT: The lodgepole chipmunk is found primarily in the Hudsonian, Canadian, and transition life zones, at elevations of 1500-3300 m. It is typically associated with mixed forests, but across its range this can include a number of forest types. Tree species found in habitats occupied by *T. speciosus* include lodgepole pine (*Pinus contorta*), Jeffrey pine (*P. jeffreyi*), sugar pine (*P. labertiana*), red fir (*Abies magnifica*), white fir (*A. concolor*), incense cedar (*Libocedrus decurrens*), chinquapin (*Castanopsis sempervirens*), and oak (e.g., *Quercus kelloggii*). Although prescribed burning for forest management can significantly alter this species' habitat, it does not appear to influence the density of. *T. speciosus*.

NATURAL HISTORY: Largely granivorous (but opportunistic), the lodgepole chipmunk feeds on a variety of seeds, nuts, and fruits of trees, shrubs, forbs, and grasses; it also consumes animal material, such as bird eggs and insects (especially caterpillars). Hypogeous (underground) fungi appear to be eaten by *T. speciosus*, but not in large quantities. This chipmunk is known to scatter-hoard, and it appears to play a role in seed dispersal. Following the wind dispersal of seeds of Jeffrey pine (*Pinus jeffreyi*), *T. speciosus* regularly scatter-hoards them to sites that are better for germination and establishment than those achieved only through passive wind dispersal. The recovery of scatter-hoarded seeds appears to be largely by olfaction and is enhanced by soil moisture. *T. speciosus* is diurnal and most active in the middle of the day. It hibernates, but it is periodically active throughout the year, frequently emerging on warm winter days. The onset of hibernation varies and usually doesn't begin until heavy snows occur. The lodgepole chipmunk emerges from hibernation in March or early April. Breeding occurs one month after emergence and is later at higher elevations; the young are born in early summer. Litter sizes average four (range = 3-6); only one litter per year is produced.

Vocalizations are used for courtship and when disturbed. Details of vocalizations, including evidence for geographic variation in call arrangements, are reported in the literature. *T. speciosus* is more arboreal than most of its congeners, and it often climbs trees when disturbed or threatened. Population densities are usually at or below one animal/ha in the spring, but these then increase, sometimes substan-

Tamias speciosus. Photo courtesy Nichole Cudworth.

tially, in July when the young are weaned. Home ranges are between 1.28 and 2.6 ha, and they do not appear to differ much between the sexes. This species shows altitudinal zonation with several other congeneric species; *T. amoenus* and *T. minimus* are both found at lower elevations, whereas *T. alpinus* is found in the alpine zone above *T. speciosus*. The zone occupied by *T. speciosus* appears to be defined in part by the aggressive behavior of *T. alpinus* and *T. amoenus*; however, *T. speciosus* is reported to be aggressive toward these two species in its preferred habitat and probably competes with them under many conditions. In some parts of its range, *T. speciosus* is sympatric with *T. amoenus*, *T. quadrimaculatus*, and *T. senex*. Only coyotes (*Canis latrans*), Cooper's Hawks (*Accipiter cooperi*), and Red-Tailed Hawks (*Buteo jamaicensis*) have been reported to prey on *T. speciosus*. Parasites include eight species of fleas, two species of lice, and five tick species. *T. speciosus* is also known to carry *Yersinia pestis*, the bacterium responsible for sylvatic (bubonic) plague.

GENERAL REFERENCES: Best, Clawson, et al. 1994b; R. M. Davis et al. 2008; Izzo et al. 2005; A. W. Linzey and Hammerson 2008o; Meyer, Kelt, et al. 2007a, 2007b; Moritz et al. 2008; Pyare and Longland 2001; Sutton 1995; Vander Wall 1993a, 1993b, 1993c, 1993e, 1998, 2002.

Tamias striatus (Linnaeus, 1758)
Eastern Chipmunk

DESCRIPTION: *T. striatus* is a relatively robust member of the genus, identified by its distinct dorsal and lateral stripes. The stripes include a median dorsal stripe of brown bordered by two wider bands of gray to brown or reddish brown agouti. The lateral stripes consist of two black bands on either side of a white or whitish yellow band. Two darker stripes on either side of the whitish stripes extend from the ears to the nose. The tail is dark above and light below. The venter is white. Few other members of the genus occur within this species' range.

SIZE: Female—HB 145.9 mm; T 95.0 mm; Mass 93.9 g.
Male—HB 148.6 mm; T 87.7 mm; Mass 101.0 g.
Sex not stated—HB 150.0 mm; T 93.2 mm; Mass 242.1 g.

DISTRIBUTION: The eastern chipmunk's range extends from southern Manitoba east to Nova Scotia (Canada), and southward to Louisiana, Mississippi, Alabama, and Georgia (USA). It is found east to the Atlantic coast in the northeastern USA; however, in the southern end of its range, it is absent from much of the coastal plain of North and South Carolina, Georgia, Alabama, and Louisiana. *T. striatus* was introduced to Newfoundland.

GEOGRAPHIC VARIATION: Eleven subspecies are recognized.

- *T. s. striatus*—southern Ohio to Georgia, along the Atlantic seaboard (USA). See description above.
- *T. s. doorsiensis*—Wisconsin (USA). This is a pale form, similar to *T. s. peninsulae*, but *T. s. doorsiensis* has brighter and more conspicuous postauricular patches; the dorsal pelage is grayer; and the tail is more frosted with white.
- *T. s. fisheri*—lower Hudson Valley in New York state, south to Virginia and eastern Kentucky, and west to Ohio (USA). This form is paler and grayer than *T. s. striatus*.
- *T. s. griseus*—west of the Great Lakes, in the upper Mississippi Valley (USA). This form is larger and grayer than *T. s. striatus*, and with a more subdued color pattern.
- *T. s. lysteri*—Ontario (Canada) and Michigan (USA) east to southeastern Canada and south to northern New York state (USA). This form is paler than *T. s. striatus*, with a bright yellowish red rump.
- *T. s. ohioensis*—Ohio. This form is very dark and dull-colored.
- *T. s. peninsulae*—Wisconsin and Michigan (USA). This form is slightly smaller than *T. s. griseus*, with a very pale and coppery color.
- *T. s. pipilans*—Louisiana (USA). This form is the largest and richest-colored form. The cheeks, the sides, and the flanks are buffy colored.

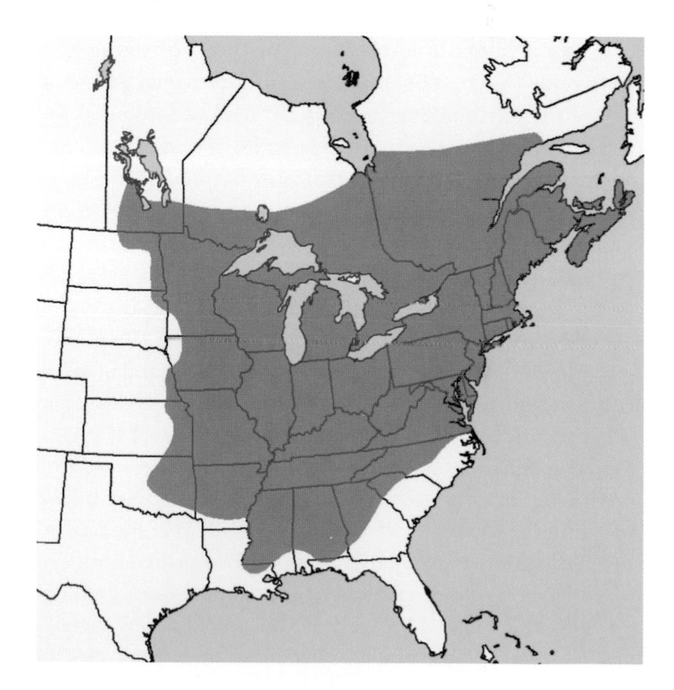

T. s. quebecensis—Québec (Canada). This form is similar to *T. s. griseus*, but *T. s. quebecensis* is smaller and darker; the gray dorsal areas lack the buffy tones of *T. s. griseus*. The white dorsal stripes do not extend to the base of the tail.

T. s. rufescens—Geauga County, Ohio (USA). This form has a relatively short tail and a brilliant red coloration. The rump is reddish tawny, the flanks and the thighs are bright ochraceous buff, and the feet are apricot buff.

T. s. venustus—extreme southwestern limits of the species' range. The dorsal stripes are shorter, and the colors are brighter and more intense than in *T. s. striatus*.

CONSERVATION: IUCN status—least concern. Population trend—stable.

HABITAT: *T. striatus* occupies a variety of specific habitat types, but it is generally associated with deciduous forests in eastern North America. Its presence and abundance are typically determined by the availability of potential burrows, tree stumps and downed woody debris, and mast-producing trees. Downed woody debris appears to be especially important for movement in areas of more open canopy, presumably because of predation risks. Although this species is strongly dependent on oaks, especially white oaks (*Quercus alba*), population densities and habitat selection in some forests may rely even more on the abundance of red maples (*Acer rubrum*). Densities of *T. striatus* in one study showed little variation in even-aged stands of northern hardwoods that varied from sapling age to mature forests. Although eastern chipmunks are sometimes more abundant in mature forests than in fragmented stands, this species appears to be relatively resilient to forest fragmentation. For example, it readily moves across open clearcuts of approximately 250 m, and the microhabitat surrounding its burrows does not differ in fragmented versus continuous forest. These chipmunks do, however, show more vigilance in forested corridors of fragmented landscapes than in continuous forests. Such increased vigilance in more open areas may reduce the time for maintenance activities (e.g., foraging).

NATURAL HISTORY: The eastern chipmunk is largely granivorous, feeding on seeds (such as those of conifers and maples), fruits, and nuts, especially acorns (*Quercus*). This species also frequently includes animal material in its diet, and it is considered a major predator of the young of ground-nesting birds, especially in years when acorns or other mast is low. Additional animals reported in this species' diet include arthropods, amphibians, snakes, and small mammals. Both epigeous (aboveground) and hypogeous (underground) fungi are also consumed when available. *T. striatus* can be considered a central-place forager, collecting large quantities of seeds and then transporting them to the nest. For example, we observed more than 122 maple samaras in the cheek pouches of a single roadkilled chipmunk. More profitable food items may be carried farther than those of lower usefulness. Although the eastern chipmunk is classified as a larder-hoarder, it will frequently cache food in both scatter-hoards and the larder(s) within its burrow. Underground larders are often filled with numerous acorns (sometimes upward of a hundred). Its scatter-hoarding behavior may contribute to seed dispersal under some circumstances, but not to the extent that is attributable to the eastern gray squirrel (*Sciurus carolinensis*). Captive individuals in ambient conditions are estimated to consume a mean of 32.7–35.7 kcal of energy/day. Body temperature is typically maintained between 36°C and 40°C, depending on the ambient temperature. Heat loss is difficult when the ambient temperature is above 36°C. The eastern chipmunk periodically reduces its body temperature and metabolism during the winter period to overcome cold temperatures and food shortages. These torpor bouts usually last one to five days. In contrast with many other ground-dwelling sciurids, energy reserves for *T. striatus* are stored primarily in the form of a larder rather than as body fat. Experimental studies show that an increase in the quality of the larder-hoard—especially the concentration of polyunsaturated fatty acids—reduces the depth

Tamias striatus. Photo courtesy Doris Potter / Focus on Nature Tours, Inc.

and the duration (the latter by more than 50%) of these periods of torpor. Such fats are critical for maintaining phospholipid membranes and preventing the depletion of body fat at low temperatures. Although eastern chipmunks also show a reduction in their aboveground activity in summer, it is not known whether this is due to physiological adaptations, behavioral adaptations, or a combination of the two. Regardless, it appears that food supplementation during this period has little effect on activity levels.

These chipmunks frequently vocalize, and they respond to the alarm calls of neighbors by sitting on their hind legs and engaging in vigilance behavior. Four different calls (in the frequency range of 1-11 kHz) are produced. Experimental playback of alarm calls results in increased vigilance and a reduction in foraging efficiency. There are two annual mating periods: late February to early April, and again in late June to July. The precise time of estrus is not clear, and estimates range from less than 1 day up to 10 days. As many as eight males will pursue a female in estrus. Dominant males are usually successful at mating when competitors are few, but when numerous males are pursuing a female, subordinates often successfully mate. One male may mate several times, and one female may mate with several males. Gestation is 31-32 days; litter sizes are usually four to five (but as high as 15); and the young are less than 5.0 g at birth. Both males and females disperse, although males disperse farther from the natal area and never establish home ranges near the mother, whereas females often do. It is suggested that this male-biased dispersal reduces both inbreeding and mate competition. The lifespan of *T. striatus* is usually 2-3 years, but it can be as long as 6 years. The eastern chipmunk is generally solitary, except when a mother is with her young. This species usually defends an area with a small radius (~ 15 m) around its burrow system. Most displays, however, are stereotypic; chases are short; and agonistic encounters are infrequent. The eastern chipmunk is diurnal and concentrates much of its activity in the middle of the day. Patterns of female and male activity may differ. Home ranges are relatively small, but they vary seasonally, geographically, and with habitat. Home range estimates are anywhere from as low as 0.05 ha to 0.60 ha, yet animals will frequently take longer forays (more than 100 m) to collect food for their larders. Eastern chipmunks are most active around the nest. *T. striatus* is capable of homing up to 500 m, although the probability of a successful return drops off at about 200 m. Experimental manipulation of food in one study resulted in a temporary reduction in range size, which then returned to normal when the food was depleted. However, in a similar study, no change was detected. There is evidence that eastern chipmunks maintain larger home ranges in association with a greater basal area of mast-producing trees, suggesting the possibility of these animals anticipating higher food supplies.

Few definitive measures of the densities and demographics of *T. striatus* are available. In one study, demographic measures (age classes, reproductive measures, and body mass) showed no differences between fragmented and uncut forests, further supporting the idea that *T. striatus* is highly resilient with regard to forest management and fragmentation. However, in another study, densities were negatively correlated with the distance from sources and with woodlot size, suggesting that isolation, rather than patch size, is most important in determining this species' distribution in fragmented landscapes. Numerous animals prey on eastern chipmunks, such as hawks, owls, snakes, and a variety of carnivorous mammals, including house cats. A diverse array of ectoparasites is reported from *T. striatus*. They include at least 2 lice, 12 ticks and mites, and 13 species of fleas. Experimental studies on a sucking louse (*Hoplopleura erratica*) indicate that the transfer of lice occurs primarily during the mating season and is usually from male chipmunks to females. Probably because of its frequent consumption of intermediate hosts, the eastern chipmunk is itself host to a number of endoparasites, including 4 protozoans, 5 cestodes, 2 trematodes, 5 nematodes, and 1 acanthocephalan. In late summer and autumn throughout its range, *T. striatus* is also frequently infected with larvae of the botfly (*Cuterebra*). However, long-term studies of botfly infections on chipmunks fail to show any effect on the reproduction and survival of chipmunks. The eastern chipmunk may also serve as one of several mammalian reservoir hosts for the human pathogen *Cryptosporidium parvum*, as well as a host for the larvae and nymphs of *Ixodes scapularis* (= *I. dammini*), the tick that transmits *Borrelia burgdorferi*, the cause of Lyme disease. Laboratory studies suggest that *T. striatus* may also be involved in the life cycle and ecology of the West Nile virus. Numerous fossil records of *T. striatus* are available; they indicate that during cooler periods of the Quaternary, suitable forests and the eastern chipmunk's distribution extended up to 600 km farther to the southwest than they do today.

GENERAL REFERENCES: J. F. Anderson et al. 1985; Baack and Switzer 2000; Bertolino 2009; Bowers 1995; Bowers and Adams-Manson 1993; Bowers et al. 1993; Bowman and Fahrig 2002; Caire and Caddell 2006; Durden 1983; Ford and Fahrig 2008; French 2000; Graham 1983; Healey and Brooks 1988; Heffner et al. 2001; Jaffe et al. 2005; D. I. King et al. 1998; Lacher and Mares 1996; Lacki et al. 1984a, 1984b; A. W. Linzey and Hammerson 2008p; Loew 1999; Mahan and Yahner 1996, 1998, 1999; J. G. A. Martin and Réale 2008; Munro et al. 2005; Perz and Le Blancq 2001; Platt et al. 2007; Reitsma et al. 1990; Reunanen and Grubb 2005; Schnurr, Canham, et al. 2004;

Slajchert et al. 1997; D. P. Snyder 1982; Thibault and Bovet 1999; White and Svendsen 1992; Wishner 1982; Wood 1993; Yahner and Mahan 1997a, 1997b; Zollner and Crane 2003.

Tamias townsendii Bachman, 1839 Townsend's Chipmunk

DESCRIPTION: Townsend's chipmunks are known for their larger size, darker anterior pelage, and grayish posterior pelage. The dorsal stripes are dark, and the tail is dark with white- or gray-tipped hairs. The ventral surface is white or cream colored.

SIZE: Female—HB 146.3 mm; T 115.9 mm; Mass 76.1 g.
Male—HB 141.8 mm; T 110.5 mm; Mass 70.3 g.
Sex not stated—HB 139.0 mm; T 110.9 mm; Mass 104.0 g.

DISTRIBUTION: This species occurs in southwestern British Columbia, including Vancouver Island (Canada), south through western Washington and Oregon to the Rogue River (USA).

GEOGRAPHIC VARIATION: Two subspecies are recognized.

T. t. townsendii—southwest corner of British Columbia (Canada) southward along the coast of Washington and Oregon to the Rogue River in southern Oregon (USA). The range extends inland about 200 km in Washington and 100 km inland in Oregon. This subspecies is large, and the upperparts vary from yellowish olive gray to rich yellow brown. The dark dorsal stripes are black or brownish black.

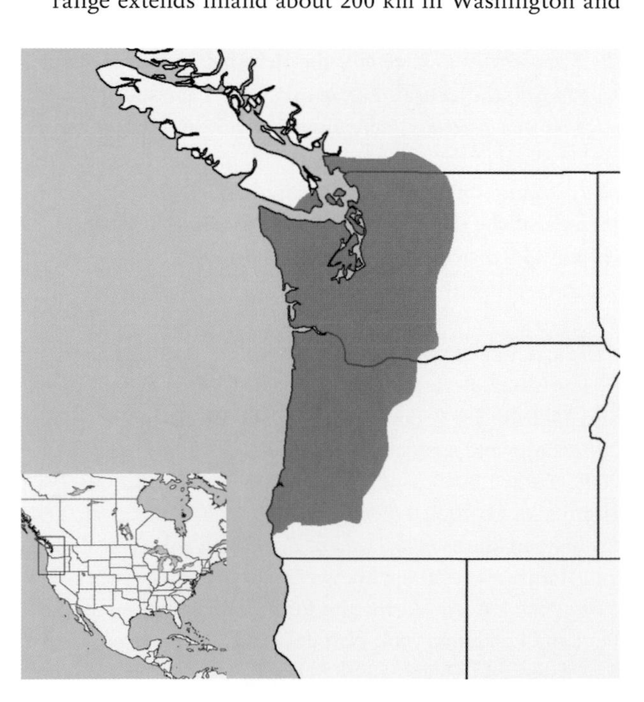

T. t. cooperi—southwestern British Columbia (Canada) southward to the Rogue River in Oregon (USA), and inland along the entire eastern edge of the range of *T. t. townsendii* for about 100 km. This form is grayer and lighter colored. The dark dorsal stripes tend to be grayer rather than browner.

CONSERVATION: IUCN status—least concern. Population trend—stable.

HABITAT: *T. townsendii* is an arboreal chipmunk found in the dense underbrush of mesic closed-canopy forests. This species is more common in upland stands; reproductive individuals, however, are more abundant in riparian forests, while juveniles appear to prefer upland habitat. Densities rise in forest clearcuts where the quantity of herbaceous plants and shrubs increases after timbering has occurred, but Townsend's chipmunk can also rely on riparian buffers following clearcutting. This species often nests on talus slopes. The abundance and size of downed woody debris influences the movement patterns—and quite possibly the survival—of Townsend's chipmunks.

NATURAL HISTORY: In cold temperate regions of the eastern Cascade Mountains (USA), this species hibernates for four to five months. There, breeding occurs in late April to early May. Hibernation varies with climate, however, and Pacific coastal animals are often active throughout the year. Gestation is 28 days, and the young emerge from the nest around early July, but some adults may remain sexually active through the summer, and lactation may occur as late as early August. Mean litter size is 3.8. The young appear to become mature adults by 3 months old, although reproduction does not occur until the following year. Details of this species' development from birth to weaning are summarized by D. A. Sutton. Although Townsend's chipmunks are diurnal and active throughout the day, clear peaks of activity occur in midday. *T. townsendii* regularly feeds on seeds, fruits, tubers, and some vegetation, and it regularly larder-hoards hazelnuts, acorns, and conifer seeds. This species is also a mycophagist, consuming both epigeous (aboveground) and hypogeous (underground) fungi; the latter may increase in diversity and standing crop size with certain forest management practices (e.g., retention of larger live, dead, and fallen trees) or stand age. The most common genera of hypogeous fungi consumed are *Melanogaster* and *Rhizopogon*, as indicated by spores in the animal's feces.

In the Pacific Northwest, *T. townsendii*, along with *Glauco-*

Tamias townsendii. Photo courtesy A. Coke Smith, www.coke smithphototravel.com.

mys sabrinus and *Tamiasciurus,* are part of an important sciurid community whose members collectively serve as important indicators of overall forest health. These species are most abundant in mature or primary forests where seed production is high, or in secondary forests managed by the variable-density thinning technique, coupled with the retention of old live, dead, and fallen trees, which together imitate the complexity of older forests. However, although some studies have reported higher densities in later-stage forests, at least one noted that only the chipmunk's mass increases, whereas density, sex ratios, and the distance moved do not differ, possibly owing to the mild climates and high primary productivity in Pacific coastal forests (e.g., the Oregon Coast Range). On southern Vancouver Island, *T. t. townsendii* often consumes or hoards conifer seeds (lodgepole pine [*Pinus contorta*] and white spruce [*Picea glauca*]) infected with the fungus *Caloscypha fulgens,* which appears to arrest seed germination and rotting and may, in turn, facilitate long-term storage of these seeds in larder-hoards. *T. t. townsendii* may also disperse this fungus, which is common in the conifer litter; however, the full extent of their relationship is not known and deserves further attention. Population densities average about 2.6 animals/ha in one study and, in another, 0.6 -1.1 animals/ha in primary stands. Home range sizes average 0.8 ha. This species may be trap prone; hence estimates based on live-trapping methods should be evaluated with caution. *T. townsendii* is often sympatric with *T. amoenus*. Both are often aggressive, but *T. townsendii* appears to negatively influence both the habitat use and reproduction of *T. amoenus*. The only parasites reported from *T. townsendii* are three species of *Eimeria* and the botfly (*Cuterebra*).

GENERAL REFERENCES: Carey 2001; Carey, Colgan, et al. 2002; Carey and Wilson 2001; E. C. Cole et al. 1998; Colgan and Claridge 2002; Hammond and Anthony 2006; Hayes et al. 1995; Lidicker 1999; A. W. Linzey and Hammerson 2008q; North et al. 1997; Sutton 1993; Waldien et al. 2006.

Tamias umbrinus (J. A. Allen, 1890) Uinta Chipmunk

DESCRIPTION: Uinta chipmunks have a generally brownish dorsum with white longitudinal stripes. This species has

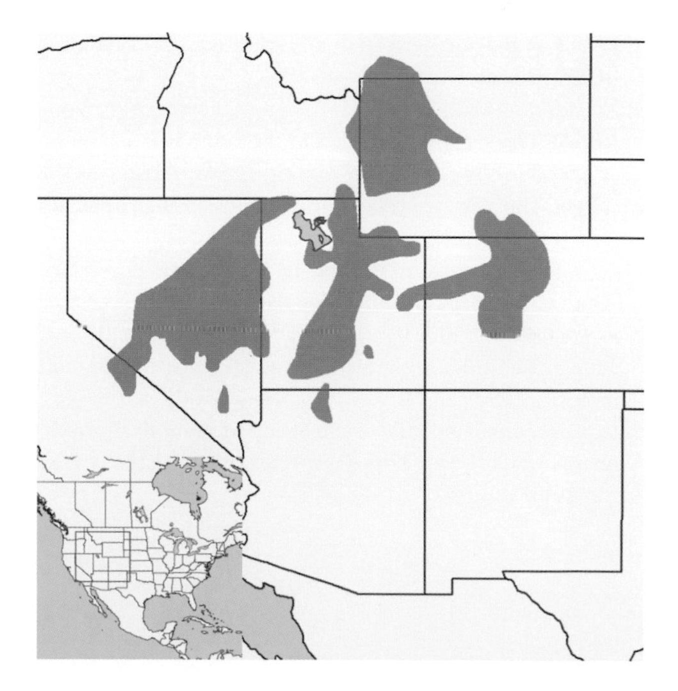

Tamias umbrinus. Photo courtesy Kelly McNulty, 2010.

two white stripes on the face, a brown to cinnamon stripe through the eyes, and a grayish head with white to cream postauricular patches. The venter is white to cream. The tail is grayish on the upper surface and orange to reddish below.

SIZE: Female—HB 125.7 mm.
Male—HB 121.8 mm.
Both sexes—T 89-119 mm; Mass 51-74 g.

DISTRIBUTION: *T. umbrinus* is found from eastcentral California through Nevada; extreme southwestern Montana and eastern Idaho to northwestern Wyoming; Utah; southeastern Wyoming and northern Colorado; and northern Arizona (USA).

GEOGRAPHIC VARIATION: Seven subspecies are recognized.

T. u. umbrinus—northern Utah and southwestern Wyoming (USA). See description above.
T. u. adsitus—southwestern Utah and northern Arizona (USA). The sides are darker and the rump is less brown.
T. u. fremonti—northwest Wyoming (USA) portion of the range. The sides are yellow, and the postauricular patches are grayish white.
T. u. inyoensis—eastern California, Nevada, and western Utah (USA). This is a relatively dark form.
T. u. montanus—primarily in Colorado (USA), with a disjunct range. The sides are gray, and the feet a cinnamon buff color.
T. u. nevadensis—endemic to the Sheep Mountains of southern Nevada (USA). This subspecies has a relatively gray dorsum.
T. u. sedulus—endemic to the Henry Mountains of southeastern Utah (USA). The dorsum is more reddish brown, the sides are yellowish, and the ventral portion of the tail is ochraceous orange.

CONSERVATION: IUCN status—least concern. Population trend—stable. *T. u. nevadensis* is considered a sensitive species in Nevada (USA).

HABITAT: Uinta chipmunks inhabit coniferous forests, particularly those dominated by pine (*Pinus*), fir (*Abies*), and spruce (*Picea*), often at the highest elevations when other chipmunk species overlap with it in distribution.

NATURAL HISTORY: This species is diurnal. Uinta chipmunks probably do not hibernate and can be active year-round in favorable weather; individuals will remain in nests during poor weather and may enter short bouts of torpor. *T. umbrinus* begins breeding in the spring. Nests are located under rocks, in underground burrows, and in tree cavities or the old arboreal nests of other species. This species is more arboreal than many chipmunks, and Uinta chipmunks frequently climb shrubs and trees in search of food and to nest. *T. umbrinus* is an herbivore, principally eating the seeds and fruits of trees, but it also incorporates a diversity of fungi into its diet and will eat insects, bird eggs, and other animal matter. Like many chipmunks, seeds are the principal component of Uinta chipmunk diets; they are carried in cheek pouches, placed in underground burrows, and cached for overwinter use. The overwinter survival rate for Uinta chipmunks is often extremely poor, with local extirpation being common. Due to the modest size of this chipmunk, its predators include a variety of raptors, mustelids, felids, and canids. *T. umbrinus* often retreats to the protective cover of shrubs or trees when under threat. Uinta chipmunks are territorial, and they aggressively defend areas from all other species of chipmunks. Tail movements are most often horizontal when an animal is excited. This chipmunk possesses a loud high-pitched alarm call and vocalizes frequently when confined. Uinta chipmunks are not trapped or hunted; however, they are occasionally a nuisance near homes or in picnic areas.

GENERAL REFERENCES: Bergstrom 1992; Bergstrom and Hoffmann 1991; Levenson 1990.

Urocitellus Obolenskij, 1927

This genus contains 12 species of ground squirrels.

Urocitellus armatus Kennicott, 1863
Uinta Ground Squirrel

DESCRIPTION: The Uinta ground squirrel can be identified by its brown to black tail that is mostly black on top, pale on the edges, and gray on the underside.

SIZE: Female—HB 216.7 mm; T 69.2 mm; Mass 347.3 g.
Male—HB 223.5 mm; T 71.8 mm; Mass 394.9 g.
Sex not stated—HB 225.0 mm; T 65.0 mm; Mass 344.6 g.

DISTRIBUTION: This species' range extends from southeast Idaho and southern Montana to western Wyoming and southcentral Utah (USA).

GEOGRAPHIC VARIATION: None.

CONSERVATION: IUCN status—least concern. Population trend—no information.

HABITAT: The Uinta ground squirrel occupies large open areas, often near timberline. Found in high valleys, this species inhabits dry meadows and agricultural and pasture lands. In western Wyoming, *U. armatus* is also reported to occur in shrub-steppe communities consisting of sagebrush (*Artemisia*) and rabbitbrush (*Chrysothamnus*). The shrubs may be used for shading.

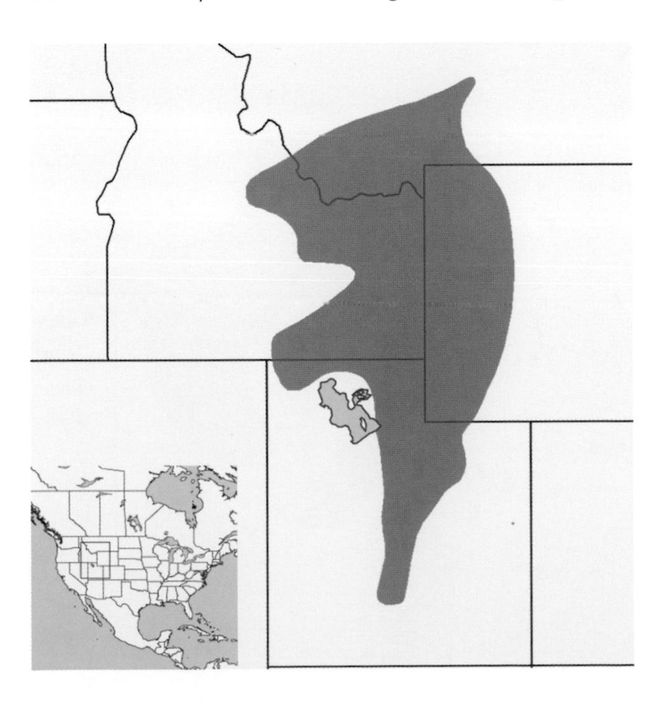

NATURAL HISTORY: *U. armatus* is herbivorous, with grasses comprising the majority of its diet; however, strong seasonal variation is reported. Immediately following emergence, 90 percent of the animal's diet consists of grass leaves, but this food component later drops when the leaves of forbs and grass seeds are added. Sagebrush (*Artemisia*) leaves, earthworms, and roots are also consumed. The mean daily energy budget of the Uinta ground squirrel is 45 kcal/day, and its metabolic rate is 500-570 ml O2/hour at 10°C. This species' arousal from hibernation and emergence from the burrow vary with conditions. One study reported emergence around the beginning of April, in which adult males emerged first, adult females followed, and then yearlings appeared. All individuals begin hibernation before 1 September, but they may immerge as early as late July. Adults tend to enter hibernation before juveniles. Males exhibit territorial behavior soon after emergence, scent-marking the ground by rubbing their cheeks on it. Female distribution, however, often causes males to shift their territory. Breeding occurs within 30 days of emergence. Aggression between males—which includes fighting, biting, and some vocalizations—is common during the breeding season. Females are in estrus for less than one day, usually two to three days after emergence from the burrow; mating occurs within the burrow. Yearling males usually do not mate, whereas yearling females can mate but may experience reproductive failure. The Uinta ground squirrel produces only one litter per year, following a gestation of 23-26 days. Litter sizes range from four to six, usually less than the initial number of embryos. The young are born around 1 May and typically emerge about 22 days later. After mating, females become aggressive and intolerant until their young are weaned. These squirrels have been observed to engage in cannibalism as well as in infanticide.

The Uinta ground squirrel's vocalizations, of which six are reported, are involved in agonistic behavior, and alarm calls are used to warn of predators. Ground predators, other than snakes, elicit a "churr" call, whereas aerial predators engender a distinctly different "chirp" call. Survival rates are reported to be directly related to body mass, which allows an animal to remain alive during hibernation. As a re-

***Urocitellus armatus*.** Photo courtesy Christopher M. Milensky.

sult, in *U. armatus* juvenile survival (about 30%) is observed to be lower than adult survival (60%), and juvenile survival is higher when the population density is lowest. During the first two years, when female body mass is correlated with age, female survival exceeds that of males, but this pattern can change in older Uinta ground squirrels. Body mass is better than age for predicting litter size in this species. Densities have been reported to be as high as 28 yearlings and adults/ha. Juvenile males regularly disperse from colonies, whereas females typically reside near their natal burrow. Raptors, weasels (*Mustela*), badgers (*Taxidea taxus*), and coyotes (*Canus latrans*) are reported to prey on Uinta ground squirrels. *U. armatus* is host to one species of *Tricomonas*, at least one coccidian (*Eimeria*), five species of fleas (*Neopsylla inopina, Opisocrostis tuberculatus, Oropsylla idahoensis, Thrassis pandorae,* and *T. francisi*), and a nematode (*Heligmosomoides polygyrus*).

GENERAL REFERENCES: Eshelman and Sonnemann 2000; Hannon et al. 2006; A. W. Linzey and Hammerson 2008b; Madan et al. 2001; Oli et al. 2001.

Urocitellus beldingi Merriam, 1888
Belding's Ground Squirrel

DESCRIPTION: *U. beldingi* is a medium-sized ground squirrel with a reddish brown dorsum and a cinnamon-colored belly. The ears and legs are small, compared with other ground squirrels; and the tail is short, flat, bushy, and reddish below. The tip of the tail has banded tricolor hairs (red, black, and white). Belding's ground squirrel is distinguished from other species of ground squirrels by the reddish ventral surface of its tail and the lack of stripes or spots on its back. *U. beldingi* is most closely related to *U. armatus.*

SIZE: Female—HB 208.6 mm; T 63.1 m; Mass 265.2 g.
Male—HB 204.7 mm; T 64.9 mm; Mass 228.6 g.
Sex not stated—HB 197.5 mm; T 67.3 mm; Mass 287.4 g.

DISTRIBUTION: This species ranges from western Oregon and southwestern Idaho south to northeastern California, northern Nevada, and northwestern Utah (USA).

GEOGRAPHIC VARIATION: Three subspecies are recognized.

U. b. beldingi—northcentral California and extreme western Nevada (USA). The reddish brown dorsum is darkest in this subspecies.

U. b. creber—eastern Oregon, northeastern California, and

northwestern Nevada (USA). The dorsal surface is lightest in this subspecies.

U. b. oregonus—eastern third of the species' range. The dorsum is buffy gray, with a faint wash of pale brown on the dorsum and head.

CONSERVATION: IUCN status—least concern. Population trend—stable.

HABITAT: Belding's ground squirrel usually is found in grassy meadows, sagebrush (*Artemisia*) communities, or a combination of the two, at relatively high elevations in alpine and subalpine zones. It also frequently occupies agricultural fields. This species' preferred habitat includes open mesic vegetation, often with a nearby water source. *U. beldingi* appears to be limited by the availability of standing water or vegetation with a substantial water content.

NATURAL HISTORY: Belding's ground squirrel is a generalist herbivore, feeding primarily on grasses and forbs, but it also consumes seeds later in the season as it prepares for hibernation. The time invested in feeding increases through the season in this diurnal ground squirrel. Animal matter is eaten when available. Visual cues are important in foraging, and the young appear to learn food preferences from their mothers. Estimates of density vary from 1.2 to 304 animals/ha, and density appears to be limited by the availability of hibernacula. Males hibernate alone, and females often do so in groups. Hibernation lasts longest for adult males; it is estimated to be 280 days at higher elevations and shorter at lower elevations. Males emerge in the early spring and are sexually active for about a month. Females emerge one to two weeks after males, become sexually active within five days, and are collectively receptive for about two weeks. Yearlings are the final animals to emerge. Although the mating season lasts a few weeks, individual females are receptive for only up to six hours on a single day.

Copulation occurs aboveground, and multiple paternity of single litters appears to be common. Gestation lasts between 23 and 28 days; mean litter sizes (based on embryo counts) differ among sites but range between 5.8 and 7.4 for adults and considerably less (4.8) for yearlings. Growth rates for the young are similar to those of other species of *Spermophilus* and *Urocitellus*. Females and their female kin defend small territories around the burrow until the young are weaned. Females are philopatric; juvenile males and recently mated adult males disperse. Natal dispersal, as well as related behavior that helps these squirrels prepare for dispersal, appears to be controlled by body condition (fat content) and mass, and it may be influenced by endogenous time-keeping cues. Juveniles regularly engage in play behavior, which may improve their motor skills and increase their reproductive potential later in life. Infanticide is common in Belding's ground squirrels. It can result in the deaths of as many as 30 percent of all juveniles, and it is most commonly committed by yearling males (for food) and by adult females soon after moving to a new colony.

***Urocitellus beldingi*.** Photo courtesy Phil Myers, Animal Diversity Web, animaldiversity.org.

Alarm calling is common in this species and most likely evolved through kin selection. Two types of alarm calls are produced: a short whistle that appears to be associated with higher risks and faster predators, which usually causes neighbors to retreat to their burrows; and a longer trill that generally results in heightened vigilance. The development of alarm calls in juveniles depends on their early experience and is highly plastic. Young Belding's ground squirrels are able to discriminate between alarm calls within five days of emergence, but their responses to such calls are continually refined for the next several weeks. Whistle calls appear to develop first. Belding's ground squirrels produce scents from both oral and dorsal apocrine glands that are used for the identification of close kin and unfamiliar relatives. Such odors vary in a linear fashion with heritability, and therefore permit the detection of distant relatives, although *U. beldingi* only display nepotistic behavior toward close kin. Littermate preferences are indirectly influenced by the presence (but not by the interaction) of the mother. Mortality results from heavy snows, especially when they occur after emergence. Predators of Belding's ground squirrels include a wide range of mammalian carnivores, rattlesnakes (*Crotalus viridis*), and even Clark's Nutcrackers (*Nucifraga columbiana*). Ectoparasites such as fleas, ticks, mites, and lice are common, but endoparasites (e.g., *Entamoeba*) are limited because of this species' herbaceous diet. *U. beldingi* is sympatric with *U. columbianus*. Belding's ground squirrels can cause significant damage to agricultural crops; yield losses in alfalfa fields, for example, can be as high as 45 percent.

GENERAL REFERENCES: Anthony 1928; Bushberg and Holmes 1985; Chappell et al. 1995; Dobson and Jones 1985; Duncan and Jenkins 1998; Holekamp 1986; W. G. Holmes 1984; W. G. Holmes and Mateo 1998; S. H. Jenkins and Eshelman 1984; A. W. Linzey and Hammerson 2008c; Mateo 1996, 2002, 2003, 2006a, 2006b; Mateo and Holmes 1997, 1999; Mateo and Johnston 2000; McCowan and Hooper 2002; Nunes, Duniec, et al. 1999; Nunes, Ha, et al. 1998; Nunes and Holekamp 1996; Nunes, Muecke, Anthony, et al. 1999; Nunes, Muecke, and Holekamp 2002; Nunes, Muecke, Lancaster, et al. 2004; Peacock and Jenkins 1988; Trombulak 1989; Verts and Costain 1988; Whisson et al. 1999.

Urocitellus brunneus (A. H. Howell, 1928) Idaho Ground Squirrel

DESCRIPTION: The Idaho ground squirrel is recognized by its small head and body; short tail; distinct eye ring; and brownish dorsal pelage, with white spots. The nose, the legs, and the ventral surface of the tail are yellow pink to orange. The pinnae are larger and the pelage darker than those of both *U. townsendii* and *U. washingtoni*. The spotting on the dorsal pelage in *U. brunneus* is more distinct, and the lateral line less distinct, than those of *U. townsendii*. Compared with *U. washingtoni*, the dorsal spots are smaller and the lateral line less distinct. *U. brunneus* is distinguished from other congeners (*U. armatus*, *U. beldingi*, and *U. elegans*) by its smaller size.

SIZE: Female—HB 173.6 mm; T 54.7 mm; Mass 116.8 g.
Male—HB 183.4 mm; T 55.1 mm.
Sex not stated—HB 179.0 mm; T 54.0 mm; Mass 205.0 g.

DISTRIBUTION: Endemic to westcentral Idaho (USA), *U. brunneus* is found in only three isolated areas: north of the Payette River to the Cuddy and Hitt mountains; between the Seven Devils and Cuddy mountains; and east of the West Mountains.

GEOGRAPHIC VARIATION: Two subspecies are recognized.

U. b. brunneus—northern portion of the species' range, in the Seven Devils Mountains and to the southeast, east of the West Mountains (USA). The dorsal pelage is dark gray with a reddish wash.

U. b. endemicus—southern terminus of the species' range, in Washington, Payette, and Gem counties in Idaho (USA). This subspecies occurs at lower elevations than *U. b. brunneus*. The dorsal pelage is grayer than that of *U. b. brunneus*; the base of the tail and the legs are brown.

CONSERVATION: IUCN status—endangered. Population trend—decreasing. *U. b. brunneus* is considered highly threatened, and in 2000 it was formally listed as threatened by the U.S. Fish and Wildlife Service under the Endangered Species Act.

HABITAT: The Idaho ground squirrel is typically found in meadows composed of forbs and grasses. In the northern part of the species' range, its preferred habitat is reported to be open conifer stands with a dense herbaceous understory. Loss of such habitat has resulted from logging and fire suppression, which both contribute to this species' decline.

Urocitellus brunneus endemicus. Photo courtesy Eric Yensen.

NATURAL HISTORY: Food habits have only been studied in *U. b. brunneus*. Its diet, similar to that of *U. columbianus*, includes grasses, leaves, flowers, roots, bulbs, and some insects and fungi. Three types of burrows are produced: for hibernation, for rearing their young, and shallow auxiliary burrows. The Idaho ground squirrel exhibits sexual dimorphism; body size and cranial dimensions are about 2.4-3.3 percent larger for males than for females. The breeding season lasts 12-13 days. Detailed studies have been conducted on the breeding behavior of *U. b. brunneus*. Males emerge two weeks prior to females. Within two days of the females' emergence, mature males court, guard, and copulate with females and then continue to guard females if the density of receptive females is low. Larger males can often displace smaller males, and paternity analyses show that the males that were the last to guard a female and did so the longest fathered 66-100 percent of the young. Mating occurs belowground, and copulatory plugs are produced after mating. One litter is produced annually; average litter size at weaning is 5.2 ± 1.4 (n = 59). Littermates emerge between late March and early April in *U. b. endemicus*. In contrast, emergence of the young is shifted about five to six weeks later in *U. b. brunneus*. Aboveground activity in *U. b. brunneus* is limited, lasting from late March to early August, and it is usually not more than four months for each individual.

U. brunneus appears to be excluded from its preferred habitat by *U. columbianus*. One simple alarm call is used when this species is threatened. Two mammals (long-tailed weasels [*Mustela frenata*] and badgers [*Taxidea taxus*]) and several raptors (Prairie Falcons [*Falco mexicanus*], Cooper's Hawks [*Accipiter cooperi*], Goshawks [*Accipiter gentilis*], Red-Tailed Hawks [*Buteo jamaicensis*], and Northern Harriers [*Circus cyaneus*]) are reported to prey on Idaho ground squirrels. Parasites include numerous ticks, fleas, mites, and lice, although *U. brunneus*—and *U. b. brunneus* in particular—are reported to have fewer ectoparasites than other species of ground squirrels, possibly owing to their fragmented and isolated populations. Detailed demographic studies of a population of *U. b. brunneus* from 1987 to 1989 showed a drop from 272 individuals to 10. The population's decline was due to the loss of older breeding females, which was attributable to a reduction in the food supply, especially of seeds, which are important for hibernation. These food shortages were ultimately caused by habitat alterations, due to fire suppression and, in turn, changes in plant-species composition. Genetic studies of 11 populations of *U. b. brunneus* indicate considerable variation in their genetic structure (due to genetic drift) and a lack of gene flow (due to significant habitat fragmentation and isolation). Four neighboring populations showed no differentiation, while seven others were each distinct. The primary factors contributing to the decline of this subspecies are habitat loss as a result of fire suppression, timbering, the replacement of native grasslands by exotic invasives, and the conversion of native steppe-brush habitat

Urocitellus brunneus brunneus. Photo courtesy Roblyn Stitt.

to agriculture. Between 2004 and 2009, however, habitat restoration and translocations have resulted in an increase in population size.

GENERAL REFERENCES: Gavin et al. 1999; Sherman 1989; Sherman and Runge 2002; Yensen 1991; Yensen and Sherman 1997; Yensen, Hammerson, et al. 2008.

Urocitellus canus Merriam, 1898
Merriam's Ground Squirrel

Urocitellus canus. Photo courtesy Ronn Altig.

DESCRIPTION: This is a small ground squirrel, dark gray in color with no visible stripes or spots. The ears are short, and the tail is short and narrow. The dorsal pelage, the cheeks, and the hind legs all have a pinkish or buffy wash. The ventral surface is whitish. The tail is gray above and cinnamon below. *U. canus* is distinguished from most other ground squirrels, except *U. mollis* and *U. townsendii*, by the absence of dorsal stripes or spots and its smaller ears. Compared with *U. mollis*, *U. canus* has a smaller and broader skull and rostrum.

SIZE: Female—HB 160.2 mm; T 40.8 mm.
Male—HB 153.3 mm; T 38.4 mm.
Sex not stated—HB 171.1 mm; T 36.3 mm; Mass 154.0 g.

DISTRIBUTION: Merriam's ground squirrel is found in the USA, from eastern Oregon to the western bank of the Snake River in westcentral Idaho and southward to the northeastern and northwestern corners of California and Nevada, respectively. It is absent from the southeast and northeast corners of Oregon.

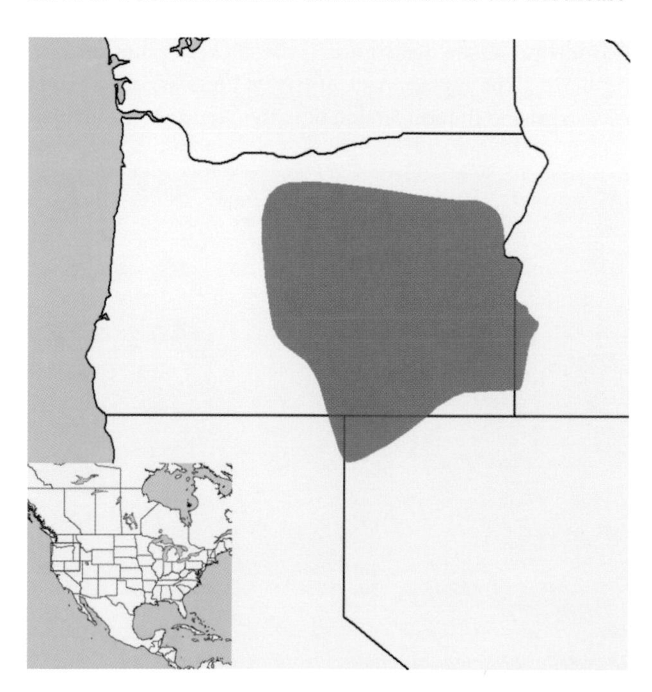

GEOGRAPHIC VARIATION: Two subspecies are recognized.

U. c. canus—throughout most of the species' range. This subspecies' skull and body size are slightly smaller than those of *U. c. vigilis*.

U. c. vigilis—from eastern Oregon to the western bank of the Snake River in westcentral Idaho (USA). It is more hoary white than *U. c. canus*.

CONSERVATION: IUCN status—least concern. Population trend—no information.

HABITAT: Merriam's ground squirrel occurs in areas of nonforested desert chaparral (upper Sonoran life zone) in eastern Oregon, as well as in the riparian zones, pastures, and disturbed open areas of the western bank of the Snake River. Its desert habitat is often associated with big sagebrush (*Artemisia tridentata*), western juniper (*Juniperus occidentalis*), and greasewood (*Sarcobatus vermiculatus*). *U. canus* is often found in conjunction with irrigation ditches and agricultural fields, where it can cause significant damage.

NATURAL HISTORY: Like many other species in the genus, *U. canus* feeds on vegetation, fruits, roots, and seeds, but it also includes animal material, such as insects, in its diet. *U. c. vigilis* in particular is noted for causing damage to alfalfa fields. Population densities are changeable, due to the food supply, disease, predators, and overwinter survival, but demographic studies are not available. Densities usually vary annually between about 20 and 50 animals/ha before and after the young are weaned. *U. canus* is diurnal, but adults may be more crepuscular than juveniles. Like other species of ground squirrels, adults enter hibernation first, sometimes as early as mid-July but usually by August. Adults emerge in early March. No precise data on the length of gestation and lactation are available. An annual litter appears in late April or early May, estimated to vary between 5 and 10 individuals. Numerous birds of prey, carnivorous mammals, and snakes are either assumed or observed to prey on *U. canus*. Little else is known about the ecology of this species. Populations of Merriam's ground squirrels are highly fragmented and appear to be declining, as a result of their elimination as pests by the farming community.

GENERAL REFERENCES: F. R. Cole and Wilson 2009; Helgen et al. 2009; Yensen and Hammerson 2008; Yensen and Sherman 2003.

Urocitellus columbianus (Ord, 1815) Columbian Ground Squirrel

DESCRIPTION: Noted for its large and robust size and its dense pelage, the Columbian ground squirrel is also distinguished by the cinnamon buff color of its dorsal pelage. The tail is dark above and lighter below. The neck is light gray, the eye ring is a pale buff, and the sides are light gray or buff.

SIZE: Female—HB 247.8 mm; T 84.3 mm; Mass 441.4 g.
Male—HB 258.0 mm; T 101.3 mm; Mass 490.1 g.
Sex not stated—HB 269.5 mm; T 98.0 mm; Mass 576.0 g.

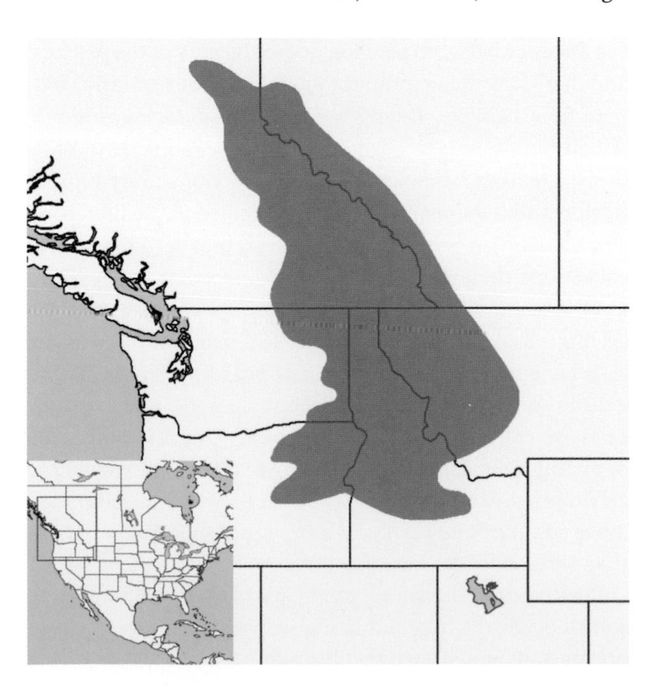

DISTRIBUTION: The Columbian ground squirrel is found across southeastern British Columbia and the southwestern edge of Alberta (Canada) southward to eastern Washington, northeastern Oregon, central Idaho, and western Montana (USA). Models of climate change suggest that alterations in vegetation will most likely result in a reduction in the geographic range of *U. columbianus* in the future.

GEOGRAPHIC VARIATION: Two subspecies are recognized.

U. c. columbianus—most of the species' range, except in northeastern Oregon. See description above.
U. c. ruficaudus—northeastern Oregon (USA). This subspecies is distinguished from *U. c. columbianus* by its more tawny throat and face, darker legs and feet, and broader skull.

CONSERVATION: IUCN status—least concern. Population trend—stable.

HABITAT: The Columbian ground squirrel occupies open habitats, sometimes in association with open forest stands, but usually in meadows of the Hudsonian, Canadian, and upper transition life zones. This species frequently occurs in wet alpine and subalpine meadows, where flooding is sometimes common. *U. columbianus* is also found on agricultural fields and rangelands.

NATURAL HISTORY: The diet of the Columbian ground squirrel consists primarily of flowers, bulbs, fruits, and seeds. This species competes with cattle and sheep for forage and is therefore the subject of strict control efforts in some parts of its range. *U. columbianus* hibernates for approximately 8.5 months and is only active during the remaining 3.5 months. The animals periodically arouse from torpor, depending on the ambient temperature. Males emerge from hibernation in full reproductive condition, and females enter estrus within a few days of emergence. Breeding continues for about 21 days thereafter. However, females that did not successfully breed during this three-week period have been observed to reenter estrus and breed a few weeks later. Yearling females will sometimes forego reproduction because of low body mass at the time of emergence from their first hibernation. Primary factors influencing age at first reproduction appear to be body mass, environmental conditions, and whether the mother is present through the first 2 years of life. Lower reproduction in older females appears to be related to senescence. The timing of the breeding season varies more at higher elevations. In general, it appears that several life-history traits, physiology, and the overall body mass of an animal all show considerable phenotypic plasticity across elevations, whereas structural size

Urocitellus columbianus. Photo courtesy Phil Myers, Animal Diversity Web, animaldiversity.org.

does not. Mean litter sizes vary from 2.1 to 4.2 in the wild and from 3.0 to 4.6 in captivity; and the range of litter sizes (estimated from embryos and placental scars) is from 2.7 to 7.0. Litter sizes are negatively correlated with both elevation and latitude. Gestation is 24 days. In *U. columbianus*, the young are born hairless and weigh between 6.8 and 11.4 g. Full adult weight is not reached until the second year. The young enter hibernation at a lower mass (only 60% of adult mass) than similar species, and they show a higher survival rate during hibernation, even though overwinter survival appears to increase with an animal's body mass. Sex ratios are often skewed toward females. Compared with several congeners, the overall life expectancy for *U. columbianus* is greater.

Population densities from diverse locations range from 24.7/ha to 61.7/ha, and juvenile densities have been reported to vary from 4.6 to 20.7/ha. Home ranges are estimated to average 4200 m^2 for males but only 1000 m^2 for females. Male home ranges overlap considerably, but males are distinctly territorial around the core areas in which they defend breeding females. This male aggression, however, declines after the breeding season. Adult females are also territorial, which may help protect juveniles until emergence. Females establish territories close to their natal burrows and show a considerable degree of site fidelity for several years. It is postulated that this may increase patterns of familiarity and social interactions, but it also heightens inbreeding, especially between daughters and fathers. *U. columbianus* is highly social and appears to develop social bonds earlier than other species. Both empirical and theoretical evidence suggest that the Columbian ground squirrels' social system functions as a social network (society) that is influenced by some individuals more than others. Cross-fostering experiments indicate that social interactions (recognition of kin and agonistic behavior) are based on experience in the nest and therefore are not directly determined by the degree of relatedness, as was previously proposed. Infanticide is reported; in one study it was as high as 7.6 percent of all juveniles (and 12.5% of all litters) over a two-year period. Infanticide in Columbian ground squirrels is generally not between close relatives. Dispersal between colonies is much higher in males than in females. This appears to be due to a higher level of adult aggression toward young males and a greater acceptance of daughters by adult females. There is little evidence that patterns of dispersal are specifically driven by the risk of depression from inbreeding. Dispersal distances in one study were usually less than 4 km, with a maximum of 8.5 km. Hence it is likely that migration patterns, rather than dispersal per se, accounted for the low genetic differentiation observed between populations across 183 km in this same study. Studies of patch colonization, however, indicate that patch occupancy is related to the distance between patches, not to the area of the patches, and that Columbian ground squirrels do not regularly colonize new habitats. Details on locomotion, self-grooming, allogrooming (which occurs but is rare), agonistic interactions, greeting behavior, basking behavior, courtship behavior, and scent-marking are all described in the literature. The Columbian ground squirrel uses a broader range of vocalizations than is typically observed in other species (e.g., *U. armatus, U. elegans, U. richardsonii*). Its vocalizations following copulation are nearly identical in structure to alarm calls but appear to function in mate guarding. These mating calls do not seem to provoke the same vigilance response as alarm calls, yet the animals typically react to playbacks of mating calls outside the breeding season as they would to alarm calls. Alarm calls in response to avian predators are more complex and involve a more rapid succession of calls than those used for ground predators.

Mortality in the young can be particularly high if heavy snows occur during the mating season. A nine-year study, however, demonstrated that the number of young surviving their first hibernation was positively associated with the amount of winter snow and the time their mothers stayed alive past weaning, and negatively linked to summer temperatures and overall population density. Similarly, population regulation has been shown experimentally to be correlated with the availability of food; supplemental feeding results in population increases by as much as 500 percent, due primarily to increased reproduction (i.e., fertility rates), the survival of juveniles, and age at first maturity. Yet there is observational evidence to suggest that the digestibility of food, rather than its quantity, is the more important proximal nutritional variable influencing juvenile survival. Forbs, for example, are 30-40 percent higher in digestibility than grasses. This suggests a tradeoff between reproduction and survival in *U. columbianus*. Mortality in males is reported to be higher during the mating season, and females that wean young are more likely to die than those that lose litters. Likewise, life-history models assume a tradeoff between early and later reproduction; experimental manipulation of litter sizes in one population, however, failed to find evidence of any short-term relationship between current and subsequent reproduction in *U. columbianus*. At least one study suggests that if such a tradeoff exists, it may be dependent on population size. Seven mammalian and three avian species are reported to prey on Columbian ground squirrels. This species distinguishes between rattlesnakes (*Crotalus viridis*) and gopher snakes (*Pituophis melanoleucus*) and shows more cautious behavior toward the former than the latter. Experimental exposure to a potential predator (domesticated dogs) results in higher cortisol levels in female ground squirrels, unless the female is lactating, when cortisol levels are normally low. Ectoparasites reported from the Columbian ground squirrel include at least one tick, four mites, two lice, and three flea species. Endoparasites consist of only one species of *Trypanosoma* and three species of *Eimeria*. *U. columbianus* is a potential reservoir host for organisms causing Rocky Mountain spotted fever, sylvatic (bubonic) plague, and encephalitis. Experimental removal of ectoparasites from female squirrels resulted in better body condition during lactation and at weaning, as well as a larger litter size, suggesting a significant cost from an animal's parasite load.

GENERAL REFERENCES: R. P. Bennett 1999a; Broussard, Dobson, et al. 2005; Broussard, Risch, et al. 2005; Dobson 1992, 1994, 1995; Dobson and Oli 2001; Dobson, Risch, et al. 1999; Elliot and Flinders 1991; Festa-Bianchet and King 1991; Hare and Murie 1992, 1996; Hubbs et al. 2000; K. Johnston and Schmitz 1997; W. King et al. 1991; Manno 2008; Manno et al. 2007; Menkens and Boyce 1993; Neuhaus 2000a, 2000b, 2003, 2006; Neuhaus, Bennett, et al. 1999; Neuhaus, Broussard, et al. 2004; Neuhaus and Pelletier 2001; Stevens 1998; Towers and Coss 1991; Weddell 1991.

Urocitellus elegans Kennicott, 1863
Wyoming Ground Squirrel

DESCRIPTION: *U. elegans* is a relatively small to medium-sized ground squirrel, with big ears and a long tail. The upperparts are mixed gray, buffy, and dusky and are indistinctly mottled. The sides of the head, neck, and body are grayish, and are brownest dorsally. The belly is pinkish, cinnamon, or buffy; the ventral surface of the tail is brownish or buffy. This species is smaller than *U. columbianus* and approximately the same size as *U. armatus* and *U. beldingi*, which are both sympatric with *U. elegans*. The Wyoming ground squirrel's tail is longer and its ventral surface is more buffy and less reddish than that of *U. beldingi*; and its ventral surface is more buffy and less gray than that of *U. armatus*. *U. elegans* is also sympatric with *U. richardsonii* and *U. townsendii* in some parts of its range, but it is distinguished from the former by its consistently smaller body size and cranial measurements, and from the latter by its larger size and more cinnamon-colored venter.

SIZE: Female—HB 206.7 mm; T 73.1 mm; Mass 284.3 g.
Male—HB 204.8 mm; T 72.7 mm; Mass 329.9 g.
Sex not stated—HB 216.0 mm; T 73.5 mm; Mass 311.0 g.

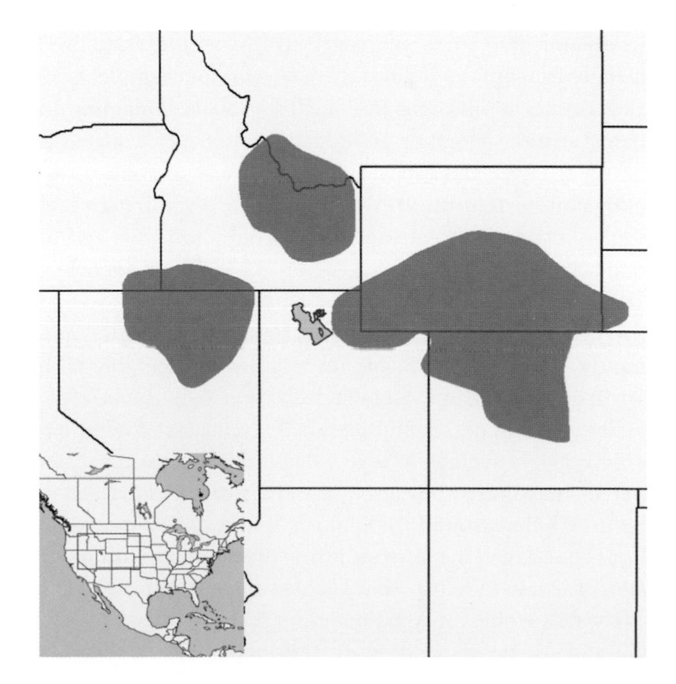

DISTRIBUTION: This species is found in parts of southeastern Oregon, Nevada, and southcentral Idaho; eastern Idaho to southwestern Montana; and northeastern Utah to southern Wyoming, central Colorado, and southwestern Nebraska (USA). The distribution of *U. elegans,* however, consists of three disjunct geographic ranges: one for each of the three subspecies.

GEOGRAPHIC VARIATION: Three subspecies, formerly grouped under *U. richardsonii,* are recognized.

U. e. elegans—southern half of Wyoming into northern and western Colorado, but also into southwestern Nebraska, northeastern Utah, and southeastern Idaho (USA). See description above. This is the smallest and darkest form, with a relatively small hindfoot.

U. e. aureus—central Idaho to southwestern Montana (USA). The underparts and the ventral surface of the tail are ochraceous buff. The general color is lighter than that of *U. e. elegans.*

U. e. nevadensis—originally located in one contiguous area of northeastern Nevada, southeastern Oregon, and southwestern Idaho (USA). It may now be extinct in Oregon; it is represented by only one population in Idaho and just a few others in Nevada. The upperparts have less brown and more gray, and the underparts are darker buff.

CONSERVATION: IUCN status—least concern. Population trend—no information.

HABITAT: The Wyoming ground squirrel is generally found in upland grasslands and sagebrush (*Artemisia*) meadows, usually in montane regions. *U. e. elegans,* for example, typically occurs at elevations from 1500 m to above timberline. In many areas of Montana and Idaho, however, *U. e. aureus* is found at lower elevations, possibly because of competitive exclusion by *U. armatus.* When possible, Wyoming ground squirrels usually select sites with well-drained soils and/or talus slopes, where the construction of burrows is possible.

NATURAL HISTORY: The diet of *U. elegans* consists predominantly of grasses and forbs, but it varies considerably with availability and can even include carrion and insects (especially grasshoppers [orthopterans]). Estimates of adult energy budgets average 35.5 kcal/day, with approximately 24 percent expended aboveground and the balance assumed to be used belowground. Individuals spend approximately 21 hours per day in the burrow during the nonhibernating season, probably to reduce the heat load during the hotter parts of the day. Home range estimates for males average between 0.2 and 0.4 ha, depending on the method of calculation. Densities have been reported to be as low as 0.2/ha and as high as 48/ha. Evidence of competitive interactions with other ground squirrels suggests that in parts of its range, *U. elegans* may be excluded from its preferred habitat by *U. armatus,* and that elsewhere *U. elegans* may exclude *Callospermophilus lateralis. U. elegans* may also interact with other species (*Ochotona princeps* [pikas] and *Marmota flaviventris*), affecting the latitudinal distribution of these species. Winter survival is reported to vary from about 55 to 100 percent for all individuals; it differs between males and females and tends to be lower for juveniles. Survival has been noted to be lower in summer, ranging between 31 and 75 percent in two studies, with juveniles again showing the lowest rates of survival. Observed sex ratios, especially of juveniles, vary considerably through the summer, both because of differences in arousal times and the timing of reentry into hibernation, and because of male-biased dispersal.

Urocitellus elegans. Photo courtesy Sean P. Maher.

U. elegans is diurnal, and its activity is bimodal, with peaks occurring in midmorning and evening. Vocalizations include a combination of five sounds but consist primarily of two major calls: short and long "chirps." The structure and frequency of its vocalizations are distinguishable from those of *U. richardsonii*, and hybrids between the two species produce calls that are intermediate between those of their parents. Males usually emerge a week before females. Within five days of emergence, females enter estrus for no more than 24 hours. Litter sizes in the wild average 5.9, which is relatively high for their adult body size. The young emerge from the burrow 4-5 weeks after birth. *U. elegans* is relatively asocial. Wyoming ground squirrels live in a colony but reside individually in a burrow. Their social structure consists primarily of clusters of related females and juveniles, usually organized in family units. Males disperse after weaning. Agonistic behavior is relatively common and limited primarily to the reproductive period. Adults hibernate first (in late July), whereas juveniles tend to enter hibernation as late as September. The heaviest individuals—often adult males—emerge first, usually by early April. Although laboratory studies suggest that food consumption, fat deposition, and hibernation are controlled by endogenous factors, hibernation is also influenced by temperature and photoperiod. Predators of Wyoming ground squirrels include a minimum of eight species of mammalian carnivores and two species of raptors—Goshawks (*Accipiter gentiles*) and Red-Tailed Hawks (*Buteo jamaicensis*)—with ground predators exerting a particularly significant effect on the survival of *U. elegans*. At least four flea species and three tick species are reported from Wyoming ground squirrels. Endoparasites include at least six coccidians (*Eimeria*), with patterns of prevalence nearly the same as those for *U. richardsonii*. *U. elegans* is affected by sylvatic (bubonic) plague, which can cause severe mortality in its populations, as well as Colorado tick-fever virus, to which this ground squirrel is reported to develop some resistance. Because of its potential threat to humans, livestock, and even agricultural crops, *U. elegans* has been targeted for intensive removal efforts.

GENERAL REFERENCES: Seville and Stanton 1993; Yensen, Mabee, et al. 2008; Zegers 1984.

Urocitellus mollis (Kennicott, 1863)
Piute Ground Squirrel

DESCRIPTION: Piute ground squirrels have a uniform pale smoke gray dorsum suffused with pinkish buff. The cheeks and the hind limbs are washed with red to rust. The venter is white to cream, washed with pinkish buff. The tail is grizzled smoke gray above, with a cinnamon cast to the underside. This species was formerly considered part of *U. townsendii*, but it differs in having a chromosome count of $2n = 36$.

SIZE: Female—Mass 121 g (82-164 g).

Male—Mass 154 g (107-205 g).

Both sexes—HB 213 mm (201-233 mm); T 52 mm (44-61 mm).

DISTRIBUTION: This species is found in the southeast corner of Oregon and the Snake River Valley (Idaho), southward through extreme eastcentral California, Nevada, and western Utah (USA).

GEOGRAPHIC VARIATION: Three subspecies are recognized.

U. m. mollis—California, Oregon, southern Idaho, Utah, and Nevada portions of the species' range (USA). It is intermediate in size among the three subspecies.

U. m. artemesiae—Snake River plain, north of the Snake River in central Idaho (USA). This is a small subspecies.

U. m. idahoensis—westcentral Idaho, northeast of the Snake River (USA). This form is larger, and the pelage tends to be more dappled. The tail is longer, broader, and darker.

CONSERVATION: IUCN status—least concern. Population trend—no information.

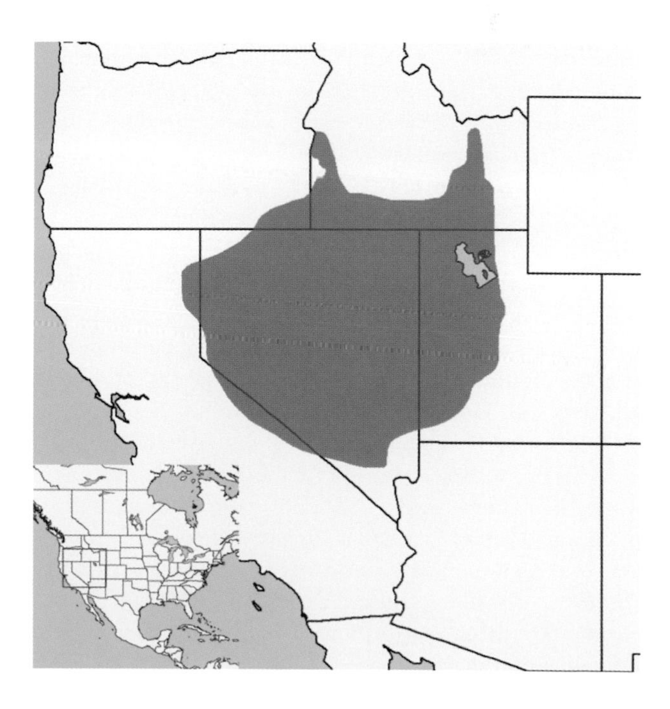

Urocitellus mollis. Photo courtesy Dean Draper.

HABITAT: Piute ground squirrels inhabit high desert sagebrush (*Artemisia*) steppes in well-drained soils with native plants, often near springs and irrigated fields.

NATURAL HISTORY: This species is diurnal. Piute ground squirrels begin hibernation in May (males) to July (females, young), and this lasts until late January to February; males emerge two to three weeks before females. Breeding occurs from late January to March; however, reproduction can be foregone during a drought. After a gestation of 24 days, a litter of 5-10 young is born within a nest chamber in the burrow. Males mature after 1 or 2 years; females typically breed as yearlings. Only one litter per year is produced. Burrow systems can be extensive, with many short escape burrows. Maternity burrows and hibernation burrows may contain multiple openings and nest chambers, and these burrows may penetrate more deeply (to about 1 m). Piute ground squirrels can form colonies, but individuals in the colonies live separately, except for mothers and their litters. Mean home range size is 0.14 ha. *U. mollis* is primarily herbivorous, but these ground squirrels will consume insects and other animal matter on occasion. Their diet includes shoots, leaves, flowers, and the seeds of grasses, forbs, and shrubs; they often feed on crops. They forage on the ground but will climb shrubs for food. Fecundity is high, but adult survivorship is low, averaging 30 percent per year. The principal predators of *U. mollis* are badgers (*Taxidea taxus*) and weasels (*Mustela*), coyotes (*Canis latrans*), raptors, and Northern Ravens (*Corvus corax*). The alarm call of this species is a short high-pitched "squeak," often uttered before retreat to the burrow or while standing upright and remaining vigilant. Piute ground squirrels are a traditional food of local Native American tribes; low levels of hunting or trapping seem to have little impact on *U. mollis* populations. It is considered a pest in many areas and is poisoned or shot.

GENERAL REFERENCES: O'Hare et al. 2006; Rickart 1986; Sharpe and Van Horne 1999; Steenhof et al. 2006.

Urocitellus parryii (Richardson, 1825)
Arctic Ground Squirrel

DESCRIPTION: Arctic ground squirrels have a dorsum that is reddish brown to cinnamon to fuscous, with white to buff spots. The head is more tawny. The eyelids have a white to cream ring. The sides, the feet, and the legs are buff to tawny to cinnamon, becoming gray in winter. The venter is white to straw yellow to cinnamon buff. The upper side of the tail is ochraceous tawny, cinnamon, or cinnamon buff, grizzled with black; the underside is tawny to russet.

SIZE: Both sexes—TL 332-495 mm; T 77-153 mm; Mass 530-816 g.

DISTRIBUTION: This species is found from northeast Yakutia (= Sakha), Magaden, Chukotka, and Kamchatka Krai (Russia) to Alaska (USA) and northern Canada.

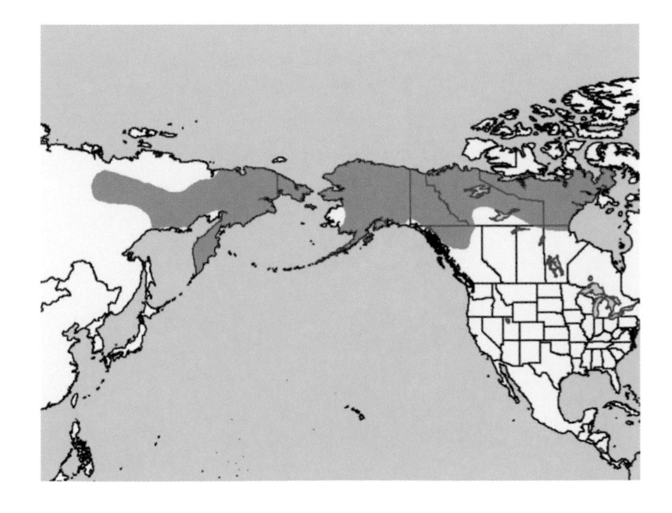

GEOGRAPHIC VARIATION: Ten subspecies are recognized.

U. p. parryii—northern coast of the Yukon and the Northwest Territories (Canada) and extreme northeast Alaska (USA). The upperparts are yellowish brown and dusky, with gray spots. The top of the head and the sides of the neck are a rich reddish brown to yellowish brown. The terminal part of the tail is black.

U. p. ablusus—western and southwestern coast of Alaska (USA). This subspecies has a brown and weakly spotted dorsum.

U. p. kennicottii—northwestern coast of Alaska (USA). This subspecies is a pale form.

U. p. kodiacensis—endemic to Kodiak Island, in southwestern Alaska (USA). This subspecies is heavily spotted with smallish flecks that are sometimes faint.

U. p. leucostictus—northeastern Siberia (Russia). The top of the head is a deep rufous. The dorsum is yellowish brown and heavily spotted with white.

U. p. lyratus—endemic to St. Lawrence Island, Alaska (USA). This is a pale and grayish brown subspecies.

U. p. nebulicola—endemic to the Shumagin Islands of southwestern Alaska (USA). This is a diminutive brown and weakly spotted form.

U. p. osgoodi—south of the Brooks Range in central Alaska (USA). This is the largest subspecies, with a very red belly.

U. p. plesius—southeastern Alaska (USA), Yukon, Northwest Territories, Nunavut, and northern British Columbia (Canada). This is a diminutive and pale gray subspecies.

U. p. stejnegeri—Kamchatka (Russia). The dorsum lacks a fulvous suffusion, and the tail is more extensively black.

CONSERVATION: IUCN status—least concern. Population trend—no information.

HABITAT: Artic ground squirrels live in open arctic tundra, forest meadows, and meadow-steppe habitats, particularly meadows with talus in alpine and subalpine zones in the mountains. They can be found near human settlements.

NATURAL HISTORY: This species is diurnal. Arctic ground squirrels hibernate for up to seven months in burrows. Adult females enter hibernation in September, followed by males a few weeks later in October. Juveniles immerge by the end of October. In March or April, males emerge to establish large territories; females emerge in another two to three weeks and establish their home ranges within a male's territory. A male defends access to several females within his territory. Mating occurs during a two-week period soon after the females emerge, involving all individuals that are more than 1 year old; more than 75 percent of females produce a litter each year. Although females mate with one to four males, the first male sires the offspring in more than 90 percent of the cases; the territory's resident male is most often the first male to mate. After a gestation of 25 days, a litter of six to eight (maximum of 14) young is born within the burrow. Juveniles emerge after 6 weeks, typically in late June or early July, and weaning occurs at about 8 weeks of age. Colonies are composed of a complex system of burrows. Burrows tend to be shallow (less than 1 m deep); permanent burrows have several entrances and nest chambers, compared with short and shallow escape burrows. Primarily herbivorous, *U. parryii* feeds on leaves, shoots, flowers, seeds, and the berries of many tundra grasses, sedges, and forbs, as well as on fungi, mosses, and lichens. Arctic ground squirrels are opportunistic and will eat insects and small vertebrates; infanticide is known. Males, in particular, collect seeds in their cheek pouches late in the active season to build substantial underground caches that will provide a food source to last until the following spring. Principal predators include ermines (*Mustela erminea*), foxes, wolverines (*Gulo gulo*), wolves (*Canis lupus*), grizzly bears (*Ursus arctos horribilis*), raptors, gulls, and jaegers. Annual survival rates for adults and juvenile females average from 25 to 40 percent, with the vast majority of deaths occurring over the winter. *U. parryii* has distinct alarm vocalizations for terrestrial predators (a "cheek-chick" call) and for aerial predators (a shrill whistle). Arctic ground squirrels are hunted primarily for their pelts; they can be a nuisance around human structures.

Urocitellus parryii. Photo courtesy Nichole Cudworth.

GENERAL REFERENCES: Boonstra et al. 2001; Eddingsaas et al. 2004; Gillis, Hik, et al. 2005; Gillis, Morrison, et al. 2005; Karels et al. 2000; Lacey and Wieczorek 2001; McLean 1982.

Urocitellus richardsonii (Sabine, 1822) Richardson's Ground Squirrel

DESCRIPTION: Richardson's ground squirrels have a fuscous cinnamon to gray buff dorsum. The head and the neck are ochraceous to cinnamon buff. The eye ring is white to buff. The snout is often light gray. The lower sides and the venter can be clay to buff to cinnamon. The tail is grizzled charcoal to black on the top and cinnamon buff to clay below.

SIZE: Female—TL 291 mm (264-318 mm); T 70 mm (55-82 mm); Mass 120-590 g.

Male—TL 307 mm (283-337 mm); T 75 mm (65-88 mm); Mass 290-745 g.

DISTRIBUTION: This species is found in the northern Great Plains: in southern Alberta, southern Saskatchewan, and southern Manitoba (Canada) south to Montana, North Dakota, northeastern South Dakota, western Minnesota, and extreme northwestern Iowa (USA).

GEOGRAPHIC VARIATION: None.

CONSERVATION: IUCN status—least concern. Population trend—stable.

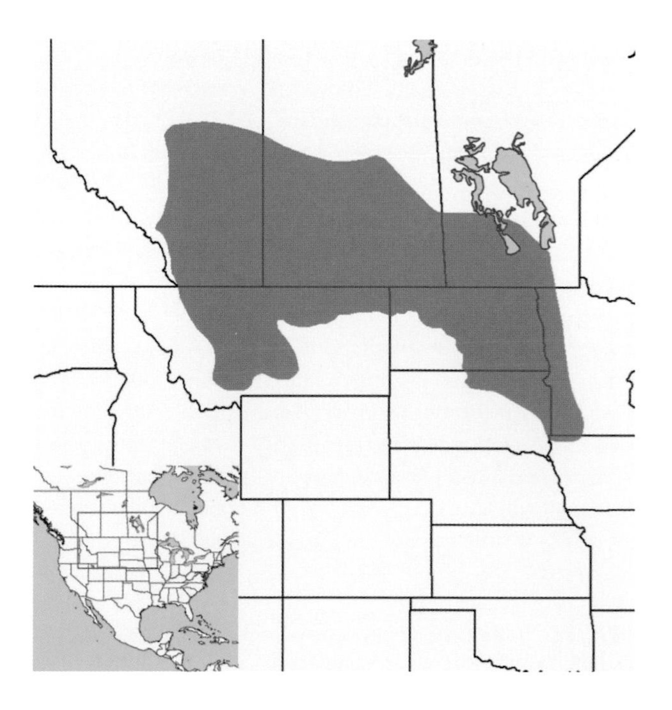

Urocitellus richardsonii. Photo courtesy Takeo Kawamichi.

HABITAT: *U. richardsonii* inhabits grasslands, pastures, fields, and croplands.

NATURAL HISTORY: Richardson's ground squirrels are diurnal, and they hibernate in burrows for six to eight months. Adult males immerge in June, followed by adult females soon after; juvenile females enter hibernation in August, whereas juvenile males remain aboveground and reach adult size as late as October. Emergence begins in March or April, with adult males appearing aboveground first, and adult females about two to three weeks later. Mating occurs three to five days after the emergence of the females. Male-male competition is intense during the two- to three-week breeding season, and many males die as a result of injuries and lost fat reserves. Females mate with several males, and a copulatory plug forms after each mating. After a gestation of 22-23 days, a litter of 3-11 is born within a grass-lined nest chamber in the burrow. The young emerge from the natal burrow at 28-30 days of age, are weaned within one to two weeks afterward, and are sexually mature at 1 year. Males disperse from their natal area during their first year of life; females remain in or near their natal area. Related females within these kin clusters treat each other amicably, but they chase unrelated individuals. Individuals hibernate alone, but they may do so in the same burrow system. Burrows consist of extensive networks of passages and entrances, with multiple chambers that extend to a depth of 1 m. Males will store seeds within the nest chamber to use in preparation for the next spring's mating season. The rigors of the mating season result in male lifespans that rarely exceed 2 years, whereas females live up to 6 years. Principal predators of this species are mustelids, canids, raptors, and snakes. The alarm calls of *U. richardsonii* are a long shrill burst that elicits vigilance for terrestrial predators, and a short "chirp" that causes the animals to escape to cover in response to aerial predators. The number of callers in a population indicates the severity of the potential threat. Richardson's ground squirrels will eat cereal grains; thus they are considered a pest and are poisoned in some agricultural areas.

GENERAL REFERENCES: Dobson, Michener, et al. 2007; Michener and Koeppl 1985; Michener and McLean 1996; Swenson 1981.

Urocitellus townsendii (Bachman, 1839) Townsend's Ground Squirrel

DESCRIPTION: Townsend's ground squirrels have a uniform pale smoke gray back suffused with pinkish buff. The cheeks and the hind limbs are washed with red to rust. The venter is white to cream, washed with pinkish buff. The tail is grizzled smoke gray above, with a cinnamon cast to the underside.

SIZE: Female—Mass 125 g.
Male—Mass 174 g.
Both sexes—HB 212 mm (200-232 mm); T 46 mm (32-54 mm).

DISTRIBUTION: *U. townsendii* occurs in southcentral Washington (USA), north and west of the Columbia River.

Urocitellus townsendii. Photo courtesy Jane Abel.

GEOGRAPHIC VARIATION: Two subspecies are recognized.

U. t. townsendii—Washington, south of the Yakima River and north of the Columbia River (USA). This subspecies has a chromosome number of $2n = 38$.
U. t. nancyae—Washington, north of the Yakima River and west of the Columbia River (USA). This subspecies has a slightly lighter dorsum, and a chromosome number of $2n = 36$.

CONSERVATION: IUCN status—vulnerable. Population trend—decreasing.

HABITAT: Townsend's ground squirrels inhabit high desert sagebrush (*Artemisia*) regions and cultivated fields.

NATURAL HISTORY: This species is diurnal. *U. townsendii* is now restricted to two subspecies found in Washington, and little ecological information is available about these populations. Two clear karyotypic forms, once considered subspecies of *U. townsendii* (*U. canus* and *U. mollis*), are much better known. Townsend's ground squirrels hibernate from late May (males) or June (females and young) until late January to February. Breeding occurs soon after emergence. A single litter of 4-16 young (to judge from embryo counts) is born within the burrow in March. Juveniles emerge from the burrow about 4 weeks later, in late March or

April. *U. townsendii* is primarily herbivorous but will consume insects. Its diet includes shoots, leaves, flowers, and the seeds of grasses, forbs, and shrubs; these squirrels often also feed on crops. Their principal predators are badgers (*Taxidea taxus*); others probably are coyotes (*Canis latrans*), raptors, and Northern Ravens (*Corvus corax*). *U. townsendii* is considered a pest and is poisoned or shot in many areas of its range.

GENERAL REFERENCES: O'Hare et al. 2006; Rickart 1999.

Urocitellus undulatus (Pallas, 1778)
Long-Tailed Ground Squirrel

DESCRIPTION: The dorsum is straw yellow to russet to ochre, with white to grayish spots; it is often suffused with gray. The head is generally darker brown on top; the cheeks and the lower jaw are straw yellow. The sides are reddish. The venter is straw yellow to orange to rust. The tail is grizzled gray to black on the upper side, with a frosting of white near the tip; the underside of the tail is rust colored near the base.

SIZE: Both sexes—HB 210-315 mm; T 100-140 mm; Mass 250-580 g.

DISTRIBUTION: *U. undulatus* is found in eastern Kazakhstan, southern Siberia and Transbaikal (Russia), northern Mongolia, and Xinjiang and Heilongjiang (China). Another population is found in Yakutia (= Sakha) in northeastern Siberia (Russia).

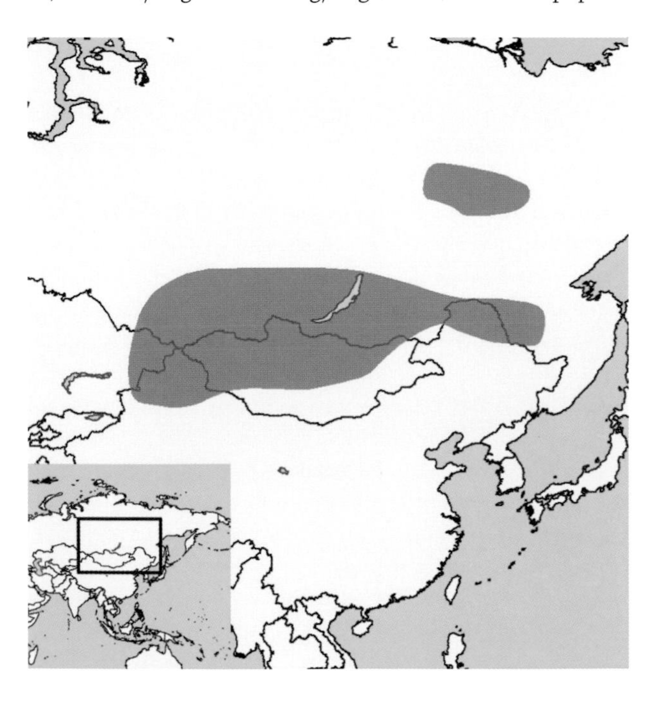

Urocitellus undulatus. Photo courtesy Andrey Tchabovsky.

GEOGRAPHIC VARIATION: Six subspecies are recognized.

U. u. undulatus—western and southwestern Transbaikal region (Russia).
U. u. eversmanni—Altai Mountains (Kazakhstan, Russia, Mongolia, and China). This subspecies is smallish but brightly colored, with intense rust and orange tones.
U. u. jacutensis—far northeast isolates of the distribution (Siberia). This subspecies is large and pale.
U. u. menzbieri—eastern portion of the species' range, along the Amur River from Russia into China. This is the largest of the subspecies and very pale.
U. u. stramineus—southwestern part of the species' range, including western Mongolia, northwestern China, and eastern Kazakhstan. This form is extremely pale, with the underparts a light orange yellow.
U. u. transbaikalicus—eastern Transbaikal region (Russia). This form is intermediate to extreme in size. The head is colored a bright cinnamon buffy clay.

CONSERVATION: IUCN status—least concern. Population trend—stable.

HABITAT: Long-tailed ground squirrels inhabit semidesert, grass, and mountain steppes; alpine meadows; savannas; and the areas along rivers.

NATURAL HISTORY: This species is diurnal. Long-tailed ground squirrels hibernate in burrows from September (males) or October to March or April; males emerge one to two weeks before females. For individuals more than 1 year old, copulation occurs soon after female emergence. After a 30-day gestation, a litter of three to nine young is born in the burrow. The young begin to leave the burrow at about 4-5 weeks of age. Only one litter is produced each year, but greater than 90 percent of females often reproduce. A single male and several females traverse and defend a group home range of about 0.16 ha; males scent-mark throughout their territories. Dense colonies have numerous conspicuous burrows, with large soil mounds demarcating the entrances. Burrows are complex and often have multiple entrances, a substantial length (up to 15 m), and depths of up to 3 m. *U. undulatus* is an herbivore and feeds primarily on green shoots, leaves, flowers, and seeds; however, it will eat insects and animal matter opportunistically. Long-tailed ground squirrels will collect food in their cheek pouches and carry it back to the burrow for an overwinter food supply. This species' primary predators are felids, foxes, wolves (*Canis lupus*), mustelids, and raptors. Alarm vocalizations are high-pitched but relatively soft. Long-tailed ground squirrels formerly were hunted in large numbers for their pelts; however, this use has nearly ended. *U. undulatus* is considered a local pest when found near agricultural areas.

GENERAL REFERENCES: Badmaev 2008; Ognev 1963; Ricankova et al. 2006; A. T. Smith and Xie 2008.

Urocitellus washingtoni (A. H. Howell, 1938) Washington Ground Squirrel

DESCRIPTION: Washington ground squirrels have a pale smoke gray dorsum washed with a pinkish buff, with small cream to buff flecks. The venter is grayish white washed with pinkish buff, and the feet are white to buff. The tail is grizzled gray, suffused with cinnamon above and with a pinkish cinnamon underside; the tail is frosted with white to buff.

SIZE: Male—Mass 158 g in late February, 257 g in late May. Both sexes—HB 185-245 mm; T 32-65 mm.

DISTRIBUTION: This species is found in eastern Washington and northeastern Oregon (USA).

GEOGRAPHIC VARIATION: None.

CONSERVATION: IUCN status—near threatened. Population trend—decreasing. *U. washingtoni* is a candidate species for protection under the U. S. Endangered Species Act. This species is endangered in Oregon (USA).

HABITAT: Washington ground squirrels occupy shrub-steppe habitats of the Columbia River Basin and are most abundant in areas of extensive grass cover on deep soils with a high clay content.

NATURAL HISTORY: *U. washingtoni* is diurnal. After accumulating fat stores, adults enter estivation that grades into hibernation from late May to early June; juveniles enter estivation one to two months later. Adults emerge in January to mid-March. Females breed within a few days of emergence. After a gestation of 23-30 days they give birth to young in February and March, in a nest chamber in inconspicuous underground burrows. The young emerge in late March or April and reach adult size by late May to early June. Female coalitions appear to form within colonies, and group members are highly social. Juvenile males typically disperse 880 m (on average), with a maximum dispersal distance of 3.5 km. Annual mortality rates exceed 60 percent, and maximum lifespan is less than 5 years. Badgers (*Taxidea taxus*), coyotes (*Canis latrans*), raptors, Northern Ravens (*Corvus corax*), and snakes are probably this species' principal predators. Major threats to conservation of *U. washingtoni* are the conversion of grasslands and sagebrush (*Artemisia*) steppes

Urocitellus washingtoni. Photo courtesy Jodie Delavan.

to agriculture and urban development, in addition to the resultant habitat fragmentation. Historically, recreational shooting and poisoning (to reduce conflict in agricultural lands) occurred. Long delays in decisions regarding legal protection have also been a major conservation issue. Washington ground squirrels are herbivores and feed heavily on succulent grasses and herbs (including cultivated plants); their diet also includes roots, bulbs, seeds, insects, and other animal matter. These ground squirrels are not hunted or trapped; however, they have been considered a pest in agricultural lands.

GENERAL REFERENCES: Betts 1990, 1999; Greene et al. 2009; Rickart and Yensen 1991.

Xerospermophilus Merriam, 1892

This genus contains four species.

Xerospermophilus mohavensis (Merriam, 1889)
Mohave Ground Squirrel

DESCRIPTION: Mohave ground squirrels are a nearly uniform light brown to pale drab on the dorsum, with a conspicuous absence of stripes or spots. There is often a suffusion of cinnamon about the head. The feet are pale buff to cinnamon. The venter is white to cream. The tail is short but broad, with a fuscous dorsal surface and a white to cream underside; the entire tail is frosted with white to cream.

SIZE: Both sexes—HB 223 mm (210-230 mm); T 66 mm (57-72 mm); Mass 70-300 g.

DISTRIBUTION: This squirrel is found in the northwestern Mohave Desert and the Owens Valley of southern California (USA).

Xerospermophilus mohavensis. Photo courtesy Philip Leitner.

GEOGRAPHIC VARIATION: None.

CONSERVATION: IUCN status—vulnerable. Population trend—decreasing.

HABITAT: Mojave ground squirrels inhabit deserts with abundant annual herbaceous vegetation on sandy or gravelly friable soils. Creosote bush (*Larrea*) associations are favored.

NATURAL HISTORY: *X. mohavensis* is diurnal. Mohave ground squirrels hibernate in burrows from midsummer (August) until February (or as late as May in the north), during which period fat stores provide their energy source. Males emerge about two weeks prior to females and establish home ranges; mating occurs in burrows soon after emergence from hibernation, usually in February and March, and females locate their nest burrows within their home range. After a gestation of 29-30 days, a litter of four to nine young is born in the burrow during late March or early April, except in years of significant drought, when no reproduction occurs. Burrows are typically found on the edges of the home range (sometimes more than 250 m from its center), and individuals use multiple burrows for temporary cover, maternity, and hibernation. Burrows are plugged each evening, probably to minimize predator access. Mohave ground squirrels remain solitary outside of the breeding season; aggression in the form of chases occurs among individuals. Male home ranges (6.74 ha) are much larger than those of females (0.73 ha), although sexual size dimorphism does not exist. *X. mohavensis* is omnivorous and feeds on the shoots, leaves, flowers, and seeds of green grasses, forbs, shrubs, cacti, and yucca; this species also consumes considerable numbers of insects and other animal matter as available. These items are collected in cheek pouches and eaten with the squirrel sitting upright on its haunches, often on a promontory. Populations fluctuate dramatically and are extirpated in marginal areas during extended droughts. Recolonization occurs from adjacent areas, and long-distance dispersal appears important for this species' persistence on the landscape, given local drought-related extirpation. Juveniles are known to move as far as 6.2 km, with males (averaging 1.5 km) moving farther than females (averaging 0.5 km). Their primary predators are raptors, snakes, badgers (*Taxidea taxus*), coyotes (*Canis latrans*), and bobcats (*Lynx rufus*). Individuals are extremely docile and tolerate approach by humans. When threatened, Mohave ground squirrels rarely run to burrows, but rather drop close to the ground and rely on camouflage. Their alarm call is a high-pitched "peep," with a raspy quality. Urban sprawl constitutes a threat to the persistence of *X. mohavensis*, due to loss of habitat and the purposeful removal of *X. tereticaudus* by poisoning.

GENERAL REFERENCES: Best 1995i; Brooks and Matchett 2002; Gall and Zembal 1980; Hafner 1992; Hafner and Yates 1983; J. H. Harris and Leitner 2004, 2005; Laabs 1998.

Xerospermophilus perotensis (Merriam, 1893) Perote Ground Squirrel

DESCRIPTION: Perote ground squirrels have a grizzled yellowish brown dorsum, with a series of thin short incomplete black lines appearing toward the posterior; faint buff flecks or spots are visible in new pelage. Their white eyelids are distinctive. The feet, the lower limbs, and the venter are a pale buff. The upper side of the tail is grizzled yellowish brown, with black becoming prominent toward the tip; the underside is ochraceous buff with a clear black line evident near the tip.

SIZE: Both sexes—HB 250 mm (243-261 mm); T 71 mm (57-78 mm).

DISTRIBUTION: This species is restricted to 2500 km^2 in Veracruz and Puebla in central México.

GEOGRAPHIC VARIATION: None.

CONSERVATION: IUCN status—endangered. Population trend—decreasing. *X. perotensis* is federally listed as threatened in México.

HABITAT: Perote ground squirrels live in high-elevation arid plains, open scrub, and rocky slopes, at elevations from 2200 to 2700 m.

Xerospermophilus perotensis. Photo courtesy Juan Cruzado.

NATURAL HISTORY: This squirrel is diurnal. Perote ground squirrels hibernate in burrows for a relatively short three to four months each year, with individuals active for about nine months (from March to November). Adult males emerge before adult females and yearlings. Mating occurs in April and May, and pregnant females are found in the population from June to August. A litter of three to five is born in a burrow and begins emerging in mid-July. Most adults have initiated hibernation by late September or early October; juveniles may remain aboveground until November. Their principal predators probably include long-tailed weasels (*Mustela frenata*) and domestic dogs. Perote squirrels are not hunted or trapped. The major threats to conservation of this species are habitat loss and fragmentation, due to the encroachment of agriculture and grazing; however, *X. perotensis* does not frequent agricultural fields. Only 30 percent of their potential habitat remains untransformed. Perote ground squirrels in many areas (82% of known localities) only remain in small fragments along railways.

GENERAL REFERENCES: Best and Ceballos 1995; Sánchez-Cordero et al. 2005; Valdez and Ceballos 1997, 2003.

Xerospermophilus spilosoma (Bennett, 1833) Spotted Ground Squirrel

DESCRIPTION: *X. spilosoma* have a gray to brown to cinnamon to reddish dorsum, with scattered small white to buff spots of variable intensity. Spotted ground squirrels exhibit considerable color variation, especially in their dorsal pelage. The venter is whitish to buff to cinnamon or pinkish buff. The tail resembles the dorsum, but with fuscous black toward the tip and a cinnamon core on the underside.

SIZE: Both sexes—TL 185-253 mm; T 55-92 mm; Mass 100-200 g.

DISTRIBUTION: *X. spilosoma* is found from southwestern South Dakota south to southern Texas and west to Arizona (USA), and from northcentral to central México.

GEOGRAPHIC VARIATION: Thirteen subspecies are recognized.

X. s. spilosoma—Durango (México). See description above.
X. s. altiplanensis—high plains of westcentral Chihuahua

(México). The dorsum of this subspecies is dark and yellowish.

X. s. ammophilus—extreme northern Chihuahua (México). This is an extremely pale form found on the sand dunes south of the city of Juárez.

X. s. annectens—southern Texas and the Rio Grande Valley (USA). This subspecies exhibits no distinctive coloration, but it has a slightly longer skull that is narrower at the base and broader at the interorbital region.

X. s. bavicorensis—endemic to the basin of the Laguna de Babícora in westcentral Chihuahua (México). This is a large subspecies that has an unusually blackish pelage.

X. s. cabrerai—San Luis Potosí and vicinity, in central México. This is a largish form, with a darker dorsum.

X. s. canescens—southcentral New Mexico southeast through Texas (USA) into Coahuila (México). This subspecies has a reddish hue, with spots that are larger and more numerous on the posterior dorsum; the apex of the tail is dark.

X. s. cryptospilotus—Four Corners area of Arizona, Utah, Colorado, and New Mexico (USA). This is a pale and more reddish form, with distinct small spotting on the dorsum.

X. s. marginatus—southeastern Colorado, southwestern Kansas, New Mexico, western Texas, and the panhandle of Oklahoma (USA). This subspecies has a reduced number of dorsal spots; the dorsum is pale and gray, tinged with red.

X. s. obsoletus—southeastern Wyoming, northeastern Colorado, western Nebraska, and northwestern Kansas (USA). This typically drab form has less distinct and sometimes obsolete spots on the dorsum.

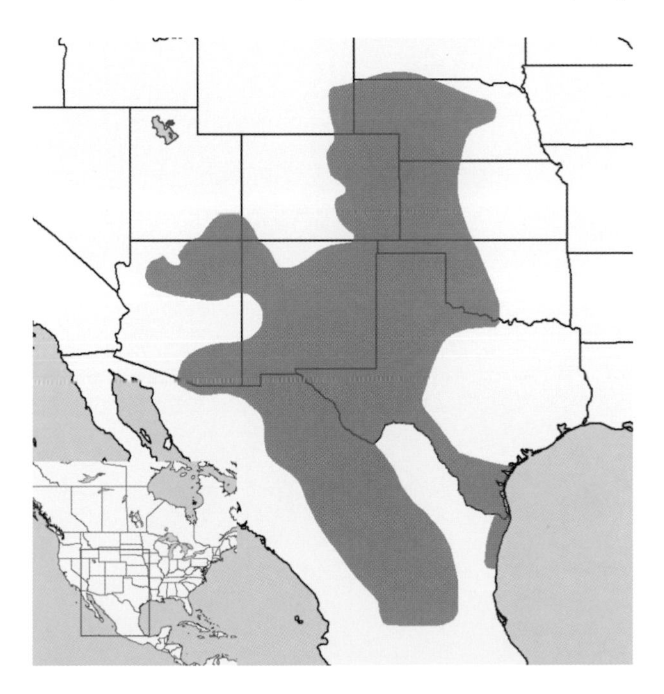

Xerospermophilus spilosoma. Photo courtesy Randall D. Babb.

X. s. oricolus—coastal Tamaulipas, in northeastern México. This form is distinctly spotted, with near cinnamon buff on the hind feet.

X. s. pallescens—central México. This subspecies is darker on the posterior dorsum, with finer spots.

X. s. pratensis—northern Arizona (USA). This is a smallish short-tailed form, with a darker dorsum and tip of the tail.

CONSERVATION: IUCN status—least concern. Population trend—stable.

HABITAT: Spotted ground squirrels are found in deep sandy soils with sparse vegetation, including desert grasslands, desert scrub, heavily grazed pastures, sand dunes, coastal plains, and grassy urban parks, cemeteries, and schoolyards.

NATURAL HISTORY: *X. spilosoma* is diurnal. Spotted ground squirrels hibernate in burrows for seven to eight months, but they may stay active or only enter short bouts of torpor during mild winters. Yearlings are the first to emerge, followed by older adult males and, finally, adult females (after about two to three weeks). Mating takes place when the females appear. After a gestation of 28 days, a litter of five to eight young is born in a burrow; the young emerge 3-4 weeks later, and are weaned after an additional three weeks. Home range sizes are relatively large for both males (1.02-4.86 ha) and females (0.50-1.55 ha). Adult males immerge in July and August, followed closely by adult females; juveniles often remain active until late September. *X. spilosoma* adults are aggressive during the breeding season but become more tolerant of conspecifics outside of that time period. Burrow densities and populations are typically low, probably reflective of resource availability and the low levels of sociality in this species. Burrows seem to be relatively simple, and they penetrate to about 1 m belowground. Spotted ground squirrels will feed on leaves, shoots, flowers, and the seeds of grasses, forbs, small shrubs, and cacti; they will also consume insects and small vertebrates. Most of their active time aboveground consists of foraging and feeding. Overwinter food caches are not common. Their predators are primarily mustelids, canids, raptors, and snakes. *X. spilosoma* is seldom hunted or trapped, but it is sometimes removed by shooting or poisoning. This species can persist in overgrazed grasslands.

GENERAL REFERENCES: Fitzgerald and Streubel 1978; Livoreil and Baudoin 1996; Livoreil et al. 1993; Mandier and Gouat 1996; Millán-Peña 1998; Stangl and Goetze 1991.

Xerospermophilus tereticaudus (Baird, 1858) Round-Tailed Ground Squirrel

DESCRIPTION: Round-tailed ground squirrels have a pale gray to pinkish buff to tawny dorsum, with the sides, the feet, and the venter white to pale buff. The cheeks are white to pale clay. The short round tail has a black tip; the tail is the color of the dorsum above, and drab to buff to cinnamon below.

SIZE: Both sexes—HB 204-278 mm; T 60-112 mm; Mass 110-170 g.

DISTRIBUTION: *X. tereticaudus* is found in the deserts of southeastern California, southern Nevada, and western Arizona (USA), and northeast Baja California and Sonora (México).

GEOGRAPHIC VARIATION: Four subspecies are recognized.

X. t. tereticaudus—southern Nevada and southeastern California (USA), and northeastern Baja California (México). See description above.

X. t. apricus—isolated in the Valle de la Trinidad, at higher elevations between mountain ranges, in northern Baja California (México). This form is slightly darker and browner.

X. t. chlorus—Coachella Valley, in southern California (USA). This subspecies has a drabber and less pink dorsum.

X. t. neglectus—southern and western Arizona (USA), and northwestern Sonora (México). This is a darker and shorter-tailed form.

CONSERVATION: IUCN status—least concern. Population trend—stable. *X. t. chlorus* is a subspecies of special concern in California (USA).

HABITAT: Round-tailed ground squirrels occur in sandy low flat deserts, commonly in communities dominated by mesquite (*Prosopis*) and creosote bush (*Larrea*). This species can also be found in urban areas, cemeteries, and parks.

NATURAL HISTORY: *X. tereticaudus* is diurnal. Round-tailed ground squirrels can stay active year-round in some locations, but they typically enter torpor in burrows. Torpor lasts for four to six months in the case of adult females (September-March), and for seven to nine months for adult males (May-January) in some areas. Males emerge first and, in some populations, defend territories; females follow two to four weeks later, and breeding occurs soon after their

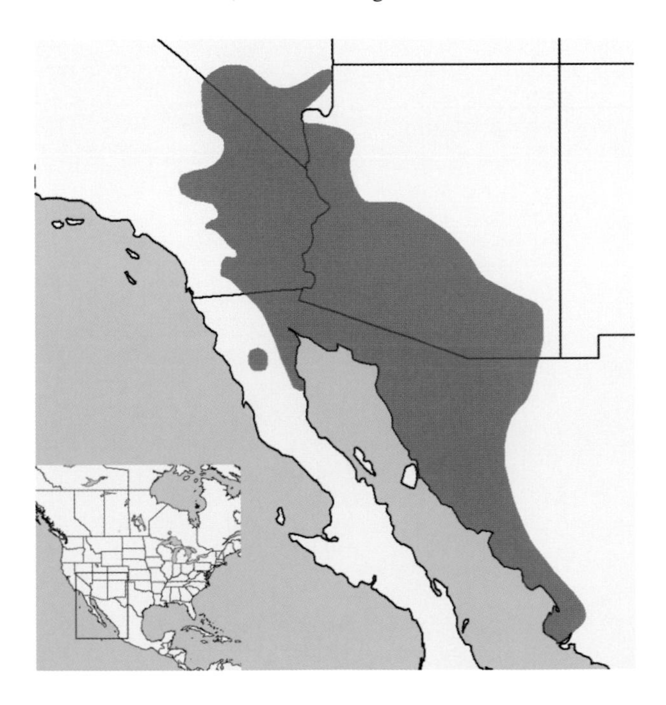

Xerospermophilus tereticaudus. Photo courtesy Randall D. Babb.

emergence. After a gestation of 26–35 days, a litter averaging 4.9–6.5 (range = 1–13) young is born within a burrow; this is the only litter produced during the year. Multiple paternity is common, and rises with an increase in litter size. Males are capable of mating with multiple females, and in multiple years. Reproduction is well timed to capitalize on the short period of winter rains that occur in the Mohave and Sonoran deserts. Pregnant females can be found from March to June. The young are weaned by 5–6 weeks, just one to two weeks after emerging from the natal burrow. Natal dispersal is male biased, with a few females remaining in or near their natal area after the June exodus of males. Sexual maturity is reached by 10–11 months of age. Home ranges of males and females do not differ in size, averaging 0.30 ha. Small burrow complexes are often evident. Individuals may overwinter together in burrows, although in other seasons nesting is rarely communal; this species does not form matriarchal colonies. Burrows are relatively inconspicuous and lack clear evidence of excavated soil or a mound; they are located at the base of shrubs, in rocks, or along a wash. *X. tereticaudus* is incredibly heat tolerant, capable of foraging at a temperature of 45°C. These animals use small patches of shade and burrows for short periods of time. Round-tailed ground squirrels forage on the ground and in the canopy in search of buds, flowers, leaves, and the fruits and seeds of forbs, shrubs, trees, and cacti, often collecting them in modest cheek pouches. Their principal predators are raptors, snakes, and, to a lesser extent, canids, mustelids, and felids. Under threat, *X. tereticaudus* returns to the cover of its burrow, often giving a high-pitched whistle or a series of "chirps" for aerial predators, or using foot drumming in the presence of snakes. Round-tailed ground squirrels are primarily threatened by the conversion of deserts to urban and agricultural lands, especially in the case of *X. t. chlorus*. *X. tereticaudus* is not hunted or trapped, but it is occasionally poisoned as a pest around human habitations and historic structures, due to its burrowing habits.

GENERAL REFERENCES: Ball et al. 2005; Drabek 1973; Dunford 1977a, 1977b, 1977c; Ernest and Mares 1987; Turkowski 1969; Wooden and Walsberg 2002.

APPENDIX: Representative Skulls of Each Squirrel Genus

Species are organized taxonomically. (*Biswamoyopterus* and *Eupetaurus* are not included.) Photos courtesy James Di Loreto and Lauren Helgen of the Smithsonian Institution, and Jennie Miller of Yale University.

Ratufa bicolor *Sciurillus pusillus*

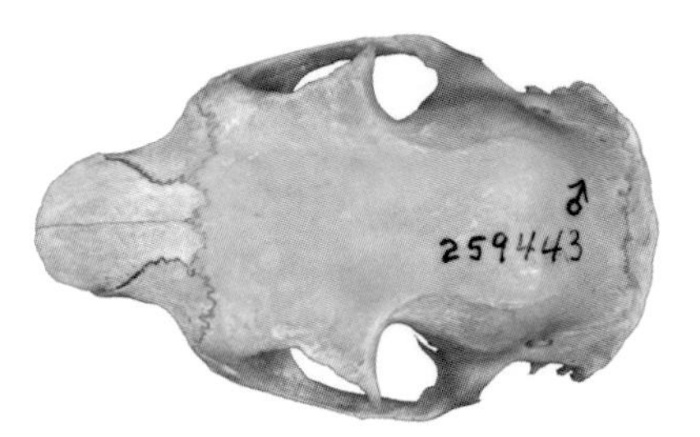

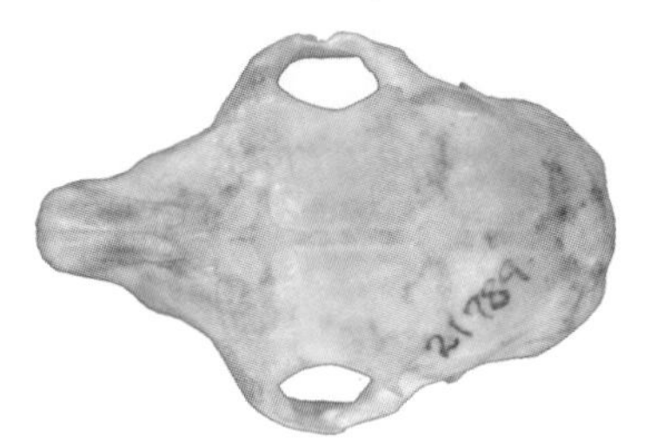

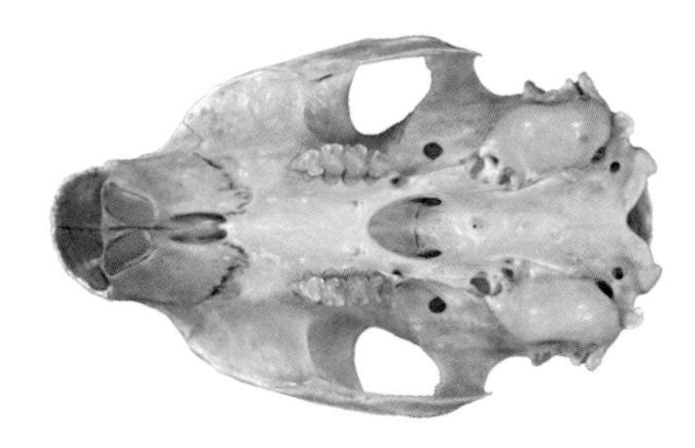

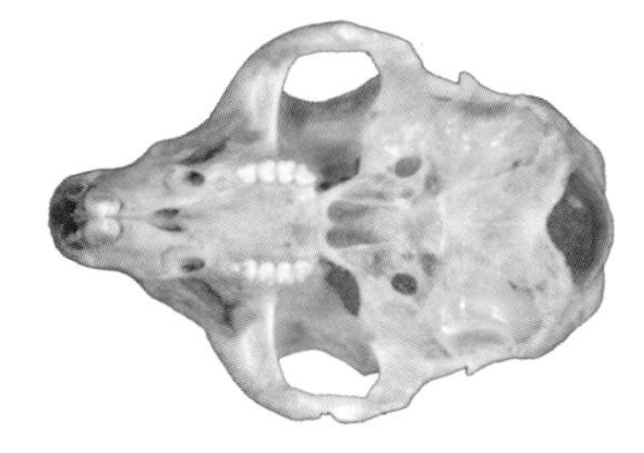

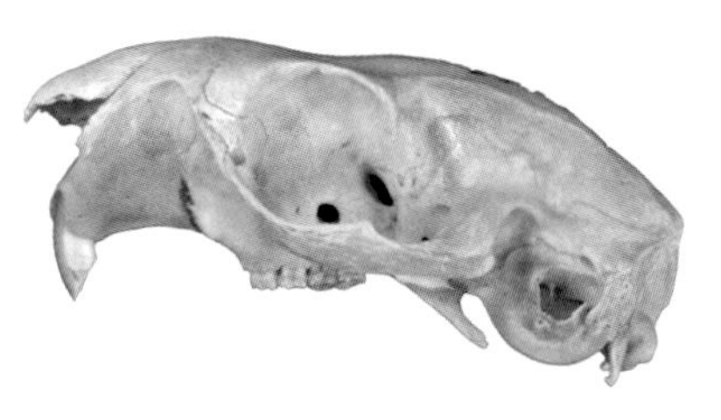

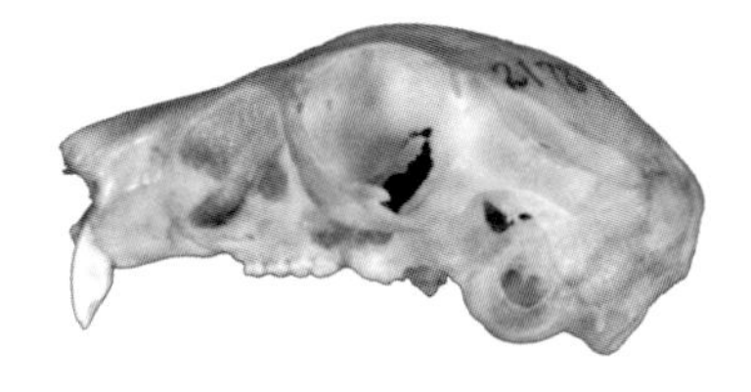

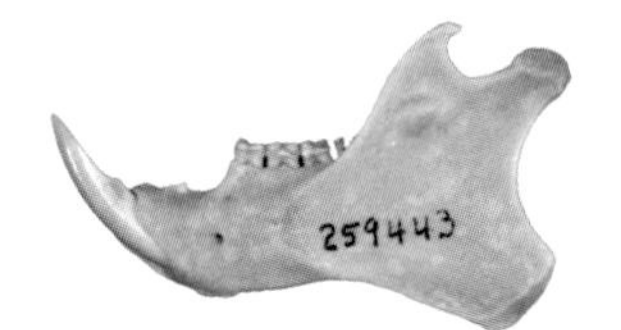

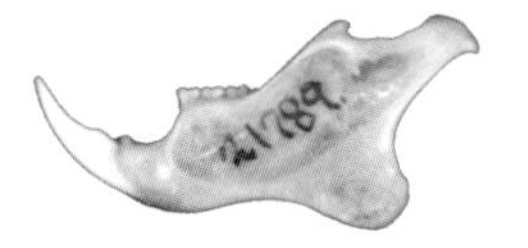

25 mm

25 mm

Microsciurus flaviventer *Rheithrosciurus macrotis*

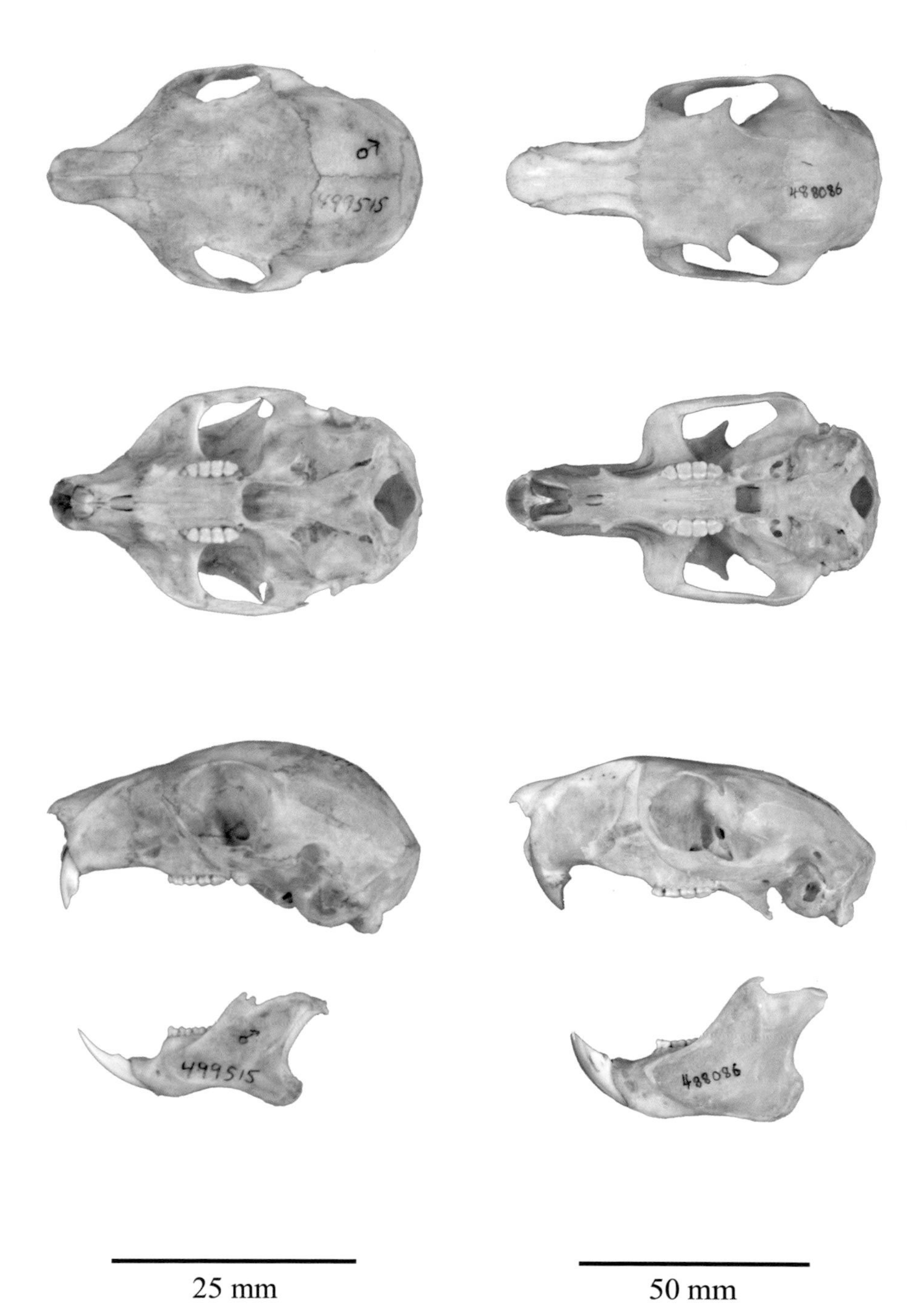

Sciurus griseus | *Syntheosciurus brochus*

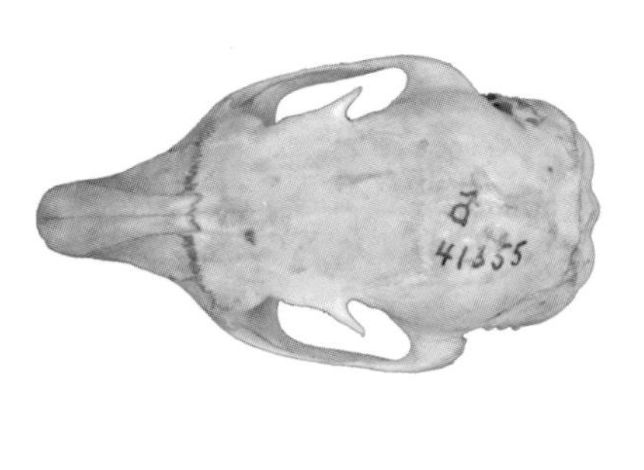

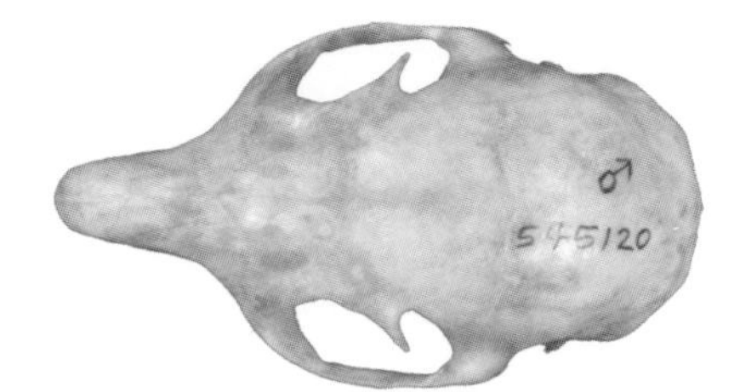

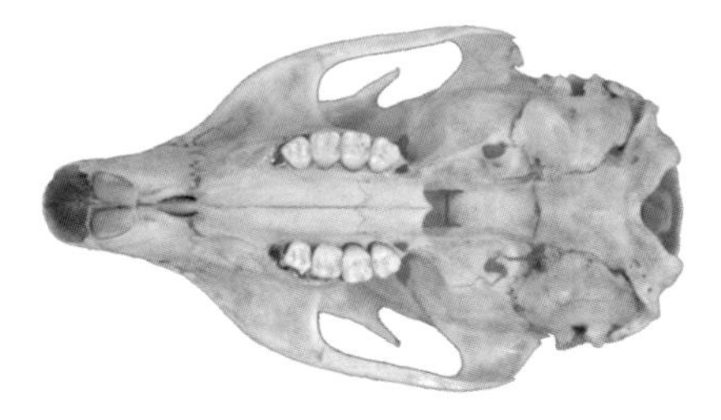

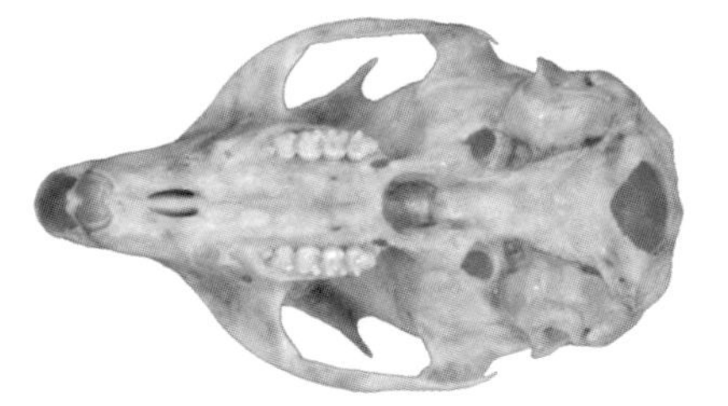

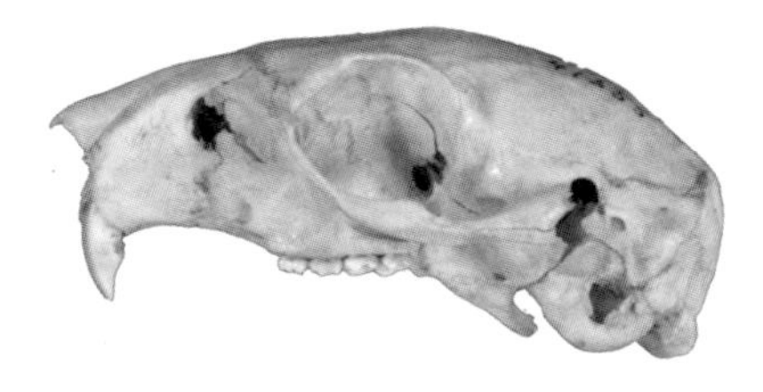

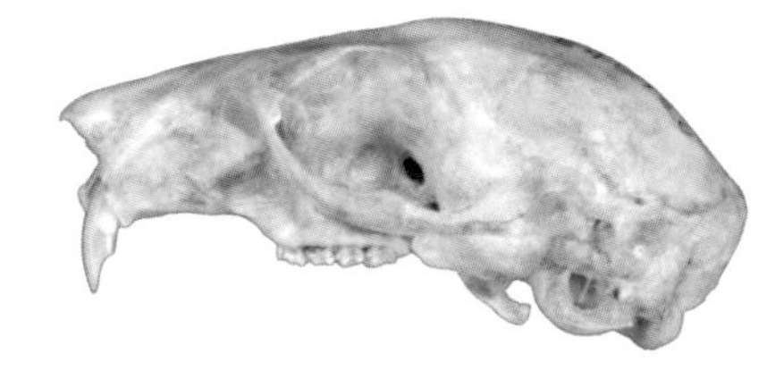

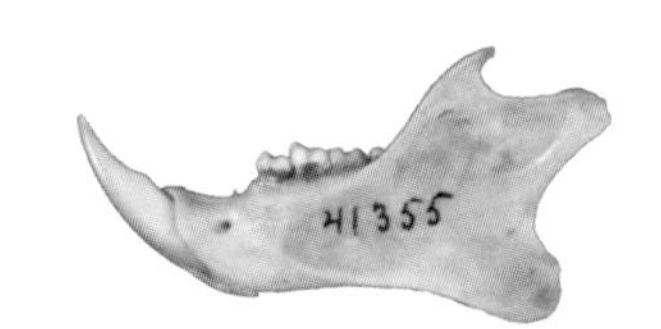

25 mm

25 mm

Tamiasciurus hudsonicus *Aeretes melanopterus*

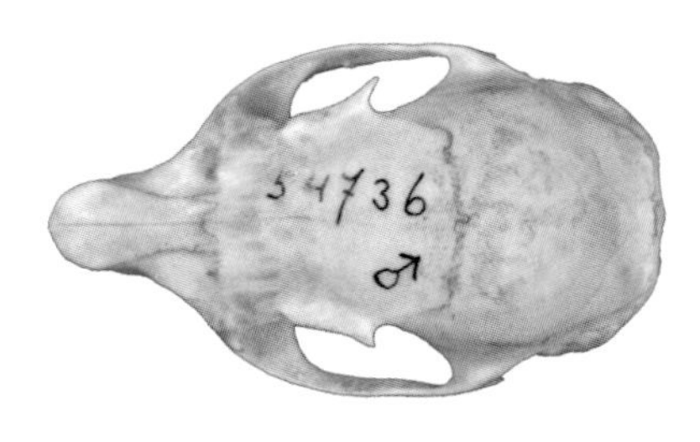

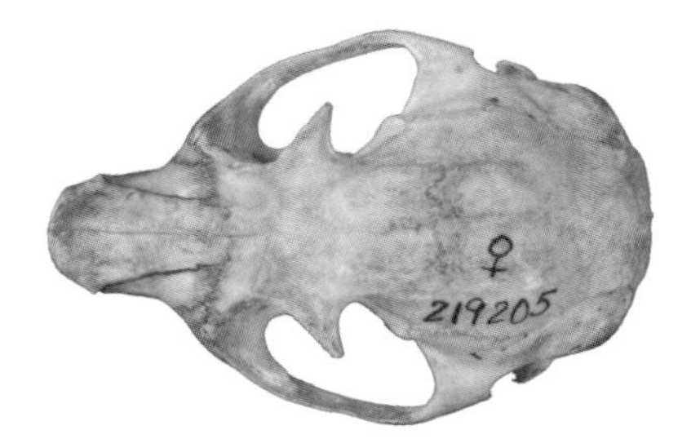

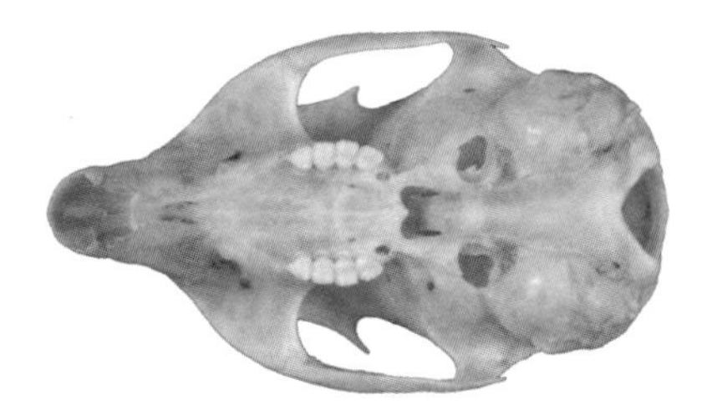

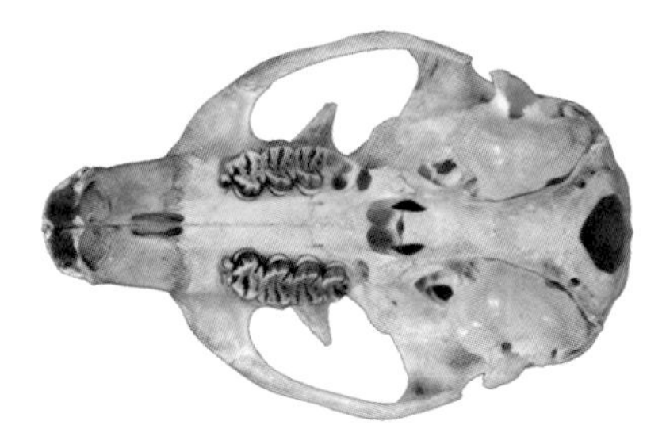

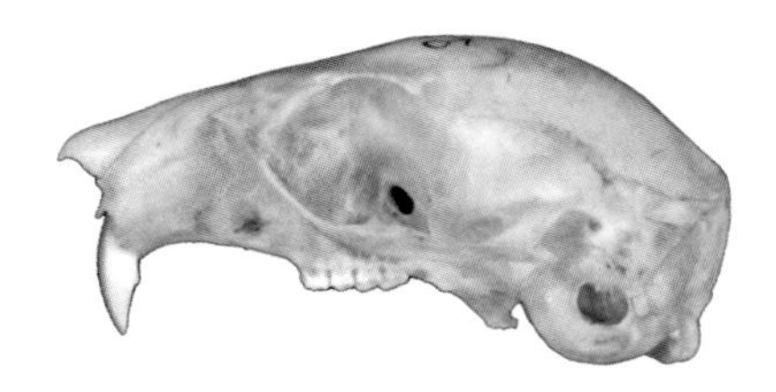

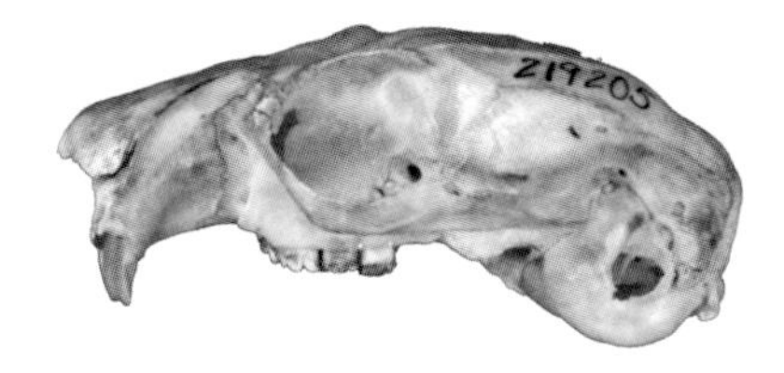

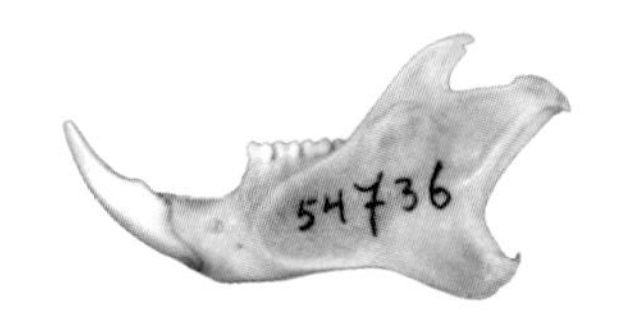

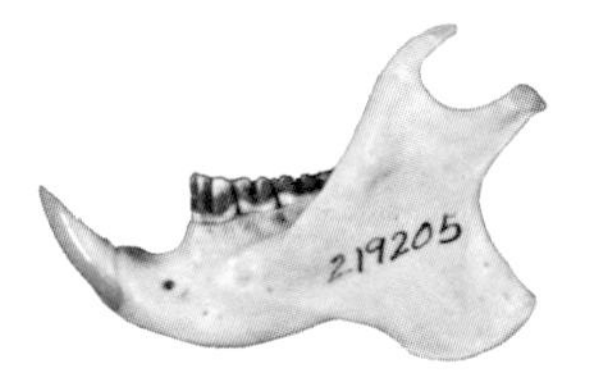

25 mm

50 mm

Aeromys tephromelas *Belomys pearsonii*

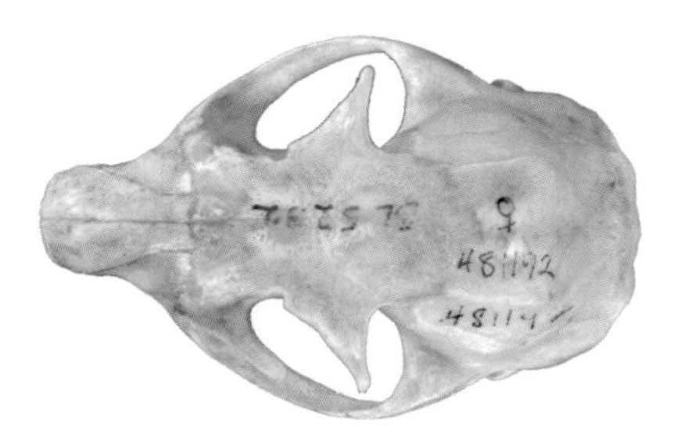

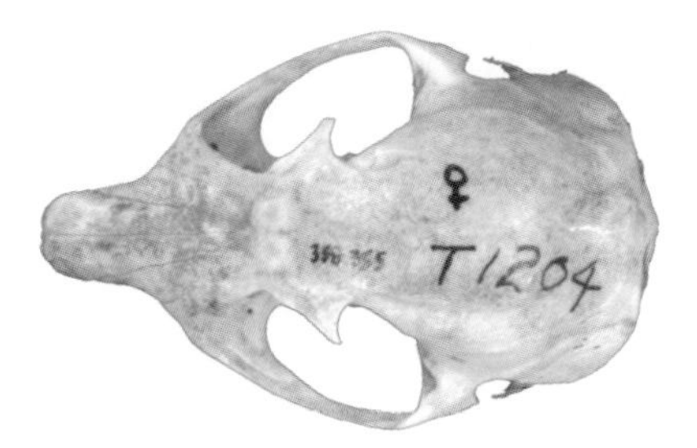

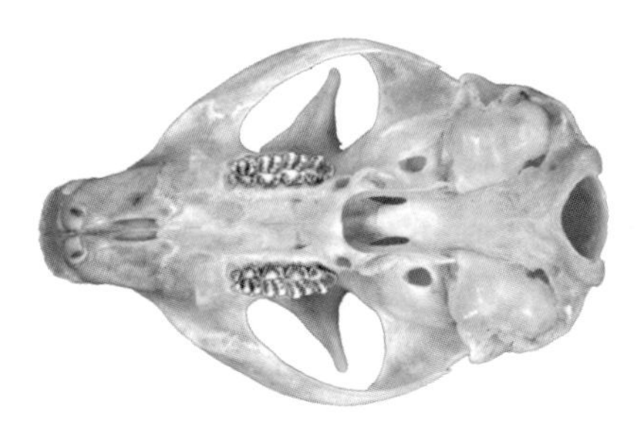

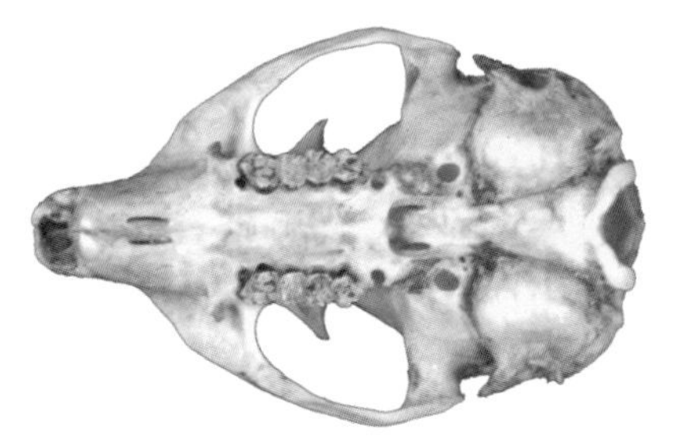

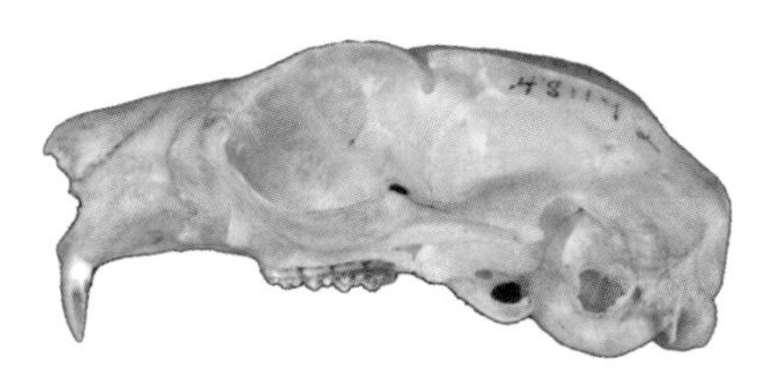

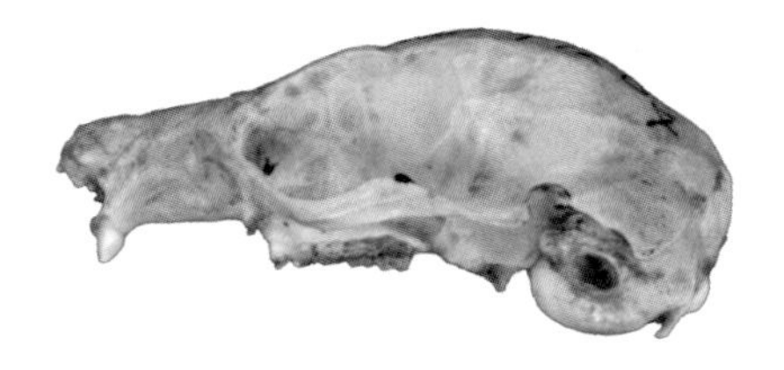

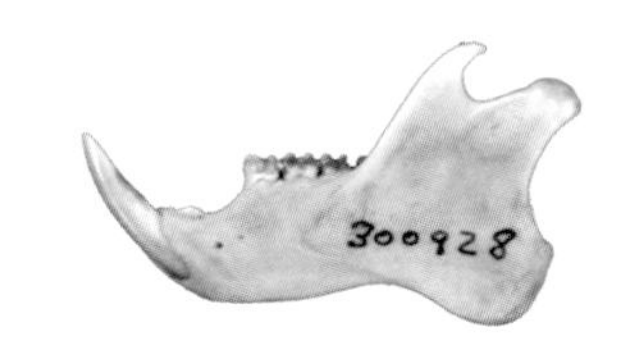

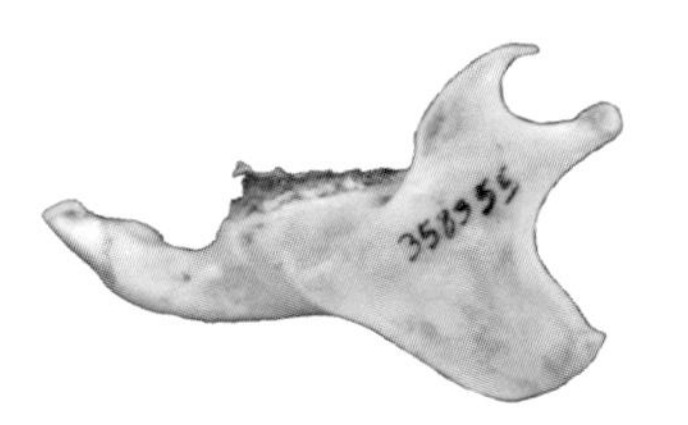

50 mm

25 mm

Eoglaucomys fimbriatus

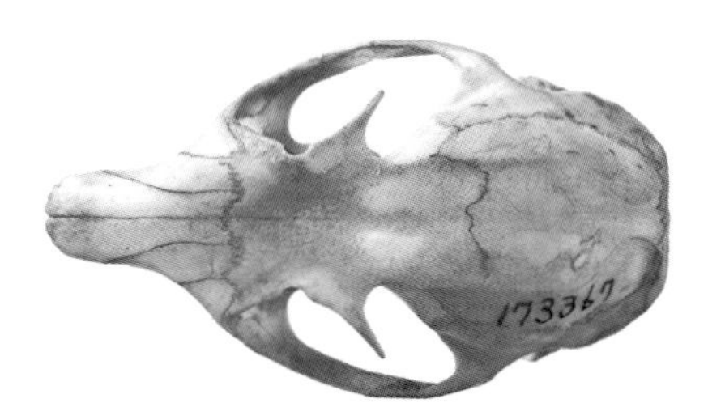

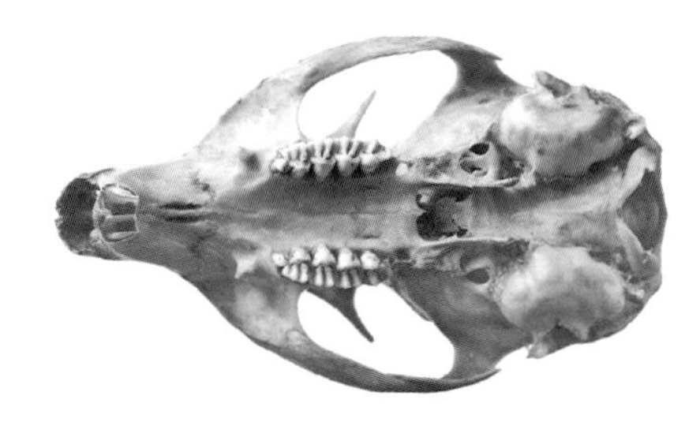

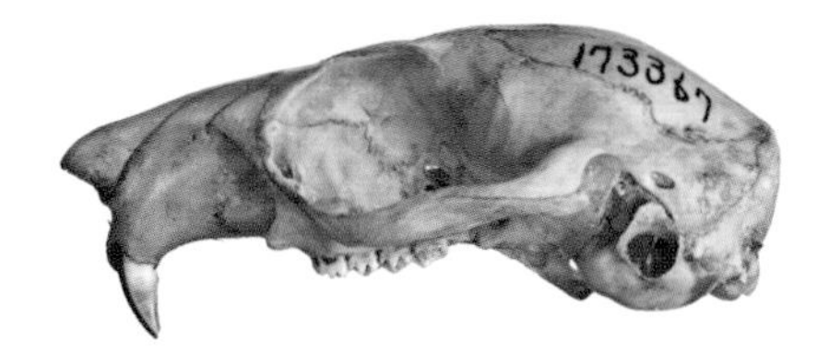

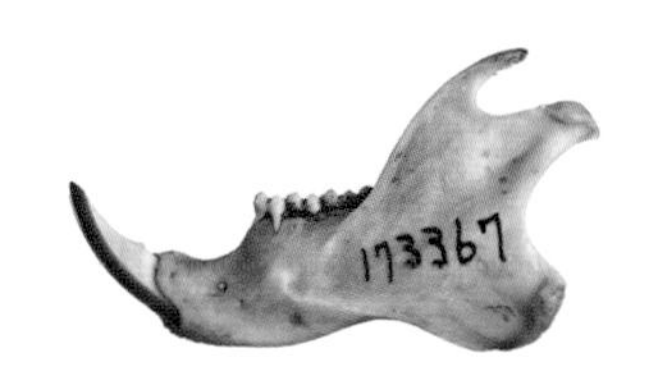

25 mm

Glaucomys sabrinus

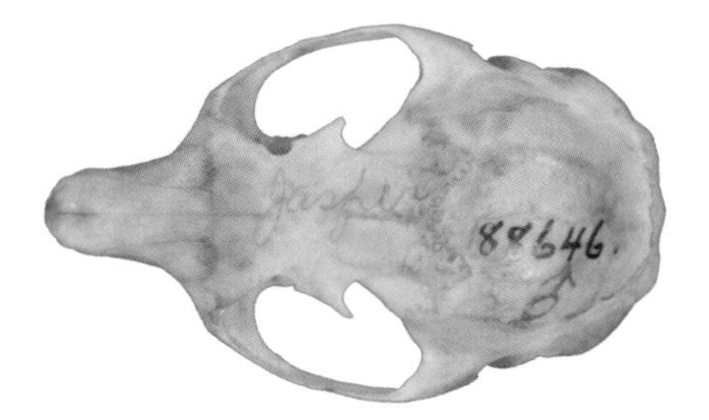

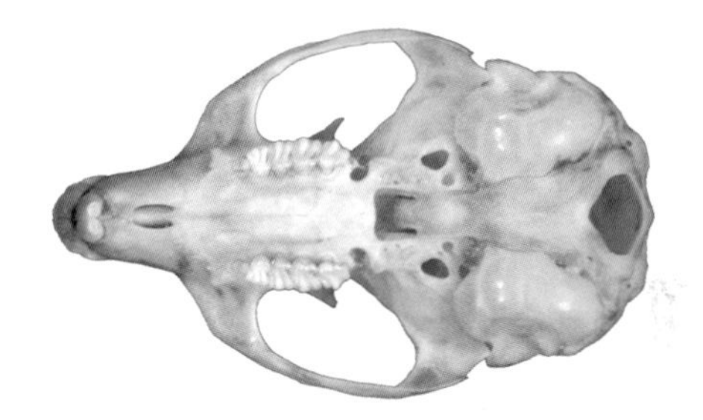

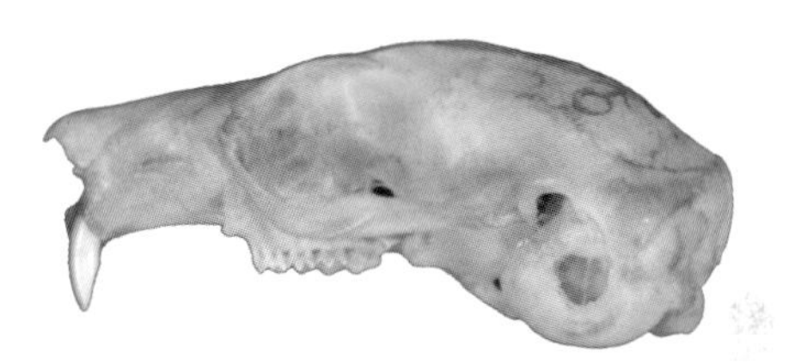

25 mm

Hylopetes spadiceus

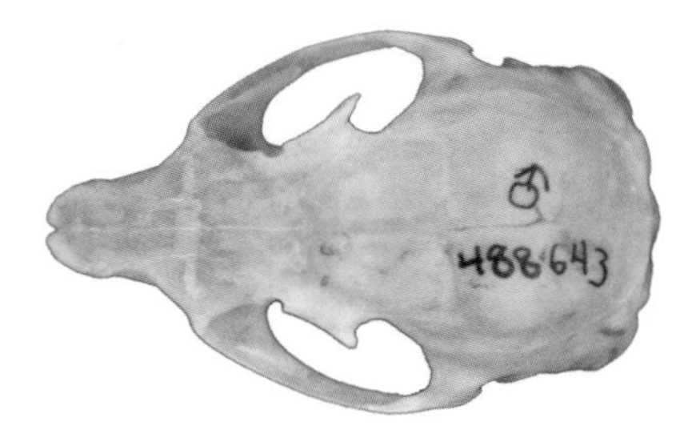

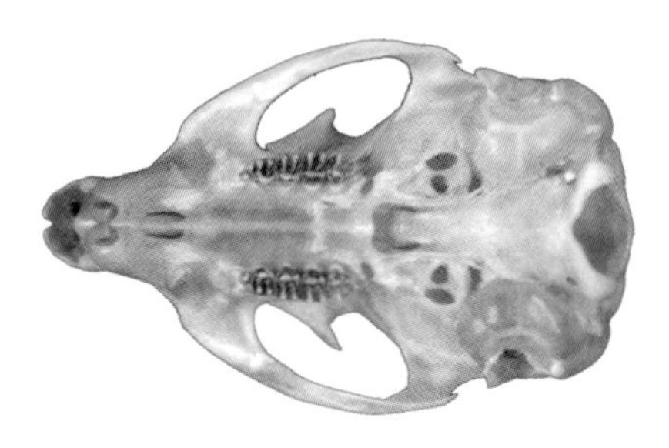

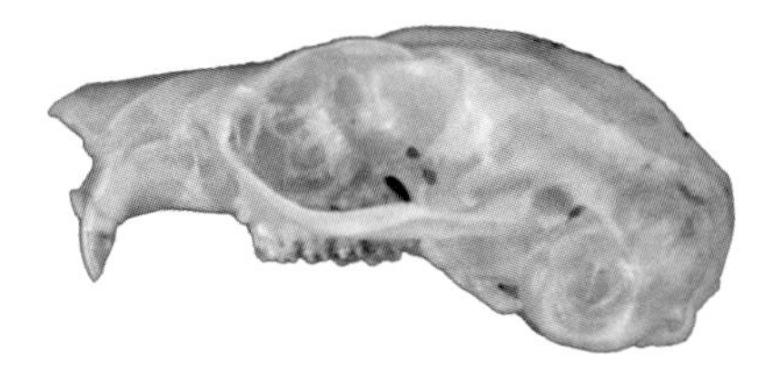

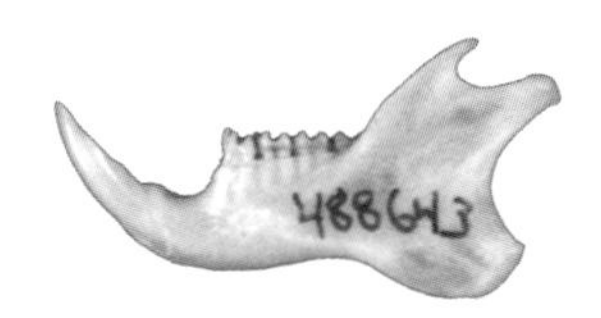

25 mm

Iomys horsfieldii

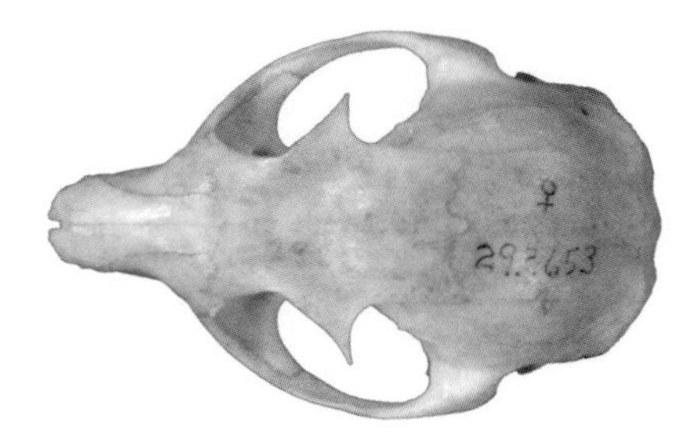

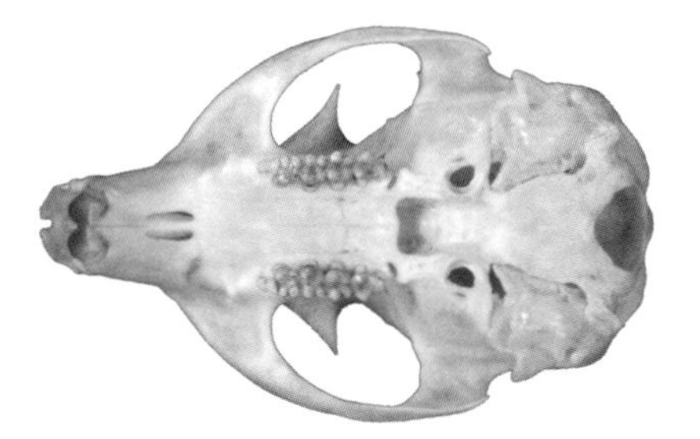

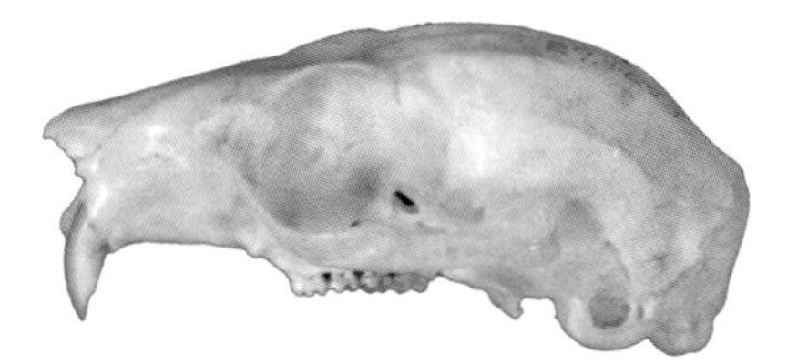

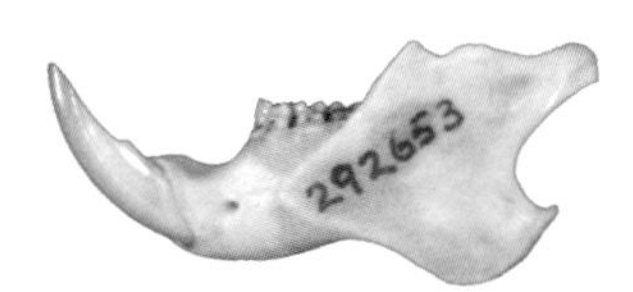

25 mm

Petaurillus kinlochii

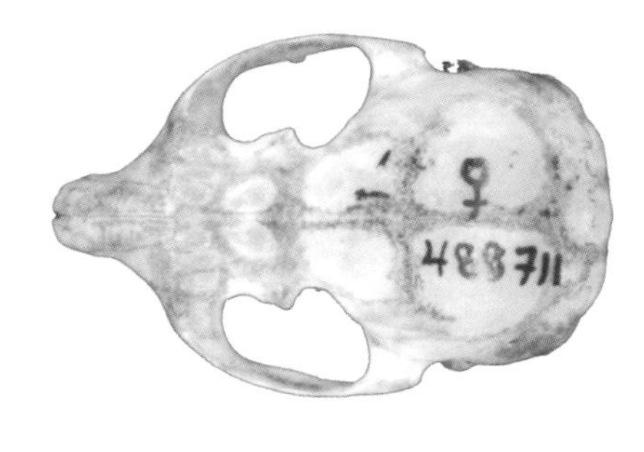

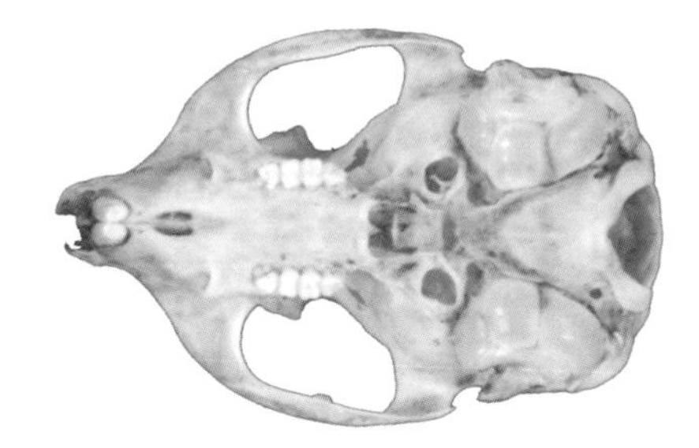

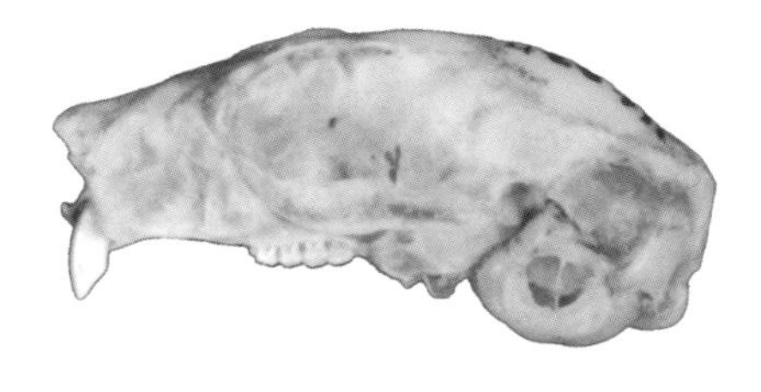

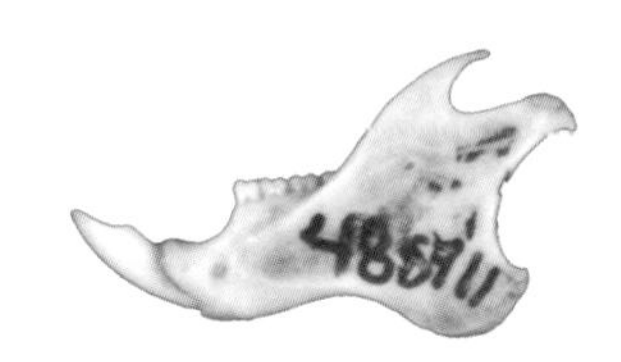

25 mm

Petaurista alborufus

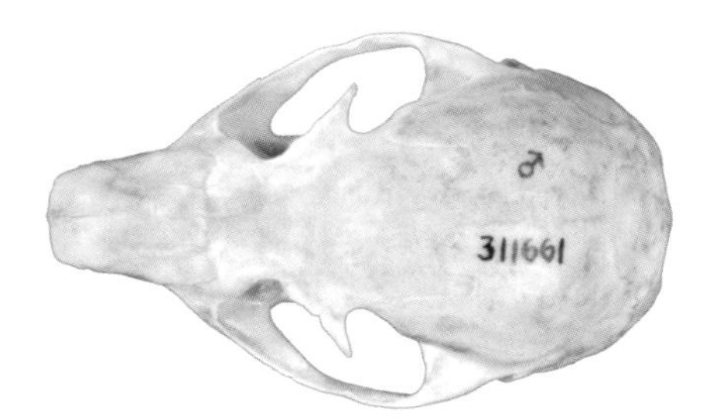

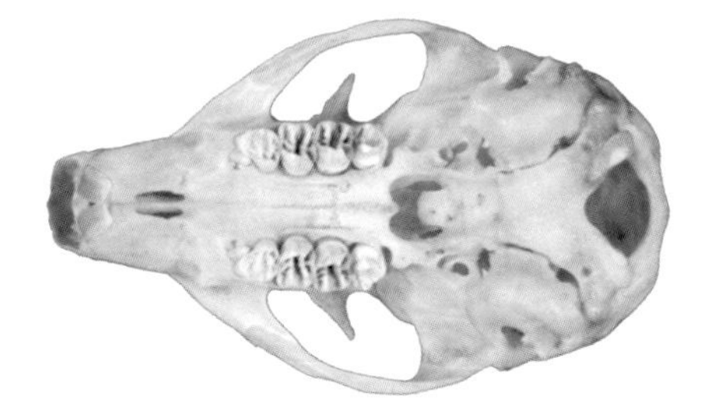

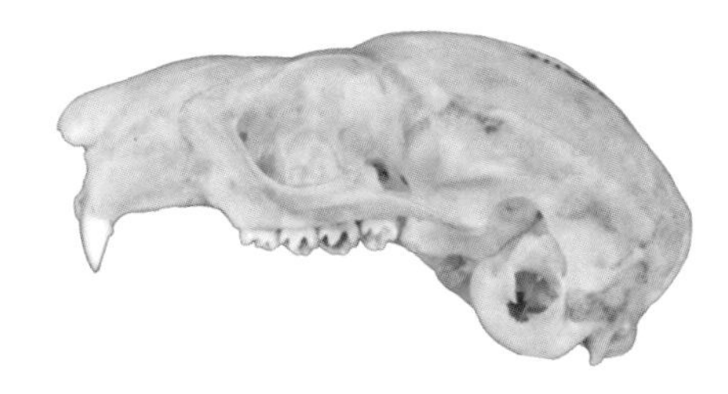

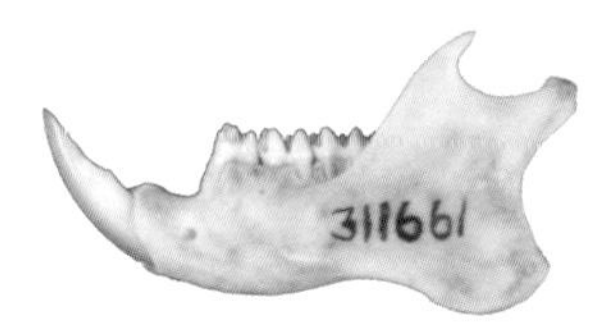

50 mm

Petinomys hageni

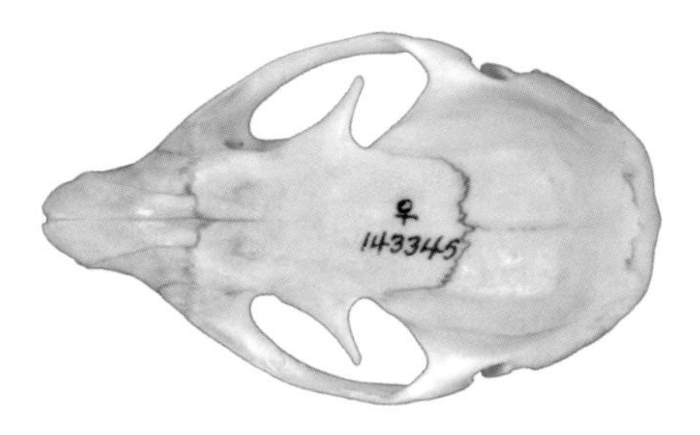

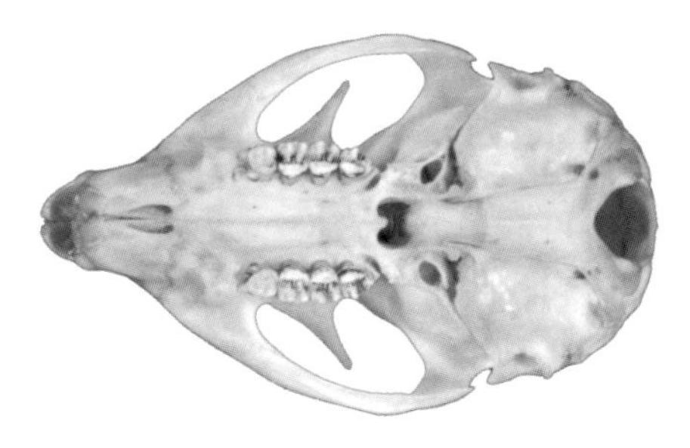

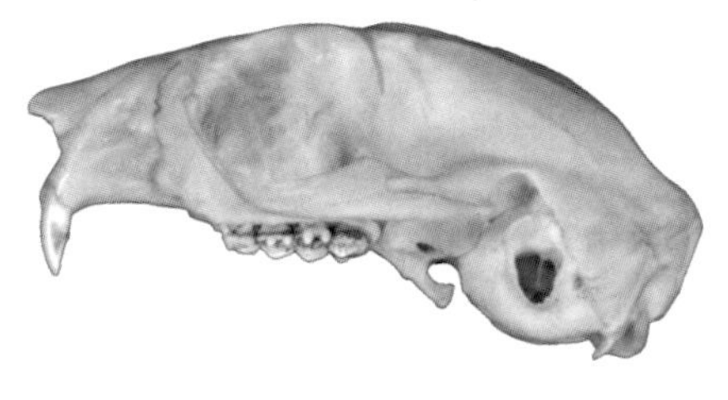

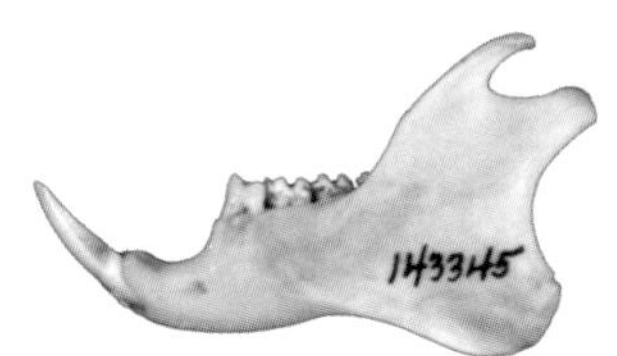

25 mm

Pteromys volans

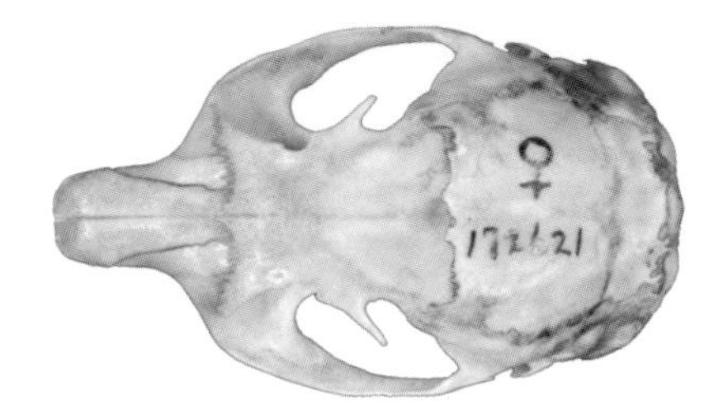

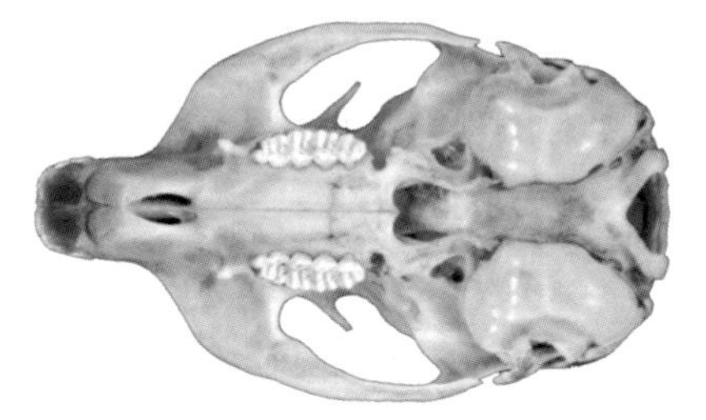

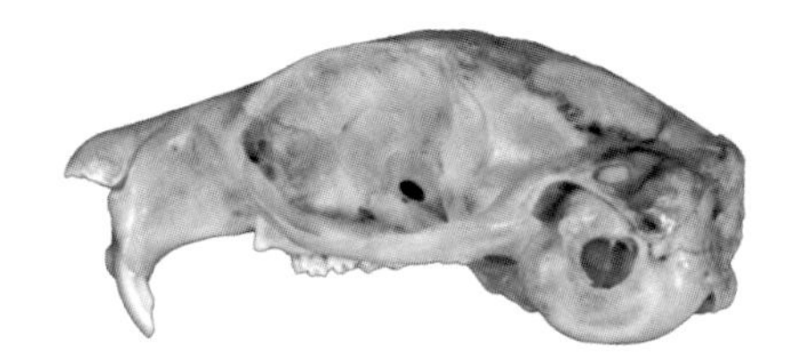

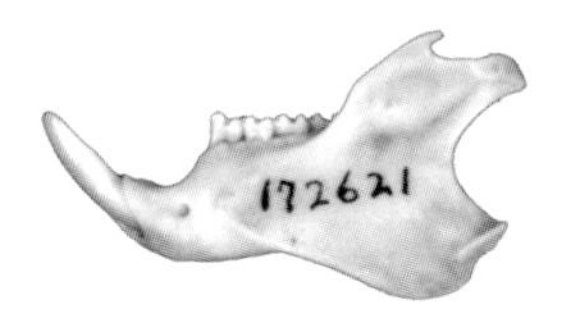

25 mm

Pteromyscus pulverulentus

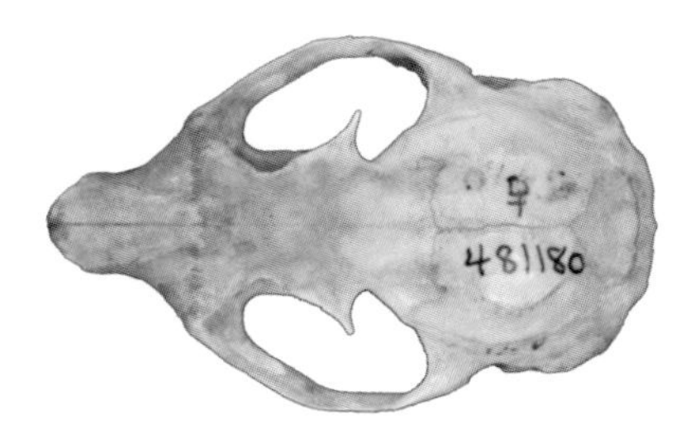

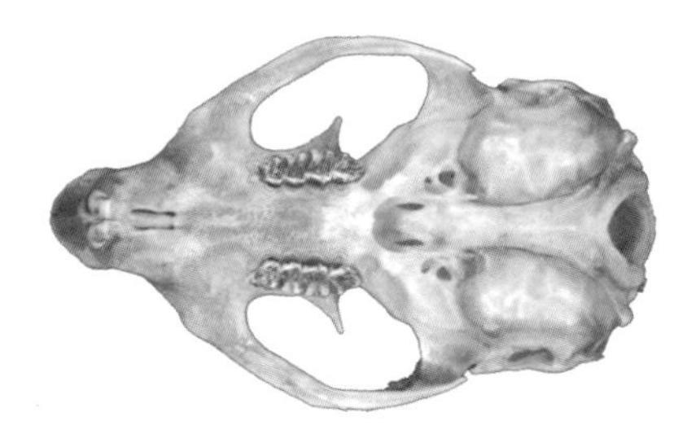

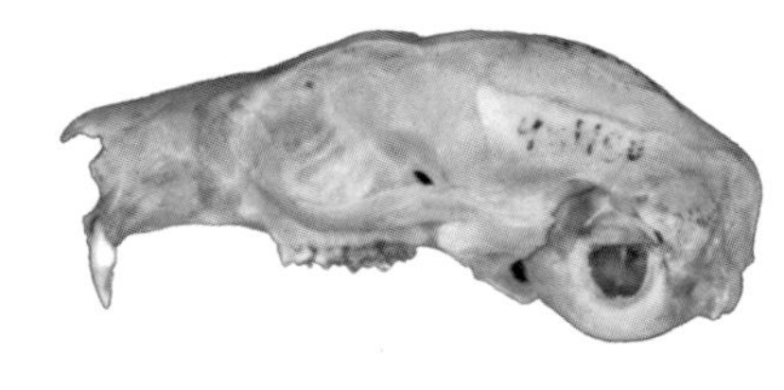

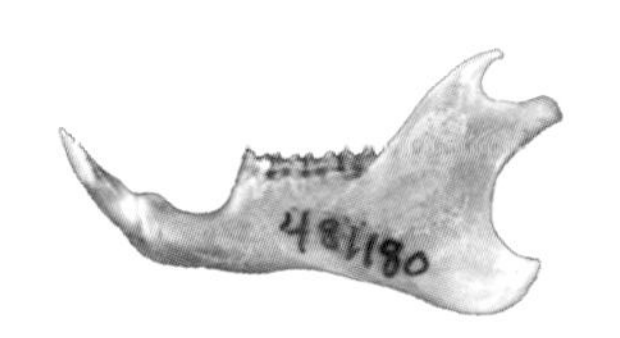

25 mm

Trogopterus xanthipes

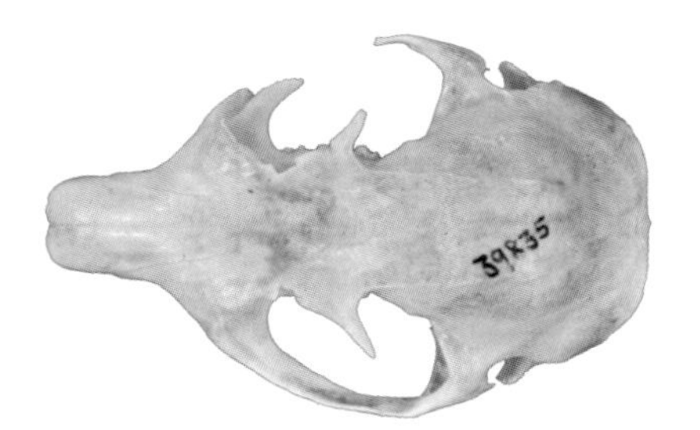

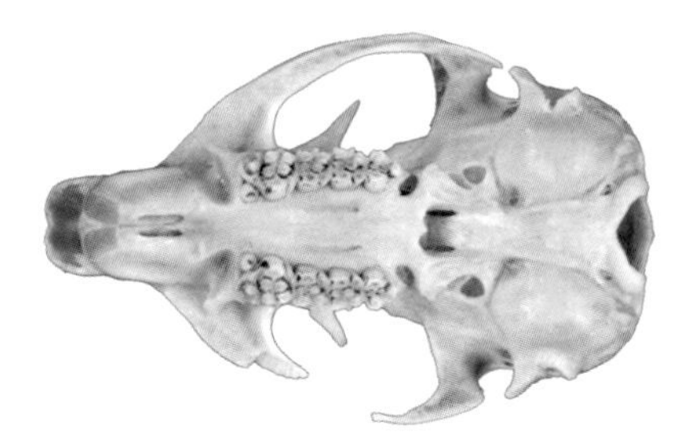

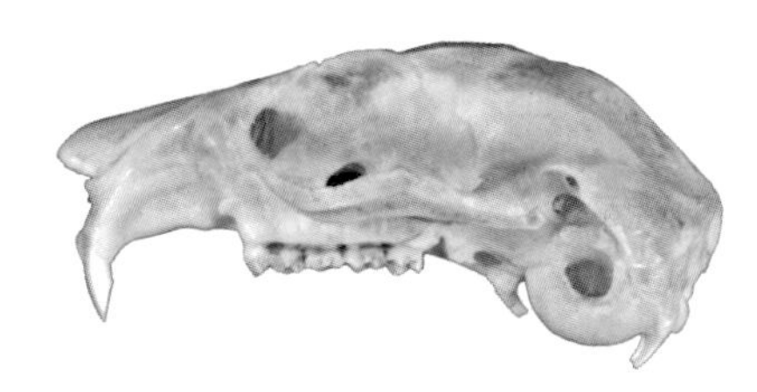

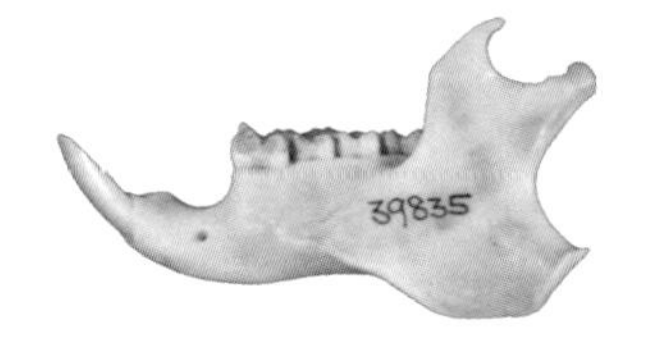

25 mm

Callosciurus prevostii | *Dremomys pernyi*

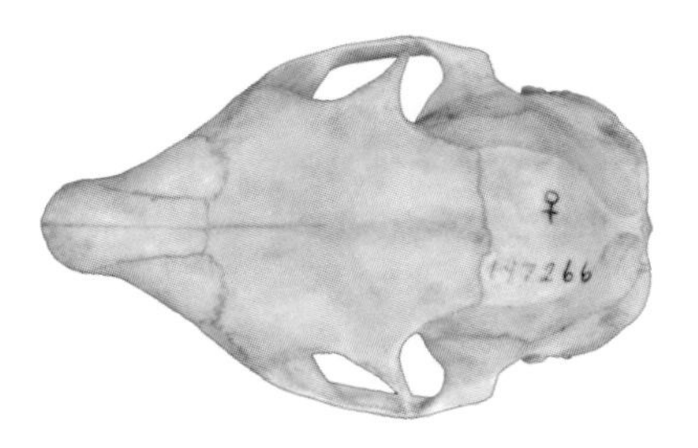

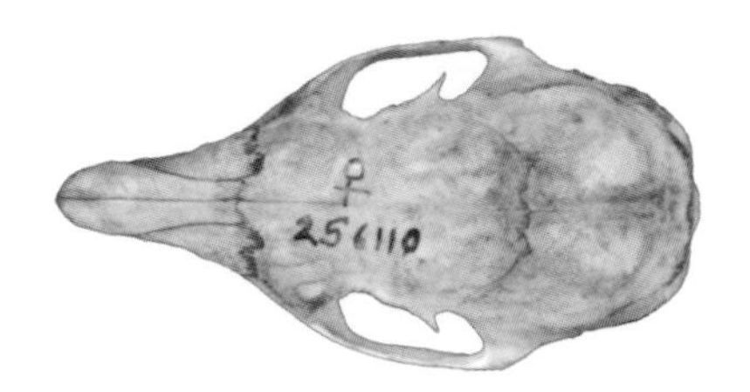

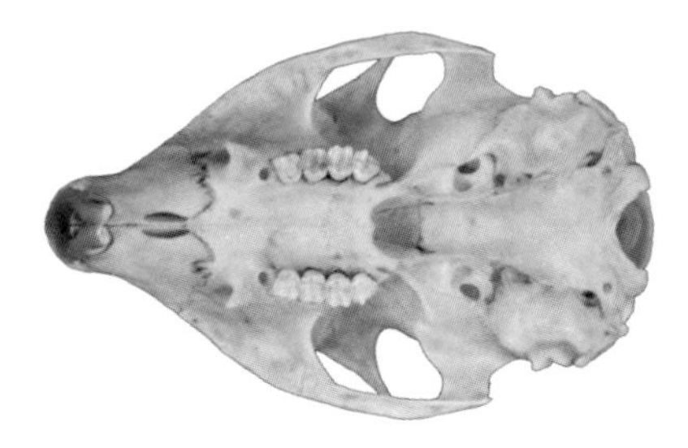

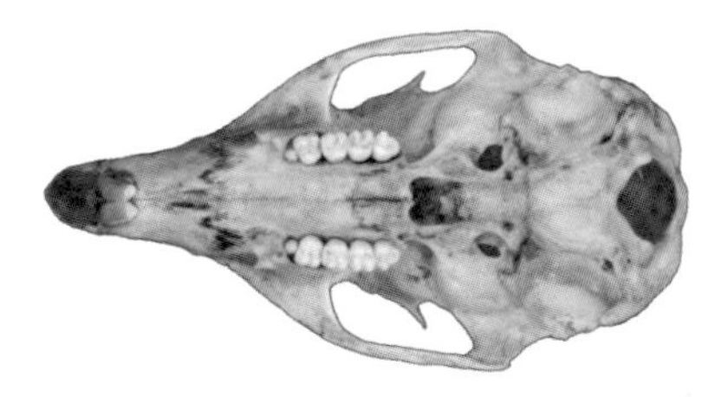

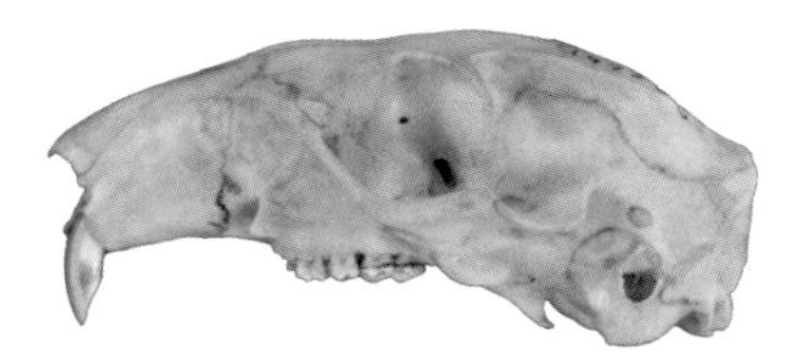

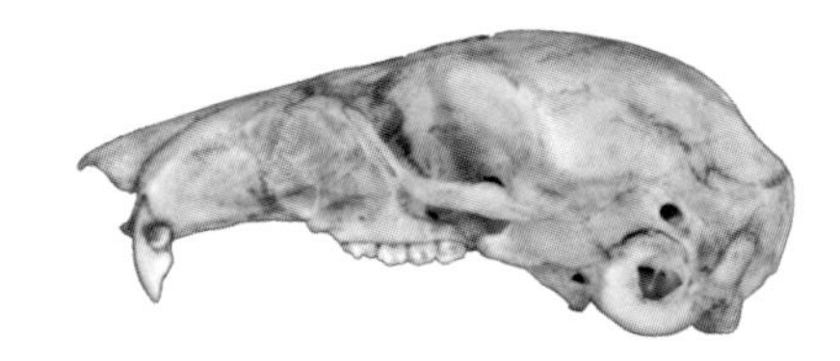

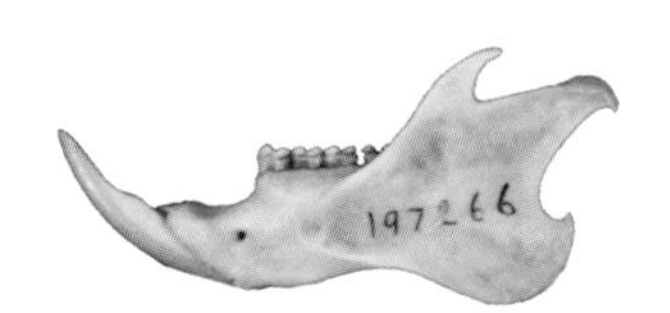

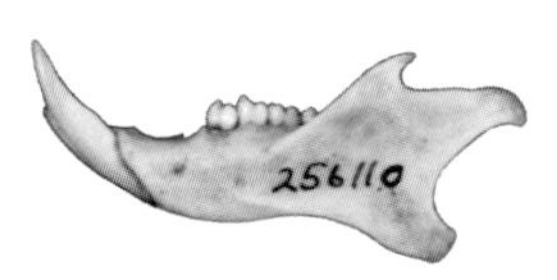

50 mm

25 mm

Exilisciurus exilis

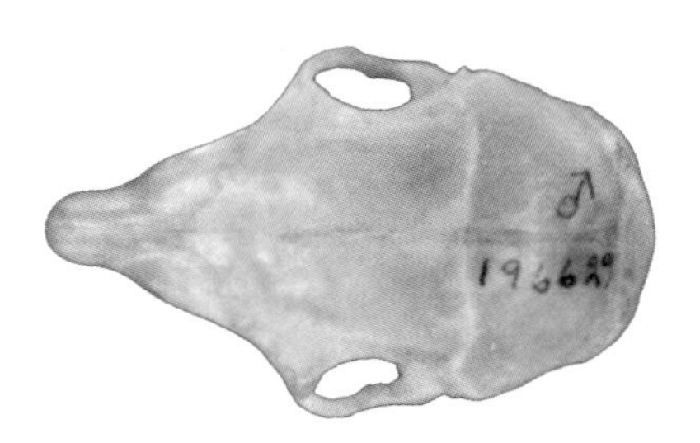

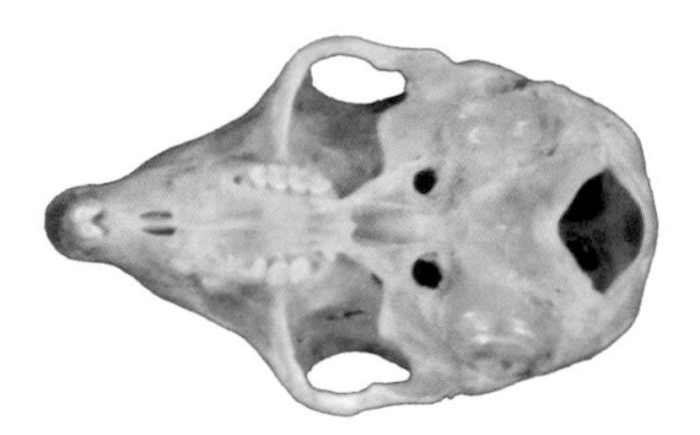

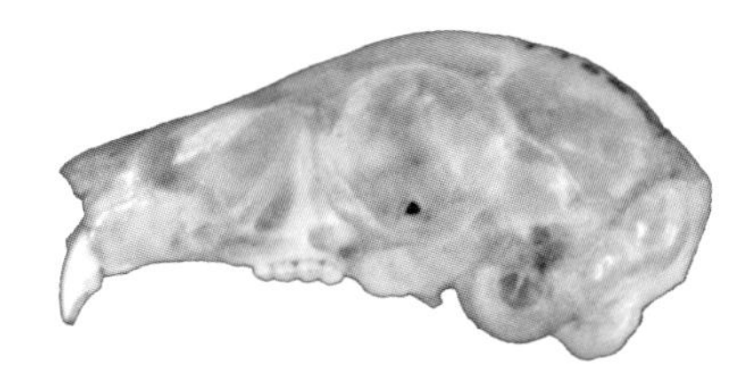

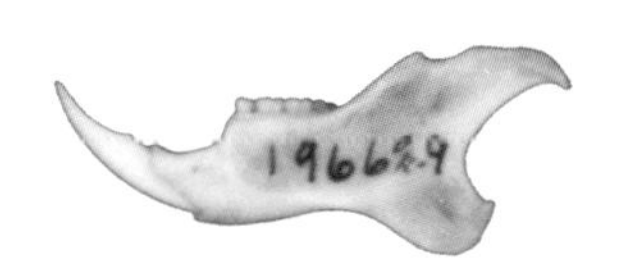

25 mm

Funambulus pennantii

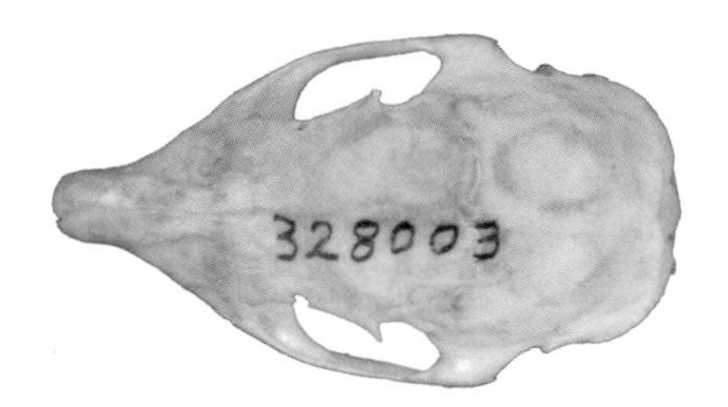

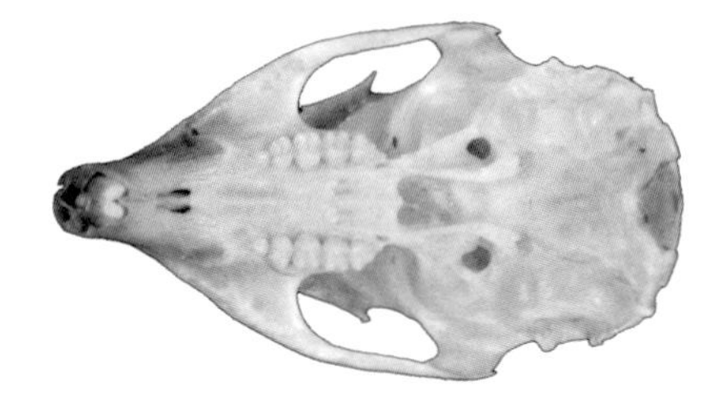

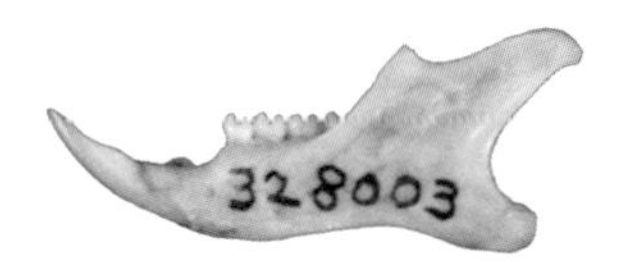

25 mm

Glyphotes simus *Hyosciurus ileile*

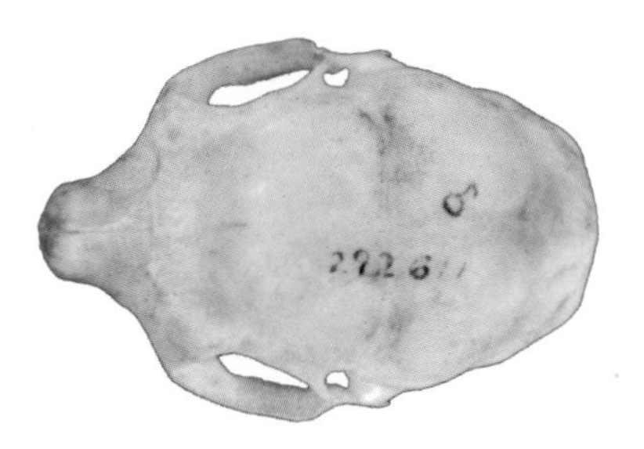

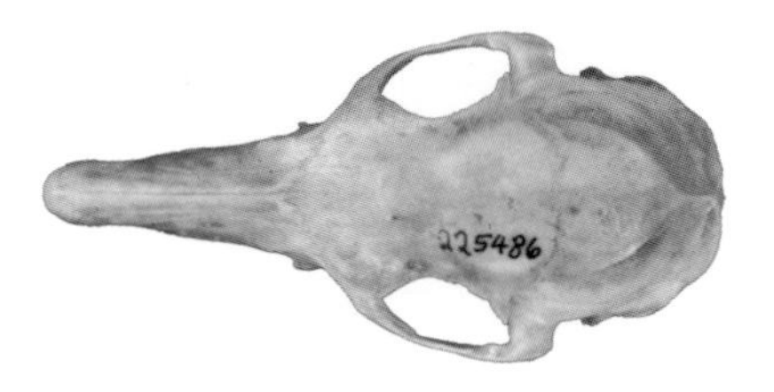

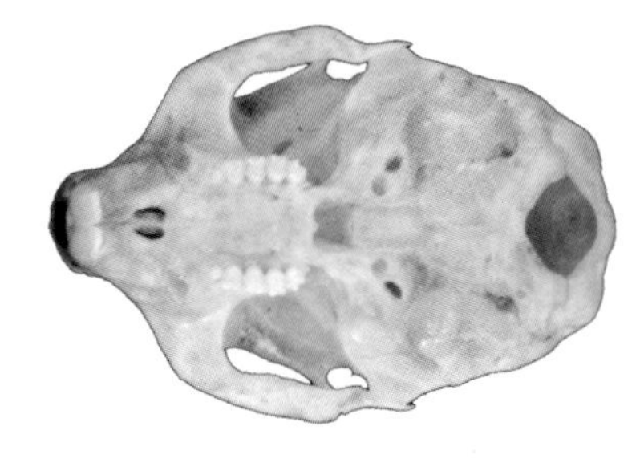

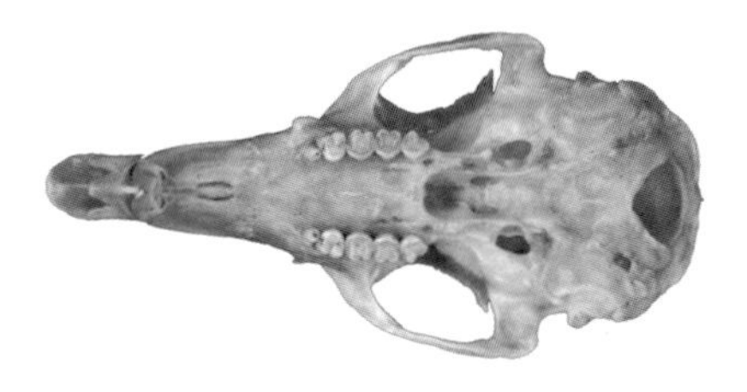

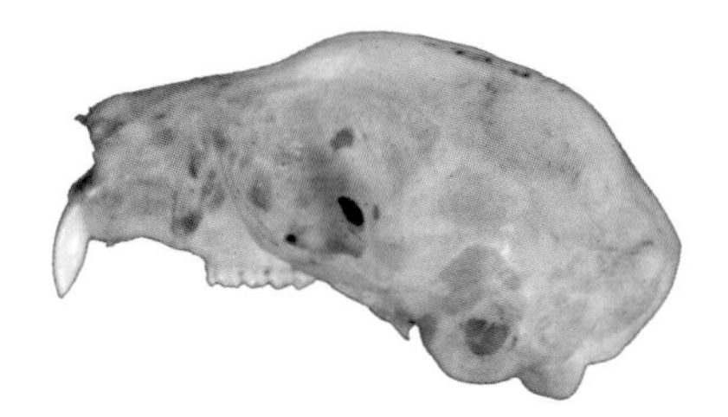

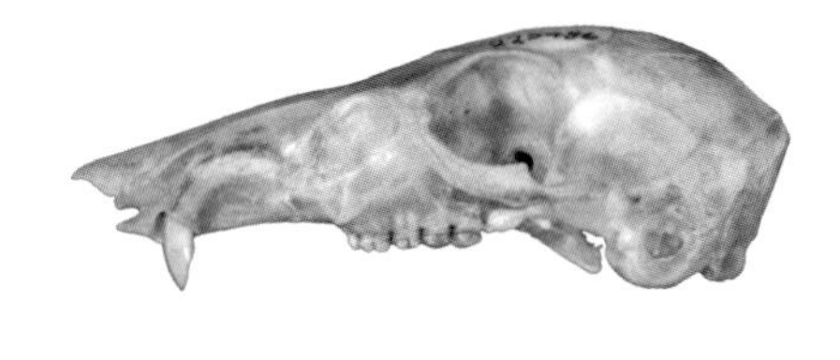

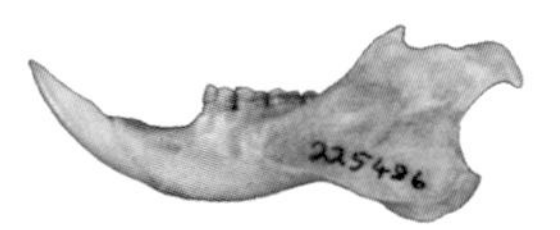

25 mm

25 mm

Lariscus insignis

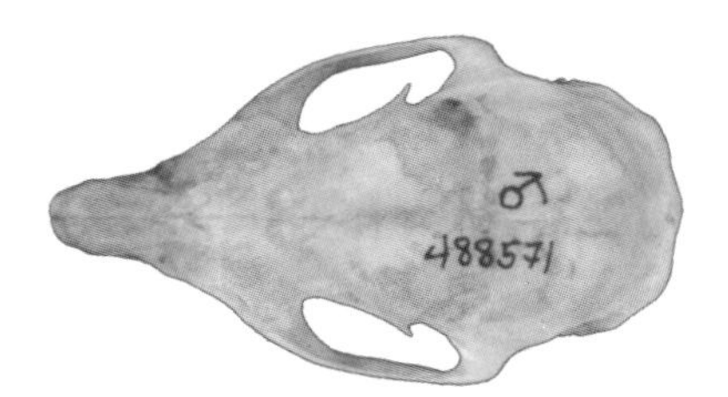

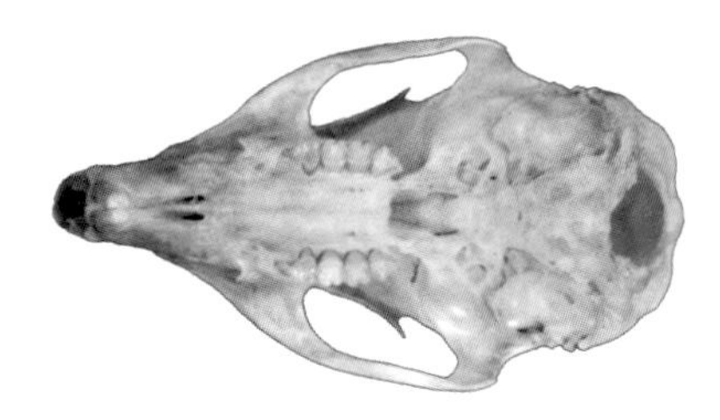

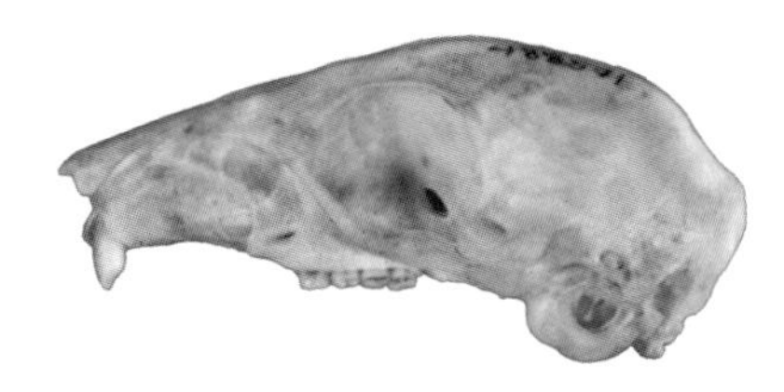

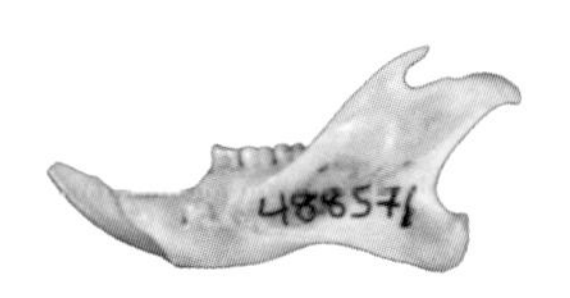

25 mm

Menetes berdmorei

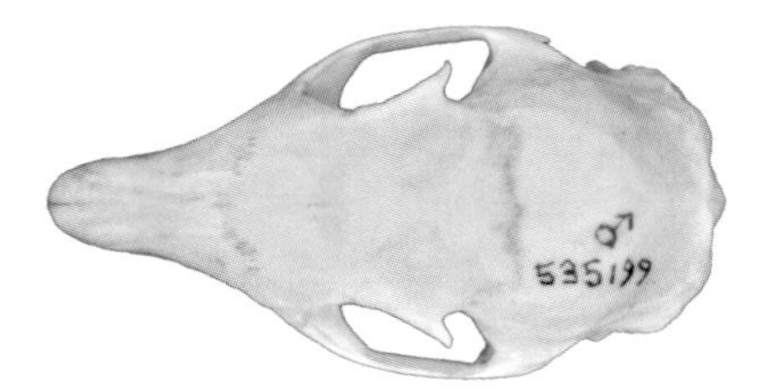

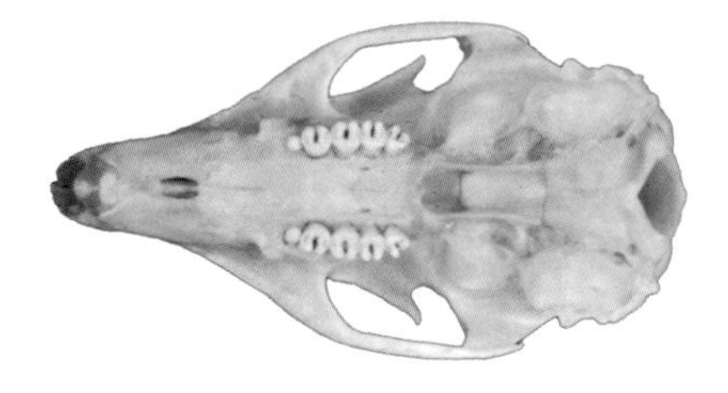

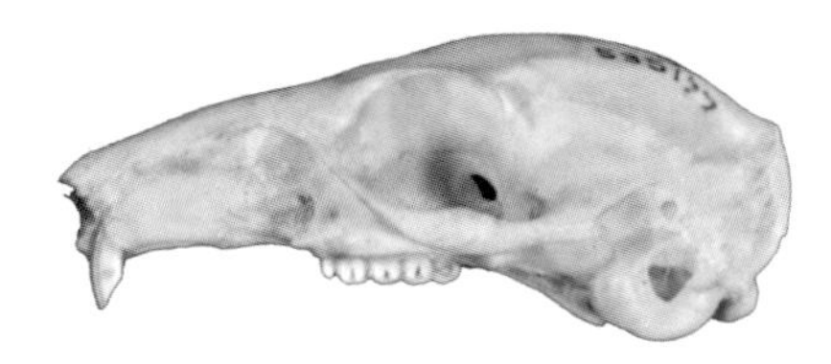

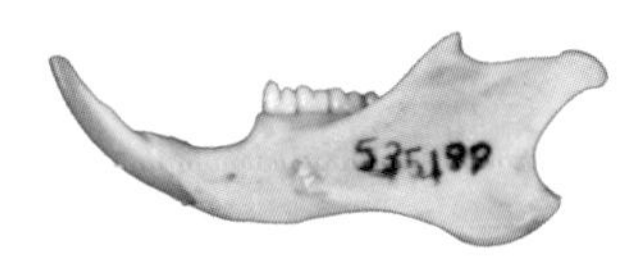

25 mm

Nannosciurus melanotis

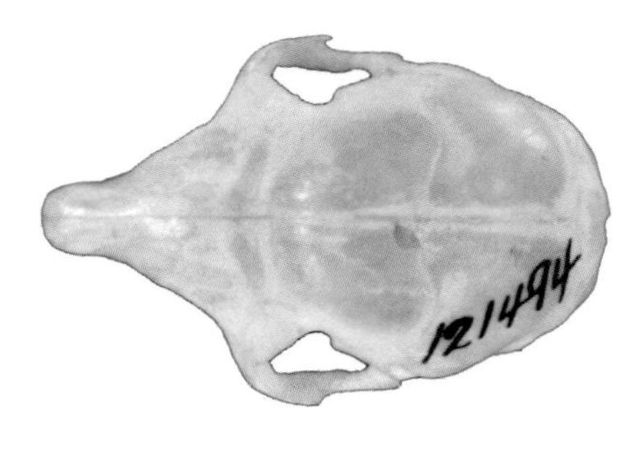

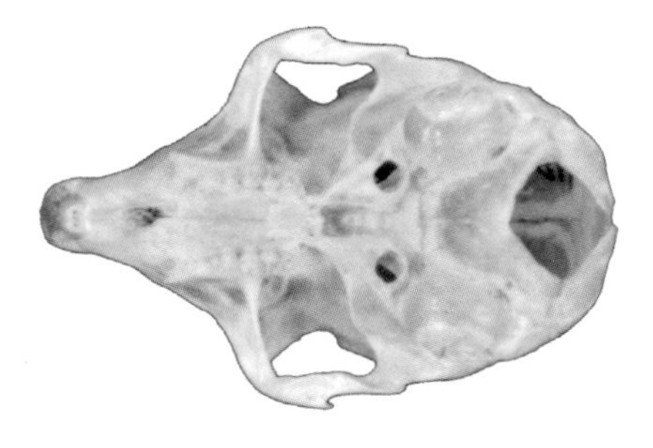

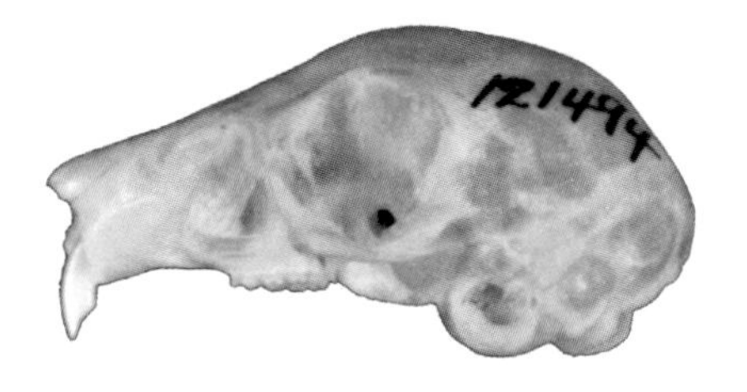

25 mm

Prosciurillus leucomus

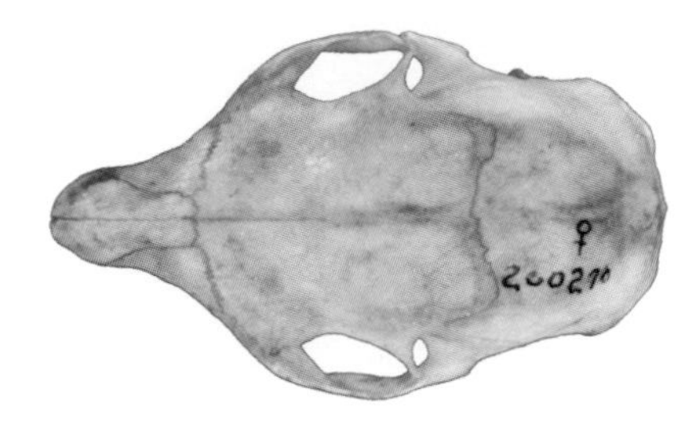

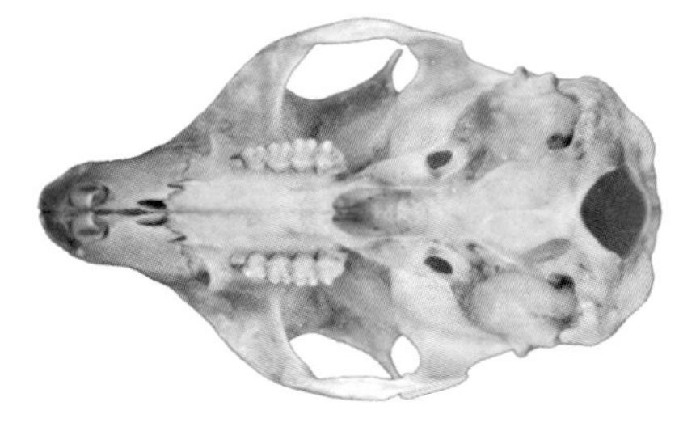

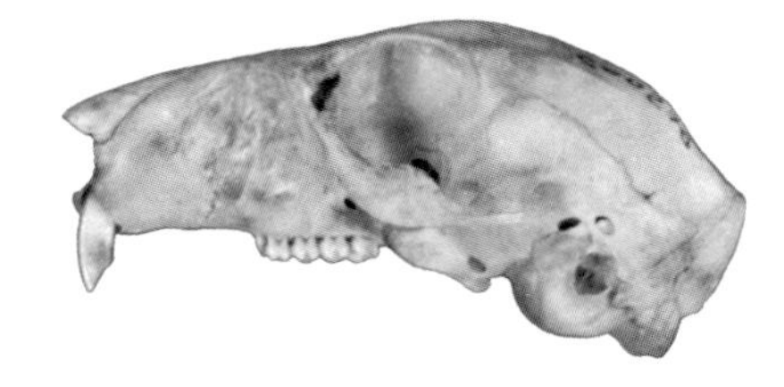

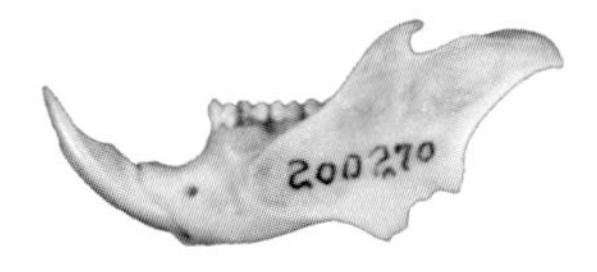

25 mm

Rhinosciurus laticaudatus *Rubrisciurus rubriventer*

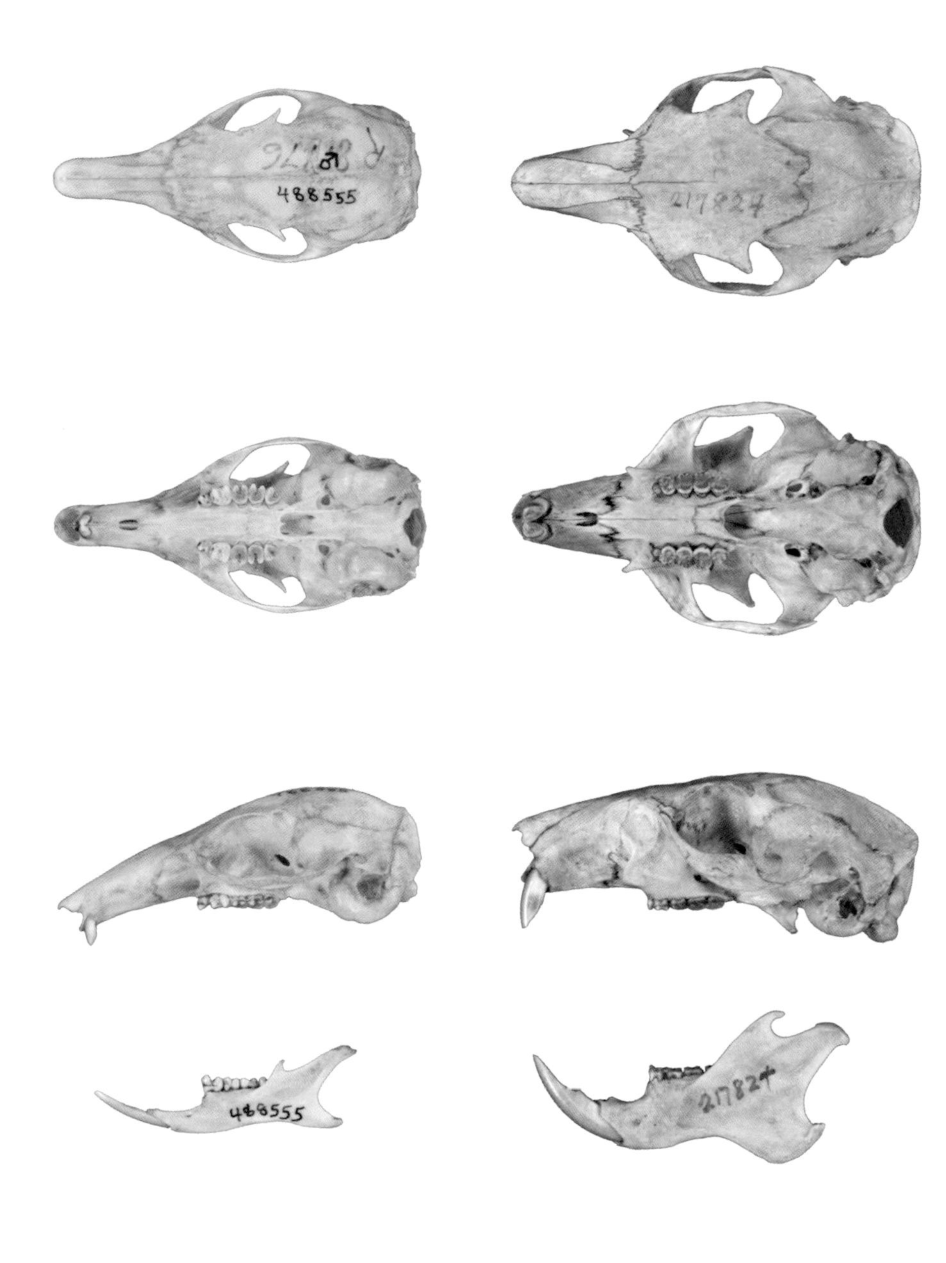

25 mm 25 mm

Sundasciurus tenuis

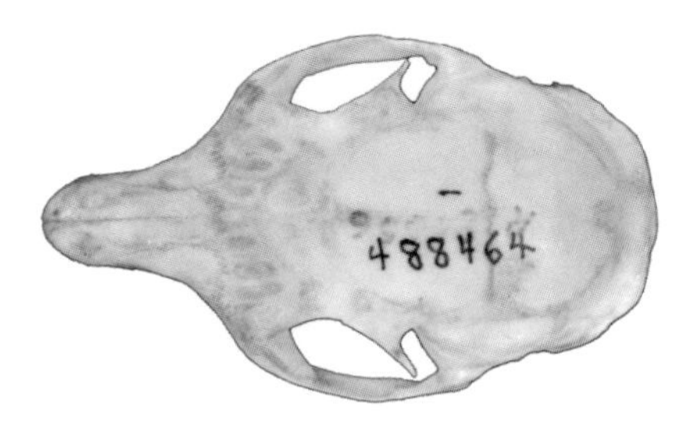

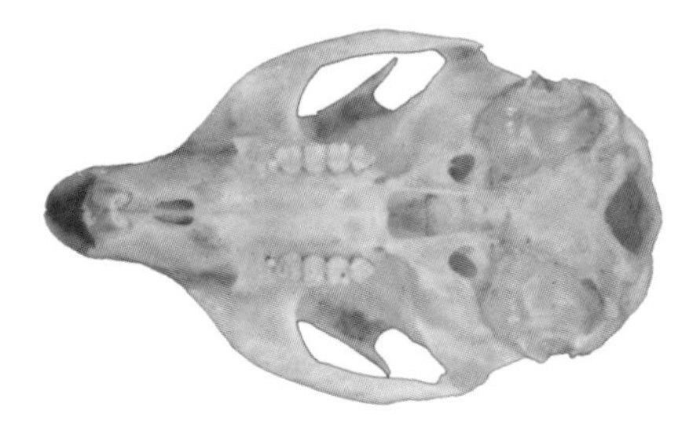

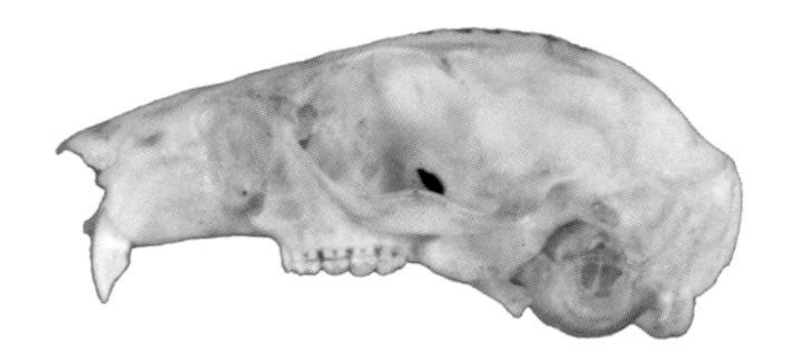

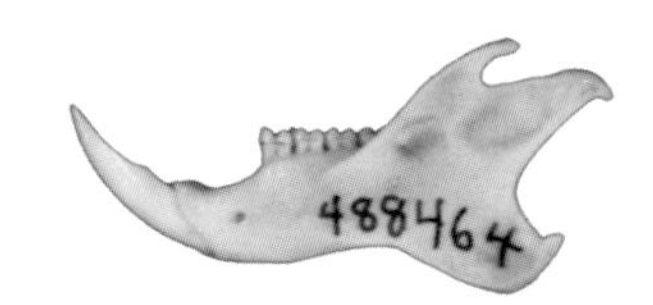

25 mm

Tamiops rodolphii

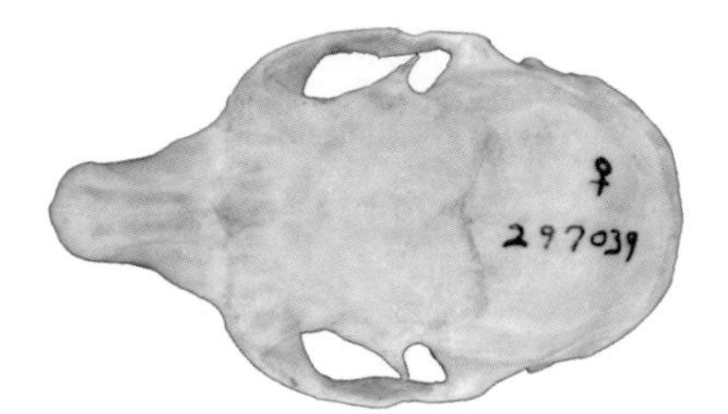

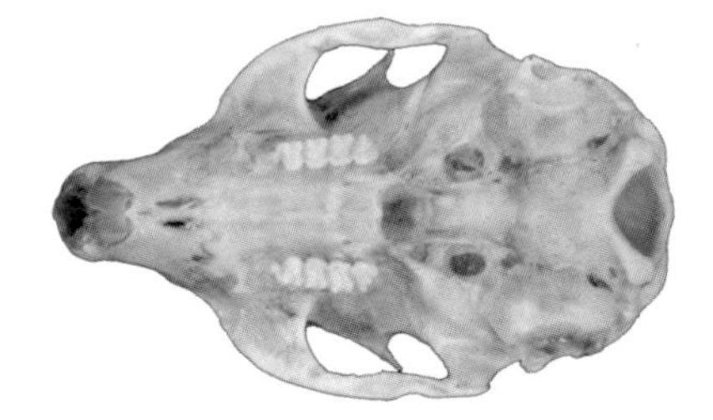

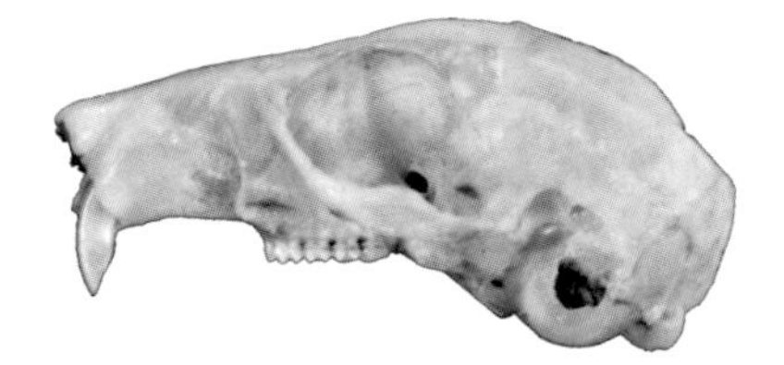

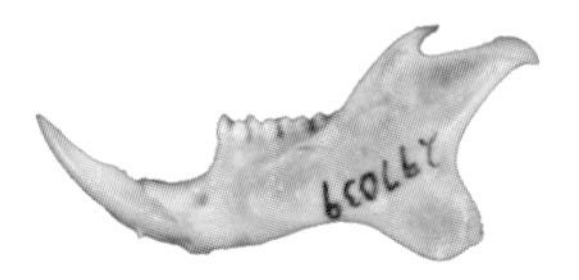

25 mm

Atlantoxerus getulus *Spermophilopsis leptodactylus*

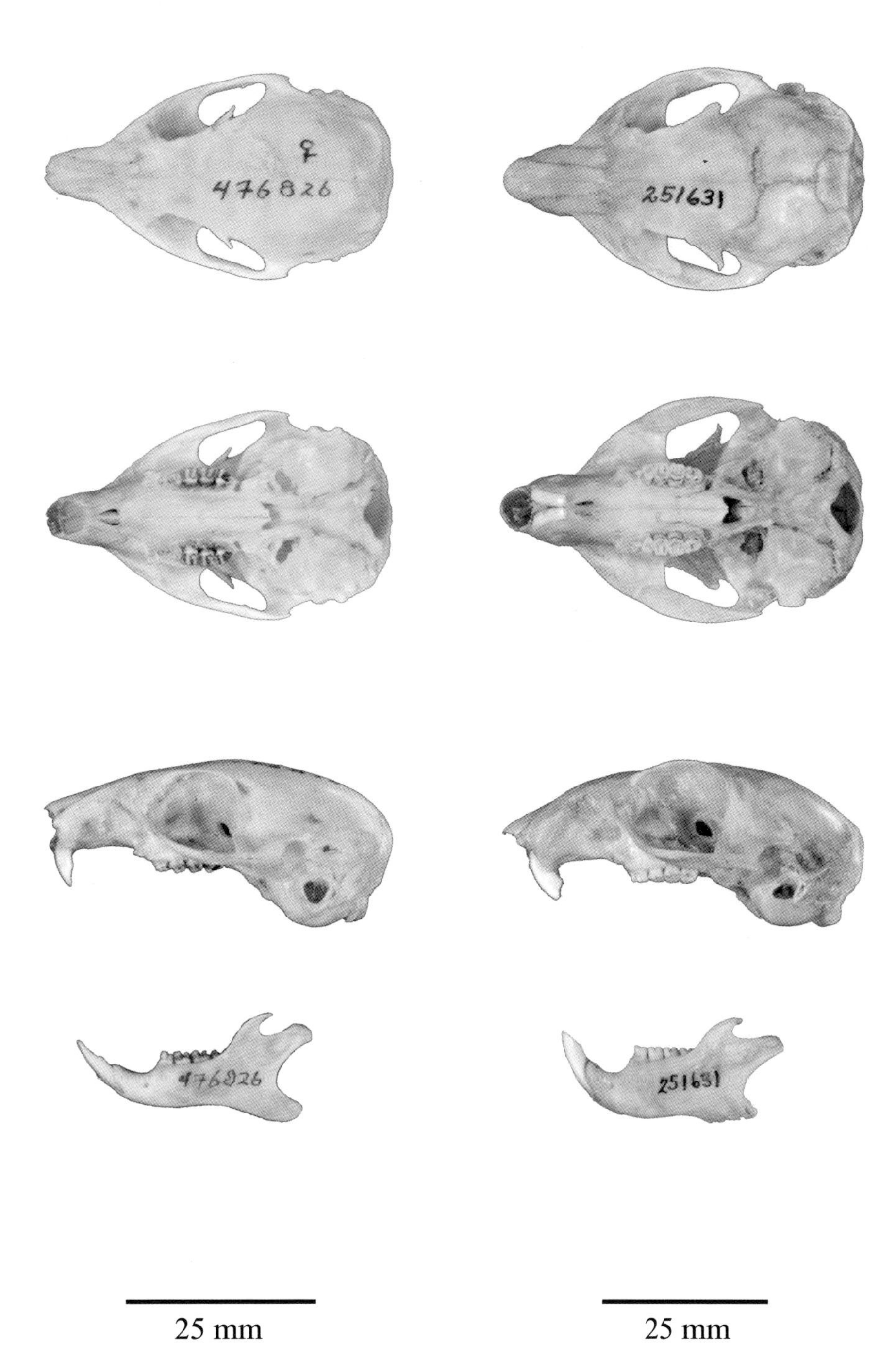

Xerus erythropus

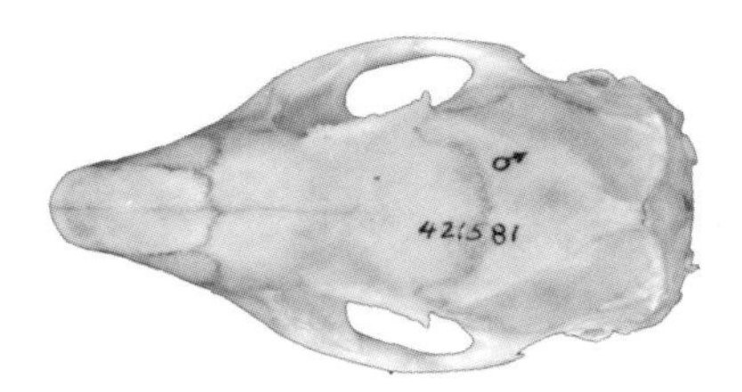

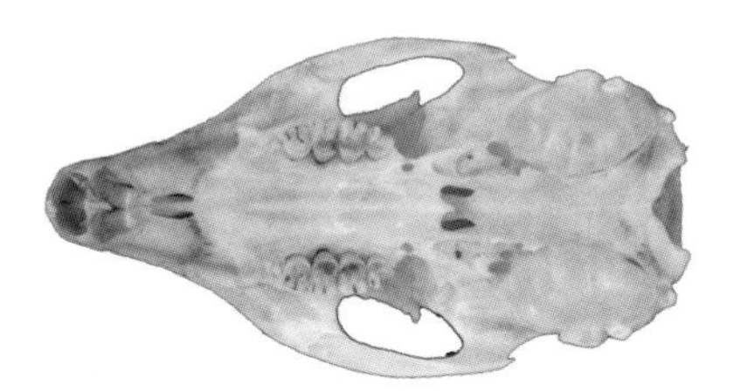

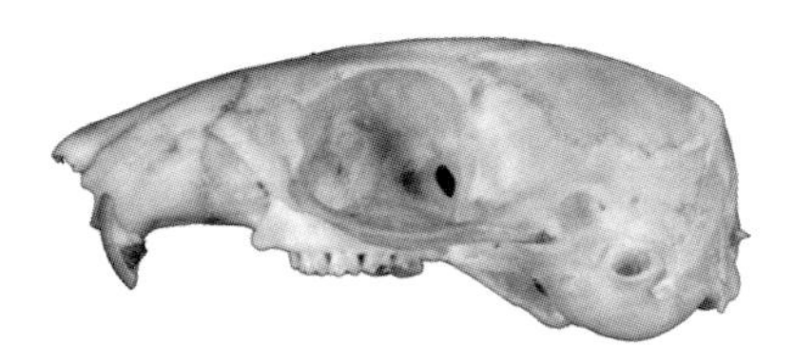

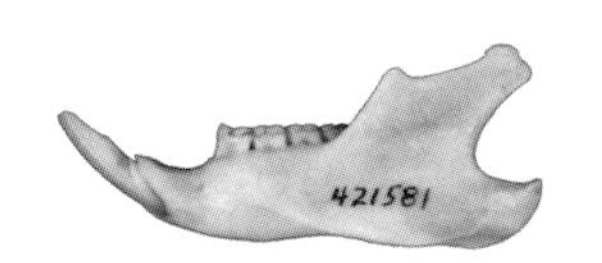

50 mm

Epixerus ebii

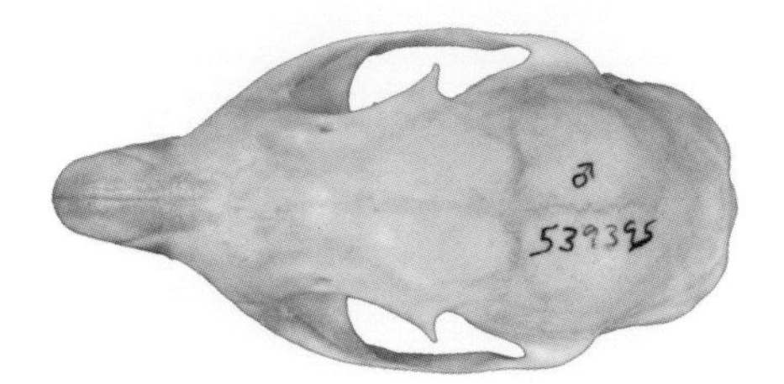

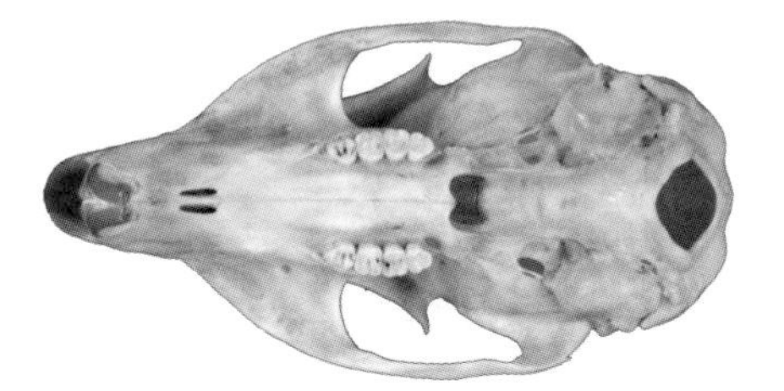

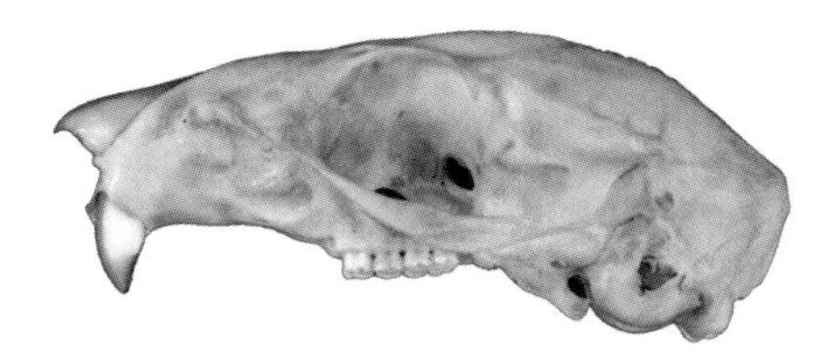

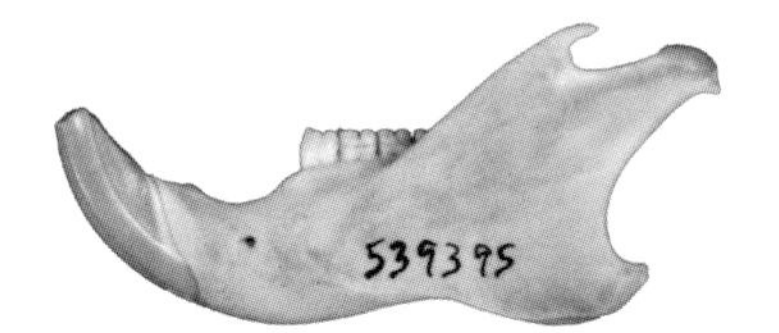

50 mm

Funisciurus anerythrus *Heliosciurus rufobrachium*

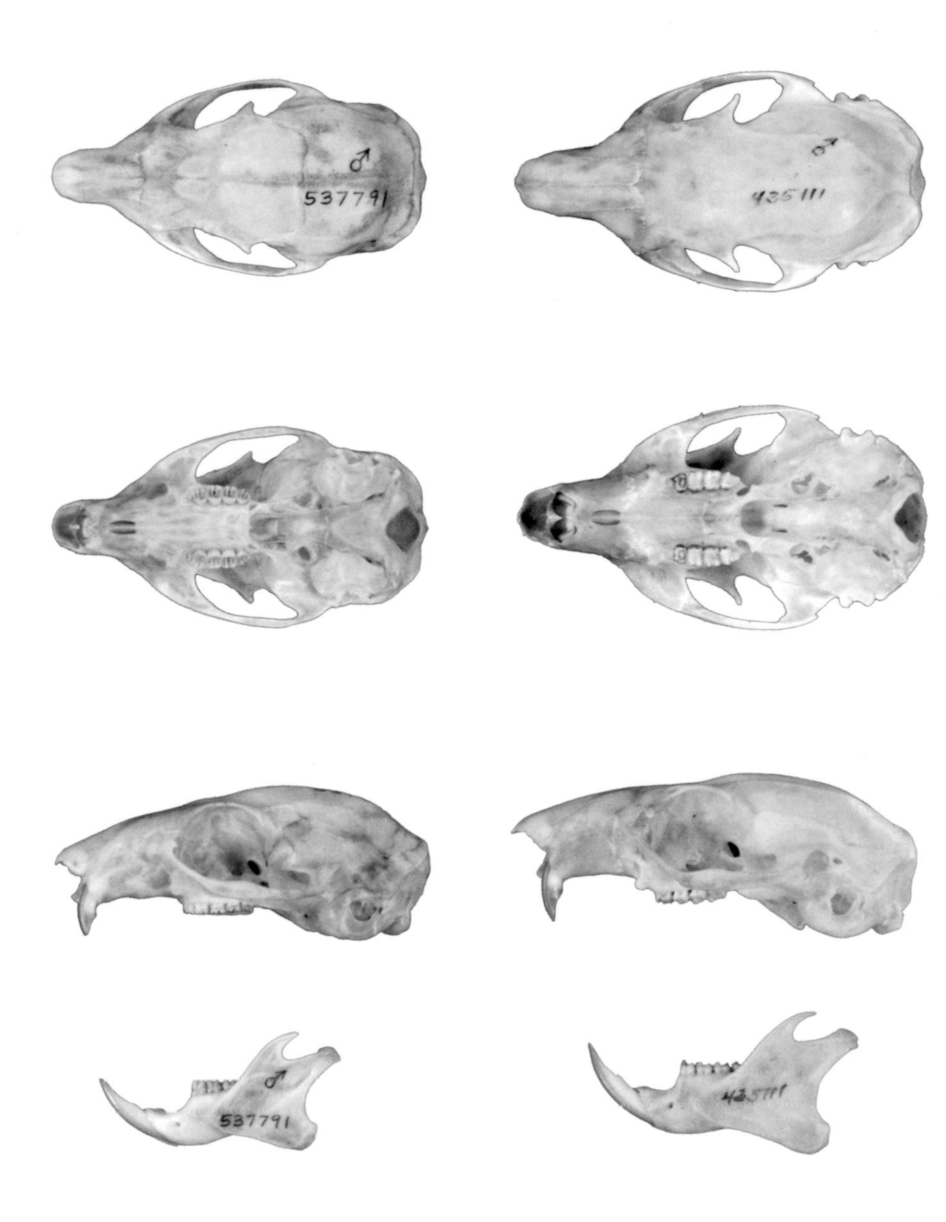

25 mm 25 mm

Myosciurus pumilio *Paraxerus cepapi*

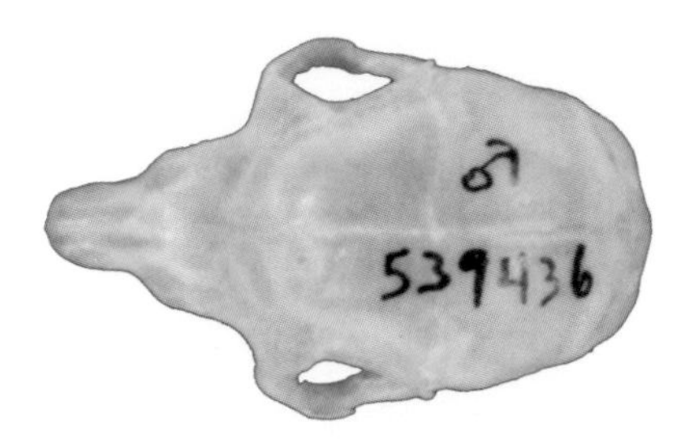

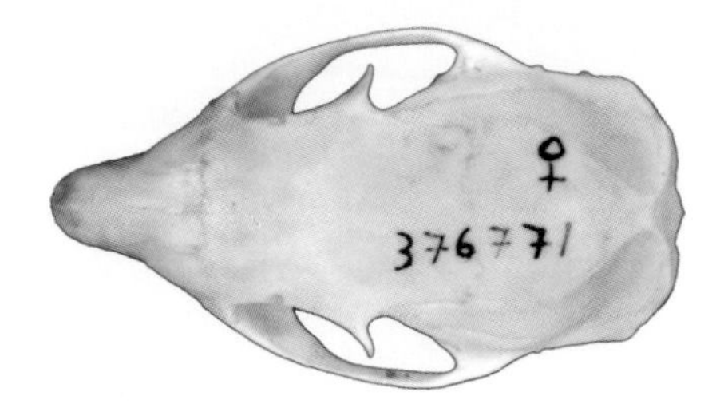

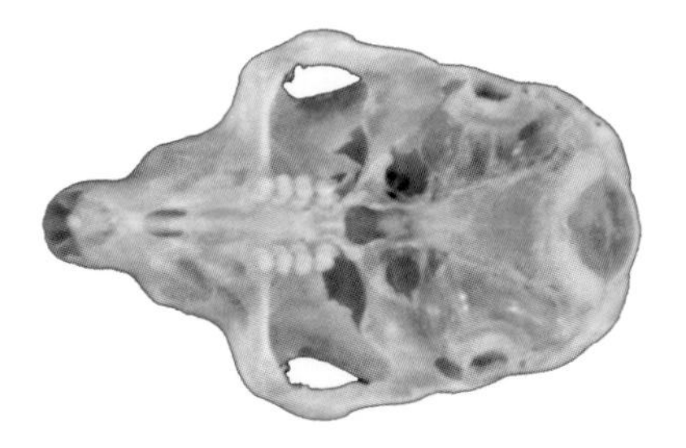

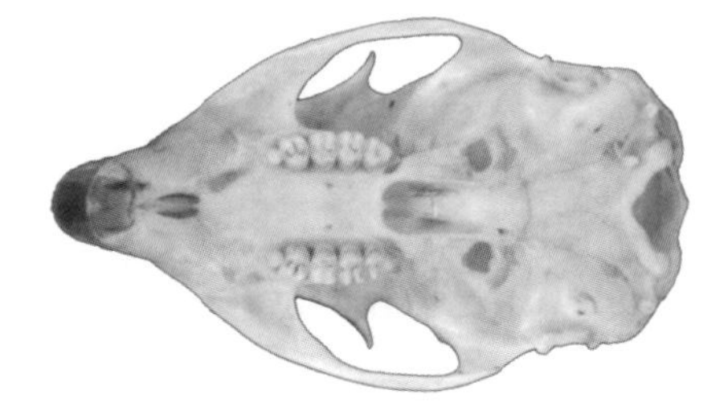

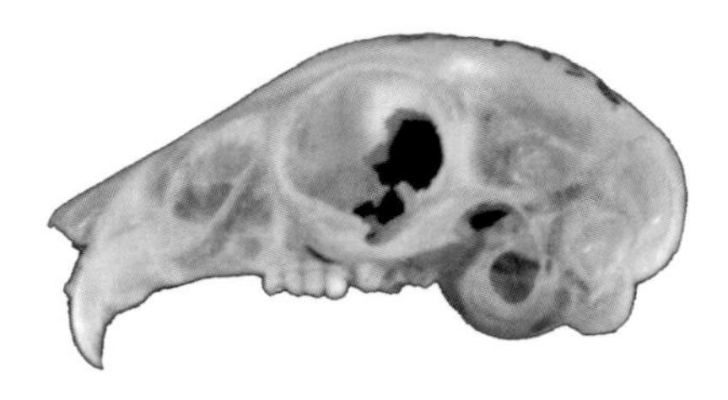

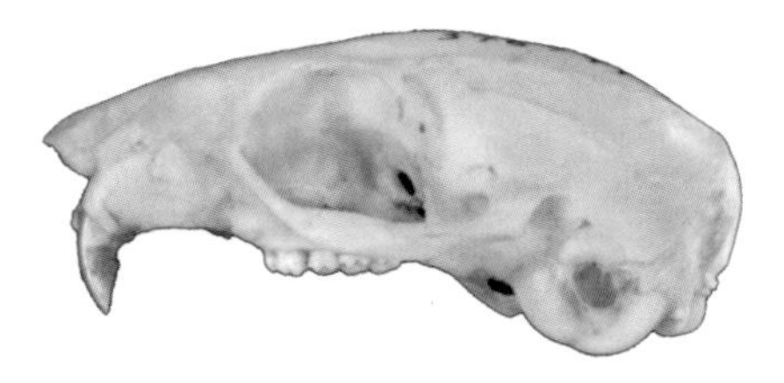

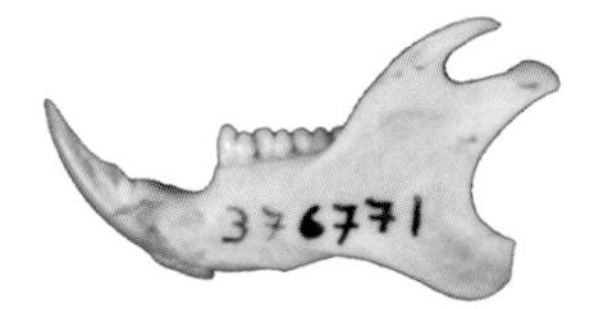

25 mm

25 mm

Protoxerus stangeri *Ammospermophilus leucurus*

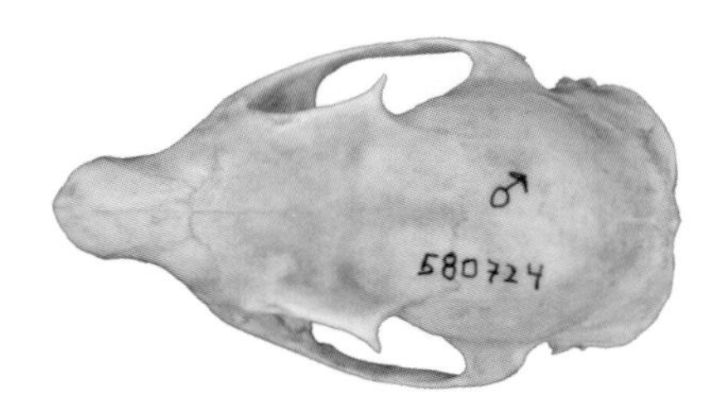

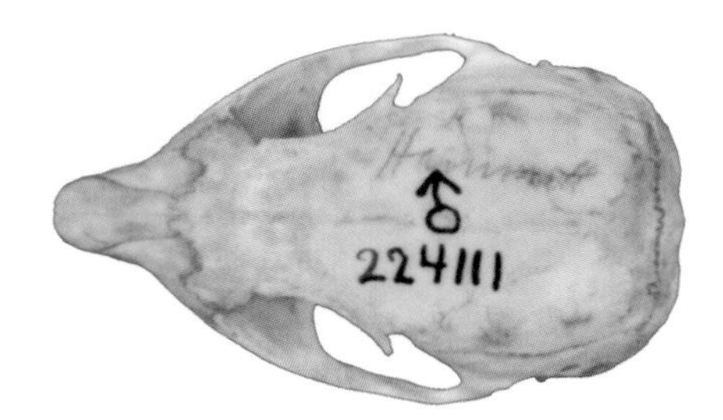

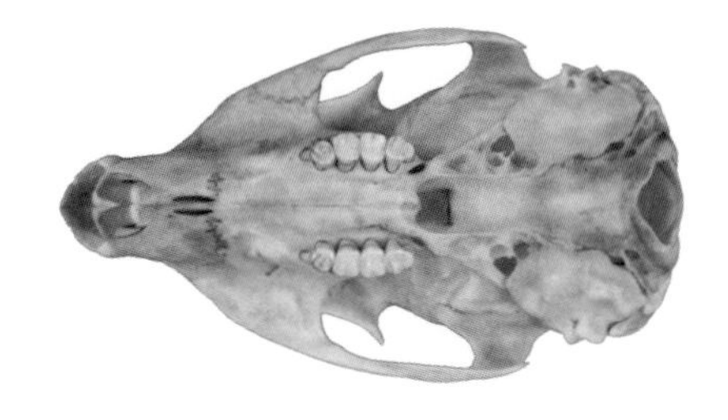

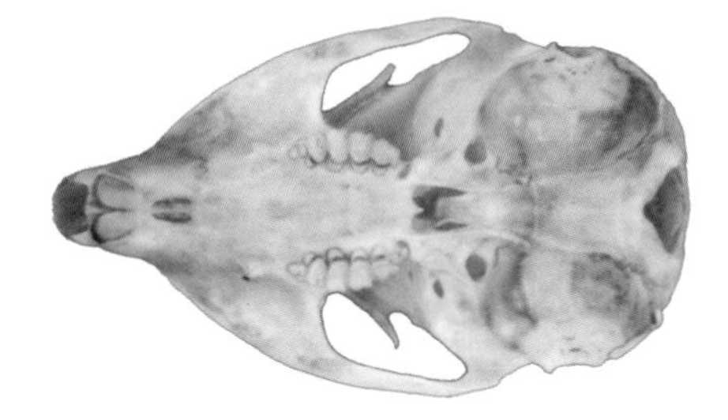

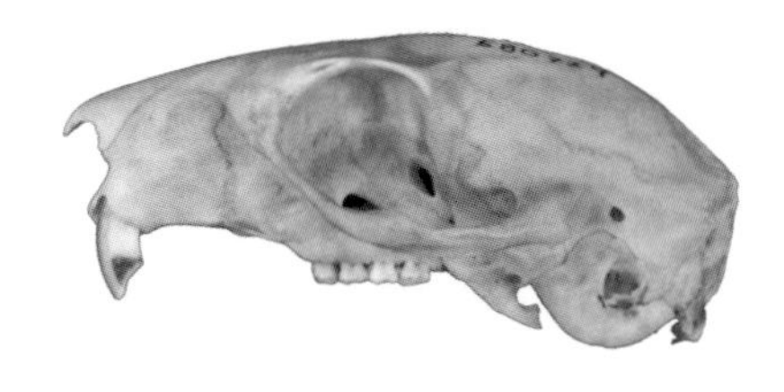

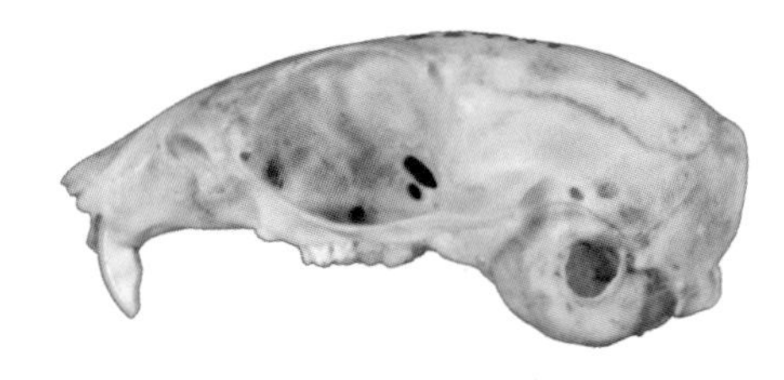

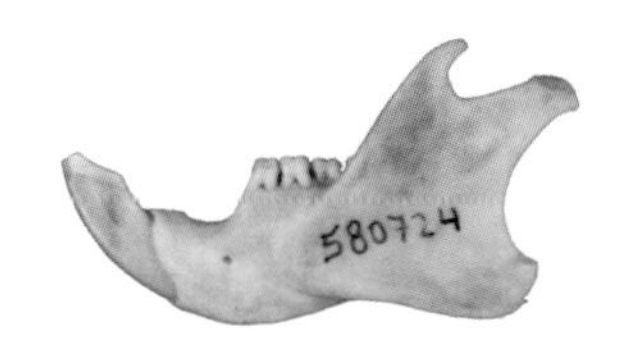

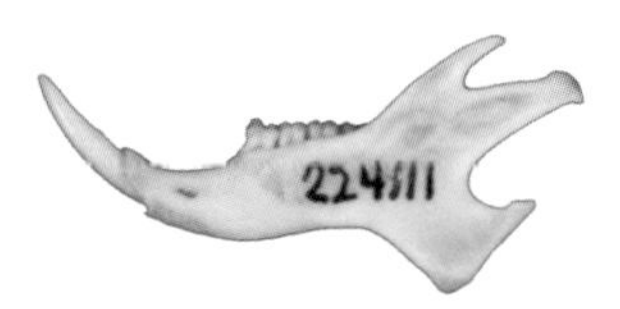

50 mm

25 mm

Callospermophilus lateralis | *Cynomys gunnisoni*

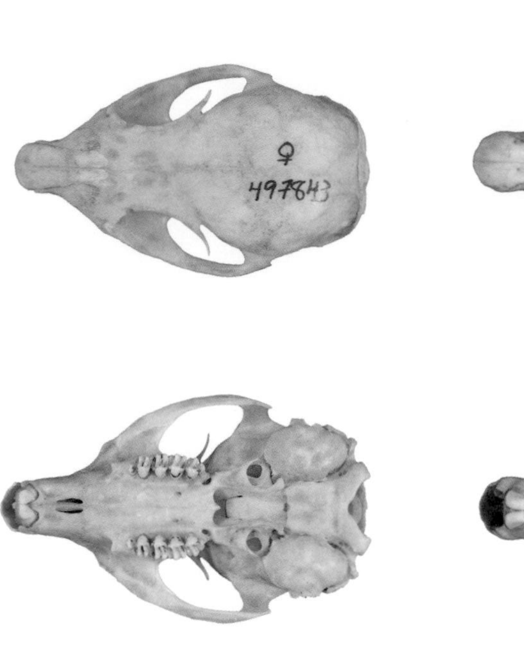

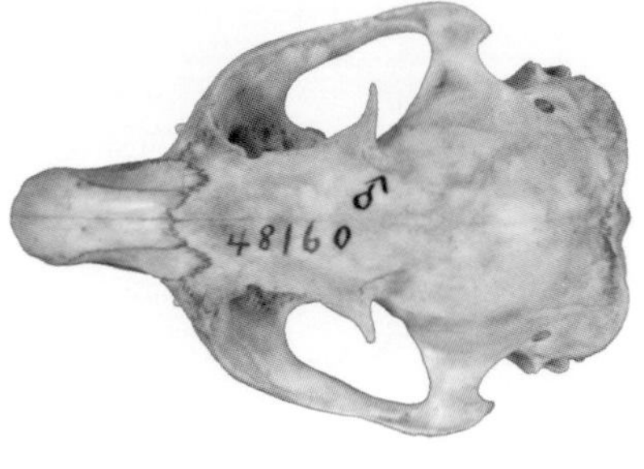

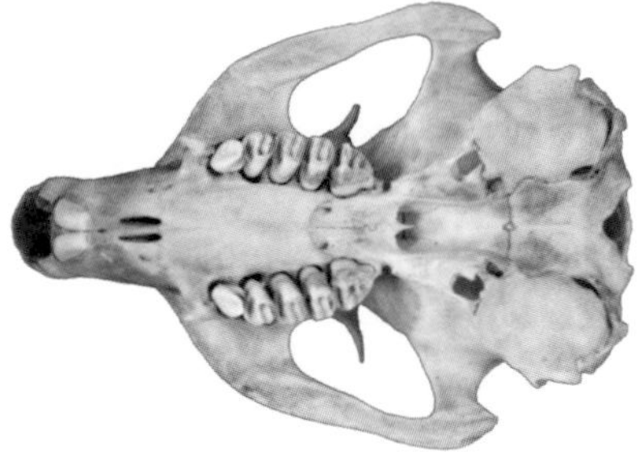

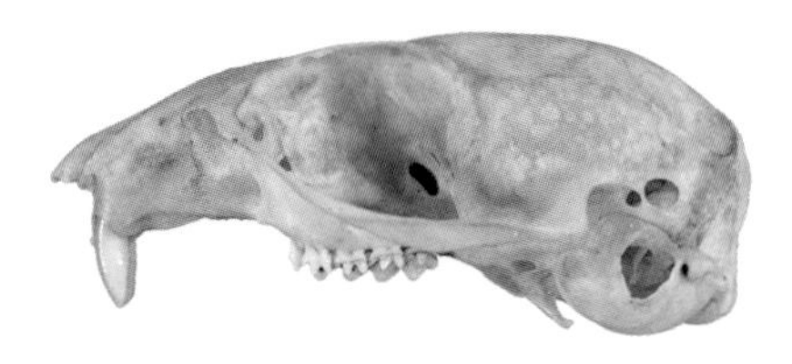

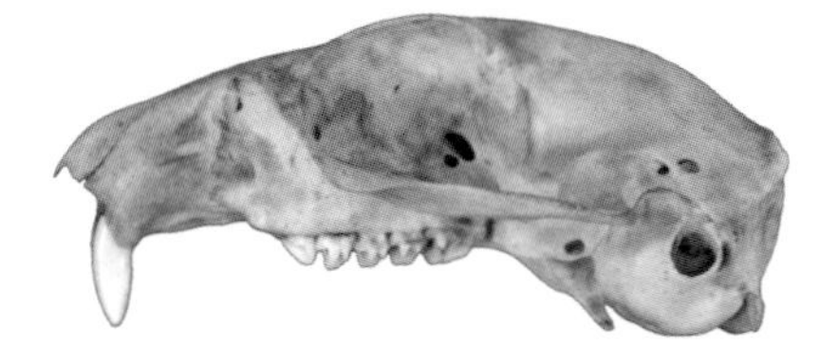

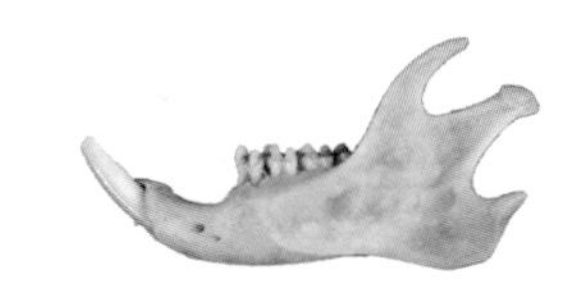

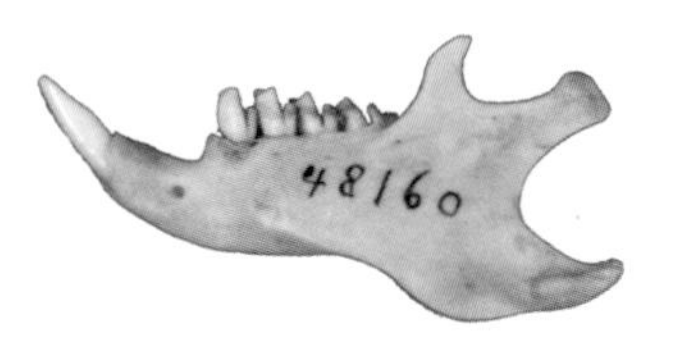

25 mm

25 mm

Ictidomys tridecemlineatus

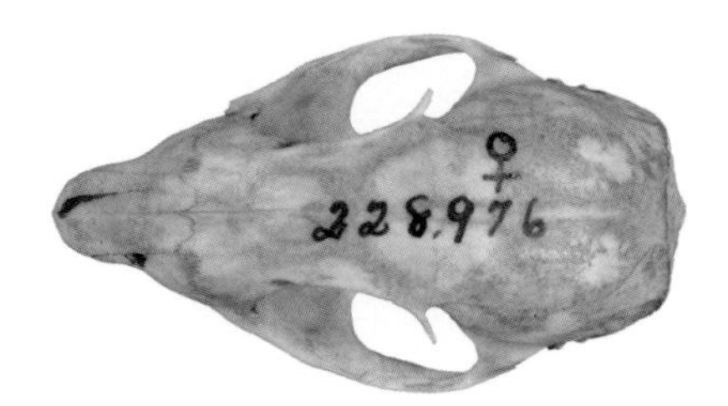

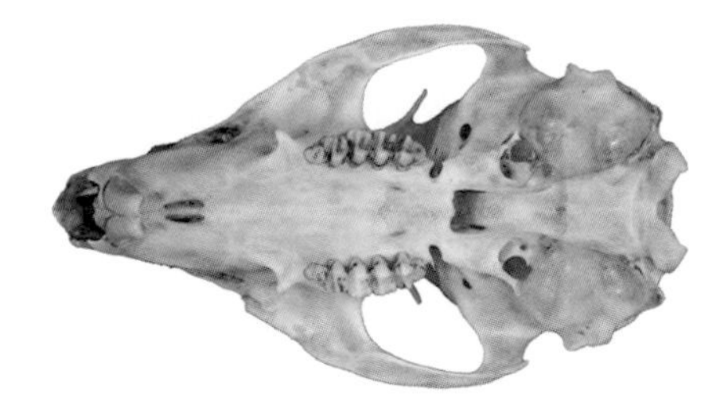

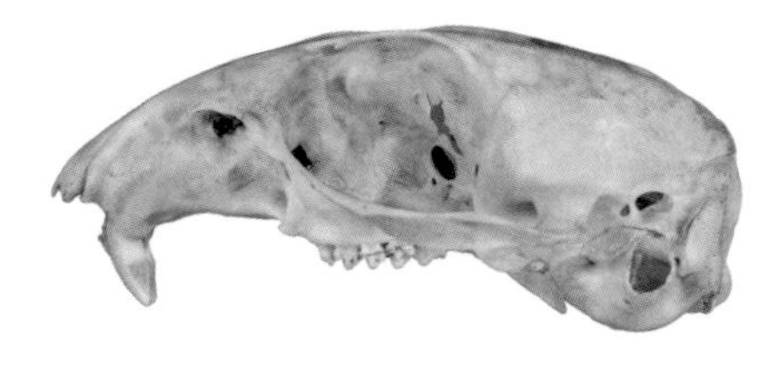

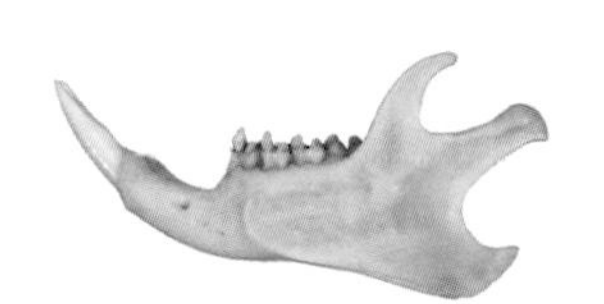

25 mm

Marmota himalayana

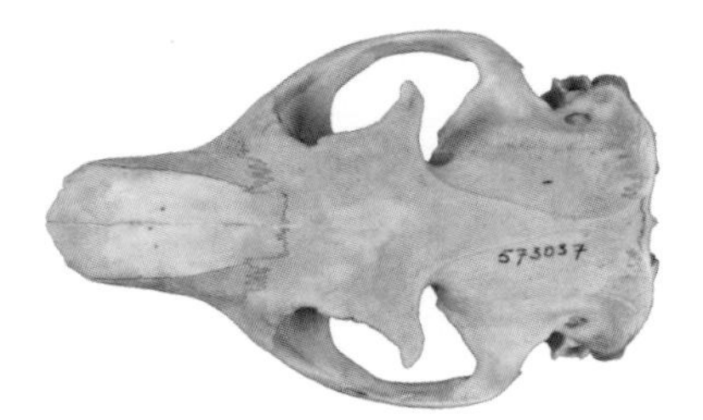

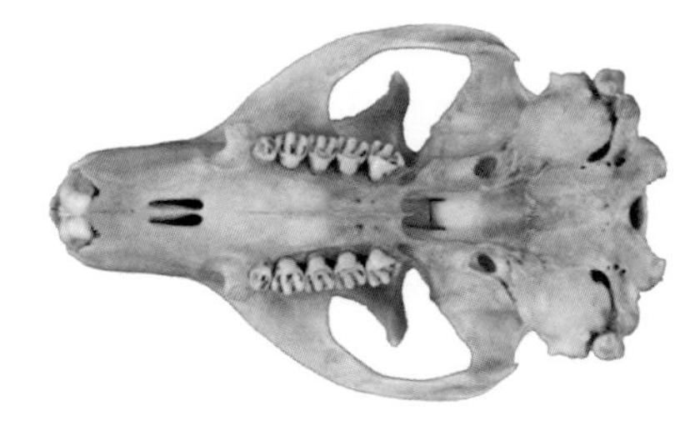

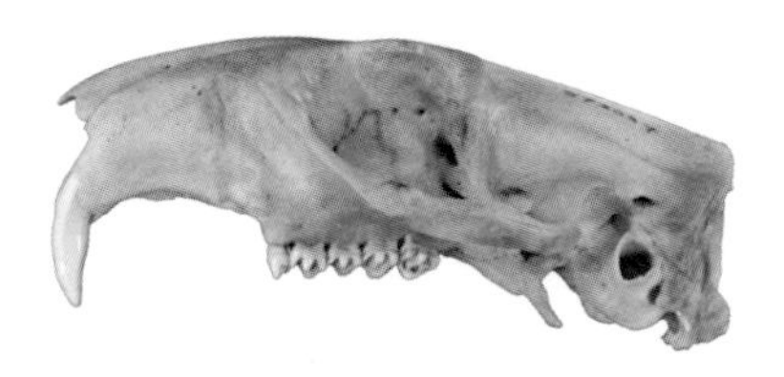

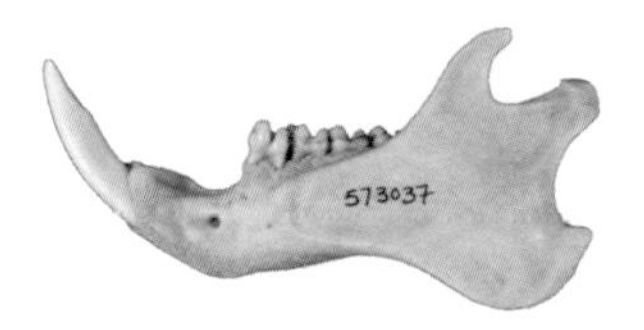

50 mm

Notocitellus annulatus *Otospermophilus variegatus*

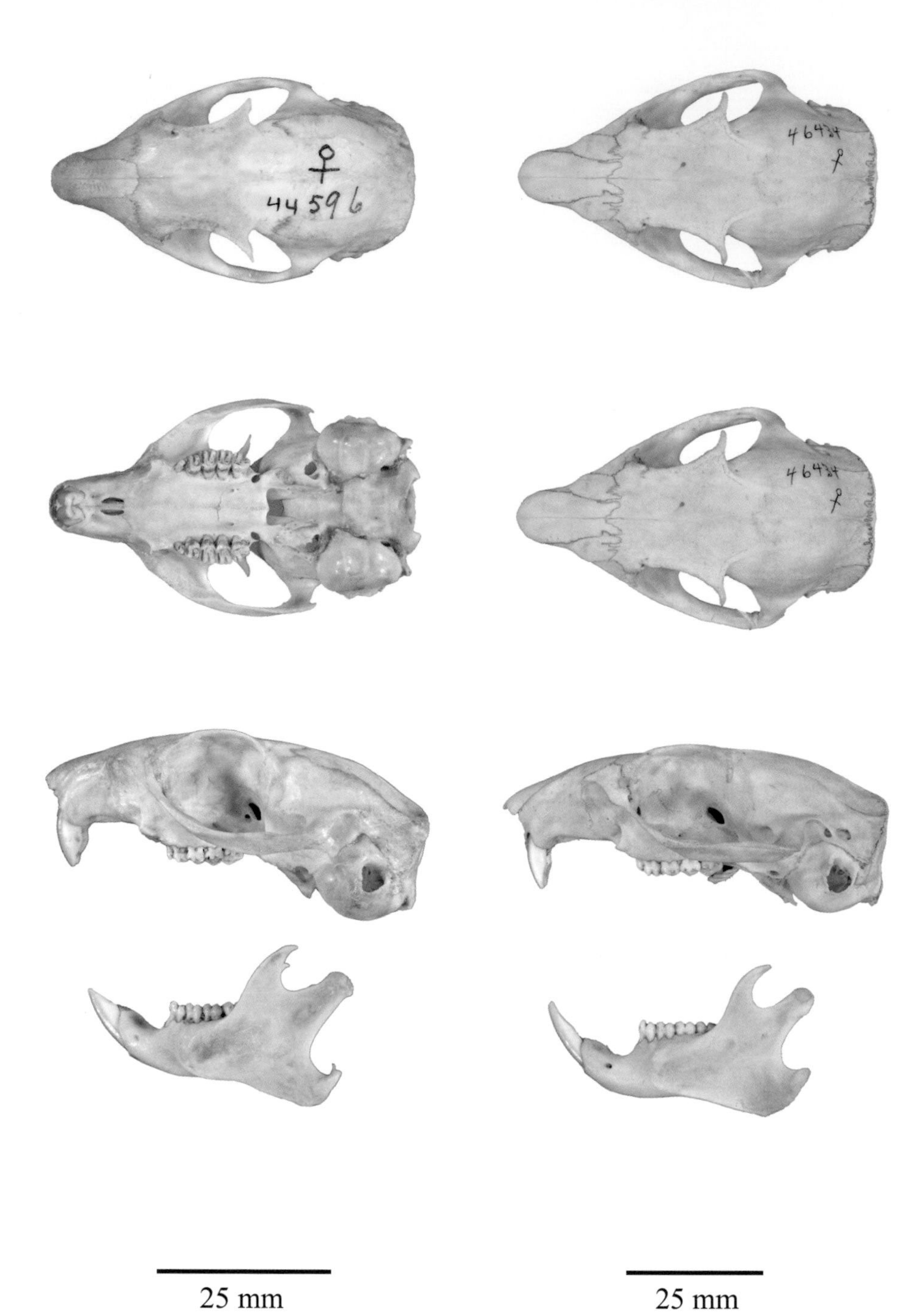

Poliocitellus franklinii

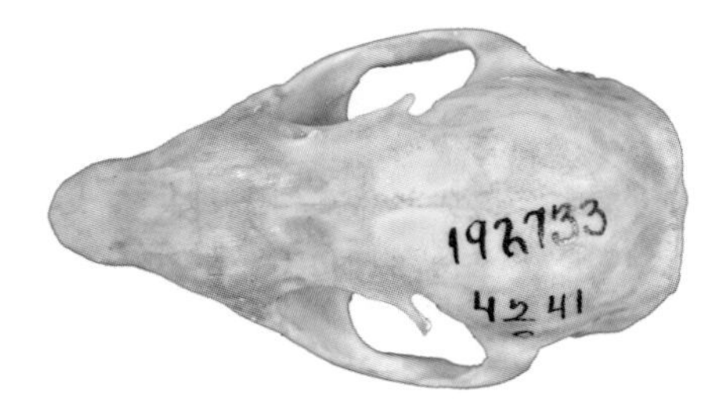

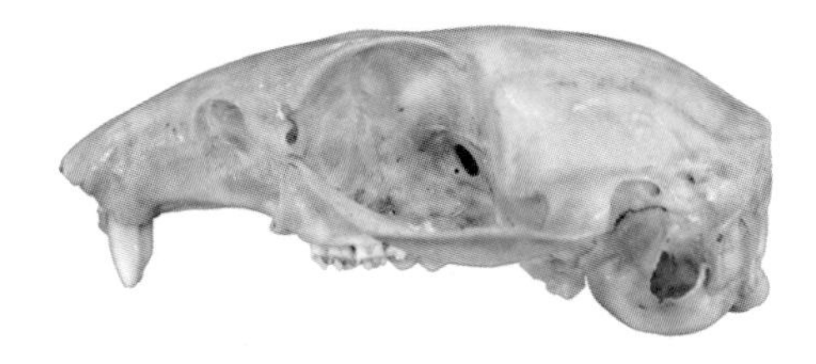

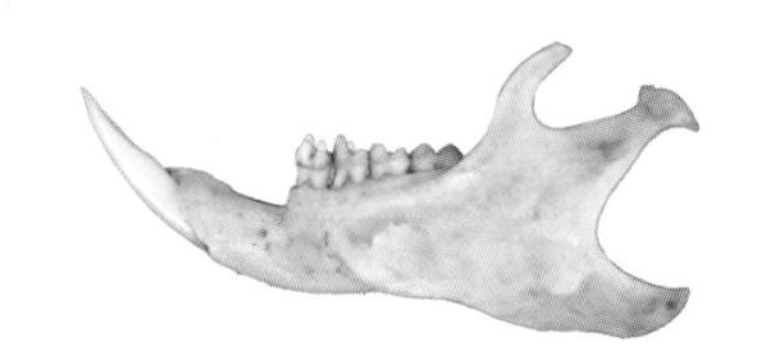

25 mm

Sciurotamias davidianus

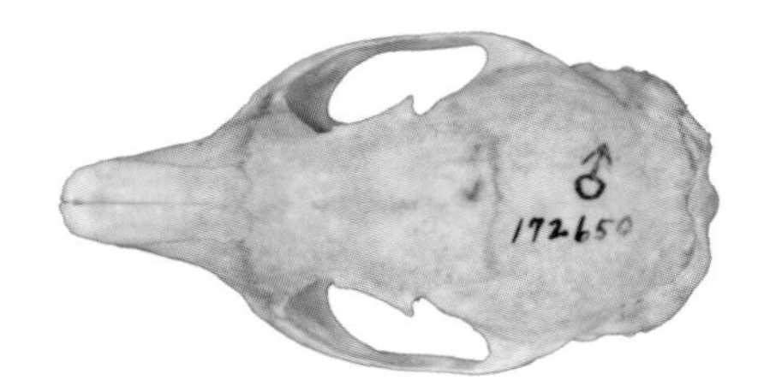

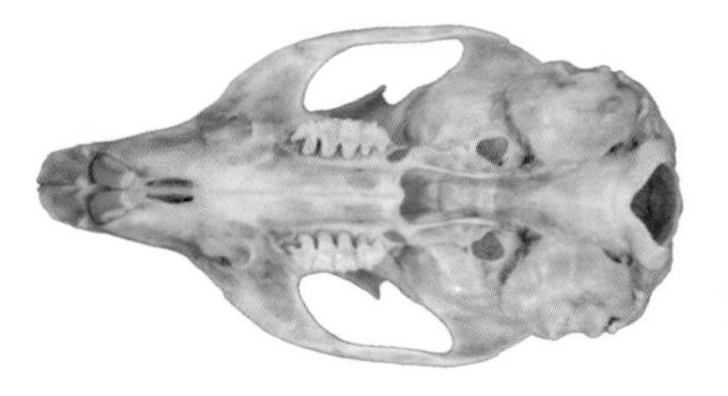

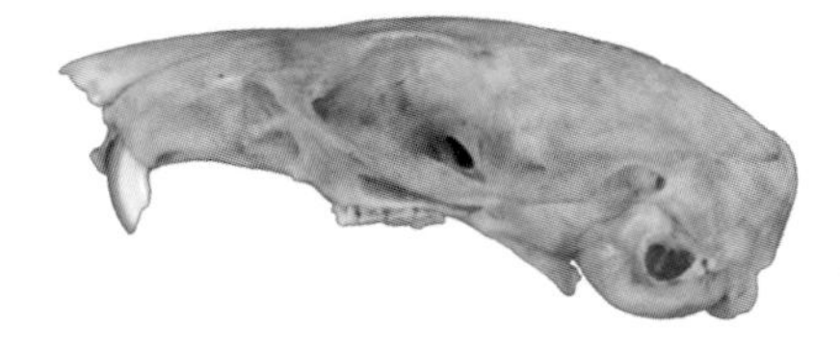

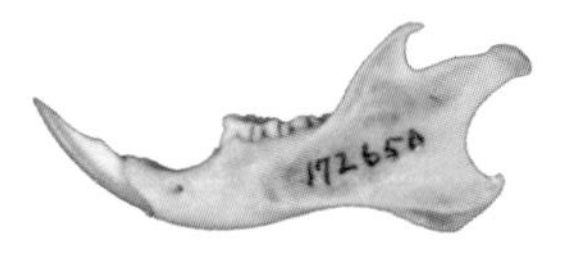

25 mm

Spermophilus citellus

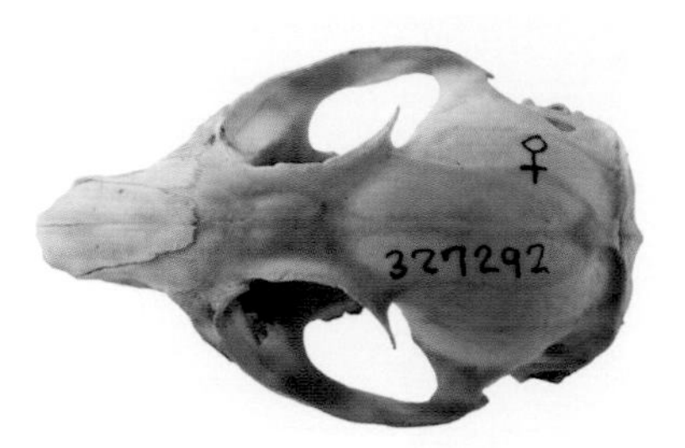

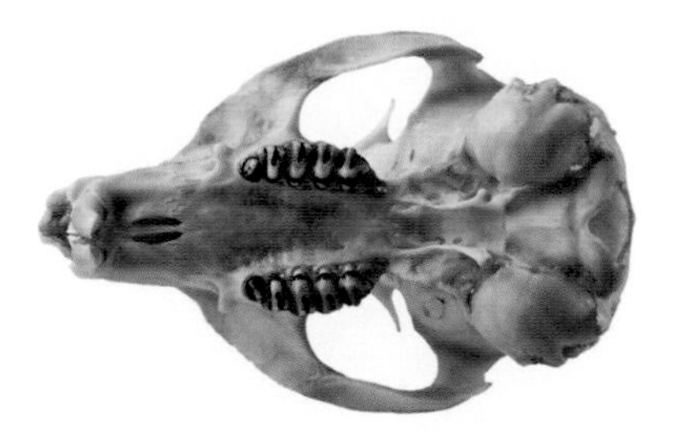

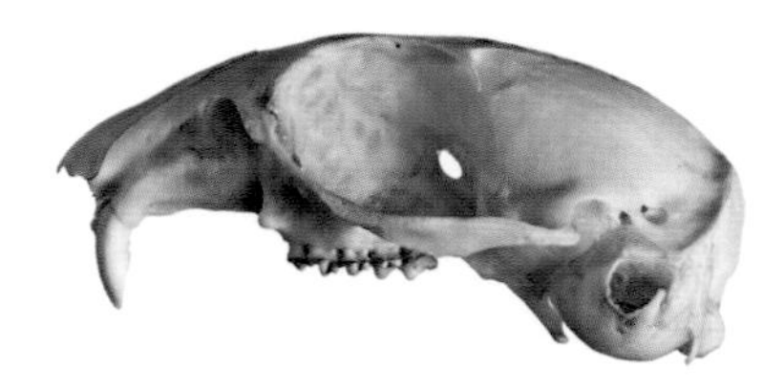

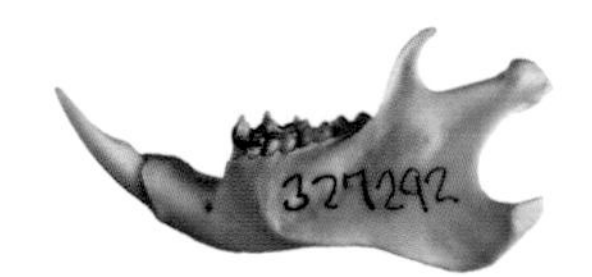

25 mm

Tamias alpinus

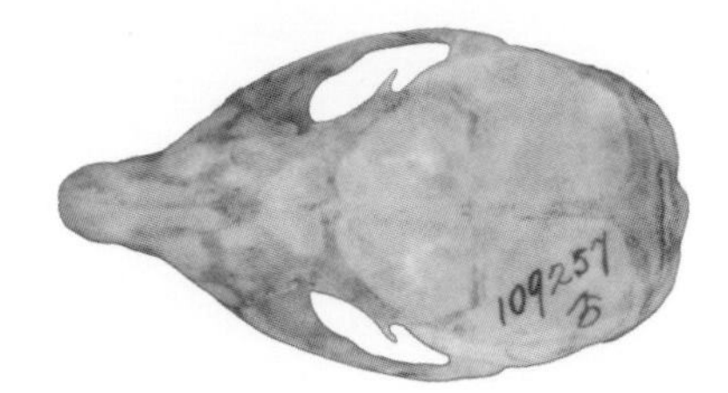

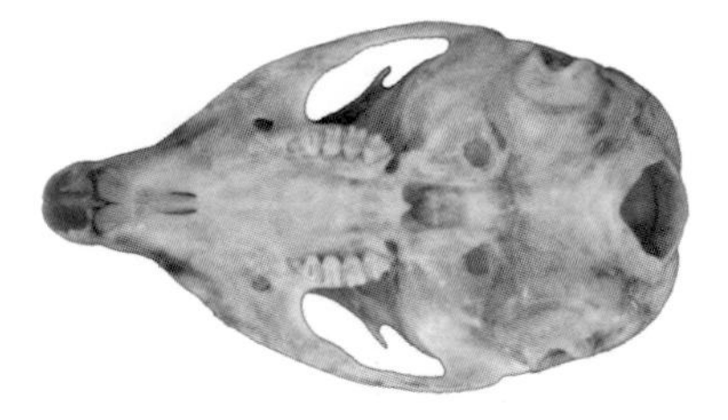

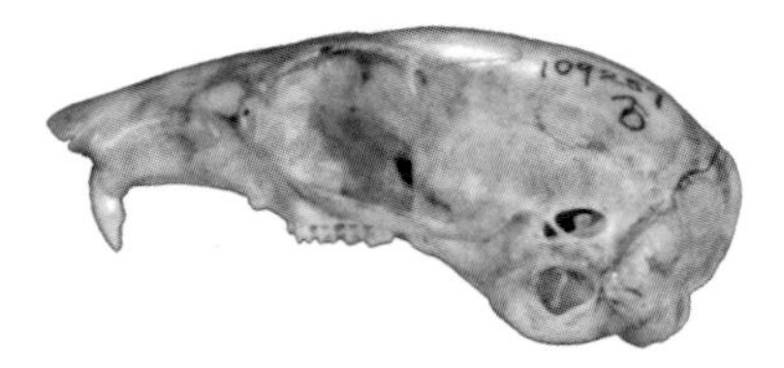

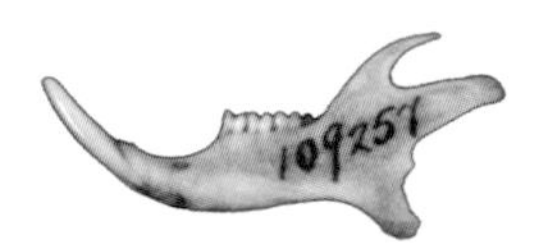

25 mm

Urocitellus columbianus *Xerospermophilus mohavensis*

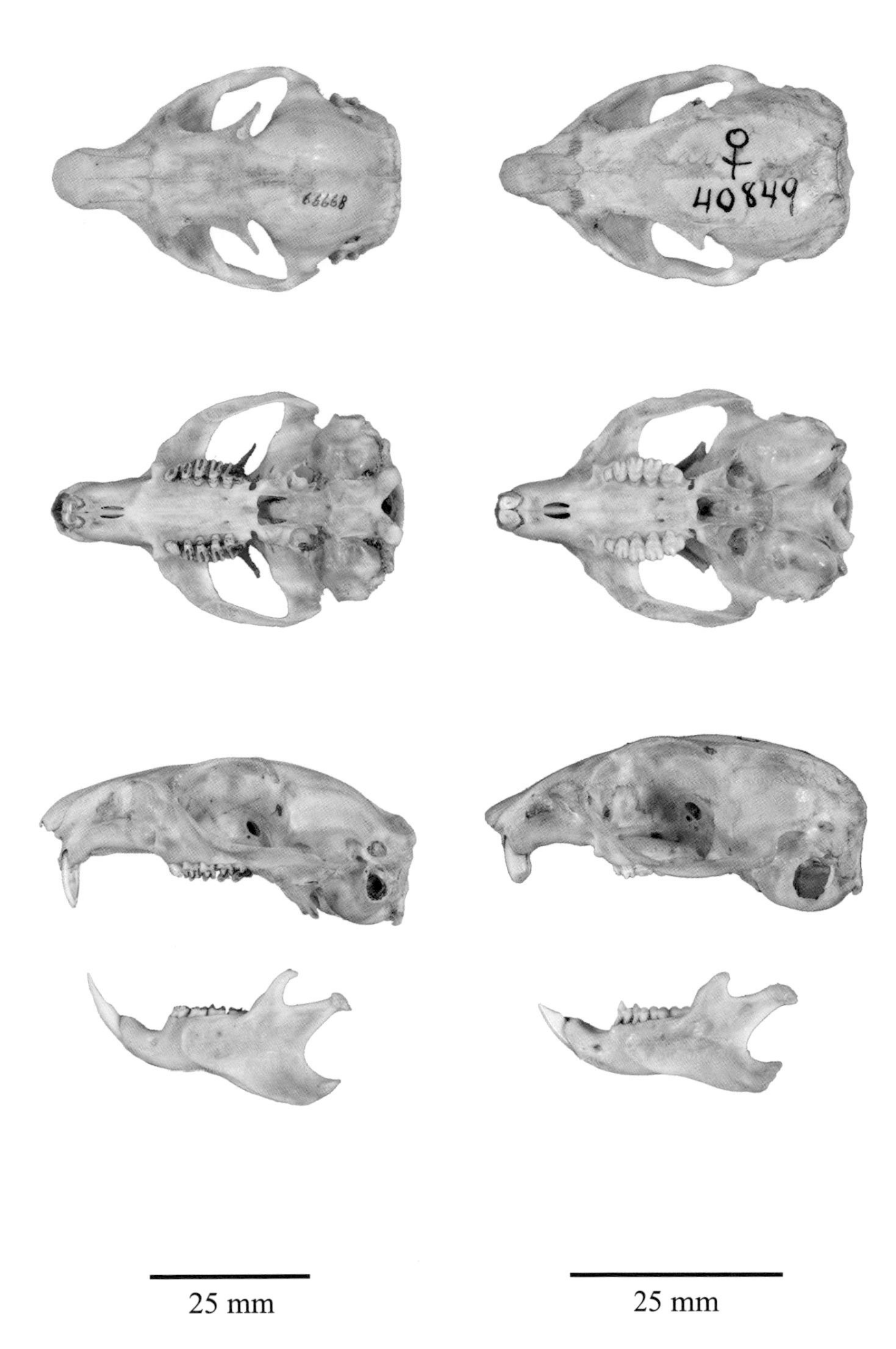

REFERENCES

Aaltonen, K., A. A. Bryant, J. A. Hostetler, and M. K. Oli. 2009. Reintroducing endangered Vancouver Island marmots: Survival and cause-specific mortality rates of captive-born versus wild-born individuals. *Biological Conservation* 142:2181–2190.

Abdullah, S. A., N. Yusoff-Rashid, and H. I. Azarae. 2001. Niche segregation among three sympatric species of squirrels inhabiting a lowland dipterocarp forest, peninsular Malaysia. *Mammal Study* 26 (2):133–144.

Adam, C. I. G. 1984. The fox squirrel in Saskatchewan. *Blue Jay* 42:241–246.

Adler, G. H., J. J. Arboledo, and B. L. Travi. 1997. Diversity and abundance of small mammals in degraded tropical dry forest of northern Colombia. *Mammalia* 61 (3):361–370.

Afanas'ev, A. V., V. S. Bazanov, M. N. Korelev, A. A. Sludskiy, and E. I. Strautman. 1953. *Zveri Kazakhstana (Les animaux du Kazakhstan)*. Alma-Ata: Académie des sciences du Kazakhstan.

Agrawal, V. C., and S. Chakraborty. 1979. Catalogue of mammals in the Zoological Survey of India: Part 1; Sciuridae. *Records of the Zoological Survey of India* 74:333–481.

Ahl, A. S. 1987. Relationship of vibrissal length and habits in the Sciuridae. *Journal of Mammalogy* 68 (4):848–853.

Airapetyants, A. E., and I. M. Fokin. 2003. Biology of European flying squirrel *Pteromys volans* L. (Rodentia: Pteromyidae) in the North-West of Russia. *Russian Journal of Theriology* 2 (2):105–113.

Albayrak, I., and A. Arslan. 2006. Contribution to the taxonomical and biological characteristics of *Sciurus anomalus* in Turkey. *Turkish Journal of Mammalogy* 30 (1):111–116.

Alberico, M., A. Cadena, J. Hernández-Camacho, and Y. Muñoz-Saba. 2000. Mamíferos (Synapsida: Theria) de Colombia. *Biota Colombiana* 1:43–75.

Alberico, M., and V. Rojas-Dias. 2002. Mamíferos de Colombia. In *Diversidad y conservación de los mamíferos Neotropicales*, ed. G. Ceballos and J. A. Simonetti. Mexico, D.F.: Comisión Nacional Para el Conocimiento y Uso de la Biodiversidad.

Allaine, D. 2000. Sociality, mating system and reproductive skew in marmots: Evidence and hypotheses. *Behavioural Processes* 51:21–34.

Allen, G. M. 1938. *The Mammals of China and Mongolia*. Vol. 11, pt. 1 of *Natural History of Central Asia*. New York: American Museum of Natural History.

———. 1939. A checklist of African mammals. *Bulletin of the Museum of Comparative Zoology* 83:1–763.

———. 1940. *The Mammals of China and Mongolia*. Vol. 11, pt. 2 of *Natural History of Central Asia*. New York: American Museum of Natural History.

Allen, G. M., and A. Loveridge. 1933. Reports on the scientific results of an expedition to the south-western highlands of Tanganyika Territory, 2: Mammals. *Bulletin of the Museum of Comparative Zoology* 75:47–140.

Allen, J. A. 1906. Mammals from the island of Hainan, China. *Bulletin of the American Museum of Natural History* 22:463–490.

———. 1914. Review of the genus *Microsciurus*. *Bulletin of the American Museum of Natural History* 33:145–165.

———. 1915. Review of the South American Sciuridae. *Bulletin of the American Museum of Natural History* 34:147–309.

Almazán-Catalán, J. A., C. Sánchez-Hernández, and M. D. L. Romero-Almaraz. 2005. Registros sobresalientes de mamíferos del Estado de Guerrero, México. *Acta Zoológica Mexicana*, n.s., 21:155–157.

Alvarenga, C. A., and S. A. Talamoni. 2006. Foraging behaviour of the Brazilian squirrel *Sciurus aestuans* (Rodentia, Sciuridae). *Acta Theriologica* 51 (1):69–74.

Álvarez-Castañeda, S. T. 2007. Systematics of the antelope ground squirrel (*Ammospermophilus*) from islands adjacent to the Baja California peninsula. *Journal of Mammalogy* 88 (5):1160–1169.

Álvarez-Castañeda, S. T., G. Arnaud, and E. Yensen. 1996. *Spermophilus atricapillus*. *Mammalian Species* 521:1–3.

Álvarez-Castañeda, S. T., I. Castro-Arellano, and T. Lacher. 2008. *Spermophilus atricapillus*. *IUCN Red List of Threatened Species*, version 2009.2.

Álvarez-Castañeda, S. T., I. Castro-Arellano, T. Lacher, and E. Vázquez. 2008a. *Spermophilus annulatus*. *IUCN Red List of Threatened Species*, version 2009.2.

———. 2008b. *Spermophilus madrensis*. *IUCN Red List of Threatened Species*, version 2009.2.

———. 2008c. *Tamias bulleri*. *IUCN Red

List of Threatened Species, 2008 [accessed 28 Oct. 2008].

———. 2008d. *Tamias durangae. IUCN Red List of Threatened Species*, version 2009.2.

Ambu, L. N., A. Tuuga, and T. P. Malim. 2009. The role of local communities in sustainable utilization of protected wildlife in Sabah. In *Biodiversity and National Development: Achievements, Opportunities, and Challenges*, ed. Y. H. Sen. Kuala Lumpur: Akademi Sains Malaysia.

Amori, G., and S. Gippoliti. 2003. A higher-taxon approach to rodent conservation priorities for the 21st century. *Animal Biodiversity and Conservation* 26 (2):1-18.

Amr, Z. S. 2000. *Mammals of Jordan*. Jordan Country Study of Biological Diversity. Amman, Jordan: United Nations Environment Programme, with funding from the Global Environment Facility.

Amtmann, E. 1966. Zur Systematik afrikanischer Streifenhörnchen der Gattung *Funisciurus*: Ein Beitrag zur Problematik klimaparalleler Variation und Phänetik. *Bonner Zoologische Beiträge* 17:1-44.

———. 1975. Family Sciuridae, pt. 6.1. In *The Mammals of Africa: An Identification Manual*, ed. J. A. J. Meester and H. W. Setzer. Washington, DC: Smithsonian Institution Press.

Anderson, J. F., R. C. Johnson, L. A. Magnarelli, and F. W. Hyde. 1985. Identification of endemic foci of Lyme disease: Isolation of *Borrelia burgdorferi* from feral rodents and ticks (*Dermacentor variabilis*). *Journal of Clinical Microbiology* 22 (1):36-38.

Anderson, S. 1997. Mammals of Bolivia: Taxonomy and distribution. *Bulletin of the American Museum of Natural History* 231:1-652.

Andō, M. 1999. Changing association of Japanese people with the Japanese giant flying squirrel, *Petaurista leucogenys*. *Mammalian Science* 78:175-179.

———. 2005. Improvement of nest box investigation techniques for study of arboreal rodents. *Mammalian Science* 91:165-176.

Andō, M., and Y. Imaizumi. 1982. Habitat utilization of the white-cheeked giant flying squirrel *Petaurista leucogenys* in a small shrine grove. *Journal of the Mammalogical Society of Japan* 9:70-81.

Andō, M., and S. Shiraishi. 1993. Gliding flight in the Japanese giant flying squirrel *Petaurista leucogenys*. *Journal of the Mammalogical Society of Japan* 18 (1):19-32.

Andō, M., S. Shiraishi, and T. A. Uchida. 1984. Field observations of the feeding behavior in the Japanese giant flying squirrel, *Petaurista leucogenys*. *Journal of the Faculty of Agriculture, Kyushu University* 28:161-175.

———. 1985a. Feeding behaviour of three species of squirrels. *Behaviour* 95:76-86.

———. 1985b. Food habits of the Japanese giant flying squirrel, *Petaurista leucogenys*. *Journal of the Faculty of Agriculture, Kyushu University* 29:189-202.

Angelici, F. M., B. Egbide, and G. C. Akani. 2001. Some new mammal records from the rainforests of south-eastern Nigeria. *Hystrix: Italian Journal of Mammalogy* 12 (1):37-43.

Angelici, F. M., and L. Luiselli. 2005. Patterns of specific diversity and population size in small mammals from arboreal and ground-dwelling guilds of a forest area in southern Nigeria. *Journal of Zoology* 265 (1):9-16.

Anonymous. 1990. The woodchuck in Saskatchewan. *Blue Jay* 48:226.

Ansell, W. F. H. 1960. *Mammals of Northern Rhodesia*. Lusaka: Government Printer.

———. 1978. *The Mammals of Zambia*. Chilanga, Zambia: National Parks & Wildlife Service.

Ansell, W. F. H., and P. D. H. Ansell. 1973. Mammals of the north-eastern montane areas of Zambia. *Puku* 7:21-69.

Ansell, W. F. H., and R. J. Dowsett. 1988. *Mammals of Malawi : An Annotated Check List and Atlas*. Zennor, St. Ives, Cornwall: Trendrine Press.

Anthony, H. E. 1928. *Field Book of North American Mammals*. New York: G. P. Putnam's Sons.

Anthony, H. E., and G. H. H. Tate. 1935. Notes on South American Mammalia: No. 1, *Sciurillus*. *American Museum Novitates* 780:1-13.

Aplin, K., and J. W. Duckworth. 2008. *Pteromyscus pulverulentus*. *IUCN Red List of Threatened Species*, version 2009.1.

Aplin, K., and D. Lunde. 2008. *Iomys horsfieldii*. *IUCN Red List of Threatened Species*, version 2009.1.

Aplin, K., D. Lunde, J. W. Duckworth, B. Lee, and R. J. Tizard. 2008. *Aeromys tephromelas*. *IUCN Red List of Threatened Species*, version 2009.1.

Arbogast, B. S. 1999. Mitochondrial DNA phylogeography of the New World flying squirrels (*Glaucomys*): Implications for Pleistocene biogeography. *Journal of Mammalogy* 80 (1):142-155.

Arbogast, B. S., R. A. Browne, P. D. Weigl, and G. J. Kenagy. 2005. Conservation genetics of endangered flying squirrels (*Glaucomys*) from the Appalachian Mountains of eastern North America. *Animal Conservation* 8:123-133.

Arenz, C. L., and D. W. Leger. 1999. Thirteen-lined ground squirrel (Sciuridae: *Spermophilus tridecemlineatus*) antipredator vigilance decreases as vigilance cost increases. *Animal Behaviour* 57:97-103.

Arita, H. T., and G. Ceballos. 1997. The mammals of Mexico: Distribution and conservation status. *Revista Mexicana de Mastozoología* 2:33-71.

Armitage, K. B. 1991. Social and population dynamics of yellow-bellied marmots: Results from long-term research. *Annual Review of Ecology and Systematics* 22:379-407.

———. 2009. Fur color diversity in marmots. *Ethology Ecology & Evolution* 21:183-194.

Armitage, K. B., and O. A. Schwartz. 2000. Social enhancement of fitness in yellow-bellied marmots. *PNAS—Proceedings of the National Academy of Sciences of the United States of America* 97:12149-12152.

Arnold, W. 1988. Social thermoregulation during hibernation in alpine marmots (*Marmota marmota*). *Journal of Comparative Physiology B—Biochemical, Systematic, and Environmental Physiology* 158:151-156.

Aschauer, A., I. E. Hoffmann, and E. Millesi. 2006. Endocrine profiles and reproductive output in European ground squirrels after unilateral ovariectomy. *Animal Reproduction Science* 92 (3-4):392-400.

Atwill, E. R., R. Phillips, M. das Graças Cabral Pereira, X. Li, and B. McCowan. 2004. Seasonal shedding of multiple *Cryptosporidium* genotypes in California ground squirrels (*Spermophilus beecheyi*). *Applied and Environmental Microbiology* 70 (11):6748-6752.

Azlan, J. M., and E. Lading. 2006. Camera trapping and conservation in Lambir Hills National Park, Sarawak. *Raffles Bulletin of Zoology* 54:469-475.

Baack, J. K., and P. V. Switzer. 2000. Alarm calls affect foraging behavior in eastern chipmunks (*Tamias striatus*, Rodentia: Sciuridae). *Ethology* 106:1057-1066.

Baba, M., T. Doi, and Y. Ono. 1982. Home range utilization and nocturnal activity of the giant flying squirrel, *Petaurista leucogenys*. *Japanese Journal of Ecology* 32:189-198.

Babu, S., and E. A. Jayson. 2009. Antipredator behaviour of large brown flying squirrel (*Petaurista philippensis*): Is this an effective census method to survey the species? *Current Science* 96 (6):772-773.

Baccus, J. T. 1978. Notes on the distribution of some mammals from Coahuila, Mexico. *Southwestern Naturalist* 23 (4):706-708.

Badmaev, B. B. 2008. Sexual differences in the annual cycle of the long-tailed ground squirrel (*Spermophilus undulatus*, Rodentia, Sciuridae) in western Transbaikalia. *Zoologichesky Zhurnal* 87:83-89.

Baillie, J., and B. Groombridge, eds. 1996. *1996 IUCN Red List of Threatened Animals*. Gland, Switzerland: IUCN.

Baker, R. H., M. W. Baker, J. D. Johnson, and R. G. Webb. 1981. New records of mammals and reptiles from northwestern Zacatecas, Mexico. *Southwestern Naturalist* 25:568-569.

Bakker, V. J., and K. Hastings. 2002. Den trees used by northern flying squirrels (*Glaucomys sabrinus*) in southeastern Alaska. *Canadian Journal of Zoology* 80 (9):1623.

Bakko, E. B., and L. N. Brown. 1967. Breeding biology of white-tailed prairie dog *Cynomys leucurus* in Wyoming. *Journal of Mammalogy* 48 (1):100-112.

Ball, L. C., P. F. Doherty, and M. W. McDonald. 2005. An occupancy modeling approach to evaluating a Palm Springs ground squirrel habitat model. *Journal of Wildlife Management* 69:894-904.

Banfield, A. W. F. 1974. *The Mammals of Canada*. Toronto: University of Toronto Press.

Bannikov, A. G. 1954. *Mlekopitaiushchie Mongolskoĭ Narodnoĭ Respubliki*. Moscow: Izdatel'stvo Akademii nauk SSSR.

Barash, D. P. 1973. The social biology of the Olympic marmot. *Animal Behaviour Monographs* 6:171-249.

———. 1989. *Marmots: Social Behavior and Ecology*. Stanford, CA: Stanford University Press.

Barragán, F., O. G. Retana, and E. J. Naranjo. 2007. The rodent trade of Tzeltal indians of Oxchuc, Chiapas, Mexico. *Human Ecology* 35:769-773.

Bartels, M. A., and D. P. Thompson. 1993. *Spermophilus lateralis*. *Mammalian Species* 440:1-8.

Bartig, J. L., T. L. Best, and S. L. Burt. 1993. *Tamias bulleri*. *Mammalian Species* 438:1-4.

Bartz, S. E., L. C. Drickamer, and M. J. C. Kearsley. 2007. Response of plant and rodent communities to removal of prairie dogs (*Cynomys gunnisoni*) in Arizona. *Journal of Arid Environments* 68 (3):422-437.

Bates, G. L. 1905. Notes on the mammals of Southern Cameroons and the Benito, 3. *Proceedings of the Zoological Society of London* 75 (1):65-88.

Baudoin, C., V. J. Sosa, and V. Serrano. 2004. Records of *Spermophilus mexicanus* (Rodentia: Sciuridae) in the Bolsón de Mapimi (Durango, Mexico) and comparison with Texan and Coahuilan forms of the *parvidens* subspecies. *Acta Zoológica Mexicana* 20:233-235.

Bayrakçi, R., A. B. Carey, and T. M. Wilson. 2001. Current status of the western gray squirrel population in the Puget Sound Trough, Washington. *Northwest Science* 75:333-341.

Becker, P., M. Leighton, and J. B. Payne. 1985. Why tropical squirrels carry seeds out of source crowns. *Journal of Tropical Ecology* 1 (2):183-186.

Bee, J. W., and E. R. Hall. 1956. *Mammals of Northern Alaska*. Miscellaneous Publication No. 8. Lawrence: University of Kansas, Museum of Natural History.

Belk, M. C., and H. D. Smith. 1991. *Ammospermophilus leucurus*. *Mammalian Species* 368:1-8.

Bendel, P. R., and J. E. Gates. 1987. Home range and microhabitat partitioning of the southern flying squirrel (*Glaucomys volans*). *Journal of Mammalogy* 68 (2):243-255.

Benkman, C. W., and A. M. Siepielski. 2004. A keystone selective agent? Pine squirrels and the frequency of serotiny in lodgepole pine. *Ecology* 85 (8):2082-2087.

Bennett, A. F., R. B. Huey, H. John-Alder, and K. A. Nagy. 1984. The parasol tail and thermoregulatory behavior of the Cape ground squirrel *Xerus inauris*. *Physiological Zoology* 57 (1):57-62.

Bennett, R. P. 1999a. Effects of food quality on growth and survival of juvenile Columbian ground squirrels (*Spermophilus columbianus*). *Canadian Journal of Zoology* 77:1555-1561.

———. 1999b. *Status of the Red-Tailed Chipmunk (*Tamias ruficaudus*) in Alberta*. Alberta Wildlife Status Report No. 19. Edmonton: Alberta Environmental Protection, Fisheries and Wildlife Management Division, and Alberta Conservation Association.

Bergstrom, B. J. 1992. Parapatry and encounter competition between chipmunk species and the hypothesized role of parasitism. *American Midland Naturalist* 128 (1):168-179.

Bergstrom, B. J., and R. S. Hoffmann. 1991. Distribution and diagnosis of 3 species of chipmunks in the Front Range of Colorado. *Southwestern Naturalist* 36 (1):14-28.

Bertolino, S. 2009. Animal trade and non-indigenous species introduction: The world-wide spread of squirrels. *Diversity and Distributions* 15 (4):701-708.

Bertolino, S., I. Currado, P. J. Mazzoglio, and G. Amori. 2000. Native and alien squirrels in Italy. *Hystrix: Italian Journal of Mammalogy* 11 (2):65-74.

Bertolino, S., and P. Genovesi. 2005. The application of the European Strategy on Invasive Alien Species: An example with introduced squirrels. *Hystrix: Italian Journal of Mammalogy* 16 (1):59-69.

Bertolino, S., P. J. Mazzoglio, M. Vaiana, and I. Currado. 2004. Activity budget and foraging behavior of introduced *Callosciurus finlaysonii* (Rodentia, Sciuridae) in Italy. *Journal of Mammalogy* 85 (2):254-259.

Best, T. L. 1993a. *Tamias ruficaudus*. *Mammalian Species* 452:1-7.

———. 1993b. *Tamias sonomae*. *Mammalian Species* 444:1-5.

———. 1994. *Tamias quadrivittatus*. *Mammalian Species* 466:1-7.

———. 1995a. *Sciurus alleni*. *Mammalian Species* 501:1-4.

———. 1995b. *Sciurus colliaei*. *Mammalian Species* 497:1-4.

———. 1995c. *Sciurus deppei*. *Mammalian Species* 505:1-5.

———. 1995d. *Sciurus nayaritensis*. *Mammalian Species* 492:1-5.

———. 1995e. *Sciurus oculatus*. *Mammalian Species* 498:1-3.

———. 1995f. *Sciurus variegatoides*. *Mammalian Species* 500:1-6.

———. 1995g. *Spermophilus adocetus*. *Mammalian Species* 504:1-4.

———. 1995h. *Spermophilus annulatus*. *Mammalian Species* 508:1-4.

———. 1995i. *Spermophilus mohavensis*. *Mammalian Species* 509:1-7.

Best, T. L., J. L. Bartig, and S. L. Burt. 1992. *Tamias canipes*. *Mammalian Species* 411:1-5.

Best, T. L., S. L. Burt, and J. L. Bartig. 1993. *Tamias durangae*. *Mammalian Species* 437:1-4.

Best, T. L., K. Caesar, A. S. Titus, and C. L. Lewis. 1990. *Ammospermophilus insularis*. *Mammalian Species* 364:1-4.

Best, T. L., and G. Ceballos. 1995. *Spermophilus perotensis*. *Mammalian Species* 507:1-3.

Best, T. L., R. G. Clawson, and J. A. Clawson. 1994a. *Tamias panamintinus*. *Mammalian Species* 468:1-7.

———. 1994b. *Tamias speciosus*. *Mammalian Species* 478:1-9.

Best, T. L., and N. J. Granai. 1994a. *Tamias merriami*. *Mammalian Species* 476:1-9.

———. 1994b. *Tamias obscurus*. *Mammalian Species* 472:1-6.

Best, T. L., C. L. Lewis, K. Caesar, and A. S. Titus. 1990. *Ammospermophilus interpres*. *Mammalian Species* 365:1-6.

Best, T. L., and S. Riedel. 1995. *Sciurus arizonensis*. *Mammalian Species* 496:1-5.

Best, T. L., H. A. Ruiz-Piña, and L. S. Leon-Paniagua. 1995. *Sciurus yucatenensis*. *Mammalian Species* 506:1-4.

Best, T. L., and H. H. Thomas. 1991. *Spermophilus madrensis*. *Mammalian Species* 378:1-2.

Best, T. L., A. S. Titus, K. Caesar, and C. L. Lewis. 1990. *Ammospermophilus harrisii*. *Mammalian Species* 366:1-7.

Best, T. L., A. S. Titus, C. L. Lewis, and K. Caesar. 1990. *Ammospermophilus nelsoni*. *Mammalian Species* 367:1-7.

Betts, B. J. 1990. Geographic distribution and habitat preferences of Washington ground squirrels (*Spermophilus washingtoni*). *Northwestern Naturalist* 71:27-37.

———. 1999. Current Status of Washington ground squirrels in Oregon and Washington. *Northwestern Naturalist* 80 (1):35-38.

Biardi, J. E., D. C. Chien, and R. G. Coss. 2006. California ground squirrel (*Spermophilus beecheyi*) defenses against rattlesnake venom digestive and hemostatic toxins. *Journal of Chemical Ecology* 32:137-154.

Biardi, J. E., R. G. Coss, and D. G. Smith. 2000. California ground squirrel (*Spermophilus beecheyi*) blood sera inhibits crotalid venom proteolytic activity. *Toxicon* 38 (5):713-721.

Bibikow, D. I. 1996. *Die Murmeltiere der Welt*. Heidelberg, Germany: Spektrum Akademischer Verlag.

Bidlack, A. L., and J. A. Cook. 2001. Reduced genetic variation in insular northern flying squirrels (*Glaucomys sabrinus*) along the North Pacific Coast. *Animal Conservation* 4 (4):283-290.

Biedrzycka, A., and M. K. Konopinski. 2008. Genetic variability and the effect of habitat fragmentation in spotted suslik *Spermophilus suslicus* populations from two different regions. *Conservation Genetics* 9 (5):1211-1221.

Biggins, D. E., and M. Y. Kosoy. 2001. Influences of introduced plague on North American mammals: Implications from ecology of plague in Asia. *Journal of Mammalogy* 82 (4):906-916.

Bihr, K. J., and R. J. Smith. 1998. Location, structure, and contents of burrows of *Spermophilus lateralis* and *Tamias minimus*, two ground-dwelling Sciurids. *Southwestern Naturalist* 43 (3):352-362.

Bishop, K. L. 2006. The relationship between 3-D kinematics and gliding performance in the southern flying squirrel, *Glaucomys volans*. *Journal of Experimental Biology* 209:689-701.

Bjarvall, A., and S. Ullstrom. 1986. *The Mammals of Britain and Europe*. Trans. D. Christie. London: Croom Helm.

Blake, B. H. 1992. Estrous calls in captive Asian chipmunks, *Tamias sibiricus*. *Journal of Mammalogy* 73 (3):597-603.

Blake, B. H., and K. E. Gillett. 1988. Estrous cycle and related aspects of reproduction in captive Asian chipmunks, *Tamias sibiricus*. *Journal of Mammalogy* 69 (3):598-603.

Blanco, J. C. 1998. *Mamíferos de España II: Cetaceos, Artidactilos, Roedores y Lagomorfos de la peninsula Iberica, Baleares y Canarias*. Barcelona: Planeta.

Blate, G. M., D. R. Peart, and M. Leighton. 1998. Post-dispersal predation on isolated seeds: A comparative study of 40 tree species in a Southeast Asian rainforest. *Oikos* 82:522-538.

Blois, J. L., R. S. Feranec, and E. A. Hadly. 2008. Environmental influences on spatial and temporal patterns of body-size variation in California ground squirrels (*Spermophilus beecheyi*). *Journal of Biogeography* 35:602-613.

Blumstein, D. T. 1999. Alarm calling in three species of marmots. *Behaviour* 136:731-757.

Blumstein, D. T., and W. Arnold. 1998. Ecology and social behavior of golden marmots (*Marmota caudata aurea*). *Journal of Mammalogy* 79:873-886.

Blumstein, D. T., J. C. Daniel, and A. A. Bryant. 2001. Anti-predator behavior of Vancouver Island marmots: Using congeners to evaluate abilities of a critically endangered mammal. *Ethology* 107:1-14.

Blyth, E. 1847. Report of the Curator, Zoology Department. *Journal of the Asiatic Society of Bengal* 16:863-880.

Bobrinskii, N. A., B. A. Kuznetzov, and A. P. Kuzyakin. 1965. *Opredelitel mlekopitaiushchikh SSSR [Classification of mammals of the USSR]*. Moscow: Prosveshchenie, Posobie dli studentov pedagogicheskikh institutov i uchiteleĭ.

Bobrinskoy, N. A., B. A. Kuznetzov, and A. P. Kuzyakin. 1944. *Mammals of USSR*. Moscow: Government Publishing Office Sovietskaya Nauka.

Bodmer, R. E. 1995. Managing Amazonian wildlife: Biological correlates of game choice by detribalized hunters. *Ecological Applications* 5:872-877.

Boellstorff, D. E., and D. H. Owings. 1995. Home range, population structure, and spatial organization of California ground squirrels. *Journal of Mammalogy* 76 (2):551-561.

Bolles, K. 1988. Evolution and variation of antipredator vocalisations of antelope squirrels, *Ammospermophilus* (Rodentia: Sciuridae). *Zeitschrift für Säugetierkunde* 53:129-147.

Bonhote, J. L. 1901. On *Sciurus notatus* and allied species. *Annals and Magazine of Natural History*, ser. 7, 7:444-455.

Bonhote, J. L., and W. W. Skeat. 1900. *On the Mammals Collected During the "Skeat Expedition" to the Malay Peninsula, 1899-1900*. London: Zoological Society of London.

Boonstra, R., C. J. McColl, and T. J. Karels. 2001. Reproduction at all costs: The adaptive stress response of male Arctic ground squirrels. *Ecology* 82 (7):1930-1946.

Booth, A. H. 1960. *Small Mammals of West Africa*. West African Nature Handbooks. London: Longmans.

Bordignon, M., and E. L. A. Monteiro-Filho. 1999. Seasonal food resources of the squirrel *Sciurus ingrami* in a secondary Araucaria forest in southern Brazil. *Studies on Neotropical Fauna and Environment* 34:137-140.

———. 2000. Behaviour and daily activity of the squirrel *Sciurus ingrami* in a secondary Araucaria forest in southern Brazil. *Canadian Journal of Zoology* 78:1732-1739.

Borges, R. M. 1990. Sexual and site differences in calcium consumption by the Malabar giant squirrel (*Ratufa indica*). *Oecologia* 85 (1):80-86.

———. 1992. A nutritional analysis of foraging in the Malabar giant squirrel (*Ratufa indica*). *Biological Journal of the Linnean Society* 47 (1):1-21.

———. 1993. Figs, Malabar giant squirrels, and fruit shortages within two tropical Indian forests. *Biotropica* 25 (2):183-190.

———. 1998. Spatiotemporal heterogeneity of food availability and dietary variation between individuals of the Malabar giant squirrel *Ratufa indica*. In *Ecology and Evolutionary Biology of Tree Squirrels*, ed. M. A. Steele, J. A. Merritt, and D. A. Zegers. Special Publication No. 6. Martinsville: Virginia Museum of Natural History.

———. 2006. Big, bigger, biggest: Why do giants occur in the Old World tropics? Paper read at the Tree and Flying Squirrel Colloquia, March, Periyar Tiger Reserve, Thekkady, Kerala, India.

Borges, R. M., B. R. G. Prasad, H. Somanathan, and S. Mali. 2006. How do foraging path statistics in the Indian giant squirrel, *Ratufa indica*, vary across timescales? A comparison of consecutive day variation with long-term variation. Paper read at the Tree and Flying Squirrel Colloquia, March, Periyar Tiger Reserve, Thekkady, Kerala, India.

Botello, F., P. Illoldi-Rangel, M. Linaje, and V. Sanchez-Cordero. 2007. New record of the rock squirrel (*Spermophilus variegatus*) in the state of Oaxaca, Mexico. *Southwestern Naturalist* 52:326-328.

Bou, J., A. Casinos, and J. Ocaña. 1987. Allometry of the limb long bones of insectivores and rodents. *Journal of Morphology* 192:113-123.

Bouchardy, C, and F. Moutou. 1989. *Observing British and European Mammals*. Trans. I. Bishop. London: British Museum (Natural History).

Bouchie, L., N. C. Bennet, T. Jackson, and J. M. Waterman. 2006. Are Cape ground squirrels (*Xerus inauris*) induced or spontaneous ovulators? *Journal of Mammalogy* 87 (1):60-66.

Bowers, M. A. 1995. Use of space and habitats by the eastern chipmunk, *Tamias striatus*. *Journal of Mammalogy* 76 (1):12-21.

Bowers, M. A., and R. H. Adams-Manson. 1993. Information and patch exploitation strategies of the eastern chipmunk, *Tamias striatus* (Rodentia: Sciuridae). *Ethology* 95 (4):299-308.

Bowers, M. A., J. L. Jefferson, and M. G. Kuebler. 1993. Variation in giving-up densities of foraging chipmunks (*Tamias striatus*) and squirrels (*Sciurus carolinensis*). *Oikos* 66 (2):229-236.

Bowles, J. B. 1975. *Distribution and Biogeography of Mammals of Iowa*. Special Publications of the Museum of Texas Tech University No. 9. Lubbock: Texas Tech Press.

Bowman, J., and L. Fahrig. 2002. Gap crossing by chipmunks: An experimental test of landscape connectivity. *Canadian Journal of Zoology* 80 (9):1556.

Bowman, J., G. L. Holloway, J. R. Malcolm, K. R. Middel, and P. J. Wilson. 2005. Northern range boundary dynamics of southern flying squirrels: Evidence of an energetic bottleneck. *Canadian Journal of Zoology—Revue Canadienne de Zoologie* 83 (11):1486-1494.

Boyd, C., T. M. Brooks, S. H. M. Butchart, G. J. Edgar, G. A. B. da Fonseca, F. Hawkins, M. Hoffmann, W. Sechrest, S. N. Stuart, and P. P. Van Dijk. 2008. Spatial scale and the conservation of threatened species. *Conservation Letters* 1:37-43.

Bradley, J. E., and J. M. Marzluff. 2003. Rodents as nest predators: Influences on predatory behavior and consequences to nesting birds. *Auk* 120 (4):1180-1187.

Brandler, O. V. 2003. On species status of the forest-steppe marmot *Marmota kastschenkoi* (Rodentia, Marmotinae). *Zoologichesky Zhurnal* 82 (12):1498-1505.

Brandler, O. V., E. A. Lyapunova, and G. G. Boeskorov. 2008. Comparative karyology of Palearctic marmots (*Marmota*, Sciuridae, Rodentia). *Mammalia* 72 (1):24-34.

Brant, J. G., and T. E. Lee. 1999. New records of mammals for Loving County, Texas. *Texas Journal of Science* 51 (4):347-350.

Briggs, J. S., and S. B. Vander Wall. 2004. Substrate type affects caching and pilferage of pine seeds by chipmunks. *Behavioral Ecology* 15 (4):666-672.

Brigido, M. D., R. Lainson, and F. T. Silveira. 2004. Blood and intestinal parasites of squirrels in Amazonian Brazil. *Memórias do Instituto Oswaldo Cruz* 99 (6):577-579.

Brooks, M. L., and J. R. Matchett. 2002. Sampling methods and trapping success trends for the Mohave ground squirrel, *Spermophilus mohavensis*. *California Fish and Game* 88 (4):165-177.

Broussard, D. R., F. S. Dobson, and J. O. Murie. 2005. The effects of capital on an income breeder: Evidence from female Columbian ground squirrels. *Canadian Journal of Zoology* 83:546-552.

Broussard, D. R., T. S. Risch, F. S. Dobson, and J. O. Murie. 2003. Senescence and age-related reproduction of female Columbian ground squirrels. *Journal of Animal Ecology* 72 (2):212-219.

Brown, D. E. 1984. *Arizona's Tree Squirrels*. Phoenix: Arizona Game and Fish Department.

Brown, L. N., and R. J. McGuire. 1975. Field ecology of the exotic Mexican red-bellied squirrel in Florida. *Journal of Mammalogy* 56:405-419.

Brugiere, D., M. C. Fleury, and M. Colyn. 2005. Structure of the squirrel community in the Forêt des Abeilles, central Gabon: Rediscovery and revalidation of *Funisciurus duchaillui* Sanborn, 1953. *Mammalia* 69 (2):223.

Bryant, A. A. 2005. Reproductive rates of wild and captive Vancouver Island marmots (*Marmota vancouverensis*). *Canadian Journal of Zoology* 83:664-673.

Buchanan, J. B., R. W. Lundquist, and K. B. Aubry. 1990. Winter populations of Douglas' squirrels in different-aged Douglas-fir forests. *Journal of Wildlife Management* 54 (4):577-581.

Buchanan, R. E., and N. E. Gibbons. 1975. *Bergey's Manual of Determinative Bacteriology*. Baltimore: Williams & Wilkins.

Buitrón-Jurado, G., and M. Tobar. 2007. Posible asociación de la ardilla enana *Microsciurus flaviventer* (Rodentia: Sciuridae) y bandadas mixtas de aves en la Amazonia ecuatoriana. *Mastozoología Neotropical* 14:235-240.

Burke da Silva, K., C. Mahan, and J. da Silva. 2002. The trill of the chase: Eastern chipmunks call to warn kin. *Journal of Mammalogy* 83 (2):546-552.

Burt, S. L., and T. L. Best. 1994. *Tamias rufus*. *Mammalian Species* 460:1-6.

Burton, R. S., and O. J. Reichman. 1999. Does immune challenge affect torpor duration? *Functional Ecology* 13 (2):232-237.

Bushberg, D. M., and W. G. Holmes. 1985. Sexual maturation in male Belding's ground squirrels: Influence of body weight. *Biology of Reproduction* 33:302-308.

Cadena, A., R. P. Anderson, and P. Rivas-Pava. 1998. Colombian mammals from the Chocoan Slopes of Nariño. *Occasional Papers of the Museum of Texas Tech University* 180:1-15.

Cai, Y. P., Q. H. Huang, H. Q. Zhao, C. L. Zhang, and H. Wang. 1992. [Effect of intraventricular injection of 6-hydroxydopamine on the initiation of hibernation in ground squirrels]. *Sheng li xue bao [Acta Physiologica Sinica]* 44 (2):175-80. [In Chinese, with English abstract.]

Caire, W., and G. M. Caddell. 2006. Westward range extension of the eastern chipmunk, *Tamias striatus* (Rodentia: Sciuridae), in central Oklahoma. *Proceedings of the Oklahoma Academy of Science* 86:91-92.

Caire, W., and C. L. Sloan. 1996. The woodchuck, *Marmota monax* (Rodentia: Sciuridae), in central Oklahoma. *Proceedings of the Oklahoma Academy of Science* 76:95.

Caire, W., J. D. Tyler, B. P. Glass, and M. A. Mares. 1989. *Mammals of Oklahoma*. Norman: University of Oklahoma Press.

Calabuig, P. 1999. Informe sobre las actuaciones realizadas para controlar la incipiente población de ardilla moruna (*Atlantoxerus getulus*) en la isla de Gran Canaria. Internal report for Cabildo de Gran Canaria, Islas Canarias.

Callahan, J. R. 1980. Taxonomic status of

Eutamias bulleri. *Southwestern Naturalist* 25 (1):1-8.

Cameron, G. N., and D. Scheel. 2001. Getting warmer: Effect of global climate change on distribution of rodents in Texas. *Journal of Mammalogy* 82 (3):652-680.

Campbell, H., A. P. Metzger, D. Spencer, S. Miller, and E. A. Wolters. 2009. Here comes the sun: Solar thermal in the Mojave Desert—Carbon reduction or loss of sequestration? Report prepared for ENSC515/520, Oregon State University, March 13.

Cao, V. S., D. T. Pham, V. M. Tran, M. T. Nguyen, G. V. Kuznetsov, and N. M. Kuljukina. 1986. Écologie des rongeurs de forêt tropicale du Vietnam. *Mammalia* 50 (3):323-328.

Carey, A. B. 1995. Sciurids in Pacific Northwest managed and old-growth forests. *Ecological Applications* 5 (3):648-661.

———. 2000. Effects of new forest management strategies on squirrel populations. *Ecological Applications* 10:248-257.

———. 2001. Experimental manipulation of spatial heterogeneity in Douglas-fir forests: Effects on squirrels. *Forest Ecology and Management* 152 (1-3):13-30.

———. 2002. Response of northern flying squirrels to supplementary dens. *Wildlife Society Bulletin* 30 (2):547-556.

Carey, A. B., W. Colgan III, J. M. Trappe, and R. Molina. 2002. Effects of forest management on truffle abundance and squirrel diets. *Northwest Science* 76 (2):148-157.

Carey, A. B., S. P. Horton, and B. L. Biswell. 1992. Northern spotted owls: Influence of prey base and landscape character. *Ecological Monographs* 62 (2):223-250.

Carey, A. B., J. Kershner, B. Biswell, and L. Dominguez de Toledo. 1999. Ecological scale and forest development: Squirrels, dietary fungi, and vascular plants in managed and unmanaged forests. *Wildlife Monographs* 142:5-71.

Carey, A. B., and S. M. Wilson. 2001. Induced spatial heterogeneity in forest canopies: Responses of small mammals. *Journal of Wildlife Management* 65 (4):1014-1027.

Caro, T. M., M. J. Kelly, N. Bol, and S. Matola. 2001. Inventorying mammals at multiple sites in the Maya Mountains of Belize. *Journal of Mammalogy* 82 (1):43-50.

Carraway, L. N., and B. J. Verts. 1994. *Sciurus griseus*. *Mammalian Species* 474:1-7.

Carvajal, A., and G. H. Adler. 2008. Seed dispersal and predation by *Proechimys semispinosus* and *Sciurus granatensis* in gaps and understorey in central Panama. *Journal of Tropical Ecology* 24:485-492.

Casher, L., R. Lane, R. Barrett, and L. Eisen. 2002. Relative importance of lizards and mammals as hosts for ixodid ticks in northern California. *Experimental and Applied Acarology* 26 (1-2):127-143.

Castro-Arellano, I., and G. Ceballos. 2005. *Spermophilus atricapillus* W. E. Bryant, 1889. In *Los mamíferos silvestres de México*, coord. G. Ceballos and G. Oliva. Mexico, D.F.: Comisión Nacional para el Conocimiento y Uso de la Biodiversidad, y Fondo de Cultura Económica.

Ceballos, G., J. Arroyo-Cabrales, and R. A. Medellín. 2002. The mammals of Mexico: Composition, distribution, and conservation status. *Occasional Papers of the Museum of Texas Tech University* 218:1-27.

Ceballos, G., E. Mellink, and L. R. Hanebury. 1993. Distribution and conservation status of prairie dogs *Cynomys mexicanus* and *Cynomys ludovicianus* in Mexico. *Biological Conservation* 63 (2):105-112.

Ceballos, G., and D. E. Wilson. 1985. *Cynomys mexicanus*. *Mammalian Species* 248:1-3.

Chabaud, A. G., and O. Bain. 1976. La lignée *Dipetalonema*: Nouvel essai de classification. *Annales de Parasitologie Humaine et Comparée* 51 (3):365-397.

Chaimanee, Y., and J. J. Jaeger. 2000. A new flying squirrel *Belomys thamkaewi* n. sp. (Mammalia: Rodentia) from the Pleistocene of West Thailand and its biogeography. *Mammalia* 64:307-318.

Chappell, M. A., G. C. Bachman, and J. P. Odell. 1995. Repeatability of maximal aerobic performance in Belding's ground squirrels, *Spermophilus beldingi*. *Functional Ecology* 9 (3):498-504.

Chasen, F. N., and C. B. Kloss. 1927. Spolia Mentawiensia—Mammals. *Proceedings of the Zoological Society of London* 97 (4):797-840.

———. 1928. On some Carnivora, Rodentia, and Insectivora principally from eastern Borneo. *Journal of the Malayan Branch of the Royal Asiatic Society* 6 (1):38-49.

Chatterjee, K., and A. Majhi. 1975. Chromosomes of the Himalayan flying squirrel, *Petaurista magnificus*. *Mammalia* 39:447-450.

Cheng, K. M., and M. C. Leung. 1997. The distribution of Cascade mantled ground squirrel, *Spermophilus saturatus*, in British Columbia. *Canadian Field-Naturalist* 111 (3):365-375.

Chiozza, F. 2008a. *Petinomys mindanensis*. *IUCN Red List of Threatened Species*, version 2009.1.

———. 2008b. *Sundasciurus davensis*. *IUCN Red List of Threatened Species*, version 2010.4.

———. 2008c. *Sundasciurus mindanensis*. *IUCN Red List of Threatened Species*, version 2010.4.

Chivers, D. J., ed. 1980. *Malayan Forest Primates: Ten Years' Study in Tropical Rain Forest*. New York: Plenum Press.

Choate, J. R., J. K. Jones Jr., and C. Jones. 1994. *Handbook of Mammals of the South-Central States*. Baton Rouge: Louisiana State University Press.

Choudhury, A. 2002. *Petaurista nobilis singhei*—First record in India and a note on its taxonomy. *Journal of the Bombay Natural History Society* 99:30-34.

Chowattukunnel, J. T., and J. H. Esslinger. 1979. A new species of *Breinlia* (*Breinlia*) (Nematoda: Filarioidea) from the South Indian flying squirrel *Petaurista philippensis* (Elliot). *Journal of Parasitology* 65 (3):375-378.

Clark, E. L. , and M. Javzansuren. 2006. *Mongolian Red List of Mammals*. London: Zoological Society of London.

Clark, E. L., J. Munkhbat, S. Dulamtseren, J. E. M. Baillie, N. Batsaikhan, S. R. B. King, R. Samiya, and M. Stubbe. 2006. *Summary Conservation Action Plans for Mongolian Mammals*. Regional Red List Series, Vol. 2. London: Zoological Society of London.

Clark, T. W. 1977. *Ecology and Ethology of the White-Tailed Prairie Dog (*Cynomys leucurus*)*. Milwaukee Public Museum Publication in Biology and Geology No. 3. Milwaukee: Milwaukee Public Museum Press.

Clark, T. W., R. S. Hoffmann, and C. F. Nadler. 1971. *Cynomys leucurus*. *Mammalian Species* 7:1-4.

Clark, T. W., and M. R. Stromberg. 1987. *Mammals in Wyoming*. Public Education Series No. 10. Lawrence: University of Kansas, Museum of Natural History.

Clawson, R. G., J. A. Clawson, and T. L. Best. 1994a. *Tamias alpinus*. *Mammalian Species* 461:1-6.

———. 1994b. *Tamias quadrimaculatus*. *Mammalian Species* 469:1-6.

Coe, M. 1972. The South Turkana Expedition: Scientific papers IX; Ecological studies of the small mammals of South Turkana. *Geographical Journal* 138 (3):316-338.

Cohas, A., N. G. Yoccoz, C. Bonenfant, B. Goosens, C. Genton, M. Galan, B. Kempenaers, and D. Allaine. 2008. The genetic similarity between pair members influences the frequency of extra-pair paternity in alpine marmots. *Animal Behaviour* 76:87–95.

Çolak, R., and S. Özkurt. 2002. Electrophoretic comparison of blood-serum proteins of *Spermophilus citellus* and *Spermophilus xanthoprymnus* (Mammalia: Rodentia) in Turkey. *Zoology in the Middle East* 25:5–8.

Cole, E. C., W. C. McComb, M. Newton, J. P. Leeming, and C. L. Chambers. 1998. Response of small mammals to clearcutting, burning, and Glyphosate application in the Oregon Coast Range. *Journal of Wildlife Management* 62 (4):1207–1216.

Cole, F. R., and D. E. Wilson. 2009. *Urocitellus canus. Mammalian Species* 834:1–8.

Colgan, W., III, and A. W. Claridge. 2002. Mycorrhizal effectiveness of *Rhizopogon* spores recovered from faecal pellets of small forest-dwelling mammals. *Mycological Research* 106 (3):314–320.

Conaway, C. H. 1968. Post-partum estrus in a Sciurid. *Journal of Mammalogy* 49 (1):158–159.

Conover, M. R., R. L. King Jr., J. E. Jimenez, and T. A. Messmer. 2005. Evaluation of supplemental feeding to reduce predation of duck nests in North Dakota. *Wildlife Society Bulletin* 33 (4):1330–1334.

Converse, S. J., W. M. Block, and G. C. White. 2006. Small mammal population and habitat responses to forest thinning and prescribed fire. *Forest Ecology and Management* 228:263–273.

Converse, S. J., G. C. White, and W. M. Block. 2006. Small mammal responses to thinning and wildfire in ponderosa pine-dominated forests of the southwestern United States. *Journal of Wildlife Management* 70:1711–1722.

Corbet, G. B. 1978. *The Mammals of the Palaearctic Region: A Taxonomic Review*. London: British Museum (Natural History) and Cornell University Press.

Corbet, G. B., and S. Harris. 1991. *The Handbook of British Mammals*, 3rd ed. Oxford: Blackwell Scientific.

Corbet, G. B., and J. E. Hill. 1980. *A World List of Mammalian Species*. London: British Museum (Natural History).

———. 1992. *The Mammals of the Indomalayan Region: A Systematic Review*. Oxford: Oxford University Press.

Corlett, R. T. 1992. The ecological transformation of Singapore, 1819–1990. *Journal of Biogeography* 19 (4):411–420.

———. 1998. Frugivory and seed dispersal by vertebrates in the Oriental (Indomalayan) region. *Biological Review* 73:413–448.

Coroiu, C., B. Kryštufek, V. Vohralík, and I. Zagorodnyuk. 2008. *Spermophilus citellus. IUCN Red List of Threatened Species*, version 2010.4.

COSEWIC [Committee on the Status of Endangered Wildlife in Canada]. 2006. *COSEWIC Assessment and Update Status Report on the Southern Flying Squirrel* Glaucomys volans *(Atlantic [Nova Scotia] Population and Great Lakes Plains Population) in Canada*. Ottawa: Committee on the Status of Endangered Wildlife in Canada.

Coss, R. G., and J. E. Biardi. 1997. Individual variation in the antisnake behavior of California ground squirrels (*Spermophilus beecheyi*). *Journal of Mammalogy* 78 (2):294–310.

Cothran, E. G. 1983. Morphologic relationships of the hybridizing ground squirrels *Spermophilus mexicanus* and *S. tridecemlineatus*. *Journal of Mammalogy* 64 (4):591–602.

Cothran, E. G., and R. L. Honeycutt. 1984. Chromosomal differentiation of hybridizing ground squirrels (*Spermophilus mexicanus* and *S. tridecemlineatus*). *Journal of Mammalogy* 65 (1):118–122.

Cowan, I. M. 1946. Notes on the distribution of the chipmunks (*Eutamias*) in southern British Columbia and the Rocky Mountains of southern Alberta with descriptions of two new races. *Proceedings of the Biological Society of Washington* 59: 107–118.

Craig, L. E., J. M. Kinsella, L. J. Lodwick, M. R. Cranfield, and J. D. Standberg. 1998. *Gondylonema macrogubernaculum* in captive African squirrels (*Xerus erythropus*) and lion-tailed macaques (*Macaca silenus*). *Journal of Zoo and Wildlife Medicine* 29 (3):331–337.

Cudworth, N. L., and J. L. Koprowski. 2010. Influences of mating strategy on space use of Arizona gray squirrels. *Journal of Mammalogy* 91 (5):1235–1241.

Cuervo-Diaz, A., J. Hernández-Camacho, and A. Cadena G. 1986. Lista actualizada de los mamíferos de Colombia: Anotaciones sobre su distribución. *Caldasia* 15 (71–75):471–501.

Cullen, L., Jr., E. R. Bodmer, and C. Valladares-Padua. 2001. Ecological consequences of hunting in Atlantic forest patches, São Paulo, Brazil. *Oryx* 35:137–144.

Cypher, B. L. 2001. Spatiotemporal variation in rodent abundance in the San Joaquin Valley, California. *Southwestern Naturalist* 46 (1):66–75.

Dalecky, A., S. Chauvet, S. Ringuet, O. Claessens, M. Larue, J. Judas, and J. F. Cosson. 2002. Large mammals on small islands: Short term effects of forest fragmentation on the large mammal fauna in French Guiana. *Revue d'Écologie—La Terre et la Vie* 8:145–164.

Dalquest, W. W. 1948. *Mammals of Washington*. Museum of Natural History Publication No. 2. Lawrence: University of Kansas, Museum of Natural History.

Dasgupta, B. 1965. Blood parasites in the Himalayan flying squirrel. *Transactions of the Royal Society of Tropical Medicine and Hygiene* 59 (6):716–717.

Dasgupta, S., B. Dasgupta, A. Chatterjee, N. C. De, J. Nandi, K. Saha, R. Roy, M. Ghosh, and G. Majumdar. 1978. *Dipetalonema laemmerli* sp. n. (Nematoda: Dipetalonematidae) from the Himalayan flying squirrel *Petaurista magnificus*. *Zeitschrift für Parasitenkunde* 155:195–198.

Davidson, A. D., and D. C. Lightfoot. 2008. Burrowing rodents increase landscape heterogeneity in a desert grassland. *Journal of Arid Environments* 72 (7):1133–1145.

Davis, D. D. 1958. *Mammals of the Kelabit Plateau, Northern Sarawak*. Chicago: Chicago Natural History Museum.

———. 1962. Mammals of the lowland rainforest of north Borneo. *Bulletin of the National Museum of Singapore* (31):1–129.

Davis, E. B., M. S. Koo, C. Conroy, J. L. Patton, and C. Moritz. 2008. The California Hotspots Project: Identifying regions of rapid diversification of mammals. *Molecular Ecology* 17 (1):120–138.

Davis, R., and S. J. Bissell. 1989. Distribution of Abert's squirrel (*Sciurus aberti*) in Colorado: Evidence for a recent expansion of range. *Southwestern Naturalist* 34 (2):306–309.

Davis, R., and D. E. Brown. 1988. Documentation of the transplanting of Abert's squirrels. *Southwestern Naturalist* 33 (4):490–492.

Davis, R. M., E. Cleugh, R. T. Smith, and C. L. Fritz. 2008. Use of a chitin synthesis inhibitor to control fleas on wild rodents important in the maintenance of plague, *Yersinia pestis*, in California. *Journal of Vector Ecology* 33 (2):278–284.

Dawaa, N. 1972. The diet of the long-tailed ground squirrel. *Scientific Journal of the National University of Mongolia* 40:53–55.

Day, F. J., and A. H. Benton. 1980. Population dynamics and coevolution of adult siphonapteran parasites of the southern flying squirrel. *American Midland Naturalist* 103:333-338.

de A. Goonatilake, W. I. L. D. P. T. S., P. O. Nameer, and S. Molur. 2008. *Funambulus layardi. IUCN Red List of Threatened Species*, version 2010.4.

de Graaff, G. 1981. *The Rodents of Southern Africa*. Durban, South Africa: Butterworths.

de Grammont, P. C., and A. Cuarón. 2008. *Spermophilus adocetus. IUCN Red List of Threatened Species*, version 2009.2.

de Grammont, P. C., A. Cuarón, and E. Vázquez. 2008. *Sciurus colliaei. IUCN Red List of Threatened Species*, version 2009.2.

de Villiers, D. J. 1986. Infanticide in the tree squirrel, *Paraxerus cepapi. South African Journal of Zoology* 21 (2):183-184.

Delany, M. J. 1975. *The Rodents of Uganda*. London: British Museum (Natural History).

Delany, M. J., and D. C. D. Happold. 1979. *Ecology of African Mammals*. Tropical Ecology Series. New York: Longman.

Demberel, J., and J. Batbold. 1991. Distribution and resources of Mongolian mammals in Mongolia. In *Epidemiological Survey of the Plague Natural Foci in the Central Asian Region*, ed. J. Batbold, Z. Adayasuren, G. Tsevegmed, and G. Erdenetsegseg. Ulaanbaatar, Mongolia: Institute of Biology, Academy of Sciences, Ulaanbaatar.

Demboski, J. R., B. K. Jacobsen, and J. A. Cook. 1998. Implications of cytochrome *b* sequence variation for biogeography and conservation of the northern flying squirrels (*Glaucomys sabrinus*) of the Alexander Archipelago, Alaska. *Canadian Journal of Zoology* 76 (9):1771-1777.

Demboski, J. R., and J. Sullivan. 2003. Extensive mtDNA variation within the yellow-pine chipmunk, *Tamias amoenus* (Rodentia: Sciuridae), and phylogeographic inferences for northwest North America. *Molecular Phylogenetics and Evolution* 26 (3):389-408.

Demirsoy, A., N. Yigit, E. Colak, M. Sozen, and A. Karatas. 2006. *Rodents of Turkiye*. Ankara: Meteksan.

den Tex, R. J., R. Thorington, J. E. Maldonado, and J. A. Leonard. 2010. Speciation dynamics in the SE Asian tropics: Putting a time perspective on the phylogeny and biogeography of Sundaland tree squirrels, *Sundasciurus. Molecular Phylogenetics and Evolution* 55 (2):711-720.

DeNatale, C. E., T. R. Burkot, B. S. Schneider, and N. S. Zeidner. 2002. Novel potential reservoirs for *Borrelia* sp. and the agent of human granulocytic ehrlichiosis in Colorado. *Journal of Wildlife Diseases* 38 (2):478-482.

Derge, K. L., and M. A. Steele. 1999. Distribution of the fox squirrel (*Sciurus niger*) in Pennsylvania. *Journal of the Pennsylvania Academy of Science* 73:43-50.

Desrochers, A., I. K. Hanski, and V. Selonen. 2003. Siberian flying squirrel responses to high- and low-contrast forest edges. *Landscape Ecology* 18 (5):543-552.

Deveaux, T. P., G. D. Schmidt, and M. Krishnasamy. 1988. Two new species of *Moniliformis* (Acanthocephala: Moniliformidae) from Malaysia. *Journal of Parasitology* 74:322-325.

Devenport, J. A., L. D. Luna, and L. D. Devenport. 2000. Placement, retrieval, and memory of caches by thirteen-lined ground squirrels. *Ethology* 106:171-183.

Ditto, A. M., and J. K. Frey. 2007. Effects of ecogeographic variables on genetic variation in montane mammals: Implications for conservation in a global warming scenario. *Journal of Biogeography* 34:1136-1149.

Dobson, F. S. 1992. Body mass, structural size, and life-history patterns of the Columbian ground squirrel. *American Naturalist* 140 (1):109-125.

———. 1994. Measures of gene flow in the Columbian ground squirrel. *Oecologia* 100 (1-2):190-195.

———. 1995. Regulation of population size: Evidence from Columbian ground squirrels. *Oecologia* 102 (1):44-51.

Dobson, F. S., and W. T. Jones. 1985. Multiple causes of dispersal. *American Naturalist* 126 (6):855-858.

Dobson, F. S., G. R. Michener, and T. S. Risch. 2007. Variation in litter size: A test of hypotheses in Richardson's ground squirrels. *Ecology* 88 (2):306-314.

Dobson, F. S., and M. K. Oli. 2001. The demographic basis of population regulation in Columbian ground squirrels. *American Naturalist* 158 (3):236-247.

Dobson, F. S., T. S. Risch, and J. O. Murie. 1999. Increasing returns in the life history of Columbian ground squirrels. *Journal of Animal Ecology* 68 (1):73-86.

Dobson, M. L., C. L. Pritchett, and J. W. Sites. 1987. Genetic variation and population structure in the cliff chipmunk, *Eutamias dorsalis*, in the Great Basin of western Utah. *Great Basin Naturalist* 47 (4):551-561.

Dolan, P. G., and D. C. Carter. 1977. *Glaucomys volans. Mammalian Species* 78:1-6.

Dorst, J., and P. Dandelot. 1970. *A Field Guide to the Larger Mammals of Africa*. Boston: Houghton Mifflin.

Dowsett, R. J., and F. Dowsett-Lemaire. 1989. Larger mammals observed in the Gotel Mts. and on the Mambilla Plateau, eastern Nigeria. In *A Preliminary Natural History Survey of Mambilla Plateau and Some Lowland Forests of Eastern Nigeria*, ed. R. J. Dowsett. Tauraco Research Report No. 1. Ely, Cambridgeshire, UK: Tauraco Press.

Drabek, C. M. 1973. Home range and daily activity of the round-tailed ground squirrel, *Spermophilus tereticaudus neglectus. American Midland Naturalist* 89 (2):287-293.

Du Chaillu, P. B. 1860. The president in the chair. *Proceedings of the Boston Society of Natural History* 7:358-364.

Dubost, G. 1968. Aperçu sur le rythme annuel de reproduction des Muridés du nord-est du Gabon. *Biologia Gabonica* 4:227-239.

Duckworth, J. W., and C. Francis. 2008a. *Petaurillus emiliae. IUCN Red List of Threatened Species*, version 2009.1.

———. 2008b. *Petaurillus hosei. IUCN Red List of Threatened Species*, version 2009.1.

Duckworth, J. W., and S. Hedges. 2008a. *Hylopetes bartelsi. IUCN Red List of Threatened Species*, version 2009.1.

———. 2008b. *Hylopetes lepidus. IUCN Red List of Threatened Species*, version 2010.4.

———. 2008c. *Hylopetes spadiceus. IUCN Red List of Threatened Species*, 2008 [accessed 28 Oct. 2008].

———. 2008d. *Lariscus hosei. IUCN Red List of Threatened Species*, version 2010.4.

Duckworth, J. W., and S. Molur. 2008. *Belomys pearsonii. IUCN Red List of Threatened Species*, version 2009.2.

Duckworth, J. W., R. J. Timmins, and K. Cozza. 1993. *A Wildlife and Habitat Survey of Phou Xang He Proposed Protected Area: Report to the Protected Areas and Wildlife Division of the National Office for Nature Conservation and Watershed Management*. Vientiane, Laos: IUCN.

Duckworth, J. W., R. J. Tizard, and S. Molur. 2008. *Hylopetes alboniger. IUCN Red List of Threatened Species*, version 2009.2.

Duncan, R. D., and S. H. Jenkins. 1998. Use of visual cues in foraging by a diurnal herbivore, Belding's ground squirrel. *Canadian Journal of Zoology* 76 (9):1766-1770.

Dunford, C. J. 1977a. Behavioral limitation of round-tailed ground squirrel density. *Ecology* 58:1254-1268.

———. 1977b. Kin selection for ground

squirrel alarm calls. *American Naturalist* 111:782-785.
———. 1977c. Social system of round-tailed ground squirrels. *Animal Behaviour* 25:885-906.
Dunn, F. L., B. L. Lim, and L. F. Yap. 1968. Endoparasite patterns in mammals of the Malayan rain forest. *Ecology* 49 (6):1179-1184.
Dupont, J. E., and D. S. Rabor. 1973. *Birds of Dinagat and Siargao, Philippines.* Nemouria (Occasional Papers of the Delaware Museum of Natural History) No. 10. Greenville: Delaware Museum of Natural History.
Durden, L. A. 1983. Sucking louse (*Hoplopleura erratica,* Insecta: Anoplura) exchange between individuals of a wild population of eastern chipmunks, *Tamias striatus,* in central Tennessee, USA. *Journal of Zoology* 201 (Sept.):117-123.
Durden, L. A., and N. E. Adams. 2005. Primary type specimens of sucking lice (Insecta: Phthiraptera: Anoplura) in the U.S. National Museum of Natural History, Smithsonian Institution. *Zootaxa* 1047:21-60.
Durden, L. A., and G. G. Musser. 1994a. The mammalian hosts of the sucking lice (Anoplura) of the world: A host-parasite list. *Bulletin of Social Vector Ecology* 19 (2):130-168.
———. 1994b. The sucking lice (Insecta, Anoplura) of the world: A taxonomic checklist with records of mammalian hosts and geographical distributions. *Bulletin of the American Museum of Natural History* 218:1-90.
Durrant, S. D. 1952. *Mammals of Utah: Taxonomy and Distribution.* Museum of Natural History Publication No. 6. Lawrence: University of Kansas, Museum of Natural History.
Eckerlin, R. P. 2005. Fleas (Siphonaptera) of the Yucatán Peninsula (Campeche, Quintana Roo, and Yucatán), Mexico. *Caribbean Journal of Science* 41 (1):152-157.
Eddingsaas, A. A., B. K. Jacobsen, E. P. Lessa, and J. A. Cook. 2004. Evolutionary history of the arctic ground squirrel (*Spermophilus parryii*) in Nearctic Beringia. *Journal of Mammalogy* 85 (4):601-610.
Edelman, A. J. 2003. *Marmota olympus. Mammalian Species* 736:1-5.
Edelman, A. J., and J. L. Koprowski. 2005. Selection of drey sites by Abert's squirrels in an introduced population. *Journal of Mammalogy* 86:1220-1226.
———. 2006. Seasonal changes in home ranges of Abert's squirrels: Impact of mating season. *Canadian Journal of Zoology* 84:404-411.
Eiler, K. C., and S. A. Banack. 2004. Variability in the alarm call of golden-mantled ground squirrels. *Journal of Mammalogy* 85:43-50.
Eisenberg, J. F. 1989. *The Northern Neotropics: Panama, Colombia, Venezuela, Guyana, Suriname, French Guiana.* Vol. 1 of *Mammals of the Neotropics.* Chicago: University of Chicago Press.
Eisenberg, J. F., and K. H. Redford. 1999. *The Central Neotropics: Ecuador, Peru, Bolivia, Brazil.* Vol. 3 of *Mammals of the Neotropics.* Chicago: University of Chicago Press.
Eisentraut, M. 1973. *Die Wirbeltierfauna von Fernando Poo und Westkamerun.* Bonn: Zoologisches Forschungsinstitut und Museum Alexander Koenig.
———. 1976. *Das Gaumenfaltenmuster der Säugetiere und seine Bedeutung für stammesgeschichtliche und taxonomische Untersuchungen.* Bonn: Zoologisches Forschungsinstitut und Museum Alexander Koenig.
Ekué, M. R. M. Personal communication.
Ellerman, J. R. 1940. *The Families and Genera of Living Rodents.* London: Printed by order of the Trustees of the British Museum.
Elliott, C. L., and J. T. Flinders. 1991. *Spermophilus columbianus. Mammalian Species* 372:1-9.
Ellis, L. S., and L. R. Maxson. 1979. Evolution of the chipmunk genera *Eutamias* and *Tamias. Journal of Mammalogy* 60 (2):331-334.
Emelianov, P. F. 1983. On the migrational activity of *Citellus musicus. Zoologichesky Zhurnal* 62:1858-1862.
Emelianov, P. F., I. K. Vagner, A. M. Karmov, E. K. Titov, V. V. Ivanovsky, and N. N. Vasiliev. 1982. Biology of *Citellus musicus* (Rodentia, Sciuridae) in the central Caucasus. *Zoologichesky Zhurnal* 61:419-425.
Emmons, L. H. 1975. Ecology and behavior of African rainforest squirrels. Ph.D. diss., Cornell University.
———. 1978. Sound communication among African rainforest squirrels. *Zeitschrift für Tierpsychologie* 47 (1):1-49.
———. 1979. Observations on litter size and development of some African rainforest squirrels. *Biotropica* 11 (3):207-213.
———. 1980. Ecology and resource partitioning among nine species of African rainforest squirrels. *Ecological Monographs* 50 (1):31-54.
———. 1984. Geographic variation in densities and diversities of non-flying mammals in Amazonia. *Biotropica* 16 (3):210-222.
———. 1987. Comparative feeding ecology of felids in a Neotropical rainforest. *Behavioral Ecology and Sociobiology* 20 (4):271-283.
Emmons, L. H., and F. Feer. 1990. *Neotropical Rainforest Mammals: A Field Guide.* Chicago: University of Chicago Press.
Emmons, L. H., J. Nais, and A. Briun. 1991. The fruit and consumers of *Rafflesia keithii* (Rafflesiaceae). *Biotropica* 23 (2):197-199.
Enders, R. K. 1953. The type locality of *Syntheosciurus brochus. Journal of Mammalogy* 34 (4):509.
———. 1980. Observations on *Syntheosciurus brochus*: Taxonomy and behavior. *Journal of Mammalogy* 61:725-727.
Endo, H., J. Kimura, T. Oshida, B. J. Stafford, W. Rerkamnuaychoke, T. Nishida, M. Sasaki, A. Hayashida, and Y. Hayashi. 2003. Geographical variation of skull morphology and its functional significances in the red-cheeked squirrel. *Journal of Veterinary Medical Science* 65 (11):1179-1183.
Ermakov, O. A., V. L. Surin, S. V. Titov, A. F. Tagiev, A. V. Luk'yanenko, and N. A. Formozov. 2002. A molecular genetic study of hybridization in four species of ground squirrels (*Spermophilus*: Rodentia, Sciuridae). *Russian Journal of Genetics* 38 (7):796-809.
Ermakov, O. A., V. L. Surin, S. V. Titov, S. S. Zborovsky, and N. A. Formozov. 2006. A search for Y-chromosomal species-specific markers and their use for hybridization analysis in ground squirrels (*Spermophilus*: Rodentia, Sciuridae). *Russian Journal of Genetics* 42 (4):538-548.
Ermakov, O. A., and S. V. Titov. 2000. Dynamics of *Spermophilus major* range boundaries in the Volga River region. *Zoologichesky Zhurnal* 79 (4):503-509.
Ermakov, O. A., S. V. Titov, A. B. Savinetsky, V. L. Surin, S. S. Zborovsky, E. A. Lyapunova, O. V. Brandler, and N. A. Formozov. 2006. Molecular genetic and paleoecological arguments for conspecificity of little (*Spermophilus pygmaeus*) and Caucasian Mountain (*S. musicus*) ground squirrels. *Zoologichesky Zhurnal* 85:1474-1483.
Ernest, K. A., and M. A. Mares. 1987. *Spermophilus tereticaudus. Mammalian Species* 274:1-9.
Escalante, T., D. Espinosa, and J. J. Morrone. 2003. Using Parsimony Analysis of Endemicity to analyze the distribution of Mexican land mammals. *Southwestern Naturalist* 48 (4):563-578.

Escalante, T., G. Rodríguez, and J. J. Morrone. 2005. The provinces of Mexican Mountain Component based on continental mammals distribution. *Revista Mexicana de Biodiversidad* 76 (2):199–205.

Eshelman, B. D., and C. S. Sonnemann. 2000. *Spermophilus armatus*. *Mammalian Species* 637:1–6.

Esselstyn, J. A., P. Widmann, and L. R. Heaney. 2004. The mammals of Palawan Island, Philippines. *Proceedings of the Biological Society of Washington* 117 (3):271–302.

Estrada, A., and R. Coates-Estrada. 1985. A preliminary study of resource overlap between howling monkeys (*Alouatta palliata*) and other arboreal mammals in the tropical rain forest of Los Tuxtlas, Mexico. *American Journal of Primatology* 9:27–37.

Everts, L. G., A. M. Strijkstra, R. A. Hut, I. E. Hoffmann, and E. Millesi. 2004. Seasonal variation in daily activity patterns of free-ranging European ground squirrels (*Spermophilus citellus*). *Chronobiology International* 21 (1):57–71.

Ewer, R. F. 1965. Food burying in the African ground squirrel, *Xerus erythropus* (E. Geoff.). *Zeitschrift für Tierpsychologie* 22 (3):321–7.

———. 1966. Juvenile behaviour in the African ground squirrel, *Xerus erythropus* (E. Geoff.). *Zeitschrift für Tierpsychologie* 23 (2):190–216.

———. 1968. *Ethology of mammals*. New York: Plenum Press.

Fa, J. E. 2000. Hunted animals in Bioko Island, West Africa: Sustainability and future. In *Hunting for Sustainability in Tropical Forests*, ed. J. G. Robinson and E. L. Bennett. New York: Columbia University Press.

Fa, J. E., and J. E. García Yuste. 2001. Commercial bushmeat hunting in the Monte Mitra forests, Equatorial Guinea: Extent and impact. *Animal Biodiversity and Conservation* 24 (1):31–52.

Fagundes, V., A. U. Christoff, R. C. Amaro-Ghilard, D. R. Scheibler, and Y. Yonenanga-Yassuda. 2003. Multiple interstitial ribosomal sites (NORs) in the Brazilian squirrel *Sciurus aestuans ingrami* (Rodentia, Sciuridae) with $2n = 40$: An overview of *Sciurus* cytogenetics. *Genetics and Molecular Biology* 26:253–257.

Falck, M. J., K. R. Wilson, and D. C. Andersen. 2003. Small mammals within riparian habitats of a regulated and unregulated aridland river. *Western North American Naturalist* 63 (1):35–42.

Faller-Menéndez, J. C., T. Urquiza-Haas, C. Chávez, S. Johnson, and G. Ceballos. 2005. Registros de mamíferos en la privada El Zapotal, en el noreste de la Península de Yucatán. *Revista Mexicana de Mastozoología* 9:128–140.

Fan, P. F., and X. L. Jiang. 2009. Predation on giant flying squirrels (*Petaurista philippensis*) by black crested gibbons (*Nomascus concolor jingdongensis*) at Mt. Wuliang, Yunnan, China. *Primates* 50 (1):45–49.

Fellowes, J. R., B. P. L. Chan, M. W. N. Lau, B. C. H. Hau, and C. S. Ng. 2002. *Report of a Rapid Biodiversity Assessment at Dapingshan Headwater Forest Nature Reserve, East Guangxi, China, 24 to 27 September 1998*. South China Forest Biodiversity Survey Report Series No. 19. Hong Kong, SAR: Kadoorie Farm and Botanic Garden.

Fellowes, J. R., B. P. L. Chan, M. W. N. Lau, S. C. Ng, and G. L. P. Siu. 2003a. *Report of Rapid Biodiversity Assessments at Damingshan National Nature Reserve, Central Guangxi, China, April and September 2000*. South China Forest Biodiversity Survey Report Series No. 34. Hong Kong, SAR: Kadoorie Farm and Botanic Garden.

———. 2003b. *Report of a Rapid Biodiversity Assessment at Dawangling Headwater Forest Nature Reserve, West Guangxi, China, August 1999*. South China Forest Biodiversity Survey Report Series No. 28. Hong Kong, SAR: Kadoorie Farm and Botanic Garden.

———. 2003c. *Report of a Rapid Biodiversity Assessment at Jiaxi Nature Reserve, Western Hainan, China, June 1999*. South China Forest Biodiversity Survey Report Series No. 25. Hong Kong, SAR: Kadoorie Farm and Botanic Garden.

———. 2003d. *Report of Rapid Biodiversity Assessments at Wuzhishan Nature Reserve, Central Hainan, China, 1999 and 2001*. South China Forest Biodiversity Survey Report Series No. 24. Hong Kong, SAR: Kadoorie Farm and Botanic Garden.

Fellowes, J. R., B. P. L. Chan, M. W. N. Lau, S. C. Ng, G. L. P. Siu, and L. K. Shing. 2003. *Report of a Rapid Biodiversity Assessment at Nanling National Nature Reserve, Northwest Guangdong, China, June to July 2000*. South China Forest Biodiversity Survey Report Series No. 29. Hong Kong, SAR: Kadoorie Farm and Botanic Garden.

Fellowes, J. R., B. P. L. Chan, S. C. Ng, and M. W. N. Lau. 2003. *Report of a Rapid Biodiversity Assessment at Diding Headwater Forest Nature Reserve, West Guangxi, China, July 1999*. South China Forest Biodiversity Survey Report Series No. 26. Hong Kong, SAR: Kadoorie Farm and Botanic Garden.

Fellowes, J. R., B. P. L. Chan, S. C. Ng, M. W. N. Lau, and G. L. P. Siu. 2002a. *Report of a Rapid Biodiversity Assessment at Diaoluoshan National Forest Park, Southeast Hainan, China, 23 to 28 May 1999*. South China Forest Biodiversity Survey Report Series No. 23. Hong Kong, SAR: Kadoorie Farm and Botanic Garden.

———. 2002b. *Report of Rapid Biodiversity Assessments at Tongtieling Forest Area and Xinglong Tropical Botanic Garden, Southeast Hainan, China, 22–23 May 1999*. South China Forest Biodiversity Survey Report Series No. 22. Hong Kong, SAR: Kadoorie Farm and Botanic Garden.

Fellowes, J. R., B. C. H. Hau, M. W. N. Lau, and S. C. Ng. 2001. *Report of Rapid Biodiversity Assessments at Bawangling National Nature Reserve and Wangxia Limestone Forest, Western Hainan, 3 to 8 April 1998*. South China Forest Biodiversity Survey Report Series No. 2. Hong Kong, SAR: Kadoorie Farm and Botanic Garden.

Fellowes, J. R., B. C. H. Hau, M. W. N. Lau, S. C. Ng, and B. P. L. Chan. 2001. *Report of Rapid Biodiversity Assessments at Jianfengling Nature Reserve, Southwest Hainan, 1998 and 2001*. South China Forest Biodiversity Survey Report Series No. 3. Hong Kong, SAR: Kadoorie Farm and Botanic Garden.

Fellowes, J. R., M. W. N. Lau, B. C. H. Hau, S. C. Ng, and B. P. L. Chan. 2002. *Report of Rapid Biodiversity Assessments at Nonggang National Nature Reserve, Southwest Guangxi, China, 19 to 27 May 1998*. South China Forest Biodiversity Survey Report Series No. 10. Hong Kong, SAR: Kadoorie Farm and Botanic Garden.

Fenner, F. 1994. Monkeypox virus. In *Virus Infections of Rodents and Lagomorphs*, ed. A. Osterhaus. New York: Elsevier Science.

Festa-Bianchet, M., and W. J. King. 1991. Effects of litter size and population dynamics on juvenile and maternal survival in Columbian ground squirrels. *Journal of Animal Ecology* 60 (3):1077–1090.

Findley, J. S., A. H. Harris, D. E. Wilson, and C. Jones. 1975. *Mammals of New Mexico*. Albuquerque: University of New Mexico Press.

Fitzgerald, J. P., C. A. Meaney, and D. M. Armstrong. 1994. *Mammals of Colorado*. [Denver:] Denver Museum of Natural History; Niwot: University Press of Colorado.

Fitzgerald, J. P., and D. P. Streubel. 1978. *Spermophilus spilosoma. Mammalian Species* 101:1-4.

Fleming, T. H. 1973. The reproductive cycles of three species of opossums and other mammals in the Panama Canal Zone. *Journal of Mammalogy* 54:439-455.

Flint, V. E., Y. D. Chugunov, and V. M. Smirin. 1965. *Mlekopitaiushchie SSSR.* Moscow: Izdatel'stvo Mysl.

Floyd, C. H. 2004. Marmot distribution and habitat associations in the Great Basin. *Western North American Naturalist* 64 (4):471-481.

Flyger, V., and J. E. Gates. 1982. Pine squirrels: *Tamiasciurus hudsonicus, T. douglasii.* In *Wild Mammals of North America: Biology, Management, and Economics*, ed. J. A. Chapman and G. A. Feldhamer. Baltimore: Johns Hopkins University Press.

Fokanov, V. A. 1966. Novye podvid surkabaibaka i zamechaniya o geogr'aficheskoi izmenchivosti *Marmota bobac* Müll. *Zoologichesky Zhurnal* 45:1862-1866.

Ford, A. T., and L. Fahrig. 2008. Movement patterns of eastern chipmunks (*Tamias striatus*) near roads. *Journal of Mammalogy* 89 (4):895-903.

Foresman, K. R. 2001. *The Wild Mammals of Montana.* Special Publication No. 12. Lawrence, KS: American Society of Mammalogists.

Forstmeier, W., and I. Weiss. 2002. Impact of nest predation by the Siberian chipmunk (*Tamias sibiricus*) on breeding success of the dusky warbler (*Phylloscopus fuscatus*). *Zoologichesky Zhurnal* 81:1367-1370.

———. 2004. Adaptive plasticity in nest-site selection in response to changing predation risk. *Oikos* 104:487-499.

Francis, C., and J. W. Duckworth. 2008a. *Petaurillus kinlochii. IUCN Red List of Threatened Species*, version 2009.1.

———. 2008b. *Petinomys sagitta. IUCN Red List of Threatened Species*, version 2009.1.

———. 2008c. *Petinomys setosus. IUCN Red List of Threatened Species*, version 2009.2.

Francis, C. M. 2008. *A Guide to the Mammals of Southeast Asia.* Princeton, NJ: Princeton University Press.

Francis, C. M., and M. Gumal. 2008a. *Petinomys genibarbis. IUCN Red List of Threatened Species*, version 2009.2.

———. 2008b. *Petinomys hageni. IUCN Red List of Threatened Species*, version 2009.1.

———. 2008c. *Petinomys lugens. IUCN Red List of Threatened Species*, version 2009.1.

———. 2008d. *Petinomys vordermanni. IUCN Red List of Threatened Species*, version 2009.2.

Frank, C. L. 1994. Polyunsaturate content and diet selection by ground squirrels (*Spermophilus lateralis*). *Ecology* 75 (2):458-463.

Frase, B. A., and R. S. Hoffmann. 1980. *Marmota flaviventris. Mammalian Species* 135:1-8.

French, A. R. 2000. Interdependency of stored food and changes in body temperature during hibernation of the eastern chipmunk, *Tamias striatus. Journal of Mammalogy* 81 (4):979-985.

Gage, K. L., and M. Y. Kosoy. 2005. Natural history of plague: Perspectives from more than a century of research. *Annual Review of Entomology* 50:505-528.

Galetti, M. 1990. Predation on the squirrel, *Sciurus aestuans*, by capuchin monkeys, *Cebus apella. Mammalia* 54 (1):152-154.

Galetti, M., M. Paschoal, and F. Pedroni. 1992. Predation on palm nuts (*Syagrus romanzoffiana*) by squirrels (*Sciurus ingrami*) in southeast Brazil. *Journal of Tropical Ecology* 8:121-123.

Galkina, L. I., D. E. Taranenko, and O. V. Brandler. 2005. To a question about species status of forest-steppe marmot *Marmota kastschenkoi* Stroganov et Judin, 1956 (Rodentia, Sciuridae). Paper read at the Fifth International Conference on Genus *Marmota*, International Marmot Network, Tashkent, Uzbekistan, August 31-September 2, 2005.

Gall, C., and R. Zembal. 1980. Observations on Mohave ground squirrels, *Spermophilus mohavensis*, in Inyo County, California. *Journal of Mammalogy* 61 (2):347-350.

Gálvez, D., and P. A. Jansen. 2007. Bruchid beetle infestation and the value of *Attalea butyracea* endocarps for neotropical rodents. *Journal of Tropical Ecology* 23:381-384.

Gangoso, L., J. Donázar, S. Scholz, C. Palacios, and F. Hiraldo. 2006. Contradiction in conservation of island ecosystems: Plants, introduced herbivores and avian scavengers in the Canary Islands. *Biodiversity and Conservation* 15 (7):2231-2248.

Gannon, W. L., and R. B. Forbes. 1995. *Tamias senex. Mammalian Species* 502:1-6.

Gannon, W. L., R. B. Forbes, and D. E. Kain. 1993. *Tamias ochrogenys. Mammalian Species* 445:1-4.

Gannon, W. L., and T. E. Lawlor. 1989. Variation of the chip vocalization of three species of Townsend chipmunks (genus *Eutamias*). *Journal of Mammalogy* 70 (4):740-753.

Gannon, W. L., and W. T. Stanley. 1991. Chip vocalization of Palmer's chipmunk (*Tamias palmeri*). *Southwestern Naturalist* 36 (3):315-317.

Garner, H. W., and J W. Bluntzer. 1975. Mammals of the Kansan-Texan boundary in Texas: Distributional records of mammals along the boundary. *Texas Journal of Science* 26:611-613.

Garrison, B. A., R. L. Wachs, and M. L. Triggs. 2005. Responses of forest squirrels to group-selection timber harvesting in the central Sierra Nevada. *California Fish and Game* 91:1-20.

Garroway, C. J., J. Bowman, T. J. Cascaden, G. L. Holloway, C. G. Mahan, J. R. Malcom, M. A. Steele, G. Turner, and P. J. Wilson. 2010. Climate change induced hybridization in flying squirrels. *Global Change Biology* 16 (1):113-121.

Gautier-Hion, A., L. H. Emmons, and G. Dubost. 1980. A comparison of the diets of three major groups of primary consumers of Gabon (primates, squirrels, and ruminants). *Oecologia* 45 (2):182-189.

Gavin, T. A., P. W. Sherman, E. Yensen, and B. May. 1999. Population genetic structure of the Northern Idaho ground squirrel (*Spermophilus brunneus brunneus*). *Journal of Mammalogy* 80 (1):156-168.

Gavish, L. 1993. Preliminary observations on the behavior and ecology of free-living populations of the subspecies *Sciurus anomalus syriacus* (golden squirrel) on Mount Hermon, Israel. *Israel Journal of Zoology* 39 (3):275-280.

Geluso, K. 2004. Westward expansion of the eastern fox squirrel (*Sciurus niger*) in northeastern New Mexico and southeastern Colorado. *Southwestern Naturalist* 49 (1):111-116.

Gerhardt, R. P., P. M. Harris, and M. A. Vásquez Marroquín. 1993. Food habits of nesting great black hawks in Tikal National Park, Guatemala. *Biotropica* 25: 349-352.

Germano, D. J., G. R. Rathbun, and L. R. Saslaw. 2001. Managing exotic grasses and conserving declining species. *Wildlife Society Bulletin* 29 (2):551-559.

Ghilardi, R., Jr., and C. J. R. Alho. 1990. Produtividade sazonal da floresta e atividade de forrageamento animal em habitat de terra firme da Amazônia. *Acta Amazonica* 20:61-76.

Ghose, R. K., and S. S. Saha. 1981. Taxonomic review of Hodgson's giant flying squirrel *Petaurista magnificus* (Hodgson) (Sciuridae: Rodentia), with description of a new species from Darjeeling district, West Bengal India. *Journal of the Bombay Natural History Society* 99:30-34.

Ghosh, S. 1981. Observation on carnivorous

habit of an Irrawaddy squirrel, *Callosciurus pygerythrus* (Geoffrey). *Journal of the Bombay Natural History Society* 77 (2):316–317.

Giacalone, J., N. Wells, and G. Willis. 1987. Observations on *Syntheosciurus brochus* in Volcán Poás National Park, Costa Rica. *Journal of Mammalogy* 68 (1):145–147.

Gillis, E. A., D. S. Hik, R. Boonstra, T. J. Karels, and C. J. Krebs. 2005. Being high is better: Effects of elevation and habitat on Arctic ground squirrel demography. *Oikos* 108:231–240.

Gillis, E. A., S. F. Morrison, G. D. Zazula, and D. S. Hik. 2005. Evidence for selective caching by Arctic ground squirrels living in alpine meadows in the Yukon. *Arctic* 58:354–360.

Gladkina, T. S., and S. V. Skalinov. 1987. Influence of modern system of farming on distribution of the little suslik (*Citellus pygmaeus* Pall.). *Soviet Journal of Ecology* 18:205–212.

Goetze, J. R., and L. L. Choate. 1987. Distributional notes on four species of Texas mammals. *Texas Journal of Science* 39: 80–381.

Golemansky, V. G., and Y. S. Koshev. 2007. Coccidian parasites (*Eucoccidia*: Eimeriidae) in European ground squirrel (*Spermophilus citellus* L., 1766) (Rodentia: Sciuridae) from Bulgaria. *Acta Zoologica Bulgarica* 59 (1):81–85.

Good, J. M., J. R. Demboski, D. W. Nagorsen, and J. Sullivan. 2003. Phylogeography and introgressive hybridization: Chipmunks (genus *Tamias*) in the northern Rocky Mountains. *Evolution* 57 (8):1900–1916.

Good, J. M., S. Hird, N. Reid, J. Sullivan, J. R. Demboski, S. J. Steppan, and T. R. Martin-Nims. 2008. Ancient hybridization and mitochondrial capture between two species of chipmunks. *Molecular Ecology* 17 (5):1313–1327.

Good, J. M., and J. Sullivan. 2001. Phylogeography of the red-tailed chipmunk (*Tamias ruficaudus*), a northern Rocky Mountain endemic. *Molecular Ecology* 10 (11):2683–2695.

Goodwin, H. T. 2009. Odontometric patterns in the radiation of extant ground-dwelling squirrels within Marmotini (Sciuridae: Xerini). *Journal of Mammalogy* 90 (4):1009–1019.

Gouat, P., and I. E. Yahyaoui. 2001. Reproductive period and group structure variety in the Barbary ground squirrel *Atlantoxerus getulus*. In *African Small Mammals: Proceedings of the 8th International Symposium on African Small Mammals, Paris, July 1999*, ed. Christiane Denys, Laurent Granjon et Alain Poule. Paris: IRD Éditions.

Graham, R. W. 1983. Paleoenvironmental implications of the quaternary distribution of the eastern chipmunk (*Tamias striatus*) in central Texas. *Quaternary Research* 21:111–114.

Grajales, K. M., R. Rodríguez-Estrella, and J. C. Hernández. 2003. Dieta estacional del coyote (*Canis latrans*) durante el periodo en el deseierto de Vizcaíno, Baja California Sur, México. *Acta Zoológica Mexicana*, n.s., 89:17–28.

Greene, E., R. G. Anthony, V. Marr, and R. Morgan. 2009. Abundance and habitat associations of Washington ground squirrels in the Columbia Basin, Oregon. *American Midland Naturalist* 162:29–42.

Gregory, S. C., W. M. Vander Haegen, W. Y. Chang, and S. D. West. 2010. Nest site selection by western gray squirrels at their northern range terminus. *Journal of Wildlife Management* 74 (1):18–25.

Grelle, C. E. V. 2003. Forest structure and vertical stratification of small mammals in a secondary Atlantic forest, southeastern Brazil. *Studies on Neotropical Fauna and Environment* 38:81–85.

Griffin, S. C., P. C. Griffin, M. L. Taper, and L. S. Mills. 2009. Marmots on the move? Dispersal in a declining montane mammal. *Journal of Mammalogy* 90:686–695.

Grijalva, M. J., and A. G. Villacis. 2009. Presence of *Rhodnius ecuadoriensis* in sylvatic habitats in the Southern Highlands (Loja Province) of Ecuador. *Journal of Medical Entomology* 46:708–711.

Grill, A., G. Amori, G. Aloise, I. Lisi, G. Tosi, L. A. Wauters, and E. Randi. 2009. Molecular phylogeography of European *Sciurus vulgaris*: Refuge within refugia? *Molecular Ecology* 18:2687–2699.

Grimshaw, J. M., N. J. Cordeiro, and C. A. H. Foley. 1995. The mammals of Kilimanjaro. *Journal of East African Natural History* 84 (2):105–139.

Grinnell, J., and T. I. Storer. 1916. *Diagnoses of Seven New Mammals from East-Central California*. University of California Publications in Zoology, Vol. 17, No. 1. Berkeley: University of California Press.

Gromov, I. M., D. I. Bibikov, N. I. Kalabukhov, and M. N. Meier. 1965. *Nazemnye belich'i (Marmotinae)*. Vol. 3, no. 2 of *Fauna SSSR, Mlekopitayushchie*. Moscow: Izdatel'stvo Nauka.

Groves, C., E. Yensen, and E. B. Hart. 1988. First specimen records of the rock squirrel (*Spermophilus variegatus*) in Idaho. *Murrelet* 69 (2):50–53.

Grubb, P. 1982. Systematics of sun-squirrels (*Heliosciurus*) in eastern Africa. *Bonner Zoologische Beiträge* 33:191–204.

———. 2001. Endemism in African rain forest mammals. In *African Rain Forest Ecology and Conservation*, ed. W. Weber, L. J. T. White, A. Vedder, and L. Naughton-Treves. New Haven, CT: Yale University Press.

———. 2008. *Paraxerus alexandri*. *IUCN Red List of Threatened Species*, version 2010.4.

Grubb, P., T. S. Jones, A. G. Davies, E. Edberg, E. D. Starin, and J. E. Hill. 1998. *Mammals of Ghana, Sierra Leone, and The Gambia*. St. Ives, Cornwall, UK: Trendine Press.

Guevara, J. A. G. 1998. Changes in habitat preferences of the tree squirrel, *Sciurus alleni*, in a pine-oak association in eastern Sierra Madre, Mexico. In *Ecology and Evolutionary Biology of Tree Squirrels*, ed. M. A. Steele, J. F. Merritt, and D. A. Zegers. Special Publication No. 6. Martinsville: Virginia Museum of Natural History.

Guichón, M. L., M. Bello, and L. Fasola. 2005. Expansión poblacional de una especie introducida en la Argentina: La ardilla de vientre rojo *Callosciurus erythraeus*. *Mastozoología Neotropical* 12:189–197.

Gunderson, A. M., B. K. Jacobsen, and L. E. Olson. 2009. Revised distribution of the Alaska marmot, *Marmota broweri*, and confirmation of parapatry with hoary marmots. *Journal of Mammalogy* 90:859–869.

Gündüz, I., M. Jaarola, C. Tez, C. Yeniyurt, P. D. Polly, and J. B. Searle. 2007a. Multigenic and morphometric differentiation of ground squirrels (*Spermophilus*, Sciuridae, Rodentia) in Turkey, with a description of a new species. *Molecular Phylogenetics and Evolution* 43 (3):916–935.

———. 2007b. *Spermophilus torosensis* Özkurt et al., 2007 (Sciuridae, Rodentia) is a subjective junior synonym of *Spermophilus taurensis* Gündüz et al., 2007, a newly described ground squirrel from the Taurus Mountains of southern Turkey. *Zootaxa* 1663:67–68.

Gür, H., and N. Barlas. 2006. Sex ratio of a population of Anatolian ground squirrels in central Anatolia, Turkey. *Acta Theriologica* 51:61–67.

Gür, H., and M. Kart Gür. 2005. Annual cycle of activity, reproduction, and body mass of Anatolian ground squir-

rels in Turkey. *Journal of Mammalogy* 86 (1):7–14.
Guralnick, R. 2007. Differential effects of past climate warming on mountain and flatland species distributions: A multispecies North American mammal assessment. *Global Ecology and Biogeography* 16:14–23.
Guzmán-Cornejo, C., R. G. Robbins, and T. M. Pérez. 2007. The *Ixodes* (Acari: Ixodidae) of Mexico: Parasite-host and host-parasite checklists. *Zootaxa* 1553:47–58.
Habibi, K. 2003. *Mammals of Afghanistan*. Coimbatore, India: Zoo Outreach Organisation, with assistance from the U.S. Fish and Wildlife Service.
Hackett, H. M., and J. F. Pagels. 2003. Nest site characteristics of the endangered northern flying squirrel (*Glaucomys sabrinus coloratus*) in southwestern Virginia. *American Midland Naturalist* 150:321–331.
Hacklander, K., E. Mostl, and W. Arnold. 2003. Reproductive suppression in female alpine marmots, *Marmota marmota*. *Animal Behaviour* 65:1133–1140.
Hadley, G. L., and K. R. Wilson. 2004a. Patterns of density and survival in small mammals in ski runs and adjacent forest patches. *Journal of Wildlife Management* 68 (2):288–298.
———. 2004b. Patterns of small mammal density and survival following ski-run development. *Journal of Mammalogy* 85:97–104.
Hafner, D. J. 1992. Speciation and persistence of a contact zone in Mojave Desert ground squirrels, subgenus *Xerospermophilus*. *Journal of Mammalogy* 73 (4):770–778.
Hafner, D. J., and T. L. Yates. 1983. Systematic status of the Mojave ground squirrel, *Spermophilus mohavensis* (subgenus *Xerospermophilus*). *Journal of Mammalogy* 64 (3):397–404.
Hafner, M. S., L. J. Barkley, and J. M. Chupasko. 1994. Evolutionary genetics of New World tree squirrels (tribe Sciurini). *Journal of Mammalogy* 75:102–109.
Haim, A., J. D. Skinner, and T. J. Robinson. 1987. Bioenergetics, thermoregulation, and urine analysis of squirrels of the genus *Xerus* from an arid environment. *South African Journal of Zoology* 22 (1):45–49.
Hall, D. S. 1991. Diet of the northern flying squirrel at Sagehen Creek, California. *Journal of Mammalogy* 72 (3):615–617.
Hall, E. R. 1946. *Mammals of Nevada*. Berkeley: University of California Press.
———. 1981. *The Mammals of North America*. New York: Wiley.
Hall, E. R. and K. R. Kelson. 1959. *Mammals of North America*, 2 vols. New York: Ronald Press.
Haltenorth, T., and H. Diller. 1980. *A Field Guide to the Mammals of Africa, Including Madagascar*. London: William Collins Sons.
Hammer, M., and E. Tatum-Hume. 2003. Expedition report: Surveying monkeys, macaws and other animals of the Peru Amazon. Sprat's Water, Suffolk, UK: Biosphere Expeditions.
Hammond, E. L., and R. G. Anthony. 2006. Mark-recapture estimates of population parameters for selected species of small mammals. *Journal of Mammalogy* 87 (3):618–627.
Hannon, M. J., S. H. Jenkins, R. L. Crabtree, and A. K. Swanson. 2006. Visibility and vigilance: Behavior and population ecology of Uinta ground squirrels (*Spermophilus armatus*) in different habitats. *Journal of Mammalogy* 87:287–295.
Hanski, I. K., and V. Selonen. 2008. Female-biased natal dispersal in the Siberian flying squirrel. *Behavioral Ecology* 20:60–67.
Hanson, M. T., and R. G. Coss. 2001a. Age differences in arousal and vigilance in California ground squirrels (*Spermophilus beecheyi*). *Developmental Psychobiology* 39 (3):199–206.
———. 2001b. Age differences in the response of California ground squirrels (*Spermophilus beecheyi*) to conspecific alarm calls. *Ethology* 107 (3):259–275.
Happold, D. C. D. 1987. *The Mammals of Nigeria*. Oxford, UK: Clarendon Press.
Hare, J. F. 2004. Kin discrimination by asocial Franklin's ground squirrels (*Spermophilus franklinii*): Is there a relationship between kin discrimination and ground squirrel sociality? *Ethology Ecology & Evolution* 16:157–169.
Hare, J. F., and J. O. Murie. 1992. Manipulation of litter size reveals no cost of reproduction in Columbian ground squirrels. *Journal of Mammalogy* 73 (2):449–454.
———. 1996. Ground squirrel sociality and the quest for the "holy grail": Does kinship influence behavioral discrimination by juvenile Columbian ground squirrels? *Behavioral Ecology* 7:76–81.
Harlow, R. F., and A. T. Doyle. 1990. Food habits of southern flying squirrels (*Glaucomys volans*) collected from red-cockaded woodpecker (*Picoides borealis*) colonies in South Carolina. *American Midland Naturalist* 124 (1):187–191.
Harris, J. H., and P. Leitner. 2004. Home-range size and use of space by adult Mohave ground squirrels, *Spermophilus mohavensis*. *Journal of Mammalogy* 85 (3):517–523.
———. 2005. Long-distance movements of juvenile Mohave ground squirrels, *Spermophilus mohavensis*. *Southwestern Naturalist* 50:188–196.
Harris, W. P., Jr. 1944. Additions and corrections to the section on Sciuridae in Ellerman's *Families and Genera of Living Rodents*. *Occasional Papers of the Museum of Zoology, University of Michigan* 484:1–24.
Harrison, D. L., and P. J. J. Bates. 1991. *The Mammals of Arabia*. Sevenoaks, Kent, UK: Harrison Zoological Museum.
Harrison, J. L. 1954. The natural food of some rats and other mammals. *Bulletin of the Raffles Museum* 25:157–165.
Harrison, R. G., S. M. Bogdanowicz, R. S. Hoffmann, E. Yensen, and P. W. Sherman. 2003. Phylogeny and evolutionary history of the ground squirrels (Rodentia: Marmotinae). *Journal of Mammalian Evolution* 10 (3):249–276.
Hart, E. B., 1992. *Tamias dorsalis*. *Mammalian Species* 399:1–6.
Haslauer, R. 2002. *Pteromyscus pulverulentus*. *Animal Diversity Web*.
Hassinger, J. D. 1973. A survey of the mammals of Afghanistan resulting from the 1965 Street Expedition (excluding bats). *Fieldiana Zoology* 60:1–195.
Haugaasen, T., and C. A. Peres. 2005. Mammal assemblage structure in Amazonian flooded and unflooded forests. *Journal of Tropical Ecology* 21:133–145.
Haukisalmi, V., and I. K. Hanski. 2007. Contrasting seasonal dynamics in fleas of the Siberian flying squirrel (*Pteromys volans*) in Finland. *Ecological Entomology* 32 (4):333–337.
Hautier, L., P. H. Fabre, and J. Michaux. 2009. Mandible shape and dwarfism in squirrels (Mammalia, Rodentia): Interaction of allometry and adaptation. *Die Naturwissenschaften* 96 (6):725–730.
Hayes, J. P., E. G. Horvath, and P. Hounihan. 1995. Townsend's chipmunk populations in Douglas-fir plantations and mature forests in the Oregon Coast Range. *Canadian Journal of Zoology* 73:67–73.
Hayman, R. W. 1950. Two new African squirrels. *Annals and Magazine of Natural History* 12 (3):263–264.
———. 1951. Notes on some Angolan mammals. *Publicações Culturais, Companhia de Diamantes de Angola (Diamang), Lisboa* 11:31–47.

Haynie, M. L., R. A. Van den Bussche, J. L. Hoogland, and D. A. Gilbert. 2003. Parentage, multiple paternity, and breeding success in Gunnison's and Utah prairie dogs. *Journal of Mammalogy* 84 (4):1244-1253.

Hayssen, V. 2008a. Patterns of body and tail length and body mass in Sciuridae. *Journal of Mammalogy* 89 (4):852-873.

———. 2008b. Reproduction within marmotine ground squirrels (Sciuridae, Xerinae, Marmotini): Patterns among genera. *Journal of Mammalogy* 89 (3):607-616.

———. 2008c. Reproductive effort in squirrels: Ecological, phylogenetic, allometric, and latitudinal patterns. *Journal of Mammalogy* 89 (3):582-606.

Hayssen, V., Ari van Tienhoven, and Ans van Tienhoven. 1993. *Asdell's Patterns of Mammalian Reproduction: A Compendium of Species-Specific Data*. Ithaca, NY: Cornell University Press.

Hazard, E. B. 1982. *The Mammals of Minnesota*. Minneapolis: University of Minnesota Press.

He, F., and J. Lin. 2006. Hoernchen gefaehrden Brutkolonien des Gelbkehlhaeherlings. *ZGAP Mitteilungen* 22 (2):7-8.

Healy, W. M., and R. T. Brooks. 1988. Small mammal abundance in northern hardwood stands in West Virginia. *Journal of Wildlife Management* 52 (3):491-496.

Heaney, L. R. 1978. Island area and body size of insular mammals: Evidence from tri-colored squirrel (*Callosciurus prevostii*) of Southeast Asia. *Evolution* 32 (1):29-44.

———. 1985. *Systematics of Oriental Pygmy Squirrels of the Genera* Exilisciurus *and* Nannosciurus *(Mammalia: Sciuridae)*. Miscellaneous Publication No. 170. Ann Arbor: University of Michigan, Museum of Zoology.

———. 2008. Biogeography of mammals in SE Asia: Estimates of rates of colonization, extinction, and speciation. *Biological Journal of the Linnean Society* 28:127-165.

Heaney, L. R., D. S. Balete, L. Dolar, A. C. Alcala, A. Dans, P. C. Gonzales, N. Ingle, M. Lepiten, W. Oliver, E. A. Rickart, B. R. Tabaranza Jr., and R. C. B. Utzurrum. 1998. A synopsis of the mammalian fauna of the Philippine Islands. *Fieldiana Zoology*, n.s., 88:1-61.

Heaney, L. R., and D. S. Rabor. 1982. Mammals of the Dinagat and Siargao Islands, Philippines. *Occasional Papers of the Museum of Zoology, University of Michigan* 699:1-30.

Heaney, L. R., B. R. Tabaranza Jr., E. A. Rickart, D. S. Balete, and N. R. Ingle. 2006. The mammals of Mt. Kitanglad Nature Park, Mindanao, Philippines. *Fieldiana Zoology*, n.s., 112:1-63.

Heaney, L. R., and R. W. Thorington Jr. 1978. Ecology of neotropical red-tailed squirrels, *Sciurus granatensis*, in Panama Canal Zone. *Journal of Mammalogy* 59 (4):846-851.

Hedges, S., J. W. Duckworth, B. Lee, and R. J. Tizard. 2008. *Lariscus insignis. IUCN Red List of Threatened Species*, version 2010.4.

Heffner, R. S., G. Koay, and H. E. Heffner. 2001. Audiograms of five species of rodents: Implications for the evolution of hearing and the perception of pitch. *Hearing Research* 157 (1-2):138-152.

Helgen, K., and K. Aplin. 2008. *Sundasciurus fraterculus. IUCN Red List of Threatened Species*, version 2010.4.

Helgen, K. M., F. R. Cole, L. E. Helgen, and D. E. Wilson. 2009. Generic revision in the Holarctic ground squirrel genus *Spermophilus. Journal of Mammalogy* 90 (2):270-305.

Henning, R. H., L. E. Deelman, R. A. Hut, E. A. Van der Zee, H. Buikema, S. A. Nelemans, H. Lip, D. De Zeeuw, S. Daan, and A. H. Epema. 2002. Normalization of aortic function during arousal episodes in the hibernating ground squirrel. *Life Sciences* 70 (17):2071-2083.

Hernandez, P. A., C. H. Graham, L. L. Master, and D. L. Albert. 2006. The effect of sample size and species characteristics on performance of different species distribution modeling methods. *Ecography* 29:773-785.

Hernandez-Camacho, J. 1960. Primitiae mastozoologicae Colombianae—I. Status taxonómico de *Sciurus pucheranii santanderensis. Caldasia* 8 (38):359-368.

Herrerías-Diego, Y., M. Quesada, K. E. Stoner, J. A. Lobo, Y. Hernández-Flores, and G. Sanchez Montoya. 2008. Effect of forest fragmentation on fruit and seed predation of the tropical dry forest tree *Ceiba aesculifolia. Biological Conservation* 141 (1):241-248.

Herron, M. D., T. A. Castoe, and C. L. Parkinson. 2004. Sciurid phylogeny and the paraphyly of Holarctic ground squirrels (*Spermophilus*). *Molecular Phylogenetics and Evolution* 31 (3):1015-1030.

Herron, M. D., and J. M. Waterman. 2004. *Xerus erythropus. Mammalian Species* 748:1-4.

Herron, M. D., J. M. Waterman, and C. L. Parkinson. 2005. Phylogeny and historical biogeography of African ground squirrels: The role of climate change in the evolution of *Xerus. Molecular Ecology* 14 (9):2773-2788.

Hershkovitz, P. 1977. *Living New World Monkeys (*Platyrrhini*)*. Chicago: University of Chicago Press.

Herzig-Straschil, B. 1978. On the biology of *Xerus inauris* (Zimmermann, 1780) (Rodentia, Sciuridae). *Zeitschrift für Säugetierkunde—International Journal of Mammalian Biology* 43 (5):262-278.

Herzig-Straschil, B., and A. Herzig. 1989. Biology of *Xerus princeps. Madoqua* 16:41-46.

Herzig-Straschil, B., A. Herzig, and H. Winkler. 1991. A morphometric analysis of the skulls of *Xerus inauris* and *Xerus princeps* (Rodentia, Sciuridae). *Zeitschrift für Säugetierkunde—International Journal of Mammalian Biology* 56 (3):177-187.

Heymann, E. W., and C. Knogge. 1997. Field observations on the Neotropical pygmy squirrel, *Sciurillus pusillus* (Rodentia: Sciuridae) in Peruvian Amazonia. *Ecotropica* 3:67-69.

Hidinger, L. A. 1996. Measuring the impacts of ecotourism on animal populations: A case study of Tikal National Park, Guatemala. *Yale F&ES Bulletin* 99:49-59.

Hilaluddin, R. Kaul, and D. Ghose. 2005. Conservation implications of wild animal biomass extractions in Northeast India. *Animal Biodiversity and Conservation* 28:169-179.

Hill, J. E. 1959. A North Bornean pygmy squirrel, *Glyphotes simus* Thomas, and its relationships. *Bulletin of the British Museum (Natural History)* 5 (9):12-266.

Hilton, C. D., and T. L. Best. 1993. *Tamias cinereicollis. Mammalian Species* 436:1-5.

Hinton, M. A. C. 1920. The subspecies of *Paraxerus flavovittis. Annals and Magazine of Natural History* 9 (5):308-312.

Hiroyuki, O. 1999. Distribution and habitat fragmentation of the Japanese giant flying squirrel *Petaurista leucogenys* in Tokyo. *Mammalian Science* 78:169-173.

Hoffmann, I. E., E. Millesi, S. Huber, L. G. Everts, and J. P. Dittami. 2003. Population dynamics of European ground squirrels (*Spermophilus citellus*) in a suburban area. *Journal of Mammalogy* 84 (2):615-626.

Hoffmann, R. S., and A. T. Smith. 2008. Family Sciuridae. In *A Guide to the Mammals of China*, ed. A. T. Smith and Y. Xie. Princeton, NJ: Princeton University Press.

Hoffmeister, D. F. 1986. *Mammals of Arizona*.

Phoenix: Arizona Game and Fish Department.

———. 1989. *Mammals of Illinois*. Urbana: University of Illinois Press.

Hoke, P., R. Demey, and A. Peal, eds. 2007. *A Rapid Biological Assessment of North Lorma, Gola, and Grebo National Forests, Liberia*. RAP Bulletin of Biological Assessment No. 44. Arlington, VA: Conservation International.

Holekamp, K. E. 1986. Proximal causes of natal dispersal in Belding's ground squirrels (*Spermophilus beldingi*). *Ecological Monographs* 56 (4):365–391.

Hollister, N. 1911. *Four New Mammals from the Canadian Rockies*. Smithsonian Miscellaneous Collections, Vol. 56, No. 26. Washington, DC: Smithsonian Institution.

———. 1913. *A Review of the Philippine Land Mammals in the United States National Museum*. Washington: Government Printing Office.

———. 1919. *East African Mammals in the United States National Museum*. United States National Museum Bulletin 99, Part 2. Washington, DC: Government Printing Office.

Holloway, G. L., and J. R. Malcolm. 2007. Nest tree use by northern and southern flying squirrels in central Ontario. *Journal of Mammalogy* 88:226–233.

Holmes, D. J., and S. N. Austad. 1994. Fly now, die later: Life-history correlates of gliding and flying in mammals. *Journal of Mammalogy* 75 (1):224–226.

Holmes, W. G. 1984. Ontogeny of dam-young recognition in captive Belding's ground squirrels (*Spermophilus beldingi*). *Journal of Comparative Psychology* 98 (3):246–256.

Holmes, W. G., and J. M. Mateo. 1998. How mothers influence the development of litter-mate preferences in Belding's ground squirrels. *Animal Behaviour* 55 (6):1555–1570.

Hoogland, J. L. 1981. The evolution of coloniality in white-tailed and black-tailed prairie dogs (Sciuridae, *Cynomys leucurus* and *C. ludovicianus*). *Ecology* 62 (1):252–272.

———. 1995. *The Black-Tailed Prairie Dog: Social Life of a Burrowing Mammal*. Chicago: University of Chicago Press.

———. 1996. *Cynomys ludovicianus*. *Mammalian Species* 535:1–10.

———. 1998. Estrus and copulation among Gunnison's prairie dogs. *Journal of Mammalogy* 79:887–897.

———. 1999. Philopatry, dispersal, and social organization of Gunnison's prairie dogs. *Journal of Mammalogy* 80:243–251.

———. 2001. Black-tailed, Gunnison's, and Utah prairie dogs all reproduce slowly. *Journal of Mammalogy* 82:917–927.

———. 2003a. Black-tailed prairie dog *Cynomys ludovicianus* and allies. In *Wild Mammals of North America: Biology, Management, and Conservation*, 2nd ed., ed. G. A. Feldhamer, B. C. Thompson, and J. A. Chapman. Baltimore: Johns Hopkins University Press.

———. 2003b. Sexual dimorphism in five species of prairie dogs. *Journal of Mammalogy* 84 (4):1254–1266.

———, ed. 2006. *Conservation of the Black-Tailed Prairie Dog: Saving North America's Western Grasslands*. Washington, DC: Island Press.

———. 2009. Nursing of own and foster offspring by Utah prairie dogs. *Behavioral Ecology and Sociobiology* 63 (11):1621–1634.

Hoogstraal, H. 1955. Notes on African *Haemaphysalis* ticks, II: The ground-squirrel parasites, *H. calcarata* Neumann, 1902, and *H. houyi* Nuttall and Warburton, 1915 (Ixodoidea, Ixodidae). *Journal of Parasitology* 41 (4):361–373.

Hoogstraal, H., and R. M. Mitchell. 1971. *Haemaphysalis (Alloceraea) aponommoides* Warburton (Ixodoidea: Ixodidae): Description of immature stages, hosts, distribution, and ecology in India, Nepal, Sikkim, and China. *Journal of Parasitology* 57 (3):635–645.

Hopewell, L. J., and L. A. Leaver. 2008. Evidence of social influences on cache-making by grey squirrels (*Sciurus carolinensis*). *Ethology* 114:1061–1068.

Hopewell, L. J., L. A. Leaver, and S. E. G. Lea. 2008. Effects of competition and food availability on travel time in scatter-hoarding gray squirrels (*Sciurus carolinensis*). *Behavioral Ecology and Sociobiology* 19 (6):1143–1149.

Hopf, H. S., G. E. J. Morley, and J. R. O. Humphries, eds. 1976. *Rodent Damage to Growing Crops and to Farm and Village Storage in Tropical and Subtropical Regions: Results of a Postal Survey 1972–73*. London: Centre for Overseas Pest Research and Tropical Products Institute.

Howell, A. H. 1915a. Description of a new genus and seven new races of flying squirrels. *Proceedings of the Biological Society of Washington* 28:109–114.

———. 1915b. Preliminary descriptions of five new chipmunks from North America. *Journal of Mammology* 6 (1):51–54.

Hubálek, Z. 1987. Geographic distribution of Bhanja virus. *Folia Parasitologica* 34 (1):77–86.

Hubbs, A. H., J. S. Millar, and J. P. Wiebe. 2000. Effect of brief exposure to a potential predator on cortisol concentrations in female Columbian ground squirrels (*Spermophilus columbianus*). *Canadian Journal of Zoology* 78 (4):578–587.

Huber, S., I. E. Hoffmann, E. Millesi, J. Dittami, and W. Arnold. 2001. Explaining the seasonal decline in litter size in European ground squirrels. *Ecography* 24 (2):205–211.

Huber, S., E. Millesi, and J. P. Dittami. 2002. Paternal effort and its relation to mating success in the European ground squirrel. *Animal Behaviour* 63 (1):157–164.

Huber, S., E. Millesi, M. Walzl, J. Dittami, and W. Arnold. 1999. Reproductive effort and costs of reproduction in female European ground squirrels. *Oecologia* 121 (1):19–24.

Huebschman, J. J. 2003. A conservation assessment of Franklin's ground squirrel (*Spermophilus franklinii* Sabine 1822): Input from natural history, morphology and genetics. Ph.D. diss., University of Nebraska-Lincoln.

Hulová, S., and F. Sedlácek. 2008. Population genetic structure of the European ground squirrel in the Czech Republic. *Conservation Genetics* 9 (3):615–625.

Hurme, E., M. Kurttila, M. Mönkkönen, T. Heinonen, and T. Pukkala. 2007. Maintenance of flying squirrel habitat and timber harvest: A site-specific spatial model in forest planning calculations. *Landscape Ecology* 22 (2):243–256.

Hurme, E., M. Mönkkönen, A. Nikula, V. Nivala, P. Reunanen, T. Heikkinen, and M. Ukkola. 2005. Building and evaluating predictive occupancy models for the Siberian flying squirrel using forest planning data. *Forest Ecology and Management* 216 (1):241.

Hurme, E., M. Mönkkönen, P. Reunanen, A. Nikula, and V. Nivala. 2008. Temporal patch occupancy dynamics of the Siberian flying squirrel in a boreal forest landscape. *Ecography* 31 (4):469–476.

Imaizumi, Y. 1961. *Coloured Illustrations of the Mammals of Japan*. Osaka, Japan: Hoikusha.

Ingles, L. G. 1965. *Mammals of the Pacific States: California, Oregon, and Washington*. Stanford, CA: Stanford University Press.

Ishii, N., and Y. Kaneko. 2008a. *Petaurista leucogenys*. *IUCN Red List of Threatened Species*, version 2009.2.

———. 2008b. *Pteromys momonga*. *IUCN Red List of Threatened Species*, version 2009.2.

Iudin, B. S, V. G. Krivosheev, and V. G. Beliaev. 1976. *Melkie mlekopitaiushchie severa Dalnego Vostoka: Fauna ėkologiia nasekomoiadnykh-Insectivora, rukokrylykh-Chiroptera, zaĭtseobraznykh-Lagomorpha, gryzunov-Rodentia i ikh parazitov*. Novosibirsk, USSR: Nauka, Sibirskoe otdelenie.

Izzo, A. D, M. Meyer, J. M. Trappe, M. North, and T. D. Bruns. 2005. Hypogeous ectomycorrhizal fungal species on roots and in small mammal diet in a mixed-conifer forest. *Forest Science* 51 (3):243–254.

Jackson, H. H. T. 1972. *Mammals of Wisconsin*. Madison: University of Wisconsin Press.

Jacobs, K. M., and D. L. Luoma. 2008. Small mammal mycophagy response to variations in green-tree retention. *Journal of Wildlife Management* 72 (8):1747–1755.

Jaffe, G., D. A. Zegers, M. A. Steele, and J. F. Merritt. 2005. Long-term patterns of botfly parasitism in *Peromyscus maniculatus*, *P. leucopus*, and *Tamias striatus*. *Journal of Mammalogy* 86:39–45.

Jannett, F. J., Jr., M. R. Broschart, L. H. Grim, and J. P. Schaberl. 2007. Northerly range extensions of mammalian species in Minnesota. *American Midland Naturalist* 158 (1):168–176.

Jardine, C., G. Appleyard, M. Kosoy, D. McColl, M. Chirino-Trejo, G. Wobeser, and F. A. Leighton. 2006. Rodent associated *Bartonella* in Saskatchewan, Canada. *Vector-Borne and Zoonotic Diseases* 5:402–409.

Jenkins, P. D., and J. E. Hill. 1982. Mammals from Siberut, Mentawai Islands. *Mammalia* 46 (2):219–224.

Jenkins, S. H., and B. D. Eshelman. 1984. *Spermophilus beldingi*. *Mammalian Species* 221:1–8.

Jeong, S. J., N. H. Kim, D. H. Kim, T. H. Kong, N. H. Ahn, T. Miyamoto, R. Higuchi, and Y. C. Kim. 2000. Hyaluronidase inhibitory active 6H-dibenzo [b,d]pyran-6-ones from the feces of *Trogopterus xanthipes*. *Planta Medica* 66:76–77.

Jernigan, K. A. 2009. Barking up the same tree: A comparison of ethnomedicine and canine ethnoveterinary medicine among the Aguaruna. *Journal of Ethnobiology and Ethnomedicine* 5:33–42.

Jessen, R., M. J. Merrick, J. L. Koprowski, and O. Ramirez. 2010. Presence of Guayaquil squirrels on the central coast of Peru: An apparent introduction. *Mammalia* 74:443–444.

Jezek, Z., and F. Fenner. 1988. *Human Monkeypox*. Monographs in Virology No. 17. Basel, Switzerland: Karger.

Jimenez-Guzman, A., and S. Guerrero-Vazquez. 1992. Fauna silvestre de Nuevo León. *Publicaciones Biologicas—FCB/UANL* 6:105–111.

Johnson, S. A., and J. Choromanski-Norris. 1992. Reduction in the eastern limit of the range of the Franklin's ground squirrel (*Spermophilus franklinii*). *American Midland Naturalist* 128 (2):325–331.

Johnson-Murray, J. L. 1977. Myology of the gliding membranes of some petauristine rodents (genera *Glaucomys*, *Pteromys*, *Petinomys*, and *Petaurista*). *Journal of Mammalogy* 58 (3):374–384.

Johnston, A. N., and R. G. Anthony. 2008. Small-mammal microhabitat associations and response to grazing in Oregon. *Journal of Wildlife Management* 72 (8):1736–1746.

Johnston, K., and O. Schmitz. 1997. Wildlife and climate change: Assessing the sensitivity of selected species to simulated doubling of atmospheric CO2. *Global Change Biology* 3 (6):531–544.

Jones, C., R. S. Hoffmann, D. W. Rice, M. D. Engstrom, R. D. Bradley, D. J. Schmidly, C. A. Jones, and R. J. Baker. 1997. Revised checklists of North American mammals north of Mexico, 1997. *Occasional Papers of the Museum of Texas Tech University* 173:1–20.

Jones, J. K., Jr., D. M. Armstrong, R. S. Hoffmann, and C. Jones. 1983. *Mammals of the Northern Great Plains*. Lincoln: University of Nebraska Press.

Jones, J. K., Jr., and H. H. Genoways. 1971. Notes on the biology of the Central American squirrel, *Sciurus richmondi*. *American Midland Naturalist* 86 (1):242–246.

———. 1975. *Sciurus richmondi*. *Mammalian Species* 53:1–2.

Jordheim, S. 1990. Woodchuck near Stewart Valley. *Blue Jay* 48:225.

Kain, D. E. 1985. The systematic status of *Eutamias ochrogenys* and *Eutamias senex* (Rodentia: Sciuridae). M.A. thesis, Humboldt State University.

Kakati, L. N., B. Ao, and V. Doulo. 2006. Indigenous knowledge of zootherapeutic use of vertebrate origin by the Ao tribe of Nagaland. *Journal of Human Ecology* 19 (3):163–167.

Kaleme, P. K., J. Bates, J. K. Peterans, M. M. Jacques, and B. R. Ndara. 2007. Small mammal diversity and habitat requirements in the Kahuzi-Biega National Park and surrounding areas, eastern Democratic Republic of the Congo. *Integrative Zoology* 2 (4):239–246.

Kaneko, K., and N. Hashimoto. 1981. Occurrence of *Yersinia enterocolitica* in wild animals. *Applied and Environmental Microbiology* 41 (3):635–638.

Karasov, W. H. 1983. Water flux and water requirement in free-living antelope ground squirrels *Ammospermophilus leucurus*. *Physiological Zoology* 56 (1):94–105.

Karels, T. J., A. E. Byrom, R. Boonstra, and C. J. Krebs. 2000. The interactive effects of food and predators on reproduction and overwinter survival of Arctic ground squirrels. *Journal of Animal Ecology* 69:235–247.

Kart Gür, M., R. Refinetti, and H. Gür. 2009. Daily rhythmicity and hibernation in the Anatolian ground squirrel under natural and laboratory conditions. *Journal of Comparative Physiology B—Biochemical, Systemic, and Environmental Physiology* 179:155–164.

Karthikeyan, S., J. N. Prasad, and B. Arun. 1992. Grizzled giant squirrel *Ratufa macroura* Thomas and Wroughton at Cauvery Valley in Karnataka. *Journal of the Bombay Natural History Society* 89:360–361.

Kashkarov, D., and L. Lein. 1927. The yellow ground squirrel of Turkestan, *Cynomys fulvus oxianus* Thomas. *Ecology* 8 (1):63–72.

Kataoka, T., and N. Tamura. 2005. Effects of habitat fragmentation on the presence of Japanese squirrels, *Sciurus lis*, in suburban forests. *Mammal Study* 30 (2):131–138.

Kaufman, G. A., and D. W. Kaufman. 2002. Woodchuck recorded in Saline and Russell counties, Kansas. *Prairie Naturalist* 34:145–147.

Kawamichi, M. 1989. Nest structure dynamics and seasonal use of nests by Siberian chipmunks (*Eutamias sibiricus*). *Journal of Mammalogy* 70 (1):44–57.

———. 1996. Ecological factors affecting annual variation in commencement of hibernation in wild chipmunks (*Tamias sibiricus*). *Journal of Mammalogy* 77 (3):731–744.

Kawamichi, T., ed. 1996. *Mammals*. Vol. 1 of *The Encyclopedia of Animals in Japan*. Tokyo: Heibonsha.

———. 1997a. The age of sexual maturity in Japanese giant flying squirrels, *Petaurista leucogenys*. *Mammal Study* 22:81–87.

———. 1997b. Seasonal changes in the diet of Japanese giant flying squirrels in relation to reproduction. *Journal of Mammalogy* 78 (1):204-212.

———. 1998. Seasonal change in the testis size of the Japanese giant flying squirrel, *Petaurista leucogenys*. *Mammal Study* 23:79-82.

———. 1999. Factors affecting dates of biannual mating in Japanese giant flying squirrels (*Petaurista leucogenys*). *Mammalian Science* 78:165-168.

Kawamichi, T., and M. Kawamichi. 1993. Gestation period and litter size of Siberian chipmunk *Eutamias sibiricus lineatus* in Hokkaido, northern Japan. *Journal of the Mammalogical Society of Japan* 18 (2):105-109.

Keane, J. J., M. L. Morrison, and D. M. Fry. 2006. Prey and weather factors associated with temporal variation in northern goshawk reproduction in the Sierra Nevada, California. *Studies in Avian Biology* 31:87-99.

Kemp, G. A., and L. B. Keith. 1970. Dynamics and regulation of red squirrel (*Tamiasciurus hudsonicus*) populations. *Ecology* 51 (5):763-779.

Kemp, G. E., O. R. Causey, H. W. Setzer, and D. L. Moore. 1974. Isolation of viruses from wild mammals in West Africa, 1966-1970. *Journal of Wildlife Diseases* 10 (3):279-293.

Key, G. E. 1985. An investigation of the pest status of the African striped ground squirrel, *Xerus erythropus* (Geoffroy), and related Sciuridae. Ph.D. diss., University of Exeter.

———. 1990a. Control of the African striped ground squirrel, *Xerus erythropus*, in Kenya. In *Proceedings of the Fourteenth Vertebrate Pest Conference, Sacramento, California, March 6-8, 1990*, ed. L. R. Davis and R. E. Marsh. Davis: University of California, Davis.

———. 1990b. Pre-harvest crop losses to the African striped ground squirrel, *Xerus erythropus*, in Kenya. *Tropical Pest Management* 36:223-229.

Kim, K. C. 1977. *Atopophthirus Emersoni*, new genus and new species (Anoplura: Hoplopleuridae) from *Petaurista elegans* (Rodentia: Sciuridae), with a key to the genera of Enderleinellinae. *Journal of Medical Entomology* 14 (4):417-420.

King, D. I., C. R. Griffin, and R. M. DeGraaf. 1998. Nest predator distribution among clearcut forest, forest edge, and forest interior in an extensively forested landscape. *Forest Ecology and Management* 104 (1-3):151-156.

King, J. A. 1955. *Social Behavior, Social Organization, and Population Dynamics in a Black-Tailed Prairie Dog Town in the Black Hills of South Dakota*. Contributions from the Laboratory of Vertebrate Biology No. 67. Ann Arbor: University of Michigan.

King, W. J., and D. Allaine. 2002. Social, maternal, and environmental influences on reproductive success in female Alpine marmots (*Marmota marmota*). *Canadian Journal of Zoology* 80:2137-2143.

King, W. J., M. Festa-Bianchet, and S. E. Hatfield. 1991. Determinants of reproductive success in female Columbian ground squirrels. *Oecologia* 86 (4):528-534.

Kingdon, J. 1971. A new race of giant squirrel (*Protoxerus stangeri cooperi*). *Uganda Journal* 35 (2):207-208.

———. 1974. *Hares and Rodents*. Vol. 2, pt. B of *East African Mammals: An Atlas of Evolution in Africa*. New York: Academic Press.

———. 1997. *The Kingdon Field Guide to African Mammals*. San Diego: Academic Press.

Kirk, E. C., P. Lemelin, M. W. Hamrick, D. M. Boyer, and J. I. Bloch. 2008. Intrinsic hand proportions of euarchontans and other mammals: Implications for the locomotor behavior of plesiadapiforms. *Journal of Human Evolution*. 55 (2):278.

Kloss, C. B. 1915. On two new squirrels from the inner Gulf of Siam. *Journal of Natural History Society of Siam* 1:157-162.

Kneeland, M. C., J. L. Koprowski, and M. C. Corse. 1995. Potential predators of Chiricahua fox squirrels (*Sciurus nayaritensis chiricahuae*). *Southwestern Naturalist* 40:340-342.

Koford, R. R. 1982. Mating system of a territorial tree squirrel (*Tamiasciurus douglasii*) in California. *Journal of Mammalogy* 63 (2):274-283.

Koh, H. S., J. X. Wang, B. K. Lee, B. G. Yang, S. W. Heo, K. H. Jang, and T. Y. Chun. 2009. A phylogroup of the Siberian chipmunk from Korea (*Tamias sibiricus barberi*) revealed from the mitochondrial DNA cytochrome *b* gene. *Biochemical Genetics* 47 (1-2):1-7.

Koh, H. S., M. Zhang, D. Bayarlkhagva, E. J. Ham, J. S. Kim, K. H. Jang, and N. J. Park. 2010. Concordant genetic distinctness of the phylogroup of the Siberian chipmunk from the Korean peninsula (*Tamias sibiricus barberi*), reexamined with nuclear DNA c-myc gene exon 2 and mtDNA control region sequences. *Biochemical Genetics* 48 (7-8):696-705.

Kolesnikov, V. V., O. V. Brandler, B. B. Badmaev, D. Zoje, and Y. Adiya. 2009. Factors that lead to a decline in numbers of Mongolian marmot populations. *Ethology Ecology & Evolution* 21:371-379.

Koprowski, J. L. 1994a. *Sciurus carolinensis*. *Mammalian Species* 480:1-9.

———. 1994b. *Sciurus niger*. *Mammalian Species* 479:1-9.

———. 1996. Natal philopatry, communal nesting, and kinship in fox squirrels and gray squirrels. *Journal of Mammalogy* 77:1006-1016.

Koprowski, J. L., and M. C. Corse. 2001. Food habits of the Chiricahua fox squirrel (*Sciurus nayaritensis chiricahuae*). *Southwestern Naturalist* 46:62-65.

———. 2005. Time budgets, activity periods, and behavior of Mexican fox squirrels. *Journal of Mammalogy* 86:947-952.

Koprowski, J. L., N. Ramos, B. S. Pasch, and C. A. Zugmeyer. 2006. Observations on the ecology of the endemic Mearns's squirrel (*Tamiasciurus mearnsi*). *Southwestern Naturalist* 51:426-430.

Koprowski, J., L. Roth, F. Reid, N. Woodman, and R. Timm. 2008. *Sciurus variegatoides*. *IUCN Red List of Threatened Species*, version 2009.2.

Koprowski, J., L. Roth, N. Woodman, J. Matson, L. Emmons, and F. Reid. 2008. *Sciurus deppei*. *IUCN Red List of Threatened Species*, version 2009.2.

Koshev, Y. S. 2008. Distribution and status of the European ground squirrel (*Spermophilus citellus*) in Bulgaria. *Lynx* 39 (2):251-261.

Kosoy, M., M. Murray, R. D. Gilmore Jr., Y. Bai, and K. L. Gage. 2003. *Bartonella* strains from ground squirrels are identical to *Bartonella washoensis* isolated from a human patient. *Journal of Clinical Microbiology* 41 (2):645-650.

Koster, J. 2008. The impact of hunting with dogs on wildlife harvests in the Bosawas Reserve, Nicaragua. *Environmental Conservation* 35 (3):211-220.

Kotler, B. P., J. S. Brown, and M. Hickey. 1999. Food storability and the foraging behavior of fox squirrels. *American Midland Naturalist* 142:77-86.

Kowalski, K., and B. Rzebik-Kowalska. 1991. *Mammals of Algeria*. Wroclaw, Poland: Polish Academy of Sciences.

Krichbaum, K., C. G. Mahan, M. A. Steele, G. Turner, and P. J. Hudson. 2010. The potential role of *Strongyloides robustus* on parasite-mediated competition between two species of flying squirrels (*Glaucomys*). *Journal of Wildlife Diseases* 46 (1):229-235.

Krivosheev, V. G. 1984. *Nazemnye mlekopitaiushchie Dalnego Vostoka SSSR*. Moscow: Izdatel'stvo Nauka.

Kuhn, H. J. 1964. *Epixerus ebii jonesi* in Liberia. *Bollettino di Zoologia* 15:157.

Kuhn, K. M., and S. B. Vander Wall. 2008. Linking summer foraging to winter survival in yellow pine chipmunks (*Tamias amoenus*). *Oecologia* 157 (2):349-360.

Kullmann, E. 1965. The mammals of Afghanistan. Special Edition, *Science* [quarterly journal of the Institute of Zoology and Parasitology, Faculty of Science, Kabul University] (August):1-58.

Kumar, A. 1998. *Biswamoyopterus biswasi* (Saha 1981) or *Ichthyophis tricolor* (Annandale 1909)? *Current Science* 75 (5):426-427.

Kuo, C., and L. Lee. 2003. Food availability and food habits of Indian giant flying squirrels (*Petaurista philippensis*) in Taiwan. *Journal of Mammalogy* 84 (4):1330-1340.

Kurland, J. A. 1973. A natural history of Kra macaques (*Macaca fascicularis* Raffles, 1821) at the Kutai Reserve, Kalimantan Timur, Indonesia. *Primates* 14:245-262.

Kuznetsov, G. V. 2006. *Mammals of Vietnam*. Moscow: KMK Scientific Press.

Kwiecinski, G. G. 1998. *Marmota monax*. *Mammalian Species* 591:1-8.

Kyle, C. J., T. J. Karels, C. S. Davis, S. Mebs, B. Clark, C. Strobeck, and D. S. Hik. 2007. Social structure and facultative mating systems of hoary marmots (*Marmota caligata*). *Molecular Ecology* 16:1245-1255.

Laabs, D. 1998. Mohave ground squirrel, *Spermophilus mohavensis*. Report by Biosearch Wildlife Surveys, regarding the West Mohave Planning Area.

Lacey, E. A., and J. R. Wieczorek. 2001. Territoriality and male reproductive success in Arctic ground squirrels. *Behavioral Ecology* 12:626-632.

Lacher, T. E., Jr., and M. A. Mares. 1996. Availability of resources and use of space in eastern chipmunks, *Tamias striatus*. *Journal of Mammalogy* 77 (3):833-849.

Lacher, T. E., Jr., M. R. Willig, and M. A. Mares. 1982. Food preference as a function of resource abundance with multiple prey types: An experimental analysis of optimal foraging theory. *American Naturalist* 120 (3):297-316.

Lacki, M. J., M. J. Gregory, and P. K. Williams. 1984a. Spatial response of an eastern chipmunk population to supplemented food. *American Midland Naturalist* 111 (2):414-416.

———. 1984b. Summer activity of *Tamias striatus* in response to supplemented food. *Journal of Mammalogy* 65 (3):521-524.

Lagaria, A., and D. Youlatos. 2006. Anatomical correlates to scratch-digging in the forelimb of European ground squirrels (*Spermophilus citellus*). *Journal of Mammalogy* 87 (3):563-570.

Lair, H. 1985. Length of gestation in the red squirrel, *Tamiasciurus hudsonicus*. *Journal of Mammalogy* 66 (4):809-810.

Lamb, T., T. R. Jones, and P. J. Wettstein. 1997. Evolutionary genetics and phylogeography of tassel-eared squirrels (*Sciurus aberti*). *Journal of Mammalogy* 78 (1):117-133.

Lambert, F. 1990. Some notes on fig-eating by arboreal mammals in Malaysia. *Primates* 31 (3):453-458.

Larsen, K. W., and S. Boutin. 1994. Movements, survival, and settlement of red squirrel (*Tamiasciurus hudsonicus*) offspring. *Ecology* 75 (1):214-223.

Laudenslayer, W. F., Jr., K. B. Buckingham, and T. A. Rado. 1995. Mammals of the California Desert. In *The California Desert: An Introduction to Natural Resources and Man's Impact*, ed. J. Latting and P. G. Rowlands. Riverside, CA: June Latting Books.

Lavers, A. J., S. D. Petersen, D. T. Stewart, and T. B. Herman. 2006. Delineating the range of a disjunct population of southern flying squirrels (*Glaucomys volans*). *American Midland Naturalist* 155 (1):188-196.

Laves, K. S., and S. C. Loeb. 1999. Effects of southern flying squirrels *Glaucomys volans* on red-cockaded woodpecker *Picoides borealis*. *Animal Conservation* 2:295-303.

Lay, D. M. 1967. A study of the mammals of Iran resulting from the Street Expedition of 1962-63. *Fieldiana Zoology* 54:1-282.

Layne, J. N. 1954. The biology of the red squirrel, *Tamiasciurus hudsonicus loquax* (Bangs), in central New York. *Ecological Monographs* 24 (3):227-268.

Lee, P. F. 1998. Body size comparison of two giant flying squirrel species in Taiwan. *Acta Zoologica Taiwanica* 9 (1):51-57.

Lee, P. F., and C. Y. Liao. 1998. Species richness patterns and research trends of flying squirrels. *Journal of Taiwan Museum* 51:1-20.

Lee, P. F., Y. S. Lin, and D. R. Progulske. 1993. Reproductive biology of the red-giant flying squirrel, *Petaurista petaurista*, in Taiwan. *Journal of Mammalogy* 74:982-989.

Lee, P. F., D. R. Progulske, and Y. S. Lin. 1986. Ecological studies on two sympatric *Petaurista* species in Taiwan. *Bulletin of the Institute of Zoology, Academia Sinica* 25:113-124.

———. 1993. Spotlight counts of giant flying squirrels (*Petaurista petaurista* and *P. alborufus*) in Taiwan. *Bulletin of the Institute of Zoology, Academia Sinica* 32 (1):54-61.

Leirs, H., J. N. Mills, J. W. Krebs, J. E. Childs, D. Akaibe, N. Woollen, G. Ludwig, C. J. Peters, and T. G. Ksiazek. 1999. Search for the *Ebola* virus reservoir in Kikwit, Democratic Republic of the Congo: Reflections on a vertebrate collection. *Journal of Infectious Diseases* 179:155-163.

Lekagul, B., and J. A. McNeely. 1977. *Mammals of Thailand*. Bangkok: Association for the Conservation of Wildlife.

Lemke, T. O., A. Cadena, R. H. Pine, and J. Hernandez-Camacho. 1982. Notes on opossums, bats, and rodents new to the fauna of Colombia. *Mammalia* 46 (2):225-234.

León, P., and S. Montiel. 2008. Wild meat use and traditional hunting practices in a rural Mayan community of the Yucatán Peninsula, Mexico. *Human Ecology* 36 (2):249-257.

Leonard, K. M., B. Pasch, and J. L. Koprowski. 2009. *Sciurus pucheranii*. *Mammalian Species* 841:1-4.

Leopold, A. S. 1959. *Wildlife of Mexico*. Berkeley: University of California Press.

Letcher, A. J., A. Purvis, S. Nee, and P. H. Harvey. 1994. Patterns of overlap in the geographic ranges of Palearctic and British mammals. *Journal of Animal Ecology* 63 (4):871-879.

Leung, M. C., and K. M. Cheng. 1997. The distribution of the Cascade mantled ground squirrel, *Spermophilus saturatus*, in British Columbia. *Canadian Field Naturalist* 111 (3):365-375.

Levenson, H. 1990. Sexual size dimorphism in chipmunks. *Journal of Mammalogy* 71:161-170.

Levenson, H., R. S. Hoffmann, C. F. Nadler, L. Deutsch, and S. D. Freeman. 1985. Systematics of the Holarctic chipmunks (*Tamias*). *Journal of Mammalogy* 66 (2):219-242.

Lewis, R. E. 1971. A new genus and species of flea from the lesser giant flying squirrel in Nepal (Siphonaptera: Ceratophyllidae). *Journal of Parasitology* 57 (6):1354-1361.

Lewis, T. L., and O. J. Rongstad. 1992. The

distribution of Franklin's ground squirrel in Wisconsin and Illinois. *Wisconsin Academy of Sciences, Arts and Letters* 80:57-62.

Li, J., and Y. Wang. 1992. Taxonomic study on subspecies of *Dremomys lokriah* (Sciuridae, Rodentia) from southwest China—Note with a new subspecies. *Zoological Research* 13:235-244.

Li, Q., R. Sun, C. Huang, Z. Wang, X. Liu, J. Hou, J. Liu, L. Cai, N. Li, S. Zhang, and Y. Wang. 2001. Cold adaptive thermogenesis in small mammals from different geographical zones of China. *Comparative Biochemistry and Physiology—Part A: Molecular & Integrative Physiology* 129 (4):949-961.

Li, T., P. C. M. O'Brien, L. Biltueva, B. Fu, J. Wang, W. Nie, M. A. Ferguson-Smith, A. S. Graphodatsky, and F. Yang. 2004. Evolution of genome organizations of squirrels (Sciuridae) revealed by cross-species chromosome painting. *Chromosome Research* 12 (4):317-335.

Liat, L. B., Y. L. Fong, and M. Krishnasamy. 1977. *Capillaria hepatica* infection of wild rodents in peninsular Malaysia. *Southeast Asian Journal of Tropical Medicine and Public Health* 8 (3):354-8.

Lidicker, W. Z., Jr. 1999. Responses of mammals to habitat edges: An overview. *Landscape Ecology* 14 (4):333-343.

Lim, B. L., K. K. P. Lim, and H. S. Yong. 1999. The terrestrial mammals of Pulau Tioman, peninsular Malaysia, with a catalogue of specimens at the Raffles Museum, National University of Singapore. *Raffles Bulletin of Zoology* 1999 (Suppl. 6):101-123.

Lin, Y. S., D. R. Progulske, P. F. Lee, and Y. T. Day. 1985. Bibliography of Petauristinae (Rodentia, Sciuridae). *Journal of Taiwan Museum* 38 (2):49-57.

Linares, O. J. 1998. *Mamíferos de Venezuela.* Caracas: Sociedad Conservacionista Audubon de Venezuela.

Linders, M. J., S. D. West, and M. H. Vander Haegen. 2004. Seasonal variability in the use of space by western gray squirrels in southcentral Washington. *Journal of Mammalogy* 85:511-516.

Linn, I., and G. Key. 1996. Use of space by the African striped ground squirrel *Xerus erythropus. Mammal Review* 26 (1):9-26.

Linnaeus, C. 1758. *Caroli Linnaei . . . Systema naturae.* Holmiae [Stockholm]: impensis direct. Laurentii Salvii.

Linzey, A. V., M. K. Clausen, and G. Hammerson. 2008. *Tamias canipes. IUCN Red List of Threatened Species,* version 2009.2.

Linzey, A. V., and G. Hammerson. 2008a. *Glaucomys sabrinus. IUCN Red List of Threatened Species,* version 2009.2.

———. 2008b. *Spermophilus armatus. IUCN Red List of Threatened Species,* version 2009.2.

———. 2008c. *Spermophilus beldingi. IUCN Red List of Threatened Species,* version 2009.2.

———. 2008d. *Spermophilus lateralis. IUCN Red List of Threatened Species,* version 2009.2.

———. 2008e. *Tamias alpinus. IUCN Red List of Threatened Species,* 2008 [accessed 28 Oct. 2008].

———. 2008f. *Tamias amoenus. IUCN Red List of Threatened Species,* version 2010.4.

———. 2008g. *Tamias minimus. IUCN Red List of Threatened Species,* version 2009.2.

———. 2008h. *Tamias ochrogenys. IUCN Red List of Threatened Species,* version 2009.2.

———. 2008i. *Tamias panamintinus. IUCN Red List of Threatened Species,* version 2009.2.

———. 2008j. *Tamias quadrimaculatus. IUCN Red List of Threatened Species,* version 2009.2.

———. 2008k. *Tamias ruficaudus. IUCN Red List of Threatened Species,* version 2009.2.

———. 2008l. *Tamias senex. IUCN Red List of Threatened Species,* version 2009.2.

———. 2008m. *Tamias siskiyou. IUCN Red List of Threatened Species,* version 2009.2.

———. 2008n. *Tamias sonomae. IUCN Red List of Threatened Species,* version 2009.2.

———. 2008o. *Tamias speciosus. IUCN Red List of Threatened Species,* version 2009.2.

———. 2008p. *Tamias striatus. IUCN Red List of Threatened Species,* version 2009.2.

———. 2008q. *Tamias townsendii. IUCN Red List of Threatened Species,* version 2009.2.

Linzey, A. V., R. Timm, S. T. Álvarez-Castañeda, I. Castro-Arellano, and T. Lacher. 2008a. *Spermophilus beecheyi. IUCN Red List of Threatened Species,* version 2009.2.

———. 2008b. *Tamias merriami. IUCN Red List of Threatened Species,* version 2009.2.

———. 2008c. *Tamias obscurus. IUCN Red List of Threatened Species,* version 2009.2.

Linzey, D. W. 1998. *The Mammals of Virginia.* Blacksburg, VA: McDonald & Woodward.

Linzey, D. W., and A. V. Linzey. 1979. Growth and development of the southern flying squirrel (*Glaucomys volans volans*). *Journal of Mammalogy* 60 (3):615-620.

Livoreil, B., and C. Baudoin. 1996. Differences in food hoarding behavior in two species of ground squirrels, *Spermophilus tridecemlineatus* and *Spermophilus spilosoma. Ethology Ecology & Evolution* 8:199-205.

Livoreil, B., P. Gouat, and C. Baudoin. 1993. A comparative study of social behavior of two sympatric ground squirrels (*Spermophilus spilosoma, S. mexicanus*). *Ethology* 93:236-246.

Loeb, S. C., and F. H. Tainter. 2000. Habitat associations of hypogeous fungi in the Southern Appalachians: Implications for the endangered northern flying squirrel (*Glaucomys sabrinus coloratus*). *American Midland Naturalist* 144 (2):286.

Loew, S. S. 1999. Sex-biased dispersal in eastern chipmunks, *Tamias striatus. Evolutionary Ecology* 13 (6):557-577.

Logan, T. M., M. L. Wilson, and J. P. Cornet. 1993. Association of ticks (Acari: Ixodoidea) with rodent burrows in northern Senegal. *Journal of Medical Entomology* 30 (4):799-801.

Lomolino, M. V., and D. R. Perault. 2001. Island biogeography and landscape ecology of mammals inhabiting fragmented, temperate rain forests. *Global Ecology and Biogeography* 10 (2):113-132.

Lomolino, M. V., and G. A. Smith. 2003. Prairie dog towns as islands: Applications of island biogeography and landscape ecology for conserving nonvolant terrestrial vertebrates. *Global Ecology and Biogeography* 12 (4):275-286.

López-Darias, M. 2006. Estudio preliminar del estado actual de la población introducida de ardilla moruna (*Atlantoxerus getulus*) en Fuerteventura. Internal report for Obra Social de La Caja de Canarias, Cabildo de Fuerteventura, and Estación Biológica de Donana, CSIC [Consejo Superior de Investigaciónes Científicas].

López-Darias, M., and J. M. Lobo. 2008. Factors affecting invasive species abundance: The Barbary ground squirrel on Fuerteventura Island, Spain. *Zoological Studies* 47 (3):268-281.

López-Darias, M., and M. Nogales. 2008. Effects of the invasive Barbary ground squirrel (*Atlantoxerus getulus*) on seed dispersal systems of insular xeric environments. *Journal of Arid Environments* 72 (6):926.

Lorenzo-Morales, J., M. López-Darias, E. Martínez-Carretero, and B. Valladares. 2007. Isolation of potentially pathogenic strains of *Acanthameoba* in wild squirrels from the Canary Islands and Morocco. *Experimental Parasitology* 117:74-79.

Lowrey, C., A. V. Linzey, and G. Hammerson. 2008. *Tamias palmeri. IUCN Red List of Threatened Species,* version 2009.2.

Lu, J. Q., and Z. B. Zhang. 2004. Effects of habitat and season on removal and hoarding of seeds of wild apricot (*Prunus armeniaca*) by small rodents. *Acta Oecologica—International Journal of Ecology* 26 (3):247-254.

———. 2008. Differentiation in seed hoarding among three sympatric rodent species in a warm temperate forest. *Integrative Zoology* 3 (2):134-142.

Luna, L. D., and T. A. Baird. 2004. Influence of density on the spatial behavior of female thirteen-lined ground squirrels, *Spermophilus tridecemlineatus*. *Southwestern Naturalist* 49:350-358.

Lundahl, S. L., and A. Olsson. 2002. Small mammal communities in the lowland rainforest of Krau Wildlife Reserve, peninsular Malaysia, with reference to human disturbance. *Malayan Nature Journal* 56:199-215.

Lunde, D. P., G. G. Musser, and N. T. Son. 2003. A survey of small mammals from Mt. Tay Con Linh II, Vietnam, with the description of a new species of *Chodsigoa* (Insectivora: Soricidae). *Mammal Study* 28:31-46.

Luo, J., and B. J. Fox. 1990. Life-table comparisons between two ground squirrels. *Journal of Mammalogy* 71 (3):364-370.

Lurz, P. W. W., J. Gurnell, and L. Magris. 2005. *Sciurus vulgaris*. *Mammalian Species* 769:1-10.

Lynch, C. D. 1983. *The Mammals of the Orange Free State*. Bloemfontein, Republic of South Africa: National Museum.

Macdonald, D. W., and P. Barrett. 1993. *Mammals of Europe*. Princeton, NJ: Princeton University Press.

MacDonald, J. A., and K. B. Storey. 2002. Purification and characterization of fructose bisphosphate aldolase from the ground squirrel, *Spermophilus lateralis*: Enzyme role in mammalian hibernation. *Archives of Biochemistry and Biophysics* 408 (2):279-285.

———. 2004. Temperature and phosphate effects on allosteric phenomena of phosphofructokinase from a hibernating ground squirrel (*Spermophilus lateralis*). *Federation of European Biochemical Societies Journal* 272 (2005):120-128.

Machado, A., and F. Domínguez. 1982. Estudio sobre la presencia de la ardilla moruna (*Atlantoxerus getulus* L.) en la isla de Fuerteventura: Su introducción, su biología y su impacto en el medio. Internal report for Instituto Nacional para la Conservación de la Naturaleza, Ministerio de Agricultura, Pesca y Alimentación.

Mackenzie, J. 1929. Notes on Berdmore's squirrel (*Menetes berdmorei*). *Journal of the Bombay Natural History Society* 33:980-981.

Mackerras, M. J. 1962. Filarial parasites (Nematoda: Filarioidea) of Australian animals. *Australian Journal of Zoology* 10:400-457.

MacKinnon, K. 1996. *Ecology of Kalimantan*. Ecology of Indonesia Series, Vol. 3. Hong Kong: Periplus Editions.

Madan, K. O., N. A. Slade, and F. S. Dobson. 2001. Effects of density reduction on Uinta ground squirrels: Analysis of life table response experiments. *Ecology* 82 (7):1921-1929.

Mahan, C. G., J. A. Bishop, M. A. Steele, G. Turner, and W. L. Meyers. 2010. Habitat characteristics and revised gap analysis for the northern flying squirrel (*Glaucomys sabrinus*), a state endangered species in Pennsylvania. *American Midland Naturalist* 164 (2):283-295.

Mahan, C. G., and R. H. Yahner. 1996. Effects of forest fragmentation on burrow-site selection by the eastern chipmunk (*Tamias striatus*). *American Midland Naturalist* 136 (2):352-357.

———. 1998. Lack of population response by eastern chipmunks (*Tamias striatus*) to forest fragmentation. *American Midland Naturalist* 140 (2):382-386.

———. 1999. Effects of forest fragmentation on behaviour patterns in the eastern chipmunk (*Tamias striatus*). *Canadian Journal of Zoology—Revue Canadienne de Zoologie* 77 (12):1991-1997.

Maher, C. R. 2009. Effects of relatedness on social interaction rates in a solitary marmot. *Animal Behaviour* 78:925-933.

Mahood, S., and T. Van Hung. 2008. *The Biodiversity of Bac Huong Hoa Nature Reserve, Quang Tri Province, Vietnam*. Conservation Report Number 35. Hanoi: BirdLife International Vietnam Programme, with financial support from the John D. and Catherine T. MacArthur Foundation.

Mandier, V., and P. Gouat. 1996. A laboratory study of social behaviour of pairs of females during the reproductive season in *Spermophilus spilosoma* and *Spermophilus mexicanus*. *Behavioural Processes* 37 (2-3):125-136.

Mann, C. S., E. Macchi, and G. Janeau. 1993. Alpine marmot (*Marmota marmota* L.). *Ibex Monograph* 1:17-30.

Manno, T. G. 2007. Why are Utah prairie dogs vigilant? *Journal of Mammalogy* 88:555-563.

———. 2008. Social networking in the Columbian ground squirrel, *Spermophilus columbianus*. *Animal Behaviour* 75 (4):1221-1228.

Manno, T. G., A. P. Nesterova, L. M. DeBarbieri, S. E. Kennedy, K. S. Wright, and F. S. Dobson. 2007. Why do male Columbian ground squirrels give a mating call? *Animal Behaviour* 74 (5):1319-1327.

Marinkelle, C. J., and R. E. Abdalla. 1978. The multiplication stages of *Trypanosoma* (*Herpetosoma*) *xeri* in the liver of the Sudanese ground squirrel *Xerus* (*Euxerus*) *erythropus*. *Journal of Wildlife Diseases* 14 (1):11-14.

Marmet, J., B. Pisanu, and J. L. Chapuis. 2009. Home range, range overlap, and site fidelity of introduced Siberian chipmunks in a suburban French forest. *European Journal of Wildlife Research* 55:497-504.

Marsh, A. C., G. Louw, and H. H. Berry. 1978. Aspects of renal physiology, nutrition, and thermoregulation in the ground squirrel *Xerus inauris*. *Madoqua* 2 (2):129-135.

Marsh, R. E. 1994. Current (1994) ground squirrel control practices in California. In *Proceedings of the Sixteenth Vertebrate Pest Conference, Santa Clara, California, February 28 and March 1-3, 1994*, ed. W. S. Halverson and A. C. Crabb. Davis: University of California, Davis.

Martin, J. G. A., and D. Réale. 2008. Temperament, risk assessment and habituation to novelty in eastern chipmunks, *Tamias striatus*. *Animal Behaviour* 75 (1):309-318.

Martin, J. M., and E. J. Heske. 2005. Juvenile dispersal of Franklin's ground squirrel (*Spermophilus franklinii*) from a prairie island. *American Midland Naturalist* 153:444-449.

Martin, J. M., E. J. Heske, and J. E. Hofmann. 2003. Franklin's ground squirrel (*Spermophilus franklinii*) in Illinois: A declining prairie mammal? *American Midland Naturalist* 150 (1):130.

Martinez, R. R., J. C. Pérez, E. E. Sánchez, and R. Campos. 1999. The antihemorrhagic factor of the Mexican ground squirrel, (*Spermophilus mexicanus*). *Toxicon* 37 (6):949-954.

Mateo, J. M. 1996. The development of alarm-call response behaviour in free-living juvenile Belding's ground squirrels. *Animal Behaviour* 52 (3):489-505.

———. 2002. Kin-recognition abilities and nepotism as a function of sociality. *Proceedings of the Royal Society of London B, Biological Sciences* 269:721-727.

———. 2003. Kin recognition in ground

squirrels and other rodents. *Journal of Mammalogy* 84 (4):1163-1181.

———. 2006a. Developmental and geographic variation in stress hormones in wild Belding's ground squirrels (*Spermophilus beldingi*). *Hormones and Behavior* 50:718-725.

———. 2006b. The nature and representation of individual recognition odours in Belding's ground squirrels. *Animal Behaviour* 71 (1):141-154.

Mateo, J. M., and W. G. Holmes. 1997. Development of alarm-call responses in Belding's ground squirrels: The role of dams. *Animal Behaviour* 54 (3):509-524.

———. 1999. Plasticity of alarm-call response development in Belding's ground squirrels (*Spermophilus beldingi*, Sciuridae). *Ethology* 105 (3):193-206.

Mateo, J. M., and R. E. Johnston. 2000. Retention of social recognition after hibernation in Belding's ground squirrels. *Animal Behaviour* 59:491-499.

Matrosova, V. A., I. A. Volodin, and E. V. Volodina. 2009. The short-term and long-term individuality in speckled ground squirrel alarm calls. *Journal of Mammalogy* 90:158-166.

Matsui, T., T. Fujino, J. Kajima, and M. Tsuji. 2000. Infectivity to experimental rodents of *Cryptosporidium parvum* oocysts from Siberian chipmunks (*Tamias sibiricus*) originated in the People's Republic of China. *Journal of Veterinary Medical Science* 62 (5):487-489.

McAllister, C. T., S. J. Upton, and B. D. Earle. 1991. *Eimeria callispermophili* and *E. morainensis* (Apicomplexa: Eimeriidae) from the Mexican ground squirrel, *Spermophilus mexicanus* (Rodentia: Sciuridae), in south central Texas, USA. *Transactions of the American Microscopical Society* 110 (1):71-74.

McAllister, C. T., S. J. Upton, J. V. Planz, and T. S. DeWalt. 1991. New host and locality records of *Coccidia* (Apicomplexa: Eimeriidae) from rodents in the southwestern and western United States. *Journal of Parasitology* 77 (6):1016-1019.

McCleery, R. A., R. R. Lopez, N. J. Silvy, and D. L. Gallant. 2008. Fox squirrel survival in urban and rural environments. *Journal of Wildlife Management* 72:133-137.

McCowan, B., and S. L. Hooper. 2002. Individual acoustic variation in Belding's ground squirrel alarm chirps in the High Sierra Nevada. *Journal of the Acoustical Society of America* 111 (3):1157-1160.

McCullough, D. A., R. K. Chesser, and R. D. Owen. 1987. Immunological systematics of prairie dogs. *Journal of Mammalogy* 68 (3):561-568.

McKeever, S. 1964. Food habits of the pine squirrel in northeastern California. *Journal of Wildlife Management* 28 (2):402-404.

McKenna, M. C. 1962. *Eupetaurus* and the living petauristine sciurids, *American Museum Novitates* 2104:1-38.

McLean, I. G. 1982. The association of female kin in the Arctic ground squirrel. *Behavioral Ecology and Sociobiology* 10 (2):91-99.

Medlin, E. C., and T. S. Risch. 2006. An experimental test of snakeskin use to deter nest predation. *Condor* 108 (4):963-965.

Medway, Lord [Galthorne Galthorne-Hardy]. 1966. Observations of the fauna of Pulau Tioman and Pulau Tulai, 2: The mammals. *Bulletin of the National Museum (Singapore)* 34:9-32.

———. 1969. *The Wild Mammals of Malaya and Offshore Islands, including Singapore*. London: Oxford University Press.

———. 1977. *Mammals of Borneo: Field Keys and an Annotated Checklist*. Monographs of the Malaysian Branch of the Royal Asiatic Society No. 7. Kuala Lumpur: Malaysian Branch of the Royal Asiatic Society.

Meijaard, E. 2003a. Mammals of south-east Asian islands and their late Pleistocene environments. *Journal of Biogeography* 30:1245-1257.

———b. Solving mammalian riddles : A reconstruction of the Tertiary and Quaternary distribution of mammals and their palaeoenvironments in island South-East Asia. Ph.D. diss., Australian National University.

Meijaard, E., and C. P. Groves. 2006. Geography of mammals and rivers in mainland southeast Asia. In *Primate Biogeography: Progress and Prospects*, ed. S. M. Lehman and J. G. Fleagle. New York: Springer.

Meijaard, E., A. C. Kitchener, and C. Smeenk. 2006. "New Bornean carnivore" is most likely a little-known flying squirrel. *Mammalian Review* 36 (4):318-324.

Meijaard, E., and D. Sheil. 2008. The persistence and conservation of Borneo's mammals in lowland rain forests managed for timber: Observations, overviews and opportunities. *Ecological Research* 23:21-34.

Mellado, M., and A. Olvera. 2008. Diets of prairie dogs (*Cynomys mexicanus*) co-existing with cattle or goats. *Mammalian Biology* 73 (1):33-39.

Mellink, E., and J. Contreras. 1993. Western gray squirrels in Baja California. *California Fish and Game* 79:169-170.

Men, X., X. Guo, W. Dong, and T. Qian. 2006. Population dynamics of *Dremomys pernyi* and *Callosciurus erythraeus* in protective and non-protective pine forests at different ages. *Zoological Research* 27 (1):29-33.

Mena-Valenzuela, P. 1998. Importancia económica de los mamíferos en tres etnias del Ecuador. In *Biología, sistemática y conservación de los mamíferos del Ecuador*, ed. S. D. Tirira. Quito: Museo de Zoologia, Centro de Biodiversidad y Ambiente, Pontifica Universidad Catolica del Ecuador.

Menkens, G. E., and M. S. Boyce. 1993. Comments on the use of time-specific and cohort life tables. *Ecology* 74 (7):2164-2168.

Mercer, J. M., and V. L. Roth. 2003. The effects of Cenozoic global change on squirrel phylogeny. *Science* 299:1568-1572.

Merritt, J. F., D. A. Zegers, and L. R. Rose. 2001. Seasonal thermogenesis of southern flying squirrels (*Glaucomys volans*). *Journal of Mammalogy* 82 (1):51-64.

Meyer, M. D., D. A. Kelt, and M. P. North. 2007a. Effects of burning and thinning on lodgepole chipmunks (*Neotamias speciosus*) in the Sierra Nevada, California. *Northwestern Naturalist* 88:61-72.

———. 2007b. Microhabitat associations of northern flying squirrels in burned and thinned forest stands of the Sierra Nevada. *American Midland Naturalist* 157 (1):202-211.

Meyer, M. D., M. P. North, and D. A. Kelt. 2005. Fungi in the diets of northern flying squirrels and lodgepole chipmunks in the Sierra Nevada. *Canadian Journal of Zoology* 83:1581-1589.

Miyao, T. 1972. Length of the large intestine of the Japanese giant flying squirrel. *Miscellaneous Notes on Mammals of Japan*, Vol. 1. Matsumoto, Japan: Mammalogical Society of Shinshu. [In Japanese.]

Michener, G. R., and J. W. Koeppl. 1985. *Spermophilus richardsonii*. *Mammalian Species* 243:1-8.

Michener, G. R., and I. G. McLean. 1996. Reproductive behaviour and operational sex ratio in Richardson's ground squirrels. *Animal Behaviour* 52:743-758.

Millán-Peña, N. 1998. Interacción social y dominancia entre dos especies de ardillas del desierto *Spermophilus spilosoma* y *Spermophilus mexicanus* de una zona árida del norte de Mexico. *Acta Zoológica Mexicana*, n.s., 73:75-87.

Miller, B., R. P. Reading, and S. Forrest. 1996. *Prairie Night: Black-Footed Ferrets and the Recovery of Endangered Species.* Washington, DC: Smithsonian Institution Press.

Miller, G. S., Jr. 1900. Mammals collected by Dr. W. L. Abbott on islands in the North China Sea. *Proceedings of the Washington Academy of Sciences* 2:203-246.

———. 1903. *Seventy New Malayan Mammals.* Washington, DC: Smithsonian Institution.

———. 1915. A new squirrel from northeastern China. *Proceedings of the Biological Society of Washington* 28:115-116.

———. 1942. Zoological results of the George Vanderbilt Sumatran Expedition, 1936-1939: Part V; Mammals collected by Frederick A. Ulmer, Jr., on Sumatra and Nias. *Proceedings of the Academy of Natural Sciences of Philadelphia* 94:107-165.

Miller, S. D., and J. F. Cully. 2001. Conservation of black-tailed prairie dogs (*Cynomys ludovicianus*). *Journal of Mammalogy* 82 (4):889-893.

Millesi, E., I. E. Hoffmann, and S. Huber. 2004. Reproductive strategies of male European sousliks (*Spermophilus citellus*) at high and low population density. *Lutra* 47 (2):75-84.

Millesi, E., I. E. Hoffmann, S. Steurer, M. Metwaly, and J. P. Dittami. 2002. Vernal changes in the behavioral and endocrine responses to GnRH application in male European ground squirrels. *Hormones and Behavior* 41 (1):51-58.

Millesi, E., S. Huber, J. Dittami, I. Hoffmann, and S. Daan. 1998. Parameters of mating effort and success in male European ground squirrels, *Spermophilus citellus. Ethology* 104 (4):298.

Millesi, E., S. Huber, L. G. Everts, and J. P. Dittami. 1999. Reproductive decisions in female European ground squirrels: Factors affecting reproductive output and maternal investment. *Ethology* 105:163-175.

Millesi, E., S. Huber, K. Pieta, M. Walzl, W. Arnold, and J. P. Dittami. 2000. Estrus and estrogen changes in mated and unmated free-living European ground squirrels. *Hormones and Behavior* 37 (3):190-197.

Millesi, E., A. M. Strijkstra, I. E. Hoffmann, J. P. Dittami, and S. Daan. 1999. Sex and age differences in mass, morphology, and annual cycle in European ground squirrels, *Spermophilus citellus. Journal of Mammalogy* 80 (1):218-231.

Millien-Parra, V., and J. J. Jaeger. 1999. Island biogeography of the Japanese terrestrial mammal assemblages: An example of a relict fauna. *Journal of Biogeography* 26:959-972.

Mishra, C., M. D. Madhusudan, and A. Datta. 2006. Mammals of the high altitudes of western Arunachal Pradesh, eastern Himalaya: An assessment of threats and conservation needs. *Oryx* 40 (1):1-7.

Mitchell, D. 2001. Spring and fall diet of the endangered West Virginia northern flying squirrel (*Glaucomys sabrinus fuscus*). *American Midland Naturalist* 146:439-443.

Mitchell, R. M. 1979. The sciurid rodents (Rodentia: Sciuridae) of Nepal. *Journal of Asian Ecology* 1:21-28.

Mitchell-Jones, A. J., G. Amori, W. Bogdanowicz, B. Krystufek, P. J. H Reijnders, F. Spitzenberger, M. Stubbe, J. B. M Thissen, V. Vohralik, and J. Zima. 1999. *The Atlas of European Mammals.* London: Academic Press.

Molur, S. 2008a. *Biswamoyopterus biswasi. IUCN Red List of Threatened Species*, version 2010.4.

———. 2008b. *Eoglaucomys fimbriatus. IUCN Red List of Threatened Species*, version 2009.2.

———. 2008c. *Petaurista magnificus. IUCN Red List of Threatened Species*, version 2009.2.

Molur, S., and P. O. Nameer. 2008. *Funambulus tristriatus. IUCN Red List of Threatened Species*, version 2010.4.

Molur, S., C. Srinivasulu, B. Srinivasulu, S. Walker, P. O. Nameer, and L. Ravikumar. 2005. *Status of South Asian Non-Volant Small Mammals: Conservation Assessment and Management Plan (C.A.M.P.) Workshop Report.* Coimbatore, India: Zoo Outreach Organisation/CBSG [Conservation Breeding Specialists Group], South Asia.

Monge, J., and L. Hilje. 2006. Feeding habits of the squirrel *Sciurus variegatoides* (Rodentia: Sciuridae) in the Nicoya Peninsula, Costa Rica. *Revista de Biología Tropical* 54 (2):681-686.

Mönkkönen, M., P. Reunanen, A. Nikula, J. Inkeröinen, and J. Forsman. 1997. Landscape characteristics associated with the occurrence of the flying squirrel *Pteromys volans* in old-growth forests of northern Finland. *Ecography* 20:634-642.

Moore, J. C. 1958a. New genera of East Indian squirrels. *American Museum Novitates* 1914:1-5.

———. 1958b. New striped tree squirrels from Burma and Thailand. *American Museum Novitates* 1879:1-6.

———. 1959. Relationships among the living squirrels of the Sciurinae. *Bulletin of the American Museum of Natural History* 118:153-206.

Moore, J. C., and G. H. H. Tate. 1965. A study of the diurnal squirrels, Sciurinae, of the Indian and Indochinese subregions. *Fieldiana Zoology* 48:1-372.

Morales, M. 1985. Aspectos biológicos de la ardilla arbórea, *Sciurus alleni*, en el municipio de Santiago, Nuevo León, México. Tesis de licenciatura, Facultad de Ciencias Biológicas [M.S. thesis], Universidad Autónama de Nuevo León.

Mori, Y., and H. Takatori. 2006. Wildlife observation activity delays the time of departure from the nest in Japanese giant flying squirrels. *Japanese Journal of Conservation Ecology* 11 (1):76-79.

Moritz, C. 2007. Final report: A re-survey of the historic Grinnell-Storer vertebrate transect in Yosemite National Park, California. Report for the Sierra Nevada Network Inventory & Monitoring Program, Sequoia & Kings Canyon National Parks.

Moritz, C., J. L. Patton, C. J. Conroy, J. L. Parra, G. C. White, and S. R. Beissinger. 2008. Impact of a century of climate change on small-mammal communities in Yosemite National Park, USA. *Science* 322 (5899):261-264.

Morris, D. W. 1996. Coexistence of specialist and generalist rodents via habitat selection. *Ecology* 77 (8):2352-2364.

———. 2005. On the roles of time, space and habitat in a boreal small mammal assemblage: Predictably stochastic assembly. *Oikos* 109 (2):223-228.

Morse, D. H. 1970. Ecological aspects of some mixed-species foraging flocks of birds. *Ecological Monographs* 40 (1):119-168.

Müller, P., and I. Vesmanis. 1971. Eine neue Subspezies von *Sciurus ingrami* (Rodentia: Sciuridae) der Insel von São Sebastião (Staat São Paulo, Braislien). *Senckenbergiana Biologica* 52 (6):377-380.

Munro, D., D. W. Thomas, and M. M. Humphries. 2005. Torpor patterns of hibernating eastern chipmunks *Tamias striatus* vary in response to the size and fatty acid composition of food hoards. *Journal of Animal Ecology* 74 (4):692-700.

Murphy, D., and D. T. Phan. 2002. *Mammal*

Observations in Cat Tien National Park, Vietnam, 2002. Cat Tien National Park Conservation Project Technical Report No. 42, S. l.: Cat Tien National Park Conservation Project.

Musser, G. G. 1968. A systematic study of the Mexican and Guatemalan gray squirrel, *Sciurus aureogaster* F. Cuvier (Rodentia: Sciuridae). Miscellaneous Publication No. 137. Ann Arbor: University of Michigan, Museum of Zoology.

Musser, G. G., L. A. Durden, M. E. Holden, and J. E. Light. 2010. Systematic review of endemic Sulawesi squirrels (Rodentia, Sciuridae), with descriptions of new species of associated sucking lice (Insecta, Anoplura), and phylogenetic and zoogeographic assessments of sciurid lice. *Bulletin of the American Museum of Natural History* 339:1-260.

Musser, G. G., and L. R. Heaney. 1992. Philippine rodents: Definitions of *Tarsomys* and *Limnomys* plus a preliminary assessment of phylogenetic patterns among native Philippine murines (Murinae, Muridae). *Bulletin of the American Museum of Natural History* 211:1-138.

Muul, I., and L. B. Liat. 1971. New locality records for some mammals of West Malaysia. *Journal of Mammalogy* 52 (2):430-437.

———. 1974. Reproductive frequency in Malaysian flying squirrels, *Hylopetes* and *Pteromyscus*. *Journal of Mammalogy* 55 (2):393-400.

Muul, I., and B. L. Lim. 1978. Comparative morphology, food habits, and ecology of some Malaysian arboreal rodents. In *The Ecology of Arboreal Folivores*, ed. G. G. Montgomery. Washington, DC: Smithsonian Institution.

Muul, I., and K. Thonglongya. 1971. Taxonomic status of *Petinomys morrisi* (Carter) and its relationship to *Petinomys setosus* (Temminck and Schlegel). *Journal of Mammalogy* 52 (2):362-369.

Muul, I., L. F. Yap, and B. L. Lim. 1973. Ecological distribution of blood parasites in some arboreal rodents. *Southeast Asian Journal of Tropical Medicine and Public Health* 4 (3):377-81.

Nagorsen, D. W. 1987. *Marmota vancouverensis*. *Mammalian Species* 270:1-5.

Nagorsen, D. W., M. A. Fraker, and Royal British Columbia Museum. 1999. *Chipmunks (*Tamias*) of the Kootenay Region, British Columbia: Results of 1997 Field Studies*. Victoria: Royal British Columbia Museum.

Nakagawa, M., H. Miguchi, K. Sato, S. Shoko, and T. Nakashizuka. 2007. Population dynamics of arboreal and terrestrial small mammals in a tropical rainforest, Sarawak, Malaysia. *Raffles Bulletin of Zoology* 55 (2):389-395.

Nakano, C., M. Ando, S. Ikeda, S. Sukemori, and Y. Kurihara. 2004. Feeding management and food preference in Japanese giant flying squirrel, *Petaurista leucogenys*. *Journal of Agricultural Science, Tokyo Nogyo Daigaku* 49 (3):150-155.

Nameer, P. O., and S. Molur. 2008a. *Funambulus palmarum*. *IUCN Red List of Threatened Species*, version 2010.4.

———. 2008b. *Funambulus pennantii*. *IUCN Red List of Threatened Species*, version 2010.4.

Nameer, P. O., S. Molur, and S. Walker. 2001. Mammals of Western Ghats: A simplistic overview. *Zoos' Print Journal* 16 (11):629-639.

Nandini, R., and N. Parthasarathy. 2008. Food habits of the Indian giant flying squirrel (*Petaurista philippensis*) in a rain forest fragment, Western Ghats. *Journal of Mammalogy* 89 (6):1550-1556.

Nel, J. A. J. 1975. Aspects of the social ethology of some Kalahari rodents. *Zeitschrift für Tierpsychologie* 37:322-331.

Nelson, E. W. 1899. Revision of the squirrels of Mexico and Central America. *Proceedings of the Washington Academy of Sciences* 1:15-110.

Neuhaus, P. 2000a. Timing of hibernation and molt in female Columbian ground squirrels. *Journal of Mammalogy* 81 (2):571-577.

———. 2000b. Weight comparisons and litter size manipulation in Columbian ground squirrels (*Spermophilus columbianus*) show evidence of costs of reproduction. *Behavioral Ecology and Sociobiology* 48 (1):75-83.

———. 2003. Parasite removal and its impact on litter size and body condition in Columbian ground squirrels (*Spermophilus columbianus*). *Proceedings of the Royal Society of London B, Biological Sciences* 270:S213-S215.

———. 2006. Causes and consequences of sex-biased dispersal in Columbian ground squirrels, *Spermophilus columbianus*. *Behaviour* 143:1013-1031.

Neuhaus, P., R. Bennett, and A. Hubbs. 1999. Effects of a late snowstorm and rain on survival and reproductive success in Columbian ground squirrels (*Spermophilus columbianus*). *Canadian Journal of Zoology* 77 (6):879-884.

Neuhaus, P., D. R. Broussard, J. O. Murie, and F. S. Dobson. 2004. Age of primiparity and implications of early reproduction on life history in female Columbian ground squirrels. *Journal of Animal Ecology* 73:36-43.

Neuhaus, P., and N. Pelletier. 2001. Mortality in relation to season, age, sex, and reproduction in Columbian ground squirrels (*Spermophilus columbianus*). *Canadian Journal of Zoology* 79:465-470.

Niethammer, J., and F. Krapp, eds. 1978. *Rodentia 1 (Sciuridae, Castoridae, Gliridae, Muridae)*. Vol. 1 of *Handbuch der Säugetiere Europas*. Wiesbaden, Germany: Akademische Verlagsgesellschaft.

Nikol'skii, A. A. 2007a. A comparative analysis of the alarm call frequency in different age rodent groups. *Zoologichesky Zhurnal* 86 (4):499-504.

———. 2007b. The influence of amplitude modulation on the structure of call spectrum in marmots (*Marmota*, Rodentia, Sciuridae). *Biology Bulletin* 34:353-360.

———. 2009. Temperature conditions in burrows of the steppe marmot, *Marmota bobak* Müller (1776), in the hibernation period. *Russian Journal of Ecology* 40:529-536.

Nikol'skii, A. A., O. A. Ermakov, and S. V. Titov. 2007. Geographical variability of the little ground squirrel (*Spermophilus pygmaeus*): A bioacoustic analysis. *Zoologichesky Zhurnal* 86:1379-1388.

Nikol'skii, A. A., and G. A. Savchenko. 1999. Structure of family groups and space use by steppe marmots (*Marmota bobak*): Preliminary results. *Vestnik Zoologii* 33:67-72.

Nikol'skii, A. A., and A. Ulak. 2006. Key factors determining the ecological niche of the Himalayan marmot, *Marmota himalayana* Hodgson (1841). *Russian Journal of Ecology* 37:46-52.

Nikolsky, A. A., N. A. Formozov, V. N. Vasiljev, and G. G. Bojeskorov. 1991. Geographical variation of the sound signal of black-capped marmot, *Marmota camtschatica* (Rodentia, Sciuridae). *Zoologichesky Zhurnal* 70:155-159.

Nitikman, L. Z. 1985. *Sciurus granatensis*. *Mammalian Species* 246:1-8.

Nogales, M., C. Nieves, J. C. Illera, D. P. Padilla, and A. Traveset. 2005. Effect of native and alien vertebrate frugivores on seed viability and germination patterns of *Rubia fruticosa* (Rubiaceae) in the eastern Canary Islands. *Functional Ecology* 19 (3):429-436.

Nogales, M., J. L. Rodríguez-Luengo, and P.

Marrero. 2006. Ecological effects and distribution of invasive non-native mammals on the Canary Islands. *Mammal Review* 36 (1):49–65.

Nor, S. M. 2001. Elevational diversity patterns of small mammals on Mount Kinabalu, Sabah, Malaysia. *Global Ecology and Biogeography* 10 (1):41–62.

Norhayati, A., A. A. Saiful, A. Shahrolnizah, B. M. Md-Zain, M. N. Shukor, H. Hazimin, and M. Nordin. 2004. Diversity and density of mammals in the peat swamp forests of the Langat Basin, Selangor, Malaysia. *Journal of Malaysian Applied Biology* 33 (2):7–17.

North, M., J. Trappe, and J. Franklin. 1997. Standing crop and animal consumption of fungal sporocarps in Pacific Northwest forests. *Ecology* 78 (5):1543–1554.

Nowak, R. M. 1999. *Walker's Mammals of the World*. Baltimore: Johns Hopkins University Press.

Nunes, S., T. R. Duniec, S. A. Schweppe, and K. E. Holekamp. 1999. Energetic and endocrine mediation of natal dispersal behavior in Belding's ground squirrels. *Hormones and Behavior* 35 (2):113–124.

Nunes, S., C. D. T. Ha, P. J. Garrett, E. M. Mueke, L. Smale, and K. E. Holekamp. 1998. Body fat and time of year interact to mediate dispersal behaviour in ground squirrels. *Animal Behaviour* 55:605–614.

Nunes, S., and K. E. Holekamp. 1996. Mass and fat influence the timing of natal dispersal in Belding's ground squirrels. *Journal of Mammalogy* 77 (3):807–817.

Nunes, S., E. M. Muecke, J. A. Anthony, and A. S. Batterbee. 1999. Endocrine and energetic mediation of play behavior in free-living Belding's ground squirrels. *Hormones and Behavior* 36 (2):153–165.

Nunes, S., E. M. Muecke, and K. Holekamp. 2002. Seasonal effects of food provisioning on body fat, insulin, and corticosterone in free-living juvenile Belding's ground squirrels (*Spermophilus beldingi*). *Canadian Journal of Zoology* 80:366–371.

Nunes, S., E. M. Muecke, L. T. Lancaster, N. A. Miller, M. A. Mueller, J. Muelhaus, and L. Castro. 2004. Functions and consequences of play behaviour in juvenile Belding's ground squirrels. *Animal Behaviour* 68 (1):27–37.

Odom, R. H., W. M. Ford, J. W. Edwards, C. W. Stihler, and J. M. Menzel. 2001. Developing a habitat model for the endangered Virginia northern flying squirrel (*Glaucomys sabrinus fuscus*) in the Allegheny Mountains of West Virginia. *Biological Conservation* 99:245–252.

O'Farrell, M. J., and W. A. Clark. 1984. Notes on the white-tailed antelope squirrel, *Ammospermophilus leucurus*, and the pinyon mouse, *Peromyscus truei*, in north central Nevada. *Great Basin Naturalist* 44:428–430.

Ognev, S. I. 1947. *Gryzuny*. Vol. 5 of *Zveri Vostochnoĭ Evropy i Severnoĭ Azii [The mammals of eastern Europe and northern Asia]* [alternate title: *The mammals of Russia (USSR) and adjacent countries*]. Moscow: Glavnoe upravlenie nauchnymi uchrezhdeniimi (Glavnauka), Gosudarstvennoe Izdatel'stvo.

———. 1963. *Rodents*. Vol. 5 of *Mammals of the U.S.S.R. and Adjacent Countries*. Trans. A. Birron and Z. S. Cole. Jerusalem: Israel Program for Scientific Translations.

O'Hare, J. R., J. D. Eisemann, and K. A. Fagerstone. 2006. Changes in taxonomic nomenclature and conservation status of ground squirrel species: Implications for pesticide labeling and use of zinc phosphide pesticide products. In *Proceedings of the Twenty-Second Vertebrate Pest Conference, Berkeley, California, March 6–9, 2006*, ed. R. M. Timm and J. M. O'Brien. Davis: University of California, Davis.

Oli, M. K. 1994. Snow leopards and blue sheep in Nepal: Densities and predator-prey ratio. *Journal of Mammalogy* 75:998–1004.

Oli, M. K., N. A. Slade, and F. S. Dobson. 2001. Effect of density reduction on Uinta ground squirrels: Analysis of life table response experiments. *Ecology* 82 (7):1921–1929.

Ong, P., B. Tabaranza, G. Rossell-Ambal, and D. Balete. 2008. *Hylopetes nigripes*. *IUCN Red List of Threatened Species*, 2008 [accessed 28 Oct. 2008].

Ormond, C. J., S. Orgeig, C. B. Daniels, and W. K. Milsom. 2003. Thermal acclimation of surfactant secretion and its regulation by adrenergic and cholinergic agonists in type II cells isolated from warm-active and torpid golden-mantled ground squirrels, *Spermophilus lateralis*. *Journal of Experimental Biology* 206:3031–3041.

Ortega, J. C. 1987. Den site selection by rock squirrel (*Spermophilus variegatus*) in southeastern Arizona. *Journal of Mammalogy* 68:792–798.

———. 1990a. Home-range size of adult rock squirrels (*Spermophilus variegatus*) in southeastern Arizona. *Journal of Mammalogy* 71 (2):171–176.

———. 1990b. Reproductive-biology of the rock squirrel (*Spermophilus variegatus*) in southeastern Arizona. *Journal of Mammalogy* 71 (3):448–457.

Osgood, W. H. 1932. Mammals of the Kelley-Roosevelts and Delacour Asiatic expeditions. *Field Museum of Natural History Publication* 312 [= *Zoological Series* 18 (10)]:193–339.

O'Shea, T. J. 1976. Home range, social behavior, and dominance relationships in the African unstriped ground squirrel, *Xerus rutilus*. *Journal of Mammalogy* 57 (3):450–460.

———. 1991. *Xerus rutilus*. *Mammalian Species* 370:1–5.

Oshida, T. 2006. Is there an evolutionary relict of the Japanese giant flying squirrel *Petaurista leucogenys* on the Asian continent? *Mammal Study* 31:69–72.

Oshida, T., A. Abramov, H. Yanagawa, and R. Masuda. 2005. Phylogeography of the Russian flying squirrel (*Pteromys volans*): Implication of refugia theory in arboreal small mammal of Eurasia. *Molecular Ecology* 14 (4):1191–1196.

Oshida, T., N. Hachiya, M. C. Yoshida, and N. Ohtaishi. 2000. Comparative anatomical note on the origin of the long accessory styliform cartilage of the Japanese giant flying squirrel, *Petaurista leucogenys*. *Mammal Study* 25:35–39.

Oshida, T., H. Hiraga, T. Nojima, and M. C. Yoshida. 2000. Anatomical and histological notes on the origin of the long accessory styliform cartilage of the Russian flying squirrel, *Pteromys volans orii*. *Mammal Study* 25:41–48.

Oshida, T., K. Ikeda, K. Yamada, and R. Masuda. 2001. Phylogeography of the Japanese giant flying squirrel, *Petaurista leucogenys*, based on mitochondrial DNA control region sequences. *Zoological Science* 18:107–114.

Oshida, T., J. K. Lee, L. K. Lin, and Y. J. Chen. 2006. Phylogeography of Pallas's squirrel in Taiwan: Geographical isolation in an arboreal small mammal. *Journal of Mammalogy* 87 (2):247–254.

Oshida, T., L. K. Lin, R. Masuda, and M. C. Yoshida. 2000. Phylogenetic relationships among Asian species of *Petaurista* (Rodentia, Sciuridae), inferred from mitochondrial cytochrome *b* gene sequences. *Zoological Science* 17:123–128.

Oshida, T., L. K. Lin, H. Yanagawa, T. Kawamichi, M. Kawamichi, and V. Cheng. 2002. Banded karyotypes of the hairy-footed flying squirrel *Belomys* (*Trogop-*

terus) pearsonii (Mammalia, Rodentia) from Taiwan. *Caryologia* 55 (3):207-211.

Oshida, T., and Y. Obara. 1993. C-band variation in the chromosomes of the Japanese giant flying squirrel, *Petaurista leucogenys*. *Journal of the Mammalogical Society of Japan* 18 (2):61-67.

Oshida, T., H. Satoh, and Y. Obara. 1991. A preliminary note on the karyotypes of giant flying squirrels *Petaurista alborufus* and *P. petaurista*. *Journal of the Mammalogical Society of Japan* 16 (2):59-69.

Oshida, T., C. M. Shafique, S. Barkati, M. Yasuda, N. A. Hussein, H. Endo, H. Yanagawa, and R. Masuda. 2004. Phylogenetic position of the small Kashmir flying squirrel, *Hylopetes fimbriatus* (= *Eoglaucomys fimbriatus*), in the subfamily Pteromyinae. *Canadian Journal of Zoology* 82:1336-1342.

Oshida, T., and M. C. Yoshida. 1994. Banded karyotype of Asiatic chipmunk, *Tamias sibiricus lineatus* Siebold. *CIS—Chromosome Information Service* 57:27-28.

———. 1998. A note on the chromosome of the white-bellied flying squirrel *Petinomys setosus* (Rodentia, Sciuridae). *Chromosome Science* 2 (2):119-121.

———. 1999. Chromosomal localization of nucleolus organizer regions in eight Asian squirrel species. *Chromosome Science* 3:55-58.

Ostroff, A., and E. J. Finck. 2003. *Spermophilus franklinii*. *Mammalian Species* 724:1-5.

Owings, D. H., R. G. Coss, D. McKernon, M. P. Rowe, and P. C. Arrowood. 2001. Snake-directed antipredator behavior of rock squirrels (*Spermophilus variegatus*): Population differences and snake-species discrimination. *Behaviour* 138:575-595.

Özkurt, S. O., M. Sozen, N. Yigit, I. Kandemir, R. Colak, M. M. Gharkheloo, and E. Colak. 2007. Taxonomic status of the genus *Spermophilus* (Mammalia: Rodentia) in Turkey and Iran with description of a new species. *Zootaxa* 1529:1-15.

Pacheco, V. 2002. Mamíferos del Perú. In *Diversidad y conservación de los mamíferos Neotropicales*, ed. G. Ceballos and J. A. Simonetti. Mexico, DF: CONABIO-UNAM [Comisión Nacional para el Conocimiento y Uso de la Biodiversidad and Universidad Nacional Autónoma de México].

Pacheco, V., R. Cadenillas, E. Salas, C. Tello, and H. Zeballos. 2009. Diversidad y endemismo de los mamíferos del Perú. *Revista Peruana de Biología* 16:5-32.

Pacheco, V., B. D. Patterson, J. L. Patton, L. H. Emmons, S. Solari, and C. F. Ascorra. 1993. List of mammal species known to occur in Manu Biosphere Reserve, Peru. *Publicaciones del Museo de Historia Natural Universidad Nacional Mayor de San Marcos* 44 (A):1-12.

Pacheco, V., S. Solari, E. Vivar, and P. Hocking. 1994. La riqueza biológica del Parque Nacional Yanachaga-Chemillen. *Magistri et Doctores* 7:3-6.

Painter, J. N., V. Selonen, and I. K. Hanski. 2004. Microsatellite loci for the Siberian flying squirrel, *Pteromys volans*. *Molecular Ecology Notes* 4 (1):119-121.

Pasch, B. S., and J. L. Koprowski. 2006a. Annual cycles in body mass and reproduction of Chiricahua fox squirrels (*Sciurus nayaritensis chiricahuae*). *Southwestern Naturalist* 51:531-535.

———. 2006b. Sex differences in space use of Chiricahua fox squirrels. *Journal of Mammalogy* 87:380-386.

Paschoal, M., and M. Galetti. 1995. Seasonal food use by the neotropical squirrel *Sciurus ingrami* in southeastern Brazil. *Biotropica* 27:268-273.

Patrick, M. J. 1991. Distribution of enteric helminths in *Glaucomys volans* L. (Sciuridae): A test for competition. *Ecology* 72 (2):755-758.

Patterson, B. D. 1984. Geographic variation and taxonomy of Colorado and Hopi chipmunks (genus *Eutamias*). *Journal of Mammalogy* 65 (3):442-456.

Patterson, B. D., and L. R. Heaney. 1987. Preliminary analysis of geographic variation in red-tailed chipmunks (*Eutamias ruficaudus*). *Journal of Mammalogy* 68 (4):782-791.

Patton, J. L. 1984. Systematic status of the large squirrels (subgenus *Urosciurus*) of the western Amazon basin. *Studies on Neotropical Fauna and Environment* 19:53-72.

Patton, J. L., M. N. F. Da Silva, and J. R. Malcolm. 2000. Mammals of the Rio Juruá and the evolutionary and ecological diversification of Amazonia. *Bulletin of the American Museum of Natural History* 244:1-306.

Pavlinov, I. I., S. V. Kruskop, A. A. Varshavskiy, and A. V. Borisenko. 2002. *Nazemnye zveri Rossii: spravochnik-opredelitel*. Moscow: Izdatel'stvo KMK.

Payne, J., and C. M. Francis. 1985. *A Field Guide to the Mammals of Borneo*. Kuala Lumpur, Malaysia: Sabah Society.

Payne, J. B. 1980. Competitors. In *Malayan Forest Primates: Ten Years' Study in Tropical Rain Forest*, ed. D. J. Chivers. New York: Plenum Press.

Payne, J. L., D. R. Young, and J. F. Pagels. 1989. Plant community characteristics associated with the endangered northern flying squirrel, *Glaucomys sabrinus*, in the southern Appalachians. *American Midland Naturalist* 121 (2):285-292.

Peacock, M. M., and S. H. Jenkins. 1988. Development of food preferences: Social learning by Belding's ground squirrels (*Spermophilus beldingi*). *Behavioral Ecology and Sociobiology* 22 (6):393-399.

Pearson, E., Y. K. Ortega, and L. F. Ruggiero. 2003. Trap-induced mass declines in small mammals: Mass as a population index. *Journal of Wildlife Management* 67 (4):684-691.

Peralvo, M., R. Sierra, K. R. Young, and C. Ulloa-Ulloa. 2007. Identification of biodiversity conservation priorities using predictive modeling: An application for the equatorial Pacific region of South America. *Biodiversity and Conservation* 16:2649-2675.

Perault, D. R., and M. V. Lomolino. 2000. Corridors and mammal community structure across a fragmented, old-growth forest landscape. *Ecological Monographs* 70 (3):401-422.

Peres, C. A., and C. Baider. 1997. Seed dispersal, spatial distribution, and population structure of Brazil nut trees (*Bertholletia excelsa*) in southeastern Amazonia. *Journal of Tropical Ecology* 13:595-616.

Perez-Orella, C., and A. I. Schulte-Hostedde. 2005. Effects of sex and body size on ectoparasite loads in the northern flying squirrel (*Glaucomys sabrinus*). *Canadian Journal of Zoology* 83:1381-1385.

Pergams, O., D. Nyberg, and G. Hammerson. 2008. *Spermophilus franklinii*. *IUCN Red List of Threatened Species*, version 2009.2.

Perz, J. F., and S. M. Le Blancq. 2001. *Cryptosporidium parvum* infection involving novel genotypes in wildlife from lower New York State. *Applied and Environmental Microbiology* 67 (3):1154-1162.

Peterson, A. T., J. Soberon, and V. Sanchez-Cordero. 1999. Conservatism of ecological niches in evolutionary time. *Science* 285 (5431):1265.

Petter, F., and M. C. Saint-Girons. 1965. Les rongeurs du Maroc. *Travaux de l'Institut Scientifique Chérifien, Série Zoologie* 31:20-31.

Phillips, W. W. A. 1980. *Manual of the Mammals of Sri Lanka*, Part 1. [Colombo]: Wildlife and Nature Protection Society of Sri Lanka.

Phuong, D. H., N. Q. Truong, N. T. Son, and N. V. Khoi. 2006. *A Photographic*

Guide to Mammals, Reptiles, and Amphibians of Phu Quoc Island, Kien Giang Province, Vietnam. Ho Chi Minh City: Ho Chi Minh City General Publishing House, Wildlife at Risk, and Phu Quoc National Park.

Piaggio, A. J., and G. S. Spicer. 2000. Molecular phylogeny of the chipmunk genus *Tamias* based on the mitochondrial cytochrome oxidase subunit II gene. *Journal of Mammalian Evolution* 7 (3):147-166.

———. 2001. Molecular phylogeny of the chipmunks inferred from mitochondrial cytochrome *b* and cytochrome oxidase II gene sequences. *Molecular Phylogenetics and Evolution* 20 (3):335-350.

Pimentel, D. S., and M. Tabarelli. 2004. Seed dispersal of the palm *Attalea oleifera* in a remnant of the Brazilian Atlantic forest. *Biotropica* 36 (1):74-84.

Pine, R. H. 1971. A review of the long-whiskered rice rat, *Oryzomys bombycinus* Goldman. *Journal of Mammalogy* 52:590-596.

Pisanu, B., L. Lebailleux, and J. L. Chapuis. 2009. Why do Siberian chipmunks (*Tamias sibiricus*, Sciuridae) introduced to French forests acquire so few intestinal parasites? *Parasitological Research* 104:709-714.

Pizzimenti, J. J., and G. D. Collier. 1978. *Cynomys parvidens. Mammalian Species* 52:1-5.

Pizzimenti, J. J., and R. S. Hoffmann. 1973. *Cynomys gunnisoni. Mammalian Species* 25:1-4.

Place, N. J. 2000. Effects of experimentally elevated testosterone on plasma glucocorticoids, body mass, and recapture rates in yellow-pine chipmunks, *Tamias amoenus. Journal of Experimental Zoology* 287:378-383.

Platt, K. B., B. J. Tucker, P. G. Halbur, S. Tiawsirisup, B. J. Blitvich, F. G. Fabiosa, L. C. Bartholomay, and W. A. Rowley. 2007. West Nile virus viremia in eastern chipmunks (*Tamias striatus*) sufficient for infecting different mosquitoes. *Emerging Infectious Diseases* 13:6.

Polyakov, A. D. 2005. Forest-steppe marmot in Kuzbass. Paper read at the Fifth International Conference on Genus *Marmota*, International Marmot Network, Tashkent, Uzbekistan, August 31-September 2, 2005.

Polyakov, A. D., and V. V. Baranova. 2007. Biology of the forest-steppe marmot in conditions of ecological pressure. *Advances in Current Natural Sciences* 2:73-76.

Popov, M. V. 1977. *Opredelitel mlekopitiushchikh Iakutii [Classification of the mammals in Yakutia].* Novosibirsk, USSR: Izdatel'stvo Nauka, Sibirskoe otdelenie.

Popov, V. A. 1960. *Mlekopitaiushchie Volzhsko-Kamskogo kraia: Nasekomoiadnye, rukokrylye, gryzuny [Mammals of the Volga-Kama region: Insectivores, chiroptera, rodents].* Kazan, USSR: Akademii nauk SSSR Kazanskii Filial.

Pozo de la Tijera, C., and J. Escobedo Cabrera. 1999. Terrestrial mammals of the Sian Ka'an Biosphere, Quintana Roo, Mexico. *Revista de Biología Tropical* 47:251-262.

Prater, S. H. 1965. *The Book of Indian Animals.* Bombay: Bombay Natural History Society and Prince of Wales Museum of Western India.

Pulawa, L. K., and G. L. Florant. 2000. The effects of caloric restrictions on the body composition and hibernation of the golden-mantled ground squirrel (*Spermophilus lateralis*). *Physiological and Biochemical Zoology* 73 (5):538-546.

Purroy, F. H., and J. M. Varela. 2003. *Guía de los mamíferos de España: Peninsula, Baleares y Canarias.* Barcelona: Lynx Edicións.

Pyare, S., and W. S. Longland. 2001. Patterns of ectomycorrhizal-fungi consumption by small mammals in remnant old-growth forests of the Sierra Nevada. *Journal of Mammalogy* 82 (3):681-689.

Qinghua, P., W. Yingxiang, and Y. Kun. 2007. *A Field Guide to the Mammals of China.* Beijing: China Forestry Publishing House.

Qumsiyeh, M. B. 1996. *Mammals of the Holy Land.* Lubbock: Texas Tech University Press.

Rabin, L. A., R. G. Coss, and D. H. Owings. 2006. The effects of wind turbines on antipredator behavior in California ground squirrels (*Spermophilus beecheyi*). *Biological Conservation* 131 (3):410-420.

Rabor, D. S. 1986. *Birds and Mammals.* Vol. 11 of *Guide to Philippine Flora and Fauna.* Quezon City: Natural Resources Management Center, Ministry of Natural Resources, and University of the Philippines.

Rahm, U. 1970a. Ecology, zoogeography, and systematics of some African forest monkeys. In *Old World Monkeys: Evolution, Systematics, and Behavior; Proceedings of a Conference on Systematics of the Old World Monkeys, Burg Wartenstein, 1969,* ed. J. R. Napier and P. H. Napier. New York: Academic Press.

———. 1970b. Note sur la reproduction des sciuridés et muridés dans la forêt equatoriale au Congo. *Revue Suisse de Zoologie* 77:635-646.

Rahm, U., and A. Christiaensen. 1963. *Les mammifères de la région occidentale du Lac Kivu. Annales Musée Royal de l'Afrique Centrale, Sciences Zoologiques,* ser. 8, 118:1-83.

Rajamani, N., S. Molur, and P. O. Nameer. 2008a. *Funambulus sublineatus. IUCN Red List of Threatened Species,* version 2010.4.

———. 2008b. *Petinomys fuscocapillus. IUCN Red List of Threatened Species,* version 2009.1.

Ramachandran, K. K. 1989. Endangered grizzled giant squirrel habitat. *Journal of the Bombay Natural History Society* 86:94-95.

Ramos-Lara, N., and F. A. Cervantes. 2007. Nest-site selection by the Mexican red-bellied squirrel (*Sciurus aureogaster*) in Michoacán, Mexico. *Journal of Mammalogy* 88:495-501.

Ransome, D. B. 2001. Population ecology and resource limitation of northern flying squirrels and Douglas' squirrels. Ph.D. diss., University of British Columbia.

Ransome, D. B., and T. P. Sullivan. 1997. Food limitation and habitat preference of *Glaucomys sabrinus* and *Tamiasciurus hudsonicus. Journal of Mammalogy* 78 (2):538-549.

———. 2002. Short-term population dynamics of *Glaucomys sabrinus* and *Tamiasciurus douglasii* in commercially thinned and unthinned stands of coastal coniferous forest. *Canadian Journal of Forest Research* 32:2043-2050.

———. 2003. Population dynamics of *Glaucomys sabrinus* and *Tamiasciurus douglasii* in old-growth and second-growth stands of coastal coniferous forest. *Canadian Journal of Forest Research* 33 (4):587.

———. 2004. Effects of food and den-site supplementation on populations of *Glaucomys sabrinus* and *Tamiasciurus douglasii. Journal of Mammalogy* 85 (2):206-215.

Raphael, M. G. 1984. Late fall breeding of the northern flying squirrel, *Glaucomys sabrinus. Journal of Mammalogy* 65 (1):138-139.

Rasmussen, N. L., and R. W. Thorington Jr. 2008. Morphological differentiation among three species of flying squirrels (genus: *Hylopetes*) from Southeast Asia. *Journal of Mammalogy* 89:1296-1309.

Rayor, L. S. 1988. Social organization and

space-use in Gunnison's prairie dog. *Behavioral Ecology and Sociobiology* 22:69–78.

Reading, R. P., H. Mix, B. Lhagvasuren, and N. Tseveenmyadag. 1998. The commercial harvest of wildlife in Dornod Aimag, Mongolia. *Journal of Wildlife Management* 62:59–71.

Redford, K. H., and J. F. Eisenberg. 1992. *The Southern Cone: Chile, Argentina, Uruguay, Paraguay*. Vol. 2 of *Mammals of the Neotropics*. Chicago: University of Chicago Press.

Refisch, J. 1998. *Les singes et les autres mammifères: Evaluation écologique intégrée de la forêt naturelle de la Lama en République du Bénin*, Volume Annexe [Appendix]. Report for the project "Promotion de l'économie forestière et du bois," ONAB [Office National du Bois], with support from KfW and GTZ [Gesellschaft für Technische Zusammenarbeit]. Cotonou, Benin: ECO Gesellschaft für sozialökologische Programmberatung.

Reid, F. A. 1997. *A Field Guide to the Mammals of Central America and Southeast Mexico*. New York: Oxford University Press.

Reid, W. D., A. Ng, R. K. Wilton, and W. K. Milsom. 1995. Characteristics of diaphragm muscle fibre types in hibernating squirrels. *Respiration Physiology* 101 (3):301–309.

Reitsma, L. R., R. T. Holmes, and T. W. Sherry. 1990. Effects of removal of red squirrels, *Tamiasciurus hudsonicus*, and eastern chipmunks, *Tamias striatus*, on nest predation in a northern hardwood forest: An artificial nest experiment. *Oikos* 57 (3):375–380.

Reunanen, P., and T. C. Grubb. 2005. Densities of eastern chipmunks (*Tamias striatus*) in farmland woodlots decline with increasing area and isolation. *American Midland Naturalist* 154 (2):433–441.

Reunanen, P., M. Mönkkönen, and A. Nikula. 2000. Managing boreal forest landscapes for flying squirrels. *Conservation Biology* 14 (1):218–226.

Reunanen, P., M. Mönkkönen, A. Nikula, E. Hurme, and V. Nivala. 2004. Assessing landscape thresholds for the Siberian flying squirrel. *Ecological Bulletins* 51:277–286.

Reunanen, P., A. Nikula, and M. Mönkkönen. 2002a. Habitat requirements of the Siberian flying squirrel in northern Finland: Comparing field survey and remote sensing data. *Annales Zoologici Fennici* 39:7–20.

———. 2002b. Regional landscape patterns and distribution of the Siberian flying squirrel *Pteromys volans* in northern Finland. *Wildlife Biology* 8:267–278.

Reunanen, P., A. Nikula, M. Mönkkönen, E. Hurme, and V. Nivala. 2002. Predicting occupancy for the Siberian flying squirrel in old-growth forest patches. *Ecological Applications* 12 (4):1188–1198.

Ricankova, V., Z. Fric, J. Chlachula, P. Stastna, A. Faltynkova, and F. Zemek. 2006. Habitat requirements of the long-tailed ground squirrel (*Spermophilus undulatus*) in the southern Altai. *Journal of Zoology* 270:1–8.

Rickart, E. A. 1986. Postnatal growth of the Piute ground squirrel. *Journal of Mammalogy* 67 (2):412–416.

———. 1999. Townsend's ground squirrel (*Spermophilus townsendii*). In *Smithsonian Book of North American Mammals*, ed. D. E. Wilson and S. Ruff. Washington, DC: Smithsonian Institution Press.

Rickart, E. A., L. R. Heaney, P. D. Heideman, and R. C. B. Utzurrum. 1993. The distribution and ecology of mammals on Leyte, Biliran, and Maripipi islands, Philippines. *Fieldiana Zoology*, n.s., 72:1–62.

Rickart, E. A., and E. Yensen. 1991. *Spermophilus washingtoni*. *Mammalian Species* 371:1–5.

Ridgway, R. 1912. *Color Standards and Color Nomenclature*. Washington, DC: published by the author.

Riley, J. 2002. Mammals on the Sangihe and Talaud islands, Indonesia, and the impact of hunting and habitat loss. *Oryx* 36 (3):288–296.

Risch, T. S., and M. J. Brady. 1996. Trap height and capture success of arboreal small mammals: Evidence from southern flying squirrels (*Glaucomys volans*). *American Midland Naturalist* 136 (2):346–351.

Risch, T. S., and S. C. Loeb. 2004. Monitoring interactions between red-cockaded woodpeckers and southern flying squirrels. In *Red-Cockaded Woodpecker: Road to Recovery*, ed. R. Costa and S. J. Daniels. Blaine, WA: Hancock House.

Rivieccio, M., B. C. Thompson, W. R. Gould, and K. G. Boykin. 2003. Habitat features and predictive habitat modeling for the Colorado chipmunk in southern New Mexico. *Western North American Naturalist* 63:479–488.

Roberts, A. 1951. *The Mammals of South Africa*. Cape Town: Trustees of "The Mammals of South Africa" Book Fund, distributed by Central News Agency.

Roberts, T. J. 1977. *The Mammals of Pakistan*. London: Ernest Benn.

———. 1997. *The Mammals of Pakistan*, rev. ed. Oxford: Oxford University Press.

Robinson, H. C., and C. B. Kloss. 1918. Results of an expedition to Korinchi Peak, Sumatra, 1: Mammals. *Journal of the Federation of Malay States Museum* 8 (2):1–81.

———. 1919. On a collection of mammals from the Bencoolen and Palembang residencies, south west Sumatra. *Journal of the Federation of Malay States Museum* 7 (4):257–291.

———. 1922. New mammals from French Indo-China and Siam. *Annals and Magazine of Natural History* 9 (9):87–99.

Robinson, H. C., and R. C. Wroughton. 1911. Notes on Indo-Malayan squirrels. *Journal of the Federation of Malay States Museum* (4):166–168.

Robinson, N. 1967. Tamed squirrels. *Nigerian Field* 32:42–46.

———. 1969. A ground squirrel (*Xerus erythropus*). *Nigerian Field* 34:136–142.

Robinson, S. S., and D. S. Lee. 1980. Recent range expansion of the groundhog, *Marmota monax*, in the Southeast (Mammalia: Rodentia). *Brimleyana* 3:43–48.

Robinson, T. L., J. D. Skinner, and A. S. Haim. 1986. Close chromosomal congruence in two species of ground squirrel: *Xerus inauris* and *X. princeps* (Rodentia: Sciuridae). *South African Journal of Zoology* 21:100–105.

Roehrs, Z. P., and H. H. Genoways. 2004. Historical biogeography of the woodchuck (*Marmota monax bunkeri*) in Nebraska and northern Kansas. *Western North American Naturalist* 64:396–402.

Rogovin, K. A. 1992. Habitat use by two species of Mongolian marmots (*Marmota sibirica* and *M. baibacina*) in a zone of sympatry. *Acta Theriologica* 37 (4):345–350.

Rogovin, K. A., and G. I. Shenbrot. 1995. Geographical ecology of Mongolian Desert rodent communities. *Journal of Biogeography* 22 (1):111–128.

Romero-Balderas, K. G., E. J. Naranjo, H. H. Morales, and R. B. Nigh. 2006. Damages caused by wild vertebrate species in corn crops at the Lacandon Forest, Chiapas, Mexico. *Interciencia* 31:276–283.

Rompola, K. H., and S. H. Anderson. 2004. Habitat of three rare species of small mammals in juniper woodlands of southwestern Wyoming. *Western North American Naturalist* 64 (1):86–92.

Root, J. J., C. H. Calisher, and B. J. Beaty. 2001. Microhabitat partitioning by two chipmunk species (*Tamias*) in western

Colorado. *Western North American Naturalist* 61 (1):114-118.

Rose, L. M. 1997. Vertebrate predation and food-sharing in *Cebus* and *Pan*. *International Journal of Primatology* 18:727-765.

———. 2001. Meat and the early human diet: Insights from Neotropical studies. In *Meat-Eating and Human Evolution*, ed. C. B. Stanford and H. T. Bunn. Oxford: Oxford University Press.

Rosenberg, D. K., and R. G. Anthony. 1992. Characteristics of northern flying squirrel populations in young second- and old-growth forests in western Oregon. *Canadian Journal of Zoology* 70:161-166.

Rosevear, D. R. 1969. *The Rodents of West Africa*. London: Trustees of the British Museum (Natural History).

Roth, J. K., and S. B. Vander Wall. 2005. Primary and secondary seed dispersal of bush chinquapin (Fagaceae) by scatterhoarding rodents. *Ecology* 86 (9):2428-2439.

Rourke, B. C., Y. Yokoyama, W. K. Milsom, and V. J. Caiozzo. 2004. Myosin isoform expression and MAFbx mRNA levels in hibernating golden-mantled ground squirrels (*Spermophilus lateralis*). *Physiological and Biochemical Zoology* 77 (4):582-593.

Ruedas, L., J. W. Duckworth, B. Lee, and R. J. Tizard. 2008a. *Aeromys thomasi*. *IUCN Red List of Threatened Species*, version 2009.1.

———. 2008b. *Hylopetes sipora*. *IUCN Red List of Threatened Species*, 2008 [accessed 28 Oct. 2008].

———. 2008c. *Hylopetes winstoni*. *IUCN Red List of Threatened Species*, version 2009.2.

———. 2008d. *Iomys sipora*. *IUCN Red List of Threatened Species*, version 2009.1.

Ruediger, J., E. A. Van der Zee, A. M. Strijkstra, A. Aschoff, S. Daan, and R. A. Hut. 2007. Dynamics in the ultrastructure of asymmetric axospinous synapses in the frontal cortex of hibernating European ground squirrels (*Spermophilus citellus*). *Synapse* 61 (5):343-352.

Rusch, D. A., and W. G. Reeder. 1978. Population ecology of Alberta red squirrels. *Ecology* 59 (2):400-420.

Ryan, L. A., and A. B. Carey. 1995. Distribution and habitat of the western gray squirrel (*Sciurus griseus*) on Ft. Lewis, Washington. *Northwest Science* 69:204-216.

Saha, S. S. 1977. A new subspecies of the flying squirrel, *Petaurista nobilis* (Gray), from Bhutan. *Proceedings of the Zoological Society of Calcutta* 28:27-29.

———. 1981. A new genus and a new species of flying squirrel (Mammalia: Rodentia: Sciuridae) from northeastern India. *Bulletin of the Zoological Survey of India* 4:331-336.

Saiful, A. A., A. H. Idris, Y. N. Rashid, N. Tamura, and F. Hayashi. 2001. Home range size of sympatric squirrel species inhabiting a lowland dipterocarp forest in Malaysia. *Biotropica* 33 (2):346-351.

Saiful, A. A., and M. Nordin. 2004. Diversity and density of diurnal squirrels in a primary hill dipterocarp forest, Malaysia. *Journal of Tropical Ecology* 20 (1):45-49.

Salafsky, N. 1993. Mammalian use of a buffer zone agroforestry system bordering Gunung Palung National Park, West Kalimantan, Indonesia. *Conservation Biology* 7:928-933.

Salazar-Bravo, J., E. Yensen, T. Tarifa, and T. L. Yates. 2002. Distributional records of Bolivian mammals. *Mastozoología Neotropical* 9:70-78.

Sanborn, C. C. 1951. Mammals from Marcapata, southeastern Peru. *Publicaciones del Museo de Historia Natural Universidad Nacional Mayor de San Marcos, Ser. A, Zoología* 6:1-26.

———. 1953a. Mammals from the departments of Cuzco and Puno, Peru. *Publicaciones del Museo de Historia Natural Universidad Nacional Mayor de San Marcos, Ser. A, Zoología* 12:1-8.

———. 1953b. Notes sur quelques mammifères de l'Afrique Équatoriale Française. *Mammalia* 17 (3):164-169.

Sánchez-Cordero, V., P. Illoldi-Rangel, M. Linaje, S. Sarkar, and A. T. Peterson. 2005. Deforestation and extant distributions of Mexican endemic mammals. *Biological Conservation* 126:465-473.

Sánchez-Cordero, V., and R. Martínez-Gallardo. 1998. Postdispersal fruit and seed removal by forest-dwelling rodents in a lowland rainforest in Mexico. *Journal of Tropical Ecology* 14 (2):139-151.

Sanders, S. D. 1983. Foraging by Douglas tree squirrels (*Tamiasciurus douglasii*: Rodentia) for conifer seeds and fungi. Ph.D. diss., University of California, Davis.

Sato, H., B. H. Al-Adhami, Y. Une, and H. Kamiya. 2007. *Trypanosoma* (*Herpetosoma*) *kuseli* sp. n. (Protozoa: Kinetoplastida) in Siberian flying squirrels (*Pteromys volans*). *Parasitology Research* 101 (2):453-61.

Sawyer, S. L., and R. K. Rose. 1985. Homing in and ecology of the southern flying squirrel *Glaucomys volans* in southeastern Virginia. *American Midland Naturalist* 113 (2):238-244.

Scheffer, V. B., J. A. Neff, F. P. Cronemiller, C. A. Kennedy, J. R. Matson, C. P. Brown, C. W. Severinghaus, J. E. Tanck, J. S. Roux, K. Stott Jr., F. W. Preston, J. R. Olive, C. V. Riley, E. P. Odum, V. S. Schantz, A. F. Halloran, G. A. Petrides, M. C. Gardner, C. H. Edmondson, and S. O. Grierson. 1948. General Notes. *Journal of Mammalogy* 29 (1):66-77.

Scheibe, J. S., K. E. Paskins, S. Ferdous, and D. Birdsill. 2007. Kinematics and functional morphology of leaping, landing, and branch use in *Glaucomys sabrinus*. *Journal of Mammalogy* 88 (4):850-861.

Schitoskey, F., and S. R. Woodmansee. 1978. Energy requirements and diet of the California ground squirrel. *Journal of Wildlife Management* 42 (2):373-382.

Schlitter, D. A. 1989. African rodents of special concern: A preliminary assessment. In *Rodents: A World Survey of Species of Conservation Concern*. ed. W. Z. Lidicker Jr. Occasional Papers of the IUCN Species Survival Commission, No. 4. Gland, Switzerland: IUCN—The World Conservation Union.

Schmidly, D. J. 1977. *The Mammals of Trans-Pecos Texas, Including Big Bend National Park and Guadalupe Mountains National Park*. College Station: Texas A&M University Press.

———. 1983. *Texas Mammals East of the Balcones Fault Zone*. College Station: Texas A&M University Press.

Schmidt, D. F., C. A. Ludwig, and M. D. Carleton. 2008. *The Smithsonian Institution African Mammal Project (1961-1972): An annotated gazetteer of collecting localities and summary of its taxonomic and geographic scope*. Smithsonian Contributions to Zoology No. 628. Washington, DC: Smithsonian Institution Press.

Schmidt, K. A. 2000. Interactions between food chemistry and predation risk in fox squirrels. *Ecology* 81 (8):2077-2085.

Schnurr, J. L., C. D. Canham, R. S. Ostfeld, and R. S. Inouye. 2004. Neighborhood analyses of small-mammal dynamics: Impacts on seed predation and seedling establishment. *Ecology* 85 (3):741-755.

Schnurr, J. L., R. S. Ostfeld, and C. D. Canham. 2002. Direct and indirect effects of masting on rodent populations and tree seed survival. *Oikos* 96 (3):402.

Schulte-Hostedde, A. I., H. L. Gibbs, and J. S. Millar. 2000. Microsatellite DNA loci suitable for parentage analysis in the yellow-pine chipmunk (*Tamias amoenus*). *Molecular Ecology* 9 (12):2180.

———. 2001. Microgeographic genetic structure in the yellow-pine chipmunk (*Tamias amoenus*). *Molecular Ecology* 10:1625–1631.

Schulte-Hostedde, A. I., and J. S. Millar. 2000. Measuring sexual size dimorphism in the yellow-pine chipmunk (*Tamias amoenus*). *Canadian Journal of Zoology—Revue Canadienne de Zoologie* 78 (5):728–733.

———. 2004. Intraspecific variation of testis size and sperm length in the yellow-pine chipmunk (*Tamias amoenus*): Implications for sperm competition and reproductive success. *Behavioral Ecology and Sociobiology* 55 (3):272–277.

Schulte-Hostedde, A. I., J. S. Millar, and H. L. Gibbs. 2004. Sexual selection and mating patterns in a mammal with female-biased sexual size dimorphism. *Behavioral Ecology* 15 (2):351–356.

Schulte-Hostedde, A. I., J. S. Millar, H. L. Gibbs, and M. Zuk. 2002. Female-biased sexual size dimorphism in the yellow-pine chipmunk (*Tamias amoenus*): Sex-specific patterns of annual reproductive success and survival. *Evolution* 56 (12):2519–2529.

Schwagmeyer, P. L. 1980. Alarm calling behavior of the 13-lined ground-squirrel, *Spermophilus tridecemlineatus*. *Behavioral Ecology and Sociobiology* 7 (3):195–200.

———. 1985. Mating competition in an asocial ground-squirrel, *Spermophilus tridecemlineatus*. *Behavioral Ecology and Sociobiology* 17 (3):291–296.

Schwanz, L. E. 2006. Annual cycle of activity, reproduction, and body mass in Mexican ground squirrels (*Spermophilus mexicanus*). *Journal of Mammalogy* 87 (6):1086–1095.

Schwartz, C. W., and E. R. Schwartz. 2001. *Wild Mammals of Missouri*. Columbia: University of Missouri Press.

Schwartz, O. A., K. B. Armitage, and D. H. Van Vuren. 1998. A 32-year demography of yellow-bellied marmots (*Marmota flaviventris*). *Journal of Zoology* 246:337–346.

Selonen, V., and I. K. Hanski. 2003. Movements of the flying squirrel *Pteromys volans* in corridors and in matrix habitat. *Ecography* 26:641–651.

———. 2004a. Habitat exploration and use in dispersing juvenile flying squirrels. *Journal of Animal Ecology* 75:1440–1449.

———. 2004b. Young flying squirrels (*Pteromys volans*) dispersing in fragmented forests. *Behavioral Ecology* 15:564–571.

Selonen, V., I. K. Hanski, and P. C. Stevens. 2001. Space use of the Siberian flying squirrel *Pteromys volans* in fragmented forest landscapes. *Ecography* 24 (5):588–600.

Semenov, Y., R. Ramousse, and M. Le Berre. 2000. Effects of ecological factors on the diurnal activity rhythm of Yakutian black-capped marmots (*Marmota camtschatica bungei*) in the Arctic. *Russian Journal of Ecology* 31:118–122.

Semenov, Y., R. Ramousse, M. Le Berre, and Y. Tutukarov. 2001. Impact of the black-capped marmot (*Marmota camtschatica bungei*) on floristic diversity of Arctic tundra in northern Siberia. *Arctic, Antarctic, and Alpine Research* 33:204–210.

Semenov, Y., R. Ramousse, M. Le Berre, V. Vassiliev, and N. Solomonov. 2001. Aboveground activity rhythm in Arctic black-capped marmot (*Marmota camtschatica bungei* Katschenko 1901) under polar day conditions. *Acta Oecologica* 22:99–107.

Servin, J., E. Chacon, N. Alonso-Perez, and C. Huxley. 2003. New records of mammals from Durango, Mexico. *Southwestern Naturalist* 48 (1):136–138.

Servin, J., V. Sánchez-Cordero, and F. A. Cervantes. 1996. First record of the Sierra Madre mantled ground squirrel (*Spermophilus madrensis*; Rodentia: Sciuridae) from Durango, Mexico. *Southwestern Naturalist* 41 (2):189–190.

Setoguchi, M. 1990. Food habits of red-bellied tree squirrels on a small island in Japan. *Journal of Mammalogy* 71 (4):570–578.

———. 1991. Nest-site selection and nest-building behavior of red-bellied tree squirrels on Tomogashima Island, Japan. *Journal of Mammalogy* 72 (1):163–170.

Setzer, H. W. 1954. A new squirrel from the Anglo-Egyptian Sudan. *Proceedings of the Biological Society of Washington* 67:87–88.

Seville, R. S., and N. L. Stanton. 1993. Eimerian guilds (Apicomplexa: Eimeriidae) in Richardson's (*Spermophilus richardsonii*) and Wyoming (*Spermophilus elegans*) ground squirrels. *Journal of Parasitology* 79 (6):973–975.

Shaffer, B. S., and J. L. J. Sanchez. 1994. Comparison of 1/8"- and 1/4"-mesh recovery of controlled samples of small- to medium-sized mammals. *American Antiquity* 59 (3):525–530.

Shafique, C. M., S. Barkati, T. Oshida, and M. Ando. 2006. Comparison of the diets between two sympatric flying squirrels in northern Pakistan. *Journal of Mammalogy* 87 (4):784–789.

Shankar-Raman, T. R., C. Mishra, and A. J. T. Johnsingh. 1995. Observations on Pallas's squirrel *Callosciurus erythraeus* Pallas and other squirrels in Mizoram, northeast India. *Journal of the Bombay Natural History Society* 92:412–415.

Shar, S., and D. Lkhagvasuren. 2008. *Spermophilus erythrogenys*. *IUCN Red List of Threatened Species*, version 2009.2.

Shar, S., D. Lkhagvasuren, H. Henttonen, T. Maran, and I. Hanski. 2008. *Pteromys volans*. *IUCN Red List of Threatened Species*, version 2009.2.

Shar, S., D. Lkhagvasuren, and A. T. Smith. 2008. *Spermophilus alashanicus*. *IUCN Red List of Threatened Species*, version 2009.2.

Sharma, S. K. 2008. Biodiversity Assessment Report: Wakal River Basin, India. Report for the Global Water for Sustainability (GLOWS) Program, Florida International University.

Sharpe, P. B., and B. Van Horne. 1999. Relationships between the thermal environment and activity of Piute ground squirrels. *Journal of Thermal Biology* 24 (4):265–278.

Sheikh, K. M., and S. Molur, eds. 2004. *Status and Red List of Pakistan's Mammals: Based on the Pakistan Mammal Conservation Assessment & Management Plan Workshop, August 18–22, 2003, [Islamabad, Pakistan]*. [Islamabad]: IUCN—Pakistan.

Shekarova, O. N., V. V. Neronov, and L. E. Savinetskaya. 2008. Speckled ground squirrel (*Spermophilus suslicus*): Current distribution, population dynamics and conservation. *Lynx* 39:317–322.

Shelley, E. L., and D. T. Blumstein. 2005. The evolution of vocal alarm communication in rodents. *Behavioral Ecology* 16 (1):169–176.

Sherman, P. W. 1989. Mate guarding as paternity insurance in Idaho ground squirrels. *Nature* 338:418–420.

Sherman, P. W., and M. C. Runge. 2002. Demography of a population collapse: The Northern Idaho ground squirrel (*Spermophilus brunneus brunneus*). *Ecology* 83 (10):2816–2831.

Shortridge, G. C. 1934a. *The Mammals of South West Africa*, vol. 1. London: Heinemann.

———. 1934b. *The Mammals of South West Africa*, vol. 2. London: Heinemann.

Shriner, W. M. 1999. Antipredator responses to a previously neutral sound by free-living adult golden-mantled ground squirrels, *Spermophilus lateralis* (Sciuridae). *Ethology* 105:747–757.

Shriner, W. M., and P. B. Stacey. 1991. Spatial relationships and dispersal patterns in the rock squirrel, *Spermophilus variegatus*. *Journal of Mammalogy* 72:601–606.

Silvius, K. M. 2002. Spatio-temporal patterns of palm endocarp use by three Amazonian forest mammals: Granivory or "grubivory"? *Journal of Tropical Ecology* 18:707–723.

Skalon, N. V., and T. N. Gagina. 2006. Forest steppe marmot and red-cheeked souslik as "flag" species for saving and recovery of Kuznetsk steppe ecosystems. In *Marmots in Anthropogenic Landscapes of Eurasia: 9th International Meeting on Marmots*. Kemerovo, Russia: EPC Grafika.

Skinner, J. D. 1990. *The Mammals of the Southern African Subregion*, 2nd ed. Pretoria, South Africa: University of Pretoria.

Skinner, J. D., and C. T. Chimimba, eds. 2005. *The Mammals of the Southern African Subregion*, 3rd ed. Cambridge; New York: Cambridge University Press.

Skurski, D. A., and J. M. Waterman. 2005. *Xerus inauris*. *Mammalian Species* 781:1–4.

Slajchert, T., U. D. Kitron, C. J. Jones, and A. Mannelli. 1997. Role of the eastern chipmunk (*Tamias striatus*) in the epizootiology of Lyme borreliosis in northwestern Illinois, USA. *Journal of Wildlife Diseases* 33 (1):40–46.

Slobodchikoff, C. N., A. Paseka, and J. L. Verdolin. 2009. Prairie dog alarm calls encode labels about predator colors. *Animal Cognition* 12 (3):435–439.

Sludskiy, A. A. 1977. [*Rodents (except the marmots, susliks, longcrawded [long-clawed] ground squirrels, gerbils, and voles)*]. Vol. 1, pt. 2 of *Mlekopitaiushchie Kazakhstana [Mammals of Kazakhstan]*. Alma-Ata, Kazakhstan: Izdatel'stvo Nauka Kazakhskoĭ SSR.

Sludskiy, A. A., S. N. Varshavsky, M. I. Ismagilov, V. I. Kapitonov, and I. G. Shubin. 1969. *[Rodents (Marmots and Ground Squirrels)]*. Vol. 1, pt. 1 of *Mlekopitaiushchie Kazakhstana [Mammals of Kazakhstan]*. Alma-Ata, Kazakhstan: Izdatel'stvo Nauka Kazakhskoĭ SSR.

Smallwood, P. D., M. A. Steele, and S. Faeth. 2001. The ultimate basis of the caching preference of rodents and the oak-dispersal syndrome: Tannins, insects, and seed germination. *American Zoologist* 41:840–851.

Smith, A. T., and C. H. Johnston. 2008a. *Aeretes melanopterus*. *IUCN Red List of Threatened Species*, version 2009.1.

———. 2008b. *Petaurista xanthotis*. *IUCN Red List of Threatened Species*, version 2009.2.

———. 2008c. *Sciurotamias davidianus*. *IUCN Red List of Threatened Species*, version 2010.4.

———. 2008d. *Sciurotamias forresti*. *IUCN Red List of Threatened Species*, version 2010.4.

———. 2008e. *Spermophilus brevicauda*. *IUCN Red List of Threatened Species*, version 2009.2.

———. 2008f. *Trogopterus xanthipes*. *IUCN Red List of Threatened Species*, version 2009.2.

Smith, A. T., and Y. Xie, eds. 2008. *A Guide to the Mammals of China*. Princeton, NJ: Princeton University Press.

Smith, C. C. 1965. Interspecific competition in the genus of tree squirrels, *Tamiasciurus*. Ph.D. diss., University of Washington.

———. 1968. The adaptive nature of social organization in the genus of tree squirrels, *Tamiasciurus*. *Ecological Monographs* 38 (1):31–63.

———. 1970. The coevolution of pine squirrels (*Tamiasciurus*) and conifers. *Ecological Monographs* 40 (3):349–371.

———. 1978. Structure and function of the vocalizations of tree squirrels (*Tamiasciurus*). *Journal of Mammalogy* 59 (4):793–808.

———. 1981. The indivisible niche of *Tamiasciurus*: An example of nonpartitioning of resources. *Ecological Monographs* 51 (3):343–363.

Smith, H. C. 1993. *Alberta Mammals: An Atlas and Guide*. Edmonton: Provincial Museum of Alberta.

Smith, W. P. 2007. Ecology of *Glaucomys sabrinus*: Habitat, demography, and community relations. *Journal of Mammalogy* 88 (4):862–881.

Smith, W. P., R. G. Anthony, J. R. Waters, and C. J. Zabel. 2003. Ecology and conservation of arboreal rodents of the Pacific Northwest. In *Mammal Community Dynamics in Western Coniferous Forests: Management and Conservation*, ed. C. J. Zabel and R. G. Anthony. New York: Cambridge University Press.

Smith, W. P., S. M. Gende, and J. V. Nichols. 2004. Ecological correlates of flying squirrel microhabitat use and density in temperate rainforests of southeastern Alaska. *Journal of Mammalogy* 85:663–674.

Smithers, R. H. N. 1971. *The Mammals of Botswana*. Salisbury: Trustees of the National Museums of Rhodesia.

———. 1983. *The Mammals of the Southern African Subregion*. Pretoria, Republic of South Africa: University of Pretoria.

Smithers, R. H. N., and J. L. P. Lobão Tello. 1976. *Check List and Atlas of the Mammals of Moçambique*. Museum Memoir No. 8. Salisbury: Trustees of the National Museums and Monuments of Rhodesia.

Smithers, R. H. N., and V. J. Wilson. 1979. *Check List and Atlas of the Mammals of Zimbabwe Rhodesia*. Museum Memoir No. 9. Salisbury, Zimbabwe Rhodesia: Trustees of the National Museums and Monuments.

Snyder, D. P. 1982. *Tamias striatus*. *Mammalian Species* 168:1–8.

Snyder, M. A. 1992. Selective herbivory by Abert's squirrel mediated by chemical variability in ponderosa pine. *Ecology* 73 (5):1730–1741.

———. 1993. Interactions between Abert's squirrel and ponderosa pine: The relationship between selective herbivory and host plant fitness. *American Naturalist* 141:866–879.

Snyman, P. S. 1940. The study and control of the vectors of rabies in South Africa. *Onderstepoort Journal of Veterinary Science and Animal Industry* 15 (1, 2):9–140.

Souza, F. L. 2000. Palm nut handling behavior by squirrels. *Ciencia e Cultura* 52:188–190.

Spiridonova, L. N., G. N. Chelomina, V. P. Starikov, V. P. Korablev, M. V. Zvirka, and E. A. Lyapunova. 2005. RAPD-PCR analysis of ground squirrels from the Tobol-Ishim interfluve: Evidence for interspecific hybridization between ground squirrel species *Spermophilus major* and *S. erythrogenys*. *Russian Journal of Genetics* 41 (9):991–1001.

Spiridonova, L. N., G. N. Chelomina, K. Tsuda, H. Yonekawa, and V. P. Starikov. 2006. Genetic evidence of extensive introgression of short-tailed ground squirrel genes in a hybridization zone of *Spermophilus major* and *S. erythrogenys*, inferred from sequencing of the mtDNA cytochrome *b* gene. *Russian Journal of Genetics* 42 (7):802–809.

Spizenberger, F., K. Bauer, A. Mayer, and E. Weiss. 2001. *Die Säugetierfauna Österreichs*. Umwelt und Wasserwirtschaft 13. Graz, Austria: Graz Bundesministerium für Land- und Forstwirtschaft.

Spratt, D. M., and G. Varughese. 1975. A taxonomic revision of filarioid nematodes from Australian marsupials. *Australian Journal of Zoology, Supplemental Series* 35:1–99.

Srere, H. K., L. C. H. Wang, and S. L. Martin. 1992. Central role for differential gene expression in mammalian hibernation. *Proceedings of the National Academy of Sciences of the United States of America* 89 (15):7119–7123.

Sridhar, H., T. R. Shankar Raman, and D.

Mudappa. 2008. Mammal persistence and abundance in tropical rainforest remnants in the southern Western Ghats, India. *Current Science* 94 (6):748–757.

Srinivasulu, C., S. Chakraborty, and M. S. Pradhan. 2004. Checklist of sciurids (Mammalia: Rodentia: Sciuridae) of South Asia. *Zoos' Print Journal* 19 (2):1351–1360.

Stafford, B. J., R. W. Thorington Jr., and T. Kawamichi. 2002. Gliding behavior of Japanese giant flying squirrels (*Petaurista leucogenys*). *Journal of Mammalogy* 83 (2):553–562.

———. 2003. Positional behavior of Japanese giant flying squirrels (*Petaurista leucogenys*). *Journal of Mammalogy* 84 (1):263–271.

Stangl, F. B., and J. R. Goetze. 1991. Comments on pelage and molt of the spotted ground squirrel, *Spermophilus spilosoma*. *Texas Journal of Science* 43:305–308.

Stanley, S. E., and R. G. Harrison. 1999. Cytochrome *b* evolution in birds and mammals: An evaluation of the avian constraint hypothesis. *Molecular Biology and Evolution* 167 (11):1575–1585.

Staples, J. F., and P. W. Hochachka. 1998. The effect of hibernation status and cold-acclimation on hepatocyte gluconeogenesis in the golden-mantled ground squirrel (*Spermophilus lateralis*). *Canadian Journal of Zoology* 76: 1734–1740.

Stapp, P. 1992. Energetic influences on the life history of *Glaucomys volans*. *Journal of Mammalogy* 73 (4):914–920.

Stapp, P., P. J. Pekins, and W. W. Mautz. 1991. Winter energy-expenditure and the distribution of the southern flying squirrels. *Canadian Journal of Zoology* 69: 2548–2555.

Steele, M. A. 1998. *Tamiasciurus hudsonicus*. *Mammalian Species* 586:1–9.

———. 1999. *Tamiasciurus douglasii*. *Mammalian Species* 630:1–8.

———. 2008. Evolutionary interactions between tree squirrels and trees: A review and synthesis. *Current Science* 95:271–276.

Steele, M. A., J. E. Carlson, P. D. Smallwood, A. B. McEuen, T. A. Contreras, and W. B. Terszaghi. 2007. Linking seed and seed shadows: A case study in the oaks (*Quercus*). In *Seed Dispersal: Theory and Its Application in a Changing World*, ed. A. J. Dennis, E. W. Schupp, R. J. Green, and D. A. Wescott. Wallingford, UK: CAB International.

Steele, M. A., S. L. Halkins, P. D. Smallwood, T. McKenna, K. Mistopolus, and M. Beam. 2008. Cache protection strategies of a scatter-hoarding rodent: Do tree squirrels engage in behavioural deception? *Animal Behaviour* 75:705–714.

Steele, M. A., and J. L. Koprowski. 2001. *North American Tree Squirrels*. Washington, DC: Smithsonian Institution Press.

Steele, M. A., S. Manierre, T. Genna, T. A. Contreras, P. D. Smallwood, and M. E. Pereira. 2006. The innate basis of food-hoarding decisions in grey squirrels: Evidence for behavioural adaptations to the oaks. *Animal Behaviour* 71:155–160.

Steele, M. A., and A. B. McEuen. 2005. Atypical acorns appear to allow escape after apical notching by tree squirrels. *American Midland Naturalist* 154:450–458.

Steele, M. A., P. D. Smallwood, A. Spunar, and E. Nelsen. 2001. The proximate basis of the oak dispersal syndrome: Can small mammals smell seed dormancy? *American Zoologist* 41:852–864.

Steele, M. A., P. D. Smallwood, W. Terzaghi, J. Carlson, T. Contreras, and A. McEuen. 2004. Oak dispersal syndromes: Do red and white oaks exhibit different dispersal strategies? *Upland Oak Ecology Symposium: History, Current Conditions, and Sustainability, Fayetteville, Arkansas, October 7–10, 2002*, ed. M. A. Spetich. General Technical Report SRS 73. Asheville, NC: Southern Research Station, U.S. Forest Service.

Steele, M. A., G. Turner, P. D. Smallwood, J. O. Wolff, and J. Radillo. 2001. Cache management by small mammals: Experimental evidence for the significance of acorn-embryo excision. *Journal of Mammalogy* 82:35–42.

Steele, M. A., L. A. Wauters, and K. W. Larsen. 2005. Selection, predation and dispersal of seeds by tree squirrels in temperate and boreal forests: Are tree squirrels keystone granivores? In *Seed Fate: Predation, Dispersal and Seedling Establishment*, ed. P.-M. Forget, J. E. Lambert, P. E. Hulme, and S. B. Vander Wall. Wallingford, UK: CAB International.

Steele, M. A., and P. D. Weigl. 1992. Energetics and patch use in the fox squirrel *Sciurus niger*: Responses to variation in prey profitability and patch density. *American Midland Naturalist* 128:156–167.

Steenhof, K., E. Yensen, M. N. Kochert, and K. L. Gage. 2006. Populations and habitat relationships of Piute ground squirrels in southwestern Idaho. *Western North American Naturalist* 66:482–491.

Steppan, S. J., M. R. Akhverdyan, E. A. Lyapunova, D. G. Fraser, N. N. Vorontsov, R. S. Hoffmann, and M. J. Braun. 1999. Molecular phylogeny of the marmots (Rodentia: Sciuridae): Tests of evolutionary and biogeographic hypotheses. *Systematic Biology* 48 (4):715–734.

Steppan, S. J., B. L. Storz, and R. S. Hoffmann. 2004. Nuclear DNA phylogeny of the squirrels (Mammalia: Rodentia) and the evolution of arboreality from c-myc and RAGI. *Molecular Phylogenetics and Evolution* 30 (3):703–719.

Stevens, S. D. 1998. High incidence of infanticide by lactating females in a population of Columbian ground squirrels (*Spermophilus columbianus*). *Canadian Journal of Zoology* 76 (6):1183–1187.

Stevenson, R. 1990. Woodchuck sightings in southeastern Saskatchewan. *Blue Jay* 48:225–226.

Stone, K. D., G. A. Heidt, W. H. Baltosser, and P. T. Caster. 1996. Factors affecting nest box use by southern flying squirrels (*Glaucomys volans*) and gray squirrels (*Sciurus carolinensis*). *American Midland Naturalist* 135 (1):9–13.

Stone, K. D., G. A. Heidt, P. T. Caster, and M. L. Kennedy. 1997. Using Geographic Information Systems to determine home range of the southern flying squirrel (*Glaucomys volans*). *American Midland Naturalist* 137 (1):106–111.

Strauss, A., I. E. Hoffmann, M. Walzl, and E. Millesi. 2009. Vaginal oestrus during the reproductive and non-reproductive period in European ground squirrels. *Animal Reproduction Science* 112 (3–4):362–370.

Strauss, A., E. Mascher, R. Palme, and E. Millesi. 2007. Sexually mature and immature yearling male European ground squirrels: A comparison of behavioral and physiological parameters. *Hormones and Behavior* 52 (5):646–652.

Strijkstra, A. M., R. A. Hut, M. C. de Wilde, J. Stieler, and E. A. Van der Zee. 2003. Hippocampal synaptophysin immunoreactivity is reduced during natural hypothermia in ground squirrels. *Neuroscience Letters* 344 (1):29–32.

Sturm, H., A. Abouchaar, R. de Bernal, and C. de Hoyos. 1970. Distribución de animales en las capas bajas de un bosque humedo tropical de la Region Carare-Opón (Santander, Colombia). *Caldasia* 10 (50):529–578.

Sullivan, R. M. 1996. Genetics, ecology, and conservation of montane populations of Colorado chipmunks (*Tamias quadrivittatus*). *Journal of Mammalogy* 77:951–975.

Sullivan, T. P., and W. Klenner. 2000. Response of northwestern chipmunks (*Tamias amoenus*) to variable habitat structure in young lodgepole pine forest. *Canadian Journal of Zoology—Revue Canadienne de Zoologie* 78 (2):283-293.

Sullivan, T. P., R. A. Lautenschlager, and R. G. Wagner. 1999. Clearcutting and burning of northern spruce-fir forests: Implications for small mammal communities. *Journal of Applied Ecology* 36 (3):327-344.

Sullivan, T. P., and D. S. Sullivan 1982. Barking damage by snowshoe hares and red squirrels in lodgepole pine stands in central British Columbia. *Canadian Journal of Forest Research* 12:443-448.

Sullivan, T. P., D. S. Sullivan, and P. M. F. Lindgren. 2000. Small mammals and stand structure in young pine, seed-tree, and old-growth forest, southwest Canada. *Ecological Applications* 10 (5):1367-1383.

Sullivan, T. P., D. S. Sullivan, P. M. F. Lindgren, and D. B. Ransome. 2005. Long-term responses of ecosystem components to stand thinning in young lodgepole pine forest, II: Diversity and population dynamics of forest floor small mammals. *Forest Ecology and Management* 205 (1-3):1-14.

Sutton, D. A. 1987. Analysis of Pacific Coast Townsend chipmunks (Rodentia: Sciuridae). *Southwestern Naturalist* 32 (3):28.

———. 1992. *Tamias amoenus*. *Mammalian Species* 390:1-8.

———. 1993. *Tamias townsendii*. *Mammalian Species* 435:1-6.

———. 1995. Problems of taxonomy and distribution in four species of chipmunks. *Journal of Mammalogy* 76 (3):843-850.

Sutton, D. A., and C. F. Nadler. 1974. Systematic revision of three Townsend chipmunks (*Eutamias townsendii*). *Southwestern Naturalist* 19 (2):199-211.

Sutton, D. A., and B. D. Patterson. 2000. Geographic variation of the western chipmunks *Tamias senex* and *T. siskiyou*, with two new subspecies from California. *Journal of Mammalogy* 81:299-316.

Swaisgood, R. R., D. H. Owings, and M. P. Rowe. 1999. Conflict and assessment in a predator-prey system: Ground squirrels versus rattlesnakes. *Animal Behaviour* 57:1033-1044.

Swaisgood, R. R., M. P. Rowe, and D. H. Owings. 1999. Assessment of rattlesnake dangerousness by California ground squirrels: Exploitation of cues from rattling sounds. *Animal Behaviour* 57 (6):1301.

Swenson, J. E. 1981. Distribution of Richardson's ground squirrel in eastern Montana. *Prairie Naturalist* 13:27-30.

Tamura, N. 1993. Role of sound communication in mating of Malaysian *Callosciurus* (Sciuridae). *Journal of Mammalogy* 74 (2):468-476.

———. 1995. Postcopulatory mate guarding by vocalization in the Formosan squirrel. *Behavioral Ecology and Sociobiology* 36 (6):377-386.

———. 2004. Effects of habitat mosaic on home range size of the Japanese squirrel, *Sciurus lis*. *Mammal Study* 29:9-14.

Tamura, N., and F. Hayashi. 2007. Five-year study of the genetic structure and demography of two subpopulations of the Japanese squirrel in a continuous forest and an isolated woodlot. *Ecological Research* 22:261-267.

———. 2008. Geographic variation in walnut seed size correlates with hoarding behaviour of two rodent species. *Ecological Research* 23:607-614.

Tamura, N., F. Hayashi, and K. Miyashita. 1988. Dominance hierarchy and mating behavior of the Formosan squirrel, *Callosciurus erythraeus thaiwanensis*. *Journal of Mammalogy* 69 (2):320-331.

Tamura, N., and H. S. Yong. 1993. Vocalizations in response to predators in three species of Malaysian *Callosciurus* (Sciuridae). *Journal of Mammalogy* 74 (3):703-714.

Tate, G. H. H. 1939. The mammals of the Guiana region. *Bulletin of the American Museum of Natural History* 76:151-229.

———. 1947. *Mammals of Eastern Asia*. New York: Macmillan.

Taylor, E. H. 1934. *Philippine Land Mammals*. Manila: Bureau of Printing.

Thapa, J., S. Molur, and P. O. Nameer. 2008. *Petaurista nobilis*. *IUCN Red List of Threatened Species*, version 2009.2.

Thayer, T. C., and S. B. Vander Wall. 2005. Interactions between Steller's jays and yellow-pine chipmunks over scatter-hoarded sugar pine seeds. *Journal of Animal Ecology* 74:365-374.

Thibault, A., and J. Bovet. 1999. Homing strategy of the eastern chipmunk, *Tamias striatus* (Mammalia: Rodentia): Validation of the critical distance model. *Ethology* 105 (1):73.

Thomas, O. 1893. Description of a new *Sciuropterus* from the Philippines. *Annals and Magazine of Natural History* 6 (12):30-31.

———. 1895. On some mammals collected by Dr. E. Modigliani in Sipora, Mentawei Islands. *Annali del Museo Civico di Storia Naturale di Genova* 14 (34):660-672.

———. 1909. Four new African squirrels. *Annals and Magazine of Natural History* 8:476-479.

———. 1916. A new genus for *Sciurus poensis* and its allies. *Annals and Magazine of Natural History* 8 (17):271-272.

———. 1920. Four new squirrels of the genus *Tamiops*. *Annals and Magazine of Natural History* 9 (5):304-308.

———. 1921. On small mammals from the Kachin Province, Northern Burma. *Journal of the Bombay Natural History Society* 27 (3):503.

———. 1926. Two new subspecies of *Callosciurus quinquestriatus*. *Annals and Magazine of Natural History*, Ser. 9, 17:639-641.

———. 1929. On mammals from the Kaoko-Veld, South-West Africa, obtained during Captain Shortridge's fifth Percy Sladen and Kaffrarian Museum Expedition. *Proceedings of the Zoological Society of London* 106:99-111.

Thomas, O., and J. Whitehead. 1898. On the mammals obtained by Mr. John Whitehead during his recent expedition to the Philippines. *Transactions of the Zoological Society of London* 14:377-412.

Thorington, K. K., J. D. Metheny, M. C. Kalcounis-Rueppell, and P. D. Weigl. 2010. Genetic relatedness in winter populations of seasonally gregarious southern flying squirrels, *Glaucomys volans*. *Journal of Mammalogy* 91 (4):897-904.

Thorington, R. W., Jr., and K. Darrow. 2000. Anatomy of the squirrel wrist: Bones, ligaments, and muscles. *Journal of Morphology* 246:85-102.

Thorington, R. W., Jr., K. Darrow, and C. G. Anderson. 1998. Wing tip anatomy and aerodynamics in flying squirrels. *Journal of Mammalogy* 79 (1):245-250.

Thorington, R. W., Jr., K. Darrow, and A. D. K. Betts. 1997. Comparative myology of the forelimb of squirrels (Sciuridae). *Journal of Morphology* 234 (2):155-182.

Thorington, R. W., Jr., and K. Ferrell. 2006. *Squirrels: The Animal Answer Guide*. Baltimore: Johns Hopkins University Press.

Thorington, R. W., Jr., and L. R. Heaney. 1981. Body proportions and gliding adaptations of flying squirrels (Petauristinae). *Journal of Mammalogy* 62 (1):101-114.

Thorington, R. W., Jr., and R. S. Hoffmann. 2005. Family Sciuridae. In *Mammal Species of the World: A Taxonomic and Geo-*

graphic Reference, 3rd ed., ed. D. E. Wilson and D. M. Reeder. Baltimore: Johns Hopkins University Press.

Thorington, R. W., Jr., A. M. Miller, and C. G. Anderson. 1998. Arboreality in tree squirrels (Sciuridae). In *Ecology and Evolutionary Biology of Tree Squirrels*, ed. M. A. Steele, J. F. Merritt, and D. A. Zegers. Special Publication No. 6. Martinsville: Virginia Museum of Natural History.

Thorington, R. W., Jr., A. L. Musante, C. G. Anderson, and K. Darrow. 1996. Validity of three genera of flying squirrels: *Eoglaucomys, Glaucomys*, and *Hylopetes*. *Journal of Mammalogy* 77 (1):69-83.

Thorington, R. W., Jr., D. Pitassy, and S. A. Jansa. 2002. Phylogenies of flying squirrels (Pteromyinae). *Journal of Mammalian Evolution* 9 (1/2):99-135.

Thorington, R. W., Jr., and E. M. Santana. 2007. How to make a flying squirrel: *Glaucomys* anatomy in phylogenetic perspective. *Journal of Mammalogy* 88 (4):882-896.

Thorington, R. W., Jr., C. E. Schennum, L. A. Pappas, and D. Pitassy. 2005. The difficulties of identifying flying squirrels (Sciuridae: Pteromyini) in the fossil record. *Journal of Vertebrate Paleontology* 25 (4):950-961.

Thorington, R. W., Jr., and B. J. Stafford. 2001. Homologies of the carpal bones in flying squirrels (Pteromyinae): A review. *Mammal Study* 26:61-68.

Thorson, J. M., R. A. Morgan, J. S. Brown, and J. E. Norman. 1998. Direct and indirect cues of predatory risk and patch use by fox squirrels and thirteen-lined ground squirrels. *Behavioral Ecology* 9:151-157.

Timm, R. M., and E. C. Birney. 1980. Mammals collected by the Menage Scientific Expedition to the Philippine Islands and Borneo, 1890-1893. *Journal of Mammalogy* 61:566-571.

Titov, S. V. 2003a. Juvenile dispersal in the colonies of *Spermophilus major* and *S. suslicus* ground squirrels. *Russian Journal of Ecology* 34:255-260.

———. 2003b. Specific features of reproductive behavior in the spotted souslik *Spermophilus suslicus* (Rodentia, Sciuridae) in the Volga River basin. *Zoologichesky Zhurnal* 82 (11):1381-1392.

———. 2004. Reproductive behavior of the russet souslik, *Spermophilus major*. *Zoologichesky Zhurnal* 83 (9):1148-1159.

Titov, S. V., A. A. Shmyrov, A. A. Kuzmin, and E. O. Ermakov. 2008. Agonistic behavior of the russet ground squirrel. *Zoologichesky Zhurnal* 87:1124-1133.

Tkachenko, V. S., S. I. Popov, and N. N. Sizonenko. 1985. Structure and distribution of burrows of the mountain souslik, *Citellus musica* (Rodentia, Cricetidae). *Zoologichesky Zhurnal* 64:1750-1753.

Tokarsky, V. A., and A. C. Valentsev. 1994. Distribution, biology, and breeding in captivity of black-capped marmot, *Marmota camtschatica* (Rodentia, Sciuridae). *Zoologichesky Zhurnal* 73:209-222.

Tong, H. 2007. *Aeretes melanopterus* (Pteromyinae, Rodentia) from Tianyuan Cave near Zhoukoudian (Choukoutien) in China. *Geobios* 40:219-230.

Tordoff, A. W., P. T. Anh, L. M. Hung, N. D. Xuan, and T. K. Phuc. 2002. A rapid bird and mammal survey of Lo Go Sa Mat Special-Use Forest and Chang Riec Protection Forest, Tay Ninh Province, Vietnam. Report for BirdLife International Vietnam Programme, the Institute of Ecology and Biological Resources, and Tay Ninh Provincial Department of Science, Technology and the Environment.

Towers, S. R., and R. G. Coss. 1991. Antisnake behavior of Columbian ground squirrels (*Spermophilus columbianus*). *Journal of Mammalogy* 72 (4):776-783.

Townsend, S. E. 2006. Burrow cluster as a sampling unit: An approach to estimate marmot activity in the eastern steppe of Mongolia. *Mongolian Journal of Biological Sciences* 4:31-36.

Townsend, S. E., and P. Zahler. 2006. Mongolian marmot crisis: Status of the Siberian marmot in the eastern steppe. *Mongolian Journal of Biological Sciences* 4:37-46.

Trevino-Villarreal, J. 1990. The annual cycle of the Mexican prairie dog (*Cynomys mexicanus*). *Occasional Papers of the Museum of Natural History, University of Kansas* 139:1-27.

Trombulak, S. C. 1987. Life history of the Cascade golden-mantled ground squirrel (*Spermophilus saturatus*). *Journal of Mammalogy* 68:544-554.

———. 1988. *Spermophilus saturatus*. *Mammalian Species* 322:1-4.

———. 1989. Running speed and body mass in Belding's ground squirrels. *Journal of Mammalogy* 70 (1):194-197.

Trufanov, G. V., and P. D. Golubev. 1982. Population structure of *Citellus musicus* in the central Caucasus. *Zoologichesky Zhurnal* 61:96-101.

Trunova, Y. E., and V. A. Lobkov. 1997. Recording hibernation features by incisor dentin of *Spermophilus suslicus* (Sciuridae, Rodentia). *Zoologichesky Zhurnal* 76:940-947.

Tsvirka, M. V., G. N. Chelomina, and V. P. Korablev. 2006. Genetic differentiation, phylogenetics, and systematics of desert ground squirrels of the subgenus *Colobotis* (*Spermophilus*, Rodentia, Sciuridae). *Zoologichesky Zhurnal* 85:629-640.

Tsvirka, M. V., L. N. Spiridonova, and V. P. Korablev. 2008. Molecular genetic relationships among East Palearctic ground squirrels of the genus *Spermophilus* (Sciuridae, Rodentia). *Russian Journal of Genetics* 44 (8):966-974.

Tsytsulina, K., N. Formozov, S. Shar, D. Lkhagvasuren, and B. Sheftel. 2008. *Tamias sibiricus*. *IUCN Red List of Threatened Species*, version 2009.2.

Tsytsulina, K., N. Formozov, and B. Sheftel. 2008. *Spermophilus fulvus*. *IUCN Red List of Threatened Species*, version 2009.2.

Tucker, M. S., D. L. Price, B. H. Kwa, and A. C. DeBaldo. 2003. Observations of *Breinlia booliati* in a new host, *Rattus rattus jalorensis*, from Kuantan, state of Pahang, Malaysia. *Journal of Parasitology* 89 (6):1220-1226.

Tumlinson, R., A. Smith, and R. Frazier. 2001. New records of the woodchuck (*Marmota monax*) from southern Arkansas. *Journal of the Arkansas Academy of Science* 55:191-192.

Turkowski, F. J. 1969. Resistance of roundtail ground squirrel (*Spermophilus tereticaudus*) to venom of scorpion (*Centruroides sculpturatus*). *Journal of Mammalogy* 50 (1):160-161.

Tyler, J. D., and S. L. Donelson. 1996. Noteworthy mammal records for western Oklahoma. *Proceedings of the Oklahoma Academy of Sciences* 76:103-104.

Ulmer, F. A. 1995. Northward extension of the range of Richmond's squirrel. *Southwestern Naturalist* 40 (4):416-418.

Umapathy, G., and A. Kumar. 2000. The occurrence of arboreal mammals in the rain forest fragments in the Anamalai Hills, South India. *Biological Conservation* 92 (3):311-319.

U. S. Fish and Wildlife Service. 1991. *Utah Prairie Dog Recovery Plan*. Denver: U. S. Fish and Wildlife Service.

———. 1998. *Recovery Plan for Upland Species of the San Joaquin Valley, California*. Portland, OR: U. S. Fish and Wildlife Service.

———. 2004. Finding for the resubmitted petition to list the black-tailed prairie

dog as threatened. *Federal Register* 69 (159):51217-51226.
Váczi, O., B. Koósz, and V. Altbäcker. 2006. Modified ambient temperature perception affects daily activity patterns in the European ground squirrel (*Spermophilus citellus*). *Journal of Mammalogy* 87 (1):54-59.
Valdez, M., and G. Ceballos. 1991. Historia natural, alimentación y reproducción de la ardilla terrestre (*Spermophilus mexicanus*) en una pradera intermontana. *Acta Zoológica Mexicana* 43:1-31.
———. 1997. Conservation of endemic mammals of Mexico: The Perote ground squirrel (*Spermophilus perotensis*). *Journal of Mammalogy* 78 (1):74-82.
———. 2003. Patrones de hibernación de ardillas de tierra (*Spermophilus mexicanus* y *S. perotensis*) en el centro de México. *Revista Mexicana de Mastozoología* 7:40-48.
Valverde, J. A. 1957. *Aves del Sahara español: Estudio ecológico del desierto*. Madrid: Instituto de Estudios Africanos, Consejo Superior de Investigaciónes Científicas.
Van Den Brink, F. H., and T. Haltenorth. 1968. *Die Säugetiere Europas Westlich des 30. Längengrades*. Hamburg: Verlag Paul Parey.
Van Heerden, J., and J. Dauth. 1987. Aspects of adaptation to an arid environment in free-living ground squirrels *Xerus inauris*. *Journal of Arid Environments* 13 (1):83-89.
Van Peenen, P. F., J. F. Duncan, and R. H. Light. 1970. Mammals of South Vietnam, I: Mammals commonly trapped during surveys. *Military Medicine* 135 (5):384-90.
Van Staalduinen, M. A., and M. J. A. Werger. 2007. Marmot disturbances in Mongolian steppe vegetation. *Journal of Arid Environments* 69:344-351.
Van Vuren, D., A. J. Kuenzi, I. Loredo, and M. L. Morrison. 1997. Translocation as a nonlethal alternative for managing California ground squirrels. *Journal of Wildlife Management* 61 (2):351-359.
Vander Haegen, W. M., M. A. Schroeder, and R. M. DeGraaf. 2002. Predation on real and artificial nests in shrubsteppe landscapes fragmented by agriculture. *Condor* 104 (3):496-506.
Vander Wall, S. B. 1993a. Cache site selection by chipmunks (*Tamias* spp.) and its influence on the effectiveness of seed dispersal in Jeffrey pine (*Pinus jeffreyi*). *Oecologia* 96 (2):246-252.
———. 1993b. Foraging success of granivorous rodents: Effects of variation of seed and soil water on olfaction. *Ecology* 79 (1):233-241.
———. 1993c. A model of caching depth: Implications for scatter hoarders and plant dispersal. *American Naturalist* 141 (2):217-232.
———. 1993d. Salivary water loss to seeds in yellow pine chipmunks and Merriam's kangaroo rats. *Ecology* 74 (5):1307-1312.
———. 1993e. Seed water content and the vulnerability of buried seeds to foraging rodents. *American Midland Naturalist* 129 (2):272-281.
———. 1995a. Dynamics of yellow pine chipmunk (*Tamias amoenus*) seed caches: Underground traffic in bitterbrush seeds. *Ecoscience* 2 (3):261-266.
———. 1995b. The effects of seed value on the caching behavior of yellow pine chipmunks. *Oikos* 74 (3):533-537.
———. 1995c. Influence of substrate water on the ability of rodents to find buried seeds. *Journal of Mammalogy* 76 (3):851-856.
———. 1995d. Sequential patterns of scatter hoarding by yellow pine chipmunks (*Tamias amoenus*). *American Midland Naturalist* 133 (2):312-321.
———. 1998. Foraging success of granivorous rodents: Effects of variation in seed and soil water on olfaction. *Ecology* 79 (1):233-241.
———. 2000. The influence of environmental conditions on cache recovery and cache pilferage by yellow pine chipmunks (*Tamias amoenus*) and deer mice (*Peromyscus maniculatus*). *Behavioral Ecology* 11 (5):544-549.
———. 2002. Masting in animal-dispersed pines facilitates seed dispersal. *Ecology* 83 (12):3508-3516.
———. 2003. Effects of seed size of wind-dispersed pines (*Pinus*) on secondary seed dispersal and the caching behavior of rodents. *Oikos* 100:25-34.
———. 2008. On the relative contributions of wind vs. animals to seed dispersal of four Sierra Nevada pines. *Ecology* 89 (7):1837-1849.
Vander Wall, S. B., M. J. Beck, J. S. Briggs, J. K. Roth, T. C. Thayer, J. L. Hollander, and J. M. Armstrong. 2003. Interspecific variation in the olfactory abilities of granivorous rodents. *Journal of Mammalogy* 84 (2):487-496.
Vander Wall, S. B., K. M. Kuhn, and J. R. Gworek. 2005. Two-phase seed dispersal: Linking the effects of frugivorous birds and seed-caching rodents. *Oecologia* 145 (2):282-287.
Vander Wall, S. B., and E. Peterson. 1996. Associative learning and the use of cache markers by yellow pine chipmunks (*Tamias amoenus*). *Southwestern Naturalist* 41 (1):88-90.
Velázquez, A., G. Bocco, F. J. Romero, and A. Pérez Vega. 2003. A landscape perspective on biodiversity conservation: The case of Central Mexico. *Mountain Research and Development* 23 (3):240-246.
———. 2004. A participatory landscape perspective on biodiversity conservation: The case of Central Mexico. In *Managing Mountain Protected Areas: Challenges and Responses for the 21st Century; Proceedings of the Mountain Protected Areas Workshop, 5th World Parks Congress, Durban, South Africa, September 2003*, ed. E. Harmon and G. L. Warboys. Colledara (Teramo), Italy: Andromeda.
Vereshchagin, N. K. 1959. *Mlekopitaiushchie Kavkaza*. Moscow: Akademii nauk SSSR.
Vernes, K. 2001. Gliding performance of the northern flying squirrel (*Glaucomys sabrinus*) in mature mixed forest of eastern Canada. *Journal of Mammalogy* 82 (4):1026-1033.
Vernes, K., S. Blois, and F. Bärlocher. 2004. Seasonal and yearly changes in consumption of hypogeous fungi by northern flying squirrels and red squirrels in old-growth forest, New Brunswick. *Canadian Journal of Zoology* 82:110-117.
Verts, B. J., and L. N. Carraway. 1998. *Land Mammals of Oregon*. Berkeley: University of California Press.
———. 2001. *Tamias minimus*. *Mammalian Species* 653:1-10.
Verts, B. J., and D. B. Costain. 1988. Changes in sex ratios of *Spermophilus beldingi* in Oregon. *Journal of Mammalogy* 69 (1):186-190.
Viljoen, S. 1975. Aspects of the ecology, reproductive physiology, and ethology of the bush squirrel, *Paraxerus cepapi cepapi* (A. Smith, 1836). M.S. thesis, University of Pretoria.
———. 1977a. Behaviour of the bush squirrel, *Paraxerus cepapi cepapi* (A. Smith, 1836). *Mammalia* 41 (2):119-166.
———. 1977b. Factors affecting breeding synchronization in an African bush squirrel (*Paraxerus cepapi cepapi*). *Journal of Reproductive Fertility* 50:125-127.
———. 1977c. Feeding habits of bush squirrel *Paraxerus cepapi cepapi* (Rodentia: Sciuridae). *Zoologica Africana* 12 (2):459-467.
———. 1978. Notes on the western striped squirrel, *Funisciurus congicus congicus* (Kuhl, 1820). *Madoqua* 11:119-128.
———. 1980. A comparative study on the

biology of two subspecies of tree squirrels, *Paraxerus palliatus tongensis*, Roberts, 1931 and *Paraxerus palliatus ornatus* (Gray, 1864) in Zululand. D.Sc. diss., University of Pretoria.

———. 1983a. Communicatory behaviour of Southern African tree squirrels, *Paraxerus palliatus ornatus, P. p. tongensis, P. c. cepapi*, and *Funisciurus congicus*. *Mammalia* 47 (4):441-462.

———. 1983b. Feeding habits and comparative feeding rates of three Southern African arboreal squirrels. *South African Journal of Zoology* 18 (4):378-387.

———. 1986. Use of space in Southern African tree squirrels. *Mammalia* 50 (3):293-310.

———. 1989. Taxonomy and historical zoogeography of the red squirrel *Paraxerus palliatus* (Peters, 1852) in the Southern African subregion (Rodentia: Sciuridae). *Annals of the Transvaal Museum* 35 (2):49-60.

———. 1997a. Striped tree squirrel. In *The Complete Book of Southern African Mammals*, ed. M. G. L. Mills and L. Hes. Cape Town, South Africa: Struik.

———. 1997b. Tree squirrel *Paraxerus cepapi*. In *The Complete Book of Southern African mammals*, ed. M. G. L. Mills and L. Hes. Cape Town, South Africa: Struik.

Volzheninov, N. N., H. S. Yeziev, and B. A. Aromov. 1986. To ecology of some mammals in western part of Gyssar Range. In *Ecology, Protection and Acclimatization of Vertebrates in Uzbekistan*. Tashkent, Uzbekistan.

Vosburgh, T. C., and L. R. Irby. 1998. Effects of recreational shooting on prairie dog colonies. *Journal of Wildlife Management* 62 (1):363-372.

Vourc'h, G., J. Marmet, M. Chassagne, S. Bord, and J. L. Chapuis. 2007. Vector-borne and zoonotic diseases. *Zoonotic Diseases* 7 (4):637-642.

Wainwright, M. 2007. *The Mammals of Costa Rica: A Natural History and Field Guide*. A Zona Tropical publication. Ithaca, NY: Cornell University Press.

Waldien, D. L., J. P. Hayes, and M. M. P. Huso. 2006. Use of downed wood by Townsend's chipmunks (*Tamias townsendii*) in western Oregon. *Journal of Mammalogy* 87 (3):454-460.

Walsberg, G. E. 2000. Small mammals in hot deserts: Some generalizations revisited. *Bioscience* 50 (2):109.

Walston, J., J. W. Duckworth, and S. Molur. 2008a. *Petaurista elegans*. *IUCN Red List of Threatened Species*, version 2009.2.

———. 2008b. *Petaurista philippensis*. *IUCN Red List of Threatened Species*, version 2010.4.

Walston, J., J. W. Duckworth, S. U. Sarker, and S. Molur. 2008. *Petaurista petaurista*. *IUCN Red List of Threatened Species*, version 2009.2.

Wampler, C. R., J. K. Frey, D. M. VanLeeuwen, J. C. Boren, and T. T. Baker. 2008. Mammals in mechanically thinned and non-thinned mixed-coniferous forest in the Sacramento Mountains, New Mexico. *Southwestern Naturalist* 53 (4):431-443.

Wang, F. 1985. Preliminary study of the ecology of *Trogopterus xanthipes*. *Acta Theriologica Sinica* 5:103-110.

Wang, J. 2000. Biodiversity: Biota and biocoenose. In *Mountain Geoecology and Sustainable Development of the Tibetan Plateau*, ed. D. Zheng, Q. Zhang, and S. Wu. Dordrecht, The Netherlands: Kluwer Academic Publishers.

Wang, S. Q., E. G. Lakatta, H. Cheng, and Z. Q. Zhou. 2002. Adaptive mechanisms of intracellular calcium homeostasis in mammalian hibernators. *Journal of Experimental Biology* 205:2957-2962.

Wang, W., K. Ma, and C. Liu. 1999. Removal and predation of *Quercus liaotungensis* acorns by animals. *Ecological Research* 14 (3):225-232.

Wang, W., H. M. Zhang, and Z. B. Zhang. 2007. Effects of predation risk on cultivated walnut (*Juglans regia*) seeds hoarding behavior by David's rock squirrel (*Sciurotamias davidianus*) in enclosure. *Acta Theriologica Sinica* 27:358-364. [In Chinese, with English abstract.]

Wang, Y. Z., Y. R. Tu, and S. Wang. 1966. Notes on some small mammals of Szechuan Province, with description of a new subspecies. *Acta Zootaxonomica Sinica* 3:85-91.

Waterhouse, G. R. 1838. Original description of *Pteromys horsfieldii*. *Proceedings of the Zoological Society of London* 1837 [1838]:87.

Waterman, J. M. 1995. The social organization of the Cape ground squirrel (*Xerus inauris*; Rodentia: Sciuridae). *Ethology* 101 (2):130.

———. 1996. Reproductive biology of a tropical, non-hibernating ground squirrel. *Journal of Mammalogy* 77 (1):134-146.

———. 1997. Why do male Cape ground squirrels live in groups? *Animal Behaviour* 53 (4):809.

———. 1998. Mating tactics of male Cape ground squirrels, *Xerus inauris*: Consequences of year-round breeding. *Animal Behaviour* 56 (2):459.

———. 2002. Delayed maturity, group fission, and the limits of group size in female Cape ground squirrels (Sciuridae: *Xerus inauris*). *Journal of Zoology* 256 (1):113-120.

Waterman, J. M., and M. D. Herron. 2004. *Xerus princeps*. *Mammalian Species* 751:1-3.

Waters, J. R., and C. J. Zabel. 1998. Abundances of small mammals in fir forests in northeastern California. *Journal of Mammalogy* 79 (4):1244-1253.

Watson, H. C. 1975. Mammals (e): Observations on the behaviour and ecology of the squirrel *Heliosciurus gambianus*. Report in *Aberdeen University Expedition to Mole National Park, Ghana*. Aberdeen, Scotland: University of Aberdeen.

Wauters, L. A., S. Bertolino, M. Adamo, S. Van Dongen, and G. Tosi. 2005. Food shortage disrupts social organization: The case of red squirrels in conifer forests. *Evolutionary Ecology* 19:375-404.

Wauters, L. A., G. Tosi, and J. Gurnell. 2002. Interspecific competition in tree squirrels: Do introduced grey squirrels (*Sciurus carolinensis*) deplete tree seeds hoarded by red squirrels (*S. vulgaris*)? *Behavioral Ecology and Sociobiology* 51:360-367.

Wauters, L. A., M. Vermeulen, S. Van Dongen, S. Bertolino, A. Molinari, G. Tosi, and E. Matthysen. 2007. Effects of spatio-temporal variation in food supply on red squirrel *Sciurus vulgaris* body size and body mass and its consequences for some fitness components. *Ecography* 30:51-65.

Webster, D. W., J. F. Parnell, and W. C. Biggs Jr. 1985. *Mammals of the Carolinas, Virginia, and Maryland*. Chapel Hill: University of North Carolina Press.

Weddell, B. J. 1991. Distribution and movements of Columbian ground squirrels (*Spermophilus columbianus* (Ord)): Are habitat patches like islands? *Journal of Biogeography* 18 (4):385-394.

Weigl, P. D. 1968. The distribution of the flying squirrels, *Glaucomys volans* and *G. sabrinus*: An evaluation of the competitive exclusion idea. Ph.D. diss., Duke University.

———. 1978. Resource overlap, interspecific interactions, and the distribution of the flying squirrels, *Glaucomys volans* and *G. sabrinus*. *American Midland Naturalist* 100 (1):83-96.

———. 2007. The northern flying squirrel (*Glaucomys sabrinus*): A conservation challenge. *Journal of Mammalogy* 88 (4):897-907.

Weigl, R. 2005. *Longevity of Mammals in*

Captivity: From the Living Collections of the World. Kleine Senckenberg-Reihe No. 48. Stuttgart: Schweizerbart.

Wells, N. M., and J. Giacalone. 1985. *Syntheosciurus brochus. Mammalian Species* 249:1–3.

Wells-Gosling, N., and L. R. Heaney. 1984. *Glaucomys sabrinus. Mammalian Species* 229:1–8.

Wenny, D. G. 1999. Two stage dispersal of *Guarea glabra* and *G. kunthiana* (Meliaceae) in Monteverde, Costa Rica. *Journal of Tropical Ecology* 15:481–496.

Werner, S. J., D. L. Nolte, and F. D. Provenza. 2005. Proximal cues of pocket gopher burrow plugging behavior: Influence of light, burrow openings, and temperature. *Physiology and Behavior* 85 (3):340–345.

Wheeler, S., and P. Myers. 2004. *Petaurista leucogenys*. Animal Diversity Web [accessed 10 Feb. 2010].

Whisson, D. A., S. B. Orloff, and D. L. Lancaster. 1999. Alfalfa yield loss from Belding's ground squirrels in northeastern California. *Wildlife Society Bulletin* 27 (1):178–183.

White, M. M., and G. E. Svendsen. 1992. Spatial-genetic structure in the eastern chipmunk, *Tamias striatus*. *Journal of Mammalogy* 73 (3):619–624.

Whitten, J. E. J. 1981. Ecological separation of three diurnal squirrels in tropical rainforest on Siberut Island, Indonesia. *Journal of Zoology* 193 (3):405–420.

Whorley, J. R., and G. J. Kenagy. 2007. Variation in reproductive patterns of antelope ground squirrels, *Ammospermophilus leucurus*, from Oregon to Baja California. *Journal of Mammalogy* 88 (6):1404–1411.

Wiewel, A. S., W. R. Clark, and M. A. Sovada. 2007. Assessing small mammal abundance with track-tube indices and mark-recapture population estimates. *Journal of Mammalogy* 88 (1):250–260.

Wilber, P. G., D. W. Duszynski, S. J. Upton, R. S. Seville, and J. O. Corliss. 1998. A revision of the taxonomy and nomenclature of the *Eimeria* spp. (Apicomplexa: Eimeriidae) from rodents in the tribe Marmotini (Sciuridae). *Systematic Parasitology* 39:113–135.

Wilson, D. E., K. M. Helgen, C. S. Yun, and B. Giman. 2006. Small mammal survey at two sites in planted forest zone, Bintulu, Sarawak. *Malayan Nature Journal* 59 (2):165–187.

Wilson, J. A., D. A. Kelt, D. H. Van Vuren, and M. L. Johnson. 2008. Population dynamics of small mammals in relation to production of cones in four types of forests in the northern Sierra Nevada, California. *Southwestern Naturalist* 53:346–356.

Winterrowd, M. F., F. S. Dobson, J. L. Hoogland, and D. W. Foltz. 2009. Social subdivision influences effective population size in the colonial-breeding black-tailed prairie dog. *Journal of Mammalogy* 90 (2):380–387.

Wishner, L. 1982. *Eastern Chipmunks: Secrets of Their Solitary Lives*. Washington, DC: Smithsonian Institution Press.

Witt, J. W. 1991. Fluctuations in the weight and trap response for *Glaucomys sabrinus* in western Oregon. *Journal of Mammalogy* 72 (3):612–615.

———. 1992. Home range and density estimates for the northern flying squirrel, *Glaucomys sabrinus*, in western Oregon. *Journal of Mammalogy* 73 (4):921–929.

Wood, M. D. 1993. The effect of profitability on caching by the eastern chipmunk (*Tamias striatus*). *American Midland Naturalist* 129 (1):139–144.

Wooden, K. M., and G. E. Walsberg. 2002. Effect of environmental temperature on body temperature and metabolic heat production in a heterothermic rodent, *Spermophilus tereticaudus*. *Journal of Experimental Biology* 205:2099–2015.

Woodman, N., N. A. Slade, R. M. Timm, and C. A. Schmidt. 1995. Mammalian community structure in lowland, tropical Peru, as determined by removal trapping. *Zoological Journal of the Linnean Society* 113:1–20.

Woodruff, D. S., and L. M. Turner. 2009. The Indochinese-Sundaic zoogeographic transition: A description and analysis of terrestrial mammal species distributions. *Journal of Biogeography* 36:803–821.

Wrigley, R. E., J. E. Dubois, and H. W. R. Copland. 1991. Distribution and ecology of six rare species of prairie rodents in Manitoba. *Canadian Field-Naturalist* 105:1–12.

Wu, X., W. Liu, X. Gao, and G. Yin. 2006. Huanglong Cave, a new Late Pleistocene hominid site in Hubei Province, China. *Chinese Science Bulletin* 51 (20):2493–2499.

Xiang, Z. F., X. C. Liang, S. Huo, and S. L. Ma. 2004. Quantitative analysis of land mammal zoogeographical regions in China and adjacent regions. *Zoological Studies* 43:142–160.

Xiao, Z., X. Gao, M. A. Steele, and Z. Zhang. 2010. Frequency-dependent selection by tree squirrels: Adaptive escape of nondormant white oaks. *Behavioral Ecology* 21 (1):169–175.

Yahner, R. H. 2003. Pine squirrels: *Tamiasciurus hudsonicus* and *T. douglasii*. In *Wild Mammals of North America: Biology, Management, and Conservation*, 2nd ed., ed. G. A. Feldhamer, B. C. Thompson, and J. A. Chapman. Baltimore: Johns Hopkins University Press.

Yahner, R. H., and C. G. Mahan. 1997a. Behavioral considerations in fragmented landscapes. *Conservation Biology* 11 (2):569–570.

———. 1997b. Effects of logging roads on depredation of artificial ground nests in a forested landscape. *Wildlife Society Bulletin* 25 (1):158–162.

Yamaguti, S. 1941. Studies on the helminth fauna of Japan, part 35: Mammalian nematodes, 2. *Japanese Journal of Zoology* 9:409–439.

Yancey, F. D., J. K. Jones Jr., and R. W. Manning. 1993. Individual and secondary sexual variation in the Mexican ground squirrel, *Spermophilus mexicanus*. *Texas Journal of Science* 45:63–68.

Yensen, E. 1991. Taxonomy and distribution of the Idaho ground squirrel *Spermophilus brunneus*. *Journal of Mammalogy* 72 (3):583–600.

Yensen, E., and G. Hammerson. 2008. *Spermophilus canus*. *IUCN Red List of Threatened Species*, version 2009.2.

Yensen, E., G. Hammerson J. Jefferson, and S. Cannings. 2008. *Spermophilus brunneus*. *IUCN Red List of Threatened Species*, version 2009.2.

Yensen, E., T. Mabee, and G. Hammerson. 2008. *Spermophilus elegans*. *IUCN Red List of Threatened Species*, version 2009.2.

Yensen, E., and P. W. Sherman. 1997. *Spermophilus brunneus*. *Mammalian Species* 560:1–5.

———. 2003. Ground squirrels (*Spermophilus* species and *Ammospermophilus* species). In *Wild Mammals of North America*, 2nd ed., ed. G. A. Feldhamer, B. C. Thompson, and J. A. Chapman. Baltimore: Johns Hopkins University Press.

Yensen, E., and M. Valdés-Alarcón. 1999. Family Sciuridae. In *Mamíferos del Noroeste de México*, ed. S. T. Alvarez-Castañeda and J. L. Patton. Baja California Sur, Mexico: Centro de Investigaciónes Biologicas del Noroeste, S.C.

Yorke, W., and P. A. Maplestone. 1926. *The Nematode Parasites of Vertebrates*. New York: Hafner.

Youlatos, D. 1999. Locomotor and postural behavior of *Sciurus igniventris* and *Micro-*

sciurus flaviventer in eastern Ecuador. *Mammalia* 63 (4):405–416.

Young, C. J., and J. K. Jones Jr. 1982. *Spermophilus mexicanus*. *Mammalian Species* 164:1–4.

Yu, F[ahong]. 2002. Systematics and biogeography of flying squirrels in the eastern and the western trans-Himalayas. Ph.D. diss., University of Florida.

Yu, Fahong, Farong Yu, P. M. McGuire, C. W. Kilpatrick, J. Pang, Y. Wang, S. Lu, and C. A. Woods. 2004. Molecular phylogeny and biogeography of woolly flying squirrel (Rodentia: Sciuridae), inferred from mitochondrial cytochrome *b* gene sequences. *Molecular Phylogenetics and Evolution* 33 (3):735–744.

Yu, Fahong, Farong Yu, J. Pang, C. W. Kilpatrick, P. M. McGuire, Y. Wang, S. Lu, and C. A. Woods. 2006. Phylogeny and biogeography of the *Petaurista philippensis* complex (Rodentia: Sciuridae), inter- and intraspecific relationships inferred from molecular and morphometric analysis. *Molecular Phylogenetics and Evolution* 38:755–766.

Zahler, P. 1996. Rediscovery of the woolly flying squirrel (*Eupetaurus cinereus*). *Journal of Mammalogy* 77:54–57.

———. 2001. The woolly flying squirrel and gliding: Does size matter? *Acta Theriologica* 46:429–435.

Zahler, P., and M. Khan. 2003. Evidence for dietary specialization on pine needles by the woolly flying squirrel (*Eupetaurus cinereus*). *Journal of Mammalogy* 84 (2):480–486.

Zapata-Ríos, G. 2001. Sustentabilidad de la cacería de subsistencia: El caso de cuatro comunidades Quichuas en la Amazonia nororiental Ecuatoriana. *Mastozoología Neotropical* 8:59–66.

Zegers, D. A. 1984. *Spermophilus elegans*. *Mammalian Species* 214:1–7.

Zervanos, S. M., C. R. Maher, J. A. Waldvogel, and G. L. Florant. 2010. Latitudinal differences in the hibernation characteristics of woodchucks (*Marmota monax*). *Physiological and Biochemical Zoology* 83:135–141.

Zhang, H. M., Y. Chen, and Z. B. Zhang. 2008. Differences of dispersal fitness of large and small acorns of Liaodong oak (*Quercus liaotungensis*) before and after seed caching by small rodents in a warm temperate forest, China. *Forest Ecology and Management* 255 (3–4):1243–1250.

Zhang, H. M., and Z. B. Zhang. 2008. Key factors affecting the capacity of David's rock squirrels (*Sciurotamias davidianus*) to discover scatter-hoarded seeds in enclosures. *Acta Entomologica Sinica* 51 (5):1017–1098. [In Chinese, with English abstract.]

Zhang, Y., S. Jin, G. Quan, S. Li, Z. Ye, F. Wang, and M. Zhang. 1997. *Distribution of Mammalian Species in China*. Beijing: China Forestry Publishing House.

Zhang, Z. B., Z. S. Xiao, and H. J. Li. 2005. Impact of small rodents on tree seeds in temperate and subtropical forests, China. In *Seed Fate: Predation, Dispersal, and Seedling Establishment*, ed. P.-M. Forget, J. E. Lambert, P. E. Hulme, and S. B. Vander Wall. Cambridge, MA: CAB International.

Zimmer, M. B., and W. K. Milsom. 2001. Effects of changing ambient temperature on metabolic, heart, and ventilation rates during steady state hibernation in golden-mantled ground squirrels (*Spermophilus lateralis*). *Physiological and Biochemical Zoology* 74 (5):714–723.

———. 2002. Ventilatory pattern and chemosensitivity in unanesthetized, hypothermic ground squirrels (*Spermophilus lateralis*). *Respiratory Physiology & Neurobiology* 133 (1–2):49–63.

Zittlau, K. A., C. S. Davis, and C. Strobeck. 2000. Characterization of microsatellite loci in northern flying squirrels (*Glaucomys sabrinus*). *Molecular Ecology* 9 (6):826–827.

Zollner, P. A., and K. J. Crane. 2003. Influence of canopy closure and shrub coverage on travel along coarse woody debris by eastern chipmunks (*Tamias striatus*). *American Midland Naturalist* 150 (1):151.

Zumpt, F. 1966. *Insecta excluding Phthiraptera*. Vol. 3 of *The Arthropod Parasites of Vertebrates in Africa South of the Sahara (Ethiopian Region)*. Publications of the South African Institute for Medical Research, Vol. 13, No. 52. Johannesburg: South African Institute for Medical Research.

Zumpt, I. F. 1968. The feeding habits of the yellow mongoose, *Cynictis penicillata*; the suricate, *Suricata suricatta*; and the Cape ground squirrel, *Xerus inauris*. *Journal of the South African Veterinary Medical Association* 39:89–91.

———. 1970. The ground squirrel. *African Wildlife* 24:115–121.

INDEX

Note: Page numbers in **boldface** indicate the starting page of the main accounts. Page numbers in *italics* indicate images of skulls. Scientific names can be searched starting with the genus, species, or subspecies name.